XXXVII Brazilian Meeting on Nuclear Physics 2014

Journal of Physics: Conference Series
Volume 630

Maresias, Sao Paulo, Brazil
8 - 12 September 2014

ISBN: 978-1-5108-0993-2
ISSN: 1742-6588

Printed from e-media with permission by:

Curran Associates, Inc.
57 Morehouse Lane
Red Hook, NY 12571

CURRAN ASSOCIATES INC. proceedings.com

Some format issues inherent in the e-media version may also appear in this print version.

Printed by Curran Associates, Inc. (2015)

Institute of Physics
Dirac House, Temple Back
Bristol BS1 6BE UK

Phone: 44 1 17 929 7481
Fax: 44 1 17 920 0979

techtracking@iop.org

Additional copies of this publication are available from:

Curran Associates, Inc.
57 Morehouse Lane
Red Hook, NY 12571 USA
Phone: 845-758-0400
Fax: 845-758-2634
Email: curran@proceedings.com
Web: www.proceedings.com

Table of contents

Volume 630

XXXVII Brazilian Meeting on Nuclear Physics
8–12 September 2014, Maresias, Sao Paulo, Brazil
Accepted papers received: 24 June 2015
Published online: 15 July 2015

Papers

Journal of Physics: Conference Series **630** (2015) 011001 doi:10.1088/1742-6596/630/1/011001

Preface

The **XXXVII Brazilian Meeting on Nuclear Physics** (or **XXXVII RTFNB 2014**) gave continuity to a long sequence of workshops held in Brazil, devoted to the study of the different aspects of nuclear physics. The meeting took place in the Maresias Beach Hotel, in the town of Maresias (state of São Paulo) from 8th to 12th September 2014. Offering gentle weather, a charming piece of green land of splendid natural beauty with beach and all amenities, the place had all the conditions for very pleasant and fruitful discussions.

The meeting involved 162 participants and attracted undergraduate and graduate students, Brazilian and South American physicists and invited speakers from overseas (USA, Italy, Spain, France, England, Switzerland, Germany and South Corea). In the program we had plenary morning sessions with review talks on recent developments in theory, computational techniques, experimentation and applications of the many aspects of nuclear physics. In the parallel sessions we had a total of 58 seminars. This volume contains 60 written contributions based on these talks and on the poster sessions. Evening talks and poster sessions gave still more insight and enlarged the scope of the scientific program. The contributed papers, representing mainly the scientific activity of young physicists, were exhibited as posters and are included in the present volume. Additional information about the meeting can be found at our website: <http:// www.sbfisica.org.br/~rtfnb/xxxvii-en/>.

Support and sponsorship came from brazilian national agencies: Conselho Nacional de Desenvolvimento Científico e Tecnológico (CNPq); Coordenação de Aperfeiçoamento de Pessoal de Nível Superior (CAPES); Fundação de Amparo à Pesquisa do Estado de São Paulo (FAPESP); Fundação de Amparo à Pesquisa do Estado do Rio de Janeiro (FAPERJ); Sociedade Brasileira de Física (SBF) and Instituto de Física da Universidade de São Paulo (IFUSP).

We honored Professor Alejandro Szanto de Toledo, who completed fifty years of scientific activities at the University of Sao Paulo (USP). He dedicated most of his life to the development of experimental nuclear physics in the country. We had a special session where his life and achievements were remembered.

The organization of the **XXXVII RTFNB 2014** was one more step in a big effort to build in our part of the world a community of physicists engaged in the difficult problems of fundamental and applied nuclear physics. The international contacts bring new knowledge, provide reference framework and stimulate collaborations that are essential for a true participation in the scientific frontier. The Editors, the Organizing Committee and the whole Brazilian community of nuclear physicists were pleased and very grateful to the visitors that were together with us during the five days in Maresias.

The Editors

Journal of Physics: Conference Series **630** (2015) 011002 doi:10.1088/1742-6596/630/1/011002

50 years in science: Alejandro Szanto de Toledo (2045-2015)

During the XXXVII RTFNB 2014, we had the pleasure to organize a tribute to Professor Alejandro Szanto de Toledo to commemorate his 50 years of dedication to nuclear physics. Named "Alexfest", it gathered friends and very prominent physicists from around the world that came to Brazil specially to render a tribute to Alex, as many friends and colleagues called him. During the whole afternoon, the invited speakers presented some of the physics topics and important contributions that Alex gave to the field of nuclear physics during his five decades of intense work. Alex is the responsible for several landmarks in the Brazilian nuclear science, especially in the Institute of Physics of the University of Sao Paulo (IFUSP), where he spent almost his entire career.

He started his professional life in the IFUSP Van de Graaf Laboratory, under Oscar Sala's supervision as an undergraduate student. His talent allowed him to give significant contributions to the machine operation since the beginning of his internship in the laboratory. He soon became the responsible for the development, construction and implementation of an ion source for the new Pelletron accelerator, project coordinated by Oscar Sala as well. This was a huge responsibility for a young master degree student. His PhD thesis was about nuclear fusion, a topic that he mastered during his career. His international reputation allowed him to become a visiting researcher at several laboratories in the US, France, Japan, Russia and Germany.

Not satisfied with the challenges that his career was imposing to him, Alex pioneered the creation of a high-energy nuclear physics group in Brazil, becoming member of the STAR (Solenoidal Tracker At Rhic) collaboration in 1995, one of the two major experiments of the RHIC accelerator (Relativistic heavy Ion Collider), located in the Brookhaven National Laboratory, USA. Later on, in 2006, he also led his group to become part of the LHC (Large Hadron Collider) experiment ALICE (A Large Ion Collider Experiment). Besides all his scientific contribution to the Brazilian nuclear physics, Alex was also an enthusiastic supporter of the Brazilian science in general. He was director of the Institute of Physics from of University of Sao Paulo and twice head of the Nuclear Physics Department of the same Institute. He was also member of the Deliberative Council of the Brazilian National Funding Agency (CNPq), the High Council of State of Sao Paulo Funding Agency (FAPESP), the Brazilian Academy of Sciences (ABS) and the Sao Paulo Academy of Sciences (ACIESP).

Just 5 months after this tribute, Alex passed away. He left many friends and admirers around the world thanks to his remarkable enthusiasm for physics and for people. He was a tireless fighter for the progress of science in our Country, never hesitating even in the most difficult times. His strength and integrity were the pillars that support everyone that had the privilege to share life moments with him. Alex will be greatly missed as a friend and as a scientist.

The Organizing Committee

Peer review statement

All papers published in this volume of *Journal of Physics: Conference Series* have been peer reviewed through processes administered by the proceedings Editors. Reviews were conducted by expert referees to the professional and scientific standards expected of a proceedings journal published by IOP Publishing.

ED-XRF analysis for Cultural Heritage: is quantitative evaluation always essential?

L Bonizzoni

Università degli Studi di Milano, Department of Physics, via Celoria 16 – 20133 Milan (Italy)

letizia.bonizzoni@mi.infn.it

Abstract. Energy Dispersive X-Ray Fluorescence (ED-XRF) is a very suitable tool for examination of Cultural Heritage materials because of its simplicity, with no requirement for any sample preparation and the possibility of operating with portable instruments, and it can probably be considered the most useful non-destructive analytical technique for ancient valuable objects of archaeological, historical or artistic interest. As regards the possibility of getting quantitative analysis in archaeometric applications, the problems arising from the limited sensitivity in detecting low Z elements, the irregular shape or the non-homogeneous composition of the sample have generated a widespread opinion that only semi-quantitative analyses are possible in XRF applications to archaeometry. In fact, this is always true for non-homogeneous samples as, typically, painting layers. On the contrary, the problems deriving from limited sensitivity in detecting matrix light elements as well as from irregular surface under analysis can be solved in most cases. Notwithstanding, working on unique and not standardized objects requires to pay attention on details and to know how to choose correct parameters and calculation algorithms to obtain reliable results. Indeed opportunities to deal with these objects are very limited and results have implication in other fields, so that each information about materials and production technique is of great interest. Two typical materials of archaeological interest showing particular features are considered - namely high corroded metallic artefacts and ceramics - revealing that, even if in cultural heritage field detailed quantitative analysis is the goal, it is not always necessary as also qualitative information by XRF spectra increase the knowledge of artefact.

1. Introduction

Although the analyses using X-rays do not represent the only application of radiation to the study of cultural heritage, those based on the detection of characteristic X-ray fluorescence are surely the most used, involving also PIXE (particle induced X emission) and SEM (scanning electron microscope). XRF acronym refers to the analysis of material obtained through X-ray fluorescence excited by X-rays; this technique, multi-elemental and non-destructive, enables in situ analysis on a high measuring points, as no actual sample collection is needed [1].

As is well known, rather complicated mathematical manipulations are required to estimate the quantitative elemental composition accurately from intensity data [2]. Two conditions simplify the problem: if the exciting radiation is monochromatic and if the object presents a flat surface so that the angle of incidence of the exciting radiation and the direction of the emitted radiation can be determined.

Yet, archaeological, historical or artistic objects in general have irregular shapes, or non-geometric reliefs on their surface, and often inhomogeneous composition, both due to degradation processes or to execution technique. During the years, various methods were developed to overcome the former difficulty [3]. Yet, hypotheses to get the equation for the quantitative determination of detected elements include also the microscopic homogeneity of samples and this means that any surface change or layering, both due to ageing or to original production, prevent us from getting reliable quantitative results.

2. Hypothesis and limits for quantitative ED-XRF analysis

The main limits to get quantitative analysis when using the energy dispersive version for X-ray spectroscopy - EDXRF - are ascribable to the limited detector sensitivity in revealing low Z elements, to the irregular shape or the non-homogeneous composition of the sample. For this reason, semi-quantitative analyses are sometimes preferred [4] in applications to archaeometry, even if the problems deriving from limited sensitivity in detecting matrix light elements as well as from irregular surface under analysis can be solved in most cases [5]. The mathematical procedures to get equations for quantitative XRF analysis show various complexity degrees depending on thickness, geometry and composition of the considered sample. Different algorithm are present in literature [2, 6] and more recently computer programs [7] allow the transformation from intensity peaks in the spectrum to concentration of detected elements. Basing on the fundamental parameter method [6], it is possible to clearly understand how to overcome the intrinsic limit of the technique.

Considering *n* elements in a sample, and supposing they all are detectable by XRF, we get

$$I_i = I_0 P \varepsilon_i G c_i \phi_1 \left[\frac{\mu^i_{ph}(E_0)}{\mu^M(E_0)\phi_1 + \mu^M(E_i)\phi_2} \right] \quad (1)$$

where:
I_i is the intensity of the considered element *i*
I_0 is the intensity of incident monochromatic radiation
P includes the atomic factors
ε_i is the detector sensitivity for the considered emission of the element *i*
G is a geometrical factor
C_i is the weight concentration of element *i*, to be detected
Φ_1 and Φ_2 are the cosecants respectively of incident and detection angle
$\mu_{ph}^i(E_0)$ is the photoelectric mass absorption coefficient at incident energy
$\mu^M(E_0)$ and $\mu^M(E_i)$ are the total mass absorption coefficients respectively for incident energy and characteristic energy of considered element *I* for the sample matrix, being M the weighed sum of the weigh concentration of all the present chemical elements, i.e. $\mu^M(E_i) = \sum_{j=1}^{n} c_j \mu^M(E_i)$.

The above formula for characteristic X-ray intensity can be easily deduced under the hypothesis of exiting monochromatic radiation, infinite thickness of the sample with respect to X-rays penetration depth, no secondary excitation effects, collimated X-ray beams and well defined angles, homogeneous sample. But while the three first circumstance have been used here only for sake of simplicity – and more complex equations can account for different situations [2, 6] – the last two are absolutely necessary to the reliability of analytical results. For samples with irregular shape, for which angles are in general not well defined, some mathematical expedients can be applied [8]. In the case of samples with coating or surface alteration, quantitative analysis is surely not possible with traditional spectrometers, preventing form the possibility to get elemental concentration for both corroded or by nature layered samples. In these cases, the quantitative analyses is almost forbidden; nonetheless, as it will be shown later, interesting information on cultural heritage materials can anyway be deducted.

What is clearly stated by equation (1) is that the intensity of emitted X-rays in a sample is a function of the concentration of the element *i* itself and of the concentration of all the elements present in the sample, through the μ^M. This last dependence is referred to as matrix effects.

Usually, to ease calculation and improve results, the ratio with a reference standard analysis, in the experimental conditions, is considered, obtaining

$$c_i = \frac{I_i}{I_i^S} \frac{\sum_{j=1}^{n} c_j(\mu_j(E_0)\phi_1 + \mu_j(E_j)\phi_2)}{\mu^S(E_0)\phi_1 + \mu^S(E_i)\phi_2} \tag{2}$$

where the summit index S refers to the same parameters of eq. 1 for the reference standard.

Being the concentrations of each chemical element in the sample present in the formula, no calibration curve is possible unless the samples in study are very similar. Indeed, this is a system of homogeneous equations that must be solved adding the closing condition

$$\sum_{j=1}^{N} c_i = 1 \tag{3}$$

Moreover, if light non detectable elements are present, equations (1) and (2) state that also detected element cannot be quantified. Evaluation of self-absorption is needed to take into account the contribution of non-detected elements; this allows to quantify the detected elements, considering the effect of the light matrix in terms of absorption of x-rays. In the cultural heritage field, this is the case, for instance, of glasses and ceramics. Many papers have been published, resolving this problem [5, 9] in various ways, yet all of them consider the intensity of Rayleigh and Compton scattered peaks to get self-absorption evaluation of the light matrix.

3. Application to Cultural Heritage materials: case studies

Notwithstanding the intrinsic limits, expressed above, of ED-XRF application on cultural heritage materials, this technique is one of the most used in the field due to its non-destructivity, rapidity, portability that allows to work also in hostile location such as on scaffoldings, platforms or places where no electricity in supplied. This is particularly valuable when artistic objects cannot be moved for practical, conservative or economic reasons and an accurate qualitative elemental characterization can however increase the knowledge of the artefact. From experience, in fact, we learn that it is possible to extract essential information already from qualitative or semi-quantitative XRF analysis whenever the quantitative evaluation of chemical elements detected is not possible in an easy way, for instance in the case of glasses or gold gildings. In this cases, a semi-quantitative analysis can be applied, based on the ratio between significant detected elements corrected by their self-absorption coefficients.

When working on cultural heritage materials, the overview on the objects has to be as large as possible, always keeping in mind the necessity of portability and non-destructivity. So, one technique alone is hardly enough to get a complete picture on the artefacts: also XRF greatly advantages from the support of some other chemical and physical techniques. The most suitable and commonly used techniques are Raman, Fourier Transform Infrared Spectroscopy (FTIR) and UV-vis-NIR spectroscopy that can give different point of view to be related [10].

In the following, two examples on different materials will be considered. The former, showing how the qualitative information get from XRF spectra are good enough to obtain important information on the artefact, giving hints to historians and archaeologists; the guess is to obtain the provenance classification of ceramics without passing through the quantitative analysis. The latter example shows how the use of a complementary analytical technique can help XRF to give good results for quantitative evaluation even if the object conditions seem to be out of reach. In both cases, the focus will be mainly on the analytical method to get the most from qualitative XRF analysis, while archaeological or historical conclusions will be left aside.

3.1. *Comparison between quantitative XRF analysis and portable XRF spectra for provenance classification of archaeological bricks*

Pottery and earthenware are the most abundant tracers in archaeological excavations: they can account for ancient commercial trades and roads, besides being indicators for technological progress; for this reason, archaeologists are highly interested in the provenance of shards. The analytical tool widely used to find out different geographical proveniences among the sets of archaeological pottery is the examination of the elemental chemical composition, mainly focused on trace elements, in association with chemometric analysis [11]. In the following, the classification results on a set of bricks from the medieval Chartusian monastery of Certosa di Pavia, in the Po valley, northern Italy, is obtained both exploiting both a traditional XRF laboratory set up (on untreated taken samples) and a portable XRF spectrometer (in situ, directly on the charterhouse walls).

The study aims at verifying if a correct classification is possible even if the quantitative determination of minor and trace elements is abandoned, allowing a faster response when a large amount of data is concerned [12]. To this aim, weight concentration was determined from laboratory XRF measurements only while entire spectra were considered for the in situ campaign. Data were then submitted to different statistical data treatments to obtain provenance classification [13].

Twenty six bricks were considered, pertaining to different periods of original building. Nondestructive EDXRF analysis in the laboratory was performed in the fracture areas of sampled fragments to avoid external contamination. A Bruker spectrometer (Artax 200) was used, equipped with a Mo anode X-ray tube with the beam collimated down to 0.65 mm in diameter. The characteristic X-ray radiation emitted by the sample is passed to a silicon drift detector. A Mo transmission filter (12.5 mm thick) has been used for incident radiation and the working conditions were 25 kV and 0.7mA with an acquisition time of 100 s for each measurement. Weight concentrations were obtained using the AXIL-QXAS software and the National Institute of Standards and Technology (NIST) standard reference material 679. Eleven elements were considered, namely K, Ca, Ti, Mn, Fe, Cu, Zn, Rb, Sr, Zr, and Pb. For each sample, three measurement points were considered to check the homogeneity of the artifact; then, elemental concentration for homogenous samples was averaged and close to 100% for multivariate analyses and classification [14]. Measurements for three samples (namely 10b, 15a, and 15c) were rejected as they turned out not to be homogeneous. Normalizing the sample concentration to 100% is necessary to eliminate the differences among samples due to different silicate presence or firing temperatures, which could induce a varying weight loss also in samples with similar raw materials. This procedure is particularly advisable whenever samples contain indefinite amounts of extraneous material. For archaeological ceramics, the basic constituent of which is clay, an extraneous substance such as crushed shell or crushed stone, called temper, could have been added to the original raw material to improve the properties of the manufactured products. Classification has been obtained by principal component analysis (PCA), exploiting the correlation matrix, which permits to compare variables on different scales standardizing data [15].

Score and loading plots for the first two principal components (captured variance is 83.7%) are shown in figure 1. We note that most of the samples form a big, not very compact ellipse-shaped group, whereas a few samples act as outliers, namely 12a and 14b (higher Ca content and lower K and Fe content), 10a and 14a (higher Ca content and lower K content), and 11d. This result will be the reference point to check the classification obtained by the spectra classification.

In situ nondestructive EDXRF analysis was carried out using Bruker's Tracer III SD portable spectrometer equipped with a Rh target X-ray tube and a silicon drift detector. Working conditions were 20 mA at 40 kV and a spot size on the sample of 4 mm radius. Measuring time was set to 100 s.

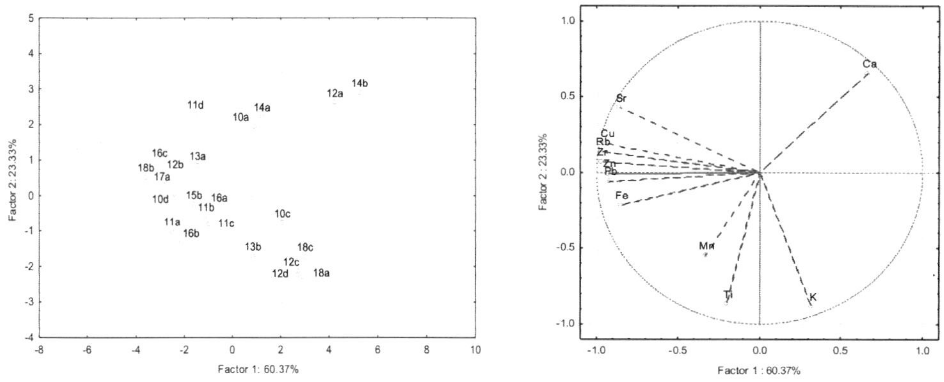

Figure 1: Score and loading plots (PC1 and PC2) from quantitative EDXRF analysis for the 23 bricks from Certosa di Pavia.

Analyses were performed directly on the monastery walls, in the same zone where samples had been previously taken, but because of the size of the instrumentation, spectra acquisition was not possible for all the sampled areas: two samples (12b and 17a) were thus left out. Our target is to obtain a correct classification directly applying multivariate analysis on the remaining 24 XRF spectra. For classification, we considered only the range of the spectra between 3 and 15 KeV (i.e. between K and Sr Ka peaks). Moreover, before performing PCA, spectra acquired on 2048 channels have been expressed in counts per second, smoothed (average on two nearest neighbors through LabVIEW software) and then rescaled so that the coherent scattering peaks have the same intensity value.

For the spectra analysis, the covariance matrix was considered [16]. In fact, the correlation matrix would standardize data, and in this case, it could confuse background noise with lower signals. On the other hand, the problem in using the covariance matrix is that the variables with the highest variance will dominate the first Principal Component. As in the samples some elements are present in a few percent while others in traces (i.e. ppm), XRF peaks in spectra highly differ in intensity and the highest peaks would dominate the first principal component, causing a loss of information. To avoid this problem, a logarithmic scale was used, getting the advantage of comparable peak heights [17].

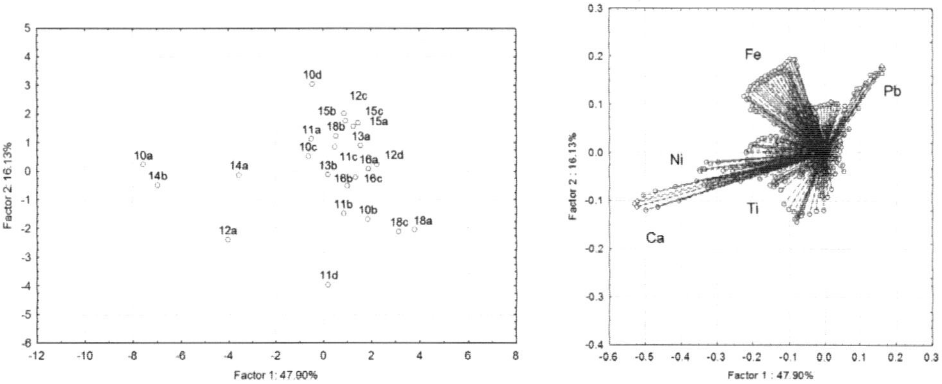

Figure 2: Score and loading plots (PC1 and PC2) from portable XRF spectra analysis for the 23 bricks from Certosa di Pavia.

The score and loading plots from PCA (covariance matrix) on the 24 in situ acquired spectra (smoothed, cut, and in logarithmic scale) is shown in figure 2. It is noticeable that a few channels (corresponding to some characteristic peaks, as indicated in the figure) weigh more on the principal components. Moreover, principal components with non-zero eigenvalues are only 23. From the PC1 and PC2 plot, it is evident that the considered samples form a compact group apart from samples 10a, 12a, 14a, and 14b, still to be distinguished on the basis of their Ca and Fe peaks, as well as sample 11d. From this results, we can say that is possible to get a reliable provenance classification of pottery avoiding quantitative analysis. Even if spot analysis can be misleading when we deal with non-homogeneous samples, we can apply multivariate analyses on entire spectra, paying attention to eventual geometry problems. This suggest to acquire more than one spectrum each sample when using handheld spectrometers.

3.2. *Getting help from other techniques: quantitative EDXRF analysis of highly corroded metal artefacts*

When dealing with highly mineralised historical objects, the quantitative EDXRF analysis cannot be applied due to the non-homogenous nature of the corrosion layer. Yet, combining EDXRF analysis with micro-Raman spectroscopy gave successful results in the identification of elements characterising the alloy of highly corroded metal samples and the patina surface and make it possible to determine the characteristic elements of the alloy [18] .

The samples considered are from the grave goods of the Royal Tomb 14 (Sipán, Peru), found in the Lambayeque region, north of the Peruvian coast, in 2007 and pertaining to the one of the most remarkable among pre Inca civilisations, the Moche culture [19]. The crafts of metalworking was highly developed in the Moche society [20]; the metal smiths developed alloys of copper with either silver or gold and some silver, used to produce hammered sheet metal to serve as raw material for the manufacture of silver- or gold-looking objects by depletion gilding and electrochemical replacement plating [21].

Most of the scientific works published about Sipan artefacts analyses [22] are dealing with well-preserved object while the two fragments analysed in this work (metal foils about 1 mm thick reported in figure 3) are highly corroded and no polishing at all was possible. Due to deterioration of the original alloy, in fact, the artefacts analysed in this work were rather fragile and could not resist hard polishing aimed at cleaning off corrosion products. They were hence characterized through XRF, using both a portable instrument and a capillary collimated spectrometer to investigate details, and micro-Raman spectroscopy, to get mineralogical data on the green corrosion patina, formed during their long burial. It was composed mainly by copper oxides and copper sulphites, such as malachite, atacamite and magnetite.

Quantitative XRF analysis was performed starting from the data acquired using the same set up (Bruker Artax 220) described in paragraph 3.1; the working conditions were 50 kV and 0.6 mA with an acquisition time of 300 s for each measurement. Detected elements were Cl, Ca, Fe, Cu, Au, Hg, Ag for each measurement point in both samples. Being Cu, Au and Ag the elements of the alloy (likely tumbaga), Fe, Ca and Cl were supposed to derive from Cu reactions with the environment. Hg presence, instead, should be explained with the tradition of painting object and corpse with cinnabar, a red pigment composed of Hg and S [23] and not linked to metal extraction techniques. Knowing that complete copper mineralisation in malachite (Cu2(OH)2CO3) would leave 57% Cu [24], quantitative analysis has been conducted bearing in mind that there are significant amounts of non-detectable elements (as C, O and H) and that and that the concentration of some elements like Fe, Al, Si, Ca and Na is increased selectively by corrosion processes in copper artefacts by comparison with relative Cu concentration with respect to the original alloy composition.

To perform quantitative analysis, only gold, copper and silver were then taken into account as they were supposed to be part of the original alloy. Setting the normalisation condition and forcing copper to be present only as atacamite and malachite the weight concentration of selected elements was evaluated.

Reliability of obtained results was proved comparing analytical results former compositional analyses on some other Moche handworks belonging to the same period [24], considering, for our samples, each measurement points as a freestanding [25]. The overall compositional data have been treated by means of cluster analysis, a multivariate method which allows to highlight groups of objects within the data set. In our application similarity between objects has been measured by Euclidean distance and the objects have been clustered by complete linkage using the statistical package SCAN (Software for Chemometric Analysis.Minitab Inc., State College, PA, USA, 1995).

Figure 3: The two sample from Sipan graves considered in the reported analyses: Sample 1Cu22-1 (on the right) and Sample 1Cu22-3 (on the left).

The dendrogram reported in figure 4 shows the similarity between the 24 measurement points on our metal sheets (code from 1 to 24) and the 28 Moche artefacts from literature [24] (codes from 25 to 53). The very large group contains all our samples which show a similarity with sub-groups from literature data higher than 90%. Two other groups are formed by objects with a high content of silver and gold and by the three objects belonging to a different historical period, respectively from left to right.

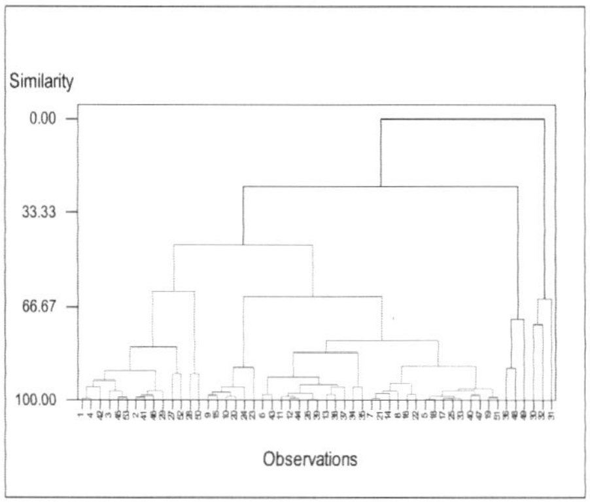

Figure 4: The two sample from Sipan graves considered in the reported analyses: Sample 1Cu22-1 (on the right) and Sample 1Cu22-3 (on the left).

This indicates that, even if our data surely lack of accuracy, they are still reliable and the result opens the possibility to perform EDXRF non-invasive measurements in situ on highly mineralised sample.

4. Conclusions

Cultural heritage field is very peculiar due to the unicity and preciousness of the artefacts involved in the analysis. For this reason, particular attention must be paid when performing scientific examinations, both when working directly on the objects and when analysing the spectra and data. Clearly, the quantitative analysis is the goal, but it is not always necessary as also reliable qualitative information increases the knowledge of artefact itself. Different experimental set-up are advisable in this cases to enlarge the amount of information obtained, such as the help of different analytical techniques that can give a larger and more complete view on the artefact.

Acknowledgments
The author is grateful to Bruker Italy for placing Tracer spectrometer at disposal. The analysis of bricks from Certosa di Pavia were performed in the TiVAl project for Lombardy cultural heritage promotion financially supported by Lombardy region.

The results presented in the paper have been performed with the cooperation and the help of Anna Galli, Mario Milazzo, Marco Martini, Marco Gondola and our valuable students.

References

[1] Mantler M and Schreiner M 2000 *X-Ray Spectrom.* **29** 3
Dran J C Calligaro T and Salomon J 2000 in *Modern Analytical Methods in Art and Archeology* (New York: Wiley)
José-Yacamàn M abnd Ascensio J A 2000 in *Modern Analytical Methods in Art and Archeology* (Wiley, New York)
[2] Lachance G R and Claisse F 1994 *Quantitative X–Ray fluorescence Analysis* (New York: Wiley)
[3] Milazzo M and Cicardi C 1998 *Archaeometry* **40**, 351
Milazzo M 2004 *Nucl. Instr. and Meth. in Phys. Res. B* **213** 683
Bonizzoni L, Maloni A and Milazzo M 2006 *X-Ray Spectrom.* **35** 390
[4] Moioli P and Seccaroni C 2000 *X-Ray Spectrom.* **29**, 48
[5] Milazzo M 2004 *Nucl. Instr. and Meth. in Phys. Res. B* **213** 683
He F and Van Espen P 1990 *NIM A* **299** 580
Wobrauschek P Halmetschlager G Zamini S Jokubonis C Trnka G and Karwowski M 2000 *X-Ray Spectrom.* **29** 25
Confalonieri L Crippa R and Milazzo M 1987 *Appli. Radiat. Isot.* **38** 139
Bui C Milazzo M and Monichino M 1987 *NIMB* **28** 88
Bonizzoni L Galli A and Milazzo M 2000 *X-ray Spectrom.* **29** 443
[6] Tertian R and Claisse F 1982 *Principles of Quantitative X-Ray Fluorescence Analysis* (London: Heyden)
[7] He F and Van Espen P 1990 *NIM A* **299** 580
Van Espen P Janssens and K Swenters I 1994 *Manual for AXIL Software*. Canberra-Packard
XRF Q 6.0 ASSING LITHOS 3000, user handbook
Solé V A Papillon E Cotte M Walter Ph and Susini J 2007 *Spectrochim. Acta Part B* **62** 63
[8] Milazzo and M Cicardi C 1997 *X-ray Spectrom.* **26** 211
Avaldi L Bui C Milazzo 1985 M *X-ray Spectrom.* **14** 159
[9] Bui C Confalonieri L and Milazzo M 1989 *International Journal of Radiation Applications and Instrumentation. Part A. Applied Radiation and Isotopes* **40** 375
[10] Rosia F Burnstockb A Van den Bergc K J Miliania C Brunetti B G and Sgamellotti A 2009 *Spectrochimica Acta A* **71** 1655
Hocquet F P Calvo del Castillo H Cervera Xicotencatl A Bourgeois C Oger C Marchal A Clar M Rakkaa S Micha E and Strivay D 2011 *Analytical and Bioanalytical Chemistry* **399** 3109
Bonizzoni L Caglio S Galli A Poldi G 2008 *Applied Physics A* **92** 203
[11] Jones R A 1986 *Greek and Cypriot Pottery* (Athens: The British School)
[12] Freitas R P Calza C Lima T A Rabello A and Lopes R T 2010 X-Ray Spectrom. 39 307
[13] Bonizzoni L Galli A Gondola and M Martini M 2013 *X-Ray Spectrom.* **42** 262
[14] Bonizzoni L Galli and A Milazzo M 2010 *X-Ray Spectrom.* **39** 346
Aruga R 1998 *Talanta* **47** 1053
[15] Marengo E Aceto M Robotti E Liparota M C Bobba M and Pantò G 2005 *Anal. Chim. Acta* **537** 359
[16] Massart D L Vandegiuste B G M Deming S N Michotte Y and Kaufman L 1988 *Data handling in science and technology vol.2* (Netherlands: Elsevier)
[17] Baxter M J Buck C E in *Modern Analytical Methods in Art and Archaeology* 2000 (New York: Wiley)

[18] Galli A Bonizzoni L Sibilia M and Martini M 2011 *X-Ray Spectrom.* **40** 74
[19] Lechtman H Erlij A and Barry E J Jr 1982 *Am. Antiq.* **47** 3
[20] Hoèrz G and Kallfass M 2000 *Mater. Character.* **45** 391
[21] Donnan C B Scott D A and Bracken T 2008 in *The Art and Archaeology of the Moche* (Eds.: S. Bourged, K. L. Jones), University of Texas Press
[22] Cesareo R, Rizzutto M A Brunetti A and Rao D V 2009 *Nucl. Instr. Meth. Phys.Res. B* **267** 2890
Cesareo R Brunetti A and Ridolfi S 2008 *X-Ray Spectrom.* **37** 309
Cesareo R Bustamante A Fabian J Calza C Dos Anjos M Lopes R T Elera C Shimada I Curayd V and Rizzutto M A 2010 *X-Ray Spectrom.* **39** 122
[23] Cesareo R Bustamante A Fabian J Calza C Dos Anjos M Lopes R T Elera C Shimada I Curayd V and Rizzutto M A 2010 *X-Ray Spectrom.* **39** 122
[24] Chapdelaine C Kennedy G and Ulceda Castillo S 2001 *Archaeometry* **43** 373
[25] Bonizzoni L Galli and A Milazzo M 2010 *X-Ray Spectrom.* **39** 346

Deformulation of a solid pharmaceutical form using computed tomography and X-ray fluorescence

J M Oliveira Junior[1], V M Balcão[2,3], M M D C Vila[2], N Aranha[1], V M H Yoshida[2], M V Chaud[2] and S Mangine Filho[1]

[1]LaFINAU – Laboratory of Applied Nuclear Physics, University of Sorocaba, Sorocaba/SP, Brazil.

[2]LaBNUS – Biomaterials and Nanotechnology Laboratory, i(bs)[2] – **i**ntelligent **bi**osensing and **bi**omolecule **s**tabilization research group, University of Sorocaba, Sorocaba/SP, Brazil.

[3]CEB - Centre of Biological Engineering, University of Minho, Braga, Portugal.

E-mail: jose.oliveira@prof.uniso.br

Abstract. Deformulation of medicines is of undeniable importance, since it can be utilized both to unravel the chemical composition of the excipients integrating a pharmaceutical formulation of a specific medicine and as an important tool to conduct morphometric studies of the formulation under study. Such strategy may be utilized in analytical studies aiming at quantifying the components of reference drugs, or in the identification of putative counterfeit pharmaceuticals. Deformulation makes use of physicochemical analysis tools to characterize, from the chemical point of view, the components integrating medicine pharmaceutical formulations and from the physical point of view, the morphological part of the pharmaceutical formulation. The techniques of computer tomography (SkyScan 1174 - Bruker microCT) and X-ray fluorescence analyses (using an X-ray source with W-anode from Hammatsu Photonics and Silicon Drift detector from Amptek) were successfully used in performing a process of deformulation of a solid pharmaceutical formulation of tablets, utilized herein as a model medicine for controlled drug release. The analytical methods used in this work, proved their effectiveness for the main goal of this study, which aimed to characterize a pharmaceutical formulation via its deconstruction.

1. Introduction

Deformulation has been increasingly practiced by pharmaceutical companies whose primary production focus is the manufacture of medicines marketed under the name of Generic. A Generic medicine is one that contains the same drug (active ingredient) in the same dose and dosage form, is administered via the same route and with the same therapeutic indication, and shall present the same biological safety as the reference drug in the country. To be interchangeable with the reference drug, the Generic product must pass and be approved in bioequivalence testing [1]. To achieve both the same therapeutic performance and safety of (so-called Reference) medicines, pharmacotechnical and technological development is required, together with significant financial investment and skilled labor. To facilitate copying into a Generic (a simpler and less expensive in investment), deformulation of the reference medicine is practiced in order to acquire knowledge about the chemical composition of the

[1] To whom any correspondence should be addressed.

compounds present in these medicines and to know the physical characteristics of the formulation, such as the mechanism of drug release, the thickness of the coating, the porosity, the morphological features, the process of drug release, etc. Deformulation enables both the decomposition of the medicine to copy and the identification of counterfeit medicines. Non-destructive analysis techniques such as computed tomography via X-ray transmission (CT) and X-ray fluorescence (XRF) are useful for deconstruction, because the former can provide the physical characteristics whereas the second provides information on the elemental constitution of the formulation, i.e. chemical composition. The results of the analyses are easily comparable and may be confronted in case of rebuttal. CT began to be used back in 2003 in the pharmaceutical field as a new tool for the nondestructive analysis of solid dosage forms, by Farber et al. [2] and Ozeki et al. [3]. The great advantage in the use of CT when compared to more traditional techniques, lies in the fact that it can be used to view not only the surface, but also the inside of objects with varying dimensions, shapes and densities, without any loss of sample. Additionally, CT enables the morphological characterization of regions inaccessible to other analytical equipments. By using CT, it is possible to determine the coating thickness of a tablet or even analyze the several layers of tablet materials subjected processing via multiple compression. Traini et al. [4] emphasize that the main analytical methods used to characterize the interior of solid dosage forms are the use of techniques of mercury porosimetry and gas adsorption. However, these researchers make it clear that, if the purpose is to conduct dynamic studies, i.e. to track changes in the internal structure of the tablets following changes in porosity during the disintegration process, the aforementioned porosimetry techniques cannot be used since they are destructive techniques preventing the use of the sample in more than one type of analysis, and recommend in these cases the use of CT. Li et al. [5] propose the use of the CT technique to study processes of drug release, using three-dimensional tomographic imaging with X-rays from synchrotron light. Oliveira Jr. et al. [6] showed, in a recent study, that it is possible to use the computed tomography technique for studies of disintegration of solid dosage forms. Scanning electron microscopy (SEM), X-ray diffraction analysis (XRD), X-ray attenuation and CT scans are analytical techniques indicated for morphological studies [7, 8, 9], whereas atomic force microscopy has been used in the analysis of the effects produced by mechanical processing upon the surface of pharmaceutical powders [10]. X-ray fluorescence analysis (XRF) [11] has been used not only to display but also to quantify the presence of trace elements in pharmaceutical formulations [12]. Computed tomography via X-ray transmission (CT) and X-ray fluorescence analysis (XRF) were utilized to perform deformulation of a solid pharmaceutical form. As a model for the study undertaken, one used a medicine in the form of coated tablets, the purpose of which is the controlled release of the drug through the increase in the osmotic pressure of the system following water absorption - Osmotic Release Oral System (OROS™). There is currently a great interest within the pharmaceutical industry in the production of these tablets, since due to expiration of the protection periods of their patents there is no longer the need to perform new clinical trials, and thus this type of medicines may be marketed immediately.

2. Materials and methods

2.1. Materials

2.1.1. Chemicals.

As model medicine, one utilized Adalat Oros™ tablets (30 mg nifedipine, Lot MS-1.0429.0001.137-9, Bayer AG, Leverkusen, NRW, Germany), acquired in a local Brazilian drug store.

2.1.2. Analytical equipment.

A computed tomographer via X-ray transmission from Bruker microCT (model SkyScan 1174, Kontich, Belgium), and an X-ray fluorescence spectrometer (XRF) from Amptek Inc. (Bedford MA, U.S.A.) were utilized in all non-destructive analyses. The analysis software utilized for processing the tomographic images were CTVox™ (version 2.6.0 r908-64bit, from Bruker microCT) and CTan™ (version 1.13.5.1-64bit, from Bruker microCT).

2.2. Experimental procedures

In this work, one utilized either dry tablets of Adalat Oros™ or tablets maintained submerged in ultrapure water during predetermined intervals of time. The tomographic images were gathered under two different conditions: (i) dry tablets, and (ii) tablets submerged in ultrapure water for a period of 1225 and 1440 min, followed by an air drying period of 48 h in a room free from dust at room temperature and in the dark. The tablets were then placed in the interior of the tomograph chamber so as to produce the tomographic images. The tomographic images were obtained using the following configurations of the tomographic system: operating voltage set at 35 kV and electric current of 800 μA. The technique employed for obtaining a tomographic image involved the acquisition of a large number of radiographs of the object, obtained by measuring the intensity of X-rays transmitted through the sample, at different angular positions. The angular step utilized was 1°, the samples were rotated 180 degrees, producing 180 radiographs (projections) per image, each containing 652×652 pixels with spatial resolution of 19.52 μm. At the outlet of the X-ray source one utilized a filter of Al with 0.25 mm thickness. Appropriate mathematical algorithms were then used to reconstruct the three-dimensional tomographic images (3D) of the object, through the appropriate composition of bi-dimensional images (2D). In the present research work, the three-dimensional images possessed 652×652×652 pixels and the same spatial resolution of the 2D images, and thus the volume of data generated for each tablet is isotropic with relation to the spatial resolution. In addition, a tomographic image of a dry tablet with a better spatial resolution was also produced, using the following experimental configurations: operating voltage set at 50 kV and electric current of 800 μA, Al filter with 0.5 mm thickness at the outlet of the X-ray source, 180° spatial rotation of the object with angular increments of 0.7°, thus generating three-dimensional images with 984×984×984 pixels and with a spatial resolution of 9.76 μm. Having all the projections (radiographs gathered at each angular position), one utilized the software NRecon™ from Bruker (version 1.6.9.4, Kontich, Belgium), which uses the algorithm of Feldkamp et al. [13] in the process of reconstruction of the tomographic images. The X-ray fluorescence equipment utilized consisted in a compact system from Amptek Inc. (Bedford MA, U.S.A.) and was composed by a silicon detector (Silicon Drift Detector) with an area of 25 mm^2 and 500 μm thickness, protected by a Beryllium window of 12.5 μm. The X-ray source utilized in excitation of samples is equipment manufactured by Hamamatsu Photonics Company, model L6731-01, that uses a target of tungsten and operates with variable voltages from 20 kV to 80 kV and maximum electric current of 100 μA. To acquire the X-ray fluorescence data one utilized the software DppMCA™ from Amptek Inc. (version 1.0.0.12, Bedford MA, U.S.A.) whereas the data gathered was analysed using the software XRF-FP™ also from Amptek Inc. (version 5.2.9). The process of excitation of samples was carried out using an Al collimator with 2-mm orifice opening at the outlet of the X-ray source. The X-ray source was configured to operate at a voltage of 25 kV and electric current of 99 μA. Due to the fact that each chemical element possesses a unique set of atomic energy levels, each chemical element also emits a unique set of characteristic X-rays when properly excited, that are characteristic of the element that emitted them. Additionally, the intensity of a line of characteristic X-rays depends on the number of corresponding atoms that have been excited in the sample. The fluorescence technique used in this study was X-ray Fluorescence by Energy Dispersion (EDXRF), also known as non-dispersive X-ray fluorescence. In this technique, the characteristic X-rays are selected and quantified by using a detection system which operates with detectors able to discriminate between X-rays of close energies. The technique of X-ray fluorescence requires a minimal sample preparation, and the measurements performed in this study were conducted by setting to 900 s the time of sample excitation. The X-ray fluorescence analyses were carried out both at the surface and interior of dry Adalat Oros™ tablets. The fluorescence system was calibrated using a reference standard sample supplied by the "National Institute of Standards and Technology – NIST" (Stainless Steel SS316), containing the chemical elements Cr, Mn, Fe, Ni, Cu and Mo at known concentrations and using samples especially prepared for this purpose, containing known concentrations of the following elements: Mg, P, Cl, K, Ca and Ti. Quali-quantitative analysis of fluorescence data was performed using the software XRF-FP™ from Amptek Inc.

3. Results and discussion

In the process of deformulation of a pharmaceutical form, the greater the detail of the physicochemical characteristics of the dosage form being studied, the easier one can reproduce it. Figure 1 shows a typical tomographic image of a tablet of Adalat Oros™. In figure 1A one can see the outside of the pharmaceutical form, in figure 1B the interior of a dried tablet can be seen after a digital cut of the image and in figures 1C and 1D the interior of the tablets after a digital cut of the images gathered following submersion of the tablets in ultrapure water for a period of 1225 and 1440 min, respectively.

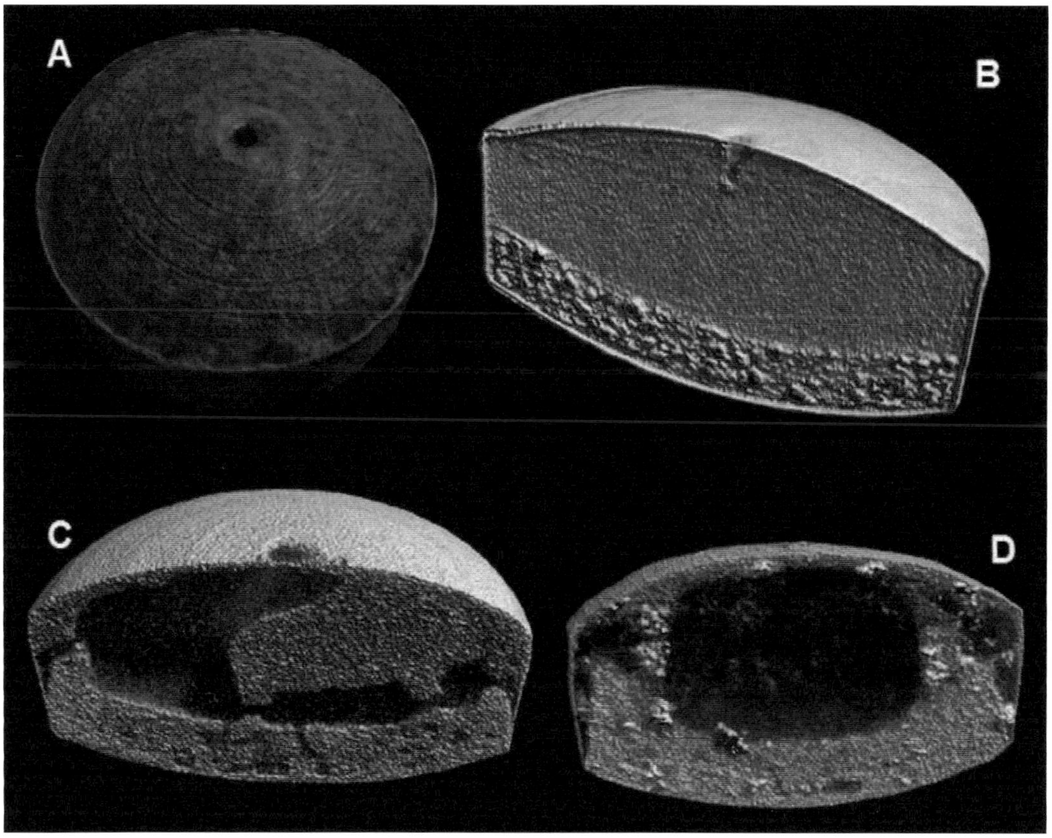

Figure 1. External part of the intact tablet of Adalat Oros™ (A), sight of its interior showing the drug (upper part) and the osmotic agent (lower part) (B), tablet of Adalat Oros™ after 1225 min (C) and 1440 min (D) of submersion in ultrapure water followed by static air drying at room temperature during 48 h prior to gathering the tomographic images.

The images in figure 1 show important details for the deformulation process of the pharmaceutical form. Through the analysis of the images in figures 1A and 1B one can obtain of morphological information of the dosage form, such as the size of the orifice through which the drug is expelled, the dimensions of the surface polymer that coats the tablet, the density and/or porosity of the drug packaged in the core of the dried pharmaceutical form, the thickness of the layer of sodium chloride crystals working as osmotic agent (osmotic pump), etc. Tomographic images like the ones displayed in figures 1C and 1D can be used for dynamic studies on the process of tablet disintegration. By calculating the initial volume of drug present in the interior of the dosage form and comparing with the volumes after each period of time under submersion in ultrapure water, it is possible to determine the rate of disintegration of the tablet. All these morphological information are susceptible to be extracted from a CT image, since materials having different densities absorb different amounts of radiation, and thus appear in the tomographic images with different shades or colours, depending on whether the images were reconstructed in shades of a single colour (8-bit) or using several colours (16-bit or 24-bit). The morphological information gathered from the dosage form under study is displayed in Table 1.

Table 1. Morphologic characteristics of a dry tablet of Adalat Oros™. The data is presented as the average of 10 measurements together with their respective standard deviations.

Morphologic characteristics	Average value	Standard deviation
Width (mm)	9.13	0.03
Height (mm)	5.3	0.5
Orifice diameter (mm)	0.48	0.04
Thickness of the coating layer (mm)	0.095	0.005
Thickness of the layer of sodium chloride crystals (mm)	1.34	0.02
Porosity of the nifedipine drug (%)	9.1	2.4
Total volume of the tablet (mm^3)	258	6
Volume of the nifedipine drug (mm^3)	164	8

Another important issue regarding deformulation is the knowledge of the elemental constituents present in the original formulation, the drug and excipients. There are many techniques of elementary analysis, but in most cases one single technique is not able to identify and/or quantify all the elemental present in the formulation, thus requiring the use of more than one analytical technique, since some of these are sensitive to organic elements, others to the presence of inorganic elements, some are destructive, some do not possess the required accuracy, and so on. In this context, we discuss the use of the technique of X-ray fluorescence (XRF) in the analysis of medicines, due to the fact that its use is relatively recent in the pharmaceutical industry, by being non-destructive, not requiring specialized technical personnel to use it, because of its relatively low cost and because it does not require any previous sample preparation. XRF, with the detectors and X-ray sources currently available on the market, can be used to quantify chemical elements present in a formulation, as long as they have an atomic number higher than that of the chemical element Magnesium (Z=12) through chemical element Fermium (Z=100), with an accuracy of approximately a few parts per million (ppm). If the fluorescence system is contained within an apparatus able to allow vacuum, it is possible to perform elemental analysis departing from the chemical element Boron (Z=5). The X-ray fluorescence analyses of tablets of Adalat Oros™ were made in both the external and internal parts of the tablets. In the inner part of the medicine, the analysis was conducted in two distinct regions, the region that concentrates the elements responsible for the osmotic pump and the region where the drug nifedipine resides. Figure 2A shows the results obtained in the elemental analysis of the external part of Adalat Oros™ tablets, with the following main chemical elements being identified: titanium ($E_{k\alpha}$ = 4.51 keV and $E_{k\beta}$ = 4.93 keV), and iron ($E_{k\alpha}$ = 6.40 keV and $E_{k\beta}$ = 7.06 keV).

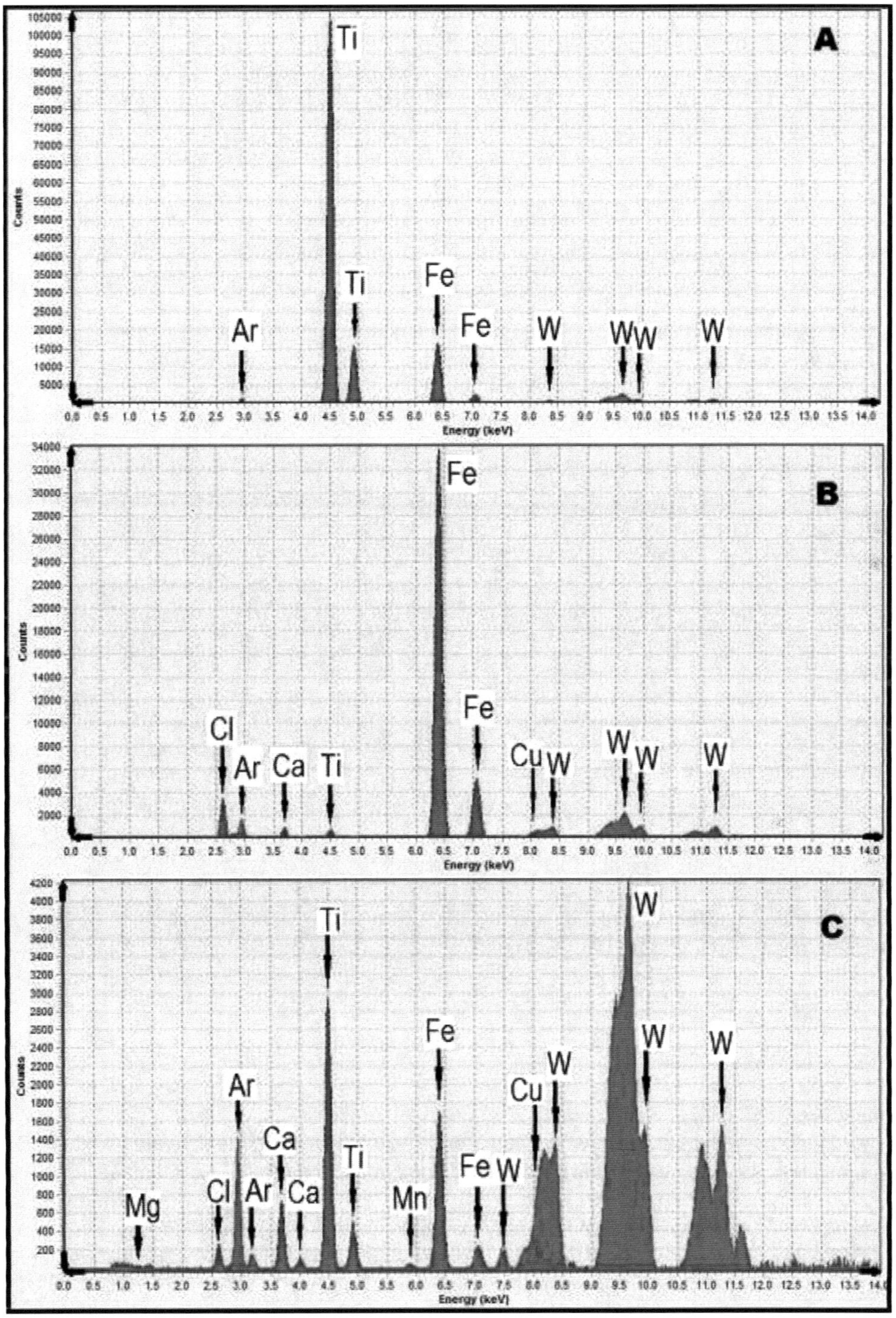

Figure 2. X-ray fluorescence spectra of the external part (A) and of the inner part of a tablet of Adalat Oros™, i.e. the region of osmotic pump (B) and the region that contains the drug nifedipine (C).

The element tungsten ($E_{L\alpha}$ = 8.40 keV, $E_{L\beta}$ = 9.67 keV, among other lines of tungsten) appears due to the target used in the X-ray source and the element argon ($E_{k\alpha}$ = 2.96 keV) appears due to its presence in the atmospheric air. Figure 2B shows the chemical elements identified in the inner part of the tablet of Adalat Oros™, of the side of the tablet where is located the osmotic pump. The main elements found in this region were: chlorine ($E_{k\alpha}$ = 2.62 keV), calcium ($E_{k\alpha}$ = 3.69 keV), titanium ($E_{k\alpha}$ = 4.51 keV), iron ($E_{k\alpha}$ = 6.40 keV and $E_{k\beta}$ = 7.06 keV) and copper ($E_{k\alpha}$ = 8.04 keV) in addition to elements argon ($E_{k\alpha}$ = 2.96 keV) and tungsten ($E_{L\alpha}$ = 8.40 keV, $E_{L\beta}$ = 9.67 keV), among other lines of tungsten). Figure 2C shows the chemical elements identified in the inner part of the tablet of Adalat Oros™, in the region that contains the drug nifedipine. The main elements found in this region were: magnesium ($E_{k\alpha}$ = 1.25 keV), chlorine ($E_{k\alpha}$ = 2.62 keV), calcium ($E_{k\alpha}$ = 3.69 keV), titanium ($E_{k\alpha}$ = 4.51 keV), manganese ($E_{k\alpha}$ = 5.89 keV), iron ($E_{k\alpha}$ = 6.40 keV and $E_{k\beta}$ = 7.06 keV) and copper ($E_{k\alpha}$ = 8.04 keV) in addition to element argon ($E_{k\alpha}$ = 2.96 keV) and tungsten ($E_{L\alpha}$ = 8.40 keV, $E_{L\beta}$ = 9.67 keV), among other lines of tungsten). Table 2 displays the amounts of each chemical element found both in the coating (external) and in the inner parts of the tablets, as percentages of the total tablet mass. According to the pharmaceutical manufacturer of Adalat Oros™ tablets, the medicine contains in addition to nifedipine the following inert components: hydroxypropyl methylcellulose (HPMC), magnesium stearate, polyethylene oxide, sodium chloride, red iron oxide, cellulose acetate, polyethylene glycol, hydroxypropyl cellulose, propylene glycol and titanium dioxide. The technique of X-ray fluorescence analysis used in this study cannot separate and quantify constituents having atomic numbers below that of magnesium element. The chemical formula of nifedipine is $C_{17}H_{18}N_2O_6$ and, since all the elements present in the molecule have an atomic number lower than Z=12, we cannot quantify them. Regarding the chemical element sodium, by having atomic number Z=11, its characteristic X-rays cannot reach the detector, and thus they do not appear in the spectra of figure 2. The chemical elements calcium, manganese and copper are contaminants, since they do not appear in the patient information leaflet provided by the pharmaceutical manufacturer. The elements iron and titanium must have originated from the red iron oxide and the titanium dioxide used in the tablet coating process. Regarding the remaining contaminants, we cannot infer about their origin, but they are probably linked to the production process. The element chlorine is present in the sodium chloride used as osmotic agent, which is the responsible for the increase in pressure inside the pharmaceutical form and concomitant expulsion of the drug (osmotic pump). The element magnesium is one of the excipients indicated by the manufacturer used in formulating the medicine. In this work, we clearly demonstrated that XRF can be used as a powerful tool for elemental analysis, and that it can be used both in qualitative identification and quantification of the elements present in a pharmaceutical formulation. It is possible to perform a quantitative analysis of light elements (Z<12), but in this case there is the need to place the experimental setup (sample, X-ray source and detector) within an evacuated chamber. The major focus of the research work undertaken was to demonstrate the potential of using the techniques of computed tomography and X-ray fluorescence analysis to perform deformulation of medicines.

4. Conclusion

By using CT images, it was possible to make a detailed physical analysis of a pharmaceutical form, including dynamic studies of the drug release process. All the morphological characteristics of the tablets were studied using three-dimensional tomographic images. The region of the tablet housing the drug, the layer used as osmotic agent and the coating characteristics including the orifice used to equilibrate the osmotic pressure and release the drug, were clearly identified. One also demonstrated the possibility of using CT for dynamic studies, by measuring the porosity. The technique of X-ray fluorescence analysis complemented the information gathered from the tomographic images, by providing identification of the elemental constituents present in the dosage form both in the coating and in the tablet matrix. It was demonstrated, how the techniques of computed tomography and X-ray fluorescence analysis can be used together to perform deformulation of a modified release dosage form.

Table 2. Concentration of the chemical elements found in the external and inner parts of tablets of Adalat Oros™, expressed in percentage of the total tablet mass.

Part of the Adalat Oros™ tablet	Chemical element	% (w/w)
Exterior – coating	Titanium	33.33 ± 0.09
	Iron	3.65 ± 0.02
Interior - side of the osmotic pump	Chlorine	10.3 ± 0.2
	Calcium	1.50 ± 0.05
	Titanium	0.54 ± 0.02
	Iron	3.80 ± 0.02
	Copper	0.024 ± 0.001
Interior - side of the drug nifedipine	Magnesium	6.4 ± 3
	Chlorine	1.5 ± 0.1
	Calcium	0.80 ± 0.03
	Titanium	1.54 ± 0.02
	Manganese	0.005 ± 0.001
	Iron	0.16 ± 0.003
	Copper	0.005 ± 0.0009

5. References

[1] Ministério da Saúde do Brasil *Lei n. 9.787 de 10 de fevereiro de 1999. Altera a Lei n. 6.360, de 23 de setembro de 1976, que dispõe sobre a vigilância sanitária, estabelece o medicamento genérico, dispõe sobre a utilização de nomes genéricos em produtos farmacêuticos e dá outras providências*. Available in http://www.planalto.gov.br/ccivil_03/leis/l9787.htm in 19/05/2014

[2] Farber L, Tardos G and Michaels J N 2003 Use of X-ray tomography to study the porosity and morphology of granules *Powder Technol.* **132** 57– 63

[3] Ozeki Y, Watanabe Y, Inoue S and Danjo K 2003 Comparison of the compression characteristics between new one-step dry-coated tablets (OSDRC) and dry-coated tablets (DC) *Int. J. Pharm.* **259** 69–77

[4] Traini D, Loreti G, Jones A S and Young P M 2008 X-ray computed microtomography for the study of modified release systems *Microsc. and Analysis* **22** 13-15

[5] Li H, Yin X, Ji J, Sun L, Shao Q, York P, Xiao T, He Y and Zhang J 2012 Microstructural investigation to the controlled release kinetics of monolith osmotic pump tablets via synchrotron radiation X-ray microtomography *Int. J. Pharm.* **427** 270-275

[6] Oliveira Jr. J M, Mangini F S, Vila M M D C and Chaud M V 2013 Using the technique of computed tomography for nondestructive analysis of pharmaceutical dosage forms *AIP Conf. Proc.* **1529** 82-85

[7] Brittain H G, Bogdanowich S J, Bugay D E, DeVincentis J, Lewen G and Newman A W 1991 Physical characterization of pharmaceutical solids *Pharm. Res.* **8** 963-973

[8] Oliveira Jr. J M and Martins A C G 2010 Non-Conventional Applications of Computerized Tomography: Analysis of Solid Dosage Forms Produced by Pharmaceutical Industry *AIP Conf. Proc.* **1245** 153-159

[9] Oliveira Jr. J M, Andréo N F, Chaud M V, Angiolucci T, Aranha N and Martins A C G 2010 Porosity measurement of solid pharmaceutical dosage forms by gamma-ray transmission *Appl. Radiat. Isot.* **68** 2223–2228

[10] Begat P, Young P M, Edge S, Kaerger S and Price R 2003 The effect of mechanical processing on surface stability of pharmaceutical powders: visualization by atomic force microscopy *J. Pharm. Sci.* **92** 611-620

[11] Bertin E P 1975 *Principles and Practice of X-Ray Spectrometric Analysis* (London: Plenum Press) p 1079

[12] Wagner M, Rostam-Khani P, Wittershagen A, Rittmeyer C, Kolbesen B O and Hoffman H 1997 Trace element determination in drugs by total-reflection X-ray fluorescence spectrometry *Spectrochim. Acta* B **52** 961-965

[13] Feldkamp L A, Davis L C and Kress J W 1984 Practical cone-beam algorithm J. *Opt. Soc. Am.* **1** 612-619

Acknowledgements

Funding by Fundação de Amparo à Pesquisa do Estado de São Paulo (FAPESP, São Paulo, Brasil) (FAPESP Ref. No. 2012/15651-4, Tomógrafo para Aplicações na Área das Ciências Farmacêutica; FAPESP Ref. No. 2013/03181-6, Project PneumoPhageKill; FAPESP Ref. No. 2012/15651-4 and FAPESP Ref. No. 2013/19300-4, Auxílio à Pesquisa - Reserva Técnica para Infra-estrutura Institucional de Pesquisa) and by FINEP – Financiadora de Estudos e Projetos (FINEP, Rio de Janeiro, Brasil) (FINEP Ref. No. 01.13.0286.00), is hereby gratefully acknowledged.

Journal of Physics: Conference Series **630** (2015) 012003 doi:10.1088/1742-6596/630/1/012003

A stopping criterion to halt iterations at the Richardson-Lucy deconvolution of radiographic images

G L Almeida[1], M I Silvani[1], E S Souza[2] and R T Lopes[2]

[1] Instituto de Engenharia Nuclear - CNEN, Caixa Postal 68550, CEP 21945-970, Rio de Janeiro - RJ, Brazil
[2] Laboratório de Instrumentação Nuclear - UFRJ, Caixa Postal 68509, CEP 21945-970, Rio de Janeiro - RJ, Brazil

E-mail: gevaldo@ien.gov.br

Abstract. Radiographic images, as any experimentally acquired ones, are affected by spoiling agents which degrade their final quality. The degradation caused by agents of systematic character, can be reduced by some kind of treatment such as an iterative deconvolution. This approach requires two parameters, namely the *system resolution* and the best *number of iterations* in order to achieve the best final image. This work proposes a novel procedure to estimate the best number of iterations, which replaces the cumbersome visual inspection by a comparison of numbers. These numbers are deduced from the image histograms, taking into account the global difference G between them for two subsequent iterations. The developed algorithm, including a Richardson-Lucy deconvolution procedure has been embodied into a Fortran program capable to plot the 1st derivative of G as the processing progresses and to stop it automatically when this derivative - within the data dispersion - reaches zero. The radiograph of a specially chosen object acquired with thermal neutrons from the Argonauta research reactor at Instituto de Engenharia Nuclear - CNEN, Rio de Janeiro, Brazil, have undergone this treatment with fair results.

Introduction

Experimentally acquired images are usually affected by agents which degrade their final quality. These spoiling agents belong to the two main branches of systematic and random families. Random uncertainties can be reduced by cutting off the associated high frequency contributions using a proper technique, such as a low-pass filtering.

An agent of pure systematic character could in principle be eliminated if one knew its features, which would then make possible to retrieve the original image. In practice this never occurs because the original matrices and the related inverted ones are, or become, ill-conditioned basically due to the arithmetic truncation precluding thus the solving of the necessary linear system of equations.

As an ideal point-source, an ideal flat surface source - i.e., a source emitting rays only perpendicularly to its surface - would produce no penumbra at all. Such a kind of source would be the ultimate one to radiography, for it, unlike the point source, would cause no image amplification. Unfortunately such a source does not exist in the real world. For neutron radiography using reactors, usual approaches to reduce the penumbra is to increase the L/D ratio or to employ collimators. The first option is not always feasible due to space constraints, while the last one yields a shadow of the collimator. Therefore, one has to face with a flat surface source emitting radiation also in other directions which do not necessarily have the same intensity as that occurring in the perpendicular one. Due to this possible anisotropy, the emission profile for every element of the source would not necessarily resemble a spherical calotte but somewhat like a bell. Borrowing a concept from X-ray diffraction, this profile can be assigned as a 3D Rocking Curve - RC, where its FWHM is a measure of the beam divergence. The cross-section of the interception of a vertical plane with the neutron beam would then define a *virtual surface source* exhibiting a certain divergence.

For a radiographic image acquired with a virtual surface source, as that available at the main port of a reactor, or by any extended source at all, the systematic error arises as an enlargement of the spatial resolution caused by the beam divergence which produces penumbrae blurring the otherwise sharp edges. This divergence has been assessed by measuring the FWHM of its related rocking curve at the main port of the Argonauta reactor where a value of 1° 16' has been found [1].

A proper way to specify this error is the *Point Spread Function* or *PSF* for short. The FWHM of this bell-shaped surface represents the resolution w of the imaging acquisition system for a given geometrical arrangement. It's important to point out that unlike the PSF, the RC doesn't change with the gap object-detector, for it is an intrinsic beam feature.

Once the *PSF* is known, the original image could theoretically be recovered, or at least improved, by applying an iterative unfolding procedure. However, the basic problem of *when* to stop the iterations remains open.

Usually, a visual inspection, which is mostly regarded as the ultimate criterion, is required to decide whether the number of performed iterations looks adequate. An excessive number not only wastes processing time, but also may degrade the image improved in previous iterations.

Since the Richardson paper in 1972 [2], the first approach employed as stopping criterion was the number of iterations. Besides χ^2 - based tests used by Lucy and others [3-5], Kolmogorov-Smirnov test and Kuhn-Tucker conditions [6] have also later on been be proposed. Another widely employed stopping criterion is the Mean Squared Error - MSE [7]. More recently Eichstädt et al [8] proposed a technique based on the well-known observation that the first iterations improve substantially the image, whereas later iterations tend to fit the noise rather than the signal.

The variety of approaches imbedded into these few examples, points out towards the difficulty to choose an ultimate criterion, a constraint arising from the intrinsically asymptotic behavior of the iteration process. Indeed, when a visual inspection is used as a tool, one cannot assert that some previous or further iterations yield poorer images than the chosen one.

In this work a novel procedure is proposed to estimate the best number of iterations, eliminating thus the cumbersomeness and processing time consumption of a pure visual inspection. No comparison is done with previous techniques but with a visual inspection, which, in spite of its subjective character, cannot be denied as the most reliable one. Its basic approach is to replace the comparison of images obtained after the iterations by a comparison of numbers, a much easier task for the human eye or a for computer program. These numbers arise from the overall difference D between the image histograms of two subsequent iterations.

As the iteration process proceeds the 1st derivative of G with regard to the number of iterations, dG/dn approaches zero until it reaches a region where the data dispersion overwhelms any further difference to zero. This is recognized by the program by checking the point where the average of the last two derivatives is smaller than half the difference between them.

When this point is reached the program halts or if instructed by the customer stores this information and continues running. The processed images around this point may be then visually analyzed - for the sake of a cross checking - in order to choose the *best* one. Usually it is hard to recognize feeble differences, if existent at all, between neighbor images and although the ultimate choice could - if desired - be done by the customer, a lot of effort is saved if a program does the hard work.

Methodology

In order to accomplish the task proposed in this work it was necessary to input an image and its spatial resolution as required by the Richardson-Lucy procedure incorporated into the developed algorithm. For this purpose a radiograph taken with a 2.23×10^5 $n.cm^{-2}.s^{-1}$ thermal neutron flux available at the main port of the *Argonauta* Reactor in *Instituto de Engenharia Nuclear - CNEN*, Brazil, has been used as original image. A neutron-sensitive imaging plate has been employed as detector and *developed* with a 0.050 mm resolution reader. A mechanical pocket chronometer, chosen for inspection due to its fine clockwork machinery has been placed directly on the imaging plate and exposed for 4 minutes.

After its *development* in the Imaging Plate Reader, the raw data is transferred to the *ImageJ* free software, which returns the pixel intensities and their coordinates. However, due to data compression, some pixel intensities and their coordinates are missed. A program developed for this purpose, *DataProc*, fills these voids with the average intensity of the neighbor pixels, weighed according to the inverse of their distances to the missing one. Once completed, the data set feeds the *Origin8* commercial software which returns the data in txt format, as required by *Deconvplot*, a developed deconvolution program embodying a Richardson-Lucy algorithm. A data checking can be done by the *DataProc* which is capable, as the *Deconvplot*, to enhance the dynamic range and plot the images on a monitor. A resume of the whole procedure is depicted in figure 1.

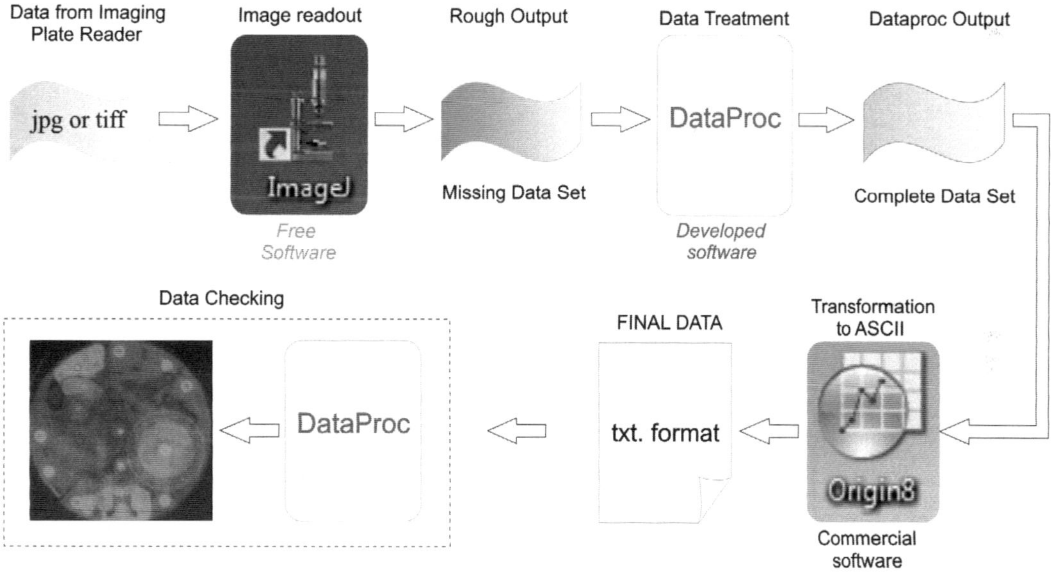

Figure 1. Prior to deconvolution the raw data arising from the Imaging Plate Reader is treated to match with the format used by the deconvolution program. For this purpose, a developed software *DataProc* is jointly used with *ImageJ* and *Origin8*.

Besides the divergence of the neutron beam, the spatial resolution depends upon the object-detector gap, as they both contribute to the penumbra that blurs the edges in the image. Since objects have a non-zero thickness, their different planes parallel to the detector exhibit different spatial resolutions.

Due to their fine granulometry, the contribution of imaging plates to the spatial resolution is not ruled by themselves but by the spot-size of the laser employed to read them. In this work, this contribution represents less than 10% of the total spatial resolution observed in the original image used as example.

Techniques have been developed to evaluate the spatial resolution [9-10], some of them based upon a *blind deconvolution* approach which do not require any information regarding the acquisition system but from the image itself. The spatial resolution for the radiograph used in this work had been early determined by another type of blind procedure [11] and is applied to carry out the deconvolution.

Figure 2 shows an actual thermal neutron radiograph of the mechanical pocket chronometer and its related histogram. It is noticeable in the histogram the existence of three pixels at the middle region, corresponding to three different gray tonality regions.

When the radiograph undergoes a deconvolution the histogram changes because an improved resolution is equivalent to an image with less blur. Since the area beneath the histogram should be maintained constant and equal to the unity - within the statistics dispersion - some grey pixels constituting the original blur have their tonalities shifted toward brighter or darker regions causing a concomitant improvement of the contrast.

This shifting process can be observed in figure 3 where the tails of the histogram after 10 iterations have been substantially raised with regard to those of the original image.

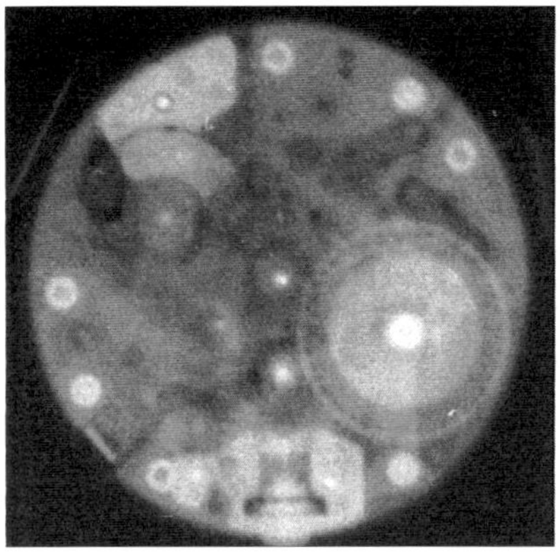

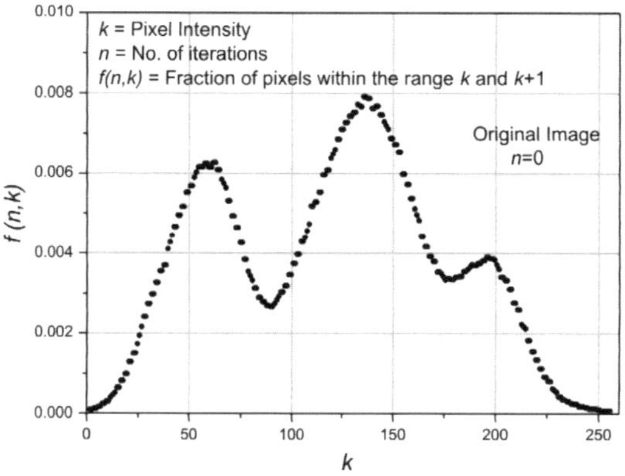

Figure 2. Thermal neutron radiograph of a pocket mechanical chronometer (left) and its related histogram (right). Bright regions correspond to high attenuating regions. The three peaks in the histogram are related to the regions where the most frequent pixel intensity occurs.

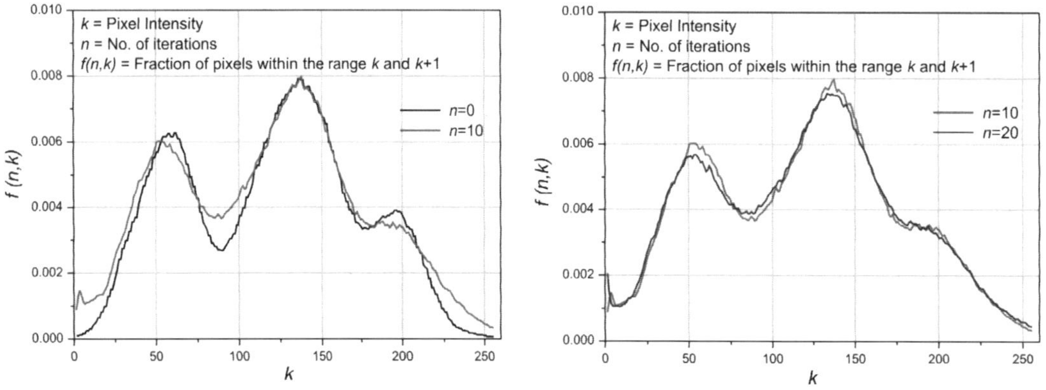

Figure 3. Impact of the iterations upon the histogram. The fraction of extreme pixels increases due to the improvement of the contrast resulting from a better resolution (left). As the iterations proceed this impact diminishes due to the asymptotic behavior of the process (right).

Journal of Physics: Conference Series **630** (2015) 012003 doi:10.1088/1742-6596/630/1/012003

As the iterations proceed the difference between two subsequent images becomes smaller, and so their histograms. This expected asymptotic behavior of the histogram is the basis for the approach proposed in this work to halt the iteration process. For this purpose, two subsequent histograms are compared by using a *global difference G(n)* between them as follows:

$$G(n) = \sum_{k=0}^{M} \left| f(n+1,k) - f(n,k) \right| \tag{1}$$

where:

n = No. of iterations.
K = Pixel intensity.
M = Maximum pixel intensity: 255 or 65,535 for jpg or tiff images respectively.
$G(n)$ = Global difference between iterations n and n+1.
$f(n,k)$ = Fraction of pixels within the interval k and k+1 for the iteration n.

The summatory in equation (1) aims at a global view of the *absolute* difference between two subsequent images avoiding thus any particular region. The use of an absolute difference precludes that deviations of different signs eliminate each other.

Figure 4 depicts the behavior of *G(n)* and its 1st derivative as the iterations progresses. Although both graphics provide the same kind of information, the derivative is more convenient and suitable to be handled by a computer program. Indeed, whereas the lower limit of *G(n)* is unknown, its asymptotic trend indicates that its 1st derivative somewhere - within the data dispersion - should reach zero.

Hence, it becomes easier for a program to choose the proper graphic scale since the upper level is previously known, and to plot *G(n)* as the subsequent iterations proceed.

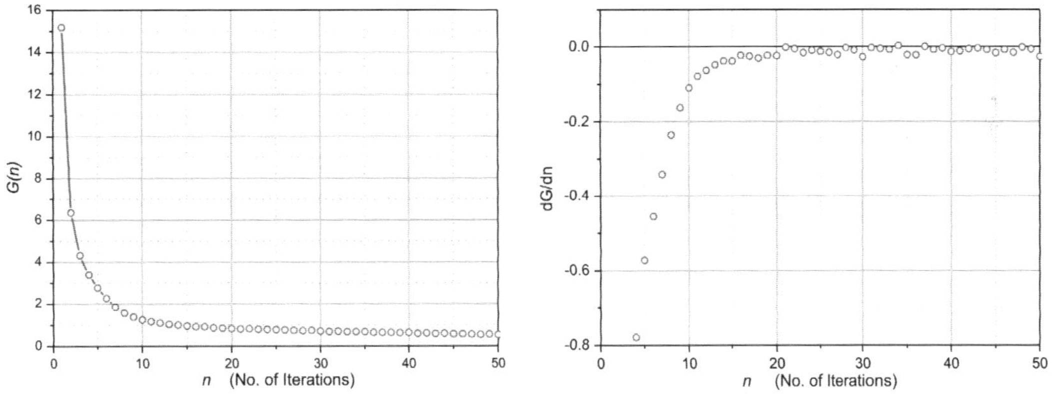

Figure 4. The *global difference G(n)* exhibits a somewhat asymptotic trend (left). The 1st derivative (right) is a more suitable approach for a computer program to analyze this behavior because its maximum is known in advance.

In order to get a closer view at the asymptotic region of the graph, a zoom can be performed by re-scaling the vertical axis. This is done by defining the first iteration that will be plotted, a procedure that will cut the points representing *dG/dn* for lower iterations, as shown in figure 4, where the lowest iteration plotted is 4.

Such a stretching of the vertical axis renders the data dispersion at the asymptotic region more visible. Due to the asymptotic behavior of *dG/dn* and its data dispersion, it somewhere would intercept or surpass the zero level. However, as a large dispersion would cause an early interception resulting in an insufficient number of iterations, it is necessary to smooth the data in order to employ a *zero level interception* as a criterion to stop the program.

The data is then smoothed by averaging the global difference as follows:

$$G_s(n) = \sum_{j=1}^{M/p} \left| H(n+1, j) - H(n, j) \right| \tag{2}$$

$$H(n, j) = p^{-1} \sum_{k=A}^{B} f(n,k) \tag{3}$$

$$\boldsymbol{A = (j\text{-}1).p} \tag{4}$$

$$\boldsymbol{B = j.p} \tag{5}$$

where:

$G_s(n)$ = Smoothed global difference between iterations *n* and *n*+1.
p = Number of points within the domain 0-M used for the smoothing.

and all other parameters and variables as formerly defined.

The 1st derivatives of the smoothed global difference ***$G_s(n)$*** for several ***p***-values are shown in figure 5, where it is possible to observe an effective smoothing of the raw data. It can be observed as well that no further improvement of the dispersion is achieved with ***p***-values beyond 8.

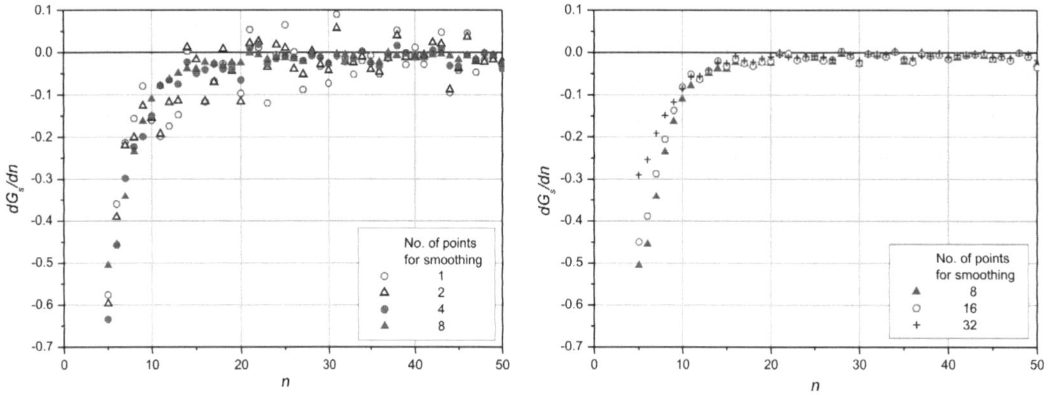

Figure 5. The first derivative of the global difference $G_s(n)$ smoothed by averaging with different number of points, shows that no noticeable improvement of data dispersion can be achieved beyond 8 points. This parameter can be chosen by the customer for the plotting as the processing progresses.

The above mentioned interception criterion can be properly applied by specifying that the average value of the two last derivatives should be greater (smaller in absolute value, for the derivatives are negative) than the half of the differences between them. This approach - which assures that the dispersion is taking into account - is equivalent to require that the last derivative should itself surpass zero, disregarding its previous value, as follows,

$$0.5.\{[dG/dn]_{n-1} + [dG/dn]_n\} \geq 0.5.\{[dG/dn]_{n-1} - [dG/dn]_n\} \tag{6}$$

$$[dG/dn]_n \geq 0 \tag{7}$$

During the processing the program displays continuously on the monitor a graph of *dG/dn* versus *n* which is reproduced in figure 6, solely to show how the display looks like. After each iteration the deconvoluted image and its histogram is recorded for further analysis, as well as the graph *dG/dn x n* when the maximal number of iterations - as previously specified by the customer - has been reached or the program has been automatically interrupted by the embedded criterion.

The 896 x 847 pixels jpg-image used in this work, has been deconvoluted with a formerly determined *point spread function* having a FWHM of 8.5 pixels, equivalent to 0.54 mm. Further details can be found elsewhere [11. The processing time required for each iteration depends upon the image size and its resolution, and in this specific case it was circa 100 seconds with a 3.6 GHz processor.

Although in this work the Richardson-Lucy procedure has been employed, images arising from any iterative process can be analyzed under the same approach.

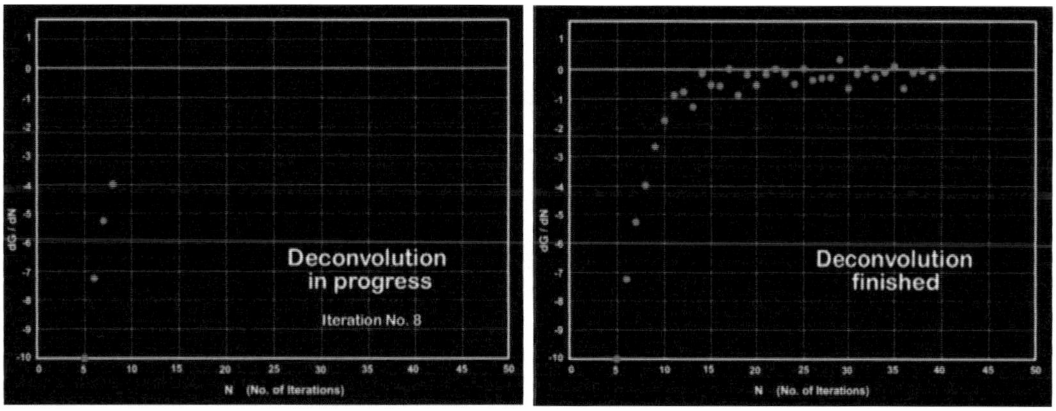

Figure 6. Example of an actual output of the program *Deconvplot* during the processing (left) and after its completion (right). Iterations can be stopped at any time by the customer or when a specified criterion has been fulfilled. All images are automatically stored in sequentially numbered files.

3. Results

A set of deconvoluted neutron radiographs of the mechanical chronometer including the original image and a conventional photograph where the metallic case has been removed is shown in figure 6. It is possible to observe a fair correlation and consistency between the quality of the images and the values of *dG/dn* of the graph. Indeed, beyond 22 iterations, where *dG/dn* surpasses zero it is hard to observe a substantial further improvement of the images.

One could eventually notice a slight improvement of the images deconvoluted with 30 and 50 iterations, but some bright spots just beginning to appear in the initial images become better defined. The acceptance of these artifacts - which could not be forecast by the *dG/dn* behavior - is a matter of trade-off between the overall image quality and their presence.

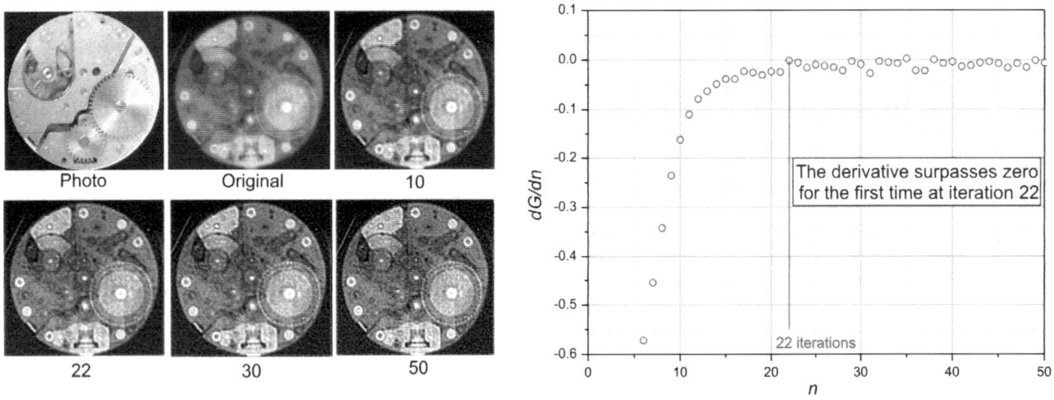

Figure 7. Original and deconvoluted neutron radiographs of a pocket chronometer. Labels refer to the number of iterations. A conventional photo with the case removed is also shown for reference. The program stopped at iteration 22 but it has been instructed to continue for the sake of cross-checking.

Conclusion

This work proposes a procedure to evaluate the best number of runs for an iterative process aiming at the improvement of images, eliminating then the cumbersomeness and time consumption of a visual inspection. Its basic idea is to replace the comparison of images by a comparison of numbers arising from the global difference G between subsequent image histograms. When the 1st derivative of G with regard to the iteration number, dG/dn, reaches zero - within the data dispersion - it means that the histograms and their related images are not substantially changing anymore. Beyond this point further iterations would not produce any additional improvement and the iterative process is interrupted.

The developed technique incorporates a Richardson-Lucy algorithm into a Fortran program which plots the behaviour of the dG/dn on a monitor as the iterations proceed and stop it automatically when a given criterion chosen by the customer is reached. This criterion can be the maximum number of iterations or the point where dG/dn surpasses zero. The soundness and performance of the technique has been verified by treating the neutron radiograph of a pocket mechanical chronometer, with fair results. In this work the Richardson-Lucy algorithm has been employed as unfolding procedure, but images improved by any iterative process can be analyzed under the same approach.

References

[1] Almeida G L, Silvani M I, Furieri C A A 2005 Evaluation of the Divergence of a Thermal Neutron Beam using a Position Sensitive Detector *Brazilian Journal of Physics* **35** No. 3B 771-74

[2] Richardson W H 1972 Bayesian-Based Iterative Method of Image Restoration *Journal of the Optical Society of America*, **62** No.1 55-59

[3] Lucy L B 1974 An iterative technique for the rectification of observed distributions *The Astronomical Journal* **19** No. 6 745-54

[4] Veklerov E, Llacer J 1987 Stopping Rule for the MLE Algorithm Based on Statistical Hypothesis Testing *IEEE Transactions on Medical Imaging*, **6** No.4 313-19

[5] Veklerov E, Llacer J and Hoffman, E J 1988 MLE reconstruction of a brain phantom using a Monte Carlo transition matrix and a statistical stopping rule *IEEE Transaction on Nuclear Science* **35** No. 1 603-7

[6] Hebert T, Leahy R and Singh M 1988 Fast MLE for SPECT using an intermediate polar representation and a stopping criterion *IEEE Transactions on Nuclear Science* **35** No. 1 615-19

[7] Laasmaa M, Vendelin M and Peterson P 2011 Application of regularized Richardson-Lucy algorithm for confocal microscopy images *Journal of Muicroscopy* **243** No.2 124-140
[8] Eichstädt S, Schmäling F, Wübbeler G, Krüger U and Elster C 2013 *16th International Congress of Metrology* 14005
[9] Marchand E W 1964 Derivation of the Point Spread Function from the Line Spread Function *Journal of the Optical Society of America*, **54** No.7 915-19
[10] Kundur D and Hatzinakos D 1996 Blind Image Deconvolution *IEEE Signal Processing Magazine*, **13** No.3 43-64
[11] Almeida G L and Silvani M I 2012 A novel algorithm for blind deconvolution applied to the improvement of radiographic images *XXXV Brazilian Workshop on Nuclear Physics AIP Conference Proceedings, São Sebastião, São Paulo Brazil* **1529** 95-99

Differences in iron concentration in whole blood of animal models using NAA

V Bahovschi , C B Zamboni, L F F Lopes Silva, S Metairon, I M M A Medeiros
Instituto de Pesquisas Energéticas e Nucleares (IPEN / CNEN - SP), Av. Professor Lineu Prestes 2242, 05508-000 São Paulo, SP- Brazil

E-mail: vanessasantos50@hotmail.com

Abstract. In this study Neutron Activation Analysis technique (NAA) was applied to determine Fe concentration in whole blood samples of several animal models such as: mice (*Mus musculus*), Golden Hamster (*Mesocricetus auratus*), Wistar rats, Albinic Rabbits of New Zealand, Golden Retriever dogs and Crioula breed horses. These results were compared with human whole blood estimation to check their similarities.

Introduction

In the health field several animal models are used for medical investigations, mainly those that involve testing of new medicines, vaccines, antibiotics, organ transplantation as well as medical diagnostic studies. While small-sized animal models such as: mice, hamsters, rats and rabbits are very convenient to perform medical investigation due to their low cost, easy handling and medico legal implications, medium and large sized animals are selected for medical diagnostic studies. The animal of choice is mainly Golden Retriever dogs, because of their physiological similarities with humans while horses (Crioula breed) are used for antivenom production in Brazil. In this study Neutron Activation Analysis technique (NAA) was applied for the determination of iron levels in the whole blood of these animal models. Iron plays important functions in blood; it acts as an indicator of a variety of anomalies and it is fundamental for the cardiac functioning and immune system [1,2]. An important aspect to be considered during these investigations (related to iron toxicity) is the need to check for similarities between the animal's and human's blood. For the development of this investigation, IAEA-A13 (Blood Animal) certified reference material was first analyzed for analytical quality control. After that, the analyses of whole blood samples of the selected animals were performed.

Experimental Procedure

The samples came from research centers in São Paulo city (Butantan and Bioscience Institute). The whole blood samples were collected from mice (*Mus musculus*), Golden Hamster (*Mesocricetus auratus*), Wistar rats, Albinic Rabbits of New Zealand, Golden Retriever dogs and Crioula breed horses. About 1-2 mL of whole blood was collected from each animal model. Aliquots of 300 μL were collect from mice, hamsters and rats and 500μL from rabbits, dogs and horses, in duplicate when biological material was available. Each aliquot of blood was then transferred to a plastic cylinder and irradiated together with the standard solution in the nuclear reactor IEA-R1 (4.0 - 4.5 MW, pool type) at IPEN (SP, Brazil). Samples of 300 µL were irradiated for 8 hours and samples of 500µL for 4 hours. Samples were gamma counted for 3 to 6 hours after a decay time of several days. A γ - spectrometer system with a semiconductor detector (HPGe) connected to an ADCAM multichannel analyzer were used to measure the induced gamma-ray activity. The Fe concentrations were obtained by using in- house software.

Results and Discussion

The results for the reference material are presented in table 1. The obtained Z-score values [3] indicated that the present results are satisfactory considering the range of certified data at the 95% ($|Z| \leq 2$) confidence level.

The Fe concentration results are summarized in table 2. To show in more detail the results in whole blood, figures 1 through 6 show all models investigated. For checking the similarities the mean values were compared to the human being estimation [4]. This comparison is shown in figure 7, where the indicative interval related to human being (in a confidence interval of 68% and 95%) was also included. The significance of differences between animal models and humans were assessed by Student's t-test. The level of statistical significance was taken as value of $p < 0.05$.

Table 1. Element concentrations obtained in the analysis of IAEA-A13 certified reference material

Element	Mean ±1 SD	Z-score
Ca, $mgkg^{-1}$	291 ± 29 286 ± 54^{CV}	0.09
Fe, mg kg^{-1}	2469 ± 138 2400 ± 144^{CV}	0.48
Mg, mg kg^{-1}	102 ± 33 99 ± 29^{CV}	0.11
K, g kg^{-1}	2.54 ± 0.24 2.50 ± 0.35^{CV}	0.10
Na, g kg^{-1}	12.90 ± 0.57 12.60 ± 1.01^{CV}	0.30
Zn, mg kg^{-1}	13.72 ± 1.21 13.00 ± 1.04^{CV}	0.21

^{cv}: certified values

Table 2. The Concentration of Fe (mgL^{-1}) in whole blood samples of animal models and Humans.

Animal model (*n*)	Mean	±1 SD	Minimum Value	Maximum Value	Range (95%)

Mice *(12)*	187	13	163	204	161 - 213
Rats *(15)*	393	42	297	492	309 - 477
Hamsters *(12)*	209	40	149	265	129 - 289
Rabbits *(14)*	363	29	163	380	305-421 301-530 [b]
Dogs *(9)*	585	17	441	605	551 - 619
Horses *(17)*	498	18	399	608	462 - 534 158-190 [c]
Humans [a]	385	28	199	474	329 – 441

[n] number of samples
[a] ref [4]
[b] ref [5]
[c] ref [6]

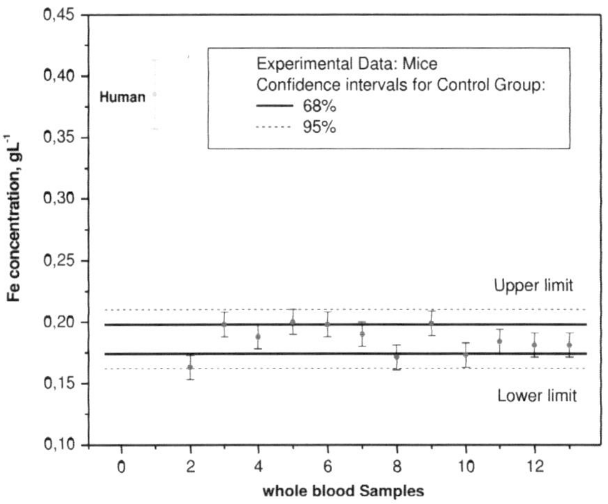

Figure 1. Fe concentration in whole blood samples of the Mice (*Mus musculus*). The human being estimation [4] was included for comparison.

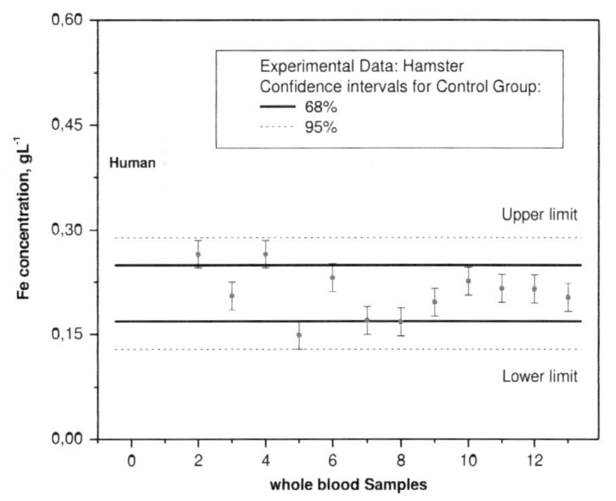

Figure 2. Fe concentration in whole blood samples of the Golden Hamster (*Mesocricetus auratus*). The human being estimation [4] was included for comparison.

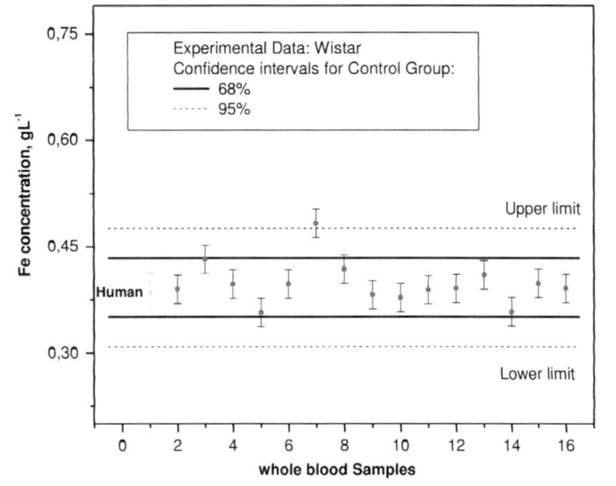

Figure 3. Fe concentration in whole blood samples of the Wistar rats. The human being estimation [4] was included for comparison.

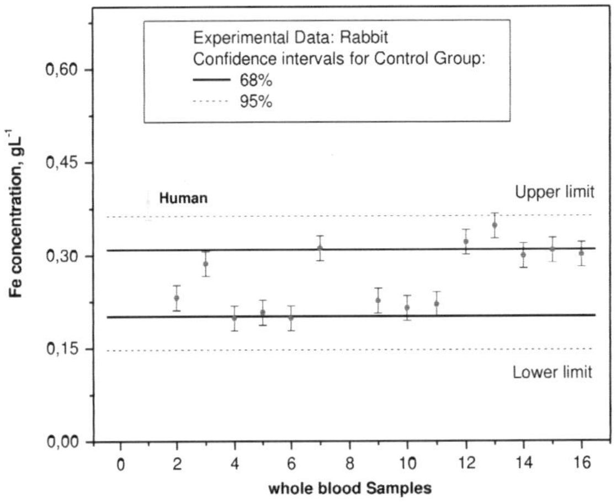

Figure 4. Fe concentration in whole blood samples of the Albinic Rabbits of New Zealand. The human being estimation [4] was included for comparison.

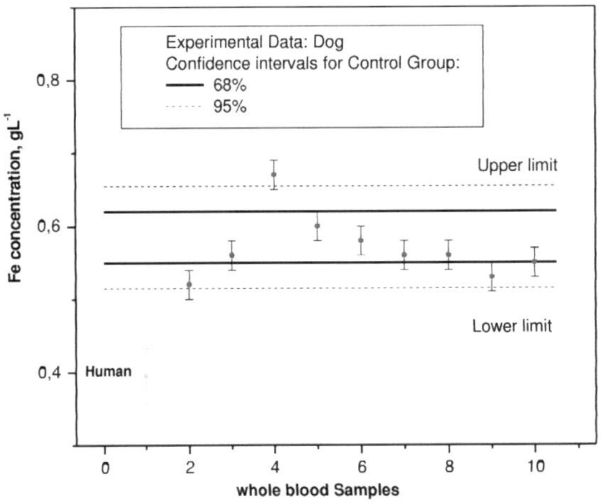

Figure 5. Fe concentration in whole blood samples of the Golden Retriever dogs. The human being estimation [4] was included for comparison.

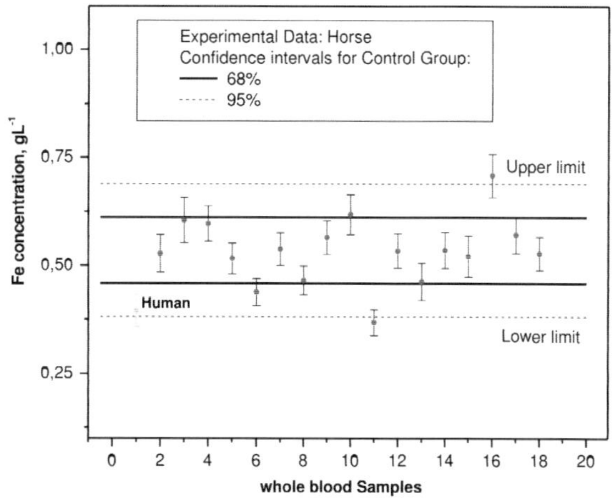

Figure 6. Fe concentration in whole blood samples of the Crioula breed horses. The human being estimation [4] was included for comparison.

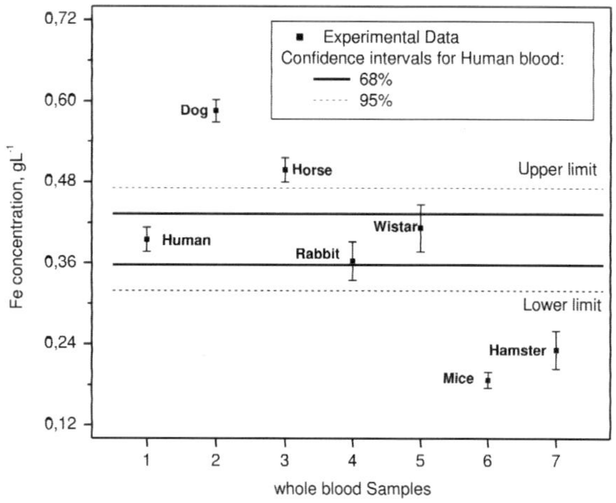

Figure 7. Fe concentration in whole blood samples of the animal models. The human being estimation [4] was included for comparison.

The results of Fe in whole blood of mice, hamster, rats and dogs are the first reference values. For rabbits there is in agreement with data from EDXRF analysis, but for horses the range determined by NAA is statistically different [6]. Regarding to the similarities with human being whole blood estimation there is a good agreement with the data from Wistar rats, Albinic Rabbits of New Zealand

and Crioula horses. However, the high Fe levels in Golden Retriever dogs, lower levels in mice (*Mus musculus*) and in Golden Hamster (*Mesocricetus auratus*) suggest that these experimental animal models are not convenient for researches even though the knowledge of Fe concentration in blood is necessary ($p < 0.05$).

Conclusion

In clinical trials where the knowledge of Fe concentration is necessary, this comparative study presents data that can help researches evaluate and conduct an appropriate medical investigation by choosing the convenient animal model. In addition, these data emphasis some differences in Fe concentration blood for mice (*Mus musculus*), Golden Hamster (*Mesocricetus auratus*), Golden Retriever dogs and Humans.

References

[1] Mahan L K, Stump S E 2010 *Krause – alimentos, nutrição e dietoterapia* 12, ed Elsevier (Rio de Janeiro, R.J.)
[2] McArdle D W, Katch I F, Katch L V 2008 *Fisiologia do exercício – Energia, Nutrição e Desempenho Humano* 6, ed Guanabara Koogan (Rio de Janeiro, R.J.)
[3] Bode P, Van Dijk C P Operational management of results in INAA utilizing a versatile system of control charts 1997 *J. Radioanal. Nucl. Chem.* **215** 87-94
[4] Lins P S, Zamboni C B, Azevedo M R A, Medeiros I M M A 2005 *Proc. of Int. Nuclear Atlantic Conf. INAC (Rio de Janeiro)*
[5] Avelar A C, Ferreira W M 2012 Composição elementar do sangue, figado, tibia e musculo de láparos de 30 dias Nova Zelândia variedade branca *Rev. Acad. Ciênc. Agrár. Ambient* **10**(2) 163-169
[6] Redígolo M M, Aguiar R O, Zamboni C B Sato I M 2013 Determination of reference interval values for inorganic elements in whole blood samples of humans and laboratory animals by X-ray fluorescence spectrometry *J. Radioanal. Nucl. Chem.* **297** 463-467

Acknowledgments

We would like to acknowledge for financial support of the Conselho Nacional de Desenvolvimento Científico e Tecnológico (CNPq).

Determination of inorganic elements in blood of mice immunized with Bothrops Snake venom using XRF and NAA

L F F Lopes da Silva[1], C B Zamboni [1], V Bahovschi[1], S Metairon[1], M F Suzuki[1], O A Sant'Anna[2], M A Rizzutto [3]

[1]Instituto de Pesquisas Energéticas e Nucleares (IPEN/CNEN-SP)
Av. Professor Lineu Prestes 2242 - 05508-000 São Paulo, SP- Brasil.
[2] Instituto Butantan
Av. Dr.Vital Brasil 1500 - 05503-900 São Paulo, SP – Brasil.
[3] Instituto de Física - Universidade de São Paulo, IFUSP, SP
Rua: do Matão Travessa R 187 - 05508-090, São Paulo, SP, Brasil.

E-mail: lais_de_fatima@ig.com.br

Abstract. In this work, mice genetically modified [H_{III} line] were immunized against different *Bothrops* snake venoms to produce anti-*Bothrops* serum (antivenom). The Neutron Activation Analysis (NAA) and Energy Dispersive X-Ray Fluorescence (EDXRF) techniques were used to evaluate Ca and Fe concentrations in blood of these immunized mice in order to establish a potential correlation between both phenotypes: antibody response and blood constituents after *Bothrops* venom administration. The results were compared with the control group (mice not immunized) and with human being estimative. These data are important for clinical screening of patients submitted to immunological therapy as well as the understanding of the envenoming mechanisms.

Introduction

In the last years we have performed investigations for checking the behaviour of ions and metal in blood of animal models as well as in antivenoms (such as, *anti-Bothrops, anti- Diphtheria, anti-Rabies and anti-Tetanus*) using Neutron Activation Analysis (NAA) technique [1-5]. The need to perform this kind of investigation in blood and also in antivenom is due to the fact that small changes (mainly excess) can lead to intoxication process to the victims of poisonous animals, mainly snakes. The main advantage to use NAA technique is the viability to analyze simultaneous elements in whole blood (for example: Br, Ca, Cl, K, Mg, Na, S) using small quantities (0.1 mL), an important condition when this biological material is scarce [1]. However, for Fe determination the amount of blood (0.1 mL) requires long neutron irradiation time (many hours) to obtain adequate uncertainty. This motivated us to check the viability of using a Portable X-Ray Fluorescence Spectrometry (PXRFS) for Fe determination in blood. Recently this portable spectrometer was used for Ca and Fe determination in our clinical research with success [6]. Now, we intend to quantify these inorganic elements in blood samples of mice immunized with a mix of different *Bothrops (B.)* species snake venoms (*B. jararaca, B. jararacussu, B. alternatus, B. moogeni, B. taeniata* and *B. neuwiedi)* using PXRFS and to compare these results with NAA procedure.

Experimental Procedure

For this study, biological samples of 20 genetically modified mice (H_{III}) were used. The blood samples were classified as a control group (CG) and immunized groups (IG) according to the mix of venoms. Several types of *Bothrops* venom *(B.)* obtained in the North of Brazil are used in the immunization procedure. The following classification was adopted:

CG: control group; mice genetically modified (H_{III}), but not immunized.
IG-1: *B.atrox* from the Brazilian Amazon in Rio Negro;
IG-2: *B.atrox* from Maranhão;
IG-3: *B. jararaca, B. jararacussu, B. alternatus, B. moogeni and B. neuwiedi;*
IG-4: *B. jararaca, B. atrox* from Rio Negro, *B. alternatus, B. moogeni and B. neuwiedi;*
IG-5: *B. taeniata* and Viperideae from Amazon;

The blood from retro–orbital venous plexus was collected from each adult male mice. A small amount of blood, 0.1 mL (±0.5%), was transferred to the filter paper and storage without refrigeration until to be used. The same procedure was used for certified standard solution preparation of Ca and Fe (Inorganic Ventures). More details about these steps are present in reference [1].

The NAA measurements were performed in the IEA-R1 nuclear reactor (4.0-4.5MW, pool type, at IPEN). Each blood sample and standards solutions of Ca and Fe, were irradiated in a thermal neutron flux and gamma counted using HPGe detector (GEM-60195, FWHM=1.92 keV for 1.33 MeV of ^{60}Co) connected to MCA (ORTEC-919E). To determine Ca, the irradiated time was 300s and the counted time was 600s. To determine Fe, the irradiated time was 8hs and the counted time was 12hs. All the quantitative analyses were performed using *ATIVAÇÃO* software and the quantitative values were determined using the certified concentration values and uncertainties of Ca (10.05 ± 0.44 μg/mL) and Fe (10.03 ± 0.54 μg/mL).

The XRF measurements were performed using a Portable X-ray Spectrometer (Amptek®) with Ag X-ray tube. The characteristic fluorescent X-rays emitted from the samples (K and L lines) were measured with a Si Drift detector (25 mm^2 x 500μm / 0.5 mil) with Be window (1.5"). The same blood sample investigated by NAA was analysed by XRF technique. All the blood samples as well as the certified standards solutions of Ca and Fe (Inorganic Ventures) were irradiated for 300s using 30 kV and 5 μA excitation. The spectra analysis was performed using WINAXIL software program and the quantitative values were determined using the same certified concentration values and uncertainties of Ca and Fe.

Results and Discussion

In figures 1 and 2 are presented the gamma ray spectrum of whole blood samples obtained by NAA using HPGe detector and in figure 3 using the PXRFS system. The element concentrations (mean value) and uncertainly propagation (combined uncertainty including contribution from all possible sources) using NAA and XRF techniques are shown in table 1. The Ca and Fe range for the control group and human being were included in this table for comparison [6, 7]. To illustrate, Ca and Fe concentrations data from XRF analyses are shown in figure 4 and 5, respectively. In these figures the confidence interval (considering ±1 and ± 2SD) for the control group (mice not immunized) was included for comparison.

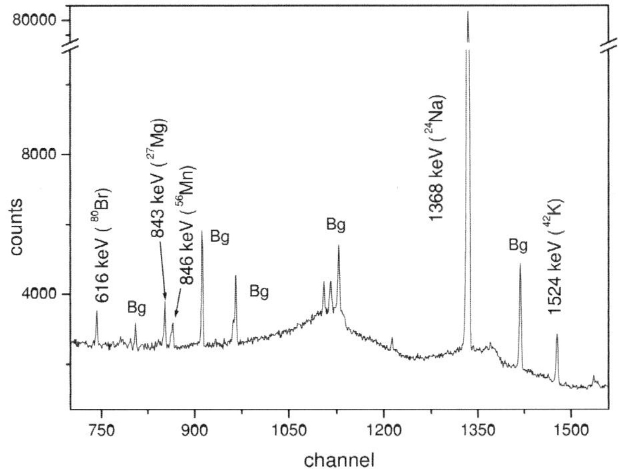

Figure 1. Partial gamma ray spectrum of whole blood obtained by NAA using HPGe detector

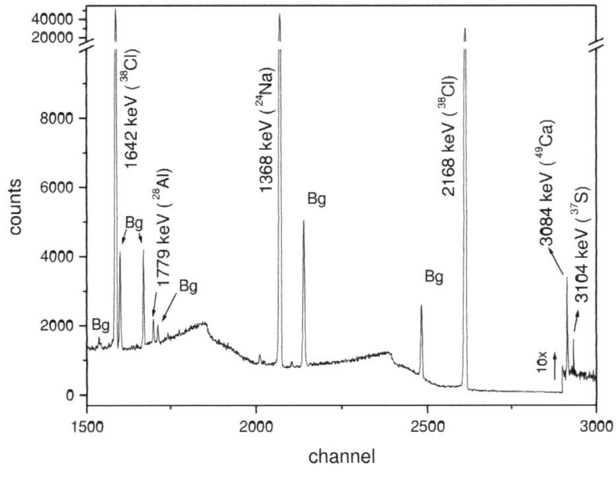

Figure 2. Partial gamma ray spectrum of whole blood obtained by NAA using HPGe detector

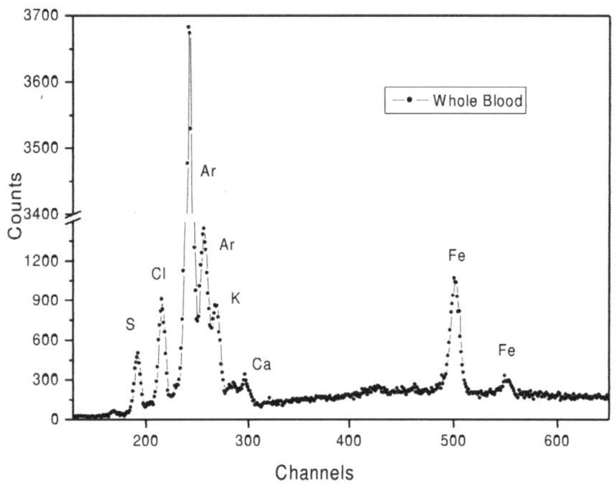

Figure 3. Spectrum of whole blood obtained by ED-XRF using PXRFS

Table 1. Element content (mean value ± uncertainty) of whole blood mice samples. The confidence interval (range) for the control group and human being were included for comparison

Groups (n)	Ca NAA	Ca XRF	Fe NAA	Fe XRF
		mgL^{-1}		
IG-1 (2)	240 ± 14	222 ± 17	390 ± 23	371 ± 26
IG-2 (3)	188 ± 11	191 ± 15	377 ± 23	362 ± 25
IG-3 (3)	201 ± 12	179 ± 14	483 ± 29	417 ± 29
IG-4 (3)	103 ± 6	118 ± 9	360 ± 24	371 ± 26
IG-5 (3)	177 ± 11	182 ± 14	354 ± 21	336 ± 23
CG [range]* (6)	[73 -355]	[48 - 356]	[304 - 477]	[268 - 376]
Human [range]*	[98 - 482][a]	nd	[277 - 513][b]	nd

n: number of samples
*confidence interval of 95%
[a] [7]
[b] [8]
nd: not determined

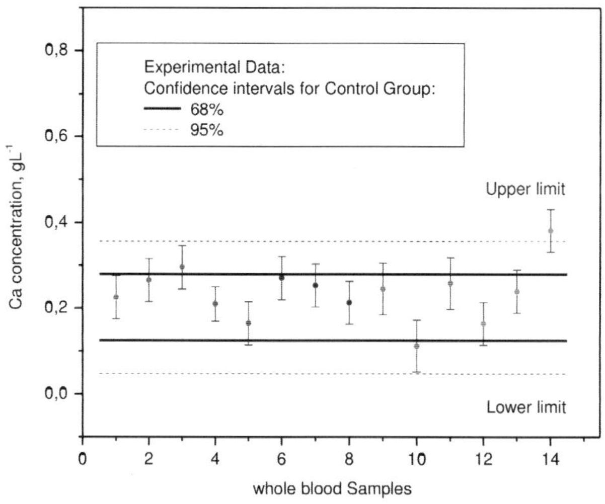

Figure 4. Ca concentration value and uncertainty in whole blood samples using PXRFS

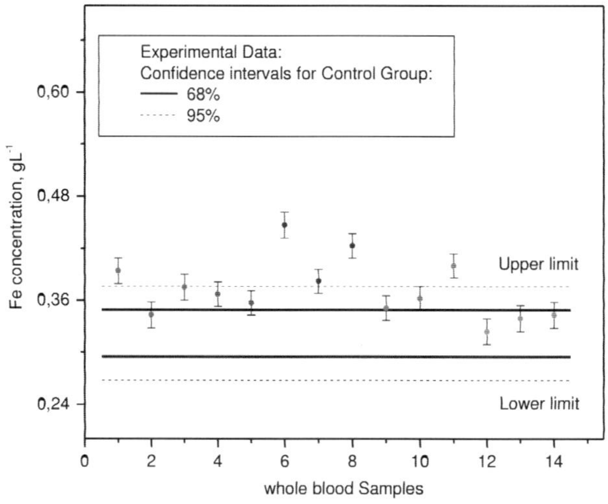

Figure 5. Fe concentration value and uncertainty in whole blood samples using PXRFS

According to table 1 the Ca data in immunized mice groups are in agreement with control group and human being estimative, while for Fe the data are close or above the upper level (considering the confidence interval of 95%). Related to the techniques (XRF and NAA) both shown to be appropriate for Ca and Fe blood analysis.

Conclusion

These data are important for clinical screening of patients were submitted to immunological therapy as well as the understanding of the envenoming mechanisms. Furthermore, considering that, in underdeveloped countries or isolated areas, when snakebites are a public health problem, as occurs in Brazil. The presented data can be also useful as markers for clinical diagnosis in snake envenoming episodes.

References

[1] Zamboni C B, Metairon S, Suzuki M F, Furtado M F, Sant'Anna O A, Tambourgi D V 2009 *J. Radioanal. Nucl. Chem.* **282** 37

[2] Baptista T S, Zamboni C B, Medeiros J A G, Freitas M G, Marcelino J R 2009 *Proc. of Int. Nuclear Atlantic Conf. INAC* (*Rio de Janeiro*)

[3] Baptista T S, Zamboni C B, Kovacs L, M G Freitas, J R Marcelino 2010 *AIP Conf. Proc. (Brazil)* vol. 1265 (Melville, New York/AIP Publishing) p 407

[4] Baptista T S, Zamboni C B, Marcelino J R 2011 *AIP Conf. Proc. (Brazil)* vol. 1351 (Melville, New York/AIP Publishing) p 332

[5] Baptista T S, Redígolo M M, Zamboni C B, Sato I M, Marcelino J R 2012 *J. Radioanal. Nucl. Chem.* **291** 399

[6] Metairon S, Zamboni C B, Suzuki M F, Rizzutto M A, Bueno Jr C R, Andrade T O, Landini V, Cangussu E B, Caetano H V A, Zatz M 2014 *http://www.globalbiotechcongress.com/home.htm*

[7] Zamboni C B, Oliveira L C, Kovacs L, Metairon S 2011 *J. Radioanal. Nucl. Chem.* **291** 389

[8] Lins P S, Zamboni C B, Azevedo M R A, Medeiros I M M A 2005 *Proc. of Int. Nuclear Atlantic Conf. INAC* (*Rio de Janeiro*)

Acknowledgments

We would like to acknowledge for financial support of the Conselho Nacional de Desenvolvimento Científico e Tecnológico (CNPq) and IPEN/CNEN-SP.

Investigation of Fe and Ca in non-stimulated human saliva using NAA

J A G de Medeiros[1,2], C B Zamboni[1], L Kovacs[1], H R Lewgoy[1]
[1]Instituto de Pesquisas Energéticas e Nucleares, IPEN - CNEN/SP
Av. Professor Lineu Prestes 2242, 05508-000 São Paulo, SP, Brasil
[2]Universidade Cidade de São Paulo – UNICID – SP
Rua Cesario Galeno, 448/475, 03071-000 – Tatuapé, São Paulo, SP, Brasil

e-mail: jose.medeiros@unicid.edu.br

Abstract. In this study we investigated non-stimulated human whole saliva of healthy subjects and patients with periodontal disease using Neutron Activation Analysis technique (NAA). The measurements were performed in the IEA-R1 nuclear reactor at IPEN-CNEN/SP. We found considerable metabolic changes mainly in Fe and Ca concentration in whole saliva of periodontal patients. These data are useful for identifying or preventing this oral disease in the Brazilian population.

1. Introduction

The use of non-stimulated human whole saliva has been presenting a significant progress in clinical testing for diagnostics of several diseases (cancer, diabetes, autoimmune disorders, cardiovascular problems, viral and bacterial infections and also in psychiatric therapy) [1]. The major advantage for using saliva in diagnosis, comparatively to serum or plasma, is the easy access and facility of collection (non - invasive) as well as the quantities that can be collected (a healthy adult can produce about 0.4 mL saliva/min). According to the latest data on oral health in Brazil (INCA- Incidence of Cancer in Brazil, 2014) [2] the prevalence of oral disease has increased in the last years and varies by regions, gender and age groups: while in the South and Southeast regions there are prevalence of oral cancer for adult men (52%) compared with all cases in the country, the North Region has higher proportion of periodontal disease index of 66 % for the age group 15 to 19 years compared with the national average of 54%. The periodontal disease begins with the accumulation and mineralization of the plaque and can lead to tooth loss [3]. Recent investigations suggest that some inorganic elements, mainly Ca a S contents in non – stimulated saliva can be biomarkers of this disease [4,5]. In this study we intend to compare the behaviour of Fe and Ca in whole saliva of healthy subjects (control group) and periodontal patients using Neutron Activation Analysis (NAA) technique. We aim to stimulate the biochemical analysis of whole saliva as a diagnostic fluid, mainly for periodontal disease that has high incidence in Brazilian subjects.

2. Experimental Procedure

The samples of non-stimulated whole saliva were collected (in duplicate) from 38 healthy subjects and 12 subjects with periodontal disease, male adults, and range from 39 to 57 years old. All participants were inhabitants of São Paulo city, where there is prevalence of this oral disease [2]. The donors made a prior rinse with distilled water, before the collection. The collection was performed at the same time of the day, near lunch time (at least with an interval of two hours fasting), spontaneously (without stimulation) directly in sterilized plastic containers in a dental office by a dentist. About 3 - 4mL of saliva was collected and lyophilized. Samples (400 ± 5% μL) and standards (aliquots of standard solutions) were irradiated for 4 h in the nuclear reactor (IEA-R1, 4.0-4.5MW, pool type) at IPEN, with a thermal neutron flux that ranged from 3 to 5.10^{12} n cm^{-2} s^{-1}. After a decay time of several days, a gamma counting of 8 hours was used to determine ^{47}Ca ($T_{1/2} \sim 4.5$ d, Eγ = 1297 keV) and ^{59}Fe ($T_{1/2} \sim$ 44 d, Eγ = 1099 keV). The element Ca was also determined by ^{49}Ca ($T_{1/2} \sim 9$ min, Eγ =3084 keV) using short time irradiation (300 s) and counting time of 1200 s. The measurements of the neutron induced activity of the sample and standard were carried out using an ORTEC GEM-60195 detector and an ORTEC 671 amplifier coupled to a MCA ORTEC 919E connected to a PC. The concentrations were performed using in-house software [6].

3. Results and Discussion

The Ca and Fe concentrations in whole saliva of healthy subjects (control group) are showed in table 1. The mean value (MV), standard deviation (SD), minimum value (Min), maximum value (Max) and reference interval (Range, for a confidence interval of 95%), are presented.

Table 1. Ca and Fe concentrations in non-stimulated whole saliva of healthy subjects.

Elements	MV	± 1SD	Min	Max	Range
gL^{-1}					(95%)
Ca	0.055	0.020	0.028	0.079	0.015 – 0.095
	0.056 [4]	0.017			
Fe	0.635	0.061	0.211	0.822	0.513 – 0.757
	0.607 [7]	0.036			

To visualize in figures 1 and 2 are showed the concentration and associated uncertainly of Ca and Fe concentrations, respectively, in patients with periodontal disease; the standard deviation (± 1 and ±2) for the control group were included for comparison.

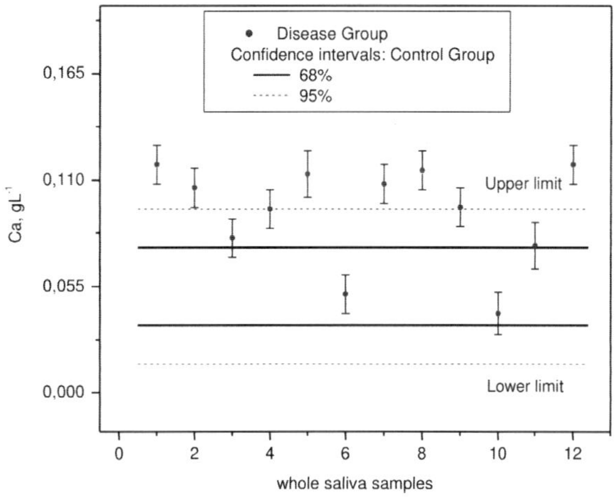

Figure 1. Ca concentrations and associated uncertainly of periodontal disease group.

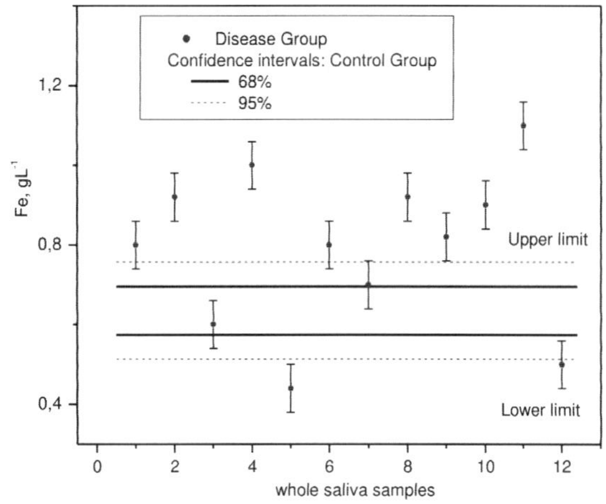

Figure 2. Fe concentrations associated uncertainly of periodontal disease group.

The results from this investigation give an indicative interval for Ca (0.015 - 0.095 g/L) in agreement with recent data [4]. Particularly, for Fe, the present result (0.513 - 0.775 g/L) is the first estimative for Brazilian population and it is in agreement with data from other population [7].

The effect of the periodontal disease (Figures 1 and 2) emphasizes the increase in Ca and Fe concentrations comparatively to healthy group. According to the Student's t-test the Ca and Fe concentration in subjects with periodontal disease are statistically different ($p < 0.05$) of health group for most of the case. Although Ca is usually used as an effective indicator of periodontal disease [1, 3-5], the similar behaviour of Fe suggests that its estimation in whole saliva can also be used as biomarker.

Another aspect to be considered is the relationship between proteins and inorganic elements. Levels of Ca and Fe are also important for keeping the health status of the tooth: while Fe present in lactoferrina protein participates in the mechanism regulating of saliva production (protecting the teeth of its demineralization), Ca present in some proteins (estaterina, histatin and proline) participates in the mechanism of production of Hydroxyapatite crystal (i. e. alterations can inhibits the growth of the tooth).

4. Conclusion

The increase of Fe in whole saliva of subjects with periodontal disease suggests that this element can be an efficient monitor of this oral disease as well as Ca. Furthermore, the data of the present investigation provide scientific basis for biomedical research of other oral diseases using the levels evaluation of these inorganic elements in human whole saliva.

Acknowledgement

The authors thank the financial support from Conselho Nacional de Desenvolvimento Científico e Tecnológico (CNPq) and Instituto de Pesquisas Energéticas e Nucleares (IPEN - CNEN/SP).

5. References

[1] Kaufman E, Lamster I B 2002 *Crit. Rev. Oral Biol. Med.* **13** 107
[2] Estimate/2014 – Cancer Incidence in Brazil http://www.inca.gov.br
[3] Schutzemberger M E, Souza R T, Petrucci R E, Machado M N, Papalexiou V, Brancher J A 2007 *RSBO.* **4(1)** 46
[4] Lewgoy H R, Zamboni C B, Metairon S, Medeiros I M M A, Medeiros J A G 2013 *J Radioanal. Nucl. Chem.* **296** 573
[5] Zamboni C B, Metairon S, Medeiros I M M A, Lewgoy H R 2013 *AIP Conf. Proc.* **1529** 70
[6] Medeiros J A G, Zamboni C B, Zahn G S, Oliveira L C, Dalaqua Jr L 2005 *39º Congresso Brasileiro de Patologia Clinica / Medicina* Laboratorial, São Paulo
[7] Zaichick V, Tsyb A, Bagirov 1995 *J. Radioanal. Nucl. Chem.* **195(1)** 123

An alternative experimental approach for subcritical configurations of the IPEN/MB-01 nuclear reactor

E Gonnelli, S M Lee, L N Pinto, H R Landim, R Diniz, R Jerez, A dos Santos

Instituto de Pesquisas Energéticas e Nucleares, IPEN/CNEN – SP, Av. Lineu Prestes 2242 CEP:05508-000, São Paulo, Brazil

Email: e.gonnelli@gmail.com

Abstract. This work presents an alternative approach for the reactivity worth experiments analysis in the IPEN/MB-01 reactor considering highly subcritical arrays. In order to reach the subcritical levels, the removal of a specific number of fuel rods is proposed. Twenty three configurations were carried out for this purpose. The control bank insertion experiment was used only as reference for the fuel rod experiment and, in addition, the control banks were maintained completely withdrawn during all the fuel rods experiment. The theoretical simulation results using the MCNP5 code and the ENDF/B-VII.0 library neutron data are in a very good agreement to experimental results.

Keywords: Subcritical Experiment; Fuel Rods; Neutron Noise; Spectral Density.
PACS: 28.50.Dr

1. Introduction

In recent years, subcritical systems have been studied in a large scale [1,2,3,4,5,6] mainly due to the development of a new technology of subcritical reactors such as Accelerator Driven Systems (ADS) [7]. It is clear that an experimental reactivity measure becomes relevant in this context. Studies that have been carried out are focused on the development of a new and precise methodology [8].

Considering all those previously presented efforts, the aim of this work was to develop an alternative experimental approach for subcritical configurations. This alternative approach is particularly interesting due to the withdrawal of fuel rods in order to maintain the subcritical states. Based on two important procedures, which were used as reference, it was possible to estimate how many fuel rods should be withdrawal to achieve a given level of negative reactivity. The first method was the k_{eff} calculation applied to nuclear multiplying systems through the MCNP5 reactor physics code. The second method was the control bank insertion, the most traditional way for reactivity variations.

Two theoretical kinetic models were employed in this work. The first one was the point reactor kinetic model in order to determine the prompt neutron decay constant (α) [9,10]. This kinetic parameter was estimated fitting the point kinetic equation to a set of APSD (Auto Power Spectral Density) experimental data [11,12] using the least-squares method.

The second model was developed by Gandini and Salvatores [1,2] which introduces the generalized reactivity (ρ_{gen}) and the subcriticality index (ζ), a function that depends on the neutron source position and magnitude. The generalized reactivity (ρ_{gen}) is the negative reactivity worth for each step of withdrawing fuel rods of this work. The proposed alternative method relies only on measured quantities, as counting rates of detectors.

2. Methodology

The proposed alternative experiment was implemented in the IPEN/MB-01 reactor and was simulated employing the MCNP5 code. Control bank experiments have an already well-known behavior [13] and, due to this fact, it was used only as reference for fuel rod experiments. The control banks method is the conventional way to study the system reactivity; controlling neutron population trough the insertion or withdrawn of control banks (simultaneously/separately) in the core active region depending on the specific range of interest. Twenty-three subcritical configurations were developed in this work.

The design of these twenty-three cases was developed employing MCNP5 based on the IPEN/MB-01 reactor geometrical structures and isotopic materials concentration. The characterization and determination of each material of the reactor is very important to obtain precise results and a good accuracy with the real experiment.

Criticality calculations performed by MCNP5 were carried out using the KCODE criticality source card. This card provides the possibility to calculate the multiplication factor, which is a ratio between two consecutive generations of neutrons and can be able to sustain a chain reaction by fission neutrons; it is characterized by k_{eff}, e.g., the eigenvalue to the neutron transport equation [10].

The way to calculate the k_{eff}, in the MCNP5, is using an estimating of the mean number of fission neutrons produced in one generation per fission neutron started. The generation can be interpreted as the life of a neutron from "birth" in fission to "death" by escape, parasitic capture, or absorption leading to fission. The computational calculation of the fission generation could be denoted by a k_{eff} cycle, which is a computed estimate of an actual fission generation [14]. The control banks k_{eff} values were used to estimate the number of fuel rods that is necessary to withdraw from the core. It was also possible calculate the fuel rods k_{eff}, and an approximated value to the subcriticality level could be achieved.

The experiments were based on the 26 x 24 fuel rods core configuration instead of the standard 28 x 26 fuel rods configuration. The total number of fuel rods for this new core configuration was 576 and provides practically zero excess of reactivity, therefore, the criticality can be reached with all the control banks withdrawn. In this way, it is possible to achieve large range of subcritical levels using this configuration. The removal of fuel rods started by the peripheral region due to its low negative reactivity worth and from the north and south faces of the core matrix. The east and west faces were left unchanged due to the presence of the experimental detectors and the external neutron source, respectively. Particularly, the number and the position of the fuel rods in the column in front of the detectors were kept unchanged to avoid any neutron flux distortion in this area and not impair the detectors measurements. The additional external neutron source of Am-Be of 1 Ci (strength $2.6x10^6$ neutrons/s) was employed in order to improve the statistics of the detectors' counts, and so the resolution of APSD´s. This neutron source was placed in the middle of the active region, between rows 14 and 15 and leaning against these two fuel rods as shown in Figure 1.

The advantage of maintaining the control banks totally withdrawn during all the fuel rods experiment is to avoid local neutron flux depression caused by control banks insertion, which depresses the neutron flux in its vicinity inducing a subsequent shadowing between the two control banks. This shadowing effect is very difficult to be taken into consideration in calculation methods.

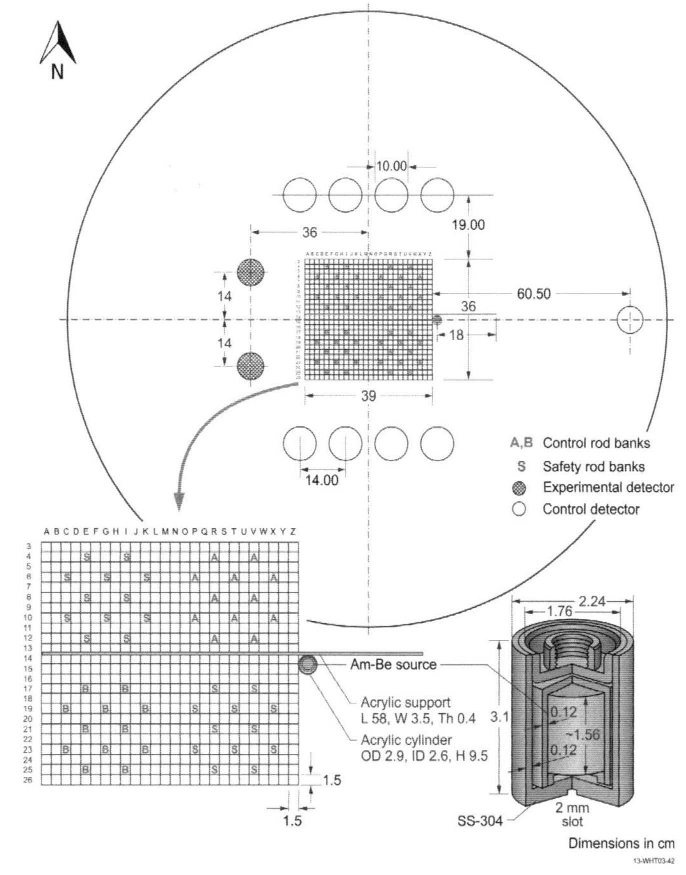

Figure 1 - Upper View of the Core of the IPEN/MB-01 Reactor. The neutron source position, the control and safety rod banks, the experimental and control detectors are represented in this figure.

The Am-Be source was installed inside of an acrylic cylinder and it was disposed in an acrylic support. Furthermore, the neutron source was used only in the experimental procedure because of the KCODE card does not allow to implement this condition during computational simulation and also subcritical systems require an external source to maintain a steady state. The first and the last core configuration can be seen, respectively, in the figures 2 and 3.

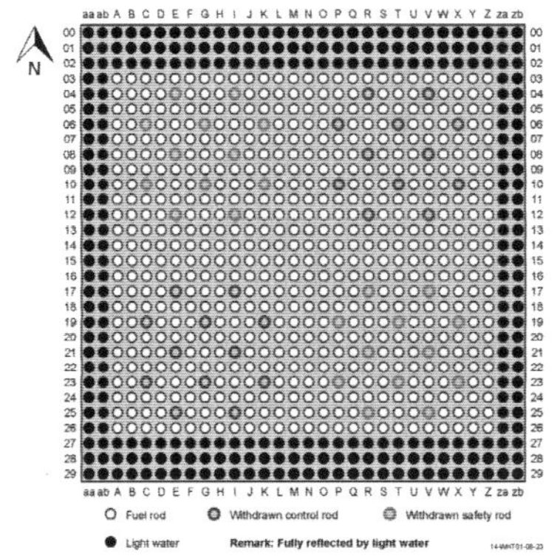

Figure 2 - Core Configuration for Case 1

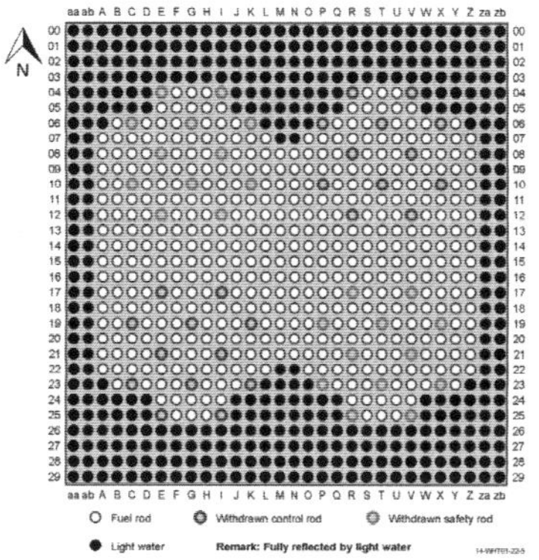

Figure 3 - Core Configuration for Case 23

Through the point reactor kinetic equations and using standard procedures [10,15], the APSD of the neutron noise was described considering the standard electronics chain, for a pulse-mode detector, and the zero-power transfer function. Moreover, the prompt neutron constant decay (α) was obtained using the following equation:

$$\Phi(\omega) = \frac{A}{\omega^2 + \alpha^2} + C \qquad (1)$$

where ω is the angular frequency, A and C are constants, being C the uncorrelated or detector noise, and Φ is the APSD to fit to the experimental data.

The generalized reactivity (ρ_{gen}) developed by Gandini and Salvatores [1,2,12] kinetic model can be expressed by the Eq. (2).

$$\rho_{\text{gen}} = (\alpha l_{eff} + \beta_{eff})(1 - P_N) \qquad (2)$$

where α is obtained by the Eq. (1), l_{eff} and β_{eff} are parameters obtained from previous works [16], and P_N is the relative power of two consecutive states By analogy with the point kinetic model, the power for a certain state is given by: [11,17]

$$P = \frac{2R^2(\gamma D) l_{eff}^{-2}}{(B)^2(\Phi^p - C)} \qquad (3)$$

where,
R is the average count-rate of the pulse mode detectors,
$\gamma$is the energy released per fission in Joule,
D is the Diven factor [17] given by: $\langle\nu(\nu-1)\rangle/\langle\nu\rangle^2$, where ν represents the number of neutrons emitted per fission, and the brackets represents the average value,
B is the prompt neutron decay constant αor its analogue, depending on the kinetic model employed,
Φ^p = mean value of the APSD on the first plateau level in V^2/Hz for current mode detector, or Counts^2/Hz for pulse mode detector,
C = mean value of the plateau of the uncorrelated noise.

Now, considering two consecutive states *a* and *b*, which *b* is more subcritical than *a*, and assuming that l_{eff} is independent of the subcriticality levels, the relative power between these two states can be written as:

$$P_N = \frac{P_b}{P_a} = \frac{R_b^2 B_a^2 (\Phi_a^p - C_a)}{R_a^2 B_b^2 (\Phi_b^p - C_b)} \qquad (4)$$

Thus, through Eq. 2 the parameter ρ_{gen} can be obtained in a purely experimental way. The multiplication factor, k_{eff}, is obtained as: $k_{eff} = 1/(1-\rho)$, where ρ is the total reactivity.

The measurements start with the two control banks completely withdrawn. Then, specified numbers of fuel rods were removed from the core in carefully chosen steps. In each of these subcritical states; the APSD, the counting rates of the detectors used, and the temperatures were measured. The inferred quantity of interest in this evaluation is k_{eff} which is obtained as 1/(1-ρ) where ρ is the total reactivity; i.e., the sum of partial ρ_{gen}.

Journal of Physics: Conference Series **630** (2015) 012007 doi:10.1088/1742-6596/630/1/012007

3. Results

Table 1 shows the MCNP5 simulated and measured values of the multiplication factor (k_{eff}) for the fuel rod experiment. It must be emphasized that all the control banks were maintained totally withdrawn from the core during the fuel rods experiment.

The MCNP5 simulations for fuel rods were run with 4050 cycles with 60000 neutron histories each. The first 50 cycles were skipped. The calculated results of configurations 1 through 21 are within 1σ range of the uncertainty of the measured value, while the one of configuration 22 is within 2σ range of the uncertainty of the measured value. The calculated result of configuration 23 is outside the 3σ range.

The uncertainty in the final value of k_{eff} for each configuration is the result of a combination of experimental uncertainties derived from acquired data and those arising from geometrical and material composition data; the latter were evaluated by computational codes.

Table 1 – The k_{eff} for the fuel rods simulation realized by the MCNP5 and its measured results.

Configuration	Fuel Rods	Empty Positions	Measured results $k_{eff} \pm \sigma_k$	MCNP5 $k_{eff} \pm \sigma_k$
1	576	276	1.0004 ± 0.0006	1.0007 ± 0.0002
2	564	288	0.9987 ± 0.0006	0.9991 ± 0.0002
3	560	292	0.9979 ± 0.0006	0.9983 ± 0.0002
4	556	296	0.9973 ± 0.0006	0.9976 ± 0.0002
5	550	302	0.9960 ± 0.0006	0.9964 ± 0.0002
6	546	306	0.9949 ± 0.0006	0.9952 ± 0.0002
7	542	310	0.9938 ± 0.0006	0.9942 ± 0.0002
8	536	316	0.9926 ± 0.0006	0.9924 ± 0.0002
9	532	320	0.9912 ± 0.0007	0.9912 ± 0.0002
10	524	328	0.9899 ± 0.0007	0.9898 ± 0.0002
11	516	336	0.9879 ± 0.0007	0.9876 ± 0.0002
12	510	342	0.9859 ± 0.0007	0.9860 ± 0.0002
13	508	344	0.9842 ± 0.0007	0.9837 ± 0.0002
14	499	353	0.9814 ± 0.0008	0.9814 ± 0.0002
15	493	359	0.9791 ± 0.0008	0.9790 ± 0.0002
16	486	366	0.9762 ± 0.0008	0.9762 ± 0.0002
17	481	371	0.9746 ± 0.0009	0.9740 ± 0.0002
18	471	381	0.9708 ± 0.0010	0.9712 ± 0.0002
19	460	392	0.9659 ± 0.0010	0.9668 ± 0.0002
20	456	396	0.9623 ± 0.0010	0.9624 ± 0.0002
21	446	406	0.9583 ± 0.0010	0.9578 ± 0.0002
22[a]	444	408	0.9548 ± 0.0010	0.9529 ± 0.0002
23[a]	444	408	0.9522 ± 0.0011	0.9475 ± 0.0002

[a] Configurations 22 and 23 have the same number of fuel rods, but the geometry is different.

The experimental uncertainties in the k_{eff} are the ones arising from the measurements of reactivity given by experimental approaches and from measurements of the average temperature given by thermocouples. The measured reactivity of the first configuration carries uncertainty components due to uncertainties of kinetic parameters used in the reactivity meter. This uncertainty was evaluated as follows.

The detector signals that gave rise to the reactivity determination were recorded. The reactivity meter was then run twice. In the first run all kinetic parameters, effective delayed neutron fraction, β_{eff}, the prompt neutron generation time, l_{eff} , the delayed neutron decay constant, λ_i, and the delayed neutron abundances, β_i, were increased by adding to their values the uncertainty on these basic parameters. The second run was the opposite. All kinetic parameters were decreased by subtracting from their values the uncertainty on these basic parameters. The reactivity meter reactivity uncertainty, only for configuration 1, was then taken to be equal to the difference of these two reactivity determinations. The reactivity uncertainty due to the uncertainty of the kinetic parameters was found to be much smaller than the statistical uncertainty of the reported reactivity (10pcm ± 3 pcm) [18].

Now, considering the subcritical configurations, the uncertainty of the measured reactivity, ρ_{gen}, already takes into account the measurement uncertainties and the uncertainty of the kinetic parameters l_{eff} and β_{eff} used in the approach. Equations 2 and 4 show that explicitly.

In the last step (configuration 22 and 23) shown in the figure 4, the difference value of -474pcm between the simulated and experimental subcritical reactivity may be seen and could be explained as the statistical resolution of the detector in this very subcritical state. In high levels of subcriticality, more sensitivity detectors should be employed.

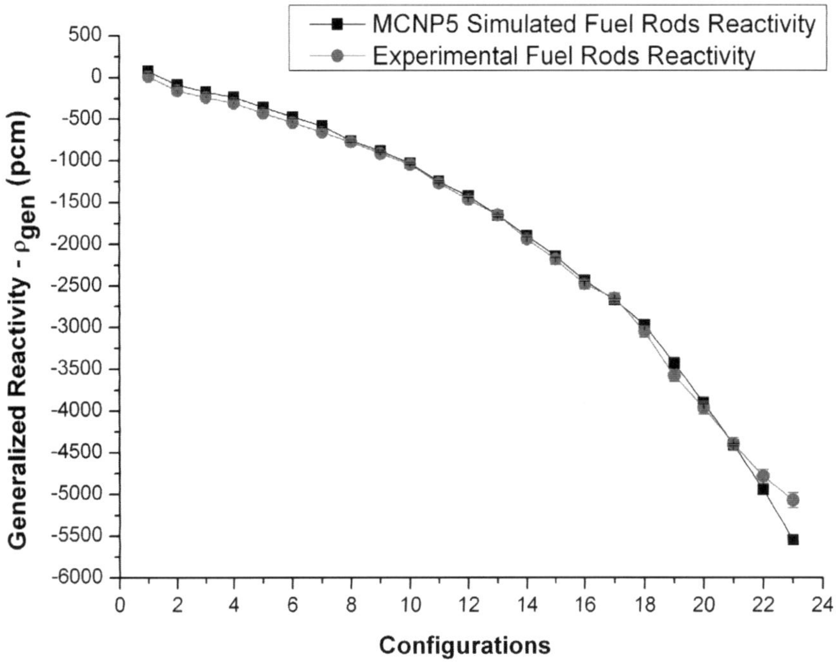

Figure 4 –The generalized reactivity (ρ_{gen}) fuel rods measured and simulated result values obtained through the Gandini and Salvarotes kinetic models and MCNP5, respectively.

4. Conclusions

The new experimental approach using fuel rods withdrawn was successfully accomplished in the IPEN/MB-01 reactor and it was possible to obtain a good accuracy of the parameter ρ_{gen}. The comparison between simulation and experiment shows good agreement, except for the configuration 23, where it is believed that the detector lost a statistically reliable response.

This work can be considered relevant in the subcritical study context since it introduces a new method to achieve subcritical reactivity levels in an alternative and well-success way.

For future work, an additional subcritical configuration could be done using boric acid diluted in the whole tank of water of the moderator/reflector. Furthermore, a study is already being done for the feasibility of use of an input logic nuclear module for data acquisition with which will be possible to make logic operations with the neutron signals from two or more detectors and improve the statistical resolution in very subcritical configurations. Also, a more complex and detailed study of the neutron flux will be developed employing the mesh tally through the MCNP5.

Acknowledgments
The authors are grateful to financial support of Fundação de Amparo à Pesquisa do Estado de São Paulo (FAPESP), Coordenação de Aperfeiçoamento de Pessoal de Nível Superior (CAPES), brazilian nuclear power company Eletronuclear. The authors would also like to thank the operational staff of the IPEN/MB-01 reactor for their professionalism and efficient operation during the course of the experiment.

References
[1] Gandini A and Salvatores M 2002 *Journal of Nuclear Science and Technology* **39** 6 pp 673-686
[2] Gandini A 2001 *Annals of Nuclear Energy* **28** pp 1193-1217
[3] Salvatores M *et al* 1996 MUSE-1: A first experiment at MASURCA to validate the physics of subcritical multiplying systems relevant to ADS *2nd ADTT Conf*
[4] Bécaresa V *et al* 2013 *Annals of Nuclear Energy* **53** pp 331-3410
[5] Fernández-Ordóñez M *et al* 2009 Reactivity monitoring of a subcritical assembly using beam-trips and current-mode fission chambers: The Yalina-Booster program *International Topical Meeting on Nuclear Research Applications and Utilization of Accelerators* **ADS/ET-04**
[6] Wright J 2005 Development and Investigation of Reactivity Measurement Methods in Subcritical Cores *Thesis for the Degree of Licentiate of Engineering* ***CTH-RF-190***
[7] Rubbia C *et al* 2004 *PHYSOR-2004* **42** 4209647
[8] Dulla S Ravetto P Rostagno M M Bianchini G Carta M and D'Angelo A 2005 *Nuclear Science and Engineering* **149** pp 88–100
[9] Lamarsh J R 1966 *Introduction to Nuclear Reactor Theory* (Addison-Wesley)
[10] Bell G I and Glasstone S 1979 *Nuclear Reactor Theory* (New York:Van Nostrand Reinhold Company)
[11] Kitamura Y Matoba M Misawa T Unesaki H Shiroya S 1999 *Journal of Nuclear Science and Engineering* 36 8 pp 653–660
[12] dos Santos A Diniz R Lee S M Jerez R 2013 *Annals of Nuclear Energy* **59** pp 243-254
[13] Pinto L N Gonnelli E dos Santos A 2014 Control rod calibration and reactivity effects at the IPEN/MB-01 reactor *AIP Conference Proceedings* **1625** pp 140-145
[14] Urbatsch T J Forster R A Prael R E and Beckman R J 1995 Estimation and interpretation of k_{eff} confidence intervals in MCNP *Los Alamos National Laboratory report* **LA–12658**
[15] Hetrick D L 1971 *Dynamics of Nuclear Reactors* (The University of Chicago Press: Chicago, IL)
[16] dos Santos A *et al* 2009 *International Handbook of Evaluated Reactor Physics Benchmark Experiments* **NEA/NSC** pp 1–142
[17] Diven B C *et al* 1956 *Phys. Rev.* **101** 1012
[18] dos Santos A *et* al 2014 *International Handbook of Evaluated Reactor Physics Benchmark Experiments* subcritical loading configurations of the IPEN/MB-01 reactor **NEA/NSC SUB-LEU-COMP-THERM-002**

Journal of Physics: Conference Series **630** (2015) 012008 doi:10.1088/1742-6596/630/1/012008

Evaluation of Microstructural Parameters of Reservoir Rocks of the Guarani Aquifer by Analysis of Images Obtained by X-Ray Microtomography

J S Fernandes[1], F A Lima[1], S F Vieira[1], P J Reis[2] and C R Appoloni[2]

[1] Instituto Federal de Educação, Ciência e Tecnologia Catarinense – IFC, Câmpus Videira SC-135, Km 125, Campo Experimental, CEP 89560-000, Videira, Santa Catarina, Brazil
[2] Laboratório de Física Nuclear Aplicada, Departamento de Física, Universidade Estadual de Londrina – UEL, CP 6001, CEP 86051-990, Londrina, Paraná, Brazil

E-mail: jaquiel.fernandes@ifc-videira.edu.br

Abstract. Microstructural parameters evaluation of porous materials, such as, rocks reservoir (water, petroleum, gas...), it is of great importance for several knowledge areas. In this context, the X-ray microtomography (μ-CT) has been showing a technical one quite useful for the analysis of such rocks (sandstone, limestone and carbonate), object of great interest of the petroleum and water industries, because it facilitates the characterization of important parameters, among them, porosity, permeability, grains or pore size distribution. The X-ray microtomography is a non-destructive method, that besides already facilitating the reuse of the samples analyzed, it also supplies images 2-D and 3-D of the sample. In this work samples of reservoir rock of the Guarani aquifer will be analyzed, given by the company of perforation of wells artesian Blue Water, in the municipal district of Videira, Santa Catarina, Brazil. The acquisition of the microtomographys data of the reservoir rocks was accomplished in a Skyscan 1172 μ-CT scanner, installed in Applied Nuclear Physics Laboratory (LFNA) in the State University of Londrina (UEL), Paraná, Brazil. In this context, this work presents the microstructural characterization of reservoir rock sample of the Guarani aquifer, analyzed for two space resolutions, 2.8 μm and 4.8 μm, where determined average porosity was 28.5 % and 21.9 %, respectively. Besides, we also determined the pore size distribution for both resolutions. Two 3-D images were generated of this sample, one for each space resolution, in which it is possible to visualize the internal structure of the same ones.

1. Introduction

The characterization of the pore space geometry and the determination of microstructural properties of the reservoir rock are extremely important for determining the capacity of a water, oil or gas reservoir. The porous rocks system has a complex geometry, involving aspects of pore shape (morphology) and the pores connection (topology). The macroscopic physical properties of these materials are strongly dependent on their microstructure. The intrinsic permeability, for example, depends on the effective porosity, the distribution of pore size and the interconnection degree between the pores. Thus, the quantitative description of the microstructure of porous materials is of fundamental importance, allowing a better understanding of how the microstructure influences physical phenomena occurring within the material, as well as the numerical determination of their microstructural properties.

The recent development of microtomograph allows characterization of porous materials with a spatial resolution between 0.5 to 5 micrometers using an X-ray tube. It makes possible to have a Synchrotron accelerator results, however, in a conventional laboratory.

The Guarani Aquifer occupies a total area of 1.2 million square kilometers and is located in the central western region of South America, being inserted within the boundaries of four countries, namely: Brazil, Argentina, Paraguay and Uruguay. Within the State of Santa Catarina it occupies an area of approximately 49,200 km^2. As the average thickness of the aquifer is 228 m, the approximate volume of the aquifer in Santa Catarina is 11,218 km^3 [1].

The City of Videira is located on the region of the aquifer. This aquifer is considered one of the largest underground freshwater reservoirs in the world, consisting basically of Botucatu sandstone formation.

The conservation of groundwater is crucial to mankind, due to the increasingly polluted and scarce of surface water present in rivers and lakes, a situation aggravated by deforestation and overuse of pesticides. Becomes therefore of the utmost importance to study the storage capacity of aquifers by population.

In this context, this article quantify the porosity of the rocks that forms the Guarani aquifer through X-ray microtomography, it made possible to estimate the amount of water the aquifer can store, because by determining the porosity of the rock and knowing the total volume of these rocks, it's possible to estimate the amount of water.

2. Material and methods

The X-ray microtomography technique is an inspection method that allows obtaining data from the internal microstructure of materials, as cross sections images from the external irradiation of the samples [2]. This technique requires the use of appropriate software for processing and analysis of images, which enable the determination of important parameters for better understanding of the material analyzed. From the cross sections images of the rocks, holds up a rendering process, obtaining three-dimensional images (3-D) [3,4,5]. These images are originally used for the characterization of properties such as pore and grains size distribution, specific surface area and connectivity of the pore system, effective porosity and permeability. An illustrative diagram of the acquisition, reconstruction and generation of image and 3-D models is shown in Figure 1.

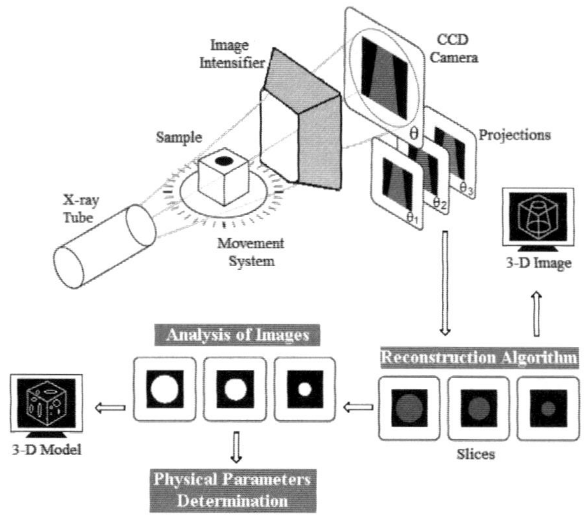

Figure 1 - Illustrative diagram of the acquisition, reconstruction and generation of image and three-dimensional models.

Furthermore, the 3-D images are used directly in numerical simulations of physical phenomena of equilibrium and transport of fluids, to determine their microstructural properties. In this procedure, fundamental physical laws are applied in the pore scale. This technique aims to characterize the geometry of the pore system and the determination of some microstructural properties.

2.1. Equipment used

For these results, a Skyscan microtomograph model 1172 was used, at the Applied Nuclear Physics Laboratory (LFNA), in the State University of Londrina (UEL), in the state of Paraná, Brazil. The microtomography images were reconstructed by NRecon software. The porosity and pore size distribution were obtained by Imago, developed by the Laboratory of Porous Media and Thermophysical Properties of Materials (LMPT) of the Department of Mechanical Engineering of the Federal University of Santa Catarina, Florianópolis, SC, in association with the Brazilian software company ESSS (Engineering Simulation and Scientific software). Figure 2 shows a picture of microtomograph Skyscan 1172.

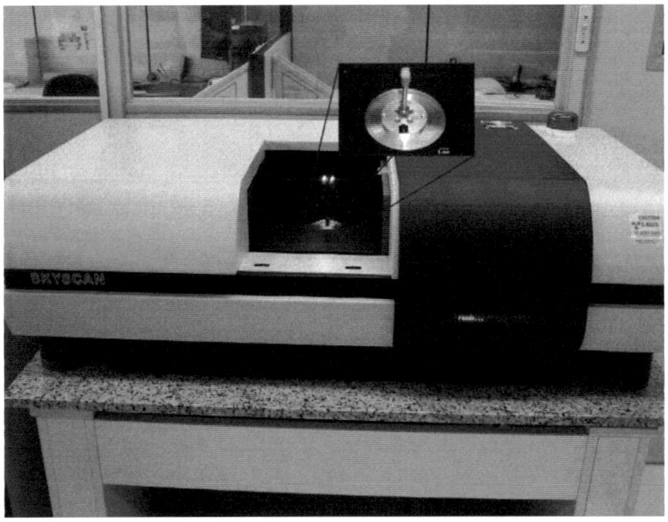

Figure 2. Skyscan 1171, X-ray source 20-100 kV , maximum power of 10 W.

The X-ray tube has a maximum power of 10 W. For this application the voltage used was 80 kV and a current of 124 μA. The sample was placed between the X-ray tube and a CCD camera of 2000 x 1048 pixels at a distance of 5.9 cm from the tube and 15.1 cm from the CCD.

2.2. Analyzed samples

The samples used were sandstones of the Botucatu formation, from two wells drilled in the region of Videira – SC, with UTM coordinates of 0551412, 7022236 for depth of 250 meters and the coordinates of 0487041, 7013997 for depth of 400 meters. The sample of sandstone collected at depth 250 m (sandstone250) has dimensions of 4 x 5 x 10 mm and the sample collected at depth of 400 m (sandstone400) has dimensions of 6 x 6 x 14 mm. The samples were rotated from 0° to 180°, with angular steps of 0.25°, totaling 1442 projections. These projections were reconstructed by NRecon software, resulting in 900 two-dimensional images for each sample. To acquire these data was used an Aluminum filter of 1 mm thick, to minimize the beam hardening effect in the sample [6].

2.3. Images processing

Figure 3 shows two longitudinal projections of the sample, 0 and 180 degrees, formed on the CCD camera. These projections are reconstructed by NRecon software using a cone beam reconstruction algorithm [7], resulting in 2-D images of the cross sections of the sample, in grayscale, as can be seen in figures 4a and 5a. After the generation of 2-D images, they are analyzed by Imago software, which is defined a region of interest (ROI) and held the binarization, transforming the image into black and white, where black represents the solid phase and white the porous layer, as can be seen in figures 4b and 5b. These images have a spatial resolution of 4.8 μm.

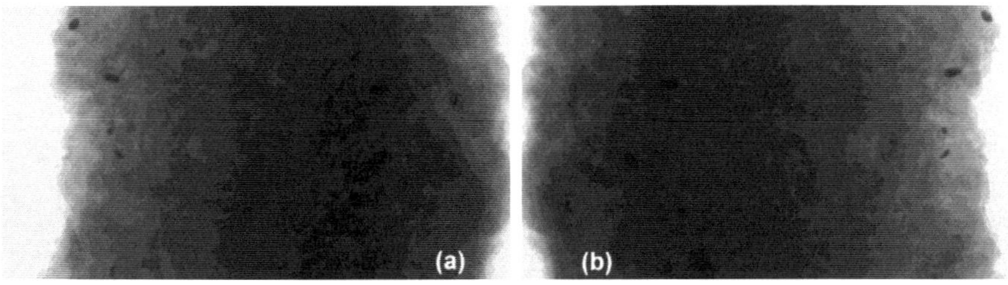

Figure 3. Projections of the Botucatu formation sample - sandstone400. (a) Image obtained at 0° position; (b) image obtained at position 180°.

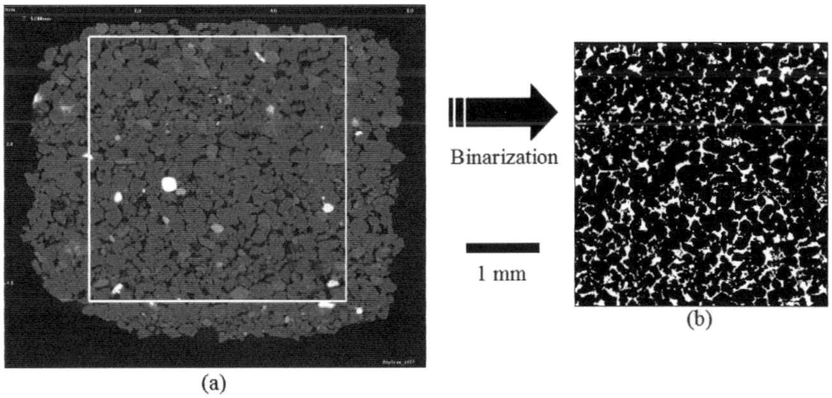

Figure 4. Images of sandstone250 rebuilt by NRecon software; (a) 2-D image in grayscale; (b) 2-D image binarized, black represents the solid phase and the white porous phase.

With 2-D images was also possible to reconstruct a 3-D image, where the CTAN software was used to reconstruct the solid phase (figure 6a and 7a) and the porous phase (figure 6b and 7b) of the sandstone250 and of sandstone400 samples, respectively.

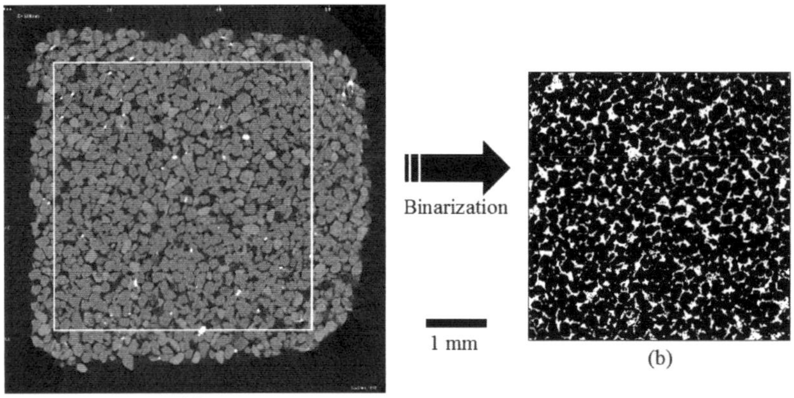

(a)

Figure 5. Images of sandstone400 rebuilt by NRecon software; (a) 2-D image in grayscale; (b) 2-D image binarized, black represents the solid phase and the white porous phase.

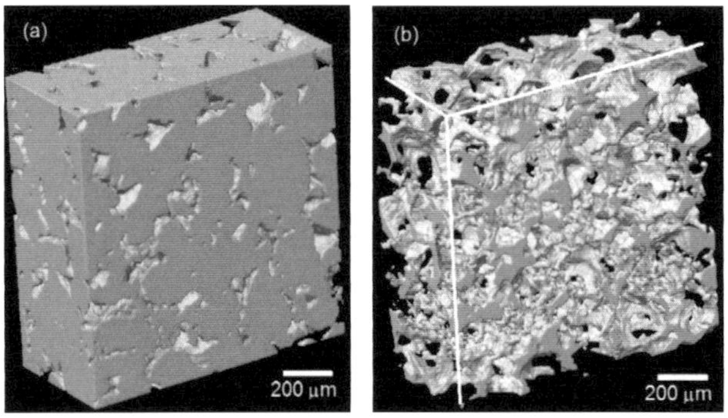

Figure 6. 3-D reconstructed volume the sandstone250; (a) solid phase reconstructed; (b) porous phase reconstructed.

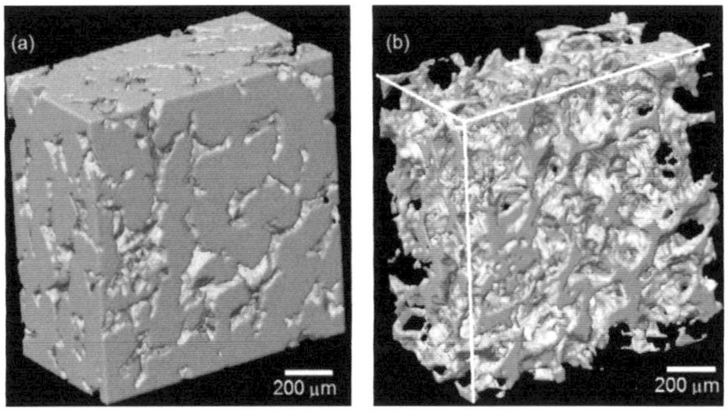

Figure 7 - 3-D reconstructed volume of sandstone400; (a) solid phase reconstructed; (b) Porous phase reconstructed.

3. Discussion and results

The porosity of the samples sandstone250 and sandstone400 obtained by Imago software for each 2-D section, are shown in the graph of figures 8 and 9, respectively, where the black lines indicate the average porosity (18.8 ± 0.6 % and 20.8 ± 0.7 %, respectively) and the red bars indicate the statistical deviation of ± 3.2 % for each section, with 95 % of confidence.

One of the advantages of microtomography methodology is the generation of 3-D images, which beyond the images presented in this article; several were generated for visualization, where the internal navigation in the samples was possible, through specific software, which allowed monitoring/mapping porous network in greater detail. It's not enough the rocks have porosity; it is necessary that their pores are interconnected and allow the water to drain.

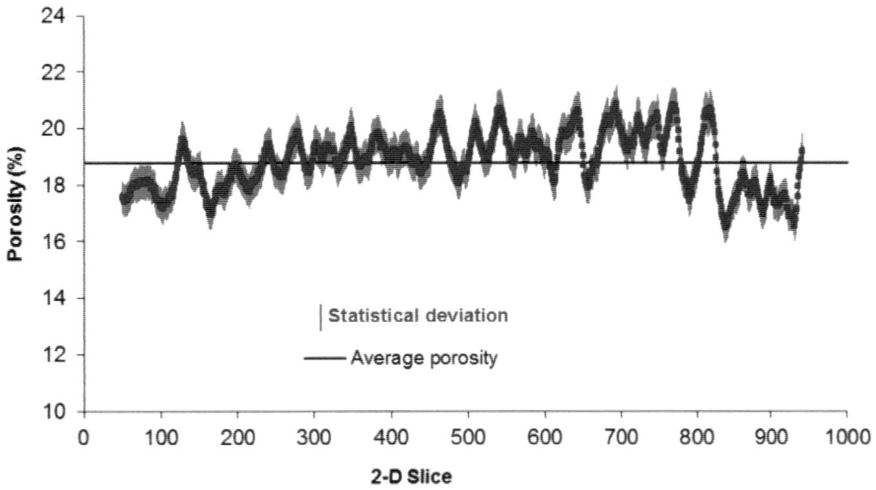

Figure 8. Porosity of 900 sections 2-D to the sample of sandstone to a depth of 250 meters and spatial resolution of 4.8 μm.

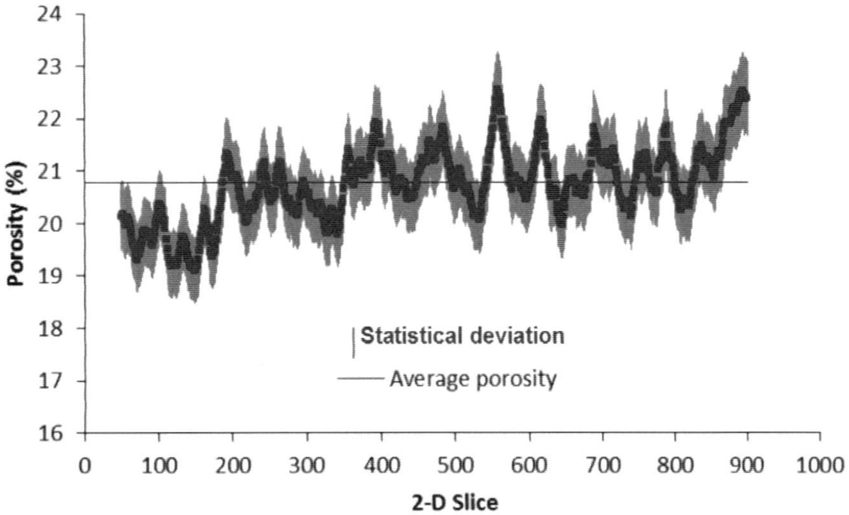

Figure 9. Porosity of 900 sections 2-D sample of sandstone to a depth of 400 meters and spatial resolution of 4.8 μm.

4. Conclusions

After analyzing the results was possible to verify that the average porosity of these rocks ranged between 18.8 and 20.8 %. As the volume of the aquifer is estimated at 272,460 km³ may suggest a volume of stored water in the range of 51,222 to 56,671 km³. However it is important to observe that this study examined only rocks of two wells, so these results should be ratified by analyzing more samples, so will be possible to have a more accurate conclusion as to the porosity of these rocks, and also its storage capacity.

Acknowledgments
This research is funded by IFC and CNPq.

References

[1] Rosa Filho E F 2003 Sistema aquífero guarani – considerações preliminares sobre a influência do arco de ponta grossa no fluxo das águas subterrânea. *Rev. Águas Subterrâneas*, **17** 91–112.
[2] Appoloni C R, Fernandes C P, Innocentini M D M and Macedo A 2004 Ceramic foams porous microstructure characterization by X-ray microtomography *Mater. Res.* **7** n. 4, 557-64.
[3] Appoloni C R, Macedo A, Fernandes C P and Philippi P C 2002 Caracterização de Microestruturas Porosas por Microtomografia de Raios *XVII Seminário Latino Americano de Análises por Técnicas de Raios X* (Brazil: São Pedro, SP) pp 419-24.
[4] Appoloni C R, Rodrigues C R O and Fernandes C P 2005 Reservoir Sandstone Porous Microstructure Characterization By High-Resolution X-Ray Microtomography *19th Int. Symp. of the Society of Core Analysts* (Toronto) pp 21-25.
[5] Appoloni C R, Fernandes C P and Rodrigues C R O 2007 X-ray microtomography study of a sandstone reservoir rock *Nucl. Instrum. Meth. A* **580**, Issue 1, 629-32.
[6] Ketcham R A and Carlson W D 2001 Acquisition, optimization and interpretation of X-ray computed tomographic imagery: applications to the geosciences *Comput. Geosci.* **27** 381–400.
[7] Feldkamp L A et al 1984 Practical cone-beam algorithm *J. Opt. Soc. Am.* **1** 612-19.

Sulfur status in long distance runners

L Kovacs[1], C Zamboni[1], T Lourenço[2] and D Macedo[2]

[1]Instituto de Pesquisas Energéticas e Nucleares (IPEN / CNEN - SP)
Av. Professor Lineu Prestes 2242, 05508-000, São Paulo, SP, Brasil.
[2]Laboratório de Bioquímica do Exercício (LABEX), Universidade de Campinas
Cidade Universitária 13083-970 - Campinas, SP, Brasil, P.O Box: 6109

E-mail: lksantos@ipen.br

Abstract. In sports medicine, sulfur plays an important role and its deficiency can cause muscle injury affecting the performance of the athletes. However, its evaluation is unusual in conventional clinical practice. In this study the sulfur levels were determined in Brazilian amateur athlete's blood using Neutron Activation Analyses (NAA) technique. Twenty six male amateur runners, age 18 to 36 years, participated of this study. The athletes had a balanced diet, without multivitamin/mineral supplements. The blood collection was performed at LABEX (Laboratório de Bioquimica do Exercício, UNICAMP-SP) and the samples were irradiated for 300 seconds in a pneumatic station in the nuclear reactor (IEA-R1, 3-4.5MW, pool type) at IPEN/CNEN-SP. The results were compared with the control group (subjects of same age but not involved with physical activities) and showed that the sulfur concentration was 44% higher in amateurs athletes than control group. These data can be considered for preparation of balanced diet, as well as contributing for proposing new protocols of clinical evaluation.

1. Introduction

Minerals are essential for metabolic and physiologic processes in the human body [1]. Among them sulfur is the third most abundant element found in human body after calcium and phosphorus and it is present in every cell [2]. A small percentage of sulfur comes in the form of inorganic sulfates and the other forms of organic sulfur presents in foods. They are important sources of sulfide in human. Sulfur plays important roles in the body, but is an element usually not analyzed in conventional clinical. According to references [3,4,5] the amount of sulfur levels in human blood ranges from 0.1 to 1.0 g/L.

Sulfur participates in the structure and activities of molecules playing important roles in growth, energy transfer mechanism, metabolism, defense, transport, and detoxification processes [2] and it is also necessary for formation of collagen (the protein found in connective tissue in our bodies). Sulfur is also present in keratin, which is necessary for the maintenance of the skin, hair, and nails, helping to give strength, shape, and hardness to these protein tissues [6].

[1]lksantos@ipen.br

Sulfur performs a number of functions in enzyme reactions and protein synthesis. It is present mainly in four common amino acids, i.e., methionine, cysteine, homocysteine and taurine (Figure 1), but only methionine and cysteine are incorporated into proteins [7]. Sulfur present in cysteine and methionine amino acids helps regulate carbohydrate metabolism, while sulfur - contained in taurine (found in bile acids) helps in the digestion process. Methionine cannot be synthesized by the body and needs to be supplied by the diet while the cysteine may, but this process requires a steady supply of sulfur [8].

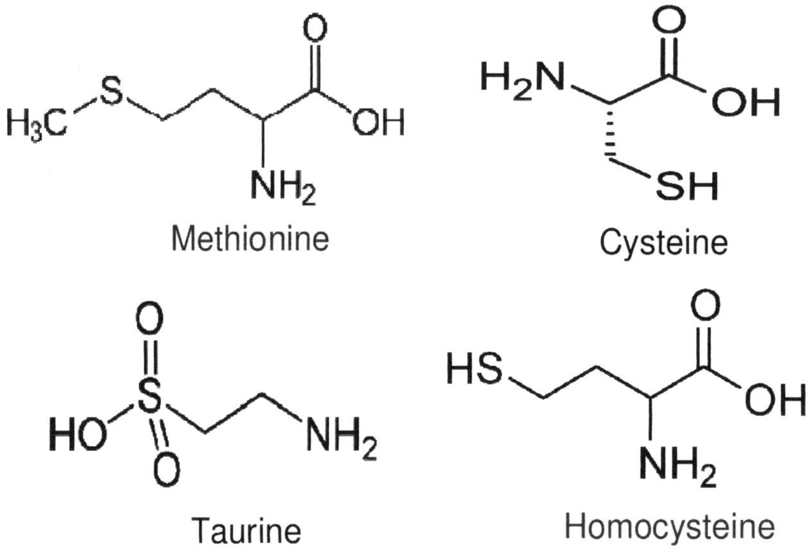

Figure 1. Structures of the sulfur-containing amino acids.

The sulfate absorption occurs in the stomach, small intestine and colon. The excess inorganic sulfur, generated as a result of liver or kidney metabolism, is excreted in urine as sulfates. There is not a recommended amount of sulfur intake but its toxicity may occur if sulfur-containing amino acids are ingested in large amounts.

In sports medicine, the sulfur intake plays a very important role: while its deficiency can cause muscle injury affecting the performance of the athletes [6], its excess can cause intoxication. However, its evaluation is unusual in conventional clinical practice.

In this study the sulfur levels were determined in blood of amateur athlete's using Neutron Activation Analyses (NAA) technique. This technique measures the total amount of sulfur present in sample independent of its chemical formula and without any pre-treatment of the sample [9]. This study is a partnership with Exercise Biochemistry Laboratory (LABEX) at University of Campinas (UNICAMP/SP, Brazil), a group of national and international reference in a medical sport field. The LABEX investigates the interrelationships metabolic that occur with physical activity monitoring the effects of different training, through tests that evaluate their clinical status, aiming to propose new evaluation protocols.

2. Materials and Methods

2.1 Collection and preparation of the samples

Twenty six male amateur athletes from LABEX, age 18 to 36 years, participated of this study. The athletes had a balanced diet, without multivitamin/mineral supplements. The blood samples were collected before (at rest) the physical training. For the blood collection, a small capillary pin

(Clinitubes®, Radiometer Copenhagen®) was inserted in the athlete's finger and exactly 50(± 0.5%) μL were dropped on to Whatman n° 41 filter paper (~2.3 cm^2) using a calibrated micropipette, and then dried for a few minutes using an infrared lamp (Figure 2). Considering the population of the study (male with 18-36 years) for the control group the whole blood samples were collected from 68 male healthy donators, with the same age and not involved with physical activities, selected from Paulista Blood Bank at São Paulo, Brazil. The sample preparation was performed as described for amateur athletes.

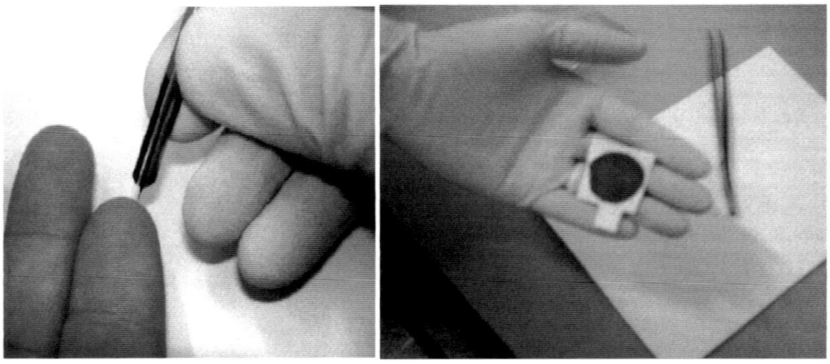

Figure 2. Blood collection and sample preparation (blood dropped in filter paper).

2.2 Experimental procedure
Blood sample and standard were irradiated for 300 s in a pneumatic station of the IEA-R1 nuclear reactor (3-4.5 MW, pool type) at IPEN. The thermal neutron flux utilized of $\sim 8 \cdot 10^{12}$ $n \cdot cm^{-2} \cdot s^{-1}$. After the irradiation, followed by a 60 s decay time, the activated materials were gamma-counted for 300 s using a high resolution HPGe detector (FWHM = 1.89 keV) connected to an ADCAM multichannel analyzer (ORTEC 919E) and a PC computer. The concentrations of sulfur were obtained using S^{36} ($T_{1/2} \sim 5$ min, $E_\gamma = 3104$ keV). The filter paper (blank) was also analyzed using the same irradiation conditions. Data reduction was done using in-house software [10]. The IAEA A-13 certified reference material was investigated for quality control.

3. Results
Partial gamma-ray whole blood spectrum is presented in Figure 3; the energy used for sulfur (3104 keV) is highlighted.

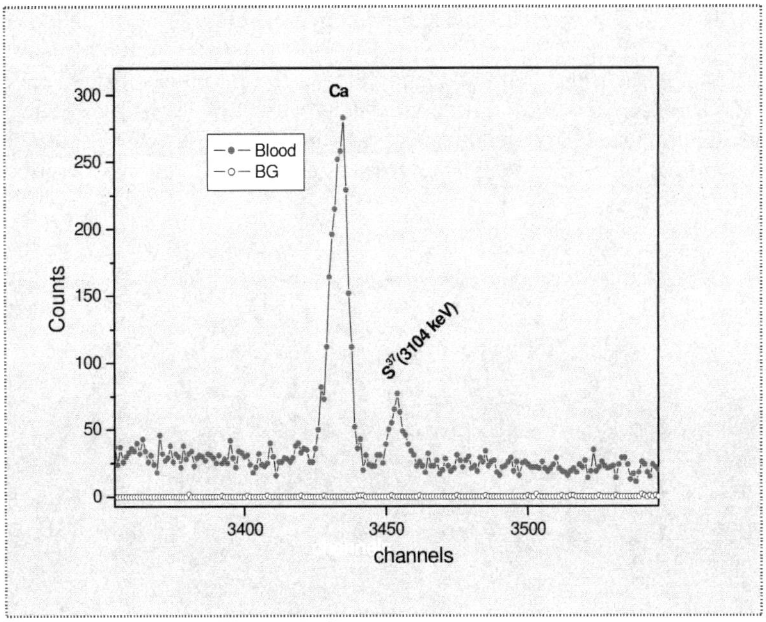

Figure 3. Gamma ray spectrum.

For analytical quality control Animal Blood (IAEA-A13) certified reference material was used. Table 1 shows the certified value for sulfur, the result obtained in this study, relative standard deviation (RSD), relative error (Er) and Z-score. The |Z score| < 2, indicated that our result was satisfactory and is within the range of certified data at the 95% confidence level. Sulfur blood concentration results, mean value, standard deviation (±1SD), median, mode, minimum value, and maximum value for amateur runner and control group are show in Table 2.

Table 1. Element concentrations obtained in the analysis of AIEA A-13 reference material.

Element gkg^{-1}	**Certified value** mean ±1 SD	**This work** mean ±1SD	**RSD** %	**Er** %	**Z- score**
S	6.50 ± 0.52	6.62 ± 0.92	13.89	1.8	0.23

Table 2. Sulfur blood concentration results, mean value, standard deviation (±1SD), median, mode, minimum value, and maximum value for amateur runner. The results for the control group were also included for comparison.

S, g/L^{-1}	Amateur Runners	Control Group
Medium value	0.65	0.45
± 1SD	0.23	0.20
Median	0.66	0.44
Mode	0.50	0.47
Minimum	0.21	0.12
Maximum	1.24	0.88

Figure 4 shows the sulfur concentration in whole blood of amateur runners (AR) and control group (CG) by Box Plot and Figure 5 shows a comparison between the individual concentrations of sulfur in whole blood of amateur runner. In this figure the indicate interval of control group, considering ± 1SD and ± 2SD were also included for comparison. According to the statistical analysis (Student *t*-test) there is a significant difference ($p < 0.05$) between these groups.

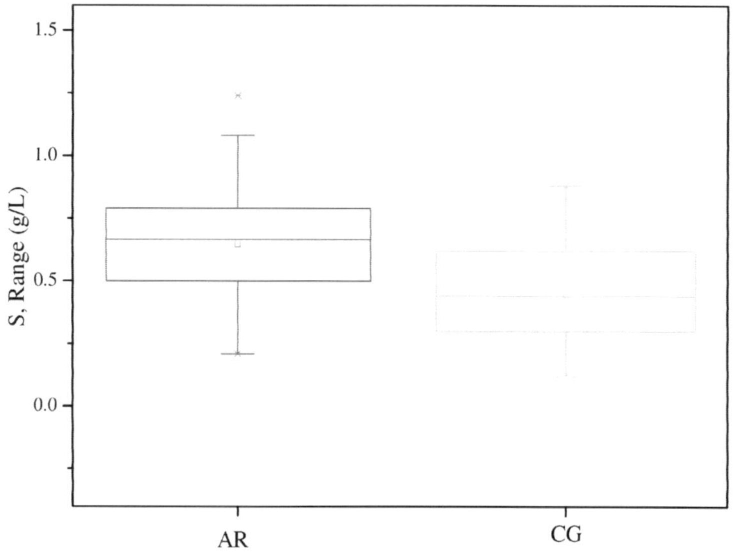

Figure 4. Sulfur concentration in blood of amateur runners (AR) and control group (CG) by Box Plot.

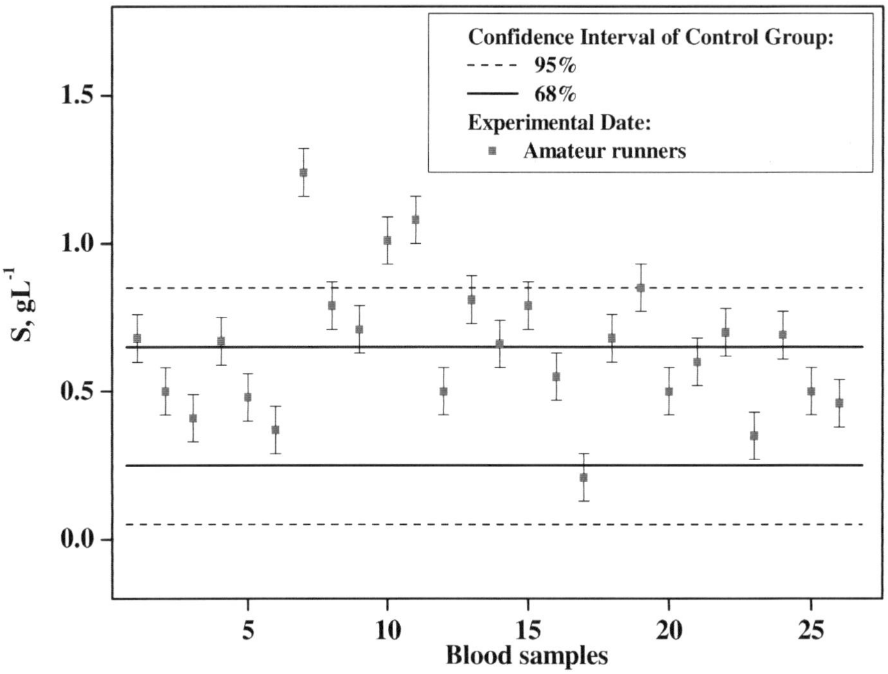

Figure 5. Sulfur concentration in blood of amateur runner.

4. Discussion

In the present study, we have demonstrated high sulfur level in the blood of amateur runners when compared to people not involved in intense physical activities. These alterations can be associated mainly with athlete's diet that is very rich in amino acids, such as, the use of amino acids in the formulations of beverages, bars cereals and supplements between other varieties of food [11].

For long-distance runners, sulfur present in some amino acids is essential, mainly the sulfur contained in methionine because it helps the body to eliminate fat and reduce the level of inflammatory histamines. Furthermore, the sulfur contained in taurine can act as a catabolic preventing the loss of muscle mass and to assist the entry of glucose into muscle.

Although all the benefits of these amino acids for the athlete the intake must be done carefully. The intake of sulfated amino acids and other sulfur compounds in excess, may contribute to the increase sulfur in the blood, which can result in intoxication.

5. Conclusion

The results showed significant differences and suggest that they can be considered for preparation of balanced diet, as well as to contribute for proposing new protocols of clinical evaluation. Moreover, these results may be useful in other areas of research such as health and nutrition.

Acknowledgments

The authors would like to acknowledge the Conselho Nacional de Desenvolvimento Científico e Tecnológico (CNPq) for the financial support.

References

[1] Maughan R J 1999 *Br Med Bull* **55**(3) 683
[2] Ingenbleek Y and Kimura H 2013 *Nutr Rev.* **71**(7) 413

[3] Denis W 1921 J Biol Chem **49** 311
[4] Mousavi-Yeganrh S et al 1984 NIM **B**(3) 364
[5] Oliveira L C and Zamboni C B 2013 AIP **1529** 73
[6] Mahan L K and Stump S E 2008 *Krause's Food & Nutrition Therapy*. (USA, Saunders/Elsevier)
[7] Brosnan J T and Brosnan M E 2006 *J Nutr* **136**(6) 16365
[8] Nimni M E et al 2007 *Nutr Metab* **4** 24
[9] Greenberg R R et al 2011 *Spectrochim. Acta B* **66** 193
[10] Medeiros J A G, Zamboni C B, Zahn G S, Oliveira L C, Dalaqua L. 2005 *Proc. 39º Congresso Brasileiro de Patologia Clinica (Brazil)* CD ROM
[11] Vally H et al 2009 *Clin Exp allergy* **39**(11) 1643

Study of a Li doped CsI scintillator crystal as a neutron detector

T Madi Filho, M C C Pereira, J R Berretta and J P N Cárdenas

Instituto de Pesquisas Energéticas e Nucleares – IPEN/CNEN-SP
Av. Prof. Lineu Prestes 2242 Cidade Universitária
CEP: 05508-000– São Paulo-SP – Brasil
tmfilho@usp.br, macoper@ipen.br, jrretta@ipen.br, ahiru@ipen.br

Corresponding author: tmfilho@usp.br

Abstract. The radiation monitoring system is an important requirement in the premises of a nuclear reactor. A variety of types of radiation (neutrons, gamma, beta and fission products) exist in a reactor, associated to the broad energy spectrum of these radiations, implying the need of detectors to be used in the reactor system and security, as well as radiation monitoring. As the neutron sources are associated to gamma radiation, it is necessary that the neutron detecting system may be capable to discriminate the gamma interference. In our work environment, there are two Nuclear Research Reactors and a neutron irradiator with two AmBe sources (592GBq of Am, each). These conditions warrant the development of new types of detectors. Due to the absence of charge in the neutron, it is necessary to use a converter material that generates radiations capable to produce signals in the detector. Materials with high cross section, like Li or B, are used for this purpose. The CsI crystal doped with ^{6}Li has been studied. The concentration of the lithium doping element (Li) studied was 10^{-3}M. The detector test was done using an AmBe source (37GBq) and gamma sources. The crystal was coupled to a photomultiplier.
Key words: neutron detector, CsI scintillator, ^{6}Li.

1. Introduction

In the design and installation of a system of measurement, it is necessary to understand the processes of interaction of radiation with the detector in order to select the one that best suits the work to be performed.

What makes neutron distinctive is the null action with electromagnetic fields, making its interaction with matter, practically, by means of a strong nuclear interaction.

Neutrons cross electronic layers without any major disruption and do not feel the Coulomb barrier that interferes with charged particles, as they approach the nucleus.

Due to the lack of electric charge, the interaction of the neutron with the atomic electric field is negligible; therefore, the interaction of neutrons with matter occurs, essentially, through direct interaction with the nuclei.

The detection of neutrons is not trivial due to lack of charge of these particles and the peculiarity of their interaction with matter. The neutron sources, also, generate gamma radiation, which may interfere with their measurement. It is necessary that the detector system may be able to discriminate these interferences.

Steven M. Grimes [1] presents the neutron detector types:

Detectors Based on Elastic Scattering, Detectors Based on Capture Reactions, Detectors Based On Fission, based on Inelastic Scattering Counters, Detectors Based on (n, p) or (n, α) Reactions.

These detectors have, internally, a converter for neutrons to produce ionizing radiation. The converters are based on nuclear reactions of the type:

Neutron + Converter → ionizing radiation

The converter is a material with a high probability of interaction with the neutron (high cross section). The cross section is a parameter that expresses the probability of neutron interaction with the target. It is dependent on the energy of the incident neutron [2].

For each of the incident neutron energy, the probability that a conversion reaction may occur depends on the cross section of the converter nuclide. The ratio between the cross sections of various nuclides used as converters, in function of the incident neutron energy, is shown in Figure 1. [2, 5]

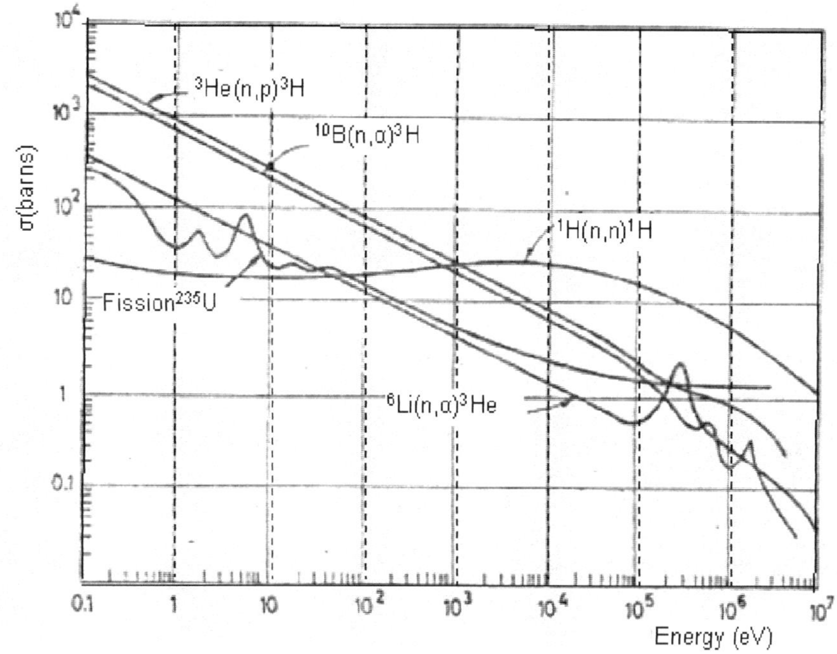

Figure 1. - Cross section of some converters as a function of incident neutron energy. (Figure adapted from Tanarro [3])

^{6}Li is an element used as a converter for detecting thermal neutrons. It features cross section of 940 barns, for thermal neutrons. The reaction of thermal neutrons with ^{6}Li used in detectors can be written as:

$$ {}^{6}_{3}Li + {}^{1}_{0}n \rightarrow {}^{3}_{1}H + {}^{4}_{2}\alpha + 4.78MeV $$

In this reaction, the emitted tritium has kinetic energy E = 2.73 MeV , while α particle is emitted with E_{α} = 2.05 MeV [5]

2. Materials and methods

Lithium doped CsI crystals were grown using the vertical Bridgman technique at the Instituto de Pesquisas Energéticas e Nucleares IPEN/CNEN-SP, [6].

In order to study the response to neutron radiation, the crystals were polished with ethylene glycol and directly coupled to the photomultiplier tube (RCA Model 8575, 21 pins), using silicon (Dow Corning), viscosity of 1.0McStokes, for the optical coupling. This procedure ensured uniform refractive index across the contact surface between the crystal and the photomultiplier. The sides of the crystal, which were not in contact with the photo-sensor, were covered with white Teflon tape to ensure good reflection of light. The electronic modules used for the processing of signals from the photomultiplier tube are shown in Fig. 2

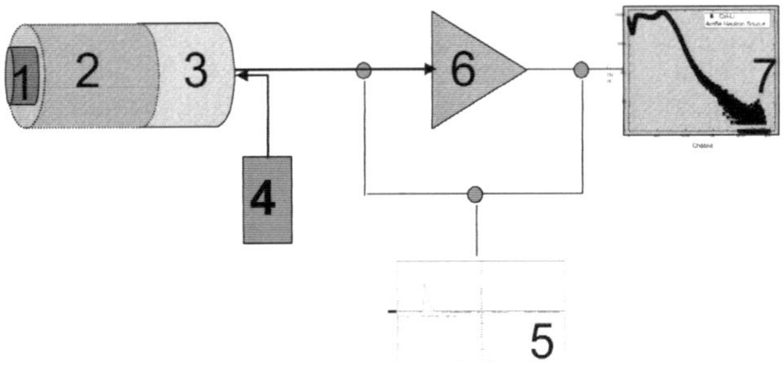

Figure 2. – Schematic of the measurement system, detector and electronics used: 1) Detector, 2) Photomultiplier, 3) Photomultiplier base, 4) HV (1900 V), 5) Oscilloscope, 6) Amplifier, 7) Multi-channel (Spectrum Master ORTEC Model 919 + computer).

The neutron radiation response of the AmBe source was measured, with pure crystal, with doped crystal and without crystal.

The AmBe reaction neutron source (α,n) was acquired in August 17,1970, with a nominal activity of $3.7 \cdot 10^{10}$ Bq (1.0Ci) of ^{241}Am and neutron emission rate (rate neutrons flux) of $2.6 \cdot 10^{6}$ n s^{-1}, calibrated in July 2, 1970. For the purpose of correcting the loss by radioactive decay, the activity was calculated for the date of the experiments, yielding a neutron emission rate of $2.42 \cdot 10^{6}$ n s^{-1} neutrons per second.

The operating voltage of the photomultiplier tube was 1900 V; the accumulation time in the counting process was 1800s. The scintillator crystals used were cut with dimensions of 20 mm diameter and 20 mm height

3. Results

Measurements to assess the ^{6}Li doped neutron detector response were performed. Therefore, a neutron source of AmBe, already described, a Co-60 source and a Cd foil of 0.5 mm thickness were used. The Co-60 source was used to verify the response in the gamma field. To verify if the photomultiplier was contributing to the neutron response, measurements with only the photomultiplier and the AmBe source were performed. Measurements with pure CsI crystal and with the crystal doped with ^{6}Li, using the neutron source, were carried out. In order to verify the response to fast and thermal neutrons, a block of 50.0 mm of paraffin was used. A Cd foil was placed around the detector and the photomultiplier to avoid the scattered neutrons contribution. Fig. 3 shows the array used in that measurements.

Fig.4 shows the results only for the photomultiplier with neutron source and for the CsI + Li crystal using the Co-60 source. In Fig 5, it is shown of measurements performed using the neutron source for a pure CsI crystal and for the crystal doped with ^{6}Li. A paraffin block to slowing down the neutron and a Cd foil to cut neutron with energy below 0.4eV, were used.

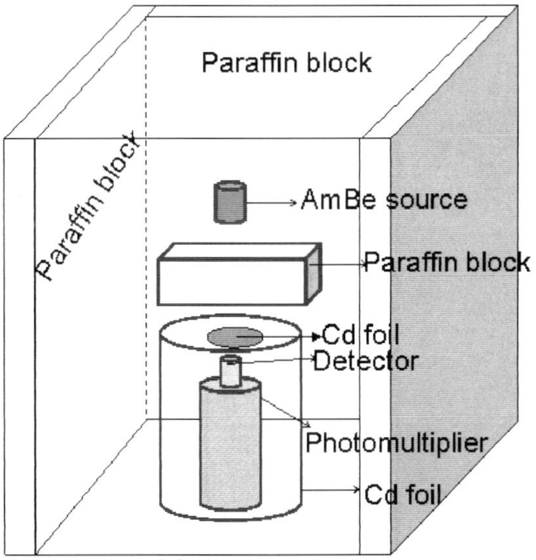

Figure 3. - Schematic of the array used for measurements

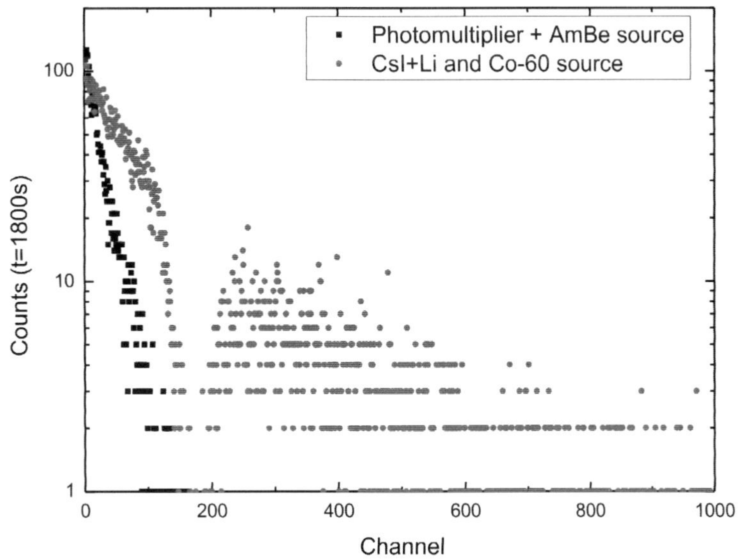

Figure 4. – Results of the measurements using neutron source and the photomultiplier without detector and CsI+Li detector with ^{60}Co source

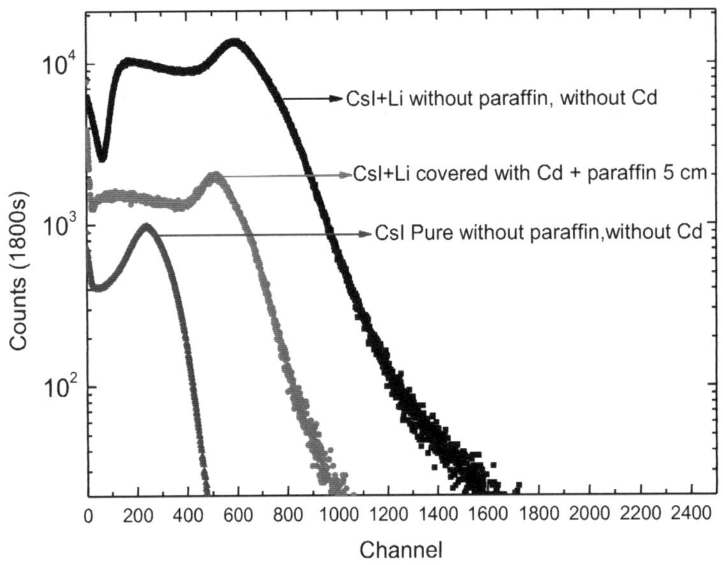

Figure 5. – Results for three different measurements conditions: (a) CsI pure and neutron source, (b) ^{6}Li doped crystal covered with paraffin and Cd foil and neutron source and (c) ^{6}Li doped crystal and neutron source.

4. Conclusions

It could be shown, experimentally that the addition of the Li to the CsI matrix resulted in crystals with promising results, when excited with neutron radiation. The crystals showed sensitive to fast and thermal neutron. Obviously, further work will have to be carried out on these materials, in particular on the concentration of dopants and crystal growth technique parameters. In our work environment can be use as area neutron detector and used in the personal dosimeter for the workers of the nuclear research reactor IEA-R1 and IPEN/MB-01.

The fast and thermal neutron were considered with energy above and below the cadmium cutoff energy (0.4 eV).

References

[1] Neutron Measurement," Steven M. Grimes, Chapter 68 in *The Measurement, Instrumentation and Sensors Handbook*, edited by John G. Webster, CRC Press, 1999

[2] CARLOS H. MESQUITA, TUFIC MADI FILHO, AND MARGARIDA M. HAMADA. *Development of Neutron Detector Using the PIN Photodiode with Polyethylene (n; p) Converter*. IEEE Transactions on Nuclear Science, Vol. 50, N. 4, August 2003

[3] TANARRO, A. S. *Instrumentación nuclear*. Servicio de Publicaciones de La J.E.N.. Madrid1970

[4] A.J. PEURRUNG. *Recent developments in neutron detection*. Nucl. Instrum. Meth. A 443 (2000).

[5] KNOLL, G. F.; *Radiation detection and measurement*. 2 ed. John Wiley&Sons. 1986

[6] Maria C. C. Pereira, Tufic M. Filho, Margarida M. Hamada, Development of crystals based on cesium iodide for measurements of gamma radiation and alpha particles. NUKLEONIKA 2009; 54(3):151–155.

Half-life determination for ^{27}Mg

G S Zahn and F A Genezini

Instituto de Pesquisas Energéticas e Nucleares - IPEN-CNEN/SP. P.O. Box 11049, São Paulo, 05422-970, Brazil

E-mail: gzahn@ipen.br

Abstract. In this work, the half-life of the short-lived magnesium radionuclide ^{27}Mg was measured by following the activity of samples after they were irradiated in the IEA-R1 reactor. An exponential decay function was then fitted to the results using the counts from a ^{60}Co source as livetime chronometer; the individual half-life values obtained for each irradiation were compiled using both the usual unweighted and σ^{-2}-weighted averages, as well as the robust averages obtained using the Normalized Residuals and the Rajeval techniques. The final half-live values obtained aren't compatible with the ENSDF compilation values, but have a similar uncertainty; analysis of the experimental literature values, all from the 50's-60's, show that further measurements should be undertaken in order to achieve a more robust consensus value for this half-life.

1. Introduction

Nuclear applications usually require a good degree of knowledge on several parameters of the nuclei involved, both regarding the safety of the experiment and the reliability of the results. For instance, in Nuclear Activation Analysis (NAA), many nuclear parameters, such as cross section, transition intensities and decay half-life, have to be well known in order to compute the results, and the uncertainties in these parameters frequently undermine the results obtained in the analyzes [1]. In the instrumental variation of NAA, which relies on the use of a well known comparator irradiated together with the samples in order to eliminate most of the nuclear parameters from the equations, the value of the decay half-life is still an important parameter and it appears inside an exponential function, so its uncertainty must be carefully assessed because it may distort the results of the whole analysis.

In the case of magnesium, only the activation of ^{26}Mg produces a radioactive isotope, ^{27}Mg; its half-life was measured in the 50's as 570.6 (18) s [2], 563.4 (18) s [3], 567.0 s [4] and 567.6 (12) s [5], and the for last time in 1970 as 567.7 (7) s [6], resulting in a compiled value of 567.5 (7) s [7]. It should be noted that most of these measurements were performed using NaI(Tl) detectors, and that in most of them the corrections for dead-time and pile-up, which can play quite a big role in these measurements [8], were not thorougly assessed.

In this work, the decay of this isotope was studied by following the activity decay for the gamma-rays peaks associated with its decay in neutron-irradiated natural magnesium samples.

2. Experimental Procedure

In the present experiment, samples were produced by cutting ~5 mg pieces of a 99.781% pure Mg foil (Reactor Experiments 1553) and packing these pieces into sealed polyethylene bags. These samples were then irradiated for 20-25 seconds in the IEA-R1 nuclear reactor under a thermal neutron flux of $\sim 5 \times 10^{12}$ n·cm^{-2}·s^{-1} using a fast pneumatic irradiation facility (sample transit time $\sim$ 10s) and then analyzed by a 20% HPGe detector coupled to a 8192-channel MCA with a source-detector distance of 9 cm. The data collection for each individual sample was made through a batch of 65 subsequent acquisitions of 60 s (realtime).

The resulting spectra were analysed using the Genie-2k software [9], which delivers reliable and accurate peak areas for standard gamma spectra [10]. The determination of the activity of ^{27}Mg was then performed using both the 843.76 keV transition ($I_\gamma = 71.8\%$) and the 1014.52 keV transition ($I_\gamma = 28.2\%$) [7] – it should be noted that a longer-lived peak was found at 846.6 keV, which could interfere with the results obtained using the 843.76 keV transition (see Fig. 1). All the samples were counted together with a ^{60}Co source, and the sum of the counts in the 1173 and 1332 keV transitions was used as a lifetime chronometer, i.e., the number of counts in each of the ^{27}Mg peaks was divided by the sum of counts in both of the ^{60}Co source's peaks to correct for losses related to pile-up and dead time. The individual uncertainties were determined by propagating, for each point, the peak area uncertainties for that specific measurement and for the ^{60}Co peaks used in the normalization; no other uncertainty was considered as a previous study already assessed that peak areas are the only relevant source of uncertainty in this type of measurement [8].

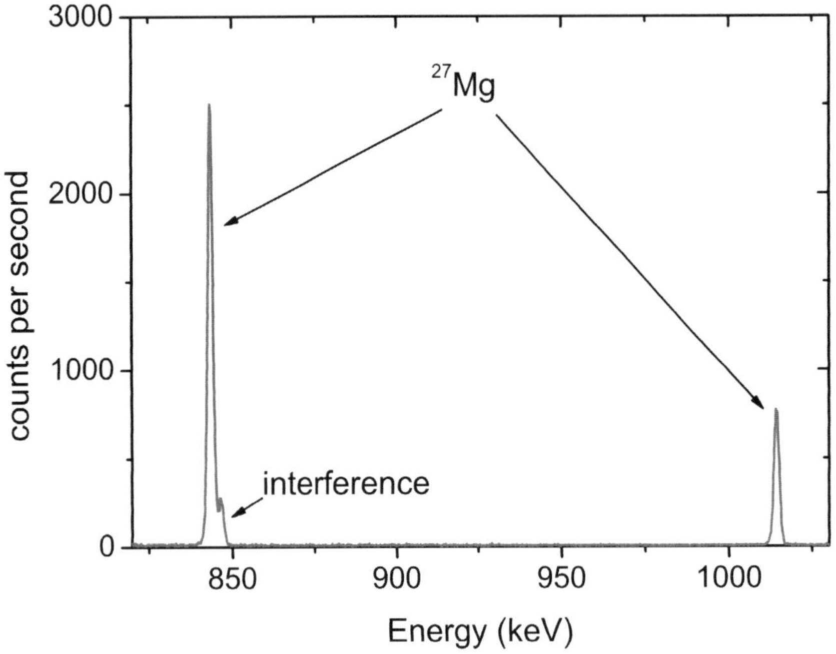

Figure 1. Sample of a spectrum obtained in the measurement of the ^{27}Mg decay, showing both peaks used in this work, together with the interferent peak at 846.6 keV.

A regular exponential decay function was fitted to the results of each of the eight individual measurements using a covariant Gauss-Marquardt routine implemented in the MatLab environment (a sample decay graph can be seen in Fig. 2), and the resulting half-life values (with the corresponding uncertainties, obtained in the fitting procedure) were then averaged using the usual arithmetic and σ^{-2}-weighted means, as well as two robust averaging techniques, *Normalized Residuals* and *Rajeval* [11].

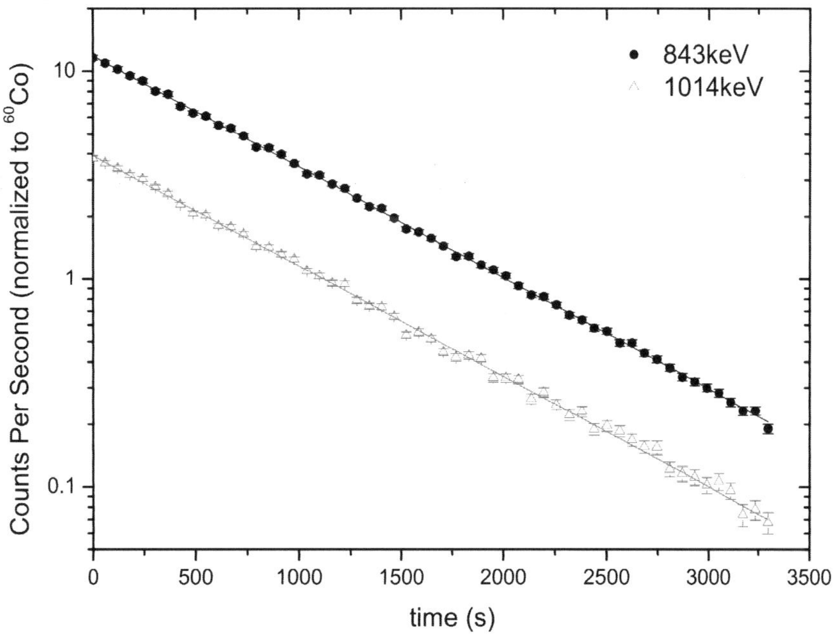

Figure 2. Sample of a activity decay graph, showing the fits for both peaks used in this work.

3. Results

The results obtained in each of the individual measurements are shown in Fig 3, together with the ENSDF compilation value; Table 1 shows the final average results using each of the transitions. The results obtained using both transitions were compatible, indicating that the interferent peak was well taken care of, and as all eight individual results were compatible in both cases, the robust averages gave the same final results as the weighted averages.

Transition (keV)	AM (s)	WM (s)	NR (s)	RT (s)	Ref. (s)
843.76	565.5 (6)	565.5(7)	564.5 (7)	564.5 (7)	567.5 (7) [7]
1014.52	564.9 (11)	565.0 (11)	565.0 (11)	565.0 (11)	

Table 1. Results of the averages for the half-lives of ^{27}Mg using each of the nuclear transitions – *AM*, *WM*, *NR* and *RT* are the arithmetic, σ^{-2}-weighted, Normalized Residuals and Rajeval averages, respectively.

As the half-life values obtained using both transitions were compatible, the most precise value of 564.5 (7) keV is adopted as the final result. This value for the half-life of ^{27}Mg is not compatible to the ENSDF compilation value of 567.5 (7) keV, with a relative error of -0.5%, and is compatible with only one of the results found in the literature – 563.4 (18) s [3]; nevertheless, all the measurements found in the literature are more than 40 years old and used precarious corrections for the pile-up and dead time losses, therefore our result indicates that further measurements should be performed in order to precisely assess the half-life of ^{27}Mg.

4. Conclusions

The result obtained for the half-life ^{27}Mg presented an uncertainty of the same magnitude as the best measurements found on the literature; the results, though, were not compatible with

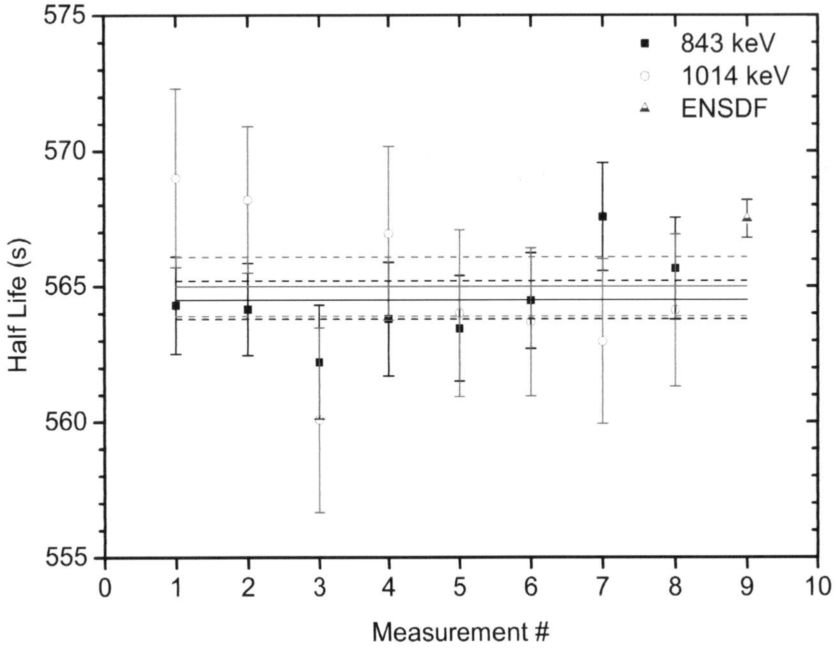

Figure 3. Results of each of the individual measurements for the half-life of ^{27}Mg together with the ENDS compilation value (in blue, from [7]); the full lines represent the result of the weighted average and the dotted lines the $1-\sigma$ interval, and the color of the lines correspond to the transition used.

most of the reported values, and are 0.5% lower than the compiled value found in [7], indicating that further measurements should be performed in order to precisely assess the value of this half life.

References

[1] Lindstrom R M, Zeisler R and Greenberg R R 2007 *Journ. Radioanal. Nucl. Chem* **271** 311–315
[2] Daniel H, Koester L and Mayer-Kuckuk T 1953 *Z. Naturforsch.* **8a** 447–448
[3] Lockett E E and Thomas R H 1953 *Nucleonics* **11** 14
[4] Sargent B W, Yaffe L and Gray A P 1953 *Can. J. Phys.* **31** 235–249
[5] Poularikas A and Fink R W 1959 *Phys. Rev.* **115** 989–992
[6] Repace J L 1970 *Radiochim. Acta* **14** 46–49
[7] Basunia M S 2011 *Nucl. Data Sheets* **112** 1875–1948
[8] Zahn G S, Oliva J W M and Genezini F A 2013 *Radiat. Phys. Chem.* **85** 70–72
[9] Canberra 1999 *Genie-2000 Spectroscopy System – operations manual* Canberra Industries USA
[10] Zahn G S, Genezini F A and Moralles M 2009 *Proceedings of the 2009 International Nuclear Atlantic Conference – INAC2009* (Rio de Janeiro, Brazil: Associação Brasileira de Energia Nuclear – Aben)
[11] Rajput M U and MacMahon T D 1992 *Nucl. Instrum. Meth. A* **312** 289–295

Journal of Physics: Conference Series **630** (2015) 012012 doi:10.1088/1742-6596/630/1/012012

A Commercial off-the-shelf pMOS Transistor as X-ray and Heavy Ion Detector

M A G Silveira[1], M A A Melo[2], V A P Aguiar[3], A Rallo[2], R B B Santos[1], N H Medina[3], N Added[3], L E Seixas Jr[4], F G Leite[1], F G Cunha[1], K H Cirne[1], R Giacomini[2], J A de Oliveira[2]

[1]Department of Physics, Centro Universitário da FEI, S.B.C, S.P., Brazil.
[2]Department of Electrical Engineering, Centro Universitário da FEI, S.B.C, S.P., Brazil.
[3]Institute of Physics, Universidade de São Paulo, S.P., Brazil.
[4]Centro de Tecnologia da Informação Renato Archer, Campinas, S.P., Brazil.

E-mail: marcilei@fei.edu.br

Abstract. Recently, p-channel metal-oxide-semiconductor (pMOS) transistors were suggested as fit for the task of detecting and quantifying ionizing radiation dose. Linearity, small detection volume, fast readout, portability, low power consumption and low radiation attenuation are some of the pMOS advantages over PIN diode and thermoluminiscent dosimeters. A hand-held measurement system using a low power commercial off-the-shelf pMOS as the sensor would have a clear advantage due to the lower cost incurred by a standard technological process. In this research work, we tested the commercial device 3N163 regarding its behaviour as an X-ray sensor, as well as its possible application as a heavy-ion detector. To study the radiation effects of X-rays, a XRD-7000 (Shimadzu) X-ray diffraction setup was used to produce 10-keV effective energy photons. Heavy ions tests involved ^{12}C, ^{16}O, ^{19}F, ^{28}Si, ^{35}Cl, ^{63}Cu and ^{107}Ag beams scattered at 15° by a 275 μg/cm^2 gold target, which provide LETs (Linear Energy Transfer) from 4 to 40 MeV/mg/cm^2. The signal readout was done using a 1 GHz oscilloscope with a 10-Gsamples/s conversion rate, high enough to permit the recording of transient pulses in the drain current. In this case, an ion can cause a current signal proportional to the ion beam used. Through this study it was found that a simple commercial pMOS device can be reliably used as a detector of X-rays as well as heavy ion detector.

1. Introduction

Radiation dosimetry is a very important area to be explored because of current needs for control and monitoring of the dose of ionizing radiation to which human beings are exposed. In the last few years, it has been suggested that electronic devices could be used as radiation monitor [1,2] since the effects of the interaction of radiation with electronic devices are relatively simple and reliable. With the widespread adoption of nuclear medicine, there is a growing demand not only for X-ray and gamma ray dosimeters, but also for good particle counters. Apart from the development of dosimeters for use in medicine, there is also a growing demand for dosimeters that can be used in environments where

[1] M.A.G Silveira, marcile@fei.edu.br.

ionizing radiation should be monitored, as in particle accelerators, satellites, avionics, skin dosimetry and others.

A dosimeter is any device that is capable of providing a reading of the measurement of the absorbed dose deposited in its sensitive volume by ionizing radiation [3]. Recently, p-channel metal-oxide-semiconductor (pMOS) transistors were suggested as fit for the task of detecting and quantifying ionizing radiation dose. Linearity, small detection volume, fast readout, easy calibration, portability, low power consumption and low radiation attenuation are some of the pMOS advantages over other electronic devices [1,2].

In order for a pMOS device be useful as a particle identifier, it is necessary that the pMOS shows a current pulse proportional to the linear energy transferred to its sensitive volume [4]. On the other hand, it is very important that the device also recovers itself quickly after any event registered, being ready to respond to a next event. This way, the device can monitor the dose of radiation imparted to it.

The physical mechanisms that are responsible for the pMOS detection principle are based on the production of electron-hole pairs along the track of the particle. Part of the generated charge accumulates in the device oxide layers, and in its oxide-semiconductor interfaces, leading to total ionizing dose (TID) effects. Besides, the charge generated in the semiconductor bulk by the passage of a single energetic ionizing particle may be collected in the terminals of the device, giving rise to single event effects (SEE) [4-7]. The pMOS transistor designed for dose measurements, the so-called RADFET (Radiation Field Effect Transistor), is manufactured with a special process aiming to produce a thick gate-oxide leading to a large detection volume (gate-oxide), which improves the sensibility of the detector [5]. In order to achieve a better understanding of the mechanisms responsible for radiation damage due to radiation effects, and check the response to radiation of a pMOS sensor, we are studying a commercial off-the-shelf pMOS - 3N163 - exposed to X-ray and heavy ion beams.

The 3N163 device is a p-channel enhancement mode MOSFET largely used in industry for several applications, presenting the following characteristics: very high input impedance, high gate breakdown voltage, ultra-low current leakage, fast switching and low capacitance. Gate oxide thickness is the main parameter controlling the sensitivity of the transistor to radiation [1-4]. Hence, this device is very appropriate in studies of radiation effects, since after irradiation its I_{DS} current is always reduced, facilitating the observation of the relationship between the accumulated dose and the change of the characteristic parameters in the device [3-6].

1.1. Radiation Effects in Electronic Devices

When electronic devices are exposed to ionizing radiation, it may experience parametric variations and even failures. The effects caused by ionizing radiation can be transient or permanent. Transient effects may provoke momentary failure, for example, changes on stored data or even peaks of electrical current that might damage the electrical circuit. These are the motivations of the growing interest in this research area, since energy absorption, carrier generation, recombination, and transport, charge trapping, and defect formation influence the effects provoked by ionizing radiation [3-7].

Total ionizing dose (TID) effects are related to the amount of energy (dose) absorbed by the material. Particles with low linear energy transfer (LET), i.e. low electronic mass stopping power, and electromagnetic radiation, tend to contribute more to these effects. When photons interact with the device oxides, they create electron-hole pairs in the material, mainly by photoelectric effect. In a typical oxide, electrons can escape swiftly to the positive electrode while holes move slowly towards the negative electrode by a hopping mechanism. The slow motion of the holes increases the probability that they are trapped by defects in the bulk of the oxide or near the oxide-semiconductor interface. This positive charge concentration may change the basic operating characteristics of the device [4, 6, 7]. In the context of radiation sensor applications, a p-channel transistor has an advantage over an n-channel transistor: because of the hole trap mechanism, the current always decreases in a p-channel transistor and its threshold voltage always become more negative, as shown in Figure 1. Therefore, these parameters vary monotonically, in stark contrast with the behavior of an n-channel transistor. In an n-channel transistor, competition between the effects of charge trapped in the oxide and charge trapped in the oxide-semiconductor interface leads to a non-monotonic variation of the electrical parameters of the device.

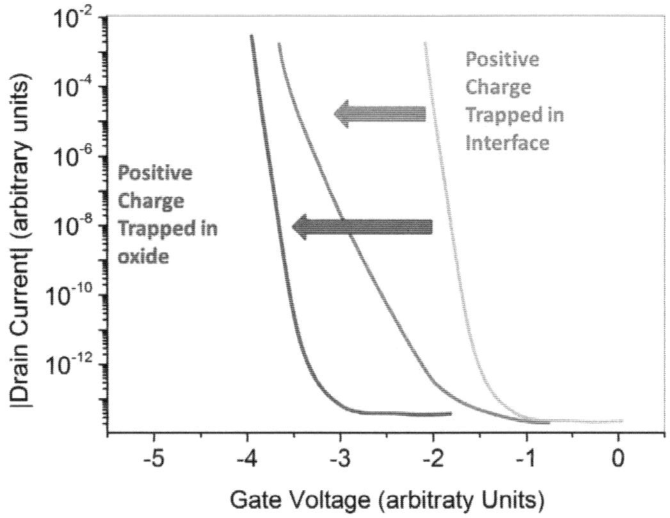

Figure 1. Effect of ionizing radiation on the behavior of the p-MOSFET when electric charges are trapped in the gate oxide and the Si/SiO_2 interface.

Single Event Effects (SEE) are induced by the travel of a single energetic particle across the active region of a semiconductor device. There is a variety of effects depending on the geometrical and physical characteristics of the device and the location in the device where most of the charge is generated, for instance. In this paper, we will concentrate on single event upset (SEU). In a SEU, the charge generated in the sensitive volume of the device is collected, leading to a current pulse of short duration. This effect begins with energy deposition that results in charge generation (electron-hole pairs) in some target medium, either owing directly to the incident particle (direct ionization) or via secondary daughter products (indirect ionization). Electrons and holes move by diffusion and drift through the semiconductor until they are collected in the terminals, giving rise to a current pulse. The electric charge collected is proportional to the linear energy transfer (LET) by the ion on the device [8]. Figure 2 shows the typical pulse generated by SEU in a pMOS.

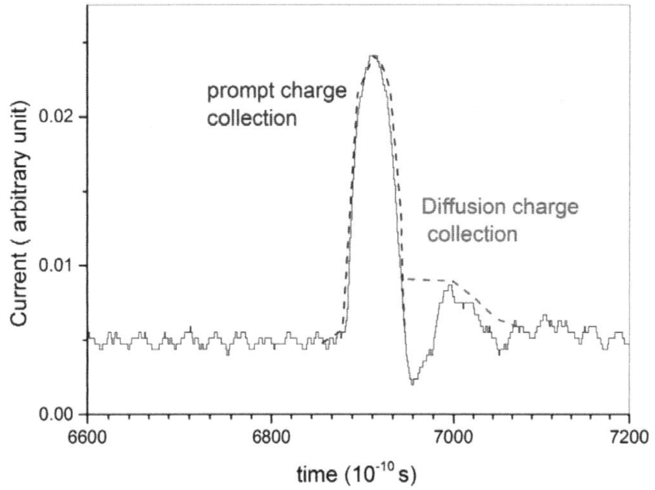

Figure 2. Current pulse of a typical SEU in pMOS.

2. Experimental Setup

In this section we present the experimental setup used in the irradiations with X-rays and with heavy ion beams as well as a brief description of the preparation of the devices used in the tests. It is worth noticing that X-rays were used for tests of the device as a dosimeter and that the tests with heavy ions probed the suitability of the device as a particle counter.

2.1. Preparation of the devices

The DUTs used in this study were 3N163 pMOS transistors, as shown in Figure 3. Preparation of the devices consisted mainly in removing the metallic layer that protects the devices, reducing any interference this could have on the absorbed dose. The removal was done mechanically. Besides, this process also allows the heavy ions to reach sensitive parts of the device.

Figure 3. Picture of the DUT 3N163.

2.2. X-ray Irradiation

The DUT was submitted to an X-ray beam of 10-keV effective energy, produced by a *XRD-7000 diffractometer* (Shimadzu) of the Centro Universitário da FEI. This diffractometer was calibrated with an ionization chamber and the X-ray dose rate in silicon was calculated using air and silicon mass attenuation coefficients. For the irradiation procedure, the dose rate was set to about 16 rad/s with an accumulated dose of 1500 krad. During the irradiation, the DUT was placed perpendicular to the X-ray beam; it was operating in the linear region, using a drain voltage $V_D = 0.1$ V, and a gate voltage V_G near the threshold voltage V_T. The V_G value was calibrated experimentally, since it depends on the transistor type, through a precision potentiometer, connected to the transistor gate terminal. Readouts were done by monitoring the voltage V_D on the drain, in order to observe possible deviations caused by the radiation effects on the transistor. These voltage values are acquired through a set of amplifiers, which wrap the signal acquisition system for the values that are within the full scale of the A/D system (0 V - 1.5 V). The information is collected and stored by an MSP 430 microcontroller (MCU), which is performed through software. Details about the bias circuit and signal measurements are shown in reference [9]. Figure 4 shows the MOSFET under test using X-ray radiation.

Figure 4. X-ray System of the Centro Universitário da FEI.

2.3. *Heavy Ion beam irradiation*

Tests on Single Event Effects were performed on the 8 MV Pelletron Accelerator of São Paulo University that provides heavy-ion beams in a wide range of masses and up to 75 MeV of energy. In tests with the 3N163 pMOS transistor for verification of their behavior as a detector of heavy ions, beams of ^{12}C, ^{16}O, ^{19}F, ^{28}Si, ^{35}Cl, ^{63}Cu and ^{107}Ag were scattered at 15° by a 275 μg.cm^{-2} gold target, which provide LETs from 4 to 40 MeV/mg/cm^2. All tests were performed in a vacuum chamber, as shown in Figure 5.

Figure 5. DUT in the vacuum chamber.

The signal readout was done using a 1 GHz oscilloscope with an ADC of 10 Gsamples/s which allowed us to register short-duration drain current pulses. Beam monitoring was done using surface barrier detectors. More details of the experimental apparatus can be seen in references [10,11].

3. Results

3.1. *The device as a dosimeter*

Figure 6 shows the monitoring of drain current as a function of the accumulated dose in the DUT 3N163. The fitted trend lines allow us to interpret this result as a calibration curve in X-ray radiation doses [9]. Besides, it provides a way to perform real-time radiation dose monitoring. The best fit found for the current it is very close to a linear relation with dose as it can be seen in the figure when comparing with the fit using a straight line.

In accordance with that reported in reference [9], through this study it is possible to observe the behaviour of a device widely used in industry when subjected to ionizing radiation of the X-ray, whereas the drain current decreases as a consequence of the cumulative dose. With these obtained results, it was possible to develop a study that suggests the use of this type of transistor as a simple, efficient and inexpensive dosimeter, since it requires less complex electronics, and can be widely used in radiation studies in various fields such as medicine, radiation protection and reliability evaluation.

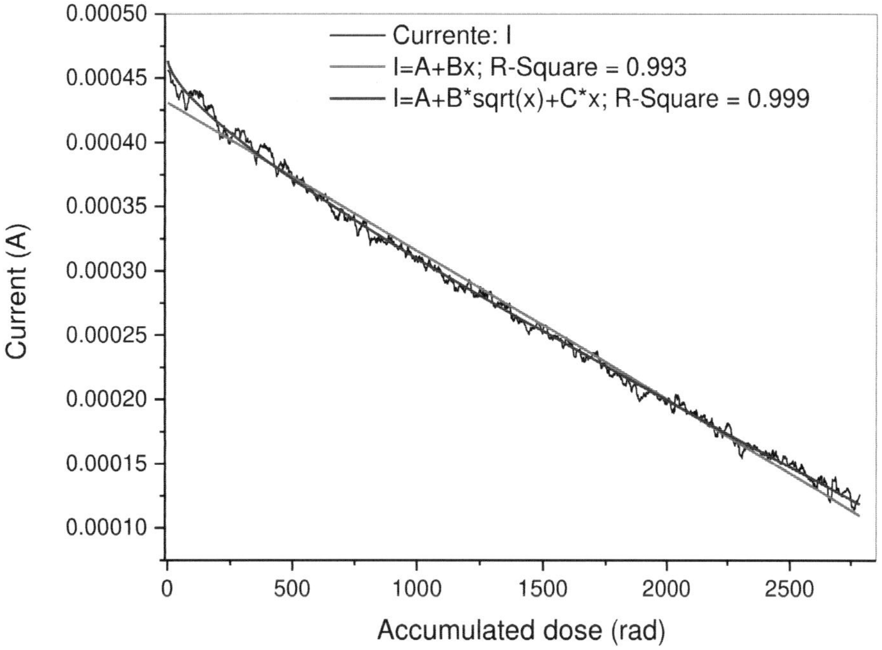

Figure 6. Linear response of the drain current as a function of the total ionizing dose accumulated in DUT 3N163.

3.2. The device as a particle counter

The charge collected on each current pulse generated by the passage of an ion through the device was determined from the analysis of the voltage pulses as a function of time. Figure 7 shows different pulses recorded by the oscilloscope for four different ion beams: ^{16}O, ^{19}F and ^{28}Si. It is possible to observe that the time duration and the shape of peak voltage vary with the ion beam. The voltage peaks related to the silicon ions are higher than the other peaks, while the peaks for oxygen are less resolved. To calculate the charge collected by the terminals of the DUT, only the first peak of each signal was considered, i.e., only the prompt charge collection was estimated in this work.

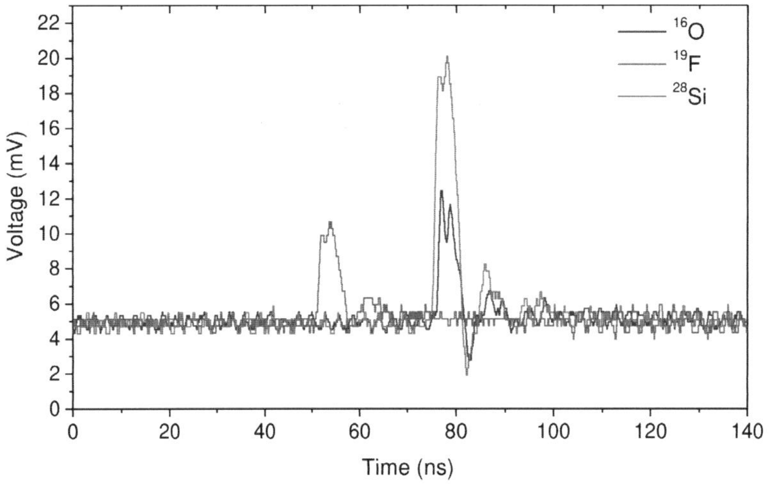

Figure 7. Voltage pulse as a function of the time, corresponding charge collected between source and drain for the O, F and Si ions in DUT 3N163.

Charge collected by the terminals of the 3N163 when exposed to different heavy ions is presented in Figure 8. It is important to note that since the device is a simple p-type CMOS transistor, the charge collected is, to a very good extent, proportional to the ion LET. In Table I, the values of energy for each ion used in this experiment are presented.

Table I – Ion beam Energies

Ion	Energy (MeV)
^{12}C	45.00 (3)
^{16}O	52.50 (3)
^{19}F	42.00 (3)
^{28}Si	66.00 (3)
^{35}Cl	75.00 (3)
^{63}Cu	82.00 (3)
^{107}Ag	97.50 (3)

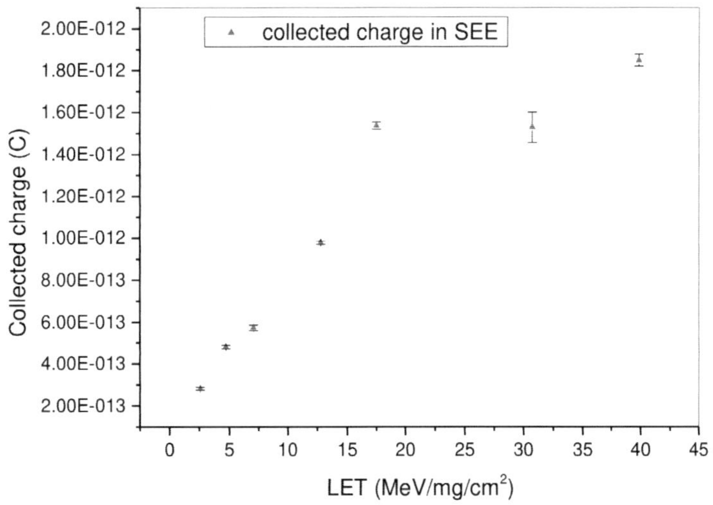

Figure 8. Collected charge as a function of the LET in DUT 3N163.

Cross section for the occurrence of an upset as a function of the incident ion LET for the 3N163 transistor is shown in Figure 9. The results indicate that ^{12}C ions produce less upset events than the higher LET, heavier ions. For a LET above 10 MeV/mg/cm^2, approximately, SEE cross section achieves a roughly constant value of $\sigma \approx 2.5x10^{-5}$ cm^2 [10]. The results indicate also there is a same proportionality factor in the number of SEU and the incident number of heavy ions from LET around 10 MeV/mg/cm^2 to LET=40 MeV/mg/cm^2, allowing to use the 3N163 pMOS transistor as a LET independent particle counter.

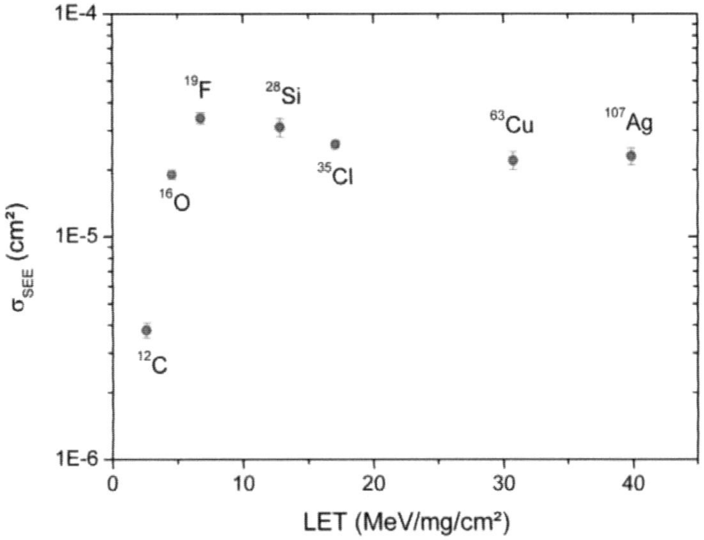

Figure 9. Cross-section as a function of the LET for DUT 3N163.

4. Conclusions

This work shows strong indications that it is possible to develop a low-cost, yet efficient, radiation monitor instrumentation using a commercial pMOS transistor. The tests performed indicate that the device is capable of measuring the total dose accumulated, and also of monitoring the number of impinging particles. The relationships between drain current and accumulated dose, and between collected charge and linear energy transfer, were observed during the tests. Additionally, a cross section for the occurrence of SEE was determined for the 3N163 transistor. New radiation tests should be performed to evaluate the behavior of this mechanism to other accumulated dose ranges.

5. References

[1] S.Martínez-García et al. 2013, IWBBIO 2013 Proceedings.
[2] L.J. Asensio et al.2006. "Evaluation of a low-cost commercial MOSFET as radiation dosimeter". Elsevier, Sensors and Actuators **A 125** 288–295.
[3] Frank Herbert Attix, 2004. "Introduction to Radiological Physics and Radiation Dosimetry". Edt: WILEY-VCH
[4] A. Johnston, World Scientific Publishing Co. Pte. Ltd., California Inst. of Tech., USA (2010).
[5] M.A. Carvajal et al., 2010."Readout techniques for linearity and resolution improvements in MOSFET dosimeters." Sensors and Actuators **A 157** 178-184.
[6] M.A.G. Silveira et al.,2012. "Performance of electronic devices submitted to X-rays and high energy proton beams", Nucl. Instr. and Methods, Physics. Research. **B**, **v.273** 135 - 138.
[7] R.D. Schrimpf, D.M. Fleetwood, 2004. "Radiation effects and soft errors in integrated circuits and electronic devices" .Edt: Singapore: World Scientific.
[8] J. Pellish, 2012. NASA Goddard Space Flight Center. NSREC, Short Course.
[9] J. A. de OLIVEIRA, 2012. IV Workshop sobre Efeitos da radiação em Componentes Eletrônicos, 2012, São José dos Campos. Proceedings of IV WERICE AEROESPACIAL. São José dos Campos: Instituto de Estudos Avançados 34 – 35.
[10] V.A.P. Aguiar et al., 2014. "Experimental Setup for Single Event Effects at the São Paulo 8UD Pelletron Accelerator", Nuclear Instruments and Meth. In Phys. Resear. B, v. 332, p. 397-400.
[11] N. H. Medina, M. A. G. Silveira, et al.,2014. "First successful SEE measurements with heavy ions in Brazil". Proceedings NSREC 2014.

Improving the quality of radiographic images acquired with conical radiation beams through divergence correction and filtering

M I Silvani[1], G L Almeida[1], R M Latini[2], A V B Bellido[2], E S Souza[3] and R T Lopes[3]

[1] Instituto de Engenharia Nuclear - CNEN, Caixa Postal 68550, CEP 21945-970, Rio de Janeiro - RJ, Brazil
[2] Instituto de Química - UFF, Outeiro de São João Baptista, s/no, Campos do Valonguinho, Centro, CEP 240020-150 Niteroi - RJ Brazil
[3] Laboratório de Instrumentação Nuclear - UFRJ, Caixa Postal 68509, CEP 21945-970, Rio de Janeiro - RJ, Brazil

E-mail: msouza@ien.gov.br

Abstract. Earlier works have shown the feasibility to correct the deformation of the attenuation map in radiographs acquired with conical radiation beams provided that the inspected object could be expressed into analytical geometry terms. This correction reduces the contribution of the main object in the radiograph, allowing thus the visualization of its otherwise concealed heterogeneities. However, the non-punctual character of the source demanded a cumbersome trial-and-error approach in order to determine the proper correction parameters for the algorithm. Within this frame, this work addresses the improvement of radiographs of specially tailored test-objects acquired with a conical beam through correction of its divergence by using the information contained in the image itself. The corrected images have afterwards undergone a filtration in the frequency domain aiming at the reduction of statistical fluctuation and noise by using a 2D Fourier transform. All radiographs have been acquired using ^{165}Dy and ^{198}Au gamma-ray sources produced at the Argonauta research reactor in Instituto de Engenharia Nuclear - CNEN, and an X-ray sensitive imaging plate as detector. The processed images exhibit features otherwise invisible in the original ones. Their processing by conventional histogram equalization carried out for comparison purposes did not succeed to detect those features.

1. Introduction

The well-known historic radiograph *Hand mit Ringen* (Hand with rings) reproduced in figure 1, taken with the newly discovered X-rays caused a high impact in Science because for the first time it was possible to visualize the bones of an *alive* creature without any intrusive procedure. However, the very 1st radiograph made by Röntgen and shown to his colleagues was a set of weights in a box [1], i.e., a potential industrial application rather than a medical one. In both cases, the new radiation made visible high attenuating materials enclosed inside low attenuating ones. If Röntgen had taken samples with an inverse configuration - a creature with an exoskeleton, instead of his wife's hand for instance - then the application of X-rays in Medicine could most likely be delayed. Indeed, the favorable ratio between the attenuation coefficients for bones and flesh - surpassing 5 for soft X-rays [2] - allowed Röntgen to exhibit a dramatic picture at the *Physik Institut* of the University of Freiburg on 1 January, 1896. As for the industrial application, its spreading as a non-destructive assay had to wait until the advent of high voltage tubes - 100 kV in 1913, 200 kV in 1922 and 1000 kV in 1931 - and of radioisotopes in 1946. Further details can be found elsewhere [3].

An inverse situation, i.e., the visualization of a low attenuating stuff enclosed within a high attenuating one, is a much more difficult task due to the feeble achievable contrast resolution, an essential parameter, as it rules the capability of the human eye to differentiate neighbor pixels as individual ones. This constraint is even strengthened when a broad angle conical beam is employed, for the apparent thickness increases with the distance between the point where the radiation hits the detector and that one directly beneath the source. This variable thickness not only deforms the attenuation map, but concomitantly increases the dynamic range of the image, limiting its further stretching, precluding thus the improvement of the contrast.

Some techniques have been developed [4-9] to overcome this hindrance but they work based upon a blind approach, i.e., using the information provided solely by the image itself. Recent works [10-11] perform their tasks by entering some physical properties of the object being radiographed and their related geometrical arrangements. Due to these utterly different approaches, [10-11] would perform better than [4-9] but the proper information should be provided as above mentioned, otherwise they couldn't work at all.

Hand mit Ringen: It was presented to Professor Ludwig Zehnder of the Physik Institut, University of Freiburg, on 1 January 1896.

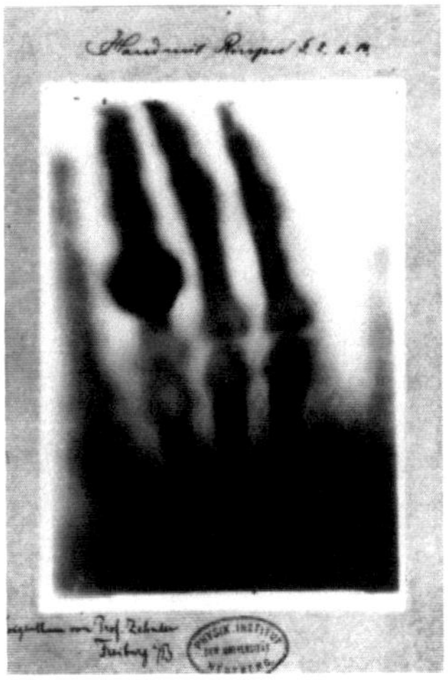

Linear Atenuation Coefficient Ratio
As inferred from Ref [2]

Energy (keV)	Bone / Muscle	Bone / Fat
20	5.62	9.30
21	5.55	9.10
22	5.47	8.87
23	5.37	8.63
24	5.26	8.36
25	5.15	8.05
26	5.01	7.71
27	4.85	7.33
28	4.66	6.89
29	4.26	6.39
30	4.14	5.80

From: Wikimedia Commons, the Free Media Repository
Source: National Aeronautics and Space Administration.

This image is in the public domain because its copyright has expired.

Figure 1. X-ray radiograph made by Wilhelm Röntgen in 1895.
The ratio of some linear attenuation coefficients are shown for reference.

This work proposes an algorithm to unveil feeble attenuating structures or heterogeneities enclosed inside high attenuating objects inspected by radiography employing conical beams. Unlike [10-11] which require a detailed description of the object, its attenuation properties, and geometrical arrangement, the proposed algorithm extracts the necessary information from the image itself. But, unlike the blind techniques, it is assumed that the image has been acquired with a conical beam and that the main object is a plate or a pipe. Besides the correction of beam attenuation and divergence, the radiographs in this work, acquired with ^{165}Dy and ^{198}Au have undergone a filtering in the frequency domain by using a Fourier transform aiming at the reduction of noise and statistical fluctuation. The grounds for this expectation are the fine results obtained in [10] for images generated without noise.

2. Methodology

Since this work addresses the detection of feeble attenuating structures concealed into high attenuating ones, two test-objects matching this condition, as sketched in figure 2, which have been earlier manufactured [11] have been radiographed with 95 and 412 keV gamma-rays from the reactor-produced ^{165}Dy and ^{198}Au respectively.

The test-object *Aluminum Plate* comprised of thin aluminum strips sandwiched between 3 mm thick aluminum plates has been exposed to a 3.89×10^8 Bq ^{165}Dy source. In this object, the thinnest 0.1 mm strips add circa 0.46% to the total attenuation factor. As for the *Iron Pipe* exposed to a 6.38×10^7 Bq ^{198}Au source, the thinnest 1 mm-thick strip adds 2.53% to that factor.

Since it is expected that thinner and farther strips from the source would be hardly detectable due to the higher attenuation factors caused by the main object, the strips have been distributed in such a way to cover a fair range of thicknesses of Al strips along the plate.

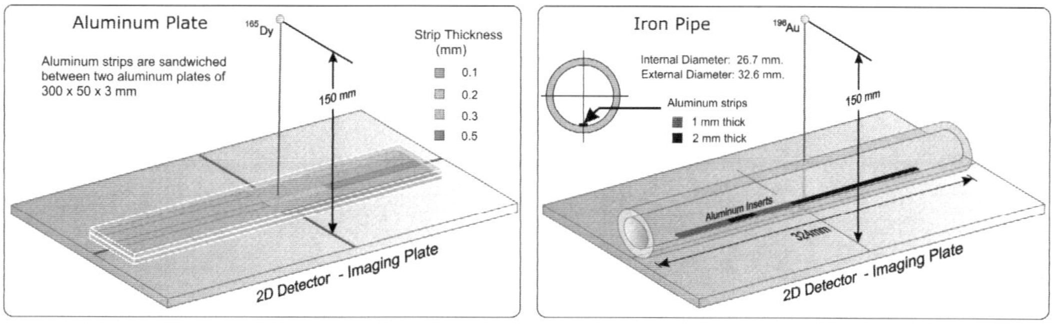

Figure 2. Geometrical arrangement used to acquire the radiographs of the specially designed test-objects with gamma-ray sources and an X-ray sensitive imaging plate.

All radiographs have been taken with an X-ray sensitive Imaging Plate - IP, and *developed* with a 0.050 mm resolution reader. Typical exposure times - ranging from a couple of hours or days for the Aluminum Plate and Iron Pipe respectively - have to be lengthened due to the limited achievable source activity produced under a 2.23×10^5 n.cm^{-2}.s^{-1} neutron flux. The expected lower efficiency of an X-ray IP when operating with higher energies certainly contributes as well to a longer exposure time, which has been chosen after some trials.

After the approach adopted in this work, the radiographs undergo a correction of divergence and attenuation with an algorithm specifically developed to perform this task, followed by a filtering in frequency domain. Both procedures have been embedded into a program written in Fortran 90 language capable to plot the images as well.

2.1. Beam divergence and attenuation correction

Unlike a parallel beam, which casts an undisturbed attenuation map of the inspected object on a 2D detector, a divergent beam not only changes this map, but also causes a secondary effect, namely an increase in the dynamic range of the acquired image. This increase precludes a further stretching of the dynamic range, making thus impossible to visualize low attenuating features, for they would be overwhelmed by the abnormally high - and wrong - attenuating factors of the main object. Indeed, the farther an object region from the source, the higher its apparent attenuation factor, since the beam would have to travel a greater distance both across the object and the air. Earlier works [10-11] suggested techniques to correct those phenomena, but they required a tedious and cumbersome trial-and-error procedure to determine the proper algorithm parameters.

This work proposes a novel algorithm which performs an automatic correction of the disturbed image by using the existent, although concealed, information contained in the image itself. It's based on the flattening of the longitudinal and transversal pixel intensity profiles, simulating thus, that one which would be exhibited by a radiograph taken with a parallel beam.

A longitudinal profile, as shown in figure 3, comprised by a sequence of transversal ones, has a convex shape with a maximum occurring at the point beneath the source. In the scheme showing the construction of the longitudinal profile as a sequence of transversal ones, they are depicted as smooth curves, but actually they are *modulated* by the heterogeneities eventually present in the main object. The baseline of the each transversal profile raises as it get closer to the source due to the higher flux causing a bending in the longitudinal profile. Due to the same reason this profile is broader at the central region.

In order to simulate a radiograph acquired with a parallel beam, both profiles should be corrected until they become flat, preserving however the *modulation*. This flattening requires a different approach for each profile type, but both demand that the source should be positioned in the vertical line passing through the center of the object to assure the symmetry required by the correction algorithm. The flattening of the longitudinal profile is carried out in two steps: first, the baseline of each transversal profile - the line connecting the pixel intensities occurring at its extremes - is raised to the same level of the baseline of the central profile, carrying with it all pixel intensities mounted on it, preserving thus the net pixel intensities with regard to the base line.

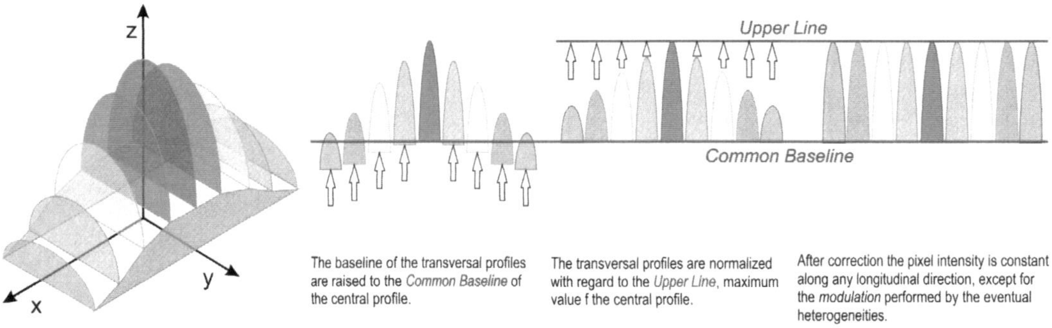

Figure 3. Typical sketch of a 2D pixel intensity profile and the procedure employed to correct its longitudinal component, comprised by a sequence of transversal ones side by side.

In the second step, all net pixel intensities measured from the henceforth *common baseline* are normalized with regard to the maximum occurring at the point beneath the source. After this process all transversal profiles would be enclosed between the *common baseline* and the *upper line* defined by the maximum pixel intensity, and the image should exhibit a homogeneous background along the longitudinal direction. It should be stressed that this homogeneity doesn't mean a loss of information of the eventual fine inner structure of the object, since it remains as irregularities *modulating* each transversal profile.

After this flattening, the image would still exhibit shadows at the bottom and upper regions, due to the convex shape of the transversal profiles which should be as well flattened without loss of information. This task is accomplished by rectification of the profile in such a way that it becomes a horizontal straight line. As each profile is unique - and unknown - a polynomial is fitted to each of them in order to obtain their individual approximate shape. The gap between the polynomial and the *upper line* at each point along the transversal profile is added to the pixel intensity at that point rectifying thus its value. In this rectification process the net pixel intensities *mounted* on the polynomial are raised to the *upper line*, flattening hence the profiles but keeping the information.

The original radiograph of the test-object *Iron Pipe* has a longitudinal profile with a large dynamic range *D* reaching 30k as shown in figure 4. Both graphs refer to the same image, but different domains for the sake of illustration. The zoomed graph on the right shows the rectification of 5 transversal profiles conducted by a 3rd degree polynomial.

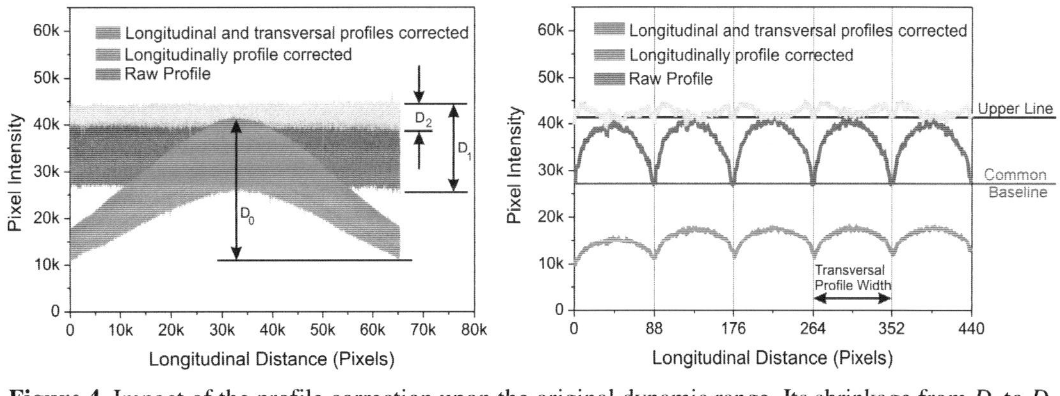

Figure 4. Impact of the profile correction upon the original dynamic range. Its shrinkage from D_0 to D_2 allows a further stretching of this parameter, making possible the visualization of otherwise invisible feeble attenuating heterogeneities.

As a tiff-type image, the pixel intensity is enclosed within the range 0 - 65,535. Hence, the high *D* value *steals* almost 50% of the potential to improve the image through stretching of the dynamic range.

A longitudinal correction yields a profile with the narrower - and better - value D_1, where the dynamic range has been reduced to about 20k. A further rectification of the transversal profiles produces an even narrower dynamic range D_2 limited to 5k, or about 8% of the total dynamic range for a tiff image. When this shallow profile undergoes a stretching of its dynamic range, the feeble attenuating features emerges from a - now diminished - low background.

2.2. Filtration of high frequencies

As shown in the scheme of figure 5, the 2nd step carried out in this work to improve the quality of the radiographic images involves the filtering in the frequency domain of their Fourier transforms, aiming at the reduction of high frequencies related to statistical fluctuation and noise. Since the Fourier transform is a widely known technique and the scheme is fairly self-explaining no further details are given.

For this purpose, a Butterworth low-pass filter, *B(u,v)*, is applied to the discrete Fourier transform *F(u,v)*, yielding the filtered transform *C(u,v)*. Examples of Fourier spectra for both transforms, as well as a map of the filter are shown for reference. The parameter L_0 - distance of a point to the center of the plane of frequencies defined as a fraction α of the semi-height of this plane - rules the frequency cutoff, while *n* specifies the abruptness of this cut.

High *n*-values would make the Butterworth filter to perform more like an ideal low-pass filter. Both parameters should be adequately chosen by the customer, a known matter of concern in the field of frequency filtering. A value n=1 has been chosen in this work because for higher values some artifacts arise as white shadows at the borders of the image. Hence, filters with a steeper roll-off, such as Chebyshev for instance have not been tested. The value of α has been chosen by applying a Butterworth high-pass filter to the filtering procedure and watching its impact on the final image. As edges - due to their steep contrast change - contain high frequency components, they have been chosen as an indicator of the point where high frequencies do not contribute to their formation anymore, but only would add noise to the image. The results in figure 6 show that beyond about 0.7 no insert is visible any more. This means that the edges of the inserts have no significant frequency component above this point, and thus, all of them may be discarded. It may also be noticed that, as expected, the granulometry becomes finer as α increases. The visually estimated minimum α-value of 0.7 has been cross-checked by carrying out filtrations around this value with a Butterworth low-pass filter as later shown in the section Results.

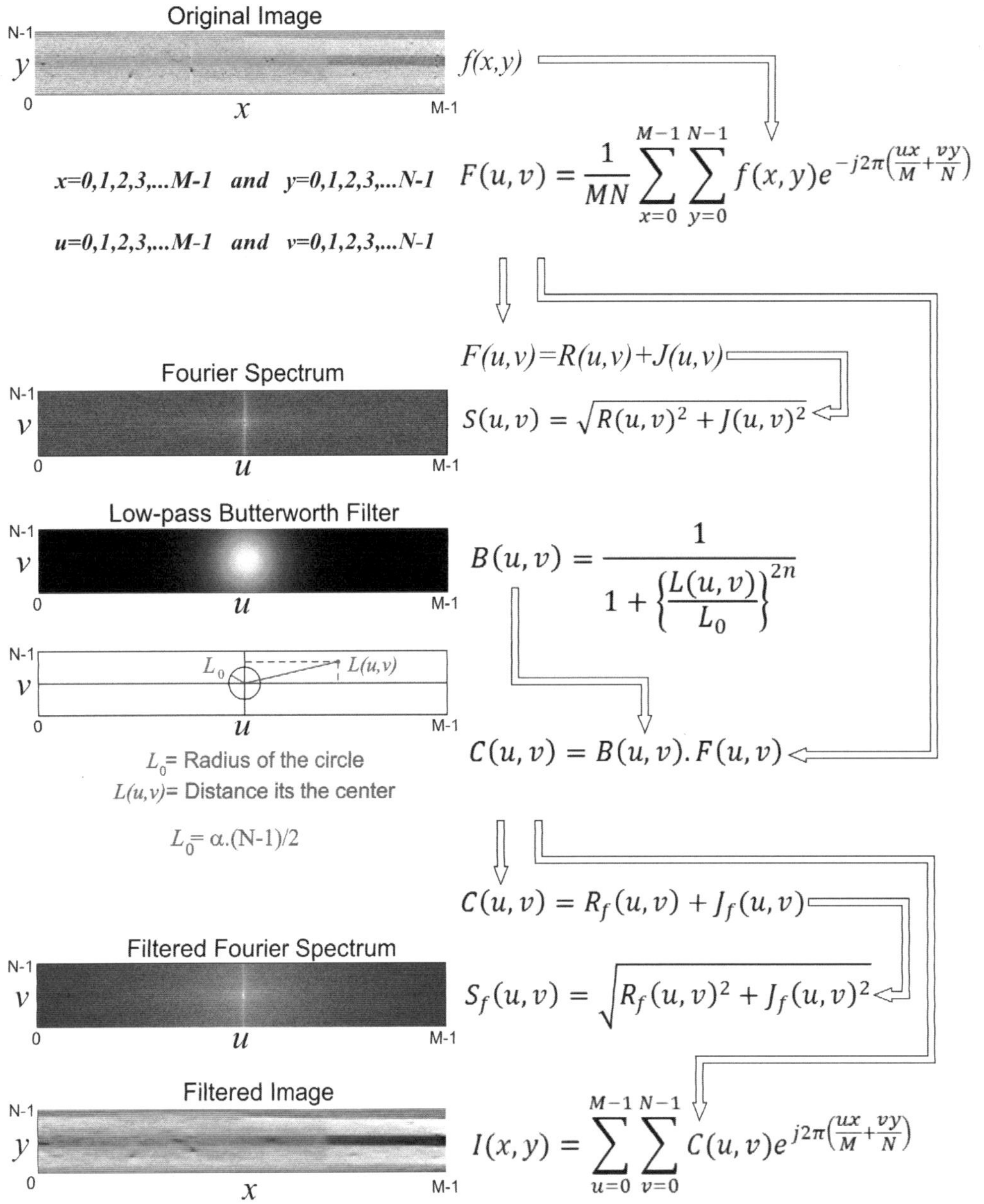

Figure 5. Scheme of the applied filtration procedure in the frequency domain by using a 2D Fourier transform and a low-pass Butterworth filter.

Once the minimum α-value is evaluated, it is used in the Butterworth low-pass-filter, and the final image is obtained after the Fourier inverse transform $I(x,y)$.

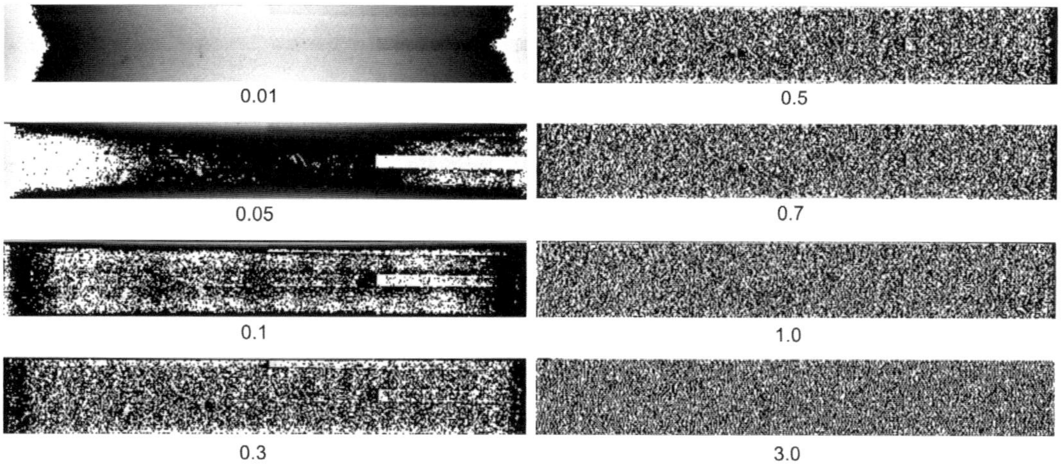

Figure 6. Filtered radiographs, acquired with 95 keV gamma-rays from ^{165}Dy, of the plate used as test-object. Labels refer to the different α-values employed to define the high-pass Butterworth filter for n=1.

3. Results

The final images for the test-objects, after they have undergone beam divergence correction and filtration, are summarized in figure 7. Combined effects of beam divergence and the related increased thickness emerge as a diffuse brighter spot at the center of the radiographs, i.e., directly beneath the source used to acquire them, surrounded by a darker region.

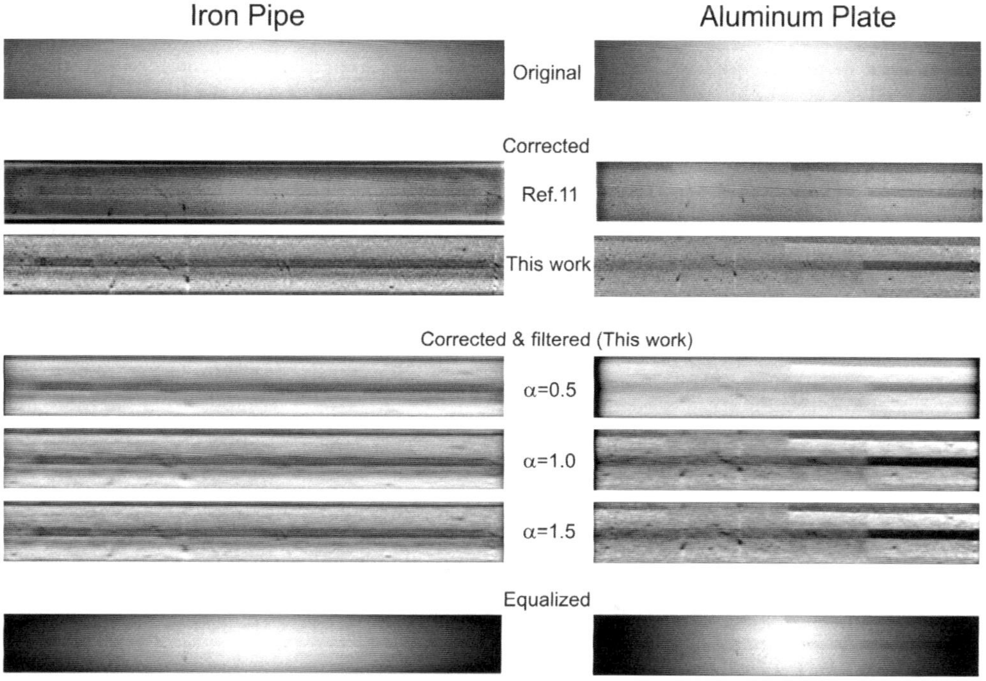

Figure 7. Original and processed radiographs. The *correction* process changes substantially the images, but a further *filtration* only slightly improves them. A value n=1 has been used for the low-pass Butterworth filter. Images from [11] and an equalized histogram are shown for comparison.

The consequent wide dynamic range makes difficult or even impossible to detect feeble differences between close attenuation factors. This drawback has been reversed by the correction algorithm allowing the retrieving of the image that would be obtained *if* the beam were parallel.

A substantial improvement can be observed between original and corrected the images. Indeed, the thin aluminium strips in both test-objects - hardly observable in the original image - can be readily identified in the corrected ones, except for the thinnest 0.1 mm-thick located near the edge of the *Aluminium Plate*. For this test-object, when exposed to the 95 keV gamma-rays from ^{165}Dy this strip adds 0.46% to the total attenuation factor. As for the *Iron Pipe* exposed to 412 keV gamma-rays from ^{198}Au, the thinnest aluminum strip, 1 mm thick, adds 2.53% to the maximum attenuation factor.

It can be observed that a filtration of the previously corrected images only causes a slight improvement of their qualities, even so, for α-values greater than 0.5. Such an outcome is consistent with the results shown in figure 6 where values below 0.7 should be discarded in order to avoid the elimination of frequencies having significant amplitude. Indeed, the design of both test-objects incorporates inserts with sharp edges, which require high frequencies for their proper representation in the frequency domain. Therefore, a low α-value eliminates them causing an image blurring.

The original images have also undergone a treatment with the conventional *histogram equalization* - a technique for contrast enhancement - for the sake of comparison with the developed algorithm. It can be observed that the achieved improvement is very poor.

4. Conclusion

An algorithm to correct the attenuation map of radiographs acquired with conical beams has been developed and its soundness and performance evaluated. It is based on the fact that for radiographs taken with conical beams, parameters such as geometrical arrangement and attenuation coefficients are concealed and embedded in the attenuation map of the final images themselves. Hence, although they cannot be easily retrieved, their overall impact can be reversed in order to emulate an image obtained with a parallel beam.

For this purpose gamma-ray radiographs of two specially tailored test-objects have been obtained with 95 and 412 keV gamma-rays from the reactor-produced isotopes ^{165}Dy and ^{198}Au respectively. The corrected images have undergone a filtration in the frequency domain after the 2D Fourier transform approach aiming at their further improvement. The corrected images were remarkably improved exhibiting features otherwise concealed in the original ones. A filtration nevertheless solely improved their quality very slightly, which emerges as a reduction of the granulometry. Both processes have been embedded into a Fortran 90 program specially written to deal with the images and to plot them as well.

References

[1] Kemp M. 1998 Röntgen's rays *Nature* 394, 25 (2 July 1998)

[2] Chiu P, Yan Y, Lin L and Chang P 1997 Attenuation of x-ray in muscle and alveolar bone during mandibular premolar periapical radiographic procedure *Kaohsiung J. Med. Sci.* **13** 631-34

[3] https://www.nde-ed.org/EducationResources/CommunityCollege/Radiography/Introduction/history.htm.

[4] Zimmerman J B, Pizer S M, Staab E V, Perry J R, Mccartney W and Brenton B C 1988 An evaluation of the effectiveness of adaptive histogram equalization for contrast enhancement *IEEE Transactions on Medical Imaging* **7** No. 4 304-12

[5] Kim J, Kim L, and Hwang S 2001 An advanced contrast enhancement using partially overlapped sub-block histogram equalization *IEEE Transactions on Circuits and Systems for Video Technology* **11** No.4 475-84

[6] Kaur M, Kaur J and Kaur J 2011 Survey of contrast enhancement techniques based on histogram equalization *International Journal of Advanced Computer Science and Applications* **2** No.7 137-41

[7] Abdullah-Al-Wadud M 2012 A modified histogram equalization for contrast enhancement preserving the small parts in images *International Journal of Computer Science and Network Security* **12** No.2, 1-4
[8] Tai S, Tsai T, Chang Y, Tsai W and Tang K 2012 Contrast enhancement through clustered histogram equalization *Research Journal of Applied Sciences Engineering and Technology* **4** No.20 3965-68
[9] Kaur R and Kaur S 2013 Contrast enhancement techniques for images - A visual analysis *International Journal of Computer Applications* **64** No.17 20-25
[10] Almeida G L, Silvani M I and Lopes R T 2013 An algorithm to unveil the inner structure of objects concealed by beam divergence in radiographic image acquisition systems *XXXVI Brazilian Workshop on Nuclear Physics* September 1-5 São Sebastião-SP Brazil
[11] Silvani M I, Almeida G L and Lopes R T 2013 Enhancing the quality of radiographic images acquired with point-like gamma-ray sources through correction of the beam divergence and attenuation *XXXVI Brazilian Workshop on Nuclear Physics* September 1-5 São Sebastião-SP Brazil

Journal of Physics: Conference Series **630** (2015) 012014 doi:10.1088/1742-6596/630/1/012014

Neutron Damage in the Plasma Chamber First Wall of the GCFTR-2 Fusion-Fission Hybrid Reactor

L N Pinto[1], E Gonnelli[1], P C R Rossi[1], T Carluccio[2], A dos Santos[1]

[1]Nuclear Engineering Center, Nuclear and Energy Research Institute- IPEN/CNEN-SP
Av. Lineu Prestes 2242 - Cidade Universitária - 05508-000 - São Paulo, SP, Brazil

[2] Centro Tecnológico da Marinha em São Paulo - CTMSP
Av. Prof. Lineu Prestes, 2468 05508-000 São Paulo, SP, Brazil

E-mail: leticiafi@gmail.com; e.gonnelli@gmail.com

Abstract. The successful development of energy-conversion machines based on either nuclear fission or fusion is completely dependent on the behaviour of the engineering materials used to construct the fuel containment and primary heat extraction systems. Such materials must be designed in order to maintain their structural integrity and dimensional stability in an environment involving high temperatures and heat fluxes, corrosive media, high stresses and intense neutron fluxes. However, despite the various others damage issues, such as the effects of plasma radiation and particle flux, the neutron flux is sufficiently energetic to displace atoms from their crystalline lattice sites. It is clear that the understanding of the neutron damage is essential for the development and safe operation of nuclear systems. Considering this context, the work presents a study of neutron damage in the Gas Cooled Fast Transmutation Reactor (GCFTR-2) driven by a Tokamak D-T fusion neutron source of 14.03 MeV. The theoretical analysis was performed by MCNP-5 and the ENDF/B-VII.1 neutron data library. A brief discussion about the determination of the radiation damage is presented, along with an analysis of the total neutron energy deposition in seven points through the material of the plasma source wall (PSW), in which was considered the HT-9 steel. The neutron flux was subdivided into three energy groups and their behaviour through the material was also examined.

Keywords: **Hybrid Reactors, Fusion Reactors, Neutron Damage, GCFTR-2, MCNP-5**
PACS: 28.52.Fa

1. Introduction

Hybrid subcritical reactors, as the Gas Cooled Fast Transmutation Reactor (GCFTR-2), or even Accelerator Driven Systems (ADS), are recently being widely studied since they would offer a range of important advantages over conventional fission reactors. The greater security due to the subcritical core is one of the strongest points in favor of hybrid reactors, along with the greater public acceptance of nuclear energy. Hybrid systems would also offer minor problems with respect to proliferation, significantly reduction of nuclear waste production and the possibility of transmutation of the existent nuclear waste. In this case, neutrons might be used to "transmute" long lived radioactive nuclei into much more short-lived ones [1,2,3,4,5,6].

In a fusion reactor, energy will be mainly produced by the fusion reaction between Deuterium (D) and Tritium (T) nuclei: $D + T \rightarrow {}^{4}He + n$. It is important to notice that this kind of system has a large neutron excess in comparison to fission reactors. Since neutrons produced by fusion reactors do not contribute for fusion reactions, the neutron leakage is complete. On the other hand, in a fission reactor the neutrons maintain the chain reaction. Despite the fact that in a fission reaction more neutrons are produced than in a fusion reaction, their leakage from the system is always kept to a minimum. In view of the high number of neutrons produced by fusion reaction without being appropriately exploited, conceiving a system in which neutrons may be utilized for other purposes is then possible. The peak of the neutron emission occurs at ~14 MeV, energy in which ^{239}Pu and ^{238}U undergo fission, enabling the utilization of fusion neutron excess as neutron source in a subcritical fission system. Therefore, a hybrid fusion-fission reactor may be seen as a coupled system; a subcritical fission reactor which has as source of neutrons a fusion reactor core [7].

1.1. The Fast Neutron Damage Problem

Damaging effects due to high-energy neutrons are considered crucial in fusion reactors, since a large number of neutrons are produced and escape from the system. In structural steels, as the one studied in this work, the main radiation damage mechanism is the displacement of atoms (DPA) from their lattice positions [8,9]. When radiation such as fast neutrons displaces atoms, vacancies are formed, which may be concentrated and create voids within the material, leading to void swelling phenomena. With the accumulation of damages, structural material may be deformed or even lose its integrity due to the high degradation of its properties. Unfortunately, currently there is no sufficient information which could be used to establish absolute damage limits for structure materials in fusion reactors. However, it is acknowledged that ferritic steels are less susceptible to displacement damage effects than austenitic steels, which led us to concentrate our study on the HT-9 steel, proved to be one of the most resistant and a strong candidate to be a structural material in fusion facilities [10,11].

This work presents a study of the displacement per atoms and the energy deposition in the first plasma source wall (PSW) composed by High-Cr martensitic HT-9 steel due to the high fast neutron flux in the Gas Cooled Fast Transmutation Reactor (GCFT-2) facility [12,13,14]. Studies of this nature are essential to evaluate which materials may be employed in fusion facilities, the thickness of material needed and also the lifetime of the structures, contributing to the understanding of these technological challenges.

2. Computational Simulation

2.1. The MCNP-5 Gas Cooled Fast Transmutation Reactor (GCFTR-2) Simulation

As the name suggests, the Gas Cooled Fast Transmutation Reactor (GCFTR-2), is a helium-cooled, fast and subcritical reactor which uses coated transuranic (TRU) fuel particles enclosed in SiC matrix pins. The GCFTR-2 was entirely reproduced using MCNP-5 simulations. The Figure 1 shows a side and an upper view scheme of the reactor.

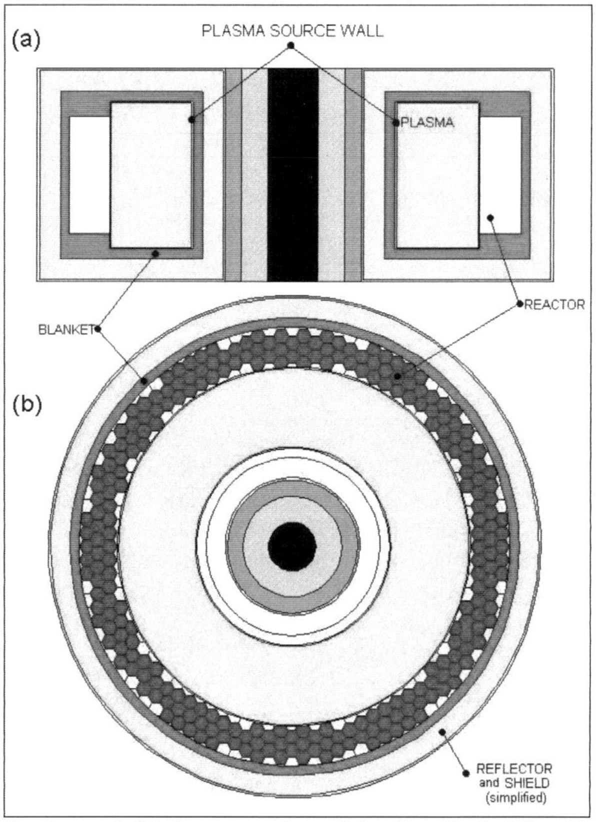

Figure 1 – (a) The side view and (b) the upper view of the Gas Cooled Fast Transmutation Reactor (GCFTR-2) simulated in the MCNP-5. In the picture, the word "reactor" corresponds to the fission core.

The PSW has a thickness of 3.5 cm and separates the fission core and plasma regions. All the calculations concerning neutron damage presented on this work were precisely made considering this wall. In order to perform those calculations coherently, a 300 ºC temperature was assumed and gas and heat production were also considered. The cross sections needed were generated by the NJOY99 Nuclear Data Processing System, broadly used when it is necessary to convert Evaluated Nuclear Data Files (ENDF) format into useful information for practical applications, as fission and fusion nuclear reactor analysis.

2.2. Fuel Element, Fuel Pin, Matrix and Triso Particles

The fuel element consists of a hexagonal arrangement of 384 fuel pins, each of them encapsulated with ODS steel, containing Triso particles enclosed by an inert matrix of SiC [14].

The Triso particles have a very interesting composition. They are constituted of five layers, the inner shell being TRU oxide and the other shells being, respectively, Zirconium carbide, Tungsten Carbide, Silicon Carbide and Tungsten carbide. In order to achieve a high burn up and no fuel failure this special configuration is required.

2.3. Determination of Radiation Damage

As previously mentioned, the DPA is a standard parameter considered for determination of radiation damage in materials. It considers the number of times in which an atom is displaced for a given neutron flux fluence. Furthermore, it is possible to extract information about the material that is under effect of the flux, and from the own flux.

The two variables that can determine the DPA rate are the number of displacements per unit volume, denoted by R divided by the atomic density of the material (HT-9 steel in this case). Equation (1) expresses the DPA rate (R_{DPA}):

$$R_{DPA} = \frac{R}{N} = \int_{Em}^{EM} \sigma_D(E_i)\Phi(E_i)dE_i \quad (1)$$

In the integral, the upper and the lower limits are denoted respectively by EM, the maximum energy of the incoming particle and Em, the minimum energy of the incoming particle. The $\Phi(E_i)$ is the energy-dependent particle flux per unit energy, and $\sigma_D(E_i)$ is the energy-dependent displacement cross-section [9,15].

Through the MCNP-5 Nuclear Reactor Physics code, the neutron flux $\Phi(E_i)$ was calculated as function of energy in the PSW and then combined with the DPA cross-section $\sigma_D(E_i)$ of the HT-9 steel. Utilizing equation (1), the computation of the R_{DPA} was possible. The achieved results are presented at the section below.

3. Results

The total neutron energy deposition in the PSW is expressed in Figure 2. It is possible to notice that the R_{DPA} (in fluence per year, FPY) presents the maximum value at the inner surface of the PSW, in the position of 0.5 cm, as expected. Due to attenuation effects, the neutron damage decreases when the R_{DPA} is calculated for other position through the PSW.

For all calculations it was considered that the neutron source strength for the GCFTR-2 was equal to 7.1×10^{19} (number per second) [13].

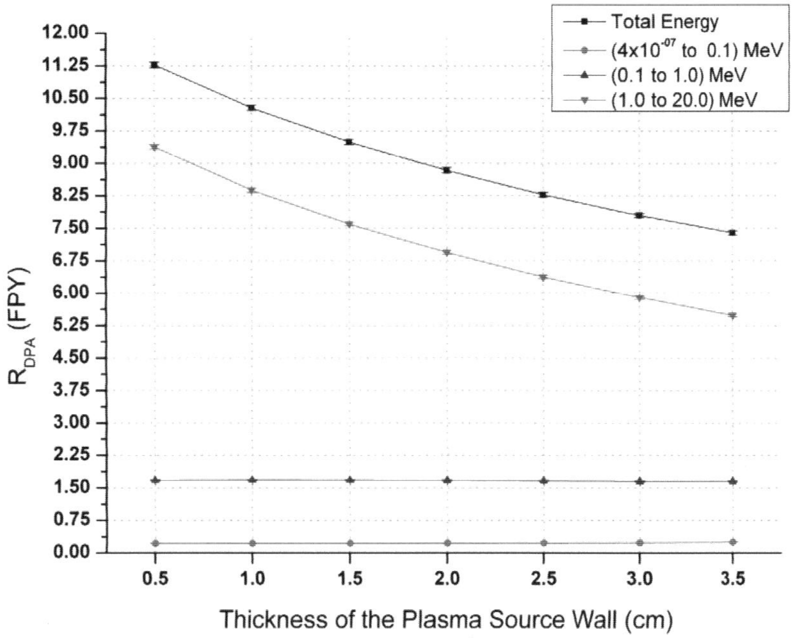

Figure 2 - Neutron energy deposition expressed in R_{DPA} (FPY) in function of seven positions within the PSW.

Besides the information concerning the total neutron energy deposition, Figure 2 also indicates the energy deposition according to the neutron energy group. Neutrons between 0.4 eV and 0.1 MeV are considered in the second epithermal energy group; between 0.1 MeV and 1 MeV in the first epithermal energy group and between 1 MeV and 20 MeV in the fast energy group [15]. Neutrons with energy up to 0.4 eV, in which thermal neutrons are included, presented no energy deposition within the material according to MCNP-5 simulations - indicating a non-interaction - and therefore are not represented in tables neither in graphs. The graph shows clearly that the major part of the neutron energy deposition was due to fast neutrons, followed by a small contribution of first epithermal group. Second epithermal group contribution was negligible.

By evaluating the energy deposition in various points through the PSW it was possible to determine which points of the material would be more affected by the radiation. The maximum energy deposition due to fast neutrons occurred at the inner points of the wall, and presented a decreasing along the material thickness. On the other hand, the first epithermal group energy deposition was practically constant, not presenting any decrease regardless of the position within the PSW.

More details may be appreciated in Tables 1 and 2, which presents, respectively, the neutron flux spectrum and the R_{DPA} for three neutron energy groups (second epithermal, first epithermal and fast) in two points (the most inner and outer) evaluated within the PSW.

Table 1- Flux spectrum for the PSW inner surface (0.5 cm) and for the outer PSW surface (3.5 cm)

Energy Group	**Energy Interval** (MeV)	**Flux Spectrum** PSW 0.5 cm surf. n / cm² / s	%f	PSW 3.5 cm surf. n / cm² / s	%f
2nd Epithermal	4E-7 to 0.1	2.17(1)E+14	30.4%	2.17(1)E+14	33.8%
1st Epithermal	0.1 to 1	2.89(2)E+14	40.5%	2.82(2)E+14	43.9%
Fast	1 to 20	2.08(1)E+14	29.1%	1.43(1)E+14	22.3%
	Total	7.14(1)E+14	100%	6.43(1)E+14	100%

Table 1 shows the neutron flux at the most inner and outer points considered within the PSW. This flux is also subdivided into three energy groups: second epithermal, first epithermal and fast. %f represents the percentage representatively correspondent of each component of the neutron flux - reminding that neutrons up to 0.4 eV were excluded from the analyses due to its non-interaction and, consequently, no energy deposition.

Table 2- R_{DPA} for the PSW inner surface (0.5 cm) and for the outer PSW surface (3.5 cm)

Energy Group	**Energy Interval** (MeV)	R_{DPA} PSW 0.5 cm surf. FPY	%f	PSW 3.5 cm surf. FPY	%f
2nd Epithermal	4E-7 to 0.1	0.221(2)	1.9%	0.253(2)	3.4%
1st Epithermal	0.1 to 1	1.67(1)	14.8%	1.66(1)	22.4%
Fast	1 to 20	9.38(5)	83.3%	5.49(3)	74.2%
	Total	11.28(7)	100%	7.39(5)	100%

Table 2 shows the R_{DPA} in fluence per year (FPY) at the most inner and outer points considered within the PSW. The R_{DPA} is also subdivided into three energy groups: second epithermal, first epithermal and fast. %f represents the percentage representatively correspondent of each component of the fluence.

By subdividing the study into three energy groups it was possible to evaluate the behaviour of neutrons within the material. Analysing the Table 1, it was possible to notice that at the most outer position within the PSW the first and second epithermal fluxes were enhanced by 3.4% each, while the fast flux presented a 6.8% of decrease. The behaviour may be explained by the thermalization of fast neutrons, which interact with the nuclei of the material.

From table 2, as seen in Figure 2, it may be observed that the larger contribution to R_{DPA} on the inner surface of the PSW (position of 0.5 cm) corresponds to the fast neutron flux (83.3% of R_{DPA}). The first and second epithermal groups correspond to 16.7% of R_{DPA}. The contribution of the second epithermal group, in this case, is found to be negligible.

The results for the neutron damage presented in Figure 2 and Table 2 are in agreement with the literature [11]. One important point that should be highlighted is that the previous studies concerning the HT-9 steel were experimental [10, 11]. Through the simulations presented in this work it was possible to achieve similar results, indicating that not only the present study contributes to the research and understanding of the irradiation effects on the HT-9 steel but also indicates that the theoretical approach may be utilized to the study of other materials with great accuracy.

4. Conclusions

The neutron damage determination was presented in this work with a special regard to the fast neutron damage problem applied to a hybrid reactor. Considering the calculated results, it was possible to confirm the predicted values of R_{DPA} in the Plasma Source Wall (PSW). The good agreement with the literature [11] indicates an excellent computational simulation of the system.

The work showed the total neutron energy deposition calculation in seven positions through the High-Cr martensitic HT-9 steel, demonstrating in which manner the energy loss may occur. Besides the total neutron energy deposition, the second epithermal, first epithermal and fast components were evaluated. As expected, according to MCNP-5 simulations, the thermal neutron energy deposition was not relevant to cause lattice damage in the steel and was not even considered in the analyses. As discussed in this work, the second epithermal group also presented negligible results.

The subdivision of the neutron flux into three energy groups was essential to demonstrate the thermalization effects. The work showed that principally neutrons belonging to the energy interval up to 0.4 eV and neutrons from the second epithermal energy group may be motive of concern from the shielding point of view, since not only they are not absorbed by the PSW but also the neutrons from the fast spectrum are thermalized, enhancing the number of low energy neutrons leaving the plasma wall.

In previous studies concerning the HT-9 steel, only mechanical tests were considered [10, 11]. In this context, computational simulations are needed to a better understanding of the experiment. The agreement between the previous results and the theoretical ones presented in this work indicates that not only the current study contributes to the research and understanding of the irradiation effects on the HT-9 steel but also indicates that the theoretical approach may be utilized to the study of other materials with great accuracy.

Studies of the same nature as the one presented are essential to evaluate which materials may be employed in fusion facilities, the thickness of material needed, the lifetime of the structures, and also exactly what kind of radiation must be shield both for biological and for instrumentation protection.

For future works, an improvement in the estimation of maximum R_{DPA} may be done by comparing other nuclear neutron data libraries, upgrading the computational code input by adding the magnets and increasing the number of the particle histories. A biological shielding study is also foreseen.

Acknowledgments

We would like to express our gratitude to IPEN/CNEN-SP, CAPES, ELETRONUCLEAR and FAPESP, institutions that made our research possible.

References

[1] Waltar A E Todd D R Tsvetkov P V 2012 *Fast Spectrum Reactors* (Springer)
[2] Nifenecker H Meplan O David S 2003 *Accelerator Driven Subcritical Reactors - Series in Fundamental and Applied Nuclear Physics* (CRC Press)
[3] Oka Y Furuta K Kondo S 1989 *Fusion Engineering and Design* **7** pp 369-376.
[4] Salvatores M *et al* 1996 MUSE-1: A first experiment at MASURCA to validate the physics of

Subcritical Multiplying Systems relevant to ADS *2nd ADTT Conf*
[5] Rubbia C *et al* 2004 *PHYSOR-2004* **42** 4209647
[6] Maiorino J R dos Santos A Pereira S A 2003 *Brazilian Journal of Physics* **33** pp 267-272
[7] Freidberg J P Kadak A C 2009 *Nature Physics* **5** pp 370-372
[8] Kulctaski G L 1974 Radiation Damage to Materials in DT Fusion Reactors *Proc. of the Fifth Conference on Plasma Physics and Controlled Nuclear Fusion Research.*
[9] Koutsky J Kocik J 1994 *Radiation Damage of Structural Materials* (Elsevier Science Ltd)
[10] Smidt F A J Malmberg P R Sprague J A Westmoreland J E 1976 *Irradiation Effects on the Microstructure and Properties of Metals* **ASTM STP 611** pp 227-241
[11] Chen Y 2013 *Nuclear Engineering and Technology* **45** pp 311–322
[12] Stacey W M *et al* 2005 *Nuclear Technology* **150** pp 162-188
[13] Stacey W M *et al* 2006 *Nuclear Technology* **156** pp 99-123
[14] Stacey W M *et al* 2007 *Nuclear Technology* **159** pp 72-105
[15] Mascitti J A Madariaga M 2011 *Science and Technology of Nuclear Installations* **ID534689** pp 1-6

Particle acceleration and radio emission for SGRs/AXPs as white dwarf pulsars

R. V. Lobato[1†], Jaziel Coelho[2], M. Malheiro[1]

[1]Departamento de Física, Instituto Tecnológico de Aeronáutica, Vila das Acácias, São José dos Campos, SP, 12228-900, Brazil

[2]Dipartimento di Fisica and ICRA, Sapienza Università di Roma, P.le Aldo Moro, 5, I-00185, Roma, Itália

E-mail: †rvlobato@ita.br

Abstract. Recently, an alternative model based on white dwarfs pulsars has been proposed to explain a class of pulsars known as Soft Gamma Repeaters (SGR) and Anomalus X-Ray Pulsars (AXP) [6][4], usually named as magnetars. In this model the magnetized white dwarfs can have surface magnetic field $B \sim 10^7 - 10^{10}$G and rotate very fast with frequencies $\omega \sim 1$ rad/s, allowing them to produce large electromagnetic (EM) potentials and generate electron-positron pairs. These EM potentials are comparable with the ones of neutron stars pulsars with strong magnetic fields. In our study we consider two possible processes associated with the particle acceleration: in one we have the pair production near to the star polar caps i.e., inside the light cylinder where magnetic-field lines are closed, on the other the creation of pair is in the Outer Magnetosphere i.e., far away of the star surface where magnetic field are open [1]. This analysis of the possibility of radio emission was done for all the 23 SGRs/AXPs of the McGill Online Magnetar Catalog [7] that contains the current information available on these sources. Our work is a first attempted to find an explanation for the puzzle why for all the SGRs/AXPs was expected radio emission, but it was observed in only four of them.

1. Introduction

Pulsars are neutrons stars with a large magnetic field, since their discovery, many kinds of pulsars are found. There is one class named as Soft Gamma Repeaters (SGRs) and Anomalus X-Ray Pulsars (AXPs) that have been intensively studied. They are known as neutron stars with large period of rotation $P \sim (2-12)$s and large spin-down $\dot{P} \sim (10^{-13} - 10^{-10})$s/s.

1.1. Canonical Spin-Powered Pulsar Model

In the framework of pulsars there is one class known as rotation-powered pulsars, a neutron star rotating uniformly with a frequency Ω, where the conversion of rotational energy into electromagnetic energy occur, i.e, the loss of rotational energy of the star provides the power. In this model we consider a dipole moment μ with a orientation α in the rotation axis. On the surface of star the magnetic field at equator is $B_s \sim \mu/R^3$ and at the poles $B_p = 2\mu/R^3$, where R is the radius of the star. The rotational energy is given by [9]:

$$E_{\rm rot} = \frac{1}{2} I\Omega^2, \tag{1}$$

where I is the moment of inertia, and the loss of rotational energy of the pulsar:

$$\dot{E}_{\rm rot} = I\Omega\dot{\Omega} + \frac{1}{2}\dot{I}\Omega^2 \approx I\Omega\dot{\Omega}. \tag{2}$$

Such configuration has time-varying dipole moment, in an infinity referential, and radiates energy with a rate

$$\dot{E}_{\rm rot} = \frac{2\mu^2\Omega^4 \sin^2\alpha}{3c^3}, \tag{3}$$

where c is the speed of light. Combining (2) and (3) we obtain the surface magnetic field at the equator as [5]:

$$B_s = B_p/2 = \left(\frac{3c^3 I}{8\pi^2 R^6} P\dot{P}\right)^{1/2}, \tag{4}$$

with $P = 2\pi/\Omega$.

2. SGRs/AXPs as White Dwarf Pulsars

Neutron stars pulsars with an extremely magnetic field $B_p \sim (10^{12} - 10^{14}\text{G})$ are known as magnetars. Currently there are 23 SRGs/AXPs [7] that are classified as magnetars or candidates. An alternative model has explained them as white dwarfs, because the magnetars modeled as neutron stars have some characteristics that are not understood very well, i.e., SRGs/AXPs are very slow rotating pulsars comparing to ordinary pulsars $P \sim (0.001 - 1)$s, spin-down rates larger than normal pulsars $\dot{P} \sim (10^{-15} - 10^{-14})$s/s, strong outburts of energies per weeks or months (for more details see [6, 4]). As show by [10], the process of release energy in a white dwarf can be explained in terms of a canonical spin-powered pulsars model, because in several aspects they are similar.
For example, if we consider a star with $M = 1.4M_\odot$ and $R = 10^6$ cm, from (4) the magnetic field at poles is:

$$B_p = 3.2 \times 10^{19}(P\dot{P})^{1/2}G, \tag{5}$$

now, in the case of a white dwarf with $M = 1.4M_\odot$ and $R = 3 \times 10^8$ cm, there is a new scale for the magnetic field at poles:

$$B_p = 4.21 \times 10^{14}(P\dot{P})^{1/2}G, \tag{6}$$

following the last consideration there are new values for the: mass density, moment of inertia, dipole moment and rotation energy. It explains a large class of problems related in considering this SGRs/AXPs as neutron stars. Furthermore, we cannot ignore that the recent astronomical observations of old SGRs (characteristic age $\sim (10^6 - 10^7)$ Myr) with low surface magnetic field, as well as the four SGRs/AXPs showing radio emission recently discussed by [3][2].

3. Models of Electromagnetic Emission

There are two models associate the pair $e^{\pm}$ production and emission of radiation in compact stars.

3.1. Polar Cap Model

The first process is the polar-cap model, where electrons are accelerated on the stellar surface emitting γ-rays by curvature radiation. In this model we consider a polar cap region. Since the corotating speed field lines cannot exceed the speed of light, the magnetic lines cannot be closed outside of the light cylinder $R_{lc} = c/\Omega$. This fact leads to the open magnetic field lines in the polar region. The potential difference along an open magnetic line given by [8]:

$$\Delta V\mathrm{max} = \frac{B_p \Omega^2 R^3}{2c^2}, \tag{7}$$

and for any potential the pairs of $e^{\pm}$ are accelerated by the Lorentz factor

$$\gamma = e\Delta V/mc^2, \tag{8}$$

where m is the mass of electron. Following [1], we determine the condition to pairs $e^{\pm}$ production in the polar cap regions of pulsars:

$$\left(\frac{e\Delta V}{mc^2}\right)\frac{\hbar}{2mcr_c}\frac{h}{r_c}\frac{B_s}{B_g} \approx \frac{1}{15}, \tag{9}$$

where h is the distance (gap) above the pulsar surface, r_c is the curvature radius of the magnetic field lines, $B_g = m^2c^3/e\hbar = 4.4 \times 10^{13}$ G and B_s the magnetic field on the surface.

3.2. Outer Gap Model

The second model of emission consider the production of the photons far away from the stellar surface, where the magnetic field lines are open. In this model, the energy is limited by a combination of curvature radiation and inverse Compton scattering. The initial inverse Compton scattering limits the energy of the primary $e^{\pm}$ to $\gamma mc^2 \sim (mc^2)^2/\hbar\omega_s$. Because the maximum energy of secondary $e^{\pm}$ is γmc^2, the highest frequency synchrotron radiation from secondary $e^{\pm}$ becomes:

$$\omega\mathrm{max} \sim \gamma \sin\theta \omega_B, \tag{10}$$

where $\omega_B = eB/mc$ and θ the angle between $e^{\pm}$ and local B. There is a critical value, where the energy spectrum can be cut [1],

$$\omega_c \sim \gamma\gamma_{\perp,c}\omega_B = \frac{\gamma_{\parallel}^3}{\omega_B^3}\left(\frac{mc^3\omega}{e^2}\right)^2. \tag{11}$$

The typical frequency of the tertiary photons is

$$\omega_s \sim \frac{1}{\omega_B^3}\left(\frac{mc^3\Omega}{e^2}\right)^2 \quad \text{valid only for } \omega_s \geq \omega_B. \tag{12}$$

From (10), (11) and (12), we can obtain the critical frequency:

$$\omega_c \approx \gamma_{\parallel}^3 \frac{m^5 c^9}{e^7}\Omega^2 B^{-3}, \tag{13}$$

and the maximum synchrotron emission frequency:

$$\omega\mathrm{max} \approx \frac{e^{15}}{\gamma_{\parallel}\hbar^2 m^9 c^{15}}\Omega^{-4}B^7. \tag{14}$$

4. Results

4.1. Polar Cap Model Applied to SGRs and AXps as White Dwarfs
From equation (9) for white dwarf with a mass $M = 1.26M_{\odot}$ and $R = 3 \times 10^8$ cm we obtain the theoretical death line:
m1 - Pure dipole:

$$4 \log B_p - 7.5 \log P = 25.84, \quad (15)$$

with $B_s = B_p$ and $r_c = (Rc/\Omega)^{1/2}$.
m2 - Magnetic field lines very curved:

$$4 \log B_p - 6.5 \log P = 19.68, \quad (16)$$

with $B_s = B_p$ and $r_c \sim R^9$ cm.
m3 - Magnetic field lines very curved on the polar cap:

$$7 \log B_p - 13 \log P = 26.06, \quad (17)$$

with $B_s = 2 \times 10^9$ G, $r_c \sim R^9$ cm and $h \sim (B_p/B_s)^{1/2} R(R\Omega/c)^{1/2}$.

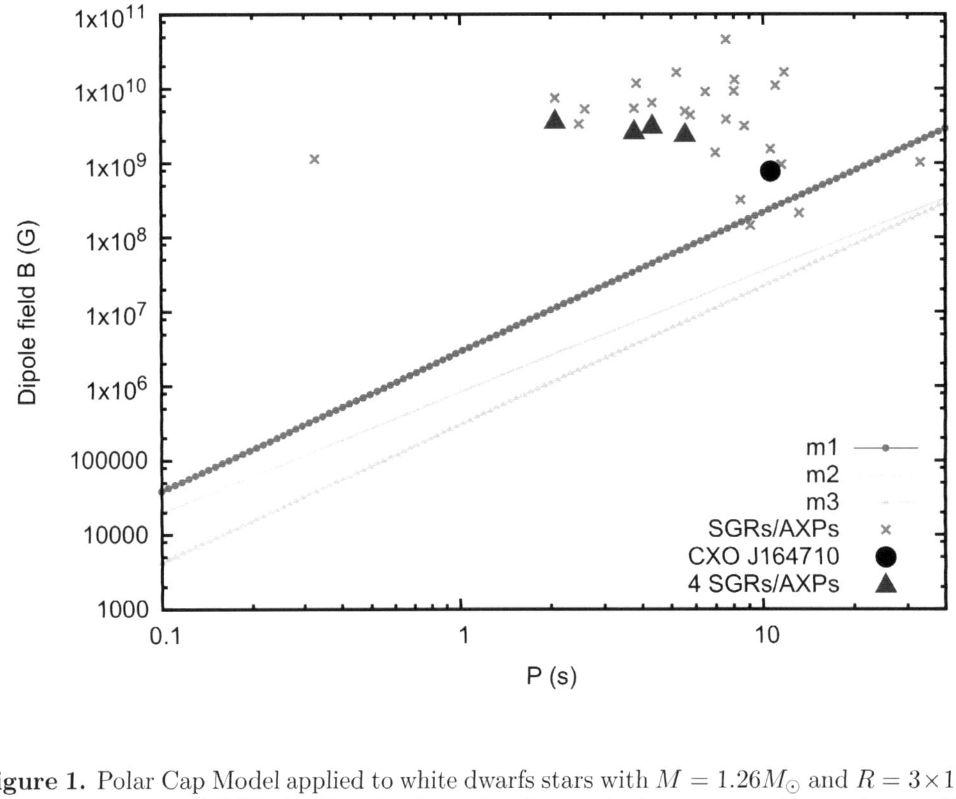

Figure 1. Polar Cap Model applied to white dwarfs stars with $M = 1.26M_{\odot}$ and $R = 3 \times 10^8$cm. The dots-cross are all SGRs/AXPs, the dots-triangle are the four sources that emit in radio, the dot-circle is a star very old.

4.2. Outer Gap Model Applied to SGRs and AXps as White Dwarfs

From (13) if we assumes that magnetic field is the B at the light cylinder

$$B = B_p R^3/r^3 = B_p(\Omega R/c)^3, \tag{18}$$

for white dwarf with a mass $M = 1.4M_\odot$ and $R = 3 \times 10^8$cm we construct:

o1:

$$5 \log B_p - 12 \log P = 34.38, \tag{19}$$

o2:

$$2 \log B_p - 5 \log P = 16.54, \tag{20}$$

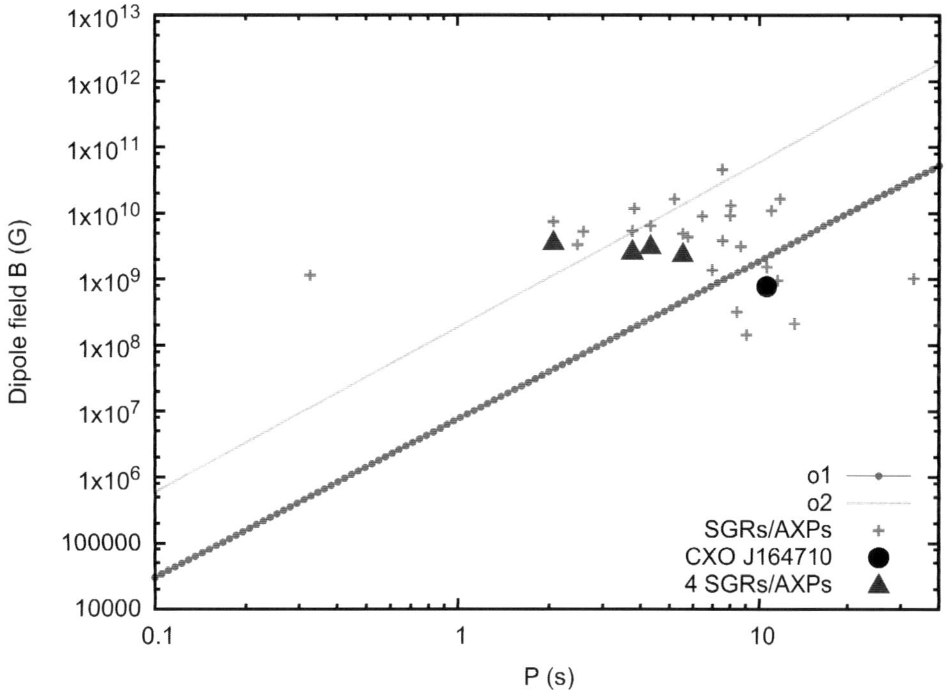

Figure 2. Outer Gap Model applied to white dwarfs stars with with $M = 1.26M_\odot$ and $R = 3 \times 10^8$cm. The dots-plus are all SGRs/AXPs, the dots-triangle are the four sources that emit in radio, the dot-circle is a star very old.

5. Conclusions

Our work is a first attempted to find an explanation for the puzzle why for all the SGRs/AXPs was expected radio emission, but it was observed only four of them.
In the polar Cap Model (figure 4.1) almost all stars should emit, however, only in four of then was observed radio emission. So there is a difference in nature of this sources, we suppose that ones (which emit) are neutron stars and others white dwarfs. For emission far from the WD star surface, using the Outer Gap Model, may explains why some of the SGRS and AXPs emit in radio and others no. As we can see in figure (4.2) all sources are very near the dead-line or below. This could explain the absence of radio emission observed in these sources.

6. Acknowledgements

The work was financially supported by CAPES (Coordenação de Aperfeiçoamento de Pessoal de Nível Superior). Acknowledgements also to McGill Pulsar Group.

References

[1] Kaiyou Chen and Malvin Ruderman. Pulsar death lines and death valley. *The Astrophysical Journal*, 402:264–270, 1993.
[2] J. G. Coelho and M. Malheiro. A different nature between the radio AXPs in comparison to the others SGRs/AXPs. *arXiv:1307.8158 [astro-ph, physics:nucl-th]*, July 2013.
[3] J. G. Coelho and M. Malheiro. Do SGRs/AXPs and radio AXPs have the same nature? *arXiv:1303.0863 [astro-ph]*, March 2013.
[4] J. G. Coelho, R. M. Marinho, M. Malheiro, R. Negreiros, D. L. Cáceres, J. A. Rueda, and R. Ruffini. Dynamical Instability of White Dwarfs and Breaking of Spherical Symmetry Under the Presence of Extreme Magnetic Fields. *The Astrophysical Journal*, 794(1):86, October 2014.
[5] Attilio Ferrari and Remo Ruffini. Theoretical Implications of the Second Time Derivative of the Period of the Pulsar NP 0532. *The Astrophysical Journal Letters*, 158:L71, November 1969.
[6] Manuel Malheiro, Jorge A Rueda, and Remo Ruffini. {SGRs} and {AXPs} as Rotation-Powered Massive White Dwarfs. *Publications of the Astronomical Society of Japan*, 64(3):56, June 2012.
[7] S. A. Olausen and V. M. Kaspi. THE McGILL MAGNETAR CATALOG. *The Astrophysical Journal Supplement Series*, 212(1):6, May 2014.
[8] M. A. Ruderman and P. G. Sutherland. Theory of pulsars - Polar caps, sparks, and coherent microwave radiation. *The Astrophysical Journal*, 196:51–72, February 1975.
[9] Stuart L. Shapiro and Saul A. Teukolsky. *Black Holes, White Dwarfs and Neutron Stars: The Physics of Compact Objects*. John Wiley & Sons, November 2008.
[10] VV Usov. Gamma-radiation generation by rotating magnetic white dwarfs. *Pisma v Astronomicheskii Zhurnal*, 14:606–609, 1988.

Large areas elemental mapping by ion beam analysis techniques

T.F. Silva, C.L. Rodrigues, J.F. Curado, P. Allegro, M.V. Moro, P.H.O.V. Campos, S.B. Santos, E.A.M. Kajiya, M.A. Rizzutto, N. Added, M.H. Tabacniks

University of São Paulo

E-mail: tfsilva@if.usp.br

Abstract. The external beam line of the Laboratory for Material Analysis with Ion Beams (LAMFI) is a versatile setup for multi-technique analysis. X-ray detectors for Particle Induced X-rays Emission (PIXE) measurements, a Gamma-ray detector for Particle Induced Gamma-ray Emission (PIGE), and a particle detector for scattering analysis, such as Rutherford Backscattering Spectrometry (RBS), were already installed. In this work, we present some results, using a large (60-cm range) XYZ computer controlled sample positioning system, completely developed and build in our laboratory. The XYZ stage was installed at the external beam line and its high spacial resolution (better than 5 μm over the full range) enables positioning the sample with high accuracy and high reproducibility. The combination of a sub-millimeter beam with the large range XYZ robotic stage is being used to produce elemental maps of large areas in samples like paintings, ceramics, stones, fossils, and all sort of samples. Due to its particular characteristics, this is a unique device in the sense of multi-technique analysis of large areas. With the continuous development of the external beam line at LAMFI, coupled to the robotic XYZ stage, it is becoming a robust and reliable option for regular analysis of trace elements ($Z > 5$) competing with the traditional in-vacuum ion-beam-analysis with the advantage of automatic rastering.

1. Introduction

Among most of the analytical techniques, there is an effort to develop imaging capabilities. This is because the correlation of the element concentration with the position on the sample can give more information on heterogeneous samples, enabling not just studies related to formation and solid phases, but also the element maps can be used as sample maps for locating interesting spots for detailed analysis. It can be found in literature many studies that produce images using XRF, XPS, EDX, PIXE, and others. But these studies are usually performed in small areas and most of them are performed in vacuum environment (which imposes restrictions on the samples that can be studied).

The external beam line in the Laboratory of Material Analysis with Ion Beams (LAMFI) is a versatile setup for multi-technique analysis using ion beams. Recently, a setup upgrade included the installation of a particle detector enabling in-air particle scattering analysis [1]. X-ray detectors for Particle Induced X-rays Emission (PIXE) measurements and Gamma-ray detector for Particle Induced Gamma-ray Emission (PIGE) measurements was already installed, and some studies using this analytical station have been reported [2-5]. With the continuous development of our external beam line, it is becoming a robust multi-analysis station for matrix and trace elements ($Z > 5$) that meets applications in disparate areas. The great advantage of using this analytical station is that it is virtually no necessary any sample preparation, what increases the broad of samples that can be studied.

Following the imaging tendency, we developed a 3D robotic system for sample positioning. The system was completely developed in our laboratory, and was coupled to the external beam line. This is a high resolution machine that enables to position the sample in the external beam line with high accuracy and high reproducibility. With the automation of this robotic device and the development of the electronic interface that establish communication between the computer that controls the positioning system and the detector electronics, we are capable to produce elemental mapping of large areas by rastering the broad beam (0.7-mm diameter) provided by the accelerator. Comparing capabilities with other similar setups for imaging using ion beam analysis techniques that can be found in other laboratories around the world, the low resolution of ours, mainly due to the broad beam, makes the observation of small details in the sample a weakness of our device. On the other hand, what makes our device unique is the ability to raster large areas, and to obtain elemental maps.

2. Methodology

The computer that controls the robotic sample holder is also able to communicate with the detector electronics and automatically record the detector spectra and the sample position coordinates. Keeping track of all this parameters it is possible to construct elemental maps using the signal collected by the detectors in a spectra corresponding to a certain element and its respective position coordinate in the sample surface. Fig. 1 illustrate the rastering process and the variables that can be chosen by the user in the software of control.

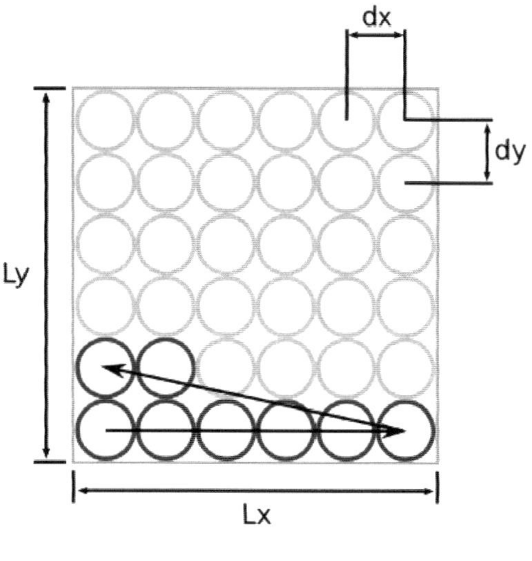

Figure 1. Illustration of the rastering process. The steps (dx and dy) and the size of the boundaries of a rectangular region (Lx and Ly) can be chosen by the user in the software of control. It is also a variable to be chosen the acquisition time in each point.

3. Results

To explore the potentiality of the rastering mode in the external beam setup to elaborate elemental maps we performed the ratering in a test sample. This sample is a structured Cu foil on top of a Ta foil. The structure of the Cu foil is composed of some holes of different sizes, in order to explore the resolution limits and the size of the minor structure that can be observed. Fig. 2 shows a picture of the test sample.

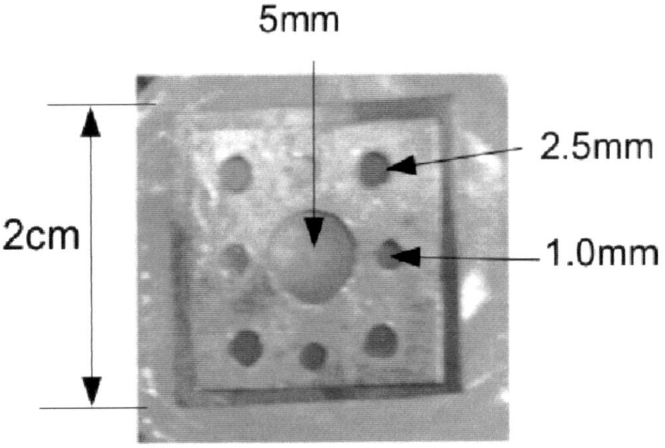

Figure 2. Picture of the test sample. The holes were fabricated with different diameters to explore the minor structure that can be observed.

For this test sample we used the robotic device to raster the broad beam with steps of 1.0 mm and measured the PIXE spectra for each point. The map obtained is presented in Fig. 3. With this measurements we were able to create maps of Cu and Ta, and despite the superposition of the K-lines of Cu and the L-lines of Ta, it is possible to observe the structures in the Cu foil.

Despite the size of the pixels that indicates the low spatial resolution of the image, we can identify all the structures of the Cu foil. The 1-mm diameter holes seems to be near the spatial resolution limit defined by the beam size.

In the beam exit at the external beam setup, there is a collimator that restricts the beam diameter to 1.0 mm. In order to exploit what is the best rastering resolution obtained, we performed a rastering of a 20.0x20.0-mm^2 area of a metallic grade. Fig. 4 shows a map of intensities for the Fe-Kα line along the measured area.

It is possible to observe that, even with the treatment of the image, the 1-mm step produced a pixelized image of the grade. A much better result is obtained using 0.7-mm step, thus this value was used as default for this beam size. To improve the spatial resolution, some other smaller collimator have been made (1.0 mm, 0.5 mm and 0.25 mm), providing the option of beam size that best suits each analysis.

The image generated in the 1.0-mm step rastering took 3 hours to be completed, while the 0.7-mm step image took 6.3 hours. Tests of movement of the equipment showed that the system takes, on average, 12 seconds between the measurement of two consecutive points with an acquisition time of 15 seconds in each spot. We estimate that 90% of this time is due to the reseting process of the detectors, meaning that there is room for improvements replacing the electronics by a faster model.

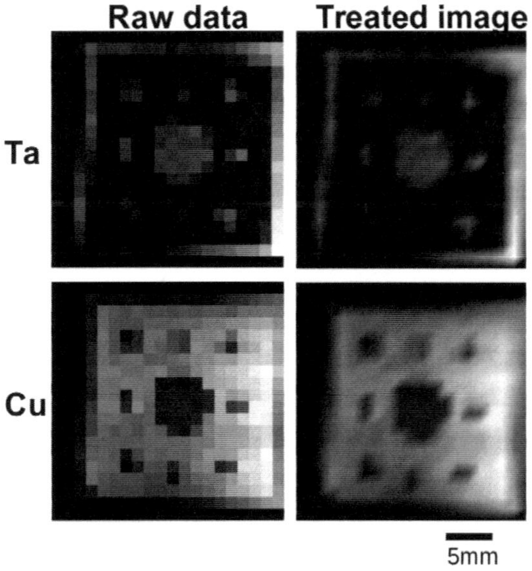

Figure 3. Picture of the test sample. The holes were fabricated with different diameters to explore the minor structure that can be observed.

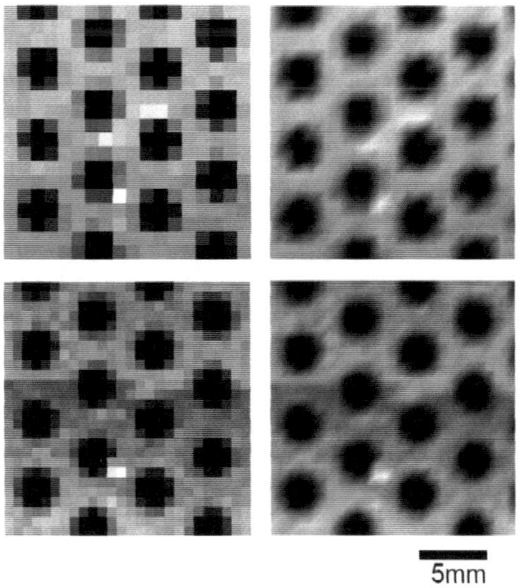

Figure 4. Resolution tests. On top images with 1.0-mm step (resulting in a 400 pixels image) and in the bottom with 0.7-mm step (resulting in a 841 pixels image). Left side raw images and in the right images with digital treatment.

4. Discussion and Conclusions

The external beam setup under development at LAMFI is becoming a versatile analysis station for ion beam measurements. As it can be seen in the theoretical efficiency curve for the x-rays detectors used for PIXE measurements (see Fig. 7), a wide range of elements can be detected. This curve was calculated using the detector efficiency provided by the manufacturer [8], the windows x-rays absorptions, and the solid angle of the detectors.

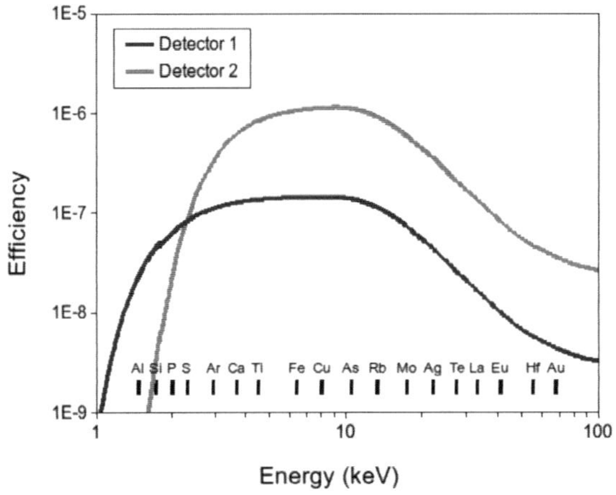

Figure 5. Theoretical relative curves of efficiency for the two x-rays detectors used for PIXE measurements at the external beam setup. Respective energies for the ka lines are marked as reference.

Also, the development of a robotic machine capable of indexing the position of PIXE measurements in a sample is showing to be very useful for production of elemental maps. The rastering of the broad beam produced by the accelerator can create maps with sub-milimeter resolution of large areas. This innovative use of a broad beam can open new perspectives of research and measurements in our laboratory.

5. References

[1] T.F. Silva, N. Added, M.V. Moro, G.F. Trindade, H.C. Santos, C.L. Rodrigues, M.A. Rizzutto, M.H. Tabacniks; In-air RBS measurements at the LAMFI external beam setup, In: *AIP Conference Proceedings*, **1625**, 160-165 (2014).

[2] R.A. Ikeoka, C.R. Appoloni, M.A. Rizzutto, T.F. Silva and A.M. Bandeira, PIXE analysis of pre-colonial pottery from Sambaqui do Panaquatira, *13th PIXE conference*, (2013).

[3] V. C. Ambiel, M.A. Rizzutto, J.F. Curado, P.H.O.V. Campos, E.A.M. Kajiya and T.F. Silva, Metallic objects belonging to the first emperor of Brazil studied with PIXE technique, *13th PIXE conference*, (2013).

[4] M.V. Moro, T.F. Silva, N. Added, G.F. Trindade, M.H. Tabacniks, Study of Boron detection limit using in-air PIGE setup at LAMFI-USP, In: *AIP Conference Proceedings*, **1625**, 120-124 (2014).

[5] H.C. Santos, N. Added, T.F. Silva, C.L. Rodrigues, External-RBS, PIXE and NRA analysis for ancient swords, *Nuc. Instr. Meth. B*, **345**, (2015), 42-47.

[6] C.M. MacRae, N.C. Wilson; Luminescence Database I: Minerals and Materials, *Microscopy and Microanalysis* **14**, (2008), 1435-8115.

[7] B.J. Luff, P.D. Townsend; 1990. Cathodoluminescence of synthetic quartz, *J. Phys. Condens. Matter*, **2** (1990), 80898097.

[8] http://www.amptek.com/products/xr-100cr-si-pin-x-ray-detector/ last access: January 22th 2015.

Journal of Physics: Conference Series **630** (2015) 012017 doi:10.1088/1742-6596/630/1/012017

Search of a systematic behaviour for the weakly bound complete fusion suppression caused by breakup

M R Cortes, P R S Gomes, J Lubian, D Mendes Jr.
Instituto de Física, Universidade Federal Fluminense, Av. Gal. Milton Tavares de Souza s/n, Niterói, R.J., Brazil

E-mail: mariane.cortess@gmail.com

Abstract. We discuss the effect of breakup on the complete and total fusion cross sections of weakly bound nuclei, both stable and radioactive, at near barrier energies. We show that there is suppression of the complete fusion of non-halo nuclei and total fusion of neutron-halo nuclei at energies above the barrier, whereas there is some enhancement at sub-barrier energies. We investigate a systematic behaviour for this effect for several weakly bound systems.

Nuclear reactions involving weakly bound nuclei, both stable and radioactive, have been widely investigated in the last years[1]. Several processes can take place after the projectile breakups. One is the incomplete fusion (ICF) in which part of the fragments is absorbed by the target. When all the fragments fuse with the target, the process is called sequential complete fusion (SDF). From the experimental point of view, the SCF cannot be distinguished from the direct complete fusion (DCF) in which the whole projectile fuses with the target without breakup. Therefore, only the complete fusion (CF) cross sections, which includes both of the DCF and SCF cross sections, can be measured. The total fusion (TF) cross section, which includes both of the DCF and SCF cross sections, $\sigma_{TF} = \sigma_{CF} + \sigma_{ICF}$. Experimentally, it is difficult to distinguish between residues from ICF and CF. Especially for light reaction systems, the excited compound nucleus emits charged particles during the process of its cooling. The residues from ICF cannot be distinguished from those from CF, and thence only the TF cross section can be measured. Particularly important is to study the effect of the breakup of those weakly bound nuclei on the complete fusion of the system at energies close to the Coulomb barrier. Recently it has been well accepted that breakup couplings suppress the complete fusion at energies close but above the Coulomb barrier and enhance fusion at sub-barrier energies. In this work we investigate a possible systematic behaviour of this suppression. The investigation of the effect of breakup on the fusion cross section can also be made using other approaches, such as the energy dependence of the optical potential in the elastic scattering of weakly bound nuclei. For tightly bound systems one observes the usual energy dependence of optical potential, namely the threshold anomaly, corresponding to the decrease of the imaginary potential when the bombarding energy decreases towards the barrier energy. This is explained by the closing of reaction channels at energies near the barrier and by the attractive polarization potential (its real part) produced by inelastic and transfer channels. For weakly bound systems a different behaviour is observed, called breakup threshold anomaly (BTA) [2]. The BTA occurs because the breakup cross section is still important at sub-barrier energies and it is possible to observe that the imaginary potential may even increase at energies close to the barrier, because breakup produces repulsive polarization potentials (real part) [3-5]. The analysis of quasi-elastic barrier distributions also shows that when the direct breakup coupling is included in the calculations, the barrier height increases and the calculations have better agreement with the data, what confirms that the real part of the polarization potential produced by the direct breakup is repulsive [6-7].

Theoretically, the most suitable calculations involving breakup are the so-called CDCC (continuum discretized coupled channel) calculations, since the breakup feeds states in the continuum. It has been shown that if one wants to describe the behaviour of elastic scattering angular distributions of weakly bound nuclei, it is essential that continuum-continuum couplings are included in the CDCC calculations [4, 8]. However, if one wants to investigate the effect of breakup on the fusion cross section, it may be more interesting to perform much simpler coupled channel calculations which do not take into account the breakup, and so the difference between data and theoretical predictions should correspond to the breakup effect.

The starting point in the comparison of fusion cross section data with theoretical predictions is the choice of a systematic bare interaction potential. We adopt the double-folding parameter-free São Paulo potential (SPP) [9, 10], using reliable nuclear densities of the nuclei involved in the collisions [11,12]. This potential has described successfully several reactions with systems in different mass ranges, including weakly bound nuclei [13,14]. In order to compare data for several systems in a single plot, it is necessary to eliminate the differences associated with trivial factors, like sizes, charges and anomalous densities. Canto *et al.* [15, 16] proposed the use of dimensionless quantities, with a method, which allows reaching a systematic understanding of this subject, since it allows the comparison of any kind of system in the same graphic. This method uses a benchmark curve, called the Universal Fusion Function (UFF), given by $F_0(x) = \ln[1 + \exp(2\pi x)]$, where $x = (E - VB) / \hbar\omega$ and $F(x) = \left(2 E_{c.m.}/\pi R_B^2\, \hbar\omega\right)\sigma_{fus}$. Here V_B, R_B and $\hbar\omega$ are the height, radius and curvature of the parabolic barrier, respectively, σ_{fus} is the fusion cross section and F(x) is called fusion function. This method was later extended for the analysis of total reaction cross section [17].

The above reduction method has a simple meaning. For systems where channel coupling effects can be neglected and the fusion cross section is well approximated by Wong's formula [18], F(x) becomes the Universal Fusion Function (UFF). This method consists of using the UFF as the benchmark curve for comparisons with fusion data. First, one evaluates the barrier parameters for the particular system under study using the SPP potential. The experimental fusion function $F_{exp}(x)$ is then determined from the experimental fusion cross section. $F_{exp}(x)$ is then compared with $F_0(x)$. However, the above described procedure has two shortcomings: The first is that Wong's approximation is not valid for light systems at sub-barrier energies. The second is that a comparison of $F_{exp}(x)$with the UFF indicates the global effect of channel coupling on the fusion cross section. In this way, breakup couplings are entangled with couplings with other bound channels. In order to single out the effects of breakup coupling and eliminate deviations arising from the inaccuracy of Wong's formula at sub-barrier energies, it is necessary to renormalize the experimental fusion function by multiplying$F_{exp}(x)$ by $F_0(x) / F_{CC}(x)$, where $F_{CC}(x)$ the fusion function with isσ_F obtained from a coupled-channel calculation including couplings to all relevant bound channels.

In Figure 1, using the complete fusion data available in the literature we have the comparison between the UFF and the renormalized functions F_{exp} fusion. For a clear view of the behaviours of the fusion functions above and below the barrier (V_B corresponds to x=0), the results are displayed in linear and logarithmic scales. The results for total fusion of stable and proton halo systems are shown in Fig. 2. At above-barrier energies, the experimental fusion functions are very close to the UFF. One concludes that the breakup process does not significantly affect the total fusion cross section above the barrier. This means that fusion of one of the fragments with the target (incomplete fusion) is comparable with fusion of the whole projectile.

Inspecting Fig. 1(a), we conclude that the data for the ^{6,7}Li +^{209}Bi [19,20] and ^{9}Be+^{208}Pb [20,21] systems at above-barrier energies are appreciably suppressed, as compared to the UFF. The suppression corresponds to multiplying the experimental fusion function by the constant factor 0.7. Actually, one can observe that the attenuation factor is slightly larger for the^{6}Li +^{209}Bi system than for ^{7}Li +^{209}Bi, as observed in Refs. [19,20], where it is pointed out that the smaller the breakup threshold energy, the larger is the complete fusion suppression. For the ^{7}Be, ^{7}Li+^{238}U systems, similar results are observed [22]. The data in figure 2 are from Refs [20, 21, 23-25].

In Fig. 3, we show total fusion functions for the^{6}He+^{209}Bi, ^{6}He+^{238}U and ^{11}Be+^{209}Bi systems. The data are from Refs. [23, 26-27]. Inspecting Fig. 3(a) we conclude that the results are rather similar to the ones for complete fusion of stable weakly bound projectiles, shown in Fig. 1(a).

We conclude that the complete fusion cross section is systematically suppressed at energies above the Coulomb barrier. The suppression is of about 30% of the complete fusion cross section. The influence of the breakup coupling on the total fusion cross section at above-barrier energies depends on the nature of the projectile. It has no appreciable effect in collisions of stable weakly bound projectiles. On the other hand, in collisions of neutron–halo nuclei, the total cross sections are suppressed by about 30%, similarly to the case of complete fusion of stable weakly bound nuclei. At sub-barrier energies, both the complete and total fusion cross sections are enhanced, owing to the coupling with the breakup channel.

It is very interesting to observe that for each weakly bound projectile, the complete fusion suppression factor is independent of the target mass [28]. An analytical relation of the suppression factor with the breakup threshold energy was recently derived by Wang *et al.* [28]. However, a physical explanation for that expression is still missing.

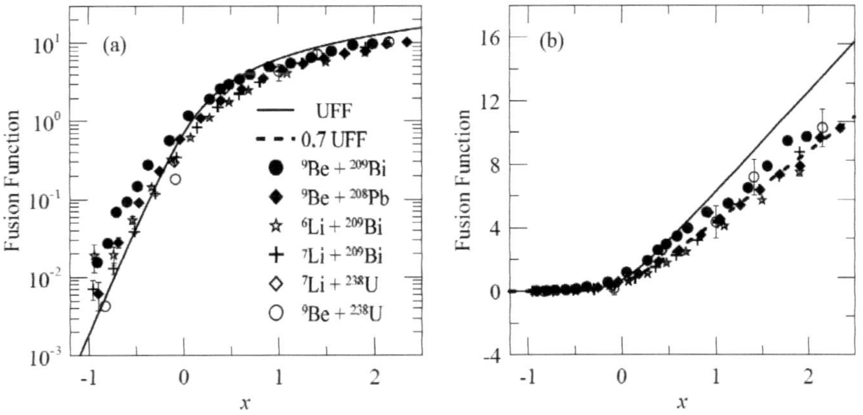

Figure 1- Comparison of the renormalized experimental complete fusion functions with the UFF, for stable and no-halo radioactive nuclei. See text for details.

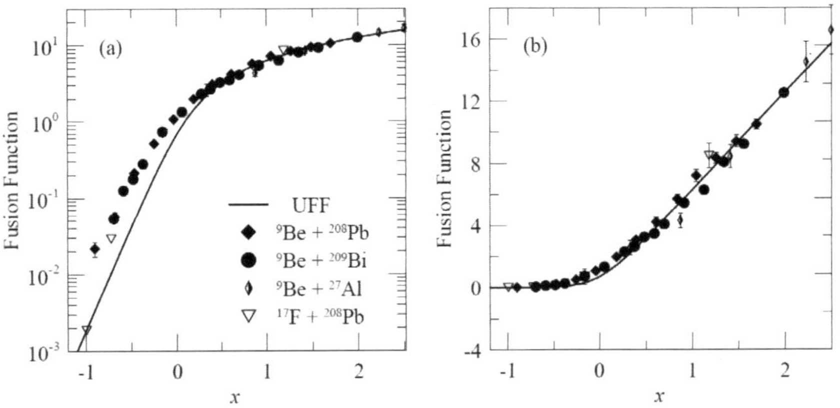

Figure 2- Comparison of the renormalized experimental total fusion functions with the UFF, for stable and no-halo radioactive nuclei. See text for details.

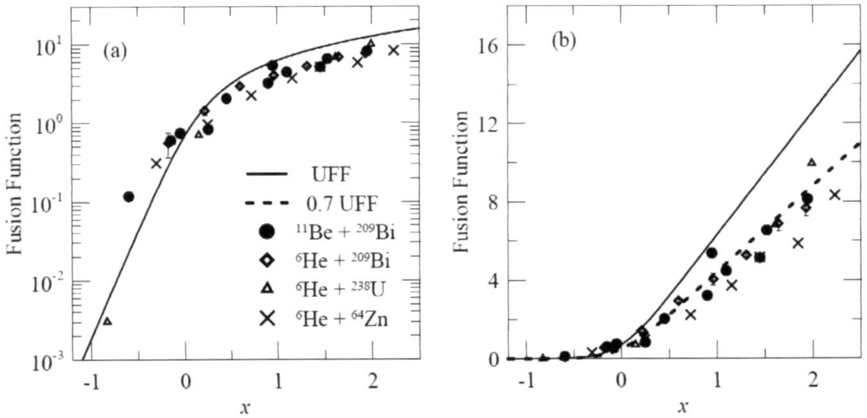

Figure 3- Comparison of the renormalized experimental total fusion functions with the UFF, for neutron-halo radioactive nuclei. See text for details.

Acknowledgements

The authors would like to thank FAPERJ, CNPq and CAPES for the partial financial support.

References

[1]. Canto L F, Gomes P R S, Donangelo R and Hussein M S 2006, *Phys. Rep.* **424** 1
[2].Hussein M S, Gomes P R S , Lubian J and Chamon L C 2006 *Phys. Rev. C***73** 044610
[3] Lubian J et al. 2007 *Nucl. Phys.A***791** 24.
[4]Canto L F, Lubian J, Gomes P R S and Hussein M S 2009 *Phys. Rev. C***80** 047601
[5]. Santra S et al. 2011 *Phys. Rev. C***83** 034616

[6] Monteiro D S et al. 2009 *Phys.Rev. C***79** 014601
[7]Lubian J, Correa T, Gomes P R S and Canto L F 2008 *Phys.Rev. C***78** 064615
[8] Lubian J et al. 2009 *Phys. Rev. C***79** 064615
[9] Chamon L C et al. 1997 *Phys. Rev.Lett.***791** 5218
[10] Chamon L C et al., 2002 *Phys. Rev. C***66** 014610
[11] Silva C P et al. 2001*Nucl. Phys.A***679** 287.
[12] Alvarez M A G et al. 2003*Nucl. Phys.A*723
[13] Crema E et al. 2005*Phys. Rev. C***72** 034610
[14] Crema E et al. 2007*Phys. Rev. C***75**037601
[15] Canto L F, Gomes P R S, Chamon L C and Crema E 2009*J. Phys. G***36** 015109
[16] Canto L F, Gomes P R S, Chamon L C and Crema E 2009 *Nucl. Phys A***821** 51
[17]. Shorto J M B et al. 2009 *Phys.Lett. B***678** 77
[18] Wong C Y 1973 *Phys.Rev.Lett.***31** 766
[19] Dasgupta M et al. 1999 *Phys.Rev.Lett.***82** 1395
[20] Dasgupta M et al. 2004 *Phys.Rev. C***70** 024606
[21] DasguptaM et al. 2002*Phys. Rev. C* **66** 041602(R).
[22] RaabeR et al. 2006*Phys.Rev. C***74** 044606.
[23] SignoriniC et al. 1997*Phys. J. A***5** 7
[24]MartíG V et al. 2005*Phys. Rev. C* **71** 027602
[25] RehmK E et al. 1998*Phys. Rev. Lett.***81** 3341
[26] KolataJ Jet al. 1998*Phys. Rev. Lett.***81**4580
[27] RaabeR et al. 2004*Nature***431** 823
[28] Wang B, Zhao J, Gomes P R S and Zhou Shan-Gui 2014 *Phys.Rev. C***90** 034612

The role of nuclear reactions in the problem of *0νββ* decay and the NUMEN project at INFN-LNS

F Cappuzzello[1,2], C Agodi[1], M Bondì[1,2], D Carbone[1], M Cavallaro[1], A Foti[1,3]
[1] INFN, Laboratori Nazionali del Sud, Italy
[2] Dipartimento di Fisica e Astronomia, Università di Catania, Italy
[3] INFN, Sezione di Catania, Italy

E-mail: cappuzzello@lns.infn.it

Abstract: An innovative technique to access the nuclear matrix elements entering the expression of the life time of the double beta decay by relevant cross sections of double charge exchange reactions is proposed. The basic point is the coincidence of the initial and final state wave-functions in the two classes of processes and the similarity of the transition operators, which in both cases present a superposition of Fermi, Gamow-Teller and rank-two tensor components with a relevant implicit momentum transfer. First pioneering experimental results obtained at the INFN-LNS laboratory for the $^{40}Ca(^{18}O,^{18}Ne)^{40}Ar$ reaction at 270 MeV, give encouraging indication on the capability of the proposed technique to access relevant quantitative information. A key aspect of the project is the use of the K800 Superconducting Cyclotron (CS) for the acceleration of the required high resolution and low emittance heavy-ion beams and of the MAGNEX large acceptance magnetic spectrometer for the detection of the ejectiles. The use of the high-order trajectory reconstruction technique, implemented in MAGNEX, allows to reach the high mass, angular and energy resolution required even at very low cross section. The LNS set-up is today an ideal one for this research even in a worldwide perspective. However a main limitation on the beam current delivered by the accelerator and the maximum rate accepted by the MAGNEX focal plane detector must be sensibly overcome in order to systematically provide accurate numbers to the neutrino physics community in all the studied cases. The upgrade of the LNS facilities in this view is part of this project.

1. A basic question in modern physics

Neutrinoless double beta decay (*0νββ*) is potentially the best resource to probe the Majorana or Dirac nature of neutrino and to extract its effective mass. Moreover, if *0νββ* is observed, it would signal that the total lepton number is not conserved. Presently, this physics case is one of the most important researches “beyond the Standard Model” and might guide the way towards a Grand Unified Theory of fundamental interactions. The *0νββ* decay rate $[T_{1/2}]^{-1}$ can be factorized as a phase-space factor G_{0v}, the nuclear matrix element (NME) M_{0v} and a term $f(m_i,U_{ei})$ containing the masses m_i and the mixing coefficients U_{ei} of the neutrino species:

$$[T_{1/2}]^{-1} = G_{0v} |M_{0v}|^2 |f(m_i,U_{ei})|^2 , \quad (1)$$

where the NME is the transition amplitude from the initial $|\varphi_i\rangle$ to the final $|\varphi_f\rangle$ nuclear state of the $\beta\beta$ process through the $0\nu\beta\beta$ decay operator:

$$|M_{0\nu}|^2 = \left|\langle\varphi_f|O^{0\nu\beta\beta}|\varphi_i\rangle\right|^2 \qquad (2)$$

Thus, if the NMEs are established with sufficient precision, the neutrino masses and the mixing coefficients can be extracted from $0\nu\beta\beta$ decay rate measurements.

The evaluation of the NMEs is presently limited to state of the art model calculations based on different methods (QRPA, shell-model, IBM etc.) [1], [2], [3], [4]. High precision experimental information from single charge-exchange, transfer reactions and electron capture are used to give constraints to the calculations [5], [6], [7], [8] [9]. However the ambiguities in the models are still too large and the constraints too loose to reach accurate values of the NMEs. Discrepancies higher than two are presently reported in literature [10]. In addition some assumptions, common to the different competing calculations, could cause overall systematic uncertainties [11].

The experimental study of nuclear transitions where the nuclear charge is changed by two units leaving the mass number unvaried, in analogy to the $\beta\beta$-decay, could give important information. The availability for the first time of valuable data on Double Charge Exchange (DCE) reactions raises the question whether they can be used toward the experimental access to $0\nu\beta\beta$ NMEs. Despite the DCE and $0\nu\beta\beta$ decay processes are mediated by different interactions, there are a number of important similarities among them:

1. Parent/daughter states of the $0\nu\beta\beta$ decay are the same as those of the target/residual nuclei in the DCE;
2. Short-range Fermi, Gamow-Teller and rank-2 tensor components are present in both the transition operators, with relative weight depending on incident energy in DCE;
3. A large linear momentum (~100 MeV/c) is available in the virtual intermediate channel in both processes. This is a crucial similarity since the $0\nu\beta\beta$ NMEs strongly depend on the momentum transfer and other processes cannot probe this feature;
4. The two processes are non-local and are characterized by two vertices localized in a pair of valence nucleons;
5. Both processes take place in the same nuclear medium, thus quenching phenomena are expected to be similar. In particular in both β-decay [4] and charge exchange reactions [12] the limited model space used in the calculations and the contribution of non-nucleonic degrees of freedom and other correlations require a renormalization of the coupling constants in the spin-isospin channel.
6. A relevant off-shell propagation through virtual intermediate channels is present in the two cases.

The description of NMEs extracted from DCE and $0\nu\beta\beta$ presents the same degree of complexity, with the advantage for DCE to be "accessible" in laboratory. However a simple relation between DCE cross sections and $\beta\beta$-decay half-lives is not trivial and needs to be explored.

2. The pioneering experiment

The ^{40}Ca(^{18}O,^{18}Ne)^{40}Ar DCE reaction was measured at the INFN-LNS in Catania (Italy) together with the competing processes: ^{40}Ca(^{18}O,^{18}F)^{40}K single charge-exchange, ^{40}Ca(^{18}O,^{20}Ne)^{38}Ar two-proton and

$^{40}Ca(^{18}O,^{16}O)^{42}Ca$ two-neutron transfer. The experiments were performed using an ^{18}O beam accelerated by the LNS Superconducting Cyclotron (CS) at 270 MeV incident energy and detecting the ejectiles by the MAGNEX large acceptance magnetic spectrometer [13] [14] [15].

This work shows for the first time experimental data on heavy-ion double charge-exchange reactions in a wide range of transferred momenta [16]. Strengths factors have been extracted under the hypothesis of a two-step charge-exchange process. Despite the approximations used in the comparison, they are reasonable within ±50%, signaling that the main physics content has been kept. A deeper investigation of DCE reactions is worthwhile in the future, studying other systems and different bombarding energies, in order to explore the systematic behavior. A rigorous treatment of the details, neglected in the present work, will be the next step toward the determination of the $0\nu\beta\beta$ NMEs.

3. The Phase2 of the NUMEN project: toward "hot" cases

The results mentioned above indicate that suitable information from heavy-ion induced DCE reactions can be extracted. In particular the determination of nuclear matrix elements for these processes seems to be at our reach and precious information toward $0\nu\beta\beta$ matrix elements can be extracted. This is at the basis of a new project at the INFN, named NUMEN (NUclear Matrix Elements of Neutrinoless double beta decay), which aims at the determination of systematic information on the nuclear matrix elements of interest for $0\nu\beta\beta$ by heavy-ion double charge exchange reactions.

The availability of the MAGNEX spectrometer for high resolution measurements of much suppressed reaction channels was essential for such a pioneering measurement. However with the present set-up it is difficult to suitably extend this research to the "hot" cases, where $\beta\beta$ decay studies are and will be concentrated. We consider that:

a) About one order of magnitude more yield would have been necessary for the reaction studied, especially at backward angles where large amounts of linear momentum (1-2 fm^{-1}) are available;

b) The $(^{18}O,^{18}Ne)$ reaction is particularly advantageous, due to the large value of both the $B[GT;^{18}Ogs(0^+) \rightarrow {}^{18}Fgs(1^+)]$ and $B[GT;^{18}Fgs(1^+) \rightarrow {}^{18}Negs(0^+)]$ strengths and to the concentration of the GT strength in the $^{18}F(1^+)$ ground state. However this reaction is of $\beta^+\beta^+$ kind, while most of the research on $0\nu\beta\beta$ is on the opposite side;

c) None of the reactions of $\beta^-\beta^-$ kind looks like as favourable as the $(^{18}O,^{18}Ne)$. For example the $(^{18}Ne,^{18}O)$ requires a radioactive beam, which cannot be available with comparable intensity. The proposed $(^{20}Ne,^{20}O)$ or the $(^{12}C,^{12}Be)$ have smaller B(GT), so a sensible reduction of the yield is foreseen in these cases;

d) In some cases gas target will be necessary, e.g. ^{136}Xe or ^{130}Xe, which are normally much thinner than solid state ones, with a consequent reduction of the collected yield;

e) In some cases the energy resolution we can provide (about half MeV) is not enough to separate the ground state from the excited states in the final nucleus. In these cases the coincident detection of γ-rays from the de-excitation of the populated states is necessary, but at the price of the collected yield.

f) A strong fragmentation of the double GT strength is known in the nuclei of interest compared to the ^{40}Ca.

Taking these considerations into account we realize that the present limit of low beam current we have experienced both for the CS accelerator and for the MAGNEX focal plane detector must be sensibly overcome. For a systematic study of the many "hot" cases of $\beta\beta$ decays an upgraded set-up, able to work with two orders of magnitude more current than the present, is thus necessary. This goal can be achieved by a substantial change in the technologies used in the beam extraction and in the detection of the ejectiles. For the accelerator the use of a stripper induced extraction is an adequate choice. More details about that are in the technical part of the project. For the spectrometer the main foreseen upgrades are:

1. The substitution of the present FPD gas tracker with a GEM tracker system;
2. The substitution of the wall of silicon pad stopping detectors with SiC detectors or similar;
3. The enhancement of the maximum magnetic rigidity;
4. The introduction of an array of detectors for measuring the coincident γ-rays.

The feasibility of DCE measurements at INFN-LNS with MAGNEX spectrometer, using the CS beams, was already demonstrated as explained before. To use this precious know-how for future application of this methodology to the relevant reaction of interest in the *0νββ* search, we need to go through a Phase2. During the Phase2 the necessary work for the upgrading of both the accelerator and MAGNEX will be carried out still preserving the access to the present facility. Due to the relevant technological challenges connected the Phase2 is foreseen to have a duration of about 3-4 years. In the meanwhile, experiments with integrated charge of tens of mC (about one order of magnitude more than collected in the pilot experiment) will be performed. These will require several weeks (4-8 depending on the case) data taking for each reaction, since thin targets (a few 10^{18} atoms/cm^2) are mandatory in order to achieve enough energy and angular resolution in the energy spectra and angular distributions. The attention will be focused on a few favourable cases with the goal to achieve conclusive results for them.

2.1 Experimental activity at Phase2

The Phase2 is crucial to allow us to optimize the experimental conditions and open a new challenging research field, carrying out an experimental investigation of few candidate nuclei for the $\beta\beta$ decay. In this framework, we propose to study the (^{18}O,^{18}Ne) reaction as a probe for the $\beta^+\beta^+$ transitions and the (^{20}Ne,^{20}O), or alternatively the (^{12}C,^{12}Be), for the $\beta^-\beta^-$. We select two systems: the ^{76}Ge-^{76}Se pair for the first class and the ^{116}Cd-^{116}Sn pair for the second. For these nuclei the ground states are resolvable from excited states by MAGNEX (being respectively 562 keV ^{76}Ge, 559 keV for ^{76}Se, 1.29 MeV for ^{116}Sn and 513 keV for ^{116}Cd) and the production technologies of the thin targets are already available at LNS. We also plan to explore the ^{130}Te and ^{106}Cd systems [17], which are candidates for *0νββ* already at our reach in terms of energy resolution and availability of thin targets. For each of them, the complete net of reactions involving the multi-step transfer processes, characterized by the same initial and final nuclei, as it is shown in Fig.1, will be studied under the same experimental conditions.

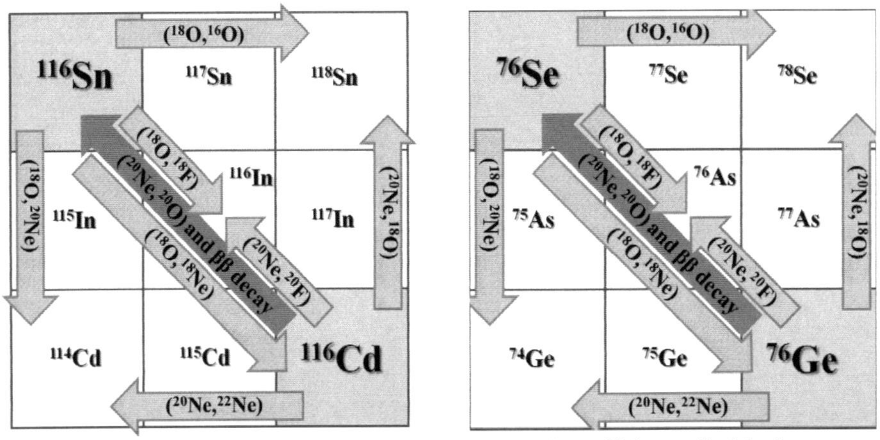

Figure 1: Scheme of the complete net of processes that will be studied in the case of the ^{116}Cd – ^{116}Sn and ^{76}Ge – ^{76}Se pairs of nuclei of interest for the $\beta\beta$ decay. Inside the arrows the reaction used to populate the final nuclei is indicated.

During the Phase2 the data reduction strategy will be optimized and the link with the theoretical physics will be strengthen, especially in the view of the construction of a "universal" framework, where $\beta\beta$-decay and DCE reactions are coherently analysed. The completion of the experimental activity of NUMEN Phase2 would represent by itself a ground-breaking result, looking forward the main goal of the proposal that has the ambition to indicate a new generation of experiments, with the challenging perspective, in the long term, to provide key information to the community to go deep insight the true nature of neutrino.

4. Phase3: The facility upgrade

Once all the building block for the upgrade of the accelerator and spectrometer facility will be ready at the LNS a Phase3, connected to the disassembling of the old set-up and re-assembling of the new will start. An estimate of about 18-24 months is considered. During this period the group will be devoted to the data analyses, to the preparation of the next experiments and test of the new detectors with Tandem beams. In addition, if necessary, experiments on single charge exchange or transfer reactions will be performed in other laboratories in order to provide possible pieces of information still lacking, e.g. measurements of B(GT) or transfer amplitudes.

5. Phase4: The experimental campaign

To perform the proposed experimental campaign, the upgrade of the CS accelerator is necessary to give high beam intensity and the upgrade of the detection system. Actually, we require a new focal plane detector, suitable to resist to high rates, and a modular gamma detector system that, together, allows us to complete the last phase of measurements, spanning among all the nuclei of interest for our studies. The Phase4 will consist of a series of experimental campaigns at high beam intensities (some pμA) and long experimental runs in order to reach in each experiment integrated charge of hundreds of mC up to C, for the experiments in coincidences, spanning all the variety of candidate isotopes, like: ^{48}Ca, ^{82}Se, ^{96}Zr, ^{100}Mo, ^{110}Pd, ^{124}Sn, ^{128}Te, ^{130}Te, ^{136}Xe, ^{148}Nd, ^{150}Nd, ^{154}Sm, ^{160}Gd, ^{198}Pt.

Journal of Physics: Conference Series **630** (2015) 012018 doi:10.1088/1742-6596/630/1/012018

Actually, once selected the optimal experimental condition for the different cases in the Phase2, with the aforementioned upgrades, the Phase4 will be devoted to collect data addressed to give, with an accurate analysis, a rigorous determination of the absolute cross sections values and their uncertainties for all the system of interest, to the challenging determination of the $0\nu\beta\beta$ decay nuclear matrix elements, that is the ambitious goal of the NUMEN project.

References

[1] Caurier E, Menendez J, Nowacki F, Poves A 2008 *Phys. Rev. Lett.* **100** 052503
[2] Suhonen J and Kortelainen M 2008 *Int. Journ. of Mod. Phys.* E **17** 1
[3] Vaquero N L, Rodriguez T R, Egido J L 2013 *Phys. Rev. Lett.* **111** 142501
[4] Barea J, Kotila J, Iachello F 2013 *Phys. Rev.* C **87** 014315
[5] Akimune H *et al* 1997 *Phys. Lett.* B **394** 23
[6] Schiffer J P *et al* 2008 *Phys. Rev. Lett.* **100** 112501
[7] Frekers D 2010 *Prog. Part. Nucl. Phys.* **64** 281
[8] Guess C J *et al* 2011 *Phys. Rev.* C **83** 064318
[9] Freeman S J and Schiffer J P 2012 *J. Phys. G: Nucl. Part. Phys.* **39** 124004
[10] Barea J, Kotila J and Iachello F 2012 *Phys. Rev. Lett.* **109** 042501
[11] *Report to the Nuclear Science Advisory Committee*, Neutrinoless Double Beta Decay, 2014
[12] Alford W P and Spicer B M 1998 *Advances in Nucl. Phys.* **24** 1
[13] Cappuzzello F, Carbone D and Cavallaro M 2011 *Nucl. Instr. and Meth.* A **638** 74
[14] Cappuzzello F *et al* 2014 *Nucl. Instr. and Meth.* A **763** 314
[15] Cavallaro M *et al* 2012 *Eur. Phys. J.* A **48:** 59
[16] Bondì M *et al* 2014 *AIP Conf. Proc.* **1595** 245
[17] Barea J, Kotila J and Iachello F 2013 *Phys. Rev.* C **87** 057301

Journal of Physics: Conference Series **630** (2015) 012019 doi:10.1088/1742-6596/630/1/012019

The d(^{9}Li,p)^{10}Li reaction as a tool to explore the ^{10}Li structure

M Cavallaro[1], M De Napoli[2], F Cappuzzello[1,3], C Agodi[1], M Bondì[1,3], D Carbone[1], A Cunsolo[1], B Davids[4], T Davinson[5], A Foti[2,3], N Galinski[4], R Kanungo[6], H Lenske[7], S E A Orrigo[8], C Ruiz[4] and A Sanetullaev[5]
[1] INFN, Laboratori Nazionali del Sud, Italy
[2] INFN, Sezione di Catania, Italy
[3] Dipartimento di Fisica e Astronomia, Università di Catania, Italy
[4] TRIUMF, Vancouver, Canada
[5] Department of Physics and Astronomy, University of Edinburgh, UK
[6] Saint Mary's University, Canada
[7] Institut fur Theoretische Physik, Universitat Gießen,
[8] Instituto de Física Corpuscular, CSIC-Universidad de Valencia, Spain

E-mail: manuela.cavallaro@lns.infn.it

Abstract. The ground and low-lying states of the unbound nucleus ^{10}Li were populated by the ^{9}Li + ^{2}H → ^{10}Li + ^{1}H reaction at 11 AMeV incident energy at the ISAC II facility (TRIUMF). In the experimental setup, the outgoing ^{9}Li at forward angles and the recoil protons at backward angles were detected and identified. This setup allows to study the ^{10}Li emitted in the crucial region at forward angles in the centre of mass.

1. Introduction

The study of the unbound system ^{10}Li is of great interest since the knowledge of its structure is a crucial ingredient in the description of the two-neutron halo nucleus ^{11}Li [1, 2]. Structure calculations for the weakly bound system ^{11}Li = ^{9}Li +2n either in the cluster model or in approaches based on the independent particle model require in fact the interaction between one neutron and the ^{9}Li core as an important input quantity. This can be deduced directly from the binding energy and states of ^{10}Li. Despite the significant amount of experimental information gathered during the last years, the properties of the ^{10}Li continuum remain unclear, to the extent that even the energy and the spin-parity of the ground state are still controversial [4-9].

In the case of N = 7 isotones, the energy and ordering of the $2s_{1/2}$, $1p_{1/2}$ and $1d_{5/2}$ orbitals can be represented as in Fig. 1, according to the excitation energy of the known single particle states. In the case of ^{15}O, ^{14}N, ^{13}C and ^{12}B ground states, it is well known that the main component of the $1d_{5/2}$ shell lies above the $2s_{1/2}$, which is in turn above the $1p_{1/2}$. In the ^{11}Be (Z = 4) ground state a shell inversion between $2s_{1/2}$ and $1p_{1/2}$ is known to appear [9,11]. If this anomaly is maintained also in the ^{10}Li (Z = 3) case, the lowest energy state in ^{10}Li is expected to have a dominant configuration with one neutron in the $2s_{1/2}$. On the other hand, if one supposes that the presence of shell inversion in ^{11}Be is related to the 2α cluster structure of Be isotopes, such a structure is not possible for ^{10}Li and the configuration of the ^{10}Li ground state is expected to be a ^{9}Li core + one neutron in the $1p_{1/2}$.

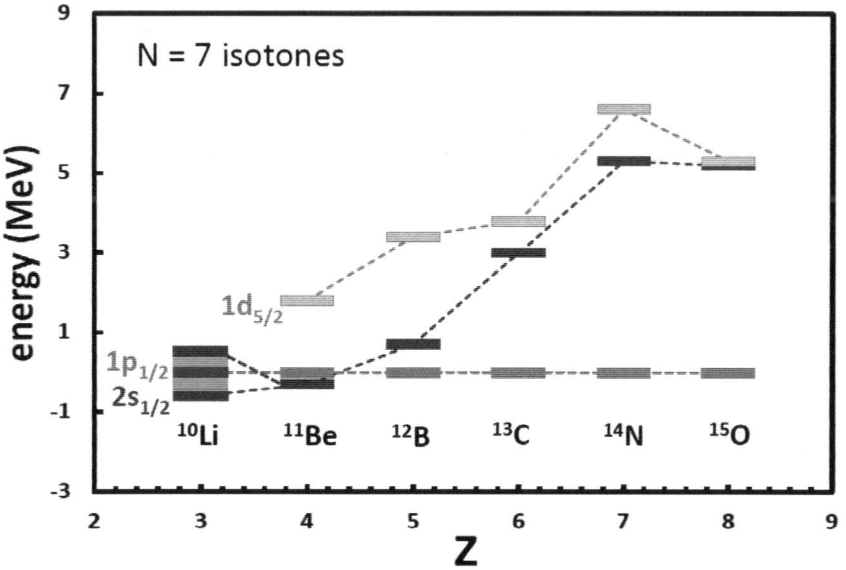

Figure 1. Relative energy of the $1d_{5/2}$ (green), $2s_{1/2}$ (blue) and $1p_{1/2}$ (red) shells for the N = 7 isotones in the region of ^{10}Li.

Transfer reactions are essential tools to probe selected components of the nuclear wave functions. Thanks to that they have been used in the past and are still crucial for nuclear spectroscopy purposes. With the advent of radioactive beam facilities, the new opportunity to explore nuclear phenomena far from the stability valley has driven a renewed interest to transfer reactions, to be studied in inverse kinematics. On one hand there is a specific attention to understand how the transfer mechanism is influenced by the reduction of the binding energy of the projectile [3,4]. On the other hand, the measured spectra could reveal unexpected features due to nucleon correlations which are beyond the mean field description of nuclear structure [5]. Examples of such phenomena have been recently found in light neutron rich nuclei such as the Li isotopes.

2. Past attempts to study the d(^{9}Li,p)^{10}Li reaction

Recently ^{10}Li has been the subject of different theoretical and challenging experimental studies [6-8], including two attempts to explore the resonant energy spectrum by the d(^{9}Li,p)^{10}Li transfer reaction at 2.35 AMeV incident energy at REX-ISOLDE and at 20 AMeV at NSCL-MSU, respectively.

In the experiment at 2.35 AMeV of Ref. [6], the energy spectrum of ^{10}Li was measured up to about 1 MeV excitation energy in an angular range between 100° and 140° in the centre of mass. Due to the low beam intensity of about 5×10^4 pps, a relatively thick target of deuterated polyethylene CD2 (660 μg/cm^2 thick) was used in order to maximize the yield, which slightly exceeded 100 events in total. The recoiling protons were detected by large area telescopes of position sensitive silicon detectors located between 18° and 80° in the laboratory. At backward laboratory angles, the energy of the protons was too low to be detected, thus excluding in the collected data the crucial region corresponding to forward angles in the centre of mass. Despite the low statistics and the not optimal angular range explored, an excitation energy resolution of about 300 keV (FWHM) was obtained and the authors could draw some conclusions about the role of the $s_{1/2}$ and $p_{1/2}$ neutron orbitals in ^{10}Li around the neutron emission threshold.

In the NSCL-MSU experiment [7], a similar experimental technique was used but at higher incident energy (20 AMeV). The beam intensity was even lower than in the REX-ISOLDE experiment (~ 7×10^3 pps) and the emittance considerably worse. The CD2 target was about 2 mg/cm^2 thick. The recoiling protons were measured at backward angles by a series of large area silicon detectors in coincidence with the ^{9}Li detected by the S800 spectrometer at forward angles. The obtained energy resolution was about 700 keV and the reconstructed ^{10}Li spectra were measured up to about 4.5 MeV excitation energy. However, due to the poor statistics (less than 100 counts in total) and energy resolution, not much was added to our understanding of ^{10}Li states.

In a recent theoretical study of the ^{10}Li low energy resonances [4], a state dependent treatment of the pairing interaction, beyond the usual BCS approximation, has proven to be necessary to reproduce the energy spectra measured in the REX-ISOLDE experiment. In addition it was shown that, despite the complications due to the resonant structure of the final states, the cross sections can be accurately described within the DWBA.

Nevertheless, due to the limitations of the existing data, many of the details predicted by the theory were not tested. For example only a measurement of the angular distribution at forward angles in the centre of mass would disentangle contributions from *s*, *p* and *d* orbitals in different portions of the energy spectrum. Also the behaviour of the observed $s_{1/2}$ and $p_{1/2}$ orbitals should be better addressed, since in both cases the coupling with the $p_{3/2}$ proton generates a doublet of ^{10}Li states, namely 1^-, 2^- for the $s_{1/2}$ and 1^+, 2^+ for the $p_{1/2}$, while experimentally there is no indication of such doublets. In addition, according to the theory, $p_{3/2}$ and $d_{5/2}$ orbitals should generate resonances above 1 MeV, but these have not been observed experimentally, since the MSU data are difficult to be reliably interpreted.

To summarize, many of the key points regarding the ^{10}Li spectrum are still very uncertain, especially from an experimental point of view. As a consequence the description of the ^{11}Li halo nucleus is quite deficient, since the binding energy strongly depends on the same n - ^{9}Li interaction responsible for the ^{10}Li structure.

3. The experiment

We studied the d(^{9}Li,p)^{10}Li reaction at 11 AMeV incident energy at the TRIUMF laboratory. At this energy the recoiling protons can be detected at backward laboratory angles, thus allowing the exploration of the crucial region at forward angles in the centre of mass.

A ^{9}Li beam, produced by the ISAC-II facility with an average current of 10^6 pps, impinged on a CD2 target, 126 μg/cm^2 thick, evaporated at the INFN-LNS laboratory (Catania). The target thickness was chosen to improve the energy resolution maintaining an acceptable count rate. The recoiling protons were detected at backward angles $127° < \theta_{LEDA} < 152°$ by the LEDA array of silicon strip detectors [12], thus allowing the study of the ^{10}Li emitted at forward angles. Protons were detected in coincidence with the ^{9}Li fragments produced from the break-up of the corresponding ^{10}Li. ^{9}Li fragments were detected and identified by a ΔE-E telescope of annular Double Sided Silicon Detectors located downstream the target. Fig. 2 shows an example of ΔE-E identification plot where the ^{9}Li locus is well identified and contoured by a solid line.

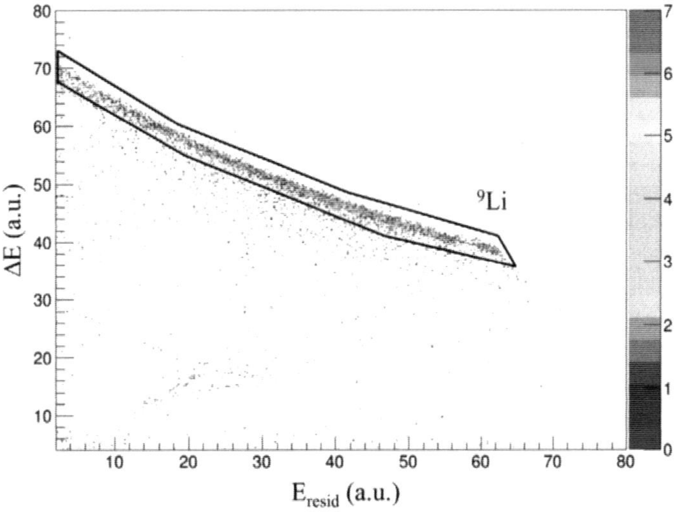

Figure 2. Energy loss (ΔE) measured by the first stage of the S2 telescope of annular silicon detectors as a function of the residual energy (E_{resid}). The graphical contour indicates the ^{9}Li events.

An additional benefit of placing the proton detectors at backward scattering angles is that no background is expected from other reaction channels. In Fig. 3 the calculated kinematics of the outgoing deuteron from for the ^{9}Li + d elastic and inelastic scattering are shown. Since there may be also a small proton contamination in the CD2 target, the ^{9}Li + proton scattering reactions are also shown in Fig. 3. In all these cases, the kinematics of the reaction does not allow to produce a light nucleus at backward scattering angles, where the LEDA detector was located.

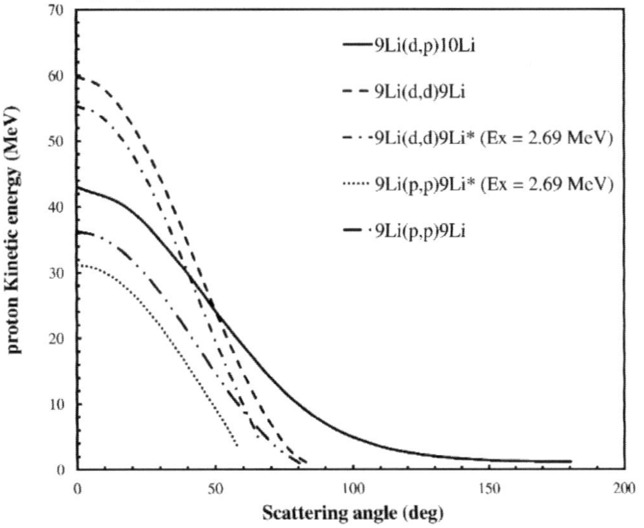

Figure 3. Energy of the emitted protons as a function of the scattering angle in the laboratory frame for the ^{9}Li(d,p)^{10}Li in inverse kinematic at 100 MeV (solid curve), in comparison with the kinematics of the lighter recoiling nuclei from competing reaction channels. The angular range covered by the LEDA detectors is $127° < \theta_{LEDA} < 152°$.

The ^{10}Li excitation energy was reconstructed with significant statistics allowing to explore its level structure. Fig. 4 shows the excitation energy spectrum obtained up to 2.5 MeV in the angular region 8.3°< θ_{CM} < 16.2°. The centroid (E_0 = 0.6 MeV) and full width at half maximum (Γ = 0.6 MeV) of the unbound ground state were deduced and also the presence of a structure at higher excitation energy at around 1.5 MeV is visible. The sharp fall down of the tail of the first peak toward zero excitation energy shows that there is no evidence in the present data of the existence of a $s_{1/2}$ low lying virtual state as supposed by [6].

The highly segmented detection system also allowed to measure the angular distributions of the observed resonances at forward angles. The analysis of the angular distribution, which is still in progress, would give some indication about the controversial question of the spin-parity of the ^{10}Li unbound ground state.

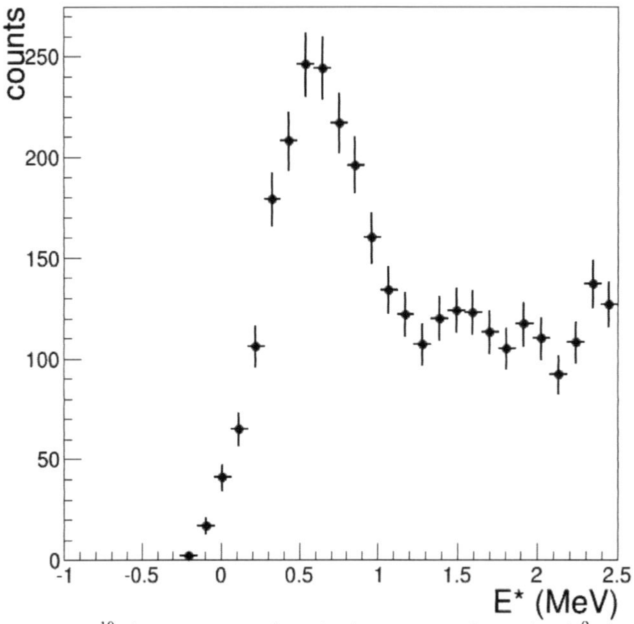

Figure 4. ^{10}Li reconstructed excitation energy from the d(^{9}Li,p)^{10}Li reaction at 11 AMeV.

[1] Tanihata I *et al* 1985 *Phys. Rev. Lett.* **55** 2676
[2] Kobayashi T *et al* 1988 *Phys. Rev. Lett.* **60** 2599
[3] Lenske H and Schrieder G 1998 *Eur. Phys. Journ. A* **2** 41
[4] Orrigo S E A and Lenske H 2009 *Phys. Lett. B* **677** 214
[5] Cappuzzello F *et al* 2012 *Phys. Lett. B* **711** 347
[6] Jeppesen H B *et al* 2006 *Phys. Lett. B.* **677** 449
[7] Santi P *et al* 2003 *Phys. Rev. C* **67** 024606
[8] Bohlen H G *et al* 1993 *Z. Phys. A* **344** 381
[9] Cappuzzello F *et al* 2004 *Nucl. Phys. A* **739** 30
[10] Nociforo C *et al* 2003 *Acta Phys. Pol. B* **34** 2387
[11] Carbone D *et al* 2014 *Acta Phys. Pol. B* **45** 431
[12] Davinson T *et al* 2000 *Nucl. Instr. And Meth. A* **454** 35

Trojan Horse Method: recent results in nuclear astrophysics

C Spitaleri[1,2], L Lamia[1,2], M Gimenez Del Santo[3], V Burjan [4], N Carlin[3], Chengbo Li[5], S Cherubini[1,2], V Crucilla[2], M Gulino[1,2], Z Hons[4], V Kroha[4], B Irgaziev[6], M La Cognata[2], J Mrazek[4], M Mukhamedzhanov[8], M G Munhoz[3], S Palmerini[2], R G Pizzone[2]. M R Puglia[2], G G Rapisarda[1,2], S Romano[1,2], L Sergi[2], Shu-Hua Zhou[7], E Somorjai[9], F A Souza[3], G Tabacaru[8], A Szanto de Toledo[3], A Tumino[2,10], Qungang Wen[11], Y Wakabayashi [12] and H Yamaguchi[12]

[1] Dipartimento di Fisica e Astronomia, Università di Catania, Catania, Italy
[2] INFN, Laboratori Nazionali del Sud, Catania, Italy
[3]Departamento de Fisica Nuclear, Universitade de Sao Paulo, Sao Paulo, Brasil
[4]Nuclear Physics Institute of ASCR, Rez, Czech Republic
[5]Beijing Radiation Center, 100875 Beijing, China
[6]GIK Institute of Engineering Sciences and Technology, Topi, Districti Swabi, K.P. Pakistan
[7]China Institute of Atomic Energy, Beijing, China
[8]Cyclotron Institute, Texas A&M University, College Station, TX 77843, USA
[9]Institute for Nuclear Research (ATOMKI), Debrecen, Hungary
[10]University of Enna, Unikore, Enna,Italy
[11],china
[12]CNS-University of Tokio, Japan

E-mail: `spitaleri@lns.infn.it`

Abstract. The accurate knowledge of thermonuclear reaction rates is important in understanding the energy generation, the neutrinos luminosity and the synthesis of elements in stars. The physical conditions under which the majority of astrophysical reactions proceed in stellar environments make it difficult or impossible to measure them under the same conditions in the laboratory. That is why different indirect techniques are being used along with direct measurements. The Trojan Horse Method (THM) is introduced as an independent technique to obtain the bare nucleus astrophysical S(E)-factor. As examples the results of recent the application of THM to the $^{2}H(^{11}B, \alpha_0{}^{8}Be)n$ and $^{2}H(^{10}B, \alpha_0{}^{7}Be)n$ reactions are presented.

1. Introduction

Big efforts have been devoted in the last years to the study of light elements abundances, being their importance related to cosmology as well as to stellar structure and evolution . In fact hints on the primordial Big Bang Nucleosynthesis (BBN) can be achieved from Li primordial abundance [1]. Moreover these studies can be a precious tool for testing and understanding the inner stellar structure, especially for what regards the mixing processes in stellar envelopes [2].

In this framework, the different nuclear processes producing or destroying lithium (Li), beryllium (Be) and boron (B) must be studied in details and an accurate knowledge of the involved nuclear cross sections are necessary. In particular we will focus our attention on one of the main destruction channels for these elements in stellar environments, the (p, α) reactions.

For this reason bare nucleus cross section measurements of the p-capture reaction at the Gamow energy (E_G) should be known with an accuracy better than 10% [3] because of their crucial role in understanding the first phases of the Universe history and the subsequent stellar evolution.

The effective cross section for stellar plasma $\sigma_{pl}(E)$, is connected to the bare nucleus cross section $\sigma_b(E)$ and to the stellar electron screening enhancement factor f_{pl} by the relation

$$\sigma_{pl}(E) = \sigma_b(E) f_{pl}(E) = \sigma_b(E) \cdot exp(\pi\eta U_{pl}/E) \tag{1}$$

where U_{pl} is the plasma potential energy and η the Sommerfeld parameter. If $\sigma_b(E)$ is measured at the ultralow energies E_G and f_{pl} is estimated within the framework of the Debye-Hückel theory, it is possible estimate from equ.(1) the $\sigma_{pl}(E)$.

In the neutron-induced reactions, the lack of a Coulomb barrier and the typical $\sigma_b(E) \sim 1/E^{1/2}$ energy dependence led to a precise knowledge of $\sigma_b(E_G)$ for many cases. Whereas the Coulomb barrier of height E_C in charged-particle induced reactions E_C causes an exponential decrease of the cross section at $E < E_C$, $\sigma_b(E) \sim exp(-2\pi\eta)$, leading to a low-energy limit of direct $\sigma_b(E)$ measurements, which is typically much larger than E_G. Owing to the strong Coulomb suppression, the behavior of the cross section at E_G is usually extrapolated from the higher energies by using the definition of the smoother astrophysical factor S(E):

$$S_b(E) = E\sigma_b(E) exp(2\pi\eta) \tag{2}$$

where $exp(2\pi\eta)$ is the inverse of the Gamow factor, which removes the dominant energy dependence of $\sigma(E)$ due to the barrier penetrability.

Although the S_b(E)-factor allows for an easier extrapolation, large uncertainties to $\sigma_b(E_G)$ may be introduced due to for instance the presence of unexpected resonances. In order to avoid the extrapolation procedure, a number of experimental solutions were proposed in direct measurements for enhancing the signal-to-noise ratio at E_G.

In recent years the availability of hight-current low-energy accelerators, such as that at the underground Laboratori Nazionali del Gran Sasso (e.g.[4]), together improved target and detection techniques have allowed us to perform $\sigma_b(E)$ measurements in some cases down to E_G or at least close to E_G.

Then in principle no $\sigma_b(E)$ extrapolation would be needed anymore for these reactions. However, the laboratory measurements at ultra low-energies suffer the presence of electron screening effects [5, 6], which exponentially enhances the measured cross section $\sigma_s(E)$ values [or equivalently the corresponding $S_s(E)$ astrophysical factor] with respect the case of bare nuclei. This is usually described in terms of an *enhancement* factor defined as [3]

$$f_{lab}(E) = \sigma_s(E)/\sigma_b(E) = exp(\pi\eta U_e/E) \tag{3}$$

In this equation U_e is the electron screening potential measured in the laboratory which is different from the U_{pl} in the stellar environment. Clearly, a precise knowledge of U_e is needed for calculating $\sigma_b(E)$ from the experimental $\sigma_s(E)$ values by using Eq.(3). In turn, the understanding of U_e may help to better understand U_{pl}, needed to calculate σ_{pl}.

Then, even in those cases for which direct cross section measurements are possible in the Gamow energy range, the bare nucleus cross section $\sigma_b(E)$ is extrapolated by high-energies direct measurements, negligibly affected by electron screening effects.

Thus, alternative methods for determining bare nucleus cross sections of astrophysical interest are needed. In this contest a number of indirect methods, e.g. the Coulomb dissociation [7], the ANC (asymptotic normalization coefficient) method [8, 9, 10], and the Trojan-horse method (THM) were developed [11, 12, 13, 14, 15, 17, 16]. The THM is a powerful tool which selects the quasi-free (QF) contribution of an appropriate three-body reaction performed at energies well above the Coulomb barrier to extract a charged particle two-body cross section at astrophysical energies free from Coulomb suppression. The THM has already been applied several times to reactions connected with fundamental astrophysical problems and here its application for studying the astrophysically relevant ^{11}B(p,α_o)^{8}Be and ^{11}B(p,α_o)^{8}Be reactions will be discussed.

2. The Trojan Horse method

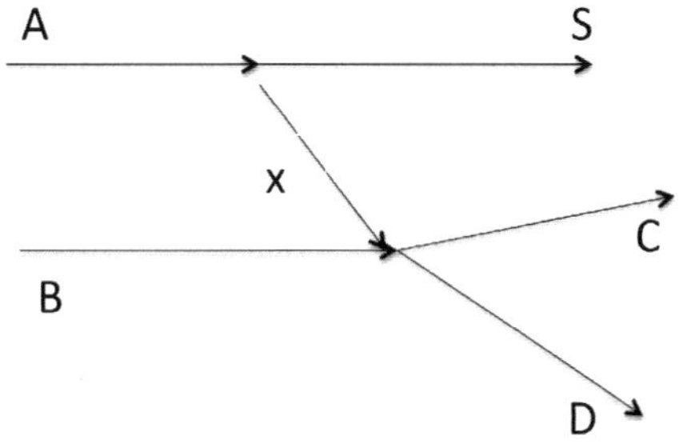

Figure 1. Feynman diagram representing the quasi-free $B(A, CD)S$ reaction.

The basic idea underlying indirect methods is the use of nuclear reaction theory to link the cross section of a reaction of astrophysical importance to the one of a different process, easier to study with present-day facilities. Indirect techniques have been developed over the past several decades to determine reaction rates that cannot be measured in the laboratory. An example is the application of the detailed balance principle to the time-reversed reaction. In the case of (α,p) reactions this approach allows one to deal with larger cross sections, thanks to the smaller Coulomb barrier for proton-induced reactions than for α-induced ones. However, only the cross section for ground state to ground state transition of the time reversed (p,α) reaction can be obtained, setting a lower limit to the total cross section.

This is a feature in common with several indirect approaches: since a different process is measured with respect to the one of astrophysical interest, different physical features could be displayed that have to be carefully inspected not to incur into systematic errors.

The THM aims at obtaining the cross section of the binary process x + B $\rightarrow$ C + D at astrophysical energies by measuring the Trojan Horse (TH) reaction (the two-body to three-body process (2 $\rightarrow$ 3 particles)) A+B $\rightarrow$ C + D+S in quasi-free (QF) kinematics regime ([17, 18, 19, 20] and references therein), where the "Trojan Horse" particle, A = (Sx), which has a dominant cluster structure, is accelerated at energies above the Coulomb barrier.

After penetrating the barrier, the TH nucleus A undergoes breakup leaving particle x (participant) to interact with target B while particle S (spectator) flies away. From the measured cross section of TH reaction, the energy dependence of the binary sub-process is determined [11, 27].

The reaction used in the THM can proceed through different reaction mechanisms. The reaction mechanisms shown schematically in Fig. 1 gives the dominant contribution to the cross section in a restricted region of the three-body phase space when the relative momentum of the fragment S and x is zero (the QF kinematical condition) or small compared to the bound state (Sx) wave number [11]. The THM has been extensively discussed in the literature [21, 22, 23, 24, 25, 26, 27, 28, 29, 30, 31, 32, 33, 36, 37, 38, 40, 41]. This method, as the Asymptotic normalization coefficient (ANC) and Coulomb dissociation (CD) methods, has a common underlying connection to nuclear reaction theory.

Since the transferred particle x in the TH reaction is virtual, its energy and momentum are not related by the on mass shell equation, thus

$$E_{xA} \neq \frac{p_{xA}^2}{2\mu_{xA}} \tag{4}$$

being the initial channel $A + B$, where A = (Sx), rather than the simple $x + A$. To increase the TH triple differential cross section, the relative kinetic energy E_{AB} in the initial channel of the TH reaction should be higher than the Coulomb barrier between particles A and B. Then, the probability to find nucleus A near B, which is given by the modulus square of the scattering wave function, is not suppressed by the Coulomb barrier, leading to a finite probability that B can be in proximity of x.

Thus, there is no additional Coulomb barrier between B and the constituent particle x of the TH nucleus B.

Moreover the THM cross section turns out to be insensitive to the electron screening effect and allows one to determine the energy dependence of the binary cross section in the region of astrophysical interest with no need of extrapolation. THM makes it possible also to retrieve independent information on the electron screening potential U_e when direct ultra-low energy measurements are available [11].

Despite being based on a relatively simple idea, the method involves a non trivial theoretical description (see [11] and references therein) and requires the careful selections of experimental conditions to guarantee that the correct reactions and the appropriate reaction mechanism are selected.

The absolute value of $S(E_{xB})$ can be found by normalization of TH data to available direct measurements at higher energies. A recent application of the THM has addressed the possibility of studying neutron induced reactions at low energies on radioactive nuclei using deuteron as source of "virtual" neutrons n, being the n-induced reactions only affected by centrifugal barrier. This idea has been successful tested with stable beams in [21, 22, 23]. Thus, the THM is currently an alternative technique for measuring cross sections for reactions related to the s, r processes in nuclear astrophysics (or nuclear structure). Furthermore the cost of production of virtual neutron targets is negligible. Indeed "virtual neutron" targets are easy to prepare from deuterated polyethylene material in any chemical laboratory. In addition, the THM cannot be applied to radiative capture reactions (p,γ), (n,γ), (α,γ).

The first attempt to deliver the TH reaction theory has been done by [16], where they introduced the surface approximation. Although the idea of the surface approximation is quite attractive, its practical realization required the surface integral formalism, recently applied to deuteron stripping reactions populating bound states and resonances [10]. The developed theory of the deuteron stripping was based on DWBA and post Continuum Discretized Coupled Channel (CDCC) formalism, the surface integral formulation of the reaction theory, and of the R-matrix method. This theory can be directly applied for analysis of the TH reactions, in which the TH nucleus is a deuteron, even if it can be also extended to other TH nuclei (see for instance [11]).

The absence of the absolute measurements allows us to consider the theory of the TH reactions in the plane wave impulse approximation (PWIA), which often predicts reasonable well the energy dependence of the three-body cross sections and it is much simpler then the DWBA and CDCC approaches.

The description of the quasi-free reaction (QFR) is very simple in the Impulse Approximation (IA) and can be represented by the Feynman diagram in Fig. 1.

In the IA, the three body reaction cross section is proportional to the cross section of the binary reaction [39] and, by using the PWIA approximation the three body reaction can be factorized into two terms corresponding to the vertices of figure 1 and is given by:

$$\frac{d^3\sigma}{d\Omega_C d\Omega_D dE_C} \propto (KF) \cdot |\phi(p_{x-S})|^2 \cdot \left[\frac{d\sigma}{d\Omega}\right]^{HOES} \tag{5}$$

where:

(i) $\left[\frac{d\sigma}{d\Omega}\right]^{HOES}$ is the half-off-energy-shell differential cross section for the binary $B(x, C)D$ reaction at the center of mass energy E_{cm} given in post-collision prescription (PCP) by the relation:

$$E_{cm} = E_{CD} - Q_{2b} \tag{6}$$

where Q_{2b} is the Q-value of the binary $x + B \rightarrow C + D$ reaction and E_{CD} is the $C - D$ relative energy;

(ii) KF is a kinematical factor containing the final state phase space factor and is a function of the masses, momenta, and angles of the outgoing particles (see [27] and ref. ther.);

(iii) $|\phi(p_{x-S})|^2$ is the Fourier transform of the radial wave function for the $(x - S)$ inter-cluster motion usually described in terms of Hankel, Eckart, and Hulthen functions depending on the $x - S$ system properties.

The applicability of the pole approximation is limited to small momenta p_{x-S}. Namely, the region where the pole diagram is expected to be predominant in the reaction mechanism was suggested to be [42]

$$0 \leq p_{x-S} \leq k_{x-S} \tag{7}$$

with

$$k_{x-S} = \sqrt{2\mu_{x-S} B_{x-S}} \tag{8}$$

where μ_{x-S} is the reduced mass and B_{x-S} is the binding energy of the system $x-S$, respectively.

If the experimental momentum distribution $|\phi(k_{x-S})|^2$ is measured, one can, in principle, extract the HOES cross-section by measuring the three body reaction and by calculation of (KF). Following the $PWIA$ the two body cross section is derived by dividing the measured three-body cross section by product of the kinematic factor and the experimental momentum distribution $(KF)|\phi(k_{x-S})_{exp}|^2$:

$$\left[\frac{d\sigma}{d\Omega}\right]^{HOES} \propto \frac{\frac{d\sigma^3}{d\Omega_C d\Omega_D dE_C}}{(KF)|\phi(k_{x-S})_{exp}|^2} \tag{9}$$

2.1. Incident energy prescriptions.

For the applicability of the IA, the incident energy E_{A-B} of the projectile must be much larger than the binding energy B_{x-S} [39] in such a way the bombarding energy E_{A-B} overcomes the Coulomb barrier $(E_{A-B})^{C.B.}$ in the entry channel as well

$$E_{A-B} > (E_{A-B})^{C.B.} \tag{10}$$

The QF kinematical conditions must be chosen in such a way that the E_{x-B} relative energy can span the astrophysical region of interest below the Coulomb barrier $(E_{x-B})^{C.B.}$ [16]

$$E_{x-B} < (E_{x-B})^{C.B}. \tag{11}$$

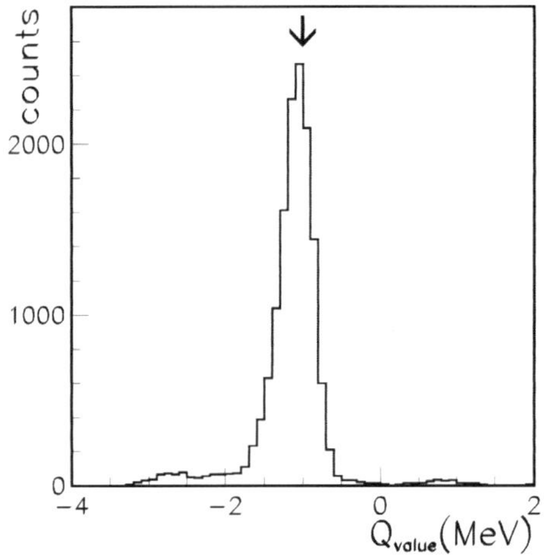

Figure 2. The experimental Q-value spectrum for the TH reaction ^{2}H(^{10}B, $\alpha_0{}^{7}$Be)n studied at INFN-LNS by using a boron beam energy of 27 MeV [12].

The commonly used THM approach differs from original idea by Baur [16], for which the initial velocity of the projectile B is compensated for by the Fermi motion of particle x. In this framework, a momentum of the order of hundreds of MeV/c could be needed. However, in the case of a Trojan Horse nucleus with a predominant $l = 0$ inter-cluster motion, these momenta populate the tail of the momentum distribution for particle x, making the separation from eventual background reaction mechanisms very critical.

Moreover the tail of the calculated momentum distribution changes depending on the theoretical approach applied, therefore a very sophisticated treatment might be required in order to get the relevant binary reaction cross section.

In order to overcome these problems, we have introduced a different approach based on the idea that the initial projectile velocity is compensated for by the binding energy of particle x inside A necessary to compensate for the high impinging momenta (equation(6)).

$$E_{cm} = E_{xB} - B_{x-S} \tag{12}$$

with E_{x-B} projectile energy in the two-body center of mass system.

Thus the relative energy of the fragments in the initial channel $x + B$ of the binary reaction can be very low and even negative. In the THM approach we fix the beam energy and vary the relative momentum $x - S$ within few tens of MeV/c; in this case the kinematics of the experiment slightly deviates from the QF condition, but one can easily cover the whole astrophysical energy region of interest.

3. Data analysis

As already mentioned, the main feature of the THM is the extraction of the astrophysical S(E)-factor at low-energies via the selection of the QF mechanism to extract the two-body cross section of interest.

Thus, particular attention is devoted to the data analysis procedure. In more detail, several steps are involved in the data analysis before the extraction of the two-body cross section of astrophysical relevance, namely:

Journal of Physics: Conference Series **630** (2015) 012020 doi:10.1088/1742-6596/630/1/012020

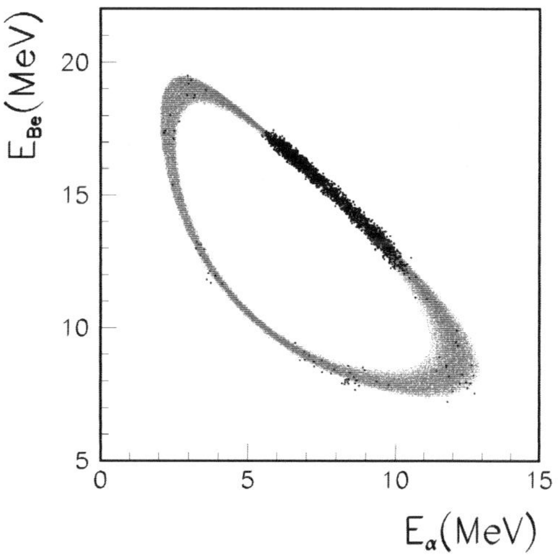

Figure 3. The experimental kinematical locus (black points) superimposed onto the simulated one (red points) for the TH reaction ^{2}H(^{10}B, $\alpha_0{}^7$Be) [12].

(i) identification of the three-body reaction channel of interest $A + B \to C + D + S$ (Fig.2 and Fig.3);
(ii) identification of the events belonging to the quasi-free (QF) reaction mechanism;
(iii) subtraction of spurious events arising from mechanisms other than the QF breakup as, for instance, sequential decay (SD);
(iv) selection of the QF reaction mechanism and related tests ;
(v) extraction of the THM two-body cross section, from the measured three-body one, in arbitrary units via Eq. (9);
(vi) correction of the THM two-body cross section via the Coulomb barrier penetration factor;
(vii) normalization high-energy direct data to obtain the bare nucleus cross section $\sigma_b(E_{CM})$, in absolute units;
(viii) validity tests;
(ix) extraction of astrophysical factor $S_b(E)$ in absolute units;
(x) determination of screening potential.

3.1. Extraction of two-body cross section from the measured three-body reaction
In the analysis, the HOES two body cross section is derived by dividing the selected three-body coincidence yield (Y) by the result of a Monte Carlo calculation for the product between the kinematical factor and the experimental momentum distribution

$$\left[\frac{d\sigma}{d\Omega}\right]^{HOES} \propto \left[\frac{Y}{(KF) \mid \phi(p_{xS}) \mid^2}\right] \tag{13}$$

where $|\phi(p_{xS})|^2$ is referred to the spectator momentum range values corresponding to those given by Eq. (1). We stress that this relation is a limit of the pole approximation applicability.

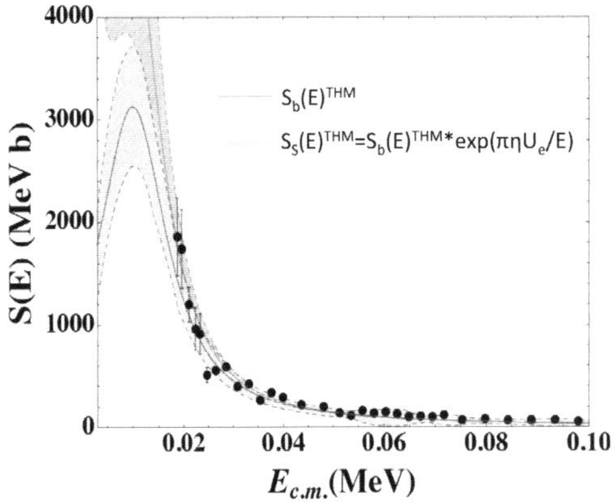

Figure 4. The TH $^{10}B(p,\alpha_0)^7Be$ $S_b{}^{THM}(E)$ factor at infinite resolution [12], together with its allowed upper and lower limits, compared with the low-energy direct data of [10]. At energies lower than 30 keV direct data are described in terms of the enhanced $S_b{}^{THM}(E)$ factor [12]. The *measured* S_b(10keV)=3127±583 (MeV b) agrees with the one *extrapolated* in [43].

This cross section needs to be normalized to the binary reaction one in order to obtain the absolute value in an energy range $\Delta(E_{CM})$ of the excitation function above the Coulomb barrier. In fact, above the Coulomb barrier, must be

$$\left[\frac{d\sigma}{d\Omega}\right]^{HOES} \propto \left[\frac{d\sigma}{d\Omega}\right]^{OES} \tag{14}$$

where the angular range $\theta(E_{CM})$ is constant in both the relations. In the THM application the reaction x+ B → C+D is induced inside the short-range nuclear field, thus the Coulomb penetration factor

$$P_l(k_{xB}r_{xB}) = \frac{k_{xB}r_{xB}}{F_l^2(k_{xB}r_{xB}) + G_l^2(k_{xB}r_{xB})} \tag{15}$$

has to be introduced in order to compare the THM cross section with the direct data from literature in the energy region below the Coulomb barrier. In Eq.(15), F_l and G_l are the regular and irregular Coulomb wave functions, k_{xB} and r_{xB} the x-B relative wave number and interaction radius, respectively. Thus, the THM cross section will be given by

$$\left[\frac{d\sigma}{d\Omega}\right]^{THM} \propto \left[\frac{Y_l}{(KF)\phi^2(p_{xS}}\right][P_l(k_{xB}r_{xB})] \tag{16}$$

being this the quantity to be compared with the OES cross section for both normalization purposes and validity test.

4. Astrophysical S-factor

The THM provides an independent measurement of the bare nucleus cross section σ_b(E) (or equivalently of the corresponding bare nucleus astrophysical factor, S_b(E)). Thus, it is possible to measure the electron screening potential [3] by comparing S_b(E) with the directly measured one with shielded nuclei. Once the two-body cross section has been extracted , the THM astrophysical S(E)-factor can be obtained according to:

$$[S_b(E)]^{THM} = E[\sigma_b(E)]^{THM}exp(2\pi\eta) \tag{17}$$

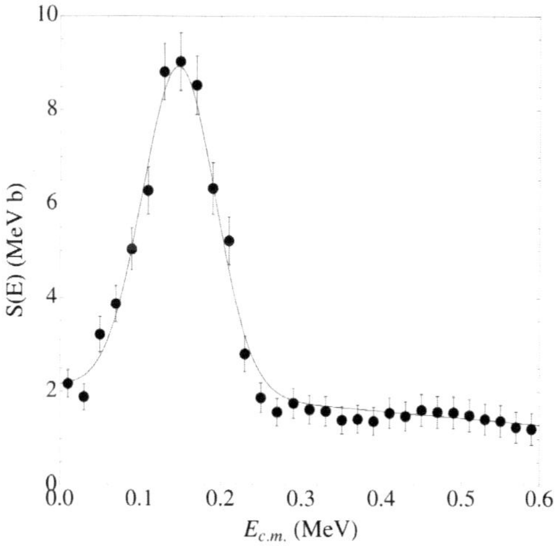

Figure 5. The TH ^{11}B(p,α_0)^{8}Be S(E)-factor (black points) compared with the one reported in NACRE ([43] and references therein), including the THM experimental resolution effects (black line). See [35] for a complete discussion.

where $\sigma_b(\mathrm{E})^{THM}$ is the THM bare nucleus cross section obtained by Eq. (18). Thus, the energy dependence of $[\mathrm{S}_b(\mathrm{E})]^{THM}$ should show the same trend of the directly measured one. Clearly, this procedure does not allow us to extract the absolute value of the astrophysical S-factor. However, the absolute scale for $\mathrm{S}_b(\mathrm{E})^{THM}$ can be obtained by normalizing the THM data to the direct $\mathrm{S(E)}^{OES}$ ones in a range of energies $\Delta\mathrm{E}^*$ where the electron screening effects are negligible, as

$$N_{abs.value} = \frac{[S(\Delta E*)]^{OES}}{[S_b(\Delta E*)]^{THM}} \tag{18}$$

In the ultra low-energy range, $\mathrm{S}_b(\mathrm{E})^{THM}$ differs from the $\mathrm{S(E)}^{OES}$ one because of the enhancing electron screening effects on direct data. Thus, THM could also provide an independent measurement of electron screening potential by simply fitting the direct data with an analytical form in which $\sigma_b(\mathrm{E})$ of Eq. (2) is replaced by $\sigma_b(\mathrm{E})^{THM}$.

5. Results

The first run of the ^{10}B(p,α_0)^{7}Be experiment was performed at the Pelletron-Linac laboratory (Departamento de Fisica Nuclear (DFN) in Sao Paulo (Brazil) [31]. The Tandem Van De Graaf accelerator provided a 27 MeV ^{10}B beam with a spot size on target of about 2 mm and intensities 1 npa. The second run was performed at the Laboratori Nazionali del Sud in Catania using the SMP Tandem Van the Graaff accelerator to produce a ^{10}B beam with an energy of 24.5 MeV [12]. The beam was accurately collimated in order to have a spot diameter of about 1.5 mm. In both cases, a $\sim$200μg/cm^2 thick deuterated polyethylene target were used. The detection set up consisted of Position Sensitive Silicon Detectors, placed on both sides of the beam direction, and of an ionization chamber (IC) for ^{7}Be identification. The obtained THM S(E)-factor is displayed in Fig.4 (solid blue line), compared with the direct measurements given in [43]. At energies lower than about 30 keV, direct data have been described in terms of the standard factor f_{enh} given in Eq.3 (see [12] for details).

The ^{11}B(p,α_0)^{8}Be experiment was performed at the Laboratori Nazionali del Sud in Catania using the SMP Tandem Van the Graaff accelerator to produce a ^{11}B beam with an energy of 27

MeV [27, 35, 36]. The beam was accurately collimated in order to have a spot diameter of about 1.5 mm. In this cases an about 150 μg/cm^2 deuterated polyethylene target were used. The detection set up consisted of Position Sensitive Detectors and Dual Position Sensitive Detectors (DPSD) placed on both sides of the beam direction. In particular the DPSD's made up of two PSD mounted one above the other with a 1 mm empty space, are specially suitable for the identification of the two α particles coming from ^{8}Be decay.

The ^{11}B(p,α_0)^{8}Be THM S(E)-factor is displayed in Fig.5 (black points), compared with the fit of direct data given in [43] smeared out for the THM experimental resolution of ~45 keV, as discussed in [35].

6. Conclusions

THM represents an useful tool for experimental nuclear astrophysics, avoiding the extrapolation procedures for extracting low-energy S(E)-factor values for astrophysical purposes. Important results have already been obtained also for pure nuclear physics (i.e. resonant reactions treatment), while new extensions of the method to further cases of interest (e.g., reactions involving neutrons and/or reactions with Radioactive Ion Beams), are currently underway.

This work has been partially supported by the Italian Ministry of University MIUR under the grant "LNS-Astrofisica Nucleare (fondi premiali)" and RFBR082838 (FIRB2008). It has been partially supported also by the grant LC 07050 of the Czech MŜMT, grant M10480902 of the Czech Academy of Science, grant LH1101 of the AMVIS project. A. M. M. acknowledges the support by U.S. Department of Energy under Grants No. DE-FG52-09NA29467, No. DE-FG02-93ER40773, and No. DE-SC0004958 and by NSF under Grant No. PHY-0852653.

7. References

[1] Fields B D 2011 *Annu. Rev. Nucl. Part. Sci.* **61** 47
[2] Boesgaard A M et al 2005 *Astronphys. J.* **621** 991
[3] Rolfs C e and Rodney W S 1988 *Cauldrons in the Cosmos*, University of Chicago Press, Chicago
[4] Bonetti R et al 1999 *Phys. Rev. Lett.* **82** 5205
[5] Assenbaum H J, Laganke K and Rolfs C 1987 *Z.Phys.* A **327** 461
[6] Strieder F, Rolfs C, Spitaleri C and Corvisiero P 2001 *Naturrwissenschaften* **88** 461
[7] Baur G, Bertulani C and Rebel H 1986 *Nucl. Phys.* A **458** 188
[8] Mukhamedzhanov A M and Tribble R E 1999 *Phys. Rev.* C **59** 3418
[9] Mukhamedzhanov A M and Gagliardi C A 2001 *Phys. Rev.* C **63** 024612
[10] Mukhamedzhanov A M et al 2008 *Phys. Rev.* C **78** 015804
[11] Tribble R E, Bertulani C, La Cognata M, Mukhamedzhanov A M and Spitaleri C 2014 *Rep. Progr. Phys.* **77** 106901
[12] Spitaleri C et al 2014 *Phys. Rev.* C **90** 035801
[13] Spitaleri C et al 2011 *Phys. Atom. Nucl.* **74** 1763
[14] Spitaleri C et al 2003 *Nucl. Phys.* A **719** 99c
[15] Typel S and Baur G 2003 *Ann. Phys.(NY)* **305** 228
[16] Baur G 1986 *Phys. Lett.* B **178** 135
[17] Spitaleri C 1991 *Problem of fundamental modern physics II* Editors Cherubini, Dal Piaz, Minetti, ISBN 981-02-0295-4 (World Scientific)
[18] Calvi G, Lattuada M, Miljanic G, Riggi F, Spitaleri C and Zadro M 1990 *Phys. Rev.* C **41** 1848
[19] Zadro M, Miljanic G, Spitaleri C, Calvi C and Riggi F 1989 *Phys. Rev.* C **40** 181
[20] Lattuada M, Riggi F, Spitaleri C, Vinciguerra D, Miljanic G, Zadro M, Jinzhang Yao 1985 *Nycl. Phys.* A **458** 493
[21] Tumino A et al 2006 *Eur. Phys. J* A **27** 243
[22] Gulino M et al 2010 *Journ. Phys.G -Nuclear and Particle Physics* **37** 1205105
[23] Gulino M et al 2013 *Phys. Rev.* **37** 1205105
[24] Cherubini S, Kondratyev V N, Lattuada M, Spitaleri C, Miljanic G, Zadro M, and Baur G 1996 *Astrophys. J.* **457** 855
[25] Spitaleri C, Aliotta M, Cherubini S, Lattuada M, Miljanic G, Romano S, Soic S, Zadro M and Zappalá R A 1999 *Phys.Rev.* C **60** 055802

[26] Spitaleri C et al 2001 *Phys.Rev.* C **63** 055801
[27] Spitaleri C et al 2004 *Phys.Rev.* C **69** 055806
[28] Tumino A et al 2005 *Eur. Phys. J.* A **25** 649
[29] La Cognata M et al 2005 *Phys. Rev.* C **72** 065802
[30] La Cognata M et al 2007 *Phys. Rev.* C **76** 065804
[31] Lamia L et al 2007 *Nucl. Phys.* A **787** 309c
[32] La Cognata M et al 2008 *J. Phys. G.- Nucl. Part. Phys.* **35** 014014
[33] La Cognata M et al 2008 *Phys. Rev. Lett.* **10** 152501
[34] La Cognata M et al 2010 *Astrophys. J.* **708** 796
[35] Lamia L et al. 2012 *J. Phys.* G **39** 015106
[36] Lamia L et al 2012 *Phys. Rev.* C **85** 025805
[37] Lamia L et al 2012 *Astron. & Astrophys.* **541** A158
[38] Lamia, L. et al 2013 *Astrophys. J.* **768** 65
[39] Chew G F and Wick G C 1952 *Phys. Rev.* C **85** 636
[40] Sergi, M.L. et al 2010 *Phys. Rev.* C **82** 032801(R)
[41] Tumino A et al 2014 *Astrophys. J.* **785** 96
[42] Shapiro I S 1967 *Interaction of High-Energy Particles with Nuclei, International School of Physics Enrico Fermi, Course 38* edited by E.O. Ericson (Academic, New York, 1967) pp. 210-251
[43] Angulo C et al (NACRE) 1999 *Nucl. Phys.* A **656** 3

Journal of Physics: Conference Series **630** (2015) 012021 doi:10.1088/1742-6596/630/1/012021

Uncertainties of optical-model parameters for the study of the threshold anomaly

D Abriola[1], A Arazi[1,2], J Testoni[1] , F Gollan[2] and G V Martí[1],

[1] Laboratorio TANDAR, Comisión Nacional de Energía Atómica, Avda. General Paz 1499, (B1650KNA) San Martín, Argentina

[2] Consejo Nacional de Investigaciones Científicas y Técnicas, Avda. Rivadavia 1917, (C1033AAJ) Buenos Aires, Argentina

E-mail: abriola@tandar.cnea.gov.ar

Abstract. In the analysis of elastic-scattering experimental data, optical-model parameters (usually, depths of real and imaginary potentials) are fitted and conclusions are drawn analyzing their variations at bombardment energies close to the Coulomb barrier (threshold anomaly). The judgement about the shape of this variation (related to the physical processes producing this anomaly) depends on these fitted values but the robustness of the conclusions strongly depends on the uncertainties with which these parameters are derived. We will show that previous published studies have not used a common criterium for the evaluation of the parameter uncertainties. In this work, a study of these uncertainties is presented, using conventional statistic tools as well as bootstrapping techniques. As case studies, these procedures are applied to re-analyze detailed elastic-scattering data for the ^{12}C + ^{208}Pb and the ^{6}Li + ^{80}Se systems.

1. Introduction

In the study of elastic scattering of atomic nuclei at low energies there has been a long-lasting interest in the so-called threshold anomaly (TA [1]). For the case of tightly bound nuclei, this phenomenon, related to the closure of reaction channels, consists in a decrease of the depth of imaginary part of the optical potential at bombarding energies below the Coulomb barrier V_B. Due to the causality-related dispersion relation (DR [1]) linking the real and imaginary parts of optical potentials, the depth of the real part also varies strongly, peaking around V_B (see Refs. [2, 3, 1]). In the case of weakly bound projectiles, the coupling to nonelastic channels (e.g. breakup) generates a repulsive polarization potential [4] that can produce either the absence of TA (*no TA* [5, 6, 7, 8]) or the so-called breakup threshold anomaly (BTA [7, 9, 8]), in which the imaginary potential increases while approaching V_B and, conversely, there is a reduction on the real part of the potential [9, 10].

For an unambiguous determination of the kind of anomaly, the variation of the optical-model parameter values as a function of the energy must be evaluated taking into account their uncertainties (hereafter called *parameter uncertainties*), which must be derived from the experimental angular distribution data and their own uncertainties (hereafter called *experimental uncertainties*). However, publications on the subject apply different, sometimes non-rigorous criteria for the estimation of the parameter uncertainties or do not consider uncertainties at all. Here, a study of the parameter uncertainties is presented, using conventional statistic tools as well as the bootstrap technique [11]. It is hoped that these methods will help to achieve a better determination of the parameters of interest in the characterization of the TA. In section 2 we discuss mathematical and physical issues that influence the determination of uncertainties when adjusting optical-model parameters using experimental angular distributions, in section 3 the $^{12}C + ^{208}Pb$ and the $^{6}Li + ^{80}Se$ systems will be reanalyzed and their uncertainties evaluated with different techniques. Finally, in section 4 we make our final remarks.

2. The determination of optical-model parameter uncertainties

In this section we will firstly consider the χ^2 test (2.1) since it is the most popular method to evaluate the optical-model parameter uncertainties, we will review the different χ^2 criteria found in the literature (2.2) and finally we will consider the two techniques that we will apply to the case studies in the next section, namely: *covariance ellipses* (2.3) and *bootstrap method* (2.4).

2.1. The χ^2 test

Let us consider a series of experimental pairs $x_i, y_i(x_i)$, where the y_i values have an uncertainty Δy_i (statistical, background substraction, etc.) which is assumed to be normally distributed. To compare these experimental data with a theoretical model that adjust the data with a function $y = f(x)$ taking into account the experimental uncertainties, we could use an adjusting program that minimizes the value of χ^2 defined as

$$\chi^2 = \sum_i^N R_i^2 = \sum_i^N \left[\frac{y_i - f(x_i)}{\Delta y_i}\right]^2, \tag{1}$$

where N is the number of experimental points and R_i are the so-called *residuals*. The adjusting program should minimize the effective distance between the experimental results y_i and their theoretical estimation $f(x_i)$. The simplest case to study is the repetition of a measurement of a physical quantity (say a life-time or the speed of light) by several laboratory groups. The theoretical model $f(x_i)$ could be as simple as the construction of the best evaluated value of that physical quantity, e.g. the arithmetic or weighted average. In this simple case the expected χ^2 should be close to $N-1$, where N is the number of measurements. If not, procedures such as the proposed by Birge [12, 13, 14] should be applied. In its original form, it applies for interlaboratory evaluations (external uncertainty) when $\chi^2/(N-1) > 1$ and consists in multiplying the experimental (internal) uncertainties by $\sqrt{\chi^2/(N-1)}$. With the increased uncertainties a new, normalized value $\chi^{2\prime}$ is obtained, here using N instead of $N-1$ for large number of points:

$$\chi^{2\prime} = \sum_{i=1}^N \frac{[y_i - f(x_i)]^2}{\Delta Y_i^2} = \sum_{i=1}^N \frac{[y_i - f(x_i)]^2}{(\sqrt{\chi^2/N}\Delta y_i)^2} = \sum_{i=1}^N \frac{[y_i - f(x_i)]^2}{\Delta y_i^2}(\frac{N}{\chi^2}) = \chi^2(\frac{N}{\chi^2}) = N \tag{2}$$

We consider now cases in which the theoretical function f has k parameters $\boldsymbol{\alpha} = (\alpha_1, ..., \alpha_k)$, denoted from now on as $f(x_i|\boldsymbol{\alpha})$, that should be adjusted to achieve the best fit of f to experimental data. It is a well-known rule [15, 11] that the uncertainty with which a single, uncorrelated parameter α_j is adjusted, can be obtained varying this parameter around its optimum value α_{j0} (while all others parameters are optimized) until the value of χ^2 increments in one unit, i.e.:

$$\chi^2 = \chi_0^2 + 1. \tag{3}$$

Here χ_0^2 denotes the minimum value of χ^2. For this rule to be valid the following conditions must apply:

- The uncertainties of raw experimental data Δy_i are properly determined.

- There are no significant systematic errors.
- The theoretical model is a true description of the data being studied.

As consequence of the previous conditions the residuals R_i will follow a normal distribution with mean value $\langle R_i \rangle = 0$ and variance $\sigma_R^2 = 1$ (we will denote this $R_i \sim \mathcal{N}(0,1)$) which implies $\chi_0^2 \sim \nu$ (although the reciprocal is not necessarily true). Here ν is the number of degrees of freedom $\nu = N - n_f$ being n_f the number of free parameters. In most of the cases $n_f = 2$ and $N \geq 20$, thus $\nu \approx N$ and no significant difference arises from the use of N instead of ν.

As we will see in section 2.2 many data sets analyzed in the area of low-energy elastic scattering do not fulfil all the conditions previously mentioned, but the rule of eq. 3 is nevertheless applied. Hence the uncertainty of the obtained parameters is underestimated. For these cases, we will extend the procedure of Birge (eq. 2) for the experimental adjustment of one or several parameters $\boldsymbol{\alpha}$ of a theoretical function $f(x_i|\boldsymbol{\alpha})$ in the following way:

$$\chi^{2\prime} = \sum_{i=1}^{N} \frac{[y_i - f(x_i|\boldsymbol{\alpha})]^2}{(\sqrt{\chi_0^2/\nu}\Delta y_i)^2} = \frac{\chi^2}{(\chi_0^2/\nu)} \tag{4}$$

For the optimum set of parameters $\boldsymbol{\alpha_0}$ we obtain

$$\chi_0^{2\prime} = \frac{\chi_0^2}{(\chi_0^2/\nu)}. \tag{5}$$

Now, with the experimental uncertainties Δy_i scaled by the Birge factor $\sqrt{\chi_0^2/\nu}$ the rule of eq. 3 can be applied:

$$\chi^{2\prime} = \chi_0^{2\prime} + 1. \tag{6}$$

This does not imply that the experimental uncertainties Δy_i are effectively changed in fact the original ones should be reported. To avoid confusions we prefer to replace in eq. 6 with eq. 4 and 5 to obtain the equivalent rule

$$\chi^2 = \chi_0^2 + \chi_0^2/\nu\,. \tag{7}$$

In the case of adjusting several (even correlated) parameters the rule of eq. 1 should be extended as $\chi^2 = \chi_0^2 + \Delta\chi^2$, where for two parameters and a 68% confidence level $\Delta\chi^2 = 2.3$. See Ref. [11] (p. 815) for a table of $\Delta\chi^2$ for different confidence levels and number of adjustable parameters. In the more general case in which $\chi_0^2/\nu > 1$ and several parameters are adjusted simultaneously eq. 7 is extended to

$$\chi^2 = \chi_0^2 + \Delta\chi^{2\prime}, \qquad \text{where} \qquad \Delta\chi^{2\prime} = \Delta\chi^2\,\chi_0^2/\nu\,. \tag{8}$$

In the study of scattering angular distributions, the experimental data (x_i, y_i) consist in angles and differential cross-sections $(\theta_i, d\sigma_i/d\Omega)$. The theoretical model $f(x_i|\boldsymbol{\alpha})$ can be either the calculated $d\sigma(\theta_i)/d\Omega$ using the phenomenological Woods-Saxon optical model or the microscopic models involving folding of nuclear densities with appropriated nucleon-nucleon potentials. In this work the São Paulo global microscopic potential (SPP [16, 17]) is used with $\alpha_1 = N_R$ and $\alpha_2 = N_I$ (depth of real and imaginary parts of the potential, respectively).

2.2. Criteria used in previous works

Table 1 presents a non-exhaustive list of experiments studying the TA, either with weakly or tightly bound projectiles. It can be seen that most studies use the standard rule $\chi^2 \leq \chi_0^2 + 1$ (eq. 3) to determine the uncertainties regardless of their χ^2/ν value . We will show in the next section that due to parameter correlations, this rule gives too small uncertainties bars, even in the best case when $\chi_0^2/\nu \approx 1$, being even worse when $\chi^2/\nu \gg 1$. It is apparent that this problem has been recognized but not clearly acknowledged and, thus, many of the different criteria listed in Table 1 attempted to increase the $\chi^2 \leq \chi_0^2 + 1$ limits. A good example of this can be seen in Ref. [18]. Since their angular distribution data for the ^{12}C + ^{208}Pb system have an average value $\langle \chi_0^2/\nu \rangle = 6$, the usual criterium $\chi^2 \leq \chi_0^2 + 1$ would have clearly been an underestimation of the parameter uncertainties and an extreme criterium $\chi^2 \leq 2\chi_0^2$ was used, which, as we will show in section 3.1, is an overestimation. In Fig. 1 we show the χ^2 vs. N_R curve corresponding to our reanalysis using SPP of the 84.9 MeV angular distribution of Ref. [18]. The optimum fit is obtained applying a combination of grid, and steepest descent searching routines and gives: $N_{R0} = 0.5738$, $N_{I0} = 0.5173$, $\chi_0^2 = 198$, $\chi_0^2/\nu = 9.9$. The covariance matrix, including the variance of each of the two parameters N_R and N_I and their covariance, is calculated using second derivatives calculated numerically around the minimum [15]. The parabolic approximation of the curve describing χ^2 as a function of N_R is shown in Fig. 1 with dashed lines. Its expression is given by

$$\chi^2 = \frac{1}{\sigma_1^2}(N_R - N_{R0})^2 + \chi_0^2, \tag{9}$$

where σ_1^2 is the variance of N_R. This variance determines the value of $|N_R - N_{R0}|$ that produces a change in one unit in the χ^2, as in eq. 10a. The solid lines in Fig. 1 represent the values corresponding to a grid on N_R while N_I is optimized. It is seen that the parabolic approximation is quite good even at several σ_1 away from the minimum. Uncertainty limits corresponding to different criteria found in the literature are:

$$\sigma_1 \rightarrow \chi^2 \leq \chi_0^2 + 1 \tag{10a}$$
$$\sigma_2 \rightarrow \chi^2 \leq \chi_0^2 + 2.3 \tag{10b}$$
$$\sigma_3 \rightarrow \chi^2 \leq \chi_0^2 + \chi_0^2/\nu \tag{10c}$$
$$\sigma_4 \rightarrow \chi^2 \leq \chi_0^2 + 2.3\chi_0^2/\nu. \tag{10d}$$

These equations are not meant to represent different confidence limits (CL) but different criteria. In the particular example of Fig. 1, σ_3 represents the 68% CL for the adjustment of the single parameter N_R, considered as uncorrelated with N_I, while σ_4 represents the 68% CL for the simultaneous adjustment of the correlated N_R and N_I. In this case, since $\chi_0^2/\nu = 9.9$, σ_1 and σ_2 do not represent 68% CL. In the parabolic approximation of eq. 9 it holds

$$\sigma_2 = \sqrt{2.3}\,\sigma_1, \quad \sigma_3 = \sqrt{\chi_0^2/\nu}\,\sigma_1, \quad \text{and} \quad \sigma_4 = \sqrt{2.3\chi_0^2/\nu}\,\sigma_1. \tag{11}$$

Perhaps, the great variety of criteria in Table 1 are due to the fact that in most cases the standard rule $\chi^2 \leq \chi_0^2 + 1$ produces too small uncertainties. This turns to be evident when the optimum values of the parameters are in effect changed by their uncertainties: the corresponding change in the angular distributions is often indistinguishable from the optimum ones.

2.3. Error ellipses

One interesting example within low-energy nuclear physics of the use of error ellipses to estimate uncertainties for correlated parameters is presented in Ref. [39]. However, this has not been applied to take into account the correlation between the optical-model parameters. In fact, none of the works listed in Table 1 uses this technique. In our case the ellipses corresponding to the different limits previously mentioned are calculated as

$$\Delta\chi^{2\prime} = (\boldsymbol{\alpha} - \boldsymbol{\alpha_0})^T C^{-1} (\boldsymbol{\alpha} - \boldsymbol{\alpha_0}), \tag{12}$$

Journal of Physics: Conference Series **630** (2015) 012021 doi:10.1088/1742-6596/630/1/012021

Table 1. Different criteria used in the calculation of parameter uncertainties in the study of the threshold anomaly.

Reference		criterium	Projectile	Target	$\langle\chi^2/N\rangle$	conclusion
Biswas 2008	[19]	$\chi_0^2+\chi_0^2/N$	^{6}Li	^{64}Ni	0.43	No TA
			^{6}Li	^{58}Ni	0.45	No conclusive
Deshmuk 2011	[20]	χ_0^2+1	^{6}Li	^{112}Sn	5.18	No TA
			^{6}Li	^{116}Sn	12.2	No TA
Fernández Niello 2007	[21]	χ_0^2+1	^{6}Li	^{27}Al	No data	No TA
Figueira 2006	[8]	χ_0^2+1	^{7}Li	^{27}Al	3.6	No TA
Figueira 2010	[22]	χ_0^2+1	^{6}Li	^{144}Sm	2.51	BTA
			^{7}Li	^{144}Sm	2.68	No TA
Figueira 2007	[23]	χ_0^2+1	^{6}Li	^{27}Al	5.22	No TA
Fimiani 2012	[24]	χ_0^2+1	^{6}Li	^{80}Se	1.05	BTA
			^{7}Li	^{80}Se	1.00	TA
García 2007	[25]	χ_0^2+N+1	^{6}He	^{209}Bi	2.2	BTA
Gomes 2004	[7]	χ_0^2+1	^{9}Be	^{27}Al	No data	No TA
Gomes 2005	[26]	$\chi_0^2+\chi_0^2/N$	^{9}Be	^{64}Zn	No data	BTA
Gómez Camacho 2007	[27]	no uncertainty bars	^{9}Be	^{64}Zn	0.61	BTA
Gómez Camacho 2008	[28]	no uncertainty bars	^{9}Be	^{144}Sm	0.61	No TA
Gómez Camacho 2010	[29]	χ_0^2+N	^{6}Li	^{58}Ni	0.61	BTA
			^{7}Be	^{58}Ni	0.05	BTA
			^{8}B	^{58}Ni	0.51/0.19	BTA
Keeley 1994	[6]	$1.15\chi_0^2$	^{6}Li	^{208}Pb	2.16	No TA
			^{7}Li	^{208}Pb	1.53	weak TA
Kumawat 2008	[30]	$1.5\chi_0^2$	^{6}Li	^{90}Zr	0.82	No TA
Lubian 2003	[31]	χ_0^2+1	^{7}Li	^{138}Ba	No data	BTA
Maciel 1999	[32]	qualitative	^{6}Li	^{138}Ba	No data	No TA
			^{7}Li	^{138}Ba	No data	TA
Nagarajan 1985	[33]	no uncertainty bars	^{16}O	^{208}Pb	No data	TA
Pakou 2003	[34]	Not given	^{6}Li	^{28}Si	No data	No TA
Pakou 2004	[35]	Not given	^{7}Li	^{28}Si	No data	No TA
Santra 2001	[18]	$2\chi_0^2$	^{12}C	^{208}Pb	6.03	TA
Souza 2007	[36]	$1.1\chi_0^2$	^{6}Li	^{59}Co	No data	TA or BTA
			^{7}Li	^{59}Co	No data	TA
Woolliscroft 2004	[37]	$1.15\chi_0^2$	^{9}Be	^{208}Pb	11	TA
Zadro 2009	[38]	χ_0^2+1	^{6}Li	^{64}Zn	1.18	No TA

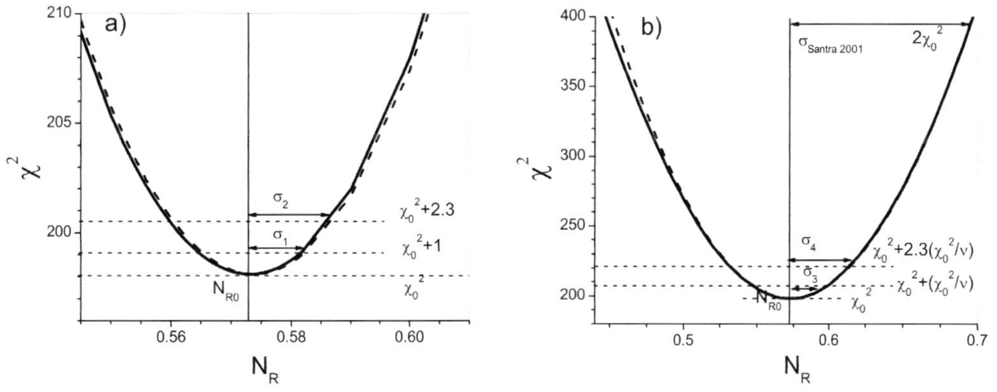

Figure 1. χ^2 as a function of N_R for ^{12}C + ^{208}Pb at 84.9 MeV (Plots a) and b) show different scales of the axes). Different criteria for the uncertainty interval of N_R are shown: $\sigma_1 \approx 0.0086$, $\sigma_2 \approx 0.013$, $\sigma_3 \approx 0.0271$, $\sigma_4 \approx 0.041$. The criterium applied in Ref. [18] gives $\sigma_{\text{Santra 2001}} \approx 0.121$.

where $\Delta\chi^{2\prime}$ is defined in eq.8, the covariance matrix C is calculated for each system and energy and $\boldsymbol{\alpha} - \boldsymbol{\alpha}_0$ represent deviations respect the optimum value of the parameters.

Figure 2 shows three of such ellipses for the ^{12}C + ^{208}Pb angular distribution at 84.9 MeV corresponding to the σ_1, σ_3 and σ_4 limits for the determination of the standard uncertainty in both parameters. The limit σ_2 should be used in cases where $(\chi^2/\nu) \approx 1$. In the cases were $(\chi^2/\nu) > 1$, σ_4 should be used. We will show the error ellipses for all energies of both cases studied in Figs. 3 and 5.

2.4. Bootstrap

Bootstrap is one of the many resampling methods designed to go beyond regular statistic test. It creates a number N_B of synthetic data sets, each of them consisting on N data points selected randomly (with reposition) from the original data set of N points (it is desirable to have $N_B \gg N$). On the average a fraction $1/e$ (about 37%) of the N points will be repeated ones (thus having more weight in the fitting procedure) and, consequently, 37% of the elements will not be included. Each synthetic data set is used to fit the parameters of interest. The standard deviation of the N_B values obtained for each parameter is a reliable estimation of its uncertainty. Hence, this procedure gauges the sensitivity of the fitted parameter to each individual data point by the simulation of N_B experiments in which the experimenter could have chosen to repeat the measurement of some of the data points at the cost of missing some others. The bootstrap method has been one of the techniques applied to the evaluation of the half-life of ^{198}Au in Ref. [40]. To the best of our knowledge, bootstrap techniques have not been applied yet to the calculation of nuclear potential parameters and their uncertainties.

In Fig. 2 we show the resulting N_R, N_I points from the adjustment of $N_B = 2200$ synthetic angular distributions. The average and variance converged for $N_B \geq 200$. It is interesting to point out that 76% of the bootstrap points are enclosed by the σ_4 ellipse.

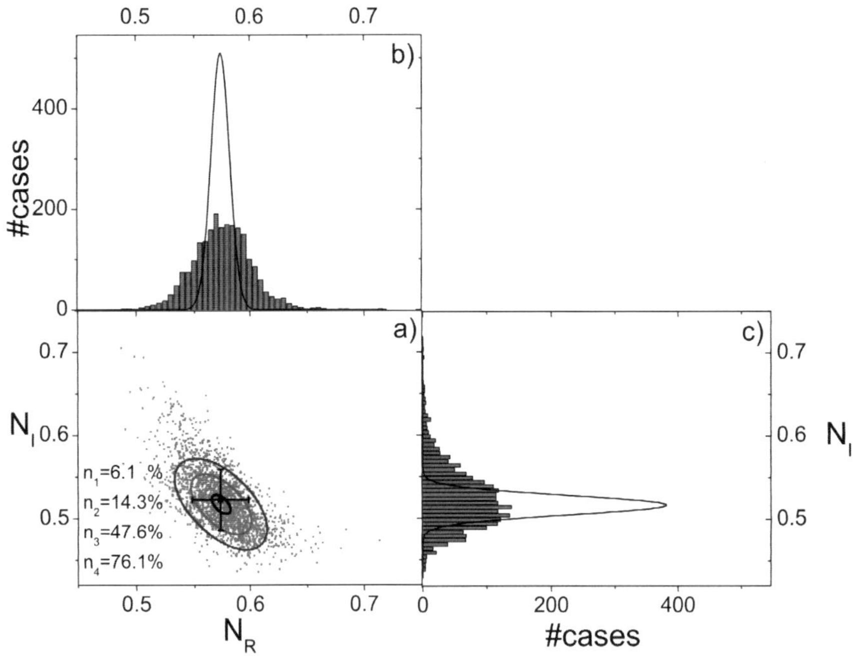

Figure 2. Plot a) shows the bootstrapped values (red points) and the ellipses described in section 2.3 (σ_2 ellipses not shown for clarity). Histograms in b) and c) are the projection of bootstrapped points on the N_R and N_I axis respectively, while the full line shows the much narrower gaussian distribution with standard deviation σ_1 ($\chi \leq \chi_0^2 + 1$) for comparison.

3. Case studies

In the first step to study the $^{12}C + {}^{208}Pb$ and the $^{6}Li + {}^{80}Se$ systems, the optimum N_R, N_I values are calculated as well as the corresponding residuals, covariance matrices and error ellipses. Since a condition of the analysis is the normality of the residuals, the Kolmogorov-Smirnov test has been applied. This statistical test evaluates the maximum difference between the empirical and reference (theoretical) cumulative distribution functions and yields the acceptance or rejection of a null hypothesis at a given confidence level [41].

In our case, we compare to two different null hypothesis: *i)* the residuals follows a normal distribution with mean value $\langle R \rangle = 0$ and variance $\sigma_R^2 = 1$, i.e. $H_{01} : R_i \sim \mathcal{N}(0, 1)$ and *ii)* the residuals follows a normal distribution but in this case with variance $\sigma_R^2 = \chi_0^2/\nu$ i.e. $H_{02} : R_i \sim \mathcal{N}(0, \chi_0^2/\nu)$.

3.1. The $^{12}C + {}^{208}Pb$ system

As expected, the residuals of the angular distributions of each individual energy, as well as the residuals corresponding to all data taken together, follow $\mathcal{N}(0, \chi_0^2/\nu)$ but do not follow $\mathcal{N}(0, 1)$. Fig. 3 shows the error ellipses corresponding to the limits σ_1, σ_3 and σ_4 of eq. 10 (σ_2 not shown for clarity) and points (N_R, N_I) representing the results from fitting bootstrapped data sets as explained in section 2.4. For all energies, $N_B \geq 300$. The numbers n_1, n_2, n_3 and n_4 indicate the number of these bootstrapped points inside each ellipse (as it has been shown in Fig. 2, expressed as a percentage of N_B). Except for the lowest and highest energies more than 50% of the bootstrapped points are enclosed inside the σ_4 ellipse. In general they extend beyond the last ellipse but keeping the same correlation. If cases c) and h) of this figure a group of points diverge

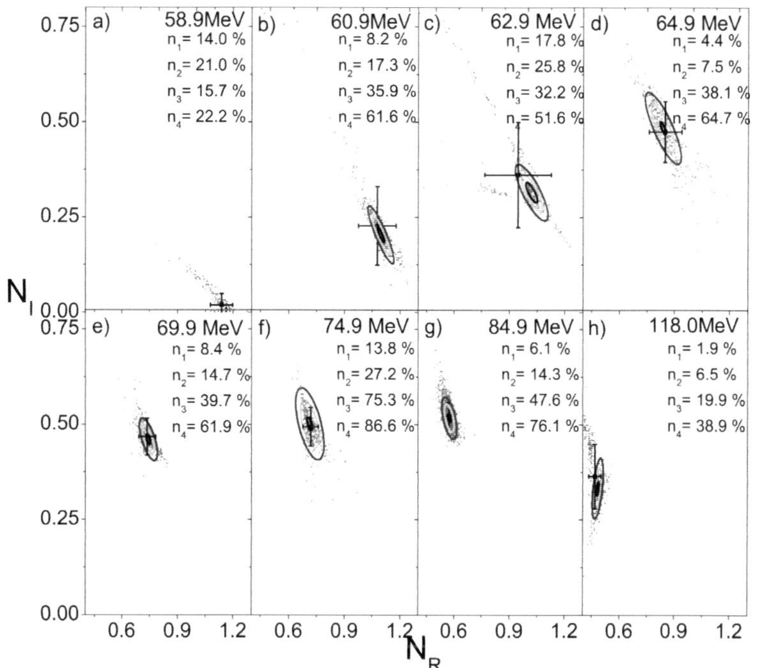

Figure 3. Distributions of N_R and N_I parameters that best adjust synthetic data sets generated through the bootstrap technique for the ^{12}C + ^{208}Pb system. The analysis based on the ellipses is described in section 3.3. The uncertainty bars indicate the medium values of the parameter distribution and their standard deviations.

from the main trend. The effect was traced to the influence of one single experimental point in both cases, been the divergent group part of a population in which that point is omitted. In Fig. 4, the parameters and their uncertainties given by the bootstrap method is compared with the ones given by Ref.[18]. It is seen that even though our assigned uncertainties are smaller the character of a regular TA is maintained.

3.2. ^{6}Li + ^{80}Se system

Since in this experiment $\chi_0^2/\nu \approx 1$ both null hypothesis almost coincide. The Kolmogorov-Smironov test indicates that residuals corresponding to all data taken together are normally distributed, and so are individual angular distributions for each energy, except for the data corresponding to 23 and 14.5 MeV. This reveals that for these two energies there is a subtle difference between experimental data and the theoretical model. Even though this difference is statistically significant, it would have remained unnoticed without this analysis. Normalization factors of around 1.6% applied to these angular distributions are enough to make them pass the test, but this may require further analysis. Figure 5 shows the error ellipses and bootstrapped points as in Fig. 2. Here, more than 43% of the bootstraped points are enclosed inside the σ_4 ellipse. The behavior of the parameters and their uncertainties given by the bootstrap limits is compared with the ones given by in Ref. [24] in Fig. 4.

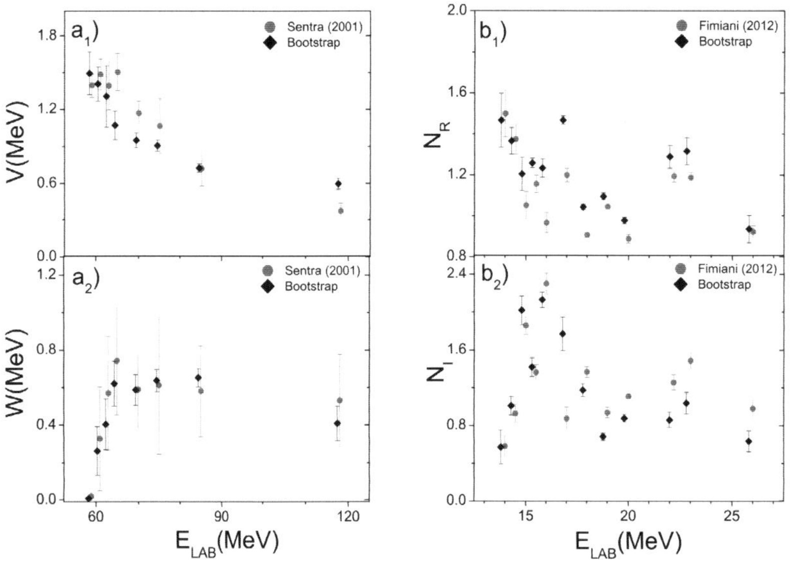

Figure 4. Comparison between the dispersion relations reported in the original works and those obtained in the present analysis with the bootstrap technique a) for the ^{12}C + ^{208}Pb system and b) for the ^{6}Li + ^{80}Se system. The bootstrap results are slightly displaced in energy for clarity.

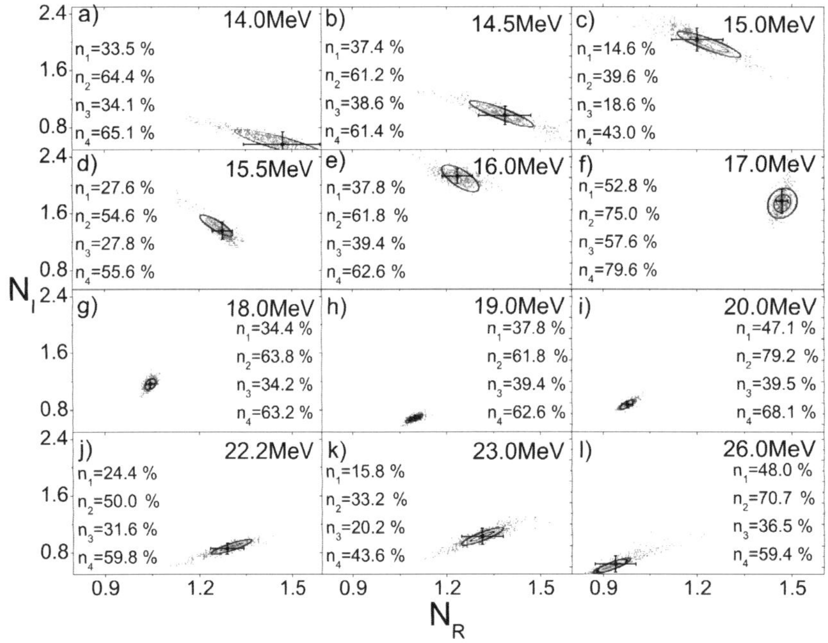

Figure 5. Distributions of N_R and N_I parameters that best adjust synthetic data sets generated by the bootstrap technique for the ^{6}Li + ^{80}Se system. The analysis based on the ellipses is described in section 3.3. The uncertainty bars indicate the medium values of the parameter distribution and their standard deviations.

4. Conclusions

We have shown that most studies use the standard rule from curve fitting: $\chi^2 \leq \chi_0^2 + 1$ (eq. 3), to determine the uncertainty in fitting parameters (usually N_R and N_I) from experimental data, regardless of their experimental value of χ_0^2/ν. This is not satisfactory even from the point of view of conventional statistic: For the cases where $\chi_0^2/\nu > 1$ that prescription is incorrect and should be replaced by $\chi^2 \leq \chi_0^2 + 2.3\chi_0^2/\nu$ of eq. 10d. Parameters uncertainties which has been already derived using the $\chi^2 \leq \chi_0^2 + 1$ recipe , can be easily scaled up (under the parabolic approximation of eq. 11) by a factor of $\sqrt{2.3\chi_0^2/\nu}$.

The bootstrap resampling method applied in this work produced a more realistic distribution of the parameters as shown in Figs. 2 b) and c). The resulting uncertainties are similar or somewhat larger than the ones obtained with the σ_4 ellipse (eq. 10d).

To judge the kind of TA it is very helpful to look at the series of 2-dim N_R vs. N_I plots with their respective uncertainty ellipses. In this way, the correlation between the parameters can be considered. In Fig. 4 our results, plotted in the conventional way, show a more reliable estimation of the parameter uncertainties than the quoted in the original works (smaller for $^{12}C + ^{208}Pb$, similar or slightly larger for $^6Li + ^{80}Se$). In these cases, however, the conclusions about the kind of TA was not modified. Further work is in progress to extend the present analysis to other systems, using bootstrap as well as other resampling techniques.

References

[1] Satchler G R 1991 *Phys. Rep.* **199** 147
[2] Abriola D *et al.*1989 *Phys. Rev.* C **39** 546
[3] Abriola D*et al.*1992 *Phys. Rev.* C **46** 244
[4] Sakuragi Y *et al.*1987 *Phys. Rev.* C **35** 2161
[5] Mahaux C *et al.* 1986 *Nucl. Phys.*A 449 354
[6] Keeley N et al. 1994 *Nucl. Phys.*A **571** 326
[7] Gomes P R S *et al.*2004 *Phys. Rev.* C **70** 054605
[8] Figueira J M *et al.* 2006 *Phys. Rev.* C **73** 054603
[9] Hussein *et al.* MS 2006 *Phys. Rev.* C **73** 044610
[10] Canto L F *et al.* 2006 *Phys. Rep.* **424** 1
[11] Press W H *et al.* 2007 Numerical Recipes (Third edition) Cambridge University Press
[12] Birge R T 1932 *Phys. Rev.* **40** 207
[13] Kacker R *et al.* 2008 *Metrologia* **45** 257
[14] Kessel R *et al.* 2011 *J. Res. Natl. Inst. Stand. Technol.* **116** 809
[15] Bevington P R and Robinson D K 2003 Data reduction and error analysis, (3rd ed.) McGraw-Hill
[16] Chamon L C *et al.* 1997 *Phys. Rev. Lett.* **79** 5218
[17] Alvarez M A G *et al.* 2003 *Nucl. Phys.* A **723** 93
[18] Santra S *et al.* 2001 *Phys. Rev.* C **64** 024602
[19] Biswas M *et al.*2008 *Nucl. Phys.* A **802**802 67
[20] Deshmuk N *et al.* 2011 *Phys. Rev.* C **83** 024607
[21] Fernández Niello J O *et al.*2007 *Nucl. Phys.* A **787** 484c
[22] Figueira J M *et al.* 2010 *Phys. Rev.* C **81** 024613
[23] Figueira J M *et al.* 2007 *Phys. Rev.* C **75** 017602
[24] Fimiani L *et al.* 2012 *Phys. Rev.* C **86** 044607
[25] García A R *et al.* 2007 *Phys. Rev.* C **76**067603
[26] Gomes P R S *et al.*2005 *Phys. Rev.* C **71** 034608
[27] Gómez Camacho A *et al.* 2007 *Phys. Rev.* C **76** 044609
[28] Gómez Camacho A *et al.* 2008 *Phys. Rev.* C **77** 054606
[29] Gómez Camacho A *et al.* 2010 *Nucl. Phys.* A **833** 158
[30] Kumawat H *et al.* 2008 *Phys. Rev.* C **78** 044617
[31] Lubian J *et al.* 2003 *Braz. J. of Physics* **33-2** 323
[32] Maciel A M M *et al.* 1999 *Phys. Rev.* C **59** 2103
[33] Nagarajan M A *et al.* 1985 *Phys. Rev. Lett.* **54** 1136
[34] Pakou A *et al.* 2003 *Phys. Lett.* B **556** 21
[35] Pakou A *et al.* 2004 *Phys. Rev.* C **69** 054602

[36] Souza F A *et al.* 2007 *Phys. Rev.* C **76** 02901
[37] Woolliscroft R J *et al.* 2004 *Phys. Rev.* C **69** 044612
[38] Zadro M *et al.* 2009 *Phys. Rev.* C **80** 064610
[39] Barbosa M D L *et al.* 2005 *Phys. Rev.* C **71** 024303
[40] Chen J *et al.* 2011 *Appl. Radiat. Isot.* **69** 1064
[41] Froedesen A G *et al.* 1979 Probability and statistics in particle physics, Universitetsforlaget

Determination of the cosmic-ray-induced neutron flux and ambient dose equivalent at flight altitude

M T Pazianotto[1], M A Cortés-Giraldo[2], C A Federico[3], O L Gonçalez[3], J M Quesada[2], B V Carlson[1]

[1]Aeronautics Technological Institute, Physics Department, São José dos Campos, Brazil
[2]University of Sevilla, Atomic, Molecular and Nuclear Physics Department, Sevilla, Spain
[3]Institute for Advanced Studies, Applied Physics Department, São José dos Campos, Brazil

E-mail: mpazianotto@gmail.com

Abstract. There is interest in modeling the atmosphere in the South Atlantic Magnetic Anomaly in order to obtain information about the cosmic-ray induced neutron spectrum and angular distribution as functions of altitude. In this work we use the Monte Carlo codes MCNPX and Geant4 to determine the cosmic-ray-induced neutron flux in the atmosphere produced by the cosmic ray protons incident on the top of the atmosphere and to estimate the ambient dose equivalent rate as function of altitude. The results present a reasonable conformity to other codes (QARM and EXPACS) based on other parameterizations.

1. Introduction

Great efforts are being made to understand the accumulated neutron dose at flight altitude in the South Atlantic Magnetic Anomaly (SAMA) [1,2], because a large part of Brazil, and therefore South America, are subject to this Anomaly. Figure 1 shows that the SAMA at 12 km altitude is located with its center point above Brazil, near Foz do Iguaçú city.

Among the many particles produced by interactions between primary cosmic rays and atmospheric atoms, neutrons are the most important component to consider in order to assess the dose deposition in the crew members of an aircraft, as well as in electronic equipment, such as on-board computers [3]. Thus, there is great interest in modeling the atmosphere and simulating the cosmic-ray induced neutron spectrum at flight altitude to develop further applications for the study of their effects on microelectronic devices [4] and carry out dosimetry studies aboard aircrafts. The purpose of this work is to determine the cosmic-ray-induced neutron (CRINS) flux in the atmosphere produced by protons using the codes MCNPX [5] and Geant4 [6] and then to estimate the ambient dose equivalent rate (ADER) through the atmosphere. In this work, we are evaluating the Geant4 by comparing it with MCNPX and other codes in order to search possible effects of the Earth's magnetic field on the neutron flux as well on the ADER in flight altitude. The magnetic field can be easily included in Geant4, but not in MCNPX v2.5 and in the other codes used in this comparison.

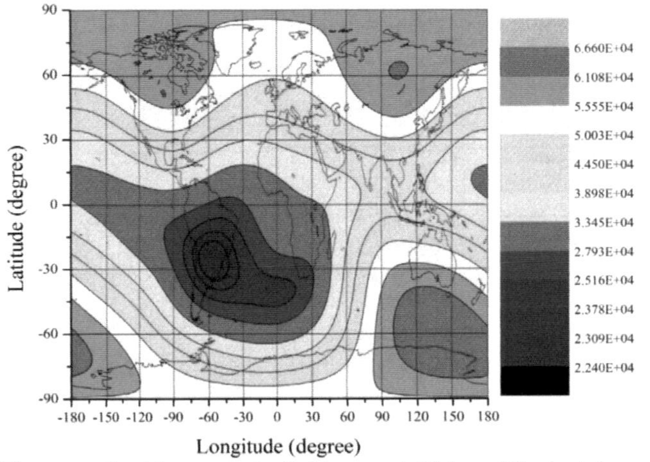

Figure 1. Earth's magnetic field map at 12 km altitude, taken from [7].

2. Methodology

2.1. Computational procedure

Figure 2 shows the geometry of the atmosphere used in the simulations. The atmosphere was modeled as a cylinder of 50 km height and 25 km radius. For the sake of reproducing the variable density of the atmosphere at different altitudes, the cylinder was divided into layers of 2.5 km thickness below 12.5 km altitude, whereas 5 km layers where used above this altitude level. The air density and composition of each layer was set according to the International Standard Atmosphere [8]; the air humidity was also modeled below 12.5 km altitude [9].

In order to reproduce an infinite medium, we used fully reflective sidewalls of the cylinder for all the particles that impinge on them. We used a primary cosmic ray flux composed only of protons at the top of the cylinder. The primary proton spectrum was obtained from QARM [10] at the geographical coordinates -23.04 S -45.15 W and at the date June 06, 2009.

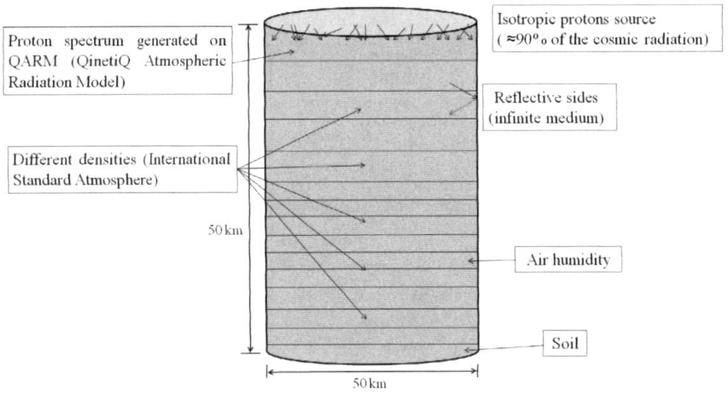

Figure 2. Schematic figure of the atmosphere modeling for cosmic ray propagation.

2.2. MCNPX physics models

In the simulations with MCNPX v2.5 the ENDF/VI nuclear data library was used for all materials. We considered the scattering matrices S(α,β) [5, 11] to correct the hydrogen cross section in water, which is important for neutron transport at low energies. Nuclear data libraries were used for neutron energies under 20 MeV. Physical models were used above this energy range according to the following parameterizations:

- Neutron and proton interaction only by elastic scattering;
- Bertini intranuclear cascade for nucleons, followed by pre-equilibrium;
- Coulomb barrier for incident charged particles;

2.3. *Geant4 physics models*

The simulations were carried out with the version 9.6.3 and 9.6.2 of the GEANT4 toolkit. For incident energies above 10 GeV, the hadronic interactions have been modeled with the Quark-gluon String Model (QGS). At lower energies, which are the most relevant in the present work, we used either the same hadronic modeling (Bertini Model) as MCNPX in the intranuclear cascade phase, followed by preequilibrium (physics list QGSP_BERT_HP), or the Binary Cascade, followed by the Exciton Pre-equilibrium model and final equilibrium de-excitation with a variety of models (Weisskopf-Ewing evaporation of nucleons and light clusters, photon emission, fission, Fermi break-up and statistical multifragmentation) as final stages (physics list QGSP_BIC_HP). For the transport of neutrons below 20 MeV, the High Precision (HP) model with G4NDL4.0 data was used.

3. Results

3.1. Comparison between MCNPX and Geant4

Figure 3 shows the integral neutron flux as a function of altitude calculated with MCNPX and Geant4. The simulations were performed using an enough events so that the statistical errors are significantly smaller than the values of the simulated data. The agreement between the MCNPX and the Geant4 simulations with the QGSP_BERT_HP physics list is certainly due the fact that the Bertini intranuclear cascade model is used in both cases. However, the agreement with the calculations done with the QGSP_BIC_HP physics list is observed only at 5 km altitude. For altitudes higher than 5 km, the QGSP_BIC_HP simulations predict a lower neutron counting rate, whereas at lower altitudes the situation is the opposite, i.e., they produce a higher neutron counting compared to those of MCNPX and the QGSP_BERT_HP Geant4 physics list.

As a preliminary analysis, the origin of this divergence could be traced back to the different neutron multiplicities of each model. At present, we are evaluating the primary proton integral flux as a function of altitude in an attempt to establish the connection between the number of neutrons produced and the protons absorbed throughout the atmosphere, which should provide us with a deeper understanding of this phenomenology.

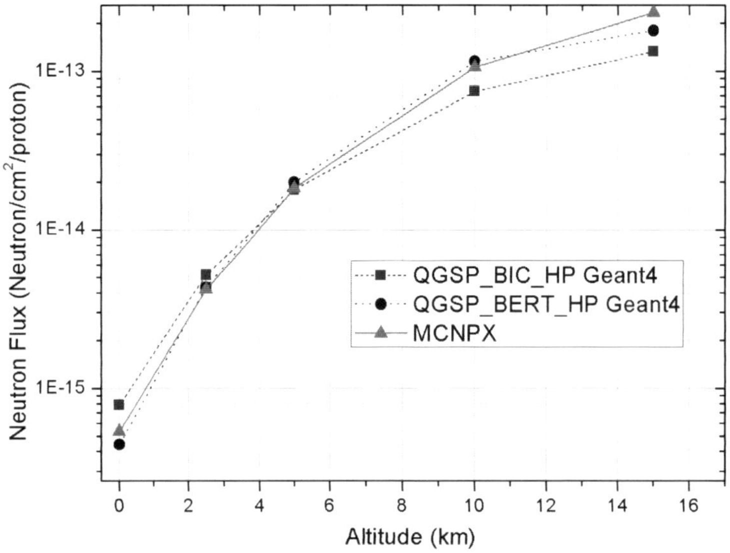

Figure 3. Integral neutron flux calculated with MCNPX and Geant4 with QGSP_BIC_HP and QGSP_BERT_HP physics lists.

Figure 4 shows the cosmic-ray induced neutron spectrum normalized per primary proton at 1 m (i.e., ground level) and 15 km altitude as simulated by MCNPX and Geant4 with the BIC and Bertini models. We observe a higher neutron production by spallation reactions at ground level with the Binary Cascade model of Geant4 than with the Bertini model. However, we have just the opposite at 15 km altitude.

In the thermal region at 1 m altitude, the neutron flux simulated with Geant4 is lower than the one simulated with MCNPX. This fact could be also explained due to the different neutron multiplicity of each model as previously mentioned.

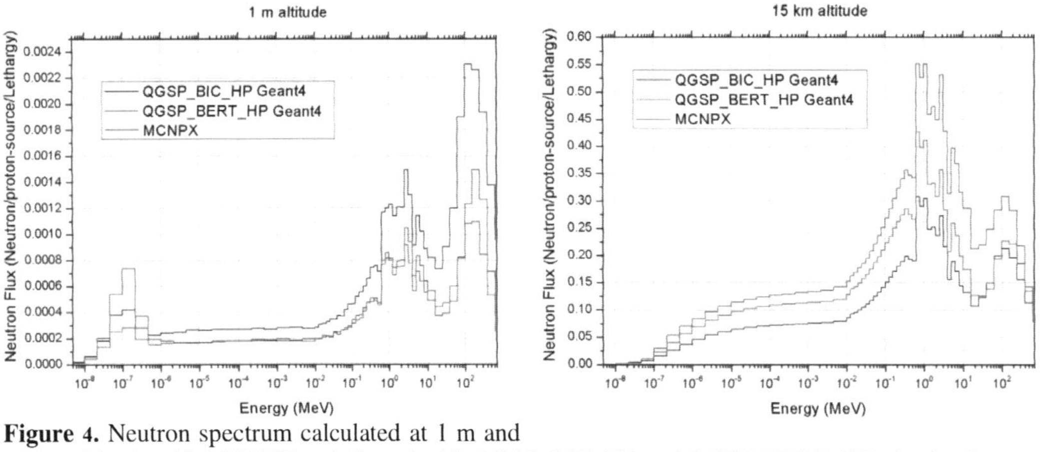

Figure 4. Neutron spectrum calculated at 1 m and 15 km altitude with MCNPX and Geant4 with QGSP_BIC_HP and QGSP_BERT_HP physics lists.

Figure 5 compares the neutron flux calculated with MCNPX, GEANT4 (using QGSP_BIC_HP) and EXPACS [12] with the experimental data from Federico et al [2] measured at 1 m altitude in São José dos Campos and other experimental results taken from the Goldhagen et al [13]. The results were normalized to 1 neutron/cm^2 integral flux in order to analyze only the energy dependence of the neutron spectra. At the thermal peak there is a wide dispersion of results, where the experimental measurement of Federico et al. [2] shows the largest amplitude and the Geant4 simulation has the smallest amplitudes. However, the results of MCNPX, EXPACS and Goldhagen somewhat agree among themselves. As well as the previous results (Figure 4), the energy dependence of the MCNPX and Geant4 simulations reproduce well the shape of the spectrum between the thermal and evaporation peaks, and these, in turn, reproduces well the experimental results of Goldhagen et al [13], but overestimate the experimental results of Federico et al [2]. However, the spallation region is clearly overestimated in the MCNPX and Geant4 calculations. The divergence between these results could be explained by the concentration of water used in the simulations, since the relative height of the peaks depends strongly on this [14]. A deeper study of this dependence is being carried out.

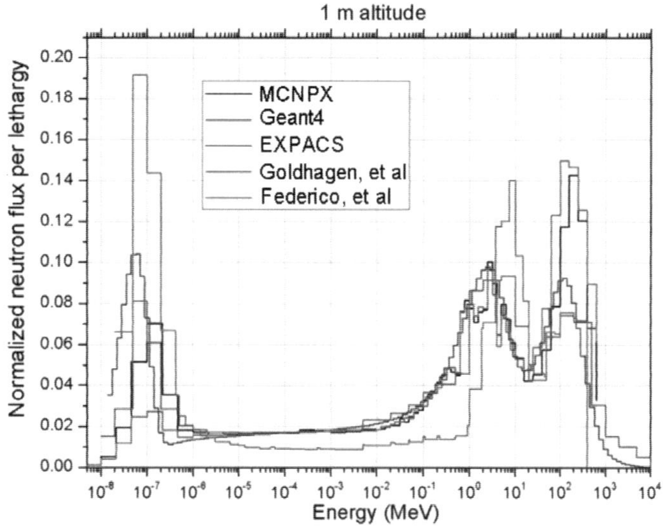

Figure 5. Energy dependence of the neutron flux at 1 m altitude calculated with EXPACS, MCNPX, GEANT4 and measured experimentally [3,12].

3.2. *Ambient dose equivalent rate as a function of altitude*

Figure 6 shows a comparison of the integral neutron flux per incident proton as a function of altitude between our Monte Carlo simulations with MCNPX and Geant4, and QARM code. The results present a reasonable conformity throughout for altitudes that ranges from 2.5 km to 15 km, but for ground level there is a divergence between our Monte Carlo simulations and QARM that could be explained due to the soil composition used on QARM code, information which could not be retrieved.

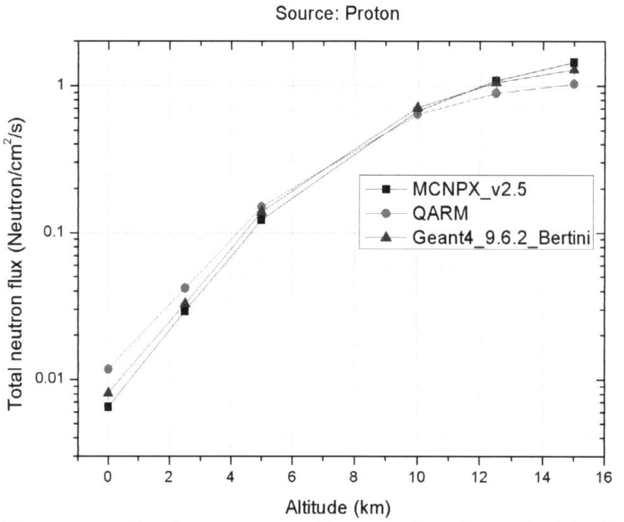

Figure 6. Absolute neutron flux as function of altitude determined by MCNPX, Geant4 and QARM.

Figure 7 compares the ambient dose equivalent rate obtained with MCNPX, GEANT4 and QARM. In order to determine the ambient dose equivalent rate, we multiplied the CRINS spectra by the integral proton flux at the top the atmosphere to obtain the CRIN in absolute value and then multiplied these spectra by a conversion factor that can be found in the ICRP74 data [15].

From this figure it can be seen that there is a remarkable agreement at 10 km altitude between the three codes. QARM gives the highest equivalent ambient dose values at altitudes lower than 10 km, whereas at higher altitudes the values predicted with both Monte Carlo codes are higher than those of QARM; nevertheless, the agreement between the calculations is quite reasonable.

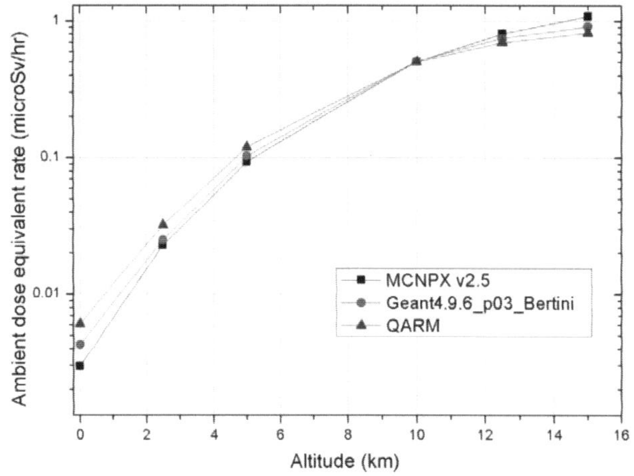

Figure 7. Ambient dose equivalent rate as function of altitudes.

As a first order approximation, we have initiated a study of the influence of Earth's magnetic field on the ambient dose equivalent rate. To do this, we have used the Geant4 code together with terrestrial

Journal of Physics: Conference Series **630** (2015) 012022 doi:10.1088/1742-6596/630/1/012022

magnetic field data from [16] at the geographical coordinates 23.04 S 45.15 W on the date June 06, 2010, when its horizontal intensity was 18,528.8 nT. Figure 8 shows the comparison of the ADER considering the Earth's magnetic field and excluding it. The magnetic field does not change the ADER for altitudes between 1 m up to 45 km, where the air density is very low compared to the ground level.

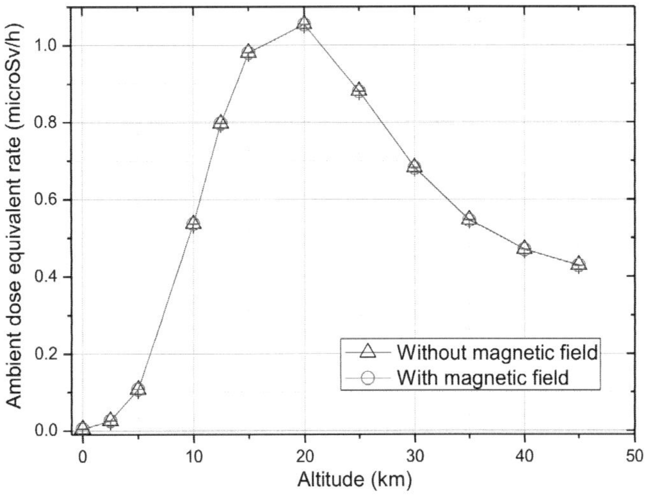

Figure 8. Ambient dose rate versus altitude calculated with GEANT4. The Earth's magnetic field was turned on and off for comparison.

4. Conclusions

The comparison between our Monte Carlo simulations with MCNPX (version 2.5) and Geant4 (version 9.6.2 and 9.6.3) and the calculations obtained with the codes QARM and EXPACS presents good general agreement, although a difference exists at 1 m altitude that could be explained by soil composition. For future work, a detailed analysis of the soil composition influence will be necessary.

As a first order approximation, the Earth's Magnetic field does not change the ambient equivalent dose rate for altitudes up to 45 km. Ongoing studies with the Geant4 toolkit and analytical calculations are being performed to quantify the influence of the Earth's magnetic field on the primary and secondary particles throughout the atmosphere and on the particle angular distributions as a function of altitude.

References

[1] Freden S C and Paulika G A 1964 Trapped protons at low altitudes in the South Atlantic magnetic anomaly *Jour. of Geophysical Research* vol 69 pp 1259-1269.

[2] Federico C A *et al* 2010 Neutron spectra measurements in south Atlantic anomaly *region Rad. Measurements* vol. 45 pp 1526-1528

[3] Prado A C M *et al* 2013 Effects of cosmic radiation on devices and embedded systems in aircrafts *2013 International Nuclear Atlantic Conference*

[4] Pereira Junior E C F *et al* 2014 The effects of total ionizing dose on the neutron SEU cross section of a 130 nm 4 Mb SRAM memory *Proc 15th Latin American Test Workshop*

[5] Pelowitz D 2005 MCNPX User's Manual Version 2.5.0 *Los Alamos National Laboratory report LA-CP-05-0369*

[6] Agostinelli S *et al* 2003 GEANT4 – a simulation toolkit *Nucl Instr Meth Phys Res A* vol 506(3) pp 250-303

[7] Space Environment Iformation System. Available at: www.spenvis.oma.be.

[8] International Organization for Standardization 1975 Standard Atmosphere ISO 2533:1975 1975

[9] Salby M L, Pielke R A, Dmowska R 1996 Fundamentals of Atmospheric Physics *International Geophysics* vol 61.

[10] Lei F, Clucas S, Dyer C, Truscott P 2004 An atmospheric radiation model based on response matrices generated by detailed Monte Carlo Simulations of cosmic ray interactions *IEEE Transactions on Nuclear Science* vol 51

[11] Mattes M and Keinert J Thermal Neutron Scattering Data for the Moderator Materials H2O, D2O and ZrHx in ENDF-6 Format and as ACE Library for MCNP(X) *Codes IAEA report INDC(NDS)-0470*

[12] Sato T EXPACS User's Manual *Japan Atomic Energy Agency*

[13] Goldhagen P *et al* 2002 Nucl Measurement of the energy spectrum of cosmic-ray induced neutrons aboard an ER-2 high-altitude airplane *Nuclear Instruments and Methods in Physics Research A* vol 476 2002

[14] Nakamura T *et al* 2005 Sequential measurements of cosmic-ray neutron spectrum and dose rate at sea level in Sendai *Japan. J. Nucl. Sci. Technol.* vol 42

[15] ICRP 1996 Conversion Coefficients for use in Radiological Protection against External Radiation ICRP Publication 74 Ann. ICRP 26 (3-4)

[16] Earth's Magnetic Field Calculators. Available at: https://www.ngdc.noaa.gov/geomag/magfield.shtml

Journal of Physics: Conference Series **630** (2015) 012023 doi:10.1088/1742-6596/630/1/012023

Angular distributions of the alpha particle production in the ^{7}Li+^{144}Sm system at near-barrier energies

P F F Carnelli[1,2,3], A Arazi[1,3], O A Capurro[1], J O Fernández Niello[1,2,3], D Martinez Heimann[1,2,3], A J Pacheco[1,3], M A Cardona[1,3], E de Barbará[1], J M Figueira[1,3], D L Hojman[1,3], G V Martí[1] and A E Negri[1,2,3]

[1] Laboratorio TANDAR, Comisión Nacional de Energía Atómica, Av. Gral. Paz 1499, B1650KNA San Martín, Buenos Aires, Argentina
[2] Instituto de Investigación e Ingeniería Ambiental, Universidad Nacional de San Martín, 25 de Mayo y Francia, B1650BWA San Martín, Buenos Aires, Argentina
[3] Consejo Nacional de Investigaciones Científicas y Técnicas, Av. Rivadavia 1917, C1033AAJ Buenos Aires, Argentina

E-mail: pcarnelli@unsam.edu.ar

Abstract. We have studied the production of alpha particles in reactions induced by ^{7}Li projectiles on a ^{144}Sm target at bombarding energies of 18, 24 and 30 MeV over the 15°-140° angular range. The purpose of the investigation has been to determine the contribution of different mechanisms in reactions that involve weakly bound projectiles. We have included in our analysis several processes that can either directly or sequentially lead to the emission of alpha particles: complete fusion, direct transfer of ^{3}H, capture breakup (incomplete fusion, sequential complete fusion) and non-capture breakup. In order to distinguish alpha particles stemming from these processes it is necessary to determine the mass and charge of the reaction products and to obtain precise measurements of their energies and scattering angles over relatively wide ranges of these variables. We have done this using a detection system consisting of an ionization chamber plus three position sensitive detectors. We present results of these measurements and a preliminary interpretation based on kinematical considerations and comparisons with predictions from a statistical model.

1. Introduction

The influence of breakup on other reaction channels (such as elastic scattering and fusion) in systems involving weakly bound projectiles at energies around the Coulomb barrier has been an extensively studied subject over the last several years (see Ref. [1] and references therein). Due to the small separation energies involved, breakup reactions are known to account in these cases for a significant part of the total cross section and in particular, weakly bound stable nuclei are a useful alternative to investigate the various aspects of this kind of processes. It is known, or it has been postulated, that the breakup of the projectile may have several outcomes: *i)* all the fragments may scatter away (non-capture breakup), *ii)* one of the fragments may be captured by the target whereas the other scatters away (incomplete fusion), or *iii)* all breakup fragments may be captured by the target (complete fusion following breakup). Even more complex processes, such as transfer reactions followed by the breakup of the remaining (projectile-like) nucleus, have also been observed. For example, for the ^{7}Li projectile, the above mentioned mechanism could consist in either neutron stripping followed by the fragmentation of ^{6}Li into a deuteron and an alpha particle, or proton pickup followed by the breakup of ^{8}Be into two alpha particles.

Non-capture breakup can be unambiguously identified through the coincident measurement of the light emitted particles. Using this method, Martinez Heimann *et al.* have recently obtained integrated and differential cross sections of non-capture breakup reactions induced by the weakly bound ^{6}Li projectile on the spherical ^{144}Sm target [2]. Shrivastava *et al.* [3] conducted a detailed study of the transfer–breakup mechanism in the ^{7}Li+^{65}Cu system. In this case the observation of large cross sections for the α+d channel was interpreted as an evidence for a two-step process that consists in direct neutron transfer followed by the breakup of ^{6}Li via the 2.186 MeV resonance in the α-d continuum. Further evidence of this mechanism has shortly afterwards been found in the ^{7}Li+^{144}Sm system [4]. More recently, an extensive overview of these and other similar complex breakup modes in reactions of ^{6}Li and ^{7}Li projectiles on heavy targets has been presented in the work of Luong *et al.* [5].

On the other hand, the experimental identification and characterization of those breakup reactions that evolve towards the subsequent capture of one or both fragments (also called capture breakup) usually presents more ambiguities. This happens because in general the fragments originated in these processes may be similar to, and may have similar energies than, the residues produced in complete fusion reactions.

In this work we present the results of angular distribution measurements of the alpha particles emitted in the ^{7}Li+^{144}Sm system at energies around the Coulomb barrier. Lithium-7 presents an alpha cluster structure in its ground state and has a quite low energy threshold for the breakup into ^{3}H+^{4}He (2.45 MeV). Consequently, it is relatively easy to excite this mode in any nuclear collision, thus giving rise to relatively large cross-section for the production of alpha particles. We analyze and discuss the energy spectra and angular distributions of these particles by means of kinematical considerations related to the relevant mechanisms, and of the predictions of a statistical evaporation model. We place particular emphasis on the experimental procedure that we have used to unfold the contributions to the energy spectra originated not only in these processes, but also those coming from the target backing.

The study hereby reported is part of an ongoing effort by our research group that includes previous investigations of the ^{6,7}Li+^{144}Sm systems. For example, Figueira *et al.* [6] reported exhaustive measurements of elastic-scattering angular distributions from which reliable values of the optical potential parameters have been obtained. In particular, for the ^{6}Li+^{144}Sm system we have also carried out studies of breakup effects through measurements of inelastic scattering cross sections [7], of the alpha-particle production at extreme backward angles [8], and of the light particles emitted in coincidence as the result of non-capture breakup [2]. The analysis and interpretation of measurements of the α-d transfer-breakup mode in ^{7}Li+^{144}Sm is still underway and we expect to publish the results soon.

It is worth mentioning that other similar works as the one presented here have been done using weakly bound stable (e.g. Refs. [9, 10]) and radioactive (e.g. Refs. [11, 12]) projectiles over a range of target masses, energies and angles.

This article is organized as follows: The experimental procedure is detailed in Sect. 2. In Sect. 3 we describe the measurements, and discuss the preliminary results. Finally, in Sect. 4 we present the conclusions and outlook of this work.

2. Experimental procedure

The experiments were performed at the 20 UD TANDAR tandem accelerator in Buenos Aires, which delivered beams of ^{7}Li projectiles with bombarding energies of 18, 24 and 30 MeV. Typical beam currents ranged from 1 to 15 pnA. The ^{144}Sm target was isotopically enriched to 96%, its thickness was 60 μg/cm^{2}, and it was evaporated onto a 20-μg/cm^{2} carbon backing.

The detection system used is a position-sensitive telescope for particle identification that consists of a segmented-anode ionization chamber filled with P10 gas, followed by an array of three silicon position-sensitive detectors (PSD). The angular acceptance of the whole device is 30°. For each particle that impinged on the detector, the charge created by the energy loss along its path through the gas was collected by means of two successive anode segments. In this way, two partial energy loss signals ΔE_1 and ΔE_2 ($\Delta E = \Delta E_1 + \Delta E_2$) were produced. Finally, the particle was detected by one of the three PSDs, which allowed the determination of its residual energy E_{res} and its position of incidence. The energy resolution of the whole system was better than 3% (FWHM/centroid ratio), which was determined by a previous characterization using several ion beams at different energies and varying the ionization gas pressure of the chamber [13]. The angular uncertainty ranged from 0.2° to 0.3°, depending on the ionization gas pressure. No appreciable difference among the three PSDs in the energy resolution as well as in the angular straggling was observed. For a more detailed explanation of the employed technique see Ref. [13].

Figure 1 shows a bi-dimensional E_{res}–ΔE spectrum for the ^{7}Li+^{144}Sm system taken at E_{lab} = 30 MeV. In this case the E_{res} signals correspond to the central PSD that covers an angular range from 40° to 50°.

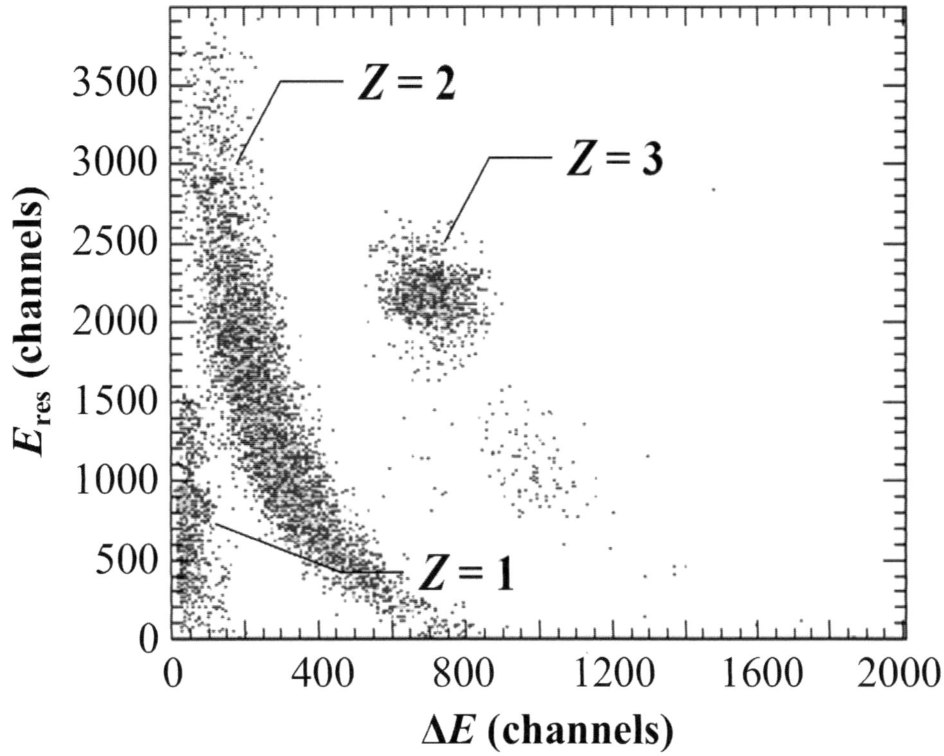

Figure 1. E_{res}–ΔE spectrum for the ^{7}Li+^{144}Sm system at E_{lab} = 30 MeV obtained with the central PSD placed at 45°.

In the data analysis procedure each PSD sensitive track was divided (by software) into bins of angular position. The effective solid angle Ω_{bin} subtended by each bin was determined using the elastic scattering of ^{16}O+^{197}Au at E_{lab} = 50 MeV. Thus, taking into account that at this energy the angular distribution of the elastically scattered ^{7}Li nuclei follows the Rutherford formula, the solid angle Ω_{bin} can be expressed as:

$$\Omega_{bin} = K\frac{N}{\delta}\left(\frac{d\sigma}{d\Omega}(\theta_{bin})\right)^{-1} \quad (1)$$

where $d\sigma/d\Omega$ (θ_{bin}) is the differential cross section for Rutherford scattering for the $^{16}O+^{197}Au$ system at the bin angular position θ_{bin}, N is the number of ^{7}Li nuclei registered in the PSD bin, δ is the effective target thickness, and K is a normalization constant. The statistical error arising from N was less than 10% and the error in the thickness δ of the ^{197}Au target was estimated to be 10% by an independent measurement, being the latter the dominant contribution to the systematic uncertainty of the measured cross sections.

The measured angular range spanned from 15° to 140°. Since the detection system covers a range of 30°, four different angular positions of the ionization chamber were required in order to complete the whole distribution (15°-45°, 45°-75°, 75°-105°, and 105°-135°). At each position, an extra run was performed moving the detector by 5° in the backward direction to cover the dead zones between the PSDs.

The intensity of the ^{7}Li beam was determined by monitoring the elastically scattered particles on the ^{144}Sm target by means of a silicon surface barrier detector placed at a polar angle θ_{mon} = 28°, small enough to ensure pure Rutherford scattering for all the projectile energies.

The irradiation time of each run was such that the number of events in the $Z = 2$ group (see Fig. 1) was of the order of 3000 (2% statistical uncertainty). However, due to very small cross sections at backward angles, such requirement was relaxed to a few hundred events and/or the solid angle of the detection system was increased through a proper adjustment of the angular bins (see next Section).

3. Results and discussion

In Fig. 2 we present the alpha-particle energy spectra for the three projectile energies (18, 24 and 30 MeV) at three different angles (θ_{bin} = 20°, 80° and 140°). In all cases, the angle θ_{bin} corresponds to the center of a PSD bin for which an angular aperture $\Delta\theta_{bin}$ = 3° was adopted by applying appropriate software cuts in the data analysis.

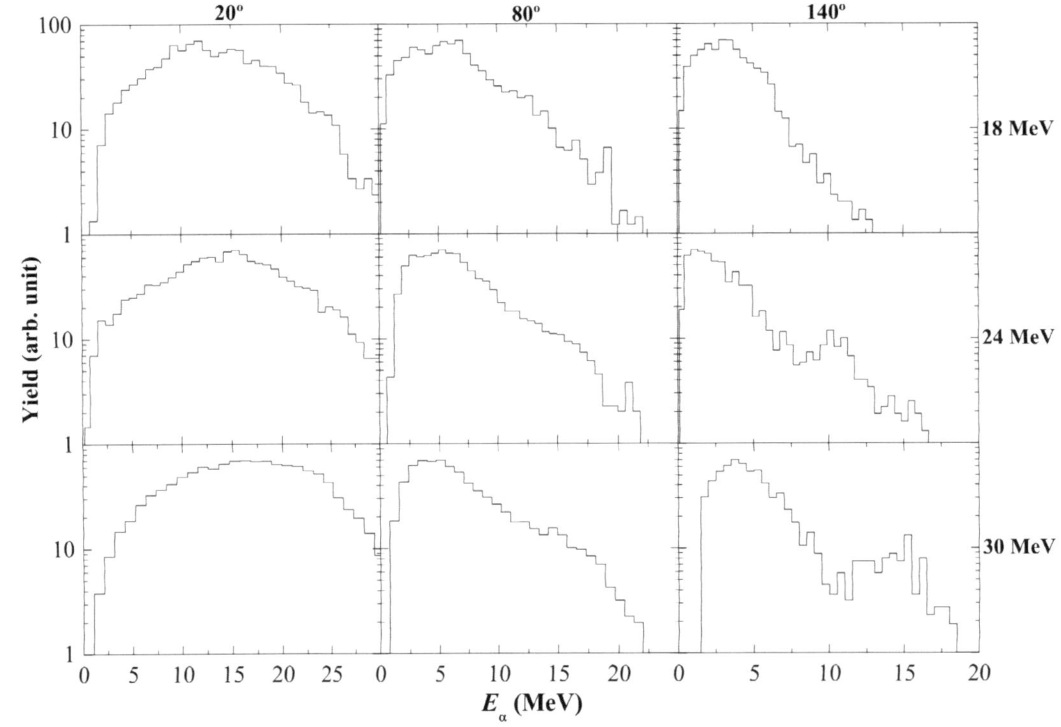

Figure 2. Energy spectra of the alpha particles emitted at θ_{bin} = 20°, 80° and 140° in the $^{7}Li+^{144}Sm$ system at E_{lab} = 18, 24 and 30 MeV. Each angle corresponds to the center of a $\Delta\theta_{bin}$ = 3° PSD bin.

At the forward angles the spectra show no structure, but at more backward angles two groups of alpha particles with distinguishable energies appear. This feature is more evident at the highest bombarding energies (24 and 30 MeV) and at the most backward angle. The energy spectra obtained at 18 MeV seem to be structureless at all angles.

In the next Sub-section we will explore the different reactions that may explain the observed structures shown in Fig. 2.

3.1. Main sources of alpha particles

There are several possible two-, three- and four-body reactions that produce alpha particles for the weakly bound projectile ^{7}Li. A list of these reactions with their respective ground-state Q-values is presented in Table 1.

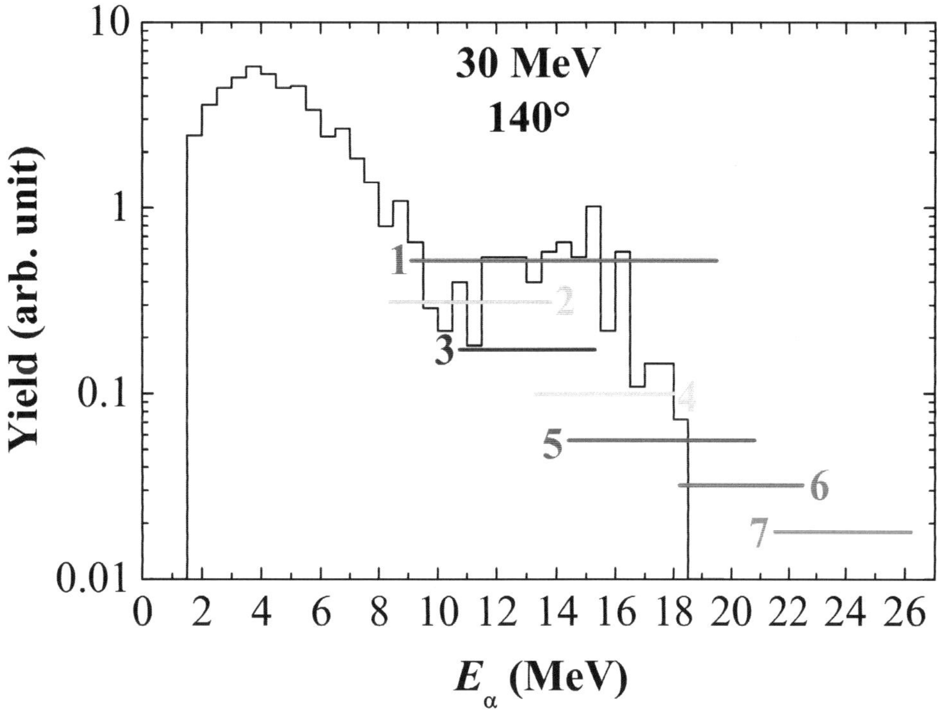

Figure 3. Energy spectrum of the alpha particles emitted in the ^{7}Li+^{144}Sm system at $E_{\rm lab}$ = 30 MeV and $\theta_{\rm bin}$ = 140.0°±1.5°. The labeled segments indicate the expected energy ranges of the alpha particles that would be emitted in reactions listed in Table 1. The main group of events, at low energies has been identified as fusion between the ^{7}Li projectile and the ^{12}C from the target backing. See text for details.

Figure 3 shows an alpha-particle energy spectrum for ^{7}Li+^{144}Sm at $E_{\rm lab}$ = 30 MeV obtained with a PSD angular bin centered at $\theta_{\rm bin}$ = 140.0°±1.5°. The labeled horizontal segments show typical energy ranges (the length represents the FWHM) of the alpha particles stemming from some of the processes listed in Table 1. For the case of incomplete fusion (channel 2 in Table 1), the mean energy has been estimated by imposing the simple model condition that the magnitudes of the asymptotic relative velocities are equal in the entrance channel (^{7}Li+^{144}Sm) and in the exit channel (α+^{147}Eu). The width has been obtained from a classical dynamical calculation using the code PLATYPUS [14]. Among the possible sources of alpha particles we have not considered the direct ^{3}H transfer to the low-lying excited states of ^{147}Eu. In fact, the alpha particles emitted in such reactions have energies close to 34

Journal of Physics: Conference Series **630** (2015) 012023 doi:10.1088/1742-6596/630/1/012023

MeV, well above the region of interest for the present study. For the processes that ultimately involve a binary breakup of the type α+X (channels 3 to 7 in Table 1) the corresponding energy distributions have been obtained with the code SUPERKIN [15]. For these cases, in the calculations we have taken into account the exact geometry of the detector as well as the dimensions of the bins used in the analysis, and we have assumed a representative uniform distribution of the center-of-mass relative energies between the breakup fragments (α and X) in the range 0 to 1 MeV. Using similar kinematical arguments, the mean energies of the alpha particles emitted in the four-body reactions (labeled 8 and 9 in Table 1, not shown in Fig. 3) have been estimated to be 14.0 MeV and 10.3 MeV, respectively.

Table 1. Reactions that produce alpha particles in the ^{7}Li+^{144}Sm system, with their respective ground state Q-values.

		Channel	Q-value (MeV)
1	^{151}Tb* → α + Y	Complete fusion (CF)	4.56
2	α + ^{147}Eu	Incomplete fusion (ICF) / t-transfer	8.06
3	α + t + ^{144}Sm	Non-capture breakup (NCBU)	-2.47
4	α + d + ^{145}Sm	n-stripping followed by breakup	-1.97
5	α + α + ^{143}Pm	p-pickup followed by breakup	11.05
6	α + n + ^{146}Eu	d-stripping followed by breakup	-0.44
7	α + p + ^{146}Sm	2n-stripping followed by breakup	4.22
8	α + n + n + ^{145}Eu	p-stripping followed by ternary breakup	-7.63
9	α + d + n + ^{144}Sm	Non-capture ternary breakup	-8.72

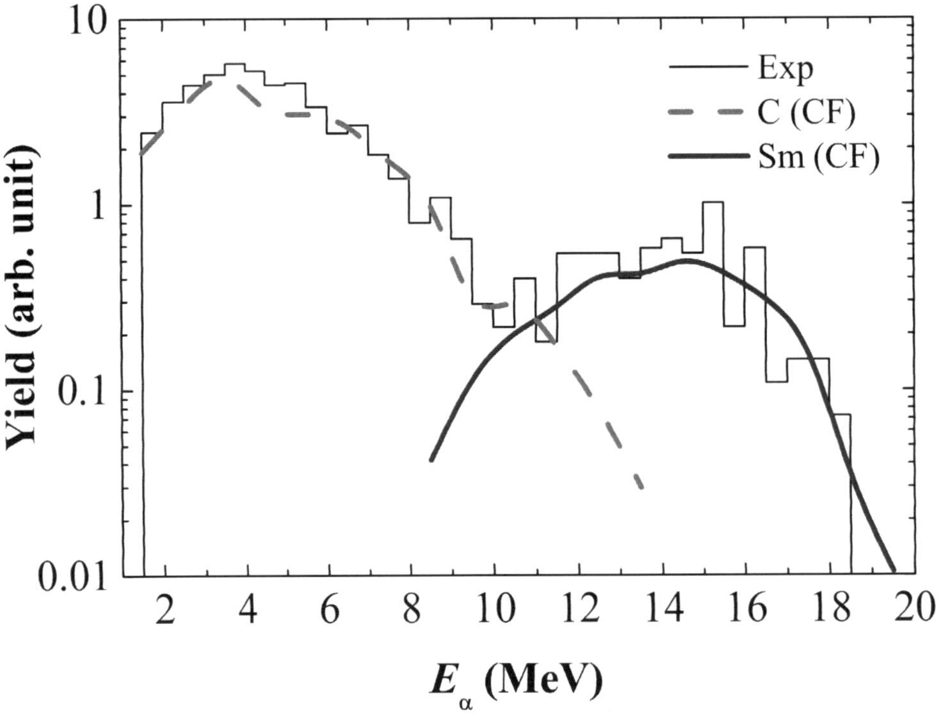

Figure 4. Alpha-particle energy spectrum for the ^{7}Li impinging on the ^{144}Sm target (and ^{12}C backing) at E_{lab} = 30 MeV and θ_{bin} = 140.0°±1.5°. The dashed and solid curves are theoretical calculations of complete fusion for the ^{7}Li+^{12}C and ^{7}Li+^{144}Sm systems, respectively.

Journal of Physics: Conference Series **630** (2015) 012023 doi:10.1088/1742-6596/630/1/012023

From this kinematical analysis it is evident that all the reactions of interest existing in the $^{7}Li+^{144}Sm$ system produce alpha particles with higher energies than the main group of events in Fig. 3. It can be seen that the expected mean energies for capture and non-capture breakup happen to be in qualitative agreement with the high-energy bump observed in most energy spectra. Alpha particles coming from direct transfer reaction that would populate the ground states or the lowest excited states of the reaction products are expected to have much higher energies and they have not been observed in the present measurements. In general, based on all the possible reactions that involve only ^{144}Sm, it is very difficult to find a quantitative explanation of the energy spectra over the whole energy range.

From these arguments we may conclude that the prominent low-energy bumps that are observed in all the spectra, most likely correspond to evaporation following the fusion of ^{7}Li projectiles with ^{12}C nuclei in the target backing. In order to further evaluate this hypothesis we have performed complete fusion-evaporation calculations for $^{7}Li+^{12}C$ (and $^{7}Li+^{144}Sm$) using the code PACE [16]. Figure 4 displays the same experimental spectrum as in Fig. 3 but now compared to the complete calculated energy distributions obtained with PACE. The excellent agreement of the calculation with the low-energy portion of the spectrum reinforces the above interpretation. In view of these results, for all the measured energy spectra we have decided to evaluate the high-energy tail of the alpha particles coming from $^{7}Li+^{12}C$ fusion and their contribution to the background over the region of interest for alpha particles coming from $^{7}Li+^{144}Sm$ reactions.

For this purpose we have resorted to auxiliary measurements with a pure ^{12}C target, which allowed us to parameterize the spectral shapes at different bombarding energies and emission angles using an asymmetric Gaussian function. In this way the ^{12}C fusion peak could be unequivocally determined in all the energy spectra and considered as background for the $^{7}Li+^{144}Sm$ data.

As an example of the procedure described above, in Fig. 5 we show background fits for $^{7}Li+^{144}Sm$ at E_{lab} = 30 MeV and at two different angles, θ_{bin} = 80° and 140° At this bombarding energy it was not possible to reliably separate the data of interest from the background below θ_{lab} ~ 80°. In the case of E_{lab} = 24 MeV, the angular limitation for a reliable separation of the alpha particles coming from reactions with ^{144}Sm from those originated in the backing was approximately $\theta_{lab} \geq 110°$. Finally, at E_{lab} = 18 MeV the background subtraction procedure could not be satisfactorily applied at any angle. Therefore, we were unable to proceed with the angular distribution analysis at this lowest bombarding energy.

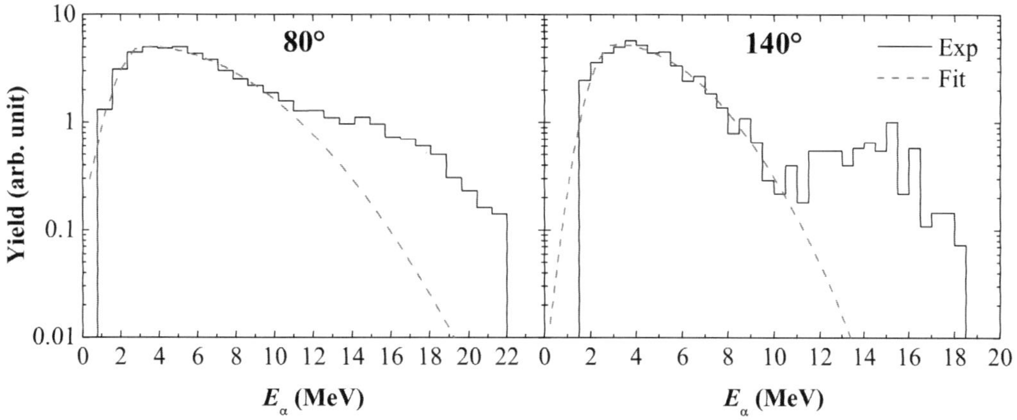

Figure 5. Fits of the $^{7}Li+^{12}C$ fusion background for E_{lab} = 30 MeV at two different angles: θ_{bin} = 80° and 140°.

3.2. Angular distributions

Angular distributions of the emitted alpha particles have been obtained using the subset of data for which the previously described background-subtraction procedure could be reliably applied. Basically, this included data taken at E_{lab} = 30 MeV for $\theta_{lab} \geq 80°$ and E_{lab} = 24 MeV for $\theta_{lab} \geq 110°$. Within these angular ranges, the energy spectra were obtained for successive angular bins of constant width $\Delta\theta_{bin}$ = 3° for both bombarding energies. By taking some measurements at overlapping angular positions of the detector, some of the angular gaps due to dead zones between adjacent PSDs could be filled out. After subtracting the background from each spectrum to obtain the alpha-particle yield that can be attributed to pure ^{7}Li+^{144}Sm reactions, the relative contribution of fusion-evaporation in comparison to the rest of the processes listed in Table 1 was evaluated. For that purpose we performed calculations with the code PACE, an example of which for E_{lab} = 30 MeV at θ_{bin} = 140° is illustrated by the solid curve in Fig. 4. In this particular case (as well as in the rest of the data; not shown) we can observe that the calculation cannot account for the total alpha production over the energy region of interest. The yield of alpha particles in excess of the prediction can therefore be interpreted as arising from the sum of capture breakup and non-capture breakup processes.

The results of the present analysis are summarized in Fig. 6, where the differential cross sections for alpha-particle production as a function of laboratory scattering angles at E_{lab} = 24 and 30 MeV are shown. The empty and full circles represent the experimental total production whereas the dashed and solid curves correspond to the theoretical calculations of the corresponding fusion-evaporation contributions. In general, it can be observed that the contribution from fusion-evaporation cannot account for the measured cross sections of alpha-particle production. At 30 MeV the discrepancy is small at the most backward angles but it increases significantly as the emission angle decreases. At 24 MeV the contribution of fusion-evaporation to the total yield seems to be even less significant and it barely explains approximately one seventh of the experimental cross section.

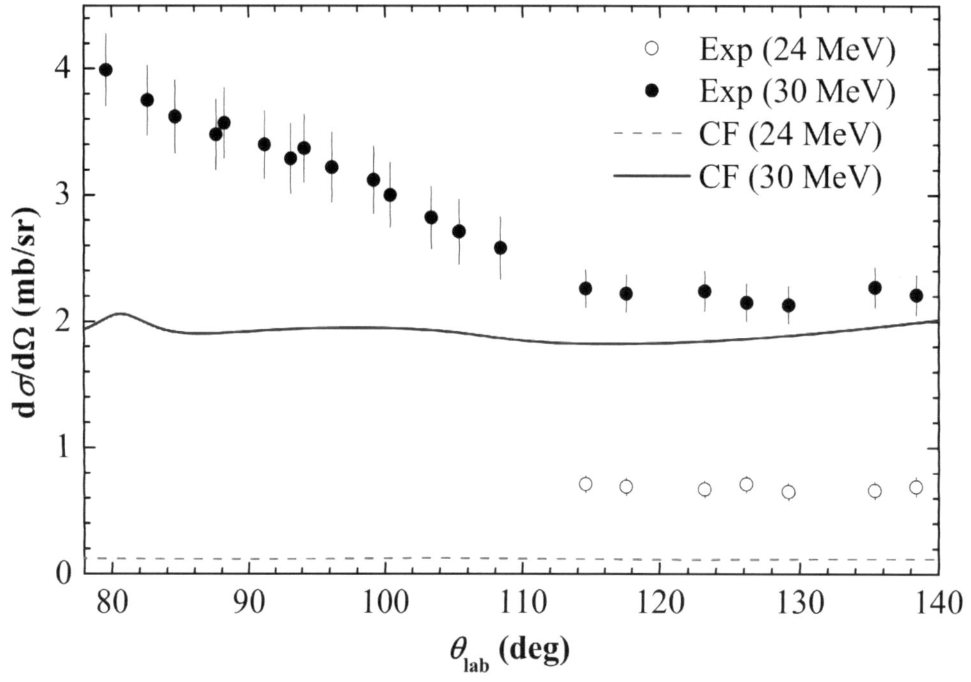

Figure 6. Angular distributions of the alpha-particle production cross section for the ^{7}Li+^{144}Sm system at laboratory energies of 24 and 30 MeV. Also shown are calculations of the corresponding fusion-evaporation contributions done with the code PACE.

Preliminary calculations based on the code PLATYPUS [14] showed us that the ICF and NCBU processes might account for these differences. The ICF has a rather uniform angular distribution in this range, but the NCBU is biased to forward angles. Hence the contribution of the NCBU process may explain the observed enhancement at angles around $\theta_{\text{lab}} \sim 80°$ in the alpha angular distribution corresponding to $E_{\text{lab}} = 30$ MeV. An assessment of this hypothesis requires a better determination of the PLATYPUS input parameters that determine the breakup probability and the overall normalization factor for the ICF and NCBU processes. We are currently investigating this possibility.

4. Conclusions and outlook

We have measured the production of alpha particles using a ^{7}Li beam with bombarding energies of 18, 24 and 30 MeV on a ^{144}Sm target over a wide range of emission angles. The procedure that we have applied for the separation of the contribution from alpha particles produced in the carbon backing allowed us to obtain significant results only for $E_{\text{lab}} = 24$ and 30 MeV at backward angles ($\theta_{\text{lab}} > 80°$). The sources of emission have been analyzed in the first place from a kinematical viewpoint, as a result of which we can establish a clear correspondence of the resulting energy spectra with fusion-evaporation and with capture (incomplete fusion) and non-capture breakup processes. No significant evidence was found of the most energetic alpha particles that would presumably originate in direct transfer reactions populating low-lying states of the transfer products. A further step in the analysis consisted in the estimate of the fusion-evaporation component by comparison with the results of calculations based on a statistical model. From the excess of the experimental cross sections over the predictions for fusion, we can evaluate the qualitative behavior of the combined breakup components as a function of bombarding energies and emission angles. At the highest measured bombarding energy (30 MeV) the results show an increasing contribution from breakup processes as we move to forward angles. At the lowest energy (24 MeV) breakup is clearly dominant over the whole measured angular range.

We are presently working on the last step of this analysis aimed to derive more quantitative information of the individual breakup components. Preliminary results obtained by application of a classical dynamical model appear to reproduce the order of magnitude of the measured cross sections and the qualitative angular dependence of the alpha-particle emission originated in breakup reactions. We expect to be able to improve the analysis from both the experimental and theoretical sides. Regarding the calculations, we are investigating the use of more reliable input parameters of the model, such as the breakup probabilities that can be obtained from a systematic evaluation of previous experiments. We are planning to incorporate the results of our recent exclusive cross-section measurements of the transfer-breakup mechanism, consisting in the coincident detection of deuterons and alpha particles in reactions induced by ^{7}Li projectiles. We also intend to apply the whole procedure described in this work to the study of other reaction systems such as ^{7}Li+^{27}Al system, which has already been measured and which analysis is currently underway.

References

[1] Canto L F, Gomes P R S, Donangelo R and Hussein M S 2006 *Phys. Rep.* **424** 1
[2] Martinez Heimann D, Pacheco A J, Capurro O A, Arazi A, Balpardo C, Cardona M A, Carnelli P F F, De Barbará E, Fernández Niello J O, Figueira J M, Hojman D L, Martí G V, Negri A E and Rodrigues Ferreira Maltez D 2014 *Phys. Rev.* C **89** 014615
[3] Shrivastava A, Navin A, Keeley N, Mahata K, Ramachandran K, Nanal V, Parkar V V, Chatterjee A and Kailas S 2006 *Phys. Lett.* B **633** 463
[4] Martinez Heimann D, Pacheco A J, Arazi A, Capurro O A, Carnelli P F F, Monteiro D S, Fernández Niello J O, Figueira J M, Fimiani L, Grinberg P, Marta H D, Martí G V, Negri A E and Testoni J E 2009 *AIP Conf. Proc.* **1098** 275
[5] Luong D H, Dasgupta M, Hinde D J, Du Rietz R, Rafiei R, Lin C J, Evers M and Diaz-Torres A 2011 *Phys. Lett.* B **695** 105
[6] Figueira J M, Fernández Niello J O, Arazi A, Capurro O A, Carnelli P F F, Fimiani L, Martí G

V, Martinez Heimann D, Negri A E, Pacheco A J, Lubian J, Monteiro D S and Gomes P R S 2010 *Phys. Rev.* C **81** 024613

[7] Woodard A E, Figueira J M, Otomar D R, Fernández Niello J O, Lubian J, Arazi A, Capurro O A, Carnelli P F F, Fimiani L, Martí G V, Martinez Heimann D, Monteiro D S, Negri A E, Pacheco A J and Gomes P R S 2012 *Nucl. Phys.* A **873** 17

[8] Capurro O A, Pacheco A J, Arazi A, Figueira J M, Martinez Heimann D and Negri A E 2011 *Nucl. Phys.* A **849** 1

[9] Kelly G R, Davis N J, Ward R P, Fulton B R, Tungate G, Keeley N, Rusek K, Bartosz E E, Cathers P D, Caussyn D D, Drummer T L and Kemper K W 2000 *Phys. Rev.* C **63** 024601

[10] Signorini C, Edifizi A, Mazzocco M, Lunardon M, Fabris D, Vitturi A, Scopel P, Soramel F, Stroe L, Prete G, Fioretto E, Cinausero M, Trotta M et al. 2003 *Phys. Rev.* C **67** 044607

[11] Kolata J J, Goldberg V Z, Lamm L O, Marino M G, O'Keeffe C J, Rogachev G, Aguilera E F, García-Martínez H, Martinez-Quiroz E, Rosales P et al. 2002 *Phys. Rev.* C **65** 054616

[12] De Faria P N, Lichtenthäler R, Pires K C C, Moro A M, Lépine-Szily A, Guimarães V, Mendes D R Jr, Arazi A, Barioni A, Morcelle V and Morais M C 2010 *Phys. Rev.* C **82** 034602

[13] Carnelli P F F, Arazi A, Fernández Niello J O, Capurro O A, Cardona M A, De Barbará E, Figueira J M, Hojman D L, Martí G V, Martinez Heimann D, Negri A E and Pacheco A J 2013 *Nucl. Inst. and Meth.* A **726** 116

[14] Diaz-Torres A 2010 *J. Phys.* G **37** 075109

[15] Martinez Heimann D, Pacheco A J and Capurro O A 2010 *Nucl. Inst. and Meth.* A **622** 642
Martinez Heimann D, Pacheco A J and Capurro O A 2012 *Nucl. Inst. and Meth.* A **694** 313

[16] Gavron A 1980 *Phys. Rev.* C **21** 230

Journal of Physics: Conference Series **630** (2015) 012024 doi:10.1088/1742-6596/630/1/012024

A cluster model for the ^{6}He+^{9}Be interaction

R. Lichtenthäler[1], K.C.C Pires[1], A. Lépine-Szily[1] and V. Morcelle[2]

[1] Instituto de Física , Universidade de São Paulo, C.P. 66318, 05314-970 São Paulo, Brazil
[2] Departamento de Física, Universidade Federal Rural do Rio de Janeiro, 23851-970, Seropédica, RJ, Brazil.

E-mail: rubens@if.usp.br

Abstract. An optical model analysis of ^{6}He+^{9}Be elastic scattering angular distributions at two energies above the Coulomb barrier is presented. A cluster model is used for the projectile-target optical potential which is decomposed into two components, the alpha-target and the 2n-target interactions. The 2n-target is adjusted to fit the data and a long range imaginary part is obtained.

1. Introduction

The study of nuclei out of the line of stability is one of the most active fields in low energy nuclear physics nowadays. Most of the research in this area has been performed in large laboratories all over the world, and at intermediate energies, from 30-200 MeV/A or higher. In São Paulo, we have an accelerator of 3-5 MeV/A, connected with the RIBRAS (Radioactive Ion Beams in Brasil) system [1], which allows the production of secondary beams of light exotic nuclei such as ^{6}He, ^{8}Li and others at low energies. Some of such light exotic nuclei are important for several reasons, for instance, in studies of the neutron halo and the 3-body structure of ^{6}He and in the possible role they could play in some astrophysical environments. The ^{6}He nucleus has a pronounced cluster structure formed by a double magic alpha core surrounded by two neutrons. Those two neutrons are bound to the $\alpha-$core by only $E_{lig} = 0.973$ MeV, forming a 3-body system, which is bound only if the core plus the two neutrons are present. Due to the small binding energy and low angular momenta of the neutrons, they form a kind of low density neutron halo, which extends over large distances from the alpha core. As a consequence the ^{6}He nucleus can be easily dissociated in the collision with a target, affecting both, the elastic scattering angular distribution and the total reaction cross section. Over the last 10 years we have developed an extensive research program in RIBRAS [2], involving mostly low energy scattering measurements of ^{6}He on targets of different masses, from light to heavy ones. This kind of study allows for instance to determine the total reaction cross sections of reactions induced by the neutron rich projectile ^{6}He on several targets and compare it with stable systems.

This contribution presents results of a new analysis of ^{6}He+^{9}Be elastic scattering angular distributions measured in RIBRAS at two energies above the Coulomb barrier, $E_{lab} = 16.2$ and 21.3 MeV [3]. The analysis has been performed in the context of a cluster optical model, where the ^{6}He+^{9}Be optical potential has been decomposed into two components, one due to the interaction of the alpha cluster with the target and another, unknown, due to the interaction between the two neutrons and the target. The latter has been adjusted to reproduce the data.

Journal of Physics: Conference Series **630** (2015) 012024 doi:10.1088/1742-6596/630/1/012024

2. The model

Scattering and reactions induced by neutron rich exotic projectiles such as ^{6}He are strongly affected by the fact that the continuum is so close to the ground state. For this reason the elastic scattering description at low energies has to take explicitly into account the effect of the coupling with the continuum. This is normally done by Continuum Discretized Coupled Channels description (CDCC) where the continuum of the projectile is discretized in bins of a given width up to a maximum excitation energy and a maximum angular momentum. The CDCC model has been succesfully applied to a number of cases [4–8] involving the scattering of ^{6}He on targets of different masses. However, for the ^{6}He+^{9}Be light system the situation is not so simple. Due to the fact that, in this case, both projectile and target are easily broken, a complete calculation would require the inclusion of both processes simultaneously, which is not possible with the present techniques. Four and three-body CDCC calculations have been presented in Ref. [9] for the ^{6}He+^{9}Be scattering considering the projectile breakup only. In the three-body CDCC description the projectile-target optical potential is obtained from a cluster model, whose ingredients are the fragment-target (α-t) and the neutron-target interactions as shown below.

$$V_{opt} = \langle \phi_{^6He} | V_{[2n+^9Be]} + V_{[\alpha+^9Be]} | \phi_{^6He} \rangle \tag{1}$$

In principle the optical potential $V_{[\alpha+^9Be]}$ can be obtained experimentally from previous α+^{9}Be scattering experiments. Such optical potential has an imaginary part, which takes into account the overall α+^{9}Be absorption, and consequently, partially includes the ^{9}Be breakup process, induced by the alpha particle. The $V_{[2n+^9Be]}$ interaction on the other hand is unknown. It is impossible to be directly measured since the di-neutron is unbound. In a four-body CDCC description it can be obtained from the known neutron-target interaction, as was done in Ref. [9].

Here we present a different approach [3]. An optical model calculation was performed but using Eq. 1 to obtain the ^{6}He+^{9}Be optical potential. It is important to make clear that the results presented here are not from a CDCC calculation, since no coupled channels calculations were performed, but a simple optical model, whose optical potential was obtained from Eq. 1. Both potentials appearing in Eq. 1 have been parameterized by a Wood-Saxon shape. The parameters for the α+^{9}Be system have been obtained from previous experiments [11] and, the $2n$+^{9}Be component was varied to best fit the angular distributions.

$\phi_{^6He}$ is the ground state ^{6}He wavefunction and was calculated in the single particle model considering the two neutrons as a single particle bound to the α core. The 2n+α binding energy was set to 1.6 MeV to reproduce 3-body effects of the ^{6}He wave function as described in Ref. [10].

The fits are shown in Figure 1 and the resulting parameters are in Table 1. In the first line of Table 1 we present the parameters of the α+^{9}Be potential taken from the literature [11]. In the second line we present the parameters of the 2n+α bound state potential used here. In the third and fourth lines the resulting 2n+^{9}Be parameters for the real and imaginary parts at the two energies analysed are presented.

The effect of the experimental angular resolution is important in this case and was taken into account by performing an average of the theoretical curves over the angular acceptance of the detectors. Its effect is displayed by the two curves (dotted and solid) in Figure 1 (see Ref. [9] for more details).

3. Conclusions

A cluster model was applied to obtain the optical potential that best reproduces the ^{6}He+^{9}Be scattering. The model considers the ^{6}He+^{9}Be interaction as the folding between the α+^{9}Be and the $2n$+^{9}Be interactions with the ground state ^{6}He wave function. The α+^{9}Be interaction was obtained from the literature and is known from previous experiments. The di-neutron-^{9}Be

Table 1. Parameters obtained from the Optical Model calculations. The calculation was performed with the folded potential (Eq. 1) using the parameters given below. Table taken from Ref. [3].

E_{lab}	Potencial	V_0 (MeV)	r_0 (fm)	a_0 (fm)	W_0 (MeV)	r_i (fm)	a_i (fm)	σ_{reac} (mb)	Ref.
	α+^{9}Be	50.00	1.85	0.55	2.50	1.85	0.55	-	[11]
	2n+α	96.06	1.90	0.25	-	-	-	-	[10]
16.2	2n+^{9}Be	61.01	1.51	0.55	20.00	1.20	1.01	1513	-
21.3	2n+^{9}Be	21.41	1.51	0.54	10.02	1.35	1.56	1944	-

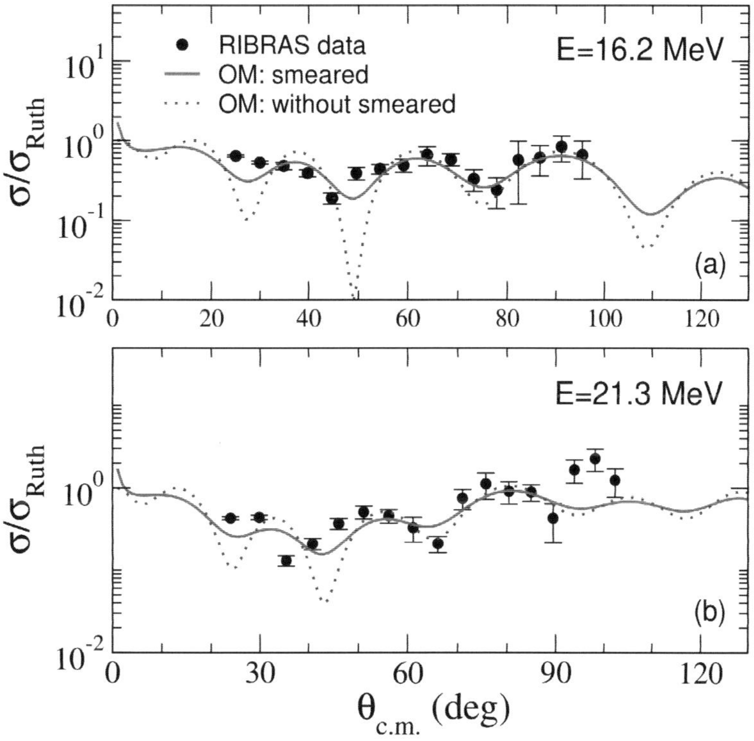

Figure 1. ^{6}He+^{9}Be Elastic angular distributions at 16.2 and 21.3 MeV compared with optical model calculations. This Figure has been taken from Ref. [3]

interaction is unknown and was adjusted to best reproduce the ^{6}He+^{9}Be angular distributions. The 2n+^{9}Be potential obtained from the present analysis shows a remarkably large imaginary diffuseness in comparison to the α+^{9}Be potential, indicating an effect of a rather extended neutron halo in the ^{6}He+^{9}Be scattering. The total reaction cross sections are also large and an enhancement has been observed with respect to the cross sections for similar stable systems.

4. Acknowledgements

The authors wish to thank the Fundação de Amparo à Pesquisa do Estado de São Paulo (FAPESP), the Conselho Nacional de Desenvolvimento Científico e Tecnológico (CNPq), the

Fundação de Amparo à Pesquisa do Estado do Rio de Janeiro (FAPERJ) and Coordenação de Aperfeiçoamento de Pessoal de Nível Superior (CAPES) for Brazilian financial support.

5. References

[1] Lichtenthäler R *et al.* 2005 *Eur. Phys. J.* A **25** 733
[2] Lépine-Szily A, Lichtenthäler R and Guimarães V 2014 *Eur. Phys. Jou.* A **50** 128
[3] Pires K C C, Lichtenthäler R, Lépine-Szily A and Morcelle V 2014 *Phys. Rev.* C 90 027605
[4] Faria P N *et al.* 2010 *Phys. Rev.* C **81** 044605
[5] Faria P N *et al.* 2010 *Phys. Rev.* C **82** 034602
[6] Mohr P *et al.* 2010 *Phys. Rev.* C **82** 044606
[7] Mohr P *et al.* 2005 *Phys. Rev.* C **71** 017601
[8] Morcelle V *et al.* 2014 *Phys. Lett.* B **732** 228
[9] Pires K C C *et al.* 2011 *Phys. Rev.* C **83** 064603
[10] Moro A M, Rusek K, Arias J M, Gómez-Camacho J and Rodríguez-Gallardo M 2007 *Phys. Rev.* C **75** 064607
[11] Perey C M and Perey F G 1976 *At. Data and Nucl. Data Tables* **17**, 1

Journal of Physics: Conference Series **630** (2015) 012025 doi:10.1088/1742-6596/630/1/012025

Coulomb-nuclear interference in the inelastic scattering of ^{6}Li on ^{76}Ge

X X Zhang[1], M R D Rodrigues[1], T Borello-Lewin[1], C L Rodrigues[1], G M Ukita[2], L R B Benevides[1], J L M Duarte[1], L B Horodynski-Matsushigue[1]

[1]Instituto de Física, Universidade de São Paulo, São Paulo, Brasil
[2]Faculdade de Psicologia, Universidade de Santo Amaro, São Paulo, Brasil

Email: xin.zhang@usp.br

Abstract. Angular distribution for the inelastic scattering of 28 MeV ^{6}Li on ^{76}Ge was measured using the São Paulo Pelletron Enge Split-pole Spectrograph facility. The coulomb-nuclear interference (CNI) analysis was applied to the first quadrupole state transition. The values of $C_L = \delta_L^C/\delta_L^N$, the ratio of charge to isoscalar deformation lengths, and of $(\delta_L^N)^2$ were extracted through the comparison of experimental and DWBA-DOMP predicted cross section. The ratio of reduced charge to isoscalar transition probabilities, B(EL) to B(ISL) respectively, are related to the square of the parameter C_L and were thus obtained with the advantage of scale uncertainties cancellation. The value of $C_2 = 1.10(2)$ obtained indicate a slight predominance of the protons relative to the neutrons for ^{76}Ge.

1. Introduction

The characteristics of excited states 2^+_1 are widely used as indicators of nuclear structure, particularly the electric reduced transition probability B(E2) is used as a measure of the collectivity of these transitions. The B(E2) is in principle sensitive only to the contribution of the charge and to quantify the contributions of neutrons is also an important ingredient to characterize the collectivity. This aim is achieved through the isoscalar (mass) reduced transition probability B(IS2) obtained from the inelastic scattering. Particularly suitable are the measurements in an incident energy that enhances coulomb-nuclear interference (CNI) and allows simultaneous extractions of B(IS2) and the ratio between electric and isoscalar reduced transition probabilities B(E2)/B(IS2) [1-3]. The ratio B(E2)/B(IS2) is related to the square of the parameter C_2. The previous CNI work, in the A~ 70 transitional mass region, considering the germanium isotopic chain [3] demonstrate an abrupt change for ^{74}Ge with $C_2 = 0.775(8)$: although for ^{70,72}Ge, values of the order of 1.0 or slightly higher were obtained. In particular, the present work pursues the study of CNI in the inelastic scattering of ^{6}Li on ^{76}Ge, in order to determine the relative contributions of protons and neutrons in the transition to the first quadrupole state.

2. Experimental setup

Measurements of the inelastic scattering of ^{6}Li on ^{76}Ge were obtained using the Pelletron Enge Split-pole Spectrograph facility. A solid-state position sensitive detector (PSD) of 500 μm thickness, and 47 x 8 mm^2 area was positioned on the focal plane. The 28.0 MeV energy ^{6}Li beam was focused after passing defining slits of 1.0 x 2.0 mm^2 on an enriched target of ^{76}Ge with 510.5 10^{15} particles/cm^2 thickness. A solid angle of 0.65 msr was used in the spectrometer entrance. Twenty-six spectra were measured at carefully chosen scattering angles in a range of $10^\circ \leq \theta_{Lab} \leq 55^\circ$, in order to characterize CNI in the angular distribution corresponding to the first quadrupole excitation.

Journal of Physics: Conference Series **630** (2015) 012025 doi:10.1088/1742-6596/630/1/012025

A surface barrier detector, with 10 mm diameter and 1000 μm thickness, was placed inside the scattering chamber at an angle of 30° relative to the ion beam. Relative normalization of the data for the various scattering angles was obtained through the total charges collected by the Faraday cup and the elastic scattering measurements from the monitor detector.

The digital pulse processing (DPP) acquisition system used in the measurement was composed by the board PCI-6133 from National Instrument with a sample of analogic inputs up to 2.5 MS/s. The analysis of the pulse shape and the use of electronic noise filters provide an important resolution improvement, which is the advantage of this acquisition system. Figure 1 shows the position spectrum along the focal plane at the scattering angle $\theta_{Lab} = 31°$. The three peaks observed on figure are associated with the elastic peak on ^{76}Ge, the 2^+_1 state inelastic peak and the elastic peak on the silicon contaminant. The energy resolution achieved was about 37 keV.

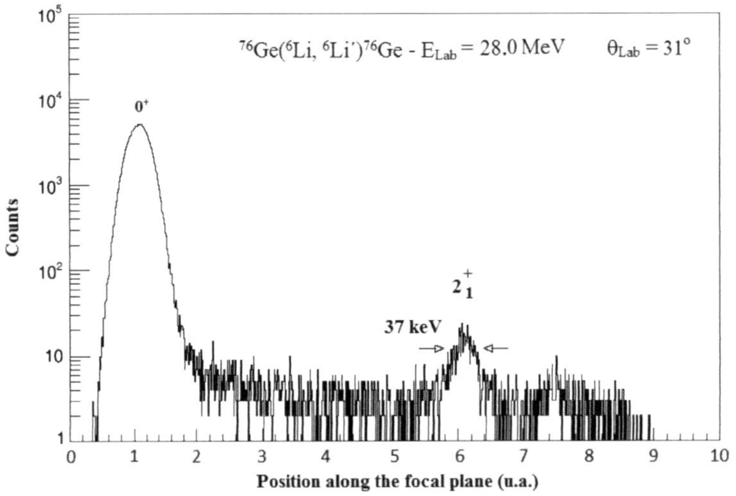

Figure 1. Position spectrum at the scattering angle $\theta_{Lab} = 31°$.

3. Data analysis and Results

The distorted wave Born approximation (DWBA) prediction using the deformed optical potential model (DOMP) approach with global optical parameters was applied. The value of the ratio between charge (δ_L^C) and mass (δ_L^N) deformation lengths C_L is obtained from the angular distribution shape. The square of mass deformation length, $(\delta_L^N)^2$, is also extracted, as a scale factor. The procedure applied for the χ^2 minimization was the iterative method of Gauss [4], extracting the correlated parameters δ_2^N and C_2.

Figure 2 illustrates the preliminary results obtained from the angular distribution of ^{76}Ge(^{6}Li, ^{6}Li')^{76}Ge to the first quadrupole excited state 2^+_1 in comparison with DWBA-DOMP prediction obtained in the fit (red solid curve). In order to illustrate the sensitivity of the method, two predictions corresponding to C = 0.78 (green dash curve) and C = 1.00 (blue dash curve) are also shown.

The results for the two correlated parameters in a preliminary data analysis are $C_2 = 1.10(2)$ and $\delta_2^N = 1.08(2)$ fm. The C_2 value obtained is slightly higher than 1.0, indicating a homogeneous excitation with a small predominant contribution of protons in the first quadrupole excited state 2^+_1 of ^{76}Ge.

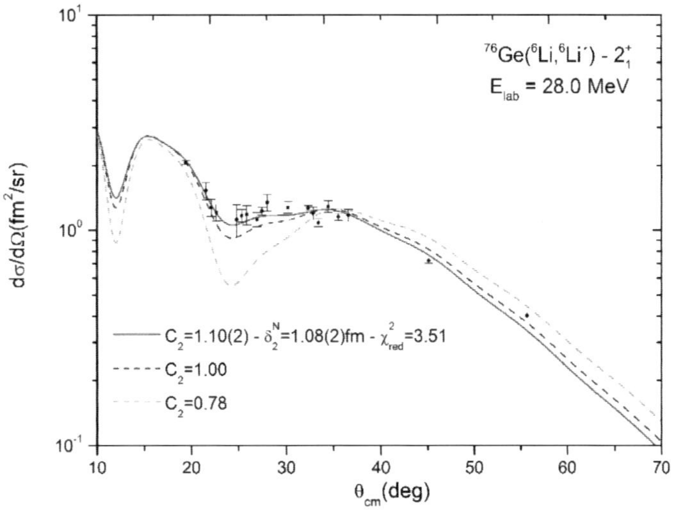

Figure 2. Experimental angular distribution in comparison with DWBA-DOMP predictions.

Acknowledgements
The author acknowledge financial support from CNEN, Comissão Nacional de Energia Nuclear.

References

[1] Ukita G M, Borello-Lewin T, Horodynski-Matsushigue L B, Duarte J L M, Gomeset L C 2001 *Phys. Rev.* C **64** 014316
[2] Duarte J L M, Ukita G M, Borello-Lewin T, Horodynski-Matsushigue L B, Gomeset L C 1997 *Phys. Rev.* C **56** 1855
[3] Barbosa M D L, Borello-Lewin T, Horodynski-Matsushigue L B, Duarte J L M, Rodrigues C L, Rodrigues M R D, Ukita G M 2005 *Phys. Rev.* C **71** 024303
[4] Bevington P R, Robinson D K 1969 *Data reduction and error analysis for the physical sciences* (New York: McGraw Hill)

Journal of Physics: Conference Series **630** (2015) 012026 doi:10.1088/1742-6596/630/1/012026

Quark matter subject to strong magnetic fields: phase diagram and applications

Débora P. Menezes, Marcus B. Pinto

Depto de Física - CFM - Universidade Federal de Santa Catarina - Florianópolis - SC - CP. 476 - CEP 88.040 - 900 - Brazil

Constança Providência, Pedro Costa, Márcio Ferreira

Centro de Física Computacional - Department of Physics - University of Coimbra - P-3004 - 516 - Coimbra - Portugal

Luis B. Castro

Departamento de Física - Universidade Federal do Maranhão - São Luís - MA - CEP 65080-805 - Brazil

E-mail: debora.p.m@ufsc.br

Abstract. In the present work we are interested in understanding various properties of quark matter subject to strong magnetic fields described by the Nambu–Jona-Lasinio model with Polyakov loop. We start by analysing the differences arising from two different vector interactions in the Lagrangian densities, at zero temperature, and apply the results to stellar matter. We then investigate the position of the critical end point for different chemical potential and density scenarios.

1. Introduction and Formalism

The study of the QCD phase diagram, when matter is subject to strong external magnetic fields has been a topic of intense investigation recently. The fact that magnetic fields can reach intensities of the order of $B \sim 10^{19}$ G or higher in heavy-ion collisions [1] and up to 10^{18} G in the center of magnetars [2] made theoretical physicists consider matter subject to magnetic field both at high temperatures and low densities and low temperatures and high densities.

In this work we study the above mentioned situations within the framework of the Nambu–Jona-Lasinio model with Polyakov Loop (PNJL) and different versions of the Nambu–Jona-Lasinio (NJL) model [3] for two commonly used parameter sets, known as HK [4] and RKH [5].

We describe quark matter subject to strong magnetic fields within the SU(3) PNJL model with vector interaction and the Lagrangian density reads:

$$\begin{aligned}\mathcal{L} &= \bar{\psi}_f \left[i\gamma_\mu D^\mu - \hat{m}_f \right] \psi_f + \mathcal{L}_{sym} + \mathcal{L}_{det} \\ &+ \mathcal{L}_{vec} + \mathcal{U}\left(\Phi, \bar{\Phi}; T\right) - \frac{1}{4} F_{\mu\nu} F^{\mu\nu}, \qquad (1)\end{aligned}$$

with

$$\mathcal{L}_{sym} = G\sum_{a=0}^{8}\left[(\bar{\psi}_f\lambda_a\psi_f)^2+(\bar{\psi}_f i\gamma_5\lambda_a\psi_f)^2\right],$$
$$\mathcal{L}_{det} = -K\left\{\det{}_f\left[\bar{\psi}_f(1+\gamma_5)\psi_f\right]+\det{}_f\left[\bar{\psi}_f(1-\gamma_5)\psi_f\right]\right\},$$

where $\psi_f = (u,d,s)^T$ represents a quark field with three flavors, $\hat{m}_c = \mathrm{diag}_f(m_u, m_d, m_s)$ is the corresponding (current) mass matrix, $\lambda_0 = \sqrt{2/3}I$ where I is the unit matrix in the three flavor space, and $0 < \lambda_a \leq 8$ denote the Gell-Mann matrices. The coupling between the magnetic field B and quarks, and between the effective gluon field and quarks is implemented *via* the covariant derivative $D^\mu = \partial^\mu - iq_f A^\mu_{EM} - iA^\mu$ where q_f represents the quark electric charge, $A^{EM}_\mu = \delta_{\mu 2}x_1 B$ is a static and constant magnetic field in the z direction and $F_{\mu\nu} = \partial_\mu A^{EM}_\nu - \partial_\nu A^{EM}_\mu$. To describe the pure gauge sector an effective potential $\mathcal{U}\left(\Phi,\bar{\Phi};T\right)$ is chosen:

$$\frac{\mathcal{U}\left(\Phi,\bar{\Phi};T\right)}{T^4} = -\frac{a\left(T\right)}{2}\bar{\Phi}\Phi + b(T)\ln\left[1-6\bar{\Phi}\Phi+4(\bar{\Phi}^3+\Phi^3)-3(\bar{\Phi}\Phi)^2\right], \tag{2}$$

where $a\left(T\right) = a_0 + a_1\left(\frac{T_0}{T}\right) + a_2\left(\frac{T_0}{T}\right)^2$, $b(T) = b_3\left(\frac{T_0}{T}\right)^3$. The standard choice of the parameters for the effective potential $\mathcal{U}$ is $a_0 = 3.51$, $a_1 = -2.47$, $a_2 = 15.2$, and $b_3 = -1.75$.

As for the vector interaction, the Lagrangian density that denotes the $U(3)_V \otimes U(3)_A$ invariant interaction is

$$\mathcal{L}_{vec} = -G_V\sum_{a=0}^{8}\left[(\bar{\psi}\gamma^\mu\lambda_a\psi)^2+(\bar{\psi}\gamma^\mu\gamma_5\lambda_a\psi)^2\right], \tag{3}$$

and a reduced Lagrangian density can be written as

$$\mathcal{L}_{vec} = -G_V(\bar{\psi}\gamma^\mu\psi)^2. \tag{4}$$

At zero temperature the PNJL model reduces to the normal NJL model.

In the SU(3) models, the above Lagrangian densities for the vector sector are not identical in a mean field approach and we discuss both cases next. We refer to the Lagrangian density given in Eq. (3) as model 1 (P1) and to the Lagrangian density given in Eq. (4) as model 2 (P2).

For our calculations, we need to evaluate the thermodynamical potential for the three flavor quark sector, $\Omega = -P = \mathcal{E} - T\mathcal{S} - \sum_f \mu_f\rho_f$ where P represents the pressure, $\mathcal{E}$ the energy density, T the temperature, $\mathcal{S}$ the entropy density, and μ_f the chemical potential of quark with flavor f. To determine the EOS for the SU(3) NJL at finite density and in the presence of a magnetic field in a mean field approximation we need to know the scalar condensates, ϕ_f, the quark number densities, ρ_f, as well as the pressure kinetic contribution from the gas of quasi-particles, θ_f. In the presence of a magnetic field all these quantities have been evaluated with great detail in [6, 7].

2. $T = 0$ - quark and stellar matter

We next restrict ourselves to the NJL model with vector interaction, disregarding the Polyakov loop, which has shown to be important at finite temperature only. The effect of the vector interaction on three flavor magnetized matter is studied for cold matter within the two different models mentioned above, a flavor dependent (P1) [8] and a flavor independent one (P2) [9].

If model P1 is considered, the pressure reads:

$$P = \theta_u + \theta_d + \theta_s - 2G_S(\phi_u^2+\phi_d^2+\phi_s^2) + 2G_V(\rho_u^2+\rho_d^2+\rho_s^2) + 4K\phi_u\phi_d\phi_s\ , \tag{5}$$

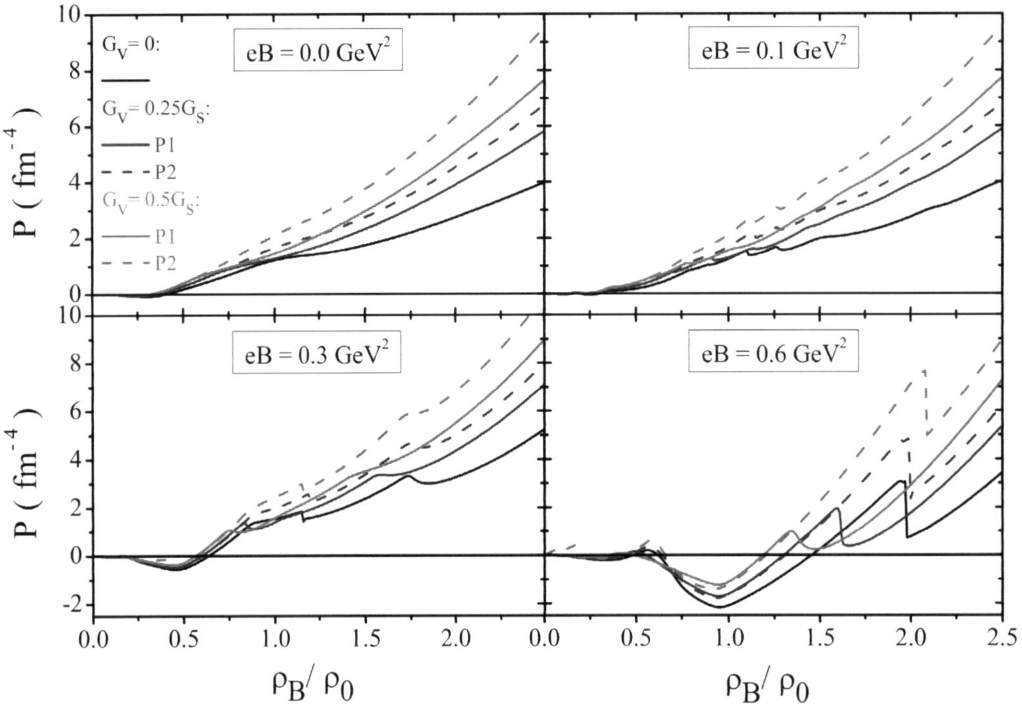

Figure 1. Pressure versus baryonic density for equal chemical potentials and models P1 and P2 for different values of G_V/G_S, and several intensities of the magnetic field. Figure taken from [10].

and the effective chemical potential, for each flavor, is given by

$$\tilde{\mu}_i = \mu_i - 4G_V\rho_i. \quad i = u, d, s \tag{6}$$

We also refer to P1 as the flavor dependent model, for the reasons that will become obvious from the analysis of our results.

If, on the other hand, model P2 is considered, the pressure becomes:

$$P = \theta_u + \theta_d + \theta_s - 2G_S(\phi_u^2 + \phi_d^2 + \phi_s^2) + G_V\rho^2 + 4K\phi_u\phi_d\phi_s \ , \tag{7}$$

where

$$\rho = \rho_u + \rho_d + \rho_s, \quad \rho_B = \rho/3, \tag{8}$$

and in this case the effective chemical potential, for each flavor, is given by

$$\tilde{\mu}_i = \mu_i - 2G_V\rho. \tag{9}$$

We next refer to P2 as the flavor independent (or flavor blind) model.

The effect of the magnetic field on the EOS can be seen in Figure 1 for the case when $\mu_u = \mu_d = \mu_s$. The van-Alphen oscillations due to the filling of the Landau levels are already seen for $eB = 0.1$ GeV^2. The softening occurring when a new Landau level starts being occupied

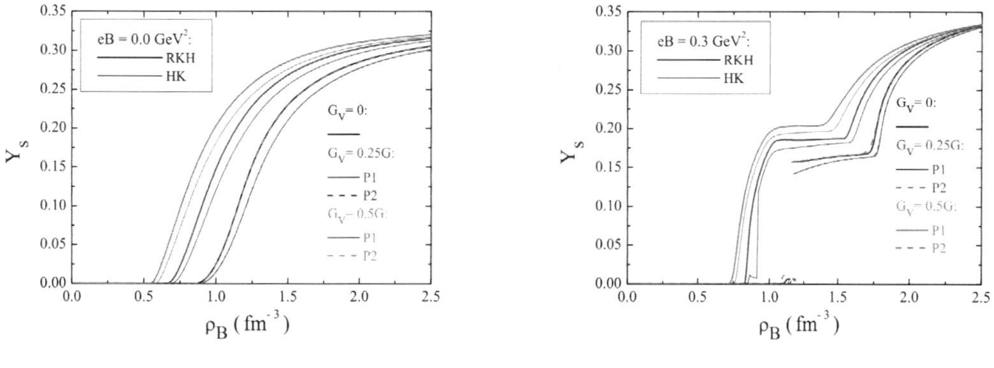

Figure 2. The strangeness fraction as a function of the baryonic density for models P1 and P2 and different values of G_V, and $B = 0$ (left panel); $eB = 0.3$ GeV^2 (right panel). Figure taken from [10].

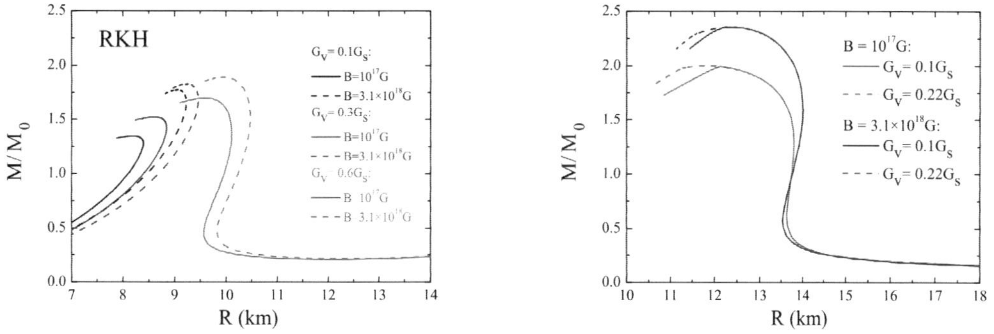

Figure 3. Mass radius curves obtained with P2 for different values of G_V, two intensities of the magnetic field, parametrization RKH for quark (left panel) and hybrid stars (right panel). Figure taken from [10].

has a strong effect at the smaller densities giving rise to a pressure that is negative within a larger range of densities.

As can be seen in Figure 2, the flavor independent vector interaction predicts a smaller strangeness content and, therefore, harder equations of state. Moreover, the strangeness content does not depend on the strength of the vector interaction. On the other hand, the flavor dependent vector interaction favors larger strangeness content the larger the vector coupling,

We now move to the study of quark and hybrid stars subject to an external magnetic field. In these cases, β-equilibrium conditions and charge neutrality are enforced. Mass-radius curves are shown in Figure 3 for both star types with the RKH parameter set. The hadronic phase in hybrid stars was obtained with the GM1 parametrization of the non-linear Walecka model [11]. A Maxwell construction was then used.

From Table 1 we can see that larger star masses are obtained for the flavor independent vector interaction and maximum masses of the order of 2 $M_\odot$ can be achieved depending on the value of the vector interaction and on the intensity of the magnetic field.

From the results displayed in Table 2, we observe that hybrid stars may bare a core containing

			HK			RKH		
			$x=0.1$	$x=0.3$	$x=0.6$	$x=0.1$	$x=0.3$	$x=0.6$
$B=0$ G	P1	M_{max} (M_0)	1.49	1.58	1.69	1.27	1.35	1.46
		R (km)	9.13	10.89	11.98	8.01	8.17	9.41
		ε_c (fm^{-4})	7.23	6.96	6.52	9.42	9.61	9.84
	P2	M_{max} (M_0)	1.56	1.72	1.91	1.35	1.54	1.74
		R (km)	9.15	10.61	11.47	8.22	8.60	9.91
		ε_c (fm^{-4})	7.35	7.37	6.92	8.71	8.58	8.09
$B=10^{17}$ G P2		M_{max} (M_0)	1.56	1.72	1.91	1.35	1.54	1.74
		R (km)	9.16	10.16	10.95	8.21	8.58	9.60
		ε_c (fm^{-4})	7.41	7.36	6.98	8.80	8.94	8.11
$B=3.1\times10^{18}$ G P2		M_{max} (M_0)	1.96	2.03	2.12	1.81	1.88	1.98
		R (km)	9.98	10.43	11.05	9.03	9.21	9.90
		ε_c (fm^{-4})	7.41	7.22	6.78	8.74	8.21	7.80

Table 1. Stellar macroscopic properties obtained from EOS of non-magnetized matter for models P1 and P2 and for magnetized matter with model P2. M_{max} is the maximum mass, R is the star radius, ε the star central energy density and $x = G_V/G_S$. Table taken from [10].

HK		M_{max} (M_0)	M_b (M_0)	R (km)	ε_c (fm^{-4})	ε (onset) (fm^{-4})	$\mu_B(\varepsilon_c)$ (MeV)	μ_B (onset) (MeV)
$B=10^{17}$ G P2	$x=0$	1.91	2.18	12.78	4.57	3.47	1360	1330
	$x=0.10$	1.99	2.30	12.14	6.27	5.05	-	1503
	$x=0.22$	2.00	2.31	11.82	5.93	7.79	1580	1726
$B=3.1\times10^{18}$ G P2	$x=0$	2.27	2.60	12.82	4.69	3.30	1324	1261
	$x=0.10$	2.35	2.70	12.34	5.29	5.59	1427	1453
	$x=0.22$	2.35	2.70	12.35	5.27	9.03	1426	1730
RKH		M_{max} (M_0)	M_b (M_0)	R (km)	ε_c (fm^{-4})	ε (onset) (fm^{-4})	$\mu_B(\varepsilon_c)$ (MeV)	μ_B (onset) (MeV)
$B=10^{17}$ G P2	$x=0$	1.97	2.26	12.48	4.29	4.28	-	1422
	$x=0.10$	2.00	2.31	11.91	7.51	5.67	-	1557
	$x=0.19$	2.00	2.31	11.83	5.91	7.83	1579	1728
$B=3.1\times10^{18}$ G P2	$x=0$	2.33	2.69	12.79	4.69	4.19	-	1335
	$x=0.10$	2.35	2.70	12.34	5.30	6.52	1428	1531
	$x=0.19$	2.35	2.70	12.34	5.30	9.05	1428	1731

Table 2. Stellar macroscopic properties obtained from EOS of magnetized hybrid stars built with GM1 and SU(3) NJL with HK and RKH parametrizations. M_{max} is the maximum gravitational mass, M_b is the maximum baryonic mass, R is the star radius, ε_c is the star central energy density, $\mu_B(\varepsilon_c)$ is the chemical potential for neutron at ε_c and μ_B(onset) is the baryonic chemical potential at the onset of the quark phase. Results taken from [10].

deconfined quarks if neither the vector interaction nor the magnetic field are too strong. The onset of the quark phase as compared with the central values for the energy density and related chemical potential show that quarks would appear only at densities/chemical potentials larger than the existing ones in the stellar core. The presence of strong magnetic fields also disfavor the existence of a quark core in hybrid stars.

Journal of Physics: Conference Series **630** (2015) 012026 doi:10.1088/1742-6596/630/1/012026

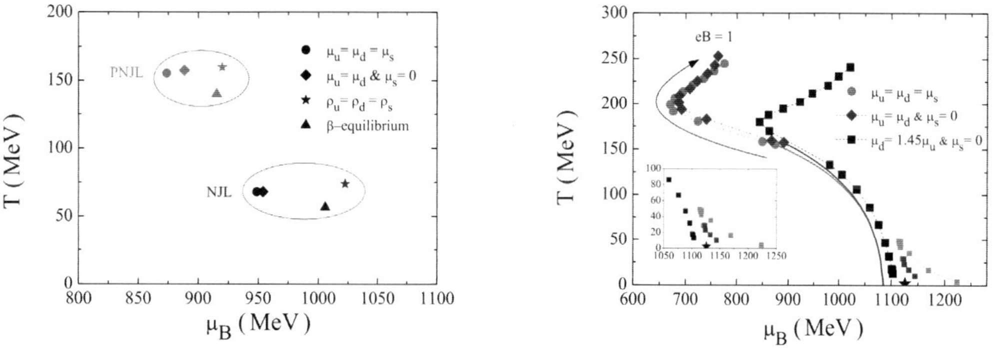

Figure 4. Left panel - Location of the CEP on a diagram T vs the baryonic chemical potential under different scenarios and models (NJL, PNJL). No external magnetic field is considered. Right panel - Effect of an external magnetic field on the CEP location within PNJL model for three different scenarios. Figure taken from [13].

3. Critical end point within the PNJL model

We move next to finite temperature and study the location of the critical end point (CEP) on the QCD phase diagram within the framework of the PNJL model and without the introduction of the vector interaction. We investigate different scenarios with respect to the isospin and strangeness content of matter, as shown in Figure 4 left for non-magnetized matter. For matter in β-equilibrium, the CEP occurs at smaller temperatures and densities. This scenario is of interest for neutron stars and confirms previous calculations that indicate that a deconfinement phase transition in the laboratory will be more easily attained with asymmetric nuclear matter [12].

As a matter of fact, by taking $\mu_s = 0$ and increasing systematically μ_d with respect to μ_u the CEP moves to smaller temperatures and larger baryonic chemical potentials [13]: less symmetric matter is less bound and, therefore, the transition to a chirally symmetric phase occurs at a smaller temperature and density than in the symmetric case ($\mu_u = \mu_d = \mu_s$). When the asymmetry is large enough, $\mu_d = 1.45\ \mu_u$, the CEP disappears.

Another interesting situation is observed when analyzing very isospin asymmetric matter subject to different intensities of the magnetic field, as seen in Figure 4 right. Starting from a scenario having an isospin asymmetry above which the CEP does not exist for a zero external magnetic field ($\mu_d = 1.45\ \mu_u$) it was shown the introduction of an external magnetic field could drive the system to a first order phase transition. The CEP occurs at very small temperatures if $eB < 0.1$ GeV^2 and, in this case, a complicated structure with several CEP at different values of (T, μ_B) are possible for the same magnetic field, because the temperature is not high enough to wash out the Landau level effects. Indeed, the magnetic field affects in a different way u and d quarks due to their different electric charge. A consequence is the possible appearance of two or more CEPs for a given magnetic field intensity. For $eB > 0.1$ GeV^2 only one CEP exists. This is an important result because it shows that an external magnetic field is able to drive a system with no CEP into a first order phase transition [13].

In the right panel of Figure 4 we presente the results for two other scenarios where a first order phase transition already occurs at zero magnetic field: $\mu_u = \mu_d = \mu_s$ and $\mu_u = \mu_d\,\&\,\mu_s = 0$. For moderate magnetic fields the trend is very similar for all scenarios: as the intensity of the magnetic field increases, the transition temperature increases and the baryonic chemical potential decreases until the critical value $eB \sim 0.4$ GeV^2. For stronger magnetic fields both T and μ_B increase.

It is also important to point out that all three scenarios presented in the right panel of Figure 4 show an inverse magnetic catalysis at finite chemical potential and zero temperature, i.e., the critical temperature decreases with increasing eB. For large values of eB this inverse magnetic catalysis tendency disappears and a magnetic catalysis takes place.

Acknowledgements: This work was partially supported by CNPq (Brazil), CAPES (Brazil) and FAPESC (Brazil) under Project 2716/2012,TR 2012000344, by Project No. PTDC/FIS/113292/2009 developed under the initiative QREN financed by the UE/FEDER through the program COMPETE "Programa Operacional Factores de Competitividade," by Grant No. SFRH/BD/51717/2011 (F.C.T.) and by NEW COMPSTAR, a COST initiative.

[1] V. Skokov, A. Y. Illarionov and V. Toneev, Int. J. Mod. Phys. A **24**, 5925 (2009).
[2] R. Duncan and C. Thompson, Astrophys. J. **392** L9 (1992); C. Kouveliotou et al., Nature (London) **393** 235 (1998).
[3] Y. Nambu and G. Jona-Lasinio, Phys. Rev. **122**, 345 (1961); **124**, 246 (1961).
[4] T. Hatsuda and T. Kunihiro, Phys. Rep. **247**, 221 (1994).
[5] P. Rehberg, S. P. Klevansky and J. Hüfner, Phys. Rev. C **53**, 410 (1996).
[6] D.P. Menezes, M. Benghi Pinto, S.S. Avancini, A. Pérez Martinez and C. Providência, Phys. Rev. C **79**, 035807 (2009).
[7] S.S. Avancini, D.P. Menezes, M.B. Pinto and C. Providência, Phys. Rev. C **80**, 065805 (2009).
[8] S. Klimt, M. Lutz, and W. Weise, Phys. Lett. **B249**, 386 (1990); M. Hanauske, L. M. Satarov, I. N. Mishustin, H. Stöcker and W. Greiner, Phys. Rev. D **64**, 043005 (2001).
[9] K. Fukushima, Phys. Rev. D **78**, 114019 (2008).
[10] Debora P. Menezes, Marcus B. Pinto, Luis R. B de Castro, Constança Providencia and Pedro Costa, Phys. Rev. C **89**, 055207 (2014).
[11] N. K. Glendenning, Compact Stars, Springer, New York (2000).
[12] R. Cavagnoli, C. Providência and D. P. Menezes, Phys. Rev. C **83**, 045201 (2011).
[13] Pedro Costa, Márcio Ferreira, Hubert Hansen, Débora P. Menezes, Constança Providência, Phys. Rev. D **89**, 056013 (2014).

Strongly coupled quark gluon plasma in a magnetic field

D. A. Fogaça, F. S. Navarra and S. M. Sanches Jr.

Instituto de Física, Universidade de São Paulo Rua do Matão Travessa R, 187, 05508-090 São Paulo, SP, Brazil

E-mail: navarra@if.usp.br

Abstract.
We derive the equation of state of a strongly coupled quark gluon plasma at finite temperature and high baryon density in a strong magnetic field. The formalism presented here is an extension of our previous works. We derive analytical expressions for the pression and energy density.

1. Introduction

After the analysis of RHIC and LHC data on heavy ion collisions, it became a consensus that the quark gluon plasma at high (but not asymptotic) temperatures is a strongly interacting system (which is now called strongly interacting quark gluon plasma or sQGP). This finding motivated a reanalysis of the previousy used equations of state (EOS), most of them based on the assumption of weak coupling between quarks and gluons. It is also reasonable to assume that the yet untested cold quark gluon plasma, which presumably exists in the core of dense stars, is also a strongly interacting system. In [1] we developed a mean field approach to the cold sQGP, which is inspired in the old relativistic mean field Walecka model. As in that model, because of the strong coupling and the large number of fermion sources, the vector boson is assumed to behave as a classical field. In contrast to the Walecka model, because of asymptotic freedom the low and high momentum modes of the gluon field have a different behavior. This justifies their separation into low ("soft") and high ("hard") momentum components already at the level of the Lagrangian. For the "hard" gluons we use the mean field approximation. The "soft" gluons are replaced by their vacuum expectation values, the in-medium gluon condensates. Using this approach we obtained the pressure and energy density of the sQGP. They both have a term similar to the bag constant usually found in the MIT bag model. Due to the strongly repulsive interaction of the vector field our EOS is much harder than the MIT one. The details of the derivation can be found in [1], where we discuss these approximations. In a subsequent paper, Ref.[2], we have applied our EOS [1] to the calculation of the structure of compact quark stars. We found that the inclusion of gluon interaction creates pressure and energy density large enough to generate stars wiht masses and radii consistent with the most recent astrophysical observations.

It is well accepted [3, 4, 5, 6, 7] that in astrophysical compact objects, such as magnetars, there is a strong magnetic field of intensity of $10^{12}\,G$ to $10^{18}\,G$. It is then natural to extend our formalism and include the effects of the magnetic field. This will be done in the next sections.

2. Equation of state

2.1. Effective Lagrangian

In what follows, we employ natural units: $\hbar = c = k_B = 1$ and the metric is given by $g_{\mu\nu} = \text{diag}(+,-,-,-)$. We first consider the homogeneous magnetic field in the Cartesian z direction:

$$\vec{B} = B\hat{z} \qquad \text{and hence} \qquad A_\mu = (0, yB, 0, 0) \tag{1}$$

Including this magnetic field in the QCD Lagrangian we obtain:

$$\mathcal{L}_{QCD} = +\sum_f \bar{\psi}_i^f \Big[i\gamma^\mu(\delta_{ij}\partial_\mu + i\delta_{ij}Q_f A_\mu - igT^a_{ij}G^a_\mu) - \delta_{ij}m_f \Big] \psi_j^f$$

$$-\frac{1}{4}F^a_{\mu\nu}F^{a\mu\nu} - \frac{1}{4}F_{\mu\nu}F^{\mu\nu} \tag{2}$$

The summation in f runs over all quark flavors, m_f is the mass of the quark of flavor f, Q_f is the charge of the quark of flavor f, i and j are the color indices of the quarks, T^a are the SU(3) generators and f^{abc} are the SU(3) antisymmetric structure constants. The object $F^{a\mu\nu} = \partial^\mu G^{a\nu} - \partial^\nu G^{a\mu} + gf^{abc}G^{b\mu}G^{c\nu}$ is the gluon field tensor and the electromagnetic Lagrangian term is $F^{\mu\nu} = \partial^\mu A^\nu - \partial^\nu A^\mu$, with A^μ given by (1).

Performing the gluon field decomposition discussed above[1]:

$$G^{a\mu} = A^{a\mu} + \alpha^{a\mu}$$

and repeating the steps described in [1] we rewrite (2) as the effective Lagrangian:

$$\mathcal{L}_0 = \frac{m_G{}^2}{2}\alpha^a_0\alpha^a_0 - \frac{B^2}{2} - \mathcal{B}_{QCD}$$

$$+\sum_f \bar{\psi}_i^f \Big\{ i\gamma^\mu \Big[\delta_{ij}\partial_\mu + i\delta_{ij}Q_f A_\mu\Big] + g_h\gamma^0 T^a_{ij}\alpha^a_0 - \delta_{ij}m_f \Big\} \psi_j^f \tag{3}$$

where $\mathcal{B}_{QCD}$ is our equivalent of the bag constant given by $\mathcal{B}_{QCD} \equiv 9\,\phi_0{}^4/136$ and m_G is the dynamical gluon mass given by $m_G{}^2 \equiv 9\,\mu_0{}^2/32$, where μ_0 is an energy scale associated with $\langle A^2\rangle$, which is the gluon condensate of dimension two [1]:

$$\langle A^2\rangle \equiv \langle g_s^2 A^{a\mu}A^{b\nu}\rangle = \langle g_s^2 A^2\rangle = -\frac{\delta^{ab}g^{\mu\nu}}{32}\mu_0{}^2 \tag{4}$$

and since $\langle g_s^2 A^2\rangle < 0$ we always have $m_G{}^2 > 0$. The constant ϕ_0 is associated with $\langle F^2\rangle$, which is the gluon condensate of dimension four [1]:

$$\mathcal{B}_{QCD} = b\phi_0^4 = \langle\frac{1}{4}F^{a\mu\nu}F^a_{\mu\nu}\rangle = \frac{\pi^2}{g_s^2}\langle F^2\rangle \tag{5}$$

In the above expression we have two QCD coupling constants given by g_s and g_h. The coupling g_s is associated to the soft gluons, while g_h is associated to the hard gluons as in [1].

2.2. Equations of motion and Landau levels

The equations of motion calculated from (3) are:

$$\Big[i\gamma^{\mu}\Big(\partial_{\mu}+iQ_fA_{\mu}\Big)+g_h\gamma^0T^a\alpha_0^a-m_f\Big]\psi^f=0 \tag{6}$$

$$m_G{}^2\alpha_0^a=-g_h\sum_f\rho_f^a=-g_h\,\rho^a \tag{7}$$

$$\partial_{\mu}F^{\mu\nu}=\sum_f Q_f(\bar{\psi}^f\gamma^{\nu}\psi^f) \tag{8}$$

The color vector current $j^{a\nu}$ is given by:

$$j^{a\nu}=\sum_f\bar{\psi}_i^f\gamma^{\nu}T_{ij}^a\psi_j^f \tag{9}$$

and its temporal component ρ^a which appears in (7) is:

$$j^{a0}=\rho^a=\sum_f\bar{\psi}_i^f\gamma^0T_{ij}^a\psi_j^f=\sum_f\psi_i^{\dagger f}T_{ij}^a\psi_j^f \tag{10}$$

In order to solve (6), we perform the steps described in [8] and write ψ_j^f as follows:

$$\psi_j^f=c_je^{-iEt}\begin{pmatrix}\Phi_f\\ \chi_f\end{pmatrix} \tag{11}$$

where c_j is the quark color vector used in some textbooks [9]:

$$c_1=\begin{pmatrix}1\\0\\0\end{pmatrix}\ \text{for red,}\ c_2=\begin{pmatrix}0\\1\\0\end{pmatrix}\ \text{for blue,}\ c_3=\begin{pmatrix}0\\0\\1\end{pmatrix}\ \text{for green} \tag{12}$$

From the above definitions it follows that:

$$c_i^{\dagger}\delta_{ij}c_j=c_1^{\dagger}c_1+c_2^{\dagger}c_2+c_3^{\dagger}c_3=3 \tag{13}$$

For future purposes we will replace the above sum by the following average:

$$c_i^{\dagger}\delta_{ij}c_j\rightarrow\frac{c_i^{\dagger}\delta_{ij}c_j}{\text{(number of quark colors)}}=\frac{c_1^{\dagger}c_1+c_2^{\dagger}c_2+c_3^{\dagger}c_3}{3}=1 \tag{14}$$

From the exact solution [8] of the Dirac equation (6) with magnetic field and hard gluon terms, we have the following expression for the eigenvalues:

$$\Big(E_{\nu}^f+g_h\mathcal{A}\Big)^2=m_f^2+k_z^2+(2\nu+1)|Q_f|B-Q_fBs \tag{15}$$

where $\nu=0,1,2,3,4,5\ldots$ and $s=+1$ or $s=-1$, for the projection *up* or *down* of the spin states, respectively. The momentum component along the magnetic field direction is given by k_z. As in our previous work [1] the constant $\mathcal{A}$ in (15) is the "algebra valued" result, $\mathcal{A}=c_i^{\dagger}T_{ij}^ac_j\alpha_0^a$ with the implicit summation over $i,j=1,2,3$ and $a=1,\ldots,8$.

With the help of (12) we are able to calculate the relation between ρ^a previously identified in (10) and the net quark density ρ. We perform the product $\rho^a\rho^a$ considering the average over the number of $SU(3)$ generators, which is 8, as follows:

$$\rho^a\rho^a = \sum_f \rho_f^a \sum_{f'} \rho_{f'}^a \longrightarrow \langle \sum_f \rho_f^a \sum_{f'} \rho_{f'}^a \rangle = \frac{1}{8} \sum_f \rho_f^a \sum_{f'} \rho_{f'}^a$$

$$= \frac{1}{8} \sum_f (\psi_i^{\dagger f} T_{ij}^a \psi_j^f) \sum_{f'} (\psi_k^{\dagger f'} T_{kl}^a \psi_l^{f'}) = \frac{1}{8} \sum_f (c_i^\dagger T_{ij}^a c_j) \psi^{\dagger f} \psi^f \sum_{f'} (c_k^\dagger T_{kl}^a c_l) \psi^{\dagger f'} \psi^{f'}$$

The result $(c_i^\dagger T_{ij}^a c_j)(c_k^\dagger T_{kl}^a c_l) = 3$ is obtained from the Gell-Mann matrices and from the color vectors (12):

$$\rho^a\rho^a = \sum_f \rho_f^a \sum_{f'} \rho_{f'}^a = \frac{3}{8} \sum_f (\psi^{\dagger f} \psi^f) \sum_{f'} (\psi^{\dagger f'} \psi^{f'})$$

As $\sum_f (\psi^{\dagger f} \psi^f) = \sum_f \rho_f = \rho$, where $f = u, d, s$ and ρ is the total net quark density, we have:

$$\rho^a\rho^a = \frac{3}{8}\rho^2 \tag{16}$$

The baryon density ρ_B is related to net quark density through:

$$\rho_B = \frac{1}{3}\rho \tag{17}$$

Returning now to (15) and rescaling the single particle energy as:

$$\tilde{E}_\nu^f \equiv E_\nu^f + g_h \mathcal{A} \tag{18}$$

we are able to rewrite (15) as:

$$\left(\tilde{E}_\nu^f\right)^2 = m_f^2 + k_z^2 + \left[2\nu + 1 - s \times sgn(Q_f)\right]|Q_f|B \tag{19}$$

where $Q_f = sgn(Q_f) \times |Q_f|$. Defining $2\nu + 1 - s \times sgn(Q_f) \equiv 2n$, (19) becomes:

$$\tilde{E}_n^{f(\pm)} = \pm\sqrt{m_f^2 + k_z^2 + 2n|Q_f|B} \tag{20}$$

and n denotes the nth Landau level. We note that, except for the rescaling (18), the equation above is the one usually found in the literature.

2.3. Thermodynamical quantities

We next follow the calculations presented in [6, 10, 11] and starting from (3) we arrive at the following thermodynamical potential:

$$\Omega = \left[-\frac{m_G{}^2}{2}\alpha_0^a\alpha_0^a + \mathcal{B}_{QCD} + \frac{B^2}{2}\right]V$$

$$+T\sum_f \sum_{\vec{k},s,n} \left\{ ln\left(1 - d_f\right) + ln\left(1 - \bar{d}_f\right)\right\} \tag{21}$$

where V is the volume, T is the temperature and the fermion distribution functions are:

$$d_f \equiv \frac{1}{1+e^{(\mathcal{E}_n^f-\nu_f)/T}} \qquad \text{and} \qquad \bar{d}_f \equiv \frac{1}{1+e^{(\mathcal{E}_n^f+\nu_f)/T}} \tag{22}$$

Using (20) in the evaluation of (21) the energy is now defined as:

$$\mathcal{E}_n^f = \sqrt{m_f^2 + k_z^2 + 2n|Q_f|B} \tag{23}$$

and the effective chemical potential for the quark f is defined as:

$$\nu_f \equiv \mu_f + g_h(c_i^\dagger T_{ij}^a c_j)\alpha_0^a \tag{24}$$

For a magnetic field pointing along the z direction, the momentum of a charged particle is restricted to discrete Landau levels [6, 12] and hence:

$$\frac{S}{(2\pi)^2}\int\int dk_x\, dk_y = \frac{S|Q_i|B}{2\pi}$$

with S being the area in the $x-y$ plane. From this last expression we have:

$$\int\int dk_x\, dk_y = 2\pi|Q_i|B \tag{25}$$

and the statistical sum becomes:

$$\frac{1}{V}\sum_{\vec{k},s,n} \longrightarrow \frac{\gamma_f}{(2\pi)^3}\sum_n \int d^3k = \frac{\gamma_f|Q_i|B}{2\pi^2}\sum_n \int_0^\infty dk_z \tag{26}$$

where γ_f is the statistical degeneracy factor of the quark f. The parallel pressure ($p_\parallel$) to the magnetic field, the magnetization (M) and perpendicular pressure ($p_\perp$) are given respectively by [6, 12]:

$$p_\parallel = -\frac{\Omega}{V} \quad , \quad M = -\frac{1}{V}\frac{\partial\Omega}{\partial B} = \frac{\partial p_\parallel}{\partial B} \quad \text{and} \quad p_\perp = p_\parallel - MB \tag{27}$$

The quark density ρ and the entropy density s read [10, 11]:

$$\rho = -\frac{1}{V}\frac{\partial\Omega}{\partial\mu_f} \qquad \text{and} \qquad s = -\frac{1}{V}\frac{\partial\Omega}{\partial T} \tag{28}$$

The energy density ε is calculated from the Gibbs relation [10, 11]:

$$\varepsilon = -p_\parallel + Ts + \sum_f \mu_f\rho_f \tag{29}$$

The evaluation of (27) to (29) with the potential (21) gives the following results:

$$p_\parallel = \frac{3{g_h}^2}{16{m_G}^2}\rho^2 - \mathcal{B}_{QCD} - \frac{B^2}{2} + \sum_f \frac{\gamma_f|Q_f|B}{2\pi^2}\sum_n \int_0^\infty dk_z \frac{{k_z}^2}{\mathcal{E}_n^f}\left(d_f + \bar{d}_f\right) \tag{30}$$

$$M = -B - T\sum_f \frac{\gamma_f|Q_f|}{2\pi^2}\sum_n \int_0^\infty dk_z \left[ln(1-d_f) + ln(1-\bar{d}_f)\right]$$

$$-\sum_f \frac{\gamma_f |Q_f| B}{2\pi^2} \sum_n \int_0^\infty dk_z \left[\frac{d_f\, n|Q_f|}{\mathcal{E}_n^f} + \frac{\bar{d}_f\, n|Q_f|}{\mathcal{E}_n^f} \right] \tag{31}$$

$$p_\perp = \frac{3{g_h}^2}{16{m_G}^2}\rho^2 - \mathcal{B}_{QCD} + \frac{B^2}{2} + \sum_f \frac{\gamma_f |Q_f| B^2}{2\pi^2} \sum_n \int_0^\infty dk_z \left[\frac{d_f\, n|Q_f|}{\mathcal{E}_n^f} + \frac{\bar{d}_f\, n|Q_f|}{\mathcal{E}_n^f} \right] \tag{32}$$

$$\rho = \sum_f \frac{\gamma_f |Q_f| B}{2\pi^2} \sum_n \int_0^\infty dk_z \left(d_f - \bar{d}_f \right) \tag{33}$$

$$s = -\sum_f \frac{\gamma_f |Q_f| B}{2\pi^2} \sum_n \int_0^\infty dk_z \left\{ d_f ln(d_f) + (1-d_f)\, ln(1-d_f) + \bar{d}_f\, ln(\bar{d}_f) + (1-\bar{d}_f)\, ln(1-\bar{d}_f) \right\} \tag{34}$$

$$\varepsilon = \frac{3{g_h}^2}{16{m_G}^2}\rho^2 + \mathcal{B}_{QCD} + \frac{B^2}{2} + \sum_f \frac{\gamma_f |Q_f| B}{2\pi^2} \sum_n \int dk_z\, \mathcal{E}_n^f (d_f + \bar{d}_f) \tag{35}$$

2.4. Zero temperature

In the zero temperature limit [6, 10, 12], applied to astrophysics, we have the distributions (22) given by:

$$d_f = \Theta(\nu_f - \mathcal{E}_n^f) \qquad \text{and} \qquad \bar{d}_f = 0 \tag{36}$$

and also [10]:

$$\lim_{T\to 0} T\, ln\left(1 - d_f\right) = (\mathcal{E}_n^f - \nu_f) \qquad \text{and} \qquad \lim_{T\to 0} T\, ln\left(1 - \bar{d}_f\right) = 0 \tag{37}$$

The quark density at zero temperature is obtained by inserting (36) into (33) [6]:

$$\rho = \sum_f \frac{\gamma_f |Q_f| B}{2\pi^2} \sum_{n=0}^{n_{max}^f} k_{z,F}^f(n) \tag{38}$$

From $\Theta(\nu_f - \mathcal{E}_n^f)$ and using (23) we have the Fermi momentum for the quark f:

$$k_{z,F}^f(n) = \sqrt{{\nu_f}^2 - m_f^2 - 2n|Q_f|B} \tag{39}$$

The summation over the Landau levels is calculated on the condition that the expression under the square root in (39) is positive, i.e., ${\nu_f}^2 \geq m_f^2 + 2n|Q_f|B$ [6]. Thus

$$n \leq n_{max}^f = int\left[({\nu_f}^2 - m_f^2)/(2|Q_f|B) \right] \tag{40}$$

where $int[a]$ denotes the integer part of a. Inserting the results (36) to (39) into (30), (31), (32) and (35) we find the energy density and the pressure components:

$$\varepsilon = \frac{3{g_h}^2}{16{m_G}^2}\rho^2 + \mathcal{B}_{QCD} + \frac{B^2}{2}$$

$$+\sum_f \frac{\gamma_f |Q_f| B}{2\pi^2} \sum_{n=0}^{n_{max}^f} \int_0^{k_{z,F}^f(n)} dk_z \sqrt{m_f^2 + k_z^2 + 2n|Q_f|B} \tag{41}$$

$$p_{\|} = \frac{3{g_h}^2}{16{m_G}^2}\rho^2 - \mathcal{B}_{QCD} - \frac{B^2}{2}$$

$$+\sum_f \frac{\gamma_f |Q_f| B}{2\pi^2} \sum_{n=0}^{n^f_{max}} \int_0^{k^f_{z,F}(n)} dk_z \frac{{k_z}^2}{\sqrt{m_f^2 + k_z^2 + 2n|Q_f|B}} \tag{42}$$

and

$$p_{\perp} = \frac{3{g_h}^2}{16{m_G}^2}\rho^2 - \mathcal{B}_{QCD} + \frac{B^2}{2}$$

$$+\sum_f \frac{\gamma_f |Q_f|^2 B^2}{2\pi^2} \sum_{n=0}^{n^f_{max}} n \int_0^{k^f_{z,F}(n)} \frac{dk_z}{\sqrt{m_f^2 + k_z^2 + 2n|Q_f|B}} \tag{43}$$

As in [6], the pressure components $p_{\|}$ and $p_{\perp}$ are different for any value of the magnetic field B. In the limit $B >> ({\nu_f}^2 - m_f^2)/2|Q_f|$ expression (40) may be rewritten as:

$$\lim_{B\to\infty} n \le \lim_{B\to\infty} n^f_{max} = \lim_{B\to\infty} int\left[\frac{\left({\nu_f}^2 - m_f^2\right)}{2|Q_f|B}\right] = 0 \tag{44}$$

which implies that only the lowest Landau level ($n = 0$) contributes. In this limit (39) becomes:

$$k^{f\,(B\to\infty)}_{z,F} = \sqrt{{\nu_f}^2 - m_f^2} \tag{45}$$

In the large magnetic field limit (41) to (43) are finally written as:

$$\varepsilon_{\;(B\to\infty)} = \frac{3{g_h}^2}{16{m_G}^2}\rho^2 + \mathcal{B}_{QCD} + \frac{B^2}{2} + \sum_f \frac{\gamma_f|Q_f|B}{2\pi^2} \int_0^{k^{f\,(B\to\infty)}_{z,F}} dk_z \sqrt{m_f^2 + k_z^2} \tag{46}$$

$$p_{\|\;(B\to\infty)} = \frac{3{g_h}^2}{16{m_G}^2}\rho^2 - \mathcal{B}_{QCD} - \frac{B^2}{2} + \sum_f \frac{\gamma_f|Q_f|B}{2\pi^2} \int_0^{k^{f\,(B\to\infty)}_{z,F}} dk_z \frac{{k_z}^2}{\sqrt{m_f^2 + k_z^2}} \tag{47}$$

$$p_{\perp\;(B\to\infty)} = \frac{3{g_h}^2}{16{m_G}^2}\rho^2 - \mathcal{B}_{QCD} + \frac{B^2}{2} \tag{48}$$

and the quark density (38) reads:

$$\rho_{\;(B\to\infty)} = \sum_f \frac{\gamma_f|Q_f|B}{2\pi^2} \sqrt{{\nu_f}^2 - m_f^2} \tag{49}$$

3. Conclusions

We have derived the equation of state of the strongly interacting QGP at finite temperature under uniform magnetic field. In all expressions the term proportional to ${g_h}^2/{m_G}^2$ is new with respect to other works in the field and comes from the EOS [1]. Special attention was given to the zero temperature case, which will be applied to stellar structure calculations.

4. Acknowledgments

This work was partially supported by the Brazilian funding agencies CAPES, CNPq and FAPESP.

5. References

[1] D. A. Fogaça and F. S. Navarra 2011 *Phys. Lett. B* **700** 236
[2] B. Franzon, D. A. Fogaça, F. S. Navarra and J. E. Horvath 2012 *Phys. Rev. D* **86** 065031
[3] D. M. Paret, J. E. Horvath and A. P. Martinez, arXiv:1407.2280 [astro-ph.HE]
[4] D. P. Menezes, M. Benghi Pinto, S. S. Avancini and C. Providência 2009 *Phys. Rev. C* **80** 065805
[5] S. S. Avancini, D. P. Menezes, Marcus B. Pinto and C. Providência 2012 *Phys. Rev. D* **85** 091901
[6] M. Strickland, V. Dexheimer and D. P. Menezes 2012 *Phys. Rev. D* **86** 125032
[7] L. Paulucci, E. J. Ferrer, V. de la Incera and J. E. Horvath 2011 *Phys. Rev. D* **83** 043009
[8] K. Bhattacharya and P. B. Pal 2004 *Pramana* **62** 1041 (arXiv:0209.053v2 [hep-ph]); K. Bhattacharya, arXiv:0705.4275v2 [hep-th].
[9] D. Griffiths 1987 *Introduction to Elementary Particles* (John Wiley & Sons Inc.) p 280.
[10] R. J. Furnstahl and Brian D. Serot 1990 *Phys. Rev. C* **41** 262
[11] D. A. Fogaça, L. G. Ferreira Filho and F. S. Navarra 2009 *Nucl. Phys. A* **819** 150
[12] S. Chakrabarty 1996 *Phys. Rev. D* **54** 1306; D. Manreza Paret and A. Perez Martinez, arXiv:1010.0909 [astro-ph.HE]; J. X. Hou, G. X. Peng, C. J. Xia and J. F. Xu, arXiv:1403.1143 [nucl-th]

The time evolution of the quark gluon plasma in the early Universe

S. M. Sanches Jr., D. A. Fogaça and F. S. Navarra

Instituto de Física, Universidade de São Paulo, Rua do Matão Travessa R, 187, 055508-090, São Paulo, SP, Brazil

E-mail: samuelsanches@usp.br

Abstract.
Our knowledge of the equation of state of the quark gluon plasma has been continuously growing due to the experimental results from heavy ion collisions and also due to the advances in lattice QCD calculations. The new findings about this state may have consequences on the time evolution of the early Universe, which can estimated by solving the Friedmann equations. The solutions of these equations give the time evolution of the energy density and also of the temperature in the beginning of the Universe. In this work we compute the time evolution of the QGP in the early Universe, comparing several equations of state, some of them based on the MIT bag model (and on its variants) and some of them based on lattice QCD calculations. Among other things, we investigate the effects of a finite baryon chemical potential in the evolution of the early Universe.

1. Introduction

In the last ten years relativistic heavy-ion collision experiments have provided us with information about the properties of matter in the early Universe (at the time when its age was about 10 microseconds and its temperature was higher than 150 MeV). It is believed that, during this period, the Universe was formed by a hot phase of deconfined quarks and gluons, i.e., a quark gluon plasma (QGP). In parallel with these experimental developments there has been a significant progress on the theoretical side, coming from the numerical simulation of finite temperature QCD on a lattice. The new findings about the nature of the QGP motivate us to investigate their consequences in the primordial Universe. This can be done by solving the Friedmann equations, which allow us to determine the precise time evolution of the thermodynamic quantities in the early Universe.

Previous works along this line and with the same motivation already exist in the literature. For a review see, e.g., [1] and for recent papers on the subject see [2, 3] and references therein. Most of these works focused on the nature of the phase transiton from the QGP to the hadron gas. There are exotic phenomena associated with the order of the phase transition. In [3] a realistic EOS was used in cosmological calculations. In this EOS the transition was actually a crossover and not a first order transition as commonly believed until recent years. The results showed a very smooth time dependence of various thermodynamic quantities and suggested indirectly that there are small chances for the observation of various exotic phenomena such as quark nuggets, strangelets, cold dark matter clumps, etc. Such phenomena are associated typically with first order phase transitions. Apart from these exotic phenomena, changing the equation of state we change the space-time evolution of the early Universe and this (specially when there is a phase transition) will change the emission of gravitational waves, as pointed out in [4] and also in [5] and [6].

In the early universe the baryon chemical potential was small (and usually neglected in cosmological calculations) but we do not know exactly how small. Moreover there may have been fluctuations in the chemical potential in the early Universe associated with the anisotropy of positively and negatively charged particles in the QGP phase, as pointed out in [7]. It is therefore interesting to estimate the effects of a non-vanishing chemical potential on the solution of the Friedmann equations.

In this work we solve numerically the Friedmann equations using some recently proposed equations of state and investigate how the results depend on details of these equations of state. We compute the time evolution of some thermodynamical quantities of the Universe, such as the energy density, pressure, temperature and sound speed. We also study the time evolution of these quantities at finite chemical potential.

2. The QGP equation of state

In the standard model of cosmology the space-time is be parametrized by the Friedmann-Lemaître-Robertson-Walker (FLRW) metric which, used in the Einstein equations, leads to the Friedmann equations, which, after some manipulation [2, 3, 8] yield the following time evolution equation:

$$-\frac{d\varepsilon}{3\sqrt{\varepsilon}\,(\varepsilon+p)}=\sqrt{\frac{8\pi G}{3}}dt \qquad (1)$$

which allows us to find the temporal evolution of the energy density ε for any given equation of state $p \equiv p(\varepsilon)$. In this section we present some equations of state which have been either phenomenologically extracted from the study of heavy ion collisions (and also of compact stars) or derived from first principles. They can be grouped into two cathegories, the first one being the "MIT family" and the second one being the "Lattice QCD family". We start presenting the MIT bag model [9] EOS and its modern variants. In what follows we use natural units $\hbar = c = k_B = 1$.

2.1. *The MIT bag model*

The energy density and pressure are given by [9, 10]:

$$\varepsilon=\frac{37\pi^2}{30}T^4+\mathcal{B} \qquad \text{and} \qquad p=\frac{37\pi^2}{90}T^4-\mathcal{B} \qquad (2)$$

where $\mathcal{B}$ is the bag constant and T is the temperature. Considering these thermodynamical quantities as functions of time, we have from (2) the following relation:

$$p[\varepsilon(t)]=\frac{1}{3}\left[\varepsilon(t)-4\mathcal{B}\right] \qquad (3)$$

A particular case, with only free gluons is described by [9, 10]:

$$\varepsilon_g=\frac{8\pi^2}{15}T^4+\mathcal{B} \qquad \text{and} \qquad p_g=\frac{8\pi^2}{45}T^4-\mathcal{B} \qquad (4)$$

and also satisfies (3).

2.2. Model 1

This model (and also the next one) was presented in [11]. The pressure and energy density as function of temperature are given by:

$$p_1 = \frac{\sigma_1}{3}T^4 - AT - \mathcal{B}_1 \qquad \text{and} \qquad \varepsilon_1 = \sigma_1 T^4 + \mathcal{B}_1 \tag{5}$$

from which we can write the following relation:

$$p_1[\varepsilon_1(t)] = \frac{1}{3}\left[\varepsilon_1(t) - 4\mathcal{B}_1\right] - A\left[\frac{\varepsilon_1(t) - \mathcal{B}_1}{\sigma_1}\right]^{1/4} \tag{6}$$

where the parameters were fixed in [11]: $\sigma_1 = 4.73$, $A = 3.94\,{T_c}^3$ and $\mathcal{B}_1 = -2.37\,{T_c}^4$. Here T_c is the critical temperature of the QGP.

2.3. Model 2

This model was also introduced in [11]. The pressure and energy density as function of temperature are given by:

$$p_2 = \frac{\sigma_2}{3}T^4 - CT^2 - \mathcal{B}_2 \qquad \text{and} \qquad \varepsilon_2 = \sigma_2 T^4 - CT^2 + \mathcal{B}_2 \tag{7}$$

which can be combined to give:

$$p_2[\varepsilon_2(t)] = \frac{1}{3\sigma_2}\left\{\left[\varepsilon_2(t) - 4\mathcal{B}_2\right] - C\left[C + \sqrt{C^2 + 4\sigma_2\left[\varepsilon_2(t) - \mathcal{B}_2\right]}\right]\right\} \tag{8}$$

and again from [11] we have: $\sigma_2 = 13.01$, $C = 6.06\,{T_c}^2$ and $\mathcal{B}_2 = -2.34\,{T_c}^4$.

2.4. Model 3

This model was proposed in [12] to study the cold and dense quark gluon plasma in the inner core of neutron stars [13]. It is derived from the QCD Lagrangian treated in the mean field approximation. Here we use its extension to the finite temperature case. We repeat all the calculations described in [12] combining the mean field approximation with the finite temperature formalism developed in [14]. After some algebra (the details will be given elsewhere) we obtain the pressure:

$$p_3 = \frac{3g^2}{16{m_g}^2}\rho^2 - \mathcal{B}_{QCD} + \sum_f \frac{\gamma_f}{6\pi^2}\int_0^\infty dk\, \frac{k^4}{\sqrt{m_f^2 + k^2}}\left(d_f + \bar{d}_f\right) + \frac{\gamma_g}{6\pi^2}\int_0^\infty dk\, k^2\,(e^{k/T} - 1)^{-1} \tag{9}$$

and the energy density:

$$\varepsilon_3 = \frac{3g^2}{16{m_g}^2}\rho^2 + \mathcal{B}_{QCD} + \sum_f \frac{\gamma_f}{2\pi^2}\int_0^\infty dk\, k^2\sqrt{m_f^2 + k^2}\left(d_f + \bar{d}_f\right) + \frac{\gamma_g}{2\pi^2}\int_0^\infty dk\, k^2\,(e^{k/T} - 1)^{-1} \tag{10}$$

where the Fermi distribution functions are given by:

$$d_f \equiv \frac{1}{1 + e^{(\mathcal{E}_f - \nu_f)/T}} \qquad \text{and} \qquad \bar{d}_f \equiv \frac{1}{1 + e^{(\mathcal{E}_f + \nu_f)/T}} \tag{11}$$

The energy of the quark of flavor f is given by $\mathcal{E}_f = \sqrt{m_f^2 + k^2}$ and ν_f is its chemical potential. The quark density is given by:

$$\rho = \sum_f \frac{\gamma_f}{(2\pi)^3} \int d^3k \Big(d_f - \bar{d}_f \Big) \tag{12}$$

For simplicity, we consider only two quark flavors, up and down, with equal masses and we work in the high temperature regime, where $T >> \nu_f$, $T >> m_f$ and $\mathcal{E}_f/T > \nu_f/T$. The statistical factors are given by $\gamma_g = 2(\text{polarizations}) \times 8(\text{colors}) = 16$ for gluons and $\gamma_f = 2(\text{spins}) \times 3(\text{colors}) = 6$ for each quark. In the high temperature limit we can solve the integrals in (9), (10) and (12) analitically [15] obtaining:

$$p_3 = \frac{3g^2}{16{m_g}^2} T^4 \mu^2 + \frac{37\pi^2}{90} T^4 - \mathcal{B}_{QCD} + \frac{1}{2} T^2 \mu^2 \tag{13}$$

and

$$\varepsilon_3 = \frac{3g^2}{16{m_g}^2} T^4 \mu^2 + \frac{37\pi^2}{30} T^4 + \mathcal{B}_{QCD} + \frac{3}{2} T^2 \mu^2 \tag{14}$$

where $\mu \equiv \nu_u = \nu_d$ is the chemical potential. Expressions (13) and (14) are the asymptotic EOS for high temperature and finite chemical potential. Setting $\mu = 0$ we recover the MIT bag model results [9, 10].

2.5. *Model 4*

The equation of state below was derived in [16], where deconfined matter in $SU(3)$ pure gauge theory was treated as an ideal gas with quasi-particle modes, which have a temperature-dependent mass given by $m(T)$. This EOS reproduces lattice results, such as the trace anomaly $(\varepsilon - 3\,p)/T^4$, as it was shown in [16]. The pressure and energy density as functions of the temperature are given by:

$$p_4 = \frac{8\,T^2\, m^2(T)}{\pi^2} K_2\left[\frac{m(T)}{T}\right] \tag{15}$$

and

$$\varepsilon_4 = \frac{8\,T^2\, m^2(T)}{\pi^2} \left\{ 3K_2\left[\frac{m(T)}{T}\right] + \left[\frac{m(T)}{T} - \left(\frac{dm}{dT}\right)\right] K_1\left[\frac{m(T)}{T}\right] \right\} \tag{16}$$

where $K_i[x]$ is the modified Bessel function of the second kind. The quasi-particle mass is given by:

$$m(T) = \frac{a}{(\frac{T}{T_0} - 1)^c} + b\,\frac{T}{T_0} \tag{17}$$

with the constants $a = 0.47\,GeV, b = 0.125\,GeV$ and $c = 0.385$ fixed in [16] to reproduce the trace anomaly. The critical temperature is $T_0 \simeq 280\,MeV$.

2.6. *Model 5*

Here we present the equation of state obtained in [17, 18] from a lattice simulation of $SU(3)$ QCD at finite temperature and chemical potential with three quark flavors (u, d and s) with equal masses and gluons. At finite chemical potential the trace anomaly reads:

$$\frac{\varepsilon_5(T,\mu) - 3\,p_5(T,\mu)}{T^4} = T\frac{\partial}{\partial T}\left[\frac{p_5(T,\mu)}{T^4}\right] + \frac{\mu^2}{T^2}\chi_2$$

$$= \frac{\varepsilon_5(T,0) - 3\,p_5(T,0)}{T^4} + \frac{\mu^2}{2T}\frac{\partial \chi_2}{\partial T} \tag{18}$$

At zero chemical potential the trace anomaly can be parametrized as [17, 18]:

$$\frac{\varepsilon_5(T,0) - 3\,p_5(T,0)}{T^4} = e^{-h_1/t - h_2/t^2} \cdot \left[h_0 + \frac{f_0 \cdot \left[tanh(f_1 \cdot t + f_2) + 1 \right]}{1 + g_1 \cdot t + g_2 \cdot t^2} \right] \tag{19}$$

and also [17]:

$$\chi_2 = e^{-h_3/t - h_4/t^2} \cdot f_3 \cdot \left[tanh(f_4 \cdot t + f_5) + 1 \right] \tag{20}$$

with $t = T/(200\,MeV)$, where $200\,MeV$ is the critical temperature. From [18] we have $h_0 = 0.1396$, $h_1 = -0.1800$, $h_2 = 0.0350$, $f_0 = 1.05$, $f_1 = 6.39$, $f_2 = -4.72$, $g_1 = -0.92$ and $g_2 = 0.57$. From [17] we have $h_3 = -0.5022$, $h_4 = 0.5950$, $f_3 = 0.1359$, $f_4 = 6.3290$ and $f_5 = -4.8303$. The pressure is calculated from (18):

$$p_5(T,\mu) = T^4 \int_0^T dT' \, \frac{e^{-h_1/t' - h_2/t'^2}}{T'} \cdot \left[h_0 + \frac{f_0 \cdot \left[tanh(f_1 \cdot t' + f_2) + 1 \right]}{1 + g_1 \cdot t' + g_2 \cdot t'^2} \right] + \frac{\chi_2}{2}\,\mu^2 T^2 \tag{21}$$

Inserting (21) into (19) we find the energy density:

$$\varepsilon_5(T,\mu) = T^4 \, e^{-h_1/t - h_2/t^2} \cdot \left[h_0 + \frac{f_0 \cdot \left[tanh(f_1 \cdot t + f_2) + 1 \right]}{1 + g_1 \cdot t + g_2 \cdot t^2} \right] + \frac{\mu^2}{2}\,T^3\,\frac{\partial \chi_2}{\partial T}$$

$$+3T^4 \int_0^T dT' \, \frac{e^{-h_1/t' - h_2/t'^2}}{T'} \cdot \left[h_0 + \frac{f_0 \cdot \left[tanh(f_1 \cdot t' + f_2) + 1 \right]}{1 + g_1 \cdot t' + g_2 \cdot t'^2} \right] + \frac{3\,\chi_2}{2\,T^2}\,\mu^2 \tag{22}$$

3. Numerical results and discussion

In this section we present the numerical solution of (1) for the different equations of state discussed above. We use the following initial condition:

$$\varepsilon_i(t_i) = 10^7 MeV/fm^3 \qquad \text{at} \qquad t_i = 10^{-9}\ s \tag{23}$$

We run the time evolution from $t_i = 10^{-9}\ s$ to $t_f = 10^{-4}\ s$. There is some uncertainty on these initial conditions but this is not important for the present study, since we are primarily interested in determining how changes in the equation of state affect the time evolution of the primordial QGP. Also, since the equations of state are different, fixing the initial energy density implies that the evolution starts at different initial temperatures for different models. In working with (1) we use the relation:

$$\frac{dT}{dt} = \frac{1}{\left(\frac{d\varepsilon}{dT}\right)} \times \frac{d\varepsilon}{dt}$$

and solve it for $T(t)$.

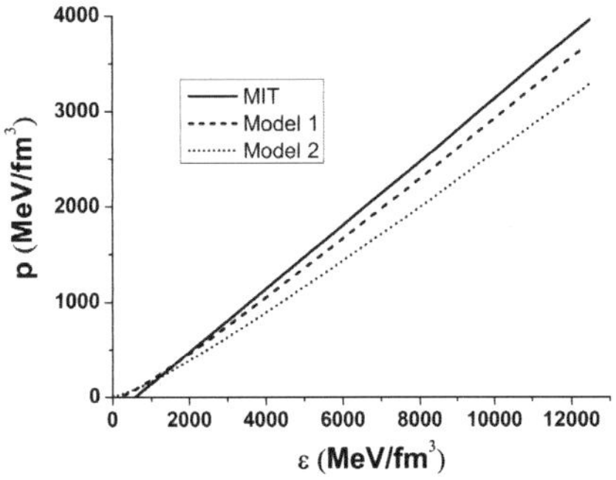

Figure 1. Equation of state of MIT-like models.

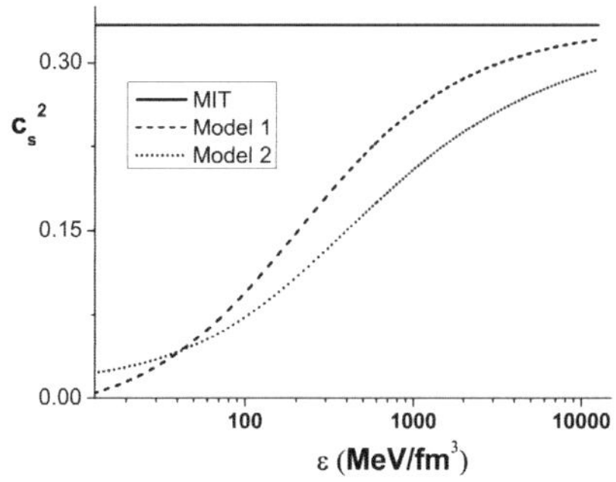

Figure 2. Speed of sound.

3.1. The MIT bag model and its variants

We start our numerical analysis by plotting in Fig. 1 the equation of state of the MIT bag model (3) and comparing it with its closest variants, models 1 and 2 at zero chemical potential, given by Eqs. (2) to (8). As a straightforward exercise we plot in Fig. 2 the sound speed given by:

$$c_s{}^2 = \frac{\partial p}{\partial \varepsilon} \tag{24}$$

For the MIT bag model we use $\mathcal{B} = 150\, MeV/fm^3$. Inserting the pressure as a function of energy density in equation (1) and using the initial condition (23) we can compute the time evolution of the energy density (Fig. 3) and of the temperature (Fig. 4). Fig. 3 shows the ratio of energy density (energy density of the model considered divided by the energy density of the MIT model) as a function of time. Comparing the models, we observe that model 1 and model 2 have a similar behavior and that, during the chosen time interval, they predict a much stronger energy dilution and a much weaker cooling than the MIT bag model. This is probably related to the critical temperature in these models, which is $T_c = 150\, MeV$. We now repeat

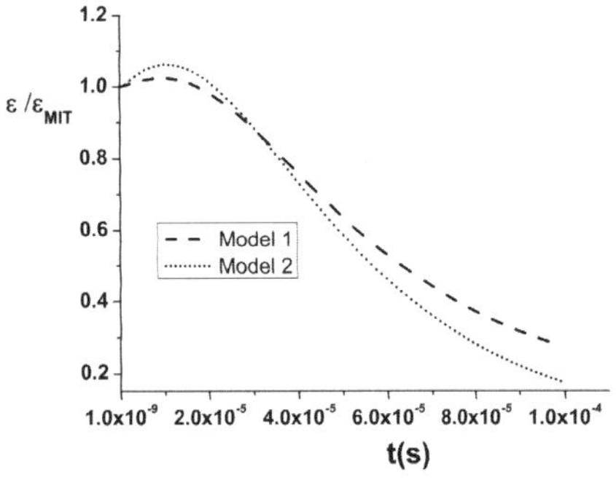

Figure 3. Energy density as a function of time.

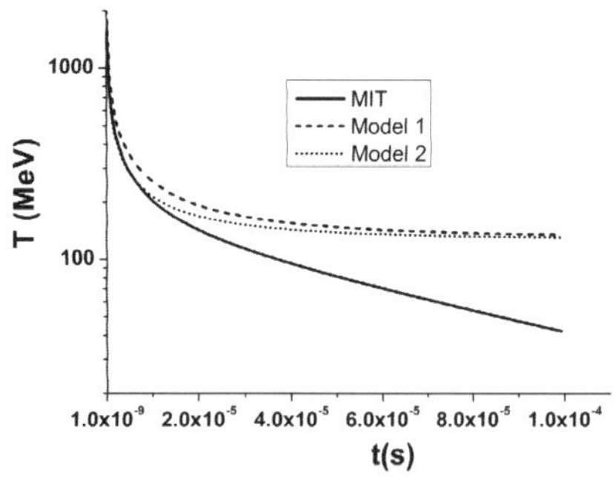

Figure 4. Temperature as a function of time.

the study performed above using the equation of state of model 3, Eqs. (13) and (14), which allow us to consider systems with finite chemical potential. Model 3 has two parameters, the dynamical gluon mass, m_G, and the QCD coupling between quarks and hard gluons, g. In Fig. 5 we show the pressure as a function of the energy density and then we calculate the sound speed, shown in Fig. 6. In most of the cases, model 3 is harder than the MIT one. Looking

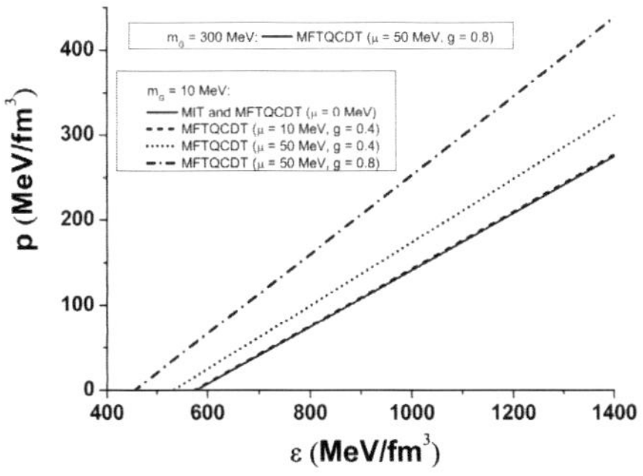

Figure 5. Pressure versus energy density.

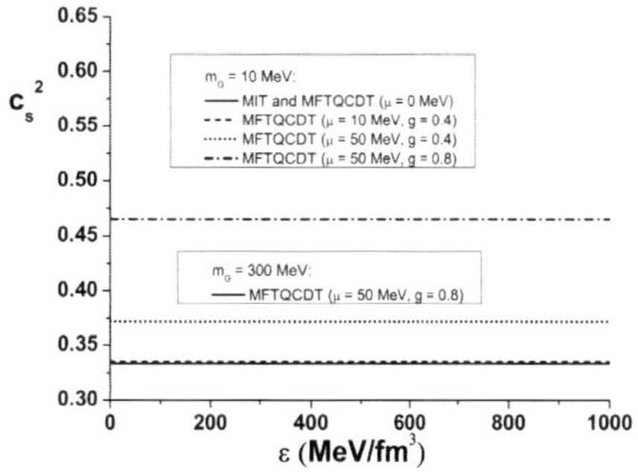

Figure 6. Sound speed.

at the expressions (13) and (14) it is easy to see qualitatively what are the effects of changing the gluon mass, the coupling constant and the chemical potential. From these expressions we can conclude that model 3 will give always a harder equation of state, except when $\mu = 0$, in which case it reproduces the MIT results. The numerical evaluation of (13) and (14), presented in Figs. 5 and 6, shows that taking a very small value for m_g, a relatively large value for g and a sizeable chemical potential we may obtain a speed of sound, which is only 50 % larger than the MIT value. The time evolution of the energy density and temperature of matter described by model 3 are shown in Fig. 7 and Fig. 8 respectively. As it could have been anticipated from the previous figures, there is only a small difference between model 3 and the MIT bag model and this difference tends to decrease with time in the case of the energy density and to remain approximately constant in the case of the temperature.

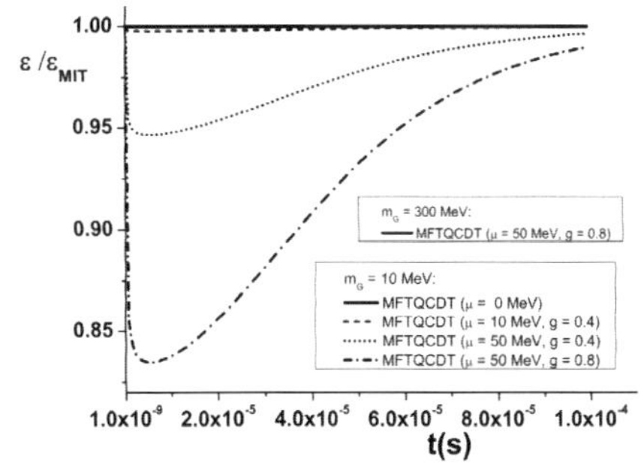

Figure 7. Time evolution of the energy density. Comparison between model 3 and the MIT bag model.

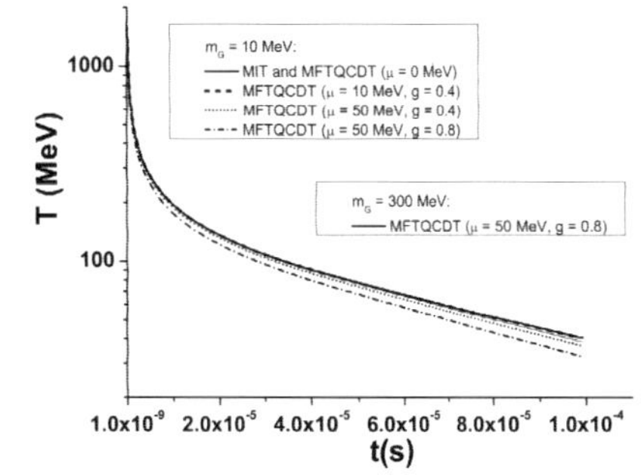

Figure 8. Time evolution of the temperature. Comparison between model 3 and the MIT bag model.

3.2. Lattice models

The two lattice models (models 4 and 5) differ mainly because in the former there are only gluons whereas in the latter quarks are included. We can hence isolate the effect of the quarks in the equation of state and in the time evolution as well. We first compare the EOS of models 4 and 5, given by (15), (16), (21) and (22), with the EOS of a gas of gluons (4) and with the MIT EOS (2). The time evolution of the energy density is shown in Fig. 9. As it can be seen there is only a small difference when we add quarks. In fact all four models have nearly the same energy density during most of the evolution and a sizeable difference appears only at very late times. In Fig. 10 we show the time evolution of the temperature calculated with the same models. It should be noted that model 4 (solid line) has a discontinuity at its critical temperature $T = 280\,MeV$. In model 5 the quark-hadron transition occurs at $T = 200\,MeV$. In the figure we can see that for similar values of the energy density the models with more degrees of freedom have a smaller temperature, as we would expect from simple thermodynamical considerations. Interestingly, models with stronger interactions have a higher temperature. In Figs. 11 and 12 we show the time evolution of the energy density and temperature computed with model 5. In the figures $\varepsilon(t)$ and $T(t)$ are scaled by the corresponding values of the energy density and temperature at zero chemical potential. From the figures we can conclude that, in the range considered, the chemical potential does not affect the time evolution. In Fig. 13 and Fig. 14 we compare the results of model 3 and model 5, the two models with which we can obtain results at finite chemical potential. Again there are only moderate differences and only at late times.

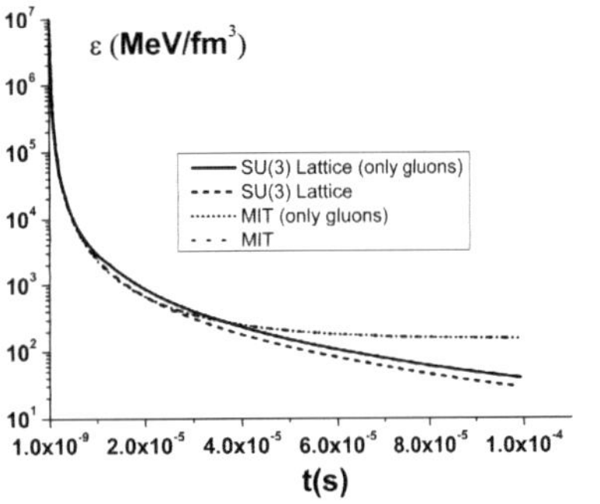

Figure 9. Time evolution of the energy density. Comparison between models 4, 5 and the MIT bag model.

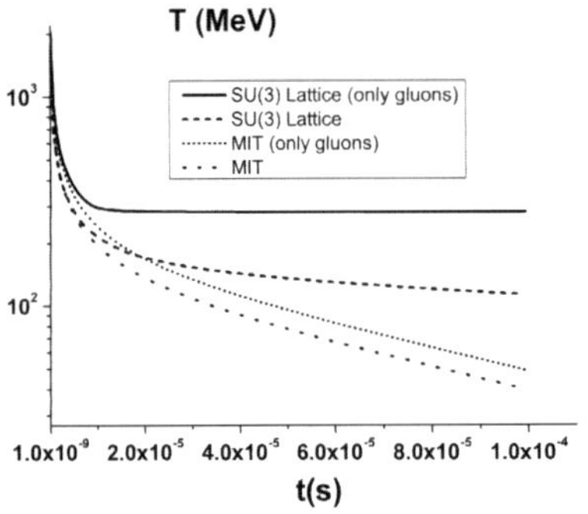

Figure 10. Time evolution of the temperature. Comparison between models 4, 5 and the MIT bag model.

4. Conclusion

As it was mentioned in the introduction, in view of the future experimental facilities for measurements relevant to cosmology, it is very interesting and very timely to identify observable effects of the QGP phase in the primordial Universe. Most of these effects have so far been associated with quark-hadron phase transition. However we think that it is also worth studying the time evolution of the QGP phase and check whether changes in the QGP equation of state induce changes in the dilution and cooling of the early Universe. These changes might, in turn, change the emission of gravitational waves or the generation of baryon number fluctuations. Taking the equation of state of the MIT bag model as a baseline, we have considered other EOS

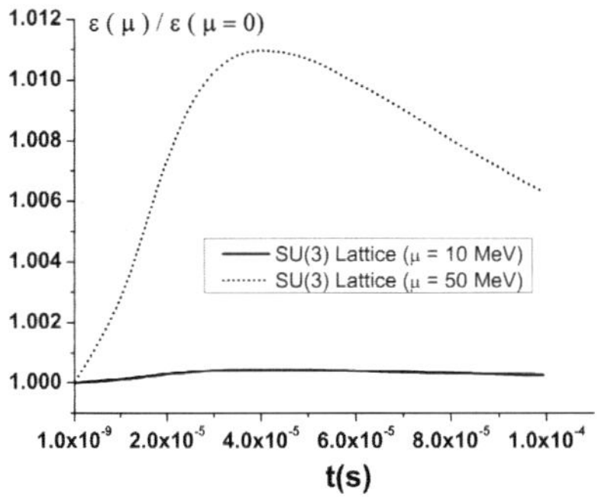

Figure 11. Time evolution of the energy density ratio calculated with model 5.

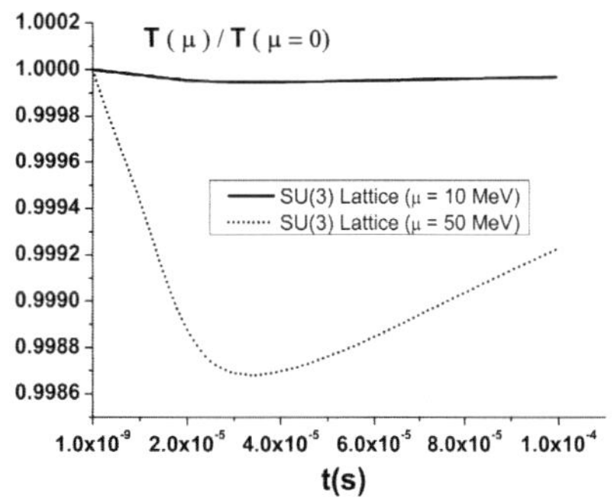

Figure 12. Time evolution of the temperature ratio calculated with model 5.

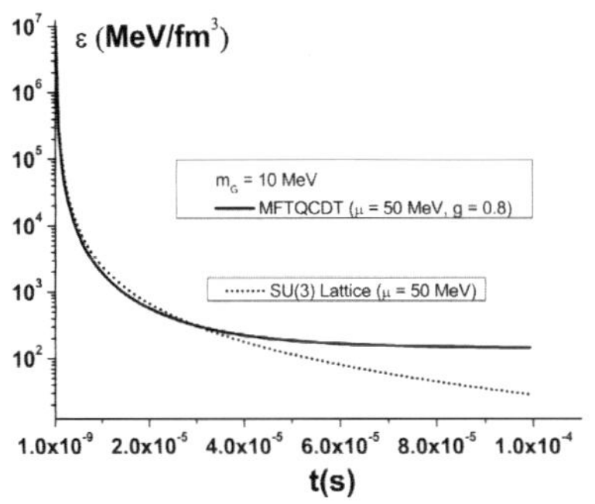

Figure 13. Time evolution of the energy density at finite chemical potential.

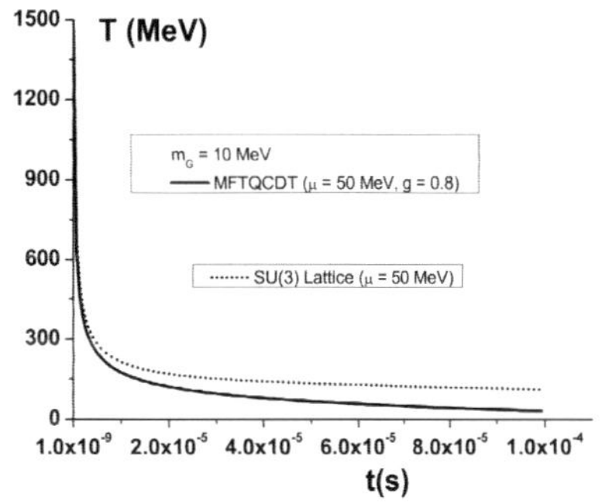

Figure 14. Time evolution of the temperature at finite chemical potential.

with different dynamical ingredients and estimated their effects on the solutions of Friedmann equations. The main conclusion is that, with the exception of results shown in Fig. 3, there are no dramatic changes. The time evolution of the energy density is only weakly sensitive to changes in the chemical potential, to changes in the degrees of freedom (addition of quarks) and to changes in the dynamical information encoded in the effective gluon mass. The temperature evolution is somewhat more sensitive to these changes, specially to the inclusion of quarks in a pure gauge theory, as it can be seen in Fig. 10. Although our results are already very suggestive, our preliminary conclusions need to be confirmed by a more complete calculation, with the inclusion of the phase transition and the evolution of the hadronic phase. Finally, we emphasize that there are still some other ingredients of the QGP phase to be considered, such as, for example the number of leptons and a strong magnetic field. Calculations along this line are in progress.

References

[1] T. Boeckel, S. Schettler and J. Schaffner-Bielich 2011 *Prog. Part. Nucl. Phys.* **66** 266; T. Boeckel and J. Schaffner-Bielich 2012 *Phys. Rev. D* 103506 and references therein.
[2] G. L. Guardo, V. Greco, M. Ruggieri arXiv:1401.7613 [hep-ph]
[3] W. Florkowski 2011 Nucl. Phys. A **853** 173
[4] V. R. C. Mouro Roque and G. Lugones, Phys. Rev. D **87**, no. 8, 083516 (2013).

[5] D. J. Schwarz and M. Stuke 2009 *JCAP* **0911**, 025; [Erratum-ibid. **1010**, E01 (2010)]
[6] D. Boyanovsky, H. J. de Vega and D. J. Schwarz 2006 *Ann. Rev. Nucl. Part. Sci.* **56**, 441; C. Schmid, D. J. Schwarz and P. Widerin, Phys. Rev. D **59**, 043517 (1999); D. J. Schwarz, Nucl. Phys. A **642**, 336 (1998).
[7] Avijeet Ray and Soma Sanyal arXiv:1309.1236 [hep-ph]
[8] U. Ornik and R. M. Weiner 1987 *Phys. Rev. D* **36** 1263
[9] A. Chodos, R. L. Jaffe, K. Johnson, C. B. Thorn and V. F. Weisskopf 1974 *Phys. Rev. D* **9** 3471
[10] D. A. Fogaça, L. G. Ferreira Filho and F. S. Navarra 2010 *Phys. Rev. C* **81** 055211
[11] V. V. Begun, M. I. Gorenstein and O. A. Mogilevsky 2012 *Phys. Atom. Nucl.* **75** 873
[12] D. A. Fogaça and F.S. Navarra 2011 *Phys. Lett. B* **700** 236
[13] B. Franzon, D. A. Fogaça, F. S. Navarra and J. E. Horvath 2012 *Phys. Rev. D* **86** 065031
[14] R. J. Furnstahl and Brian D. Serot 1990 *Phys. Rev. C* **41** 262
[15] Robert F. Tooper 1969 *The Astrophysical Journal* **156** 1075
[16] P. Castorina, D. E. Miller and H. Satz 2011 *Eur. Phys. J. C* **71** 1673
[17] S. Borsanyi, G. Endrodi, Z. Fodor, S. D. Katz, S. Krieg, C. Ratti and K. K. Szabo 2012 *JHEP* **1208** 053
[18] S. Borsanyi, Z. Fodor, C. Hoelbling, S. D. Katz, S. Krieg and K. K. Szabo 2014 *Phys. Lett. B* **730** 99

Fermions embedded in a scalar-vector kink-like smooth potential

W M Castilho[1], **A S de Castro**[2]

UNESP, Campus de Guaratinguetá, Departamento de Física e Química, 12516-410 Guaratinguetá, SP, Brazil

E-mail: [1]castilho.w@gmail.com, [2]castro@pq.cnpq.br

Abstract. The behaviour of massive fermions is analyzed with scalar and vector potentials. A continuous chiral-conjugation transformation decouples the equation for the upper component of the Dirac spinor provided the vector coupling does not exceed the scalar coupling. It is shown that a Sturm-Liouville perspective is convenient for studying scattering as well as bound states. One possible isolated solution (excluded from the Sturm-Liouville problem) corresponding to a bound state might also come into sight. For potentials with kink-like profiles, beyond the intrinsically relativistic isolated bound-state solution corresponding to the zero-mode solution of the massive Jackiw-Rebbi model in the case of no vector coupling, a finite set of bound-state solutions might appear as poles of the transmission amplitude in a strong coupling regime. It is also shown that the possible isolated bound solution disappears asymptotically as the magnitude of the scalar and vector coupling becomes the same. Furthermore, we show that due to the sizeable mass gain from the scalar background the high localization of the fermion in an extreme relativistic regime is conformable to comply with the Heisenberg uncertainty principle.

1. Introduction

The Dirac Hamiltonian with a mixing of scalar potential and time component of vector potential in a four-dimensional space-time is invariant under an SU(2) algebra when the difference between the potentials, or their sum, is a constant [1]. The near realization of these symmetries may explain degeneracies in some heavy meson spectra (spin symmetry) [2, 3] or in single-particle energy levels in nuclei (pseudospin symmetry) [3–38]. When these symmetries are realized, the energy spectrum does not depend on the spinorial structure, being identical to the spectrum of a spinless particle [39]. Despite the absence of spin effects in 1+1 dimensions, many attributes of the spin and pseudospin symmetries in four dimensions are preserved.

In a pioneering work, Jackiw and Rebbi [40] have shown that massless fermions coupled to scalar fields with kink-like profiles in 1+1 dimensions develops quantum states with fractional fermion number due to the zero-mode solution. This phenomenon has been seen in certain polymers such as polyacetylene [41–44]. Recently the complete set of solutions for the kink-like scalar potential behaving like $\tanh x/\lambda$ has been considered for massless fermions in Ref. [45], and for massive fermions in Ref. [46]. The complete set of solutions for massive fermions under the influence of a kink-like scalar potential added by the time component of a vector potential with the same functional form was considered in Refs. [47] and [48], in Ref. [47] for the background field behaving like $\operatorname{sgn} x$, and in Ref. [48] for the background behaving like $\tanh x/\lambda$.

In Refs. [47] and [48], it has been shown that the Dirac equation with a scalar potential plus a time component of vector potential of the same functional form is manageable if the vector coupling does not exceed the scalar coupling, and that the bound states for mixed scalar-vector potentials with kink-like profiles are intrinsically relativistic solutions. Furthermore, it has been shown that the fermion can be confined in a highly localized region of space under a very strong field without any chance for spontaneous pair production related to Klein's paradox.

Here we shall outline the scalar-vector mixing framework developed in Refs. [47]-[48] with the Sturm-Liouville perspective plus its isolated solution. We show that the isolated solution disappears asymptotically as one approaches the conditions for the realization of the so-called spin and pseudospin symmetries in four dimensions. After a general consideration of the Sturm-Liouville problem for an arbitrary kink-like potential, we concentrate our attention on the isolated solution which corresponds to the zero-mode solution of the Jackiw-Rebbi model in the case of a pure scalar coupling. It is shown that such an isolated solution is an intrinsically relativistic solution even if the fermion is massive. Then we use the smooth step potential $\tanh x/\lambda$ to show in detail how the additional mass acquired by the fermion from the scalar background can acquiesce highly localized states without violating the Heisenberg uncertainty principle.

2. Mixed scalar-vector interactions

Consider the Lagrangian density for a massive fermion

$$L = \bar{\Psi}\left(i\hbar c\gamma^{\mu}\partial_{\mu} - Imc^{2} - V\right)\Psi \tag{1}$$

where $\hbar$ is the constant of Planck, c is the velocity of light, I is the unit matrix, m is the mass of the free fermion and the square matrices γ^{μ} satisfy the algebra $\{\gamma^{\mu},\gamma^{\nu}\} = 2Ig^{\mu\nu}$. The spinor adjoint to Ψ is defined by $\bar{\Psi} = \Psi^{\dagger}\gamma^{0}$. In 1+1 dimensions Ψ is a 2×1 matrix and the metric tensor is $g^{\mu\nu} = \mathrm{diag}(1,-1)$. For vector and scalar interactions the matrix potential is written as

$$V = \gamma^{\mu}A_{\mu} + IV_{s} \tag{2}$$

We say that A_{μ} and V_s are the vector and scalar potentials, respectively, because the bilinear forms $\bar{\Psi}\gamma^{\mu}\Psi$ and $\bar{\Psi}I\Psi$ behave like vector and scalar quantities under a Lorentz transformation, respectively. Eq. (1) leads to the Hamiltonian form for the Dirac equation

$$i\hbar\frac{\partial\Psi}{\partial t} = H\Psi \tag{3}$$

with the Hamiltonian given as

$$H = \gamma^{5}c\left(p_{1} + \frac{A_{1}}{c}\right) + IA_{0} + \gamma^{0}\left(mc^{2} + V_{s}\right) \tag{4}$$

where $\gamma^{5} = \gamma^{0}\gamma^{1}$. Requiring $(\gamma^{\mu})^{\dagger} = \gamma^{0}\gamma^{\mu}\gamma^{0}$, one finds the continuity equation $\partial_{\mu}J^{\mu} = 0$, where the conserved current is $J^{\mu} = c\bar{\Psi}\gamma^{\mu}\Psi$. The positive-definite function $J^{0}/c = |\Psi|^{2}$ is interpreted as a position probability density and its norm is a constant of motion. This interpretation is completely satisfactory for single-particle states [49]. The space component of the vector potential can be gauged away by defining a new spinor just differing from the old by a phase factor so that we can consider $A_1 = 0$ without loss of generality.

Assuming that the potentials are time independent, one can write $\Psi(x,t) = \psi(x)\exp(-iEt/\hbar)$ in such a way that the time-independent Dirac equation becomes $H\psi = E\psi$. Meanwhile $J^{\mu} = c\bar{\psi}\gamma^{\mu}\psi$ is time independent and J^{1} is uniform.

From now on, we use an explicit representation for the 2×2 matrices γ as

$$\gamma^{0} = \sigma_{3}, \quad \gamma^{1} = i\sigma_{2} \tag{5}$$

in such a way that $\gamma^{5} = \sigma_{1}$. Here, σ_1, σ_2 and σ_3 stand for the Pauli matrices.

2.1. Nonrelativistic limit
The Lorentz nature of the potentials does no matter in a weak-coupling regime. Indeed, fermions (antifermions) are subject to the effective potential $V_s + A_0$ ($V_s - A_0$) with energy $E \approx +mc^2$ ($-mc^2$) so that a mixed potential with $A_0 = -V_s$ ($A_0 = +V_s$) is associated with free fermions (antifermions) in a nonrelativistic regime [47]-[48]. The changes of signs of A_0 and E as well as the invariance of the sign of V_s when one exchanges the roles of fermions and antifermions are justified by the charge-conjugation transformation.

2.2. Extreme relativistic limit
For potentials localized in the range λ, quantum effects appear when λ is comparable to the Compton wavelength $\lambda_C = \hbar/mc$, and relativistic quantum effects are expected when λ is of the same order or smaller than the Compton wavelength. The fermion is under an extreme relativistic regime if $\lambda << \lambda_C$. When the fermion localized in the region Δx reduces its extension (increasing the intensity of the potential, for example) then the uncertainty in the momentum must expand, in consonance with the Heisenberg uncertainty principle. Nevertheless, the maximum uncertainty in the momentum is comparable with mc requiring that is impossible to localize a fermion in a region of space less than or comparable with half of its Compton wavelength (see, for example, [50, 51]). Nevertheless, due to the mass gain granted by its interaction with the scalar-field background, those bound states could be highly localized without any chance of spontaneous pair production.

2.3. Charge conjugation
The charge-conjugation operation is accomplished by the transformation $\psi \to \sigma_1\psi^*$ followed by $A_0 \to -A_0$, $V_s \to V_s$ and $E \to -E$ [52]. As a matter of fact, A_0 distinguishes fermions from antifermions but V_s does not, and so the spectrum is symmetrical about $E = 0$ in the case of a pure scalar potential.

2.4. Chiral conjugation
The chiral-conjugation operation $\psi \to \gamma^5\psi$ (according to Ref. [53]) is followed by the changes of the signs of V_s and m, but not of A_0 and E [52]. One sees that the charge-conjugation and the chiral-conjugation operations interchange the roles of the upper and lower components of the Dirac spinor.

2.5. Continuous chiral conjugation
The unitary operator

$$U(\theta) = \exp\left(-\frac{\theta}{2}i\gamma^5\right) \tag{6}$$

where θ is a real quantity such that $0 \leq \theta \leq \pi$, allows one to write

$$h\phi = E\phi \tag{7}$$

where

$$\phi = U\psi, \quad h = UHU^{-1} \tag{8}$$

with

$$h = \sigma_1 c p_1 + IA_0 + \sigma_3\left(mc^2 + V_s\right)\cos\theta - \sigma_2\left(mc^2 + V_s\right)\sin\theta \tag{9}$$

It is instructive to note that the transformation preserves the form of the current in such a way that $J^\mu = c\bar{\phi}\gamma^\mu\phi$. An additional important feature of the continuous chiral transformation (see, e.g., [54])) induced by (6) is that it is a symmetry transformation when $m = V_s = 0$.

In terms of the upper and the lower components of the spinor ϕ, the Dirac equation decomposes into:

$$\hbar c\frac{d\phi_\pm}{dx} \pm \left(mc^2 + V_s\right)\sin\theta\,\phi_\pm = i\left[E \pm \left(mc^2+V_s\right)\cos\theta - A_0\right]\phi_\mp \tag{10}$$

Furthermore,

$$\frac{J^0}{c} = |\phi_+|^2 + |\phi_-|^2, \quad \frac{J^1}{c} = 2\mathrm{Re}\left(\phi_+^*\phi_-\right) \tag{11}$$

2.6. *Special mixing*

Choosing

$$A_0 = V_s\cos\theta \tag{12}$$

one has

$$\hbar c\frac{d\phi_+}{dx} + \left(mc^2 + V_s\right)\sin\theta\,\phi_+ \;=\; i\left(E + mc^2\cos\theta\right)\phi_- \tag{13a}$$

$$\hbar c\frac{d\phi_-}{dx} - \left(mc^2 + V_s\right)\sin\theta\,\phi_- \;=\; i\left[E - \left(mc^2 + 2V_s\right)\cos\theta\right]\phi_+ \tag{13b}$$

Note that due to the constraint represented by (12), the vector and scalar potentials have the very same functional form and the parameter θ in (6) measures the dosage of vector coupling in the vector-scalar admixture in such a way that $|V_s| \geq |A_0|$. Note also that when the mixing angle θ goes from $\pi/2-\varepsilon$ to $\pi/2+\varepsilon$ the sign of the spectrum undergoes an inversion under the charge-conjugation operation whereas the spectrum of a massless fermion is invariant under the chiral-conjugation operation. Combining charge-conjugation and chiral-conjugation operations makes the spectrum of a massless fermion to be symmetrical about $E=0$ in spite of the presence of vector potential.

We now split two classes of solutions depending on whether E is equal to or different from $-mc^2\cos\theta$.

2.7. *The class $E \neq -mc^2\cos\theta$*

For $E \neq -mc^2\cos\theta$, using the expression for ϕ_- obtained from (13a), viz.

$$\phi_- = \frac{-i}{E + mc^2\cos\theta}\left[\hbar c\frac{d\phi_+}{dx} + \left(mc^2 + V_s\right)\sin\theta\,\phi_+\right] \tag{14}$$

one finds

$$J^1 = \frac{2\hbar c^2}{E + mc^2\cos\theta}\,\mathrm{Im}\left(\phi_+^*\frac{d\phi_+}{dx}\right) \tag{15}$$

Inserting (14) into (13b) one arrives at the following second-order differential equation for ϕ_+:

$$-\frac{\hbar^2}{2}\frac{d^2\phi_+}{dx^2} + V_{\mathtt{eff}}\,\phi_+ = E_{\mathtt{eff}}\,\phi_+ \tag{16}$$

where

$$V_{\mathtt{eff}} = \frac{\sin^2\theta}{2c^2}V_s^2 + \frac{mc^2 + E\cos\theta}{c^2}V_s - \frac{\hbar\sin\theta}{2c}\frac{dV_s}{dx} \tag{17}$$

and

$$E_{\mathtt{eff}} = \frac{E^2 - m^2c^4}{2c^2} \tag{18}$$

Therefore, the solution of the relativistic problem for this class is mapped into a Sturm-Liouville problem for the upper component of the Dirac spinor. In this way one can solve the Dirac problem for determining the possible discrete or continuous eigenvalues of the system by recurring to the solution of a Schrödinger-like problem because ϕ_+ is a square-integrable function.

Notice that the effective potential in (17) behaves like V_s in the case of scalar and vector potentials of the same magnitude. For the case of a pure scalar coupling ($E \neq 0$), it is also possible to write a second-order differential equation for ϕ_- just differing from the equation for ϕ_+ in the sign of the term involving dV_s/dx, namely,

$$-\frac{\hbar^2}{2}\frac{d^2\phi_\pm}{dx^2} + \left(\frac{V_s^2}{2c^2} + mV_s \mp \frac{\hbar}{2c}\frac{dV_s}{dx}\right)\phi_\pm = E_{\mathtt{eff}}\,\phi_\pm \tag{19}$$

which is of the form of supersymmetric quantum mechanics, as has already been appreciated in the literature [55, 56].

2.8. The class $E = -mc^2\cos\theta$
Defining

$$v(x) = \int^x dy\, V_s(y) \tag{20}$$

the solutions for (13a) and (13b) with $E = -mc^2\cos\theta$ are

$$\phi_+ = N_+ \tag{21a}$$

$$\phi_- = N_- - 2\frac{i}{\hbar c}N_+\left[mc^2x + v(x)\right]\cos\theta \tag{21b}$$

for $\sin\theta = 0$, and

$$\phi_+ = N_+\exp\left\{-\frac{\sin\theta}{\hbar c}\left[mc^2x + v(x)\right]\right\} \tag{22a}$$

$$\phi_- = N_-\exp\left\{+\frac{\sin\theta}{\hbar c}\left[mc^2x + v(x)\right]\right\} + i\phi_+\cot\theta \tag{22b}$$

for $\sin\theta \neq 0$. N_+ and N_- are normalization constants. It is instructive to note that there is no solution for scattering states. Both set of solutions present a space component for the current equal to $J^1 = 2c\mathrm{Re}\left(N_+^* N_-\right)$ and a bound-state solution demands $N_+ = 0$ or $N_- = 0$, because ϕ_+ and ϕ_- are square-integrable functions vanishing as $|x| \to \infty$. It is remarkable that the eigenenergy does not depend on the magnitude of the potential but the eigenspinor does. Note also that

$$\phi_\pm = N_\pm\exp\left\{\mp\frac{1}{\hbar c}\left[mc^2x + v(x)\right]\right\} \tag{23}$$

in the case of a pure scalar coupling ($E = 0$) so that either $\phi_+ = 0$ or $\phi_- = 0$. There is no bound-state solution for $\sin\theta = 0$, and for $\sin\theta \neq 0$ the existence of a bound state solution depends on the asymptotic behaviour of $v(x)$ [57, 58].

3. Kink potentials

Now we consider a kink-like potential with the asymptotic behaviour $V_s(x) \to \pm v_0$ as $x \to \pm x_0$, with v_0 = constant and $x_0 >> \lambda$ (λ is the range of the interaction centered on the origin).

We turn our attention to scattering states for fermions with $E \neq -mc^2\cos\theta$ coming from the left. Then, ϕ for $x \to -\infty$ describes an incident wave moving to the right and a reflected wave moving to the left, and ϕ for $x \to +\infty$ describes a transmitted wave moving to the right or an evanescent wave. The upper components for scattering states are written as

$$\phi_+ = \begin{cases} A_+ e^{+ik_- x} + A_- e^{-ik_- x}, & \text{for} \quad x \to -\infty \\ \\ B_\pm e^{\pm ik_+ x}, & \text{for} \quad x \to +\infty \end{cases} \tag{24}$$

where

$$\hbar c k_\pm = \sqrt{\left(E \mp v_0\cos\theta\right)^2 - \left(mc^2 \pm v_0\right)^2} \tag{25}$$

Note that k_+ is a real number for a progressive wave and an imaginary number for an evanescent wave (k_- is a real number for scattering states). Therefore,

$$J^1_{\gtrless}(-\infty) = \frac{2\hbar c^2 k_-}{E + mc^2\cos\theta}\left(|A_\pm|^2 - |A_\mp|^2\right), \quad \text{for} \quad E \gtrless -mc^2\cos\theta \tag{26}$$

and

$$J^1_{\gtrless}(+\infty) = \pm\frac{2\hbar c^2 \mathrm{Re}\, k_+}{E + mc^2\cos\theta}|B_\pm|^2, \quad \text{for} \quad E \gtrless -mc^2\cos\theta \tag{27}$$

Note also that

$$J^1_{\gtrless}(-\infty) = J_{\mathtt{inc}} - J_{\mathtt{ref}} \tag{28}$$

and

$$J^1_{\gtrless}(+\infty) = J_{\mathtt{tran}} \tag{29}$$

where $J_{\mathtt{inc}}$, $J_{\mathtt{ref}}$ and $J_{\mathtt{tran}}$ are nonnegative quantities characterizing the incident, reflected and transmitted waves, respectively. Note also that the roles of A_+ and A_- are exchanged as the sign of $E + mc^2\cos\theta$ changes. In fact, if $E > -mc^2\cos\theta$, then $A_+e^{+ik_-x}$ ($A_-e^{-ik_-x}$) will describe the incident (reflected) wave, and $B_- = 0$. On the other hand, if $E < -mc^2\cos\theta$, then $A_-e^{-ik_-x}$ ($A_+e^{+ik_-x}$) will describe the incident (reflected) wave, and $B_+ = 0$. Therefore, the reflection and transmission amplitudes are given by

$$r_{\gtrless} = \frac{A_\mp}{A_\pm}, \quad t_{\gtrless} = \frac{B_\pm}{A_\pm}, \quad \text{for} \quad E \gtrless -mc^2\cos\theta \tag{30}$$

To determine the transmission coefficient we use the current densities $J^1_{\gtrless}(-\infty)$ and $J^1_{\gtrless}(+\infty)$. The x-independent space component of the current allows us to define the reflection and transmission coefficients as

$$R_{\gtrless} = \frac{|A_\mp|^2}{|A_\pm|^2}, \quad T_{\gtrless} = \frac{\mathrm{Re}\, k_+}{k_-}\frac{|B_\pm|^2}{|A_\pm|^2}, \quad \text{for} \quad E \gtrless -mc^2\cos\theta \tag{31}$$

Notice that $R_{\gtrless} + T_{\gtrless} = 1$ by construction. The possibility of bound states requires a solution with an asymptotic behaviour given by (24) with $k_\pm = i|k_\pm|$ and $A_+ = B_- = 0$, or $k_\pm = -i|k_\pm|$ and $A_- = B_+ = 0$, to obtain a square-integrable ϕ_+. On the other hand, if one considers the transmission amplitude t in (30) as a function of the complex variables $k_\pm$ one sees that for $k_\pm > 0$ one obtains the scattering states whereas the bound states would be obtained by the poles lying along the imaginary axis of the complex k-plane.

As for $E = -mc^2\cos\theta$, the existence of a bound-state solution requires $|v_0| > mc^2$ so that the eigenspinor behaves asymptotically as

$$\phi \sim \begin{pmatrix} 1 \\ i\cot\theta \end{pmatrix} f \tag{32}$$

for $v_0 > mc^2$, and

$$\phi \sim \begin{pmatrix} 0 \\ 1 \end{pmatrix} f \tag{33}$$

for $v_0 < -mc^2$. Here,

$$f = \exp\left\{ -\frac{\sin\theta}{\hbar c} \left[|v_0| + mc^2 \mathrm{sgn}\,(v_0 x) \right] |x| \right\} \tag{34}$$

Armed with the knowledge about asymptotic solutions and with the definition of the transmission coefficient we should proceed for searching solutions on the entire region of space. Nevertheless, we can not tell much more about the problem until the potential function is specified.

4. The smooth step potential

Now the scalar potential takes the form

$$V_s = v_0 \tanh x/\lambda \tag{35}$$

where the skew positive parameter λ is related to the range of the interaction which makes V_s to change noticeably in the interval $-\lambda < x < \lambda$, and v_0 is the height of the potential at $x = +\infty$. Notice that as $\lambda \to 0$, the case of an extreme relativistic regime, the smooth step approximates the sign potential already considered in Ref. [47].

As commented before, there is no isolated solution from the Sturm-Liouville problem when $\sin\theta = 0$, and the existence of a well-behaved isolated solution when $\sin\theta \neq 0$ makes

$$v\,(x) = \lambda v_0 \ln\left(\cosh x/\lambda\right) \tag{36}$$

and requires $|v_0| > mc^2$:

$$\phi = \begin{pmatrix} 1 \\ i\cot\theta \end{pmatrix} N_> f \tag{37}$$

for $v_0 > mc^2$, and

$$\phi = \begin{pmatrix} 0 \\ 1 \end{pmatrix} N_< f \tag{38}$$

for $v_0 < -mc^2$. Here,

$$f = \frac{\exp\left(-\alpha_1 x\right)}{\cosh^{\alpha_2} x/\lambda} \tag{39}$$

where

$$\alpha_1 = \frac{\mathrm{sgn}\,(v_0)\, mc \sin\theta}{\hbar}, \qquad \alpha_2 = \frac{\lambda |v_0| \sin\theta}{\hbar c} \tag{40}$$

The normalization condition

$$\int_{-\infty}^{+\infty} dx\, \left(|\phi_+|^2 + |\phi_-|^2 \right) = 1 \tag{41}$$

and the integral tabulated (see the formula 3.512.1, or 8.380.10, in Ref. [59])

$$\int_0^\infty dx\, \frac{\cosh 2\beta_1 x}{\cosh^{2\beta_2} \gamma x} = \frac{2^{2\beta_2}}{4\gamma} B\left(\beta_2 + \frac{\beta_1}{\gamma}, \beta_2 - \frac{\beta_1}{\gamma} \right) \tag{42}$$

where $B\,(z_1, z_2)$ is the beta function [60], allow one to determine $N_{\gtrless}$. In the way indicated one can find the position probability density [48]

$$|\phi|^2 = \frac{2f^2}{2^{2\alpha_2} \lambda B\left(\alpha_+, \alpha_-\right)} \tag{43}$$

where

$$\alpha_{\pm} = \alpha_2 \pm \lambda \alpha_1 \tag{44}$$

Therefore, a massive fermion tends to concentrate at the left (right) region when $v_0 > 0$ ($v_0 < 0$), and tends to avoid the origin more and more as $\sin\theta$ decreases. A massless fermion has a position probability density symmetric around the origin. One can see that the best localization occurs for a pure scalar coupling. In fact, the fermion becomes delocalized as $\sin\theta$ decreases.

The expectation value of x and x^2 is given by

$$< x >= -\frac{4}{2^{2\alpha_2}\lambda B\left(\alpha_+, \alpha_-\right)} \int_0^\infty dx \, \frac{x \sinh 2\alpha_1 x}{\cosh^{2\alpha_2} x/\lambda} \tag{45}$$

and

$$< x^2 >= \frac{4}{2^{2\alpha_2}\lambda B\left(\alpha_+, \alpha_-\right)} \int_0^\infty dx \, \frac{x^2 \cosh 2\alpha_1 x}{\cosh^{2\alpha_2} x/\lambda} \tag{46}$$

Defining

$$\Delta\left(\alpha\right) = \boldsymbol{\psi}\left(\alpha_+\right) - \boldsymbol{\psi}\left(\alpha_-\right) \tag{47a}$$

$$\Sigma^{(1)}\left(\alpha\right) = \boldsymbol{\psi}^{(1)}\left(\alpha_+\right) + \boldsymbol{\psi}^{(1)}\left(\alpha_-\right) \tag{47b}$$

where

$$\boldsymbol{\psi}\left(z\right) = \frac{d \ln \Gamma\left(z\right)}{dz} \tag{48}$$

is the digamma (psi) function and

$$\boldsymbol{\psi}^{(1)}\left(z\right) = \frac{d\boldsymbol{\psi}\left(z\right)}{dz} \tag{49}$$

is the trigamma function [60], printed in a boldface type to differ from the Dirac eigenspinor, and using a pair of integrals tabulated in [48], $< x >$ and $< x^2 >$ can be simplified to

$$< x >= -\frac{\lambda}{2}\Delta\left(\alpha\right) \tag{50}$$

and

$$< x^2 >= \frac{\lambda^2}{4}\Sigma^{(1)}\left(\alpha\right) + < x >^2 \tag{51}$$

and hence the fermion is confined within an interval

$$\Delta x = \sqrt{< x^2 > - < x >^2} \tag{52}$$

given by

$$\Delta x = \frac{\lambda}{2}\sqrt{\Sigma^{(1)}\left(\alpha\right)} \tag{53}$$

With the help of a few approximate formulas for the special functions in Ref. [48], one obtains the values for $< x >$ and Δx either in the case of $\sin\theta \simeq 0$ or $\lambda << \lambda_C$:

$$< x > \simeq -\mathrm{sgn}\left(v_0\right) \frac{\hbar c}{\sin\theta} \frac{mc^2}{v_0^2 - m^2c^4} \tag{54a}$$

$$\Delta x \simeq \frac{\hbar c}{\sqrt{2}\sin\theta} \frac{\sqrt{v_0^2 + m^2c^4}}{v_0^2 - m^2c^4} \tag{54b}$$

One can also see that when $\lambda >> \lambda_C$ or $|v_0| >> mc^2$

$$< x > \simeq \frac{\lambda}{2} \ln \frac{|v_0| - \text{sgn}(v_0)\, mc^2}{|v_0| + \text{sgn}(v_0)\, mc^2} \tag{55a}$$

$$\Delta x \simeq \sqrt{\frac{\lambda \hbar c}{2 \sin\theta} \frac{|v_0|}{v_0^2 - m^2 c^4}} \tag{55b}$$

Again one can see that the fermion becomes delocalized as $\sin\theta$ decreases and that the best localization occurs for a pure scalar coupling. More than this, $< x > \to -\infty$ and $\Delta x \to \infty$ as $|v_0| \to mc^2$, and besides $< x > \to 0$ and $\Delta x \to 0$ as $|v_0| \to \infty$.

Those last results show that Δx reduces its extension with rising $|v_0|$ or $\sin\theta$, or decreasing λ, in such a way that Δx can be arbitrarily small for a potential strong enough or short-range enough. The impasse related to the Heisenberg uncertainty principle can be broken by resorting to the concepts of effective mass and effective Compton wavelength. Indeed, if one defines an effective mass as

$$m_{\tt eff} = m\sqrt{1 + \left(\frac{v_0}{mc^2}\right)^2} \tag{56}$$

and an effective Compton wavelength $\lambda_{\tt eff} = \hbar / (m_{\tt eff} c)$, one will find

$$\Delta x = \frac{\sqrt{2}\lambda_{\tt eff}}{4 \sin\theta} \sqrt{\left(\alpha_+^2 + \alpha_-^2\right) \Sigma^{(1)}(\alpha)} \tag{57}$$

It follows that the high localization of fermions, related to high values of $|v_0|$ or small values of λ, never menaces the single-particle interpretation of the Dirac theory even if the fermion is massless ($m_{\tt eff} = |v_0|/c^2$). As a matter of fact, (54b) furnishes

$$(\Delta x)_{\min} \simeq \frac{\lambda_{\tt eff}}{\sqrt{2} \sin\theta} \tag{58}$$

for $|v_0| >> mc^2$ or $\lambda << \lambda_C$.

5. Final remarks

After reviewing the use of a continuous chiral transformation for solving the Dirac equation in the background of scalar and vector potential developed in Refs. [47]-[48], we have done an extension for arbitrary kink-like potentials which generalizes the Jackiw-Rebbi [40] model not only for considering massive fermions but also for taking into account an additional vector coupling. Concentrating our attention on the zero-mode solution, we have shown that, due to the sizeable additional mass acquired by the fermion resulting from its interaction with the scalar-field background, those bound states can be highly localized by very strong potentials without any chance of spontaneous pair production. This fact is convincing because the scalar is stronger than the vector coupling, and so the conditions for Klein's paradox are never reached.

Acknowledgments

This work was supported in part by means of funds provided by CNPq.

References

[1] Bell J and Ruegg H 1975 *Nucl. Phys. B* **98** 151
[2] Page P R, Goldman T and Ginocchio J N 2001 *Phys. Rev. Lett.* **86** 204
[3] Ginocchio J N 2005 *Phys. Rep.* **414** 165
[4] Ginocchio J N 1997 *Phys. Rev. Lett.* **78** 436

[5] Ginocchio J N and Leviatan A 1998 *Phys. Lett. B* **425** 1
[6] Lalazissis G A, Gambhir Y K, Maharana J P, Warke C S and Ring P 1998 *Phys. Rev. C* **58** R45
[7] Meng J, Sugawara-Tanabe K, Yamaji S, Ring P and Arima A 1998 *Phys. Rev. C* **58** R628
[8] Sugawara-Tanabe K and Arima A 1998 *Phys. Rev. C* **58** R3065
[9] Ginocchio J N 1999 *Phys. Rep.* **315** 231
[10] Marcos S, Savushkin L N, López-Quelle M and Ring P 2000 *Phys. Rev. C* **62** 054309
[11] Alberto P, Fiolhais M, Malheiro M, Delfino A and Chiapparini M 2001 *Phys. Rev. Lett.* **86** 5015
[12] Marcos S, Lpez-Quelle M, Niembro R, Savushkin L and Bernardos P 2001 *Phys. Lett. B* **513** 30
[13] Ginocchio J N and Leviatan A 2001 *Phys. Rev. Lett.* **87** 072502
[14] Alberto P, Fiolhais M, Malheiro M, Delfino A and Chiapparini M 2002 *Phys. Rev. C* **65** 034307
[15] Ginocchio J N 2002 *Phys. Rev. C* **66** 064312
[16] Ti-Sheng C, Hong-Feng L, Jie M, Shuang-Quan Z and Shan-Gui Z 2003 *Chinese Phys. Lett.* **20** 358
[17] Zhou S G, Meng J and Ring P 2003 *Phys. Rev. Lett.* **91** 262501
[18] Mao G 2003 *Phys. Rev. C* **67** 044318
[19] Lisboa R, Malheiro M, de Castro A S, Alberto P and Fiolhais M 2004 *Phys. Rev. C* **69** 024319
[20] Leviatan A 2004 *Phys. Rev. Lett.* **92** 202501
[21] Guo J Y and Sheng Z Q 2005 *Phys. Lett. A* **338** 90
[22] Alberto P, Lisboa R, Malheiro M and de Castro A S 2005 *Phys. Rev. C* **71** 034313
[23] Guo J Y, Wang R D and Fang Z 2005 *Phys. Rev. C* **72** 054319
[24] Guo J Y, Fang Z and u F 2005 *Nuc. Phys. A* **757** 411
[25] Berkdemir C 2006 *Nucl. Phys. A* **770** 32
[26] u Q and Zhu S J 2006 *Nucl. Phys. A* **768** 161
[27] He , Zhou S, Meng J, Zhao E and Scheid W 2006 *Eur. Phys. J. A* **28** 265
[28] Jolos R and oronov 2007 *Phys. Atomic Nuclei* **70** 812
[29] Chun-Yan S, Jiang-Ming Y and Jie M 2009 *Chinese Phys. Lett.* **26** 122102
[30] Liang H, Hui Long W, Meng J and an Giai N 2010 *Eur. Phys. J. A* **44** 119
[31] Lisboa R, Malheiro M, Alberto P, Fiolhais M and de Castro A S 2010 *Phys. Rev. C* **81** 064324
[32] Chun-Yan S and Jiang-Ming Y 2010 *Chinese Phys. C* **34** 1425
[33] Liang H, Zhao P, Zhang Y, Meng J and Giai N 2011 *Phys. Rev. C* **83** 041301
[34] Chun-Yan S, Jiang-Ming Y and Jie M 2011 *Chinese Phys. Lett.* **28** 092101
[35] Lu B N, Zhao E G and Zhou S G 2012 *Phys. Rev. Lett.* **109** 072501
[36] de Castro A S and Alberto P 2012 *Phys. Rev. A* **86** 032122
[37] Alberto P, de Castro A S and Malheiro M 2013 *Phys. Rev. C* **87** 031301
[38] Alberto P, Castro A, Fiolhais M, Lisboa R and Malheiro M 2014 *J. Phys.: Conf. Series* **490** 012069
[39] Alberto P, de Castro A S and Malheiro M 2007 *Phys. Rev. C* **75** 047303
[40] Jackiw R and Rebbi C 1976 *Phys. Rev. D* **13** 3398
[41] Su W P, Schrieffer J R and Heeger A J 1979 *Phys. Rev. Lett.* **42** 1698
[42] Goldstone J and Wilczek F 1981 *Phys. Rev. Lett.* **47** 986
[43] Jackiw R and Semenoff G 1983 *Phys. Rev. Lett.* **50** 439
[44] Niemi A J and Semenoff G W 1986 *Phys. Rep.* **135** 99
[45] Charmchi F and Gousheh S 2014 *Phys. Rev. D* **89** 025002
[46] Charmchi F and Gousheh S 2014 *Nucl. Phys. B* **883** 256
[47] Castilho W M and de Castro A S 2014 *Ann. Phys. (NY)* **340** 1
[48] Castilho W M and de Castro A S 2014 *Ann. Phys. (NY)* **346** 164
[49] Thaller B 1992 *The Dirac Equation (Texts and Monographs in Physics)* (Berlin: Springer)
[50] Greiner W 1990 *Relativistic quantum mechanics: wave equations* (Berlin: Springer)
[51] Strange P 1998 *Relativistic Quantum Mechanics: with applications in condensed matter and atomic physics* (Cambridge: Cambridge University Press)
[52] de Castro A S, Alberto P, Lisboa R and Malheiro M 2006 *Phys. Rev. C* **73** 054309
[53] Watanabe S 1957 *Phys. Rev.* **106** 1306
[54] Touschek B 1957 *Il Nuovo Cimento* **5** 754
[55] Cooper F, Khare A, Musto R and Wipf A 1988 *Ann. Phys. (NY)* **187** 1
[56] Nogami Y and Toyama F M 1993 *Phys. Rev. A* **47** 1708
[57] de Castro A S 2005 *Ann. Phys. (NY)* **316** 414
[58] de Castro A S and Hott M 2005 *Phys. Lett. A* **342** 53
[59] Gradshteyn I and Ryzhik I 1980 *Table of Integrals, Series and Products* 2nd ed (San Diego: Academic)
[60] Abramowitz M and Stegun I A (eds) 1965 *Handbook of Mathematical Functions: with Formulas, Graphs, and Mathematical Tables (Dover Books on Mathematics)* 9th ed (Dover: Dover Publications)

Journal of Physics: Conference Series **630** (2015) 012030 doi:10.1088/1742-6596/630/1/012030

Quantum relativistic transport theory for asymmetric nuclear matter: collective modes

Felipe dos Passos, Sidney S. Avancini and Bruno Bertolino

Departamento de Física, CFM, Universidade Federal de Santa Catarina, Florianopolis, SC, CEP 88.040-900, Brazil

E-mail: fpassos100@gmail.com

Abstract. Isospin, density and electromagnetic waves in asymmetric nuclear matter with the addition of electrons (ANM) are studied. The use of a semi classical relativistic approach based on the Wigner function and its Vlasov like equation to study ANM allowed to obtain the dispersion relations for small amplitude oscillations. The small longitudinal wave case results were equivalent with a previous work that used a somewhat different formalism called generating function. The small transverse wave results, that cannot be obtained directly using the generating function formalism, were also obtained.

1. Introduction

Relativistic phenomenological models, such as the quantum hadrodynamical model (QHD) [1], are important tools to investigate and describe nuclei and nuclear matter. With the adequate parameters, these models are effective to study systems that vary from simple nuclei, both stable and unstable [2, 3], to astrophysical objects, e.g. neutron stars and supernovae [4, 5]. Due to this success, it is evident that a full investigation of the models properties is necessary. In particular, the quantum relativistic kinetic theory, and its quantum relativistic transport equations, play an important role to obtain the non-equilibrium behavior of these models.

In order to achieve a complete quantum relativistic kinetic theory, it is necessary to use a quantum relativistic distribution function, also known as the covariant Wigner function [6]. This approach allows us to develop a more robust theory, capable of investigating phenomena in non-equilibrium systems. Besides, its deduction is very similar to the classical one, allowing a deeper understanding of the quantum effects on the theory.

There are previous works already done on instabilities, phase transitions and collective modes in asymmetric nuclear matter (ANM) where the results were obtained using various methods. For example in Ref. [7] the author used a non relativistic Landau-Fermi liquid theory. In Refs. [8, 9], the Wigner function was used in a semi classical approach for the symmetric nuclear matter and the Vlasov equation (transport equation) was obtained, but the way to manipulate the Vlasov equation was through the generating function method. The same formalism was used in Refs. [10, 11] for ANM.

The present work aims to develop a different, more straightforward, formalism. Instead of using the generating functions to solve the Vlasov equation, we use the Vlasov equation and the system's equations of state. In this way we are able to reobtain the results from Ref. [11] for longitudinal small amplitude oscillations around a equilibrium state in nuclear matter.

Furthermore, we were able to obtain results for transverse oscillations, which to our knowledge was never made before.

2. Theory Overview

We consider a system of baryons of mass M interacting with and through a isoscalar-scalar field ϕ with mass m_s, a isoscalar-vector field $\mathcal{V}^\mu$ with mass m_v and an isovector-vector field $\vec{b}^\mu$ with mass m_ρ. Furthermore we include electrons with mass m_e. The electrons and the protons interact through the electromagnetic field A^μ. The Lagrangian density reads:

$$\mathcal{L} = \sum_{i=p,n} \mathcal{L}_i + \mathcal{L}_e + \mathcal{L}_\sigma + \mathcal{L}_\omega + \mathcal{L}_\rho + \mathcal{L}_A, \tag{1}$$

where:

$$\mathcal{L}_i = \bar{\psi}_i \left[i\gamma_\mu D^\mu - M^* \right] \psi_i;$$
$$D^\mu = i\partial^\mu - g_v \mathcal{V}^\mu - \frac{g_\rho}{2}\vec{\tau}\cdot\vec{b}^\mu - e\frac{1+\tau_3}{2}A^\mu;$$
$$M^* = M - gs\phi;$$
$$\mathcal{L}_e = \bar{\psi}_e \left[\gamma_\mu (i\partial^\mu + eA^\mu) - m_e \right] \psi_e;$$
$$\mathcal{L}_\sigma = \frac{1}{2}\left(\partial_\mu\phi\partial^\mu\phi - m_s^2\phi^2 - \frac{1}{3}\kappa\phi^3 - \frac{1}{4!}\lambda\phi^4 \right);$$

$$\mathcal{L}_\omega = -\frac{1}{4}\Omega_{*\mu\nu}\Omega^{*\mu\nu} + \frac{1}{2}m_v^2 \mathcal{V}_\mu \mathcal{V}^\mu;$$
$$\mathcal{L}_\rho = -\frac{1}{4}\vec{B}_{\mu\nu}\vec{B^{\mu\nu}} + \frac{1}{2}m_\rho^2 \vec{b}^\mu \vec{b}_\mu;$$
$$\mathcal{L}_A = -\frac{1}{4}F^{\mu\nu}F_{\mu\nu};$$
$$\Omega^{*\mu\nu} = \partial^\mu \mathcal{V}^\nu - \partial^\nu \mathcal{V}^\mu;$$
$$\vec{B^{\mu\nu}} = \partial^\mu \vec{b}^\nu - \partial^\nu \vec{b}^\mu - g_\rho(\vec{b}^\mu \times \vec{b}^\nu);$$
$$F^{\mu\nu} = \partial^\mu A^\nu - \partial^\nu A^\mu;$$

We also need the Wigner function definition:

$$F(x,p)^\beta_\alpha = \frac{1}{(2\pi)^4}\int d^4R\, e^{-ip\cdot R}\left\langle \bar{\psi}^\beta(x+R/2)\psi_\alpha(x-R/2) \right\rangle, \tag{2}$$

where the so called Wigner function actually is a 4×4 matrix.

Using the field equations together with the definition of the Wigner function, and assuming the ansatz $f(x,\Pi) = 2\Theta(\Pi_0)\delta(\Pi^2 - M^{*2})\tilde{f}(x,\vec{\Pi})$ for the scalar Wigner function component, which is in complete agreement with the stationary case, it is possible to obtain the Vlasov equation (transport equation) for the system, in $O(\hbar)$:

$$\left[\Pi^\mu \partial_\mu - \Pi^\mu \Omega_{\mu k}\partial^k_\Pi + M^*(\partial^k M^*)\partial^\Pi_k \right] \tilde{f}(x,\vec{\Pi}) = 0, \tag{3}$$

where $\Omega^{\mu\nu} = \partial^\mu V^\nu - \partial^\nu V^\mu$;, $V^\mu = g_v\mathcal{V}^\mu + \frac{g_\rho}{2}\vec{\tau}\cdot\vec{b^\mu} + e\frac{1+\tau_3}{2}A^\mu$, Π^μ is the kinetic four-momentum, the greek indices represents four-vectors and the latin indices spatial three-vectors.

The process to obtain this equation is straightforward, and details con be found in Ref [12]. We can rewrite this equation as:

$$\frac{d}{dt}f(x,\vec{\Pi}) = \frac{\partial f}{\partial t} + \{f, H\} = 0, \tag{4}$$

where H is the system's Hamiltonian. This is the conservation of the absolute particle number (particles-antiparticles). Thus, the Vlasov equation represents one of the most basic conservation law in physics.

3. Small Oscillations

Solving the Vlasov equation for the most general case is impossible, but as we are interested in studying the collective modes, we should approach the equilibrium state through small oscillations. These small deviations can be described by:

$$f_i = f_{i(0)} + \delta f_i, \qquad \phi = \phi_0 + \delta\phi, \qquad V_i^{\mu} = V_{i(0)}^{\mu} + \delta V^{\mu},$$

where $f_{i(0)}$, ϕ_0 and $V_{i(0)}^{\mu}$ are the equilibrium quantities.

When we apply these near equilibrium conditions in the Vlasov equation and in the field equations, assuming again $O(\hbar)$ and considering the Fourier transform of all the quantities considered, we obtain the following set of equations:

$$\begin{cases} \delta f = \left[\delta\vec{V} - \dfrac{\left(\delta V^0 - \frac{\vec{\Pi}}{E_{\Pi}^{(0)}}\cdot\delta\vec{V} + \frac{M^{(0)*}g_s}{E_{\Pi}^{(0)}}\delta\phi\right)}{\omega - \frac{\vec{\Pi}\cdot\vec{q}}{E_{\Pi}^{(0)}}}\vec{q}\right]\cdot\nabla_{\vec{\Pi}}f(\vec{\Pi}), \\ \delta j_i^{\mu}(x) = \begin{cases} \frac{2}{(2\pi)^3}\int d^3\vec{\Pi}\ \delta f_i, & \mu = 0 \\ \frac{2}{(2\pi)^3}\int d^3\vec{\Pi}\frac{\Pi^k}{E_{\Pi}^{(0)}}\delta f_i, & \mu = k \end{cases}, \\ \delta\rho_s^i(x) = \frac{2}{(2\pi)^3}\int d^3\vec{\Pi}\left[\frac{M^{(0)*}}{E_{\Pi}^{(0)}}\delta f_i - g_s d\rho_s^{(0)i}\delta\phi\right], \\ \left[-\omega^2 + \vec{q}^2 + \tilde{m}_s^2\right]\delta\phi(\vec{q},\omega) = g_s\sum_{i=p,n}\frac{2}{(2\pi)^3}\int d^3\vec{\Pi}\frac{M^{(0)*}}{E_{\Pi}^{(0)}}\delta f_i(\vec{q},\omega,\vec{\Pi}), \\ \delta V_i^{\mu}(\omega,\vec{q}) = \sum_{j=p,n}\left[\frac{g_\omega^2}{-\omega^2+\vec{q}^2+m_\omega^2} + \frac{(g_\rho/2)^2\tau_i\tau_j}{-\omega^2+\vec{q}^2+m_\rho^2}\right]\delta j_j^{\mu}(\omega,\vec{q}), \end{cases} \tag{5}$$

where $M^{*(0)} = M - g_s\phi^{(0)}$, $E_{\pi}^{(0)} = \sqrt{\vec{\Pi}^2 + (M^{*(0)2}}$, $d\rho_s^{(0)i} = \frac{\partial}{\partial M^{*(0)}}\rho_s^{(0)i}$, $\tilde{m}_s^2 = m_s^2 + k\phi^{(0)} + \frac{\lambda}{2}\phi^{(0)2} + \frac{2}{(2\pi)^3}g_s^2\sum_{i=p,n}\int d^3\vec{\Pi}\frac{\partial}{\partial M^{(0)*}}\rho_s^{(0)i}$ and $i = p,\ n$. Again, obtaining these equations is straightforward and a more detailed derivation can be found in Ref [13].

Strictly speaking the problem is solved. This set is compatible and determined. But, as we can see, for the most general case the solution is too complex. For this reason we will work with two cases in which the system can be analytically solved up to a certain point. The electron equations can be obtained from (5) by putting $g_s = g_\omega = g_\rho = 0$, since the electronic Lagrangian does not have such terms.

3.1. Longitudinal Waves

Let us first consider the longitudinal waves for two main reasons: it is the simplest problem and it was already solved in Ref. [11], so we are able to compare results. In our case, a longitudinal wave is described by the following set:

$$\delta\vec{V}_i = \delta V_{i3}\hat{e}_3, \qquad \vec{q} = q\hat{e}_3, \qquad \omega\delta V_i^0 = q\delta V_i^3,$$

where the last equation comes from the potential conservation law and $i = p,\ n$. Inputting this new information on the set obtained previously and after long algebraic manipulations, we obtain the following equation:

$$\begin{pmatrix} 1 + F^{pp}L_p & F^{pn}L_p & C_e^{pe}L_p \\ F^{np}L_n & 1 + F^{nn}L_n & 0 \\ C_e^{ep}L_e & 0 & 1 - C_e^{ee}L_e \end{pmatrix}\begin{pmatrix} A_{\omega p} \\ A_{\omega n} \\ A_{\omega e} \end{pmatrix} = 0 \tag{6}$$

where the F's, C's are functions depending on the equilibrium densities and fields and the A's are the variables. This is the called dispersion equation for the system. This matrix is equivalent to the matrix obtained in Ref [11]. Thus the method used is equivalent to the generating function method used in that reference.

If we want non-trivial solutions for this equation, we should make the determinant of the first matrix zero. With this condition we are able to get F's, C's values and using these we get information about the system properties, such as the collective modes amplitude for both protons, neutrons and electrons. This was already investigated in [11].

3.2. Transverse Waves

We are now heading into new territory. These calculations are harder to be done due to the fact that the generating function formalism is not well adapted to work in this condition. But as we will show the formalism used in the present work proved to be effective in this case. Let us begin with the transverse wave condition:

$$\delta\vec{V_i} = \delta V_{i\perp}\hat{e}_\perp, \qquad \vec{q} = q\hat{e}_3, \qquad \vec{q}\cdot\delta\vec{V_i} = 0$$

Again, using these equations in the set obtained previously and after long algebraic manipulations, we obtain:

$$\begin{pmatrix} 1+F^{pp} & F^{pn} & C_e^{pp*} \\ F^{np} & 1+F^{nn} & C_e^{np*} \\ C_e^{ep*} & 0 & 1-C_e^{ee*} \end{pmatrix} \begin{pmatrix} B_{\omega p} \\ B_{\omega n} \\ B_{\omega e} \end{pmatrix} = 0 \tag{7}$$

where the F's, C's are functions depending on the equilibrium densities and fields and the B's are the variables.

Again if we want non-trivial solutions for this equation, we should make the determinant of the first matrix zero. Of course the problem is solved and the next step are numerical simulations. We are implementing the algorithms and we expect to have results soon.

4. Conclusions

In order to validate our formalism to solve the Vlasov Equation we were able to rederive results obtained in [11] for longitudinal waves. This shows that the formalism used in this work is equivalent to the generating functions method.

Furthermore the formalism developed here was able to solve the transverse wave collective mode, which was not treated before as seen in Ref. [11]. This couldn't be done using the generating function formalism, at least not in a direct way. Using this result we could investigate how an electromagnetic wave propagates through the nuclear matter described by the ANM Lagrangian.

Also we see that the Wigner function is well suited to work in near equilibrium $O(\hbar)$ conditions. There are various works in the literature studying near equilibrium systems using a multitude of methods and approximations, but using the covariant Wigner function we are able to get more detailed and reliable results. Besides, the formalism used in this work gives results for the mesonic collective modes that are ignored in other methods, e.g. the Landau-Fermi Liquid Theory.

The next step in this research is the numerical solution of the transverse wave modes. After we have the numerical solution, we will be able to draw conclusions from the theory and investigate if there is any regions of instability in the npe matter. We expect to solve the same problem, but including a strong background magnetic field that are speculated to exist in the core of some neutron stars, named magnetars.

Acknowledgments

We would like to thank CNPq-Conselho Nacional de Desenvolvimento Científico e Tecnológico for the finnancial support given to the authors.

References

[1] Serot B D and Walecka J D 1986 *Advances in Nuclear Physics* **16**
[2] Reinhard P G, Rufa M, Maruhn J, Greiner W and Friedrich J 1986 *Z. Phys.* A **13** 323
[3] Sugahara Y and Toki H 1994 *Nucl. Phys.* A **579** 557
[4] Shen H, Toki H, Oyamatsu K and Sumiyoshi K 1998 *Nucl. Phys.* A **637** 435
Sumiyoshi K and Toki H 1994 *Astrophys. J.* **422** 700
Menezes D P and Providencia C 2003 *Phys. Rev.* C **68** 035804
Panda P K, Menezes D P and Providencia C 2004 *Phys. Rev.* C **60** 025207
[5] Horowitz C J and Piekarewicz 2001 *Phys. Rev. Lett.* **86** 5647
[6] Hakim R 1978 *La Rivista Del Nuovo Cimento* **1** 1
Hakim R and Heivaerts J 1978 *Phys. Rev.* A **18** 1250
Hakim R 2011 *Introduction to Relativistic Statistical Mechanics: Classical and Quantum* (Singapore, World Scientific)
[7] Haensel P 1978 *Nucl. Phys.* A **301** 53
[8] Nielsen M, Providencia C and Providencia J 1991 *Phys. Rev.* C **44** 209
Nielsen M, Providencia C , Providencia J and Wang-Ru L 1994 *Mod. Phys. Lett.* A **10** 919
[9] Nielsen M, Providencia C and Providencia J 1993 *Phys. Rev.* C **47** 200
[10] Avancini S S, Brito L, Menezes D P and Providencia C 2005 *Phys. Rev.* C **71** 044323
[11] Brito L et al 2006 *Phys. Rev.* C **74** 045801
[12] Blattel B, Koch V and Mosel U 1993 *Rep. Prog. Phys.* **56** 1
[13] Passos F, 2014 *Formalismo de Vlasov covariante aplicado a modelos efetivos* (Florianópolis, Brasil: http://tede.ufsc.br/teses/PFSC0253-D.pdf)

Hyperon-meson and delta-meson coupling to protoneutron stars structure using the nonlinear Walecka model

W A Souto[1], J C T Oliveira[2], H Rodrigues[3], S B Duarte[4] and M Chiapparini[5]

[1] Governo do Estado de Roraima, Secretaria de Segurança Pública, Instituto de Criminalística, Rua José Pinheiro, 952, 69.300-000, Roraima, Brazil
[2] Departamento de Física, Universidade Federal de Roraima, Campus do Paricarana s/n, 69310-270, Roraima, Brazil
[3] Centro Federal de Educação Tecnológica do Rio de Janeiro, Av. Maracanã 229, 20271-110, Rio de Janeiro, RJ, Brazil
[4] Centro Brasileiro de Pesquisas Físicas, Rua Dr. Xavier Sigaud 150, 22290-180, Rio de Janeiro, RJ, Brazil
[5] Instituto de Física, Universidade do Estado do Rio de Janeiro, Rua São Francisco Xavier 524, 20550-900, Rio de Janeiro, RJ, Brazil

E-mail: wsouto@yahoo.com

Abstract. In this work we determine the equation of state and the population of baryons and leptons and discuss the effects of the hyperon-meson coupling constants to the formation of delta resonances in the stellar medium. We also discuss the structure of the protoneutron stars including the delta matter in their composition, and compared the results of a cooled neutron star, after escape of neutrinos. For protoneutron stars structure and composition, the neutrinos are considered trapped.

1. Introduction

The high production of delta-resonances ($\rho \geq 3\rho_0$) in the dense phase of relativistic heavy ion collisions [1-3] leads to a great interest in the study of the effects of delta matter in the structure of protoneutron stars and neutron stars. In the present work we extended the nonlinearWalecka (NLW) model to accommodate in the context of the relativistic mean field approximation the Rarita-Schwinger field for the 3/2-spin resonances. In this work the delta-rich hadronic condensate in stellar matter is studied for the hadrons in beta-equilibrium with the leptons present in the medium. The baryon sector includes the complete 1/2-spin octet and delta resonances with 3/2-spin, which interact by exchange of scalar, vector and isovector mesons.

2. The model

In sector baryonic, we consider in the calculations the 1/2-spin baryon octet (n, p, Λ^0, Σ^-, Σ^0, Σ^+, Ξ^-, Ξ^0), the 3/2-spin baryonic resonances represented by the delta-matter (Δ^-, Δ^0, Δ^+, Δ^{++}) and by Ω^- [4,5]. In the leptonic sector are considered electrons, muons and trapped neutrinos. For mean field

Journal of Physics: Conference Series **630** (2015) 012031 doi:10.1088/1742-6596/630/1/012031

approximation we have adopted the nonlinear Walecka (NLW) model with the Lagrangian densities [6,7] given by $\mathcal{L} = \mathcal{L}_F + \mathcal{L}_I$, where

$$\mathcal{L}_F = \sum_B \Psi_B (i\gamma_\mu \partial_\mu - m_B)\Psi_B + \sum_\zeta R_\zeta \nu\, (i\gamma_\mu \partial_\mu - m_\zeta)\, R\nu_\zeta$$
$$+\ 1/2\,(\partial_\mu \sigma\, \partial^\mu \sigma - m_\sigma^2 \sigma^2) - b/3\, m_N (g_{\sigma N}\sigma)^3 - c/4\,(g_{\sigma N}\sigma)^4 - 1/4\,\omega_{\mu\nu}\,\omega^{\mu\nu}$$
$$+ 1/2\, m_\omega^2\, \omega_\mu\, \omega^\mu - 1/4\, \rho_{\mu\nu} . \rho^{\mu\nu} + 1/2\, m_\rho^2\, \rho_\mu . \rho^\mu + \sum_\lambda \Psi_\lambda (i\gamma_\mu \partial_\mu - m_\lambda)\Psi_\lambda .$$

is the Lagrangian density for free baryons, electrons, muons and neutrinos ($\lambda = e^-, \mu^-, \nu_e, \nu_\mu$), and σ, ω and ρ are meson fields. In addition, the interaction Lagrangian density has the following form

$$\mathcal{L}_I = \sum_B \Psi_B\, (g_{\sigma B}\, \sigma - g_{\omega B}\, \gamma_\mu\, \omega_\mu - 1/2\, g_{\rho B}\, \gamma_\mu\, \tau . \rho^\mu)\Psi_B$$
$$+ \sum_\zeta R_\zeta \nu\, (g_{\sigma\zeta}\, \sigma - g_{\omega\zeta}\, \gamma_\mu\, \omega_\mu - 1/2\, g_{\rho\,\zeta}\, \gamma_\mu\, \tau . \rho^\mu) R\nu_\zeta .$$

In the equations above the operator Ψ_B represents the Dirac spinor describing the 1/2-spin baryon-octet, with $B = n, p, \Lambda^0, \Sigma^-, \Sigma^0, \Sigma^+, \Xi^-, \Xi^0$. The operator $R_\zeta \nu$ is the Rarita-Schwinger spinor for the 3/2-spin resonances [8], with $\zeta = \Delta^-, \Delta^0, \Delta^+, \Delta^{++}, \Omega^-$.
In this work we have used the set of coupling constants given in Ref. [9], namely $(g_{\sigma N}/m_\sigma)^2 = 9.927$ fm^2, $(g_{\omega N}/m_\omega)^2 = 4.820\, fm^2$, $(g_{\rho N}/m_\rho)^2 = 4.791\, fm^2$, $b = 0.008659$ and $c = -0.002421$.

3. The hyperon-meson coupling constants

The coupling constants between the hyperons (Λ, Σ, Ξ) and the mesons ω and ρ are fixed by using SU(6) symmetry [9], while the hyperons-σ coupling constants are constrained by the consistence of the hypernuclear potential in the nuclear matter with hypernuclear data (See Table 1) [10].

Table 1. Sets of parameters used for the hyperon-meson coupling constants.

	$x_{\omega\Lambda}$	$x_{\omega\Sigma}$	$x_{\omega\Xi}$	$x_{\sigma\Lambda}$	$x_{\sigma\Sigma}$	$x_{\sigma\Xi}$	$x_{\rho\Lambda}$	$x_{\rho\Sigma}$	$x_{\rho\Xi}$
Set 1	0.6666	0.6666	0.6666	0.6104	0.6104	0.6104	0.6104	0.6104	0.6104
Set 2	0.6666	0.6666	0.3333	0.6106	0.4046	0.3195	1.000	1.000	1.000

Set1 in Table 1 is taken from Ref. [9], where the couplings between hyperons and σ and ω mesons have to be compatible with the allowed maximum neutron star masses and with the Λ binding energy in nuclear matter. We shall assume that all hyperons in the octet have the same coupling as the Λ [9]. While Set2 is obtained from Set1 imposing SU(6) symmetry and different binding energies for hyperons in nuclear matter [10]. In Table 1 we have $x_{ij} = g_{ij}/g_{iN}$ where $i = \omega, \sigma, \rho$; $j = \Lambda, \Sigma, \Xi$ and $N = p, n$.
In addition, we use the results of the finite density QCD Sum Rules (QCDSR) to determine the possible values of delta-meson coupling constants [11], where we have taken different values for the quantities $\alpha = g_{\omega\Delta}/g_{\omega N}$, $\beta = g_{\sigma\Delta}/g_{\sigma N}$ and $\gamma = g_{\rho\Delta}/g_{\rho N}$.

4. Results and conclusions

Using the equation of state obtained with the mentioned model, we solve numerically the equation TOV (Tolman-Oppenheimer-Volkoff) [12,13] and so we obtained the values of the maximum mass of

Journal of Physics: Conference Series **630** (2015) 012031 doi:10.1088/1742-6596/630/1/012031

the star, before and after the cooling phase. In Figure 1 we show the mass-radius diagrams for the protoneutron stars (with trapped neutrinos) and neutron stars (without neutrinos), referring to the cases: QCDSR, QCDSR + Set1, and QCDSR + Set2, respectively. We conclude that the protoneutron stars after the cooling phase induced by escape of neutrinos leads to the formation of less massive neutron stars with the same baryon-meson coupling constants set in the equation of state.

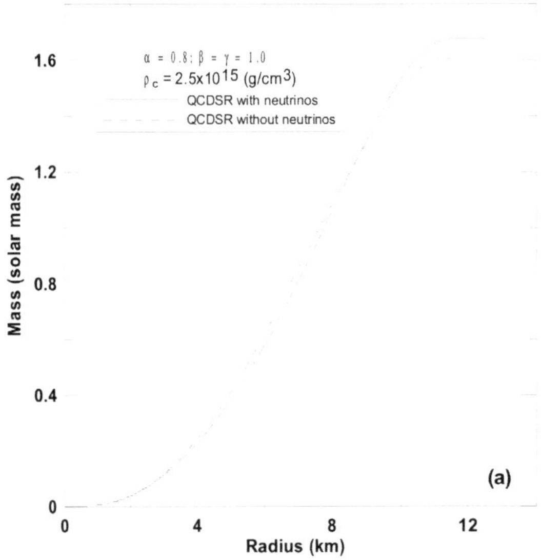

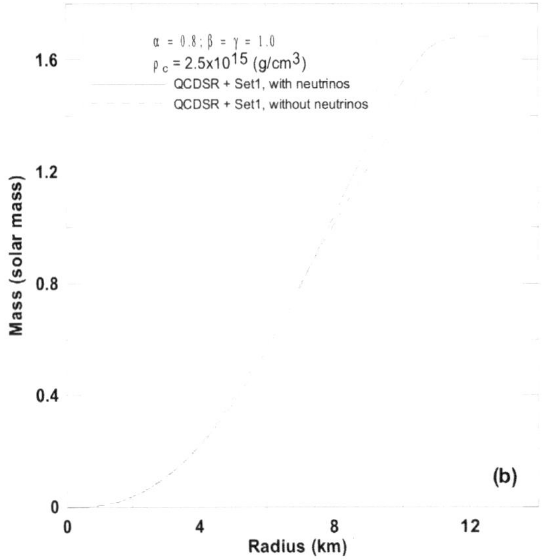

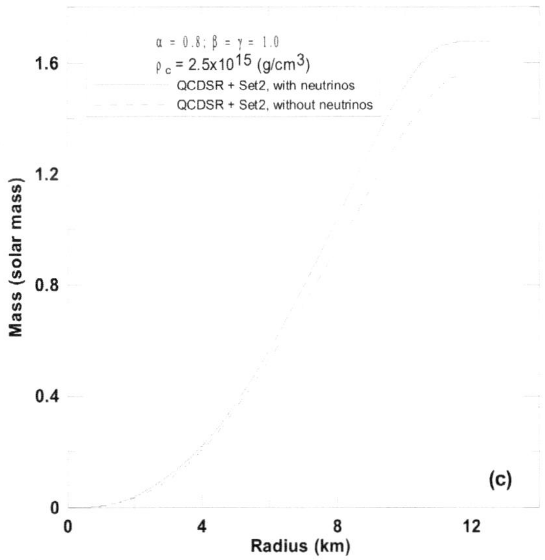

Figure 1. Mass-radius diagrams referring to the cases: QCDSR (part-a), QCDSR + Set1 (part-b), and QCDSR + Set2 (part-c), with neutrinos (solid lines) and without neutrinos (dashed lines). The figure shows the results for $\alpha = 0.8$, $\beta = \gamma = 1.0$, and central density 2.5×10^{15} (g/cm^3).

References

[1] Hjort E L et al 1997 Phys. Rev. Lett. **79** 4345
[2] Hofmann M et al 1995 Phys. Rev. C **51** 2095
[3] Hong B et al 1997 Phys. Lett. B **407** 115
[4] Oliveira J C T, Kyotoku M, Chiapparini M, Rodrigues H and Duarte S B 2000 Mod. Phys. Lett. A **15** 1529
[5] Oliveira J C T, Duarte S B, Rodrigues H, Chiapparini M and Kyotoku M 2007 Int. J. Mod. Phys. D **16** 175
[6] Walecka J D 1974 Ann. Phys. **83** 497
[7] Serot B D and Walecka J D 1986 Adv. Nucl. Phys. **16** 1
[8] Rarita W and Schwinger J 1941 Phys. Rev. **60** 61
[9] Glendenning N K and Moszkowski S A 1991 Phys. Rev. Lett. **67** 2414
[10] Chiapparini M et al 2009 Nucl. Phys. A **826** 178
[11] Kosov D S, Fuchs C, Martemyanov B V and Faessler A 1998 Phys. Lett. B **421** 37
[12] Tolman R C 1939 Phys. Rev. **55** 364
[13] Oppenheimer J R and Volkoff G M 1939 Phys. Rev. **55** 374

Journal of Physics: Conference Series **630** (2015) 012032 doi:10.1088/1742-6596/630/1/012032

Hydrodynamics applied to relativistic heavy ion collisions: influence of equations of state on pion interferometry

D S Lemos, J Carvalho and O Socolowski Jr

Instituto de Matemática, Estatística e Física, Universidade Federal do Rio Grande, Brazil

E-mail: denerslemos@gmail.com

Abstract. In this work we study the effects of a first order phase transition EoS and a lattice QCD based EoS on the pion interferometry. We use a simple hydrodynamical model with cylindrical symmetry and longitudinal boost invariance. We conclude that the results of the HBT radii are very similar for both EoS.

1. Introduction

Hydrodynamics has a long history in high-energy collisions [1]. Nowadays, it is an essential tool for describing the collective behaviour of the matter produced in relativistic heavy-ion collisions at RHIC and LHC [2, 3, 4, 5, 6, 7]. Because the hydrodynamical approach involves thermodynamical properties of the system, it is the perfect tool to study the equation of state (EoS) of the strongly interacting matter formed in the collision.

While several hydrodynamical models have successfully used equations of state with first-order phase transition between a hadronic matter and a quark-gluon plasma (QGP), results of lattice QCD have shown that the transition between these phases of matter is not of first-order, but it is of crossover type [8]. It is importante to see how the hydrodynamical evolution is affected by the use of different equations of state and how much the observables are affected in these process. Our aim in this work is to perform a study of interferometry of identical pions (HBT effect), by using a simple hydrodynamical model with two different equations of state: an EoS with first-order phase transition between a QGP and a resonance gas with excluded volume correction and a QCD EoS with a crossover transition between the QGP and the hadron phase. The hydrodynamical calculations are performed for central Au + Au collisions at RHIC energy of $\sqrt{s} = 130\text{A GeV}$, by using cylindrical symmetry and Lorentz boost invariance along the collision axis.

This work is organized as follows. In Sec. 2 we describe the hydrodynamical model used here; in Sec. 3 we show the equations of states used; in Sec. 4 we present the HBT effect; in Sec. 5 we show the numerical results; finally, conclusions are presented in Sec. 6. Much of this work has been presented in [9].

2. Hydrodynamic Evolution

In the hydrodynamical description of a relativistic heavy-ion collision, we assume that a matter in local thermal equilibrium is formed and its evolution is governed by the conservation of energy-momentum and other conserved numbers (baryonic number, strangeness, electric charge, etc.). Here, we assume that all the conserved numbers are zero. Thus, in an arbitrary coordinate system, the hydrodynamic equations are written as [10]

$$T^{\mu\nu}{}_{;\mu} = \partial_\mu T^{\mu\nu} + \Gamma^\mu{}_{\sigma\mu} T^{\sigma\nu} + \Gamma^\nu{}_{\sigma\mu} T^{\mu\sigma} = 0, \tag{1}$$

where $\Gamma^\mu{}_{\sigma\nu}$ are the Christoffel symbols. The energy-momentum tensor, $T^{\mu\nu}$, for a perfect fluid is given by

$$T^{\mu\nu} = (\varepsilon + P)\, u^\mu u^\nu - P g^{\mu\nu}, \tag{2}$$

where ε, P, u^μ, and $g^{\mu\nu}$ are, respectively, the energy density, the pressure, the fluid flow four-velocity, and the metric tensor. Here, we use a simplified approach for describing the fluid evolution: we suppose transversal and longitudinal expansion by using longitudinal boost invariance [11] and cylindrical symmetry. In this framework, it is more convenient to use hyperbolic-cilindrical coordinates, $\{\tau, \rho, \phi, \eta\}$, defined as: $\tau = (t^2 - z^2)^{1/2}$, $\rho = (x^2 + y^2)^{1/2}$, $\phi = \tan^{-1}(y/x)$, and $\eta = (1/2)\ln[(t+z)/(t-z)]$. The hydrodynamical equations (1) are explicitly written as

$$\partial_\tau T^{\tau\tau} + \partial_\rho\left[(T^{\tau\tau} + P)\, v_T\right] + (T^{\tau\tau} + P)\left(\frac{1}{\tau} + \frac{v_T}{\rho}\right) = 0, \tag{3}$$

$$\partial_\tau T^{\tau\rho} + \partial_\rho\left(T^{\tau\rho} v_T + P\right) + T^{\tau\rho}\left(\frac{1}{\tau} + \frac{v_T}{\rho}\right) = 0, \tag{4}$$

where $T^{\tau\tau} = (\varepsilon + P)\gamma_T^2 - P$ and $T^{\tau\rho} = (\varepsilon + P)\gamma_T^2 v_T$, with $\gamma_T = (1 - v_T^2)^{-1/2}$. To solve these equations, we need to provide initial conditions and an equation of state.

We assume that the hydrodynamical evolution stops on a sharply defined hypersurface characterized by a freeze-out temperature $T_{\rm fo}$. The invariant momentum distribution of observed particles is given by the Cooper-Frye formula [12]

$$E\frac{d^3N}{dp^3} = \int d\sigma_\mu p^\mu f(x, p), \tag{5}$$

where $d\sigma_\mu$ is a normal vector to the freeze-out surface and $f(x, p)$ is the equilibrium distribution function.

3. Equation of state

We consider equations of state for a system with zero net baryon density, neutral charge, and neutral strangeness. We compare two types of equations of state: EoS with first order phase transition (1ST), and EoS based on lattice QCD (QCD-s95p).

The EoS 1ST corresponds to a first order phase transition between a hadronic gas, which includes resonances up to 2 GeV and excluded volume correction [13], and a QGP, described by the MIT bag model (see Ref. [2] for details). The critical temperature used in EoS 1ST is $T_{\rm c} = 160\,{\rm MeV}$. The EoS QCD-s95p [14] is constructed by using the trace anomaly, $\Theta^\mu_{\ \mu}(T) = \varepsilon(T) - 3P(T)$, which is evaluated on lattice points. This EoS is parametrized to connect the trace anomaly of a QGP (high temperature) with the trace anomaly of a hadron gas (low temperature). The pressure is calculated by using

$$\frac{P(T)}{T^4} - \frac{P(T_{\rm low})}{T_{\rm low}^4} = \int_{T_{\rm low}}^{T} \frac{\Theta^\mu_{\ \mu}(T')}{T'^5}\, dT', \tag{6}$$

where $P(T_{\rm low})$ is the pressure of the hadronic gas. The Fig. 1 shows a comparison of the two EoS.

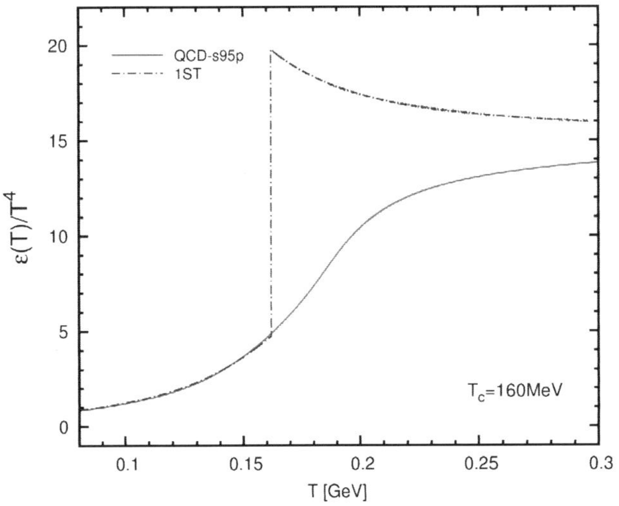

Figure 1. Phase diagram comparing the two EoS: first-order phase transition (1ST) and based on lattice QCD (QCD-s95p).

4. Pion interferometry

The interferometry of identical particles (or HBT effect) [15] is a powerfull method for probing the space-time geometry of the particle-emitting source. For two particles with four-momenta p_1 and p_2, we can define the two-particle correlation function as

$$C_2(p_1, p_2) = \frac{\mathcal{P}_2(p_1, p_2)}{\mathcal{P}_1(p_1)\mathcal{P}_1(p_2)}, \tag{7}$$

where $\mathcal{P}_1$ and $\mathcal{P}_2$ are, respectively, the one and two-particle inclusive distributions. The two-particle correlation function can be calculated from the emission function, $S(x, \mathbf{p})$, which describes the probability of a particle to be emitted from a space-time point x with momentum $\mathbf{p}$. We can write [16, 17]

$$C_2(\mathbf{q}, \mathbf{K}) = 1 + \frac{\left|\int d^4x \exp(iq \cdot x) S(x, \mathbf{K})\right|^2}{\int d^4x S(x, \mathbf{K} + \mathbf{q}/2) \int d^4x S(x, \mathbf{K} - \mathbf{q}/2)}, \tag{8}$$

where $q = p_1 - p_2$ is the relative four-momentum and $K = (p_1 + p_2)/2$ is the average four-momentum of the pair. The emission function can be written as

$$S(x, \mathbf{p}) = \int d\sigma_\mu(x') p^\mu \delta^{(4)}(x' - x) f(x', \mathbf{p}), \tag{9}$$

where $f(x, \mathbf{p})$ is the equilibrium distribution function. $\int d^4x S(x, \mathbf{p})$ coincides with the one-particle spectrum (5).

5. Numerical results

In this section, we show the results for the transverse-momentum distribution and HBT radii for pions calculated by using two equation of state, namely, 1ST and QCD-s95p. To solve the hydrodynamical equations (3) and (4), we use the method RHLLE [21] and simple initial conditions. We consider only central collisions with constant initial energy density of

Journal of Physics: Conference Series **630** (2015) 012032 doi:10.1088/1742-6596/630/1/012032

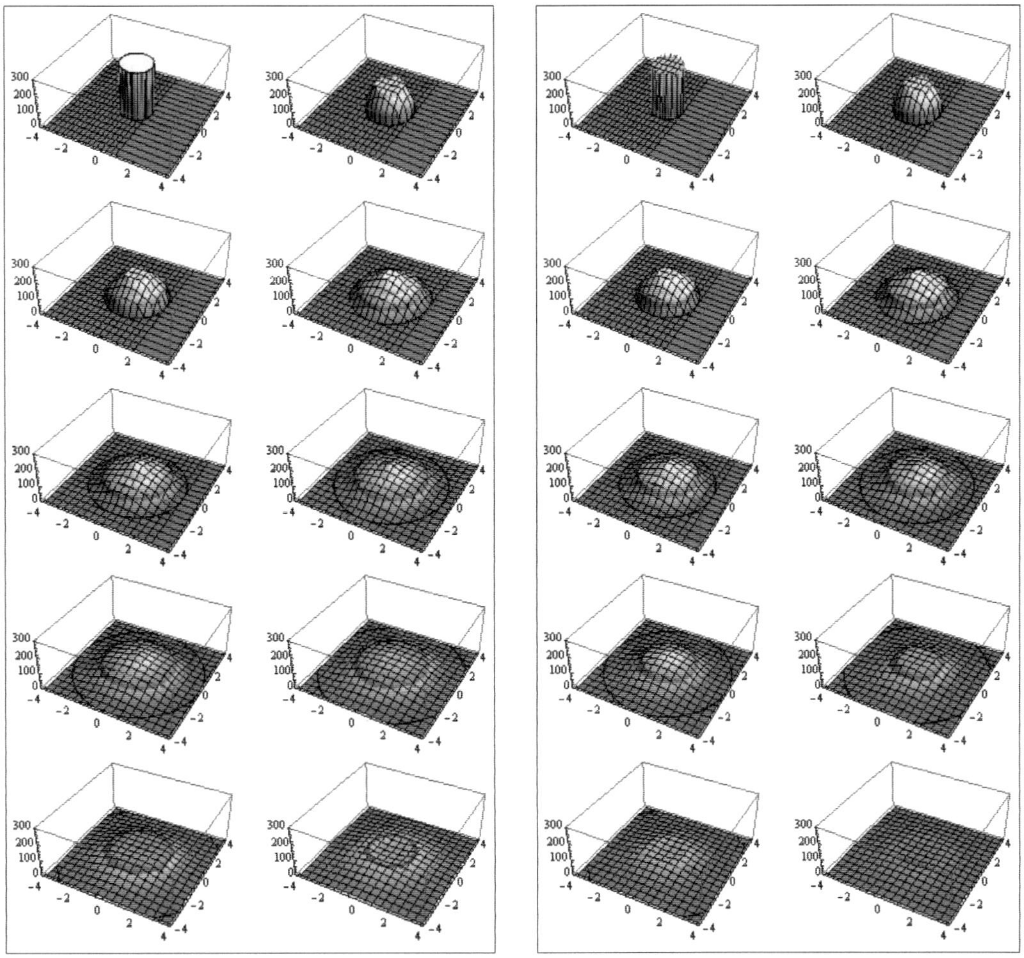

Figure 2. Temperature evolution (in MeV) in xy plane ($\eta = 0$). *Left*: EoS with first-order phase transition is used. *Right*: EoS based on lattice QCD is used. Coordinates x and y are in units of the nucleus radius.

$10\,\mathrm{GeV/fm^3}$. The initial time is fixed in 0.3 fm. We look at two scenarios: (A) without initial transverse velocity; (B) with initial transverse velocity parameterized as [22] $\tanh(\alpha\rho/R_A)$, where the parameter $\alpha = 0.1$ and the nucleus radius $R_A = 7\,\mathrm{fm}$.

5.1. Transverse-momentum distribution

The transverse-momentum spectra for pions are shown in Fig. 3 for scenario A and in Fig. 4 for scenario B. The figures show the results for the following freeze-out temperatures: 120 MeV, 130 MeV, 135 MeV, and 140 MeV. We see a small difference between the two equations of state. The p_T distribution is lightly flatter for QCD-s95p, because, in this equation of state, the pressure is not constant in the crossover transition region causing a larger acceleration of the system. This can be seen in Fig. 2 where the temperature evolution of the fluid is shown for both equations of state: 1ST on the left and QCD-s95p on the right. We see that the fluid cools faster for the QCD-s95p EoS, showing a larger acceleration. We also note that the presence of an initial transverse velocity (scenario B) gives better results when comparing with experimental data.

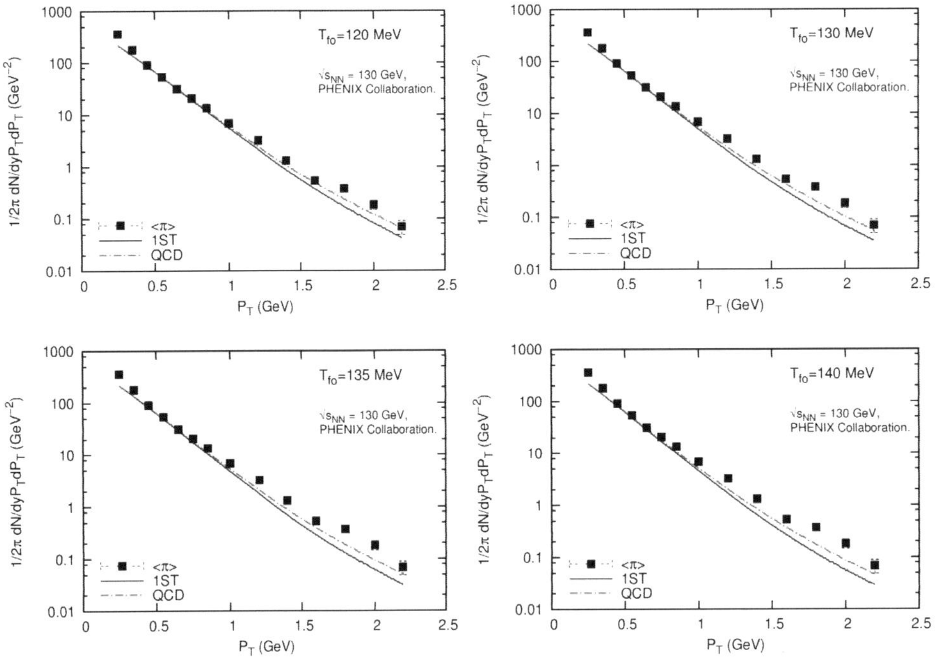

Figure 3. Transverse-momentum distributions for pions in Au+Au collisions at 130A GeV, calculated by using four differents freeze-out temperatures. Scenario without initial transverse velocity. Comparison of 1ST EoS (solid line) and QCS-s95p EoS (dashed line). Data are from PHENIX collaboration [20]; centrality 0-5%.

5.2. HBT radii

We choose the freeze-out temperature $T_{\text{fo}} = 120$ MeV to calculate the HBT radii. To compare our results with experimental data, we use the Bertsch-Pratt *out-long-side* coordinates [16] defined as:

$$q_o = \frac{\mathbf{q}_T \cdot \mathbf{K}_T}{K_T}, \quad q_\ell = q_z, \quad q_s = \frac{(\mathbf{K} \times \mathbf{q})_z}{K_T}, \tag{10}$$

and fit the correlation function (8) with the three-dimensional gaussian

$$C_2(\mathbf{q}, \mathbf{K}) = 1 + \lambda \exp\left[-R_o^2 q_o^2 - R_\ell^2 q_\ell^2 - R_s^2 q_s^2\right]. \tag{11}$$

Figures 5 and 6 show the radii R_o, R_ℓ, R_s, and the ratio R_o/R_s for the scenarios A and B, respectively, as function of K_T. The results are compared with experimental data for Au+Au collisions at 130A GeV from PHENIX [19] and STAR [18] collaborations. We can see a small difference between the equations of state, with QCD-s95p EoS giving smaller R_o, R_ℓ, and R_o/R_s. This behaviour is associated with the larger acceleration of the system for QCD-s95p EoS. The R_s radius does not show any significant difference for the two EoS. It should be noted that the

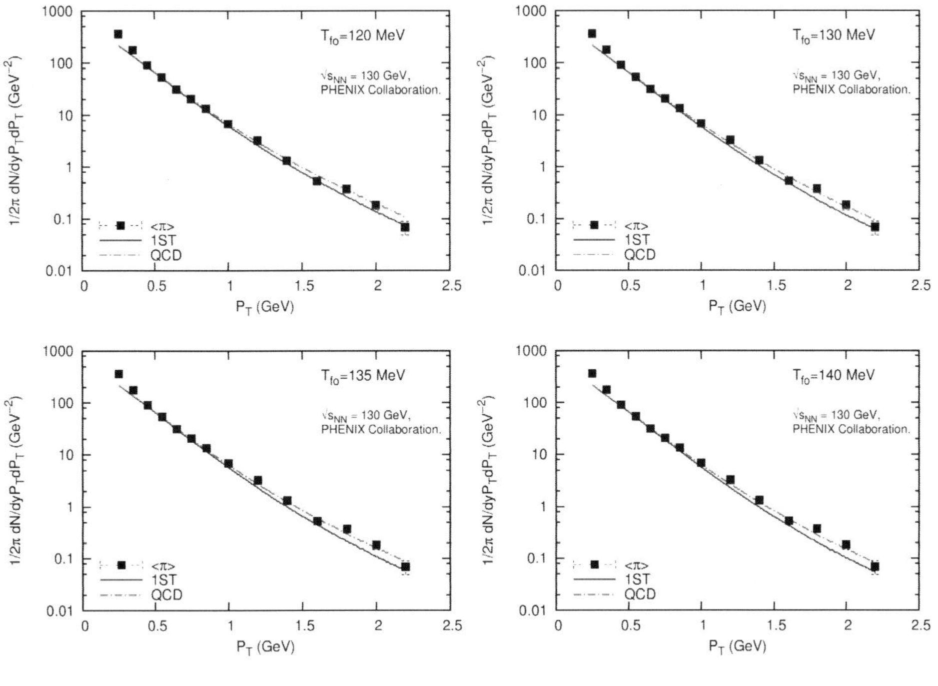

Figure 4. Transverse-momentum distributions for pions in Au+Au collisions at 130A GeV, calculated by using four differents freeze-out temperatures. Scenario with initial transverse velocity. Comparison of 1ST EoS (solid line) and QCS-s95p EoS (dashed line). Data are from PHENIX collaboration [20]; centrality 0-5%.

scenario B shows a better R_o and R_o/R_s behaviour when comparing with experimental data. In scenario A, R_o radius is approximately constant with K_T; in scenario B, it decreases with K_T (same behaviour shown by experimental data). It shows the importance of initial transverse velocity in explaining the HBT radii, as already discussed in [23].

6. Conclusions and Perspectives

In this work, we have used a simple hydrodynamical model in 1+1 dimensions for calculating both p_T distribution and HBT radii for pions. We have studied the effects of two types of equations of state. We have found that, although the use of different equations of state affects the hydrodynamical evolution, the effects on both p_T distribution and HBT radii are small. The HBT radii show a tendency to be closer to the data for QCD-s95p EoS, although our numerical results are far from the data.

For a complete understanding of the effects of the equations of state on pion interferometry a more realistic treatment should be done. It includes: hydrodynamics in 3+1 dimensions with viscous effects; event-by-event initial conditions; equations of state with non-zero baryonic chemical potential. Also, others observables should be studied. In part, some of these points are addressed in [24]. In fact, in [25], the HBT effect was studied by using the 3+1 hydrodinamical code NEXSPHERIO, but with a phenomenological parametrization of QCD EoS, that differs from the QCD-s95p EoS used here.

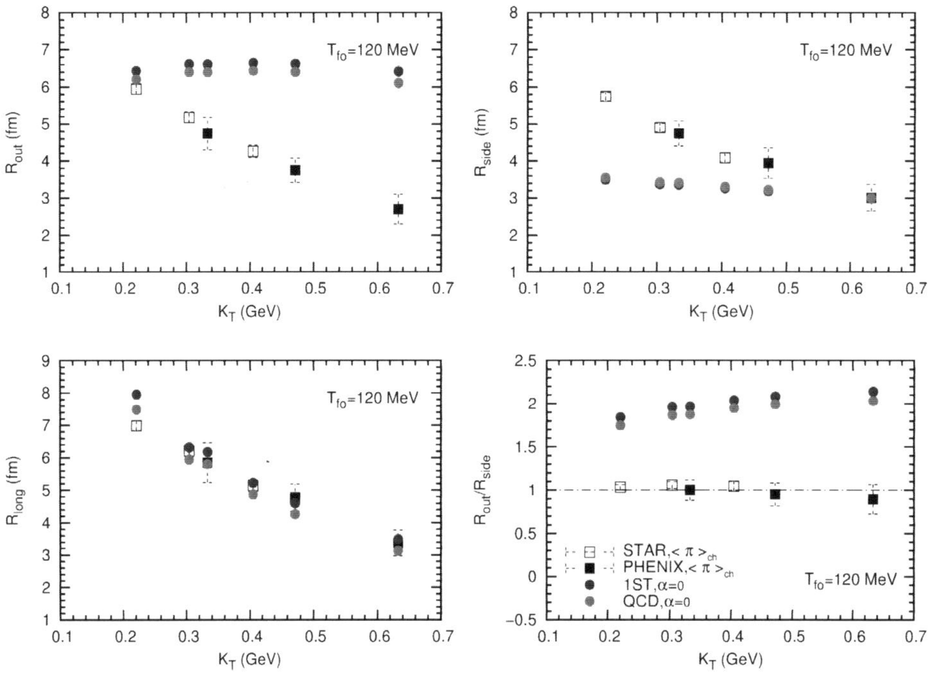

Figure 5. K_T dependence of HBT radii for pions in Au+Au collisions at 130A GeV. Scenario without initial transverse velocity and $T_{\rm fo} = 120\,{\rm MeV}$. Data are from STAR [18] and PHENIX [19]; most central collisions.

7. Acknowledgments

DSL acknowledges the financial support by Universidade Federal do Rio Grande (FURG) for participating in the XXXVII RTFNB.

References

[1] Landau L D (1953) *Izv. Akad. Nauk SSSR* **17** 51.
[2] Hama Y, Kodama T, Socolowski Jr O 2005 *Braz. J. Phys.* **35** 24.
[3] Socolowski Jr O, Grassi F, Hama Y, Kodama T (2004) *Phys. Rev. Lett.* **93** 182301.
[4] Andrade R P G, Grassi F, Hama Y, Qian W L (2011) *Nucl. Phys.* A **854** 81.
[5] Werner K, Karpenko I, Pierog T (2011) *Acta Phys. Polon. Supp.* **4** 629.
[6] Noronha-Hostler J, Denicol G S, Noronha J, Andrade R P G, Grassi F (2013) *J. Phys. Conf. Ser.* **458** 012018.
[7] Hirano T (2011) *Acta Phys. Polon.* B **42** 2811.
[8] Fodor Z, Katz S D (2002) *J. High Energy Phys.* **3** 14; Karsh F (2002) *Nucl. Phys.* A **698** 199.
[9] Carvalho J 2014 A influência das condições iniciais e das equações de estado no estudo hidrodinâmico da interferometria de píons *Master dissertation* FURG.
[10] Misner C W, Thorne K S, Wheeler J A (2000) Gravitation (Freeman).
[11] Bjorken J D (1983) *Phys. Rev.* D **27** 140.
[12] Cooper F, Frye G (1974) *Phys. Rev.* D **10** 186.
[13] Yen G D, Gorenstein M I, Greiner W, Yang S N (1997) *Phys. Rev.* C **56** 2210.
[14] Huovinen P, Petreczky P (2010) *Nucl. Phys.* A **837** 26.
[15] Boal D H, Gelbke C K, Jennings B K 1990 *Rev. Mod. Phys.* **62** 553; Wong C Y (1994) Introduction to High-Energy Heavy-Ion Collisions (World Scientific); Heinz U, Jacak B V 1999 *Ann. Rev. Nucl. Part. Sci.* **49** 529; Florkowski W (2010) Phenomenology of Ultra-Relativistic Heavy-Ion Collisions (World Scientific).

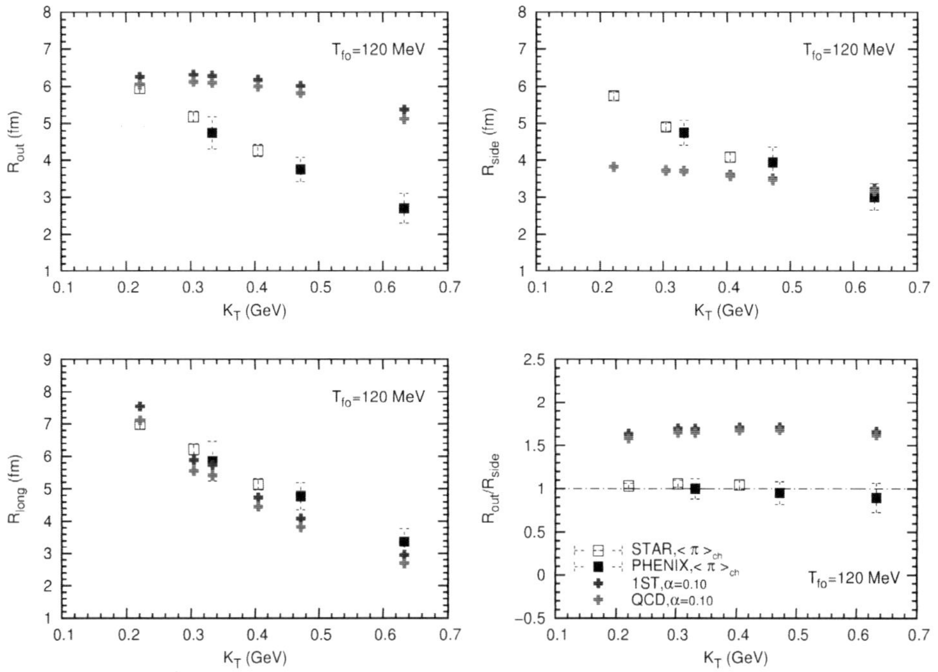

Figure 6. K_T dependence of HBT radii for pions in Au+Au collisions at 130A GeV. Scenario with initial transverse velocity and $T_{\rm fo} = 120\,{\rm MeV}$. Data are from STAR [18] and PHENIX [19]; most central collisions.

[16] Lisa M A, Pratt S, Soltz R, Wiedemann U (2005) *Ann. Rev. Nucl. Part. Sci.* **55** 357.
[17] Wiedemann U A, Heinz U (1997) *Phys. Rev.* C **56** 610.
[18] Adler C et al (2001) *Phys. Rev. Lett.* **87** 082301.
[19] Adcox K et al (2002) *Phys. Rev. Lett.* **88** 192302.
[20] Adcox K et al (2002) *Phys. Rev. Lett.* **88** 242301; Adcox K et al (2004) *Phys. Rev.* C **69** 024904.
[21] Rischke D H, Bernard S, Maruhn J A (1995) *Nucl. Phys.* A **595** 346.
[22] Karpenko Iu A, Sinyukov Yu M (2010) *Phys. Rev.* C **81** 054903.
[23] Pratt S (2009) *Phys. Rev. Lett.* **102** 232301.
[24] Dudek D M, Qian W L, Wu C, Socolowski Jr O, Padula S S, Krein G, Hama Y, Kodama T. (2014) *Preprint* arXiv:1409.0278 [nucl-th]
[25] Hama Y, Andrade R P G, Grassi F, Socolowski Jr O, Kodama T, Tavares B, Padula S S 2006 *Nucl. Phys.* A **774** 169

Journal of Physics: Conference Series **630** (2015) 012033 doi:10.1088/1742-6596/630/1/012033

Correlations between the nuclear matter symmetry energy, its slope, and curvature.

B. M. Santos[1], **M. Dutra**[2], **O. Lourenço**[3], **A. Delfino**[1]

[1]Instituto de Física, Universidade Federal Fluminense, 24210-150, Niterói, RJ, Brazil

[2]Departamento de Física e Matemática - ICT, Universidade Federal Fluminense, 28895-532 Rio das Ostras, RJ, Brazil

[3]Departamento de Ciências da Natureza, Matemática e Educação, CCA, Universidade Federal de São Carlos, 13600-970, Araras, SP, Brazil

E-mail: biank_ce@if.uff.br

Abstract. By using point-coupling versions of finite range nuclear relativistic mean field models containing cubic and quartic self interactions in the scalar field σ, a nonrelativistic limit is achieved. This approach allows for an analytical expression for the symmetry energy (J) as a function of its slope (L) in a unified form, namely, $L = 3J + f(m^*, \rho_o, B_o, K_o)$, where the quantities m^*, ρ_o, B_o and K_o are bulk parameters at the nuclear matter saturation density ρ_o. This result establishes a linear correlation between L and J which is reinforced by exact relativistic calculations we have performed. An analogous analytical correlation can also be found for J, L and the symmetry energy curvature ($K_{\rm sym}$). Based on these results, we propose a graphic constraint in $L \times J$ plane which finite range models should satisfy.

1. Introduction

The investigation on correlations between observables is an important issue in physics since the knowledge of one observable may carry information about other. Several bulk parameter quantities help understanding the nuclear matter properties. One of them is the symmetry energy $\mathcal{S}$, which can be expanded as a function of the nuclear density ρ as $\mathcal{S}(\rho) = J + Lx + \frac{1}{2}K_{\rm sym}x^2 + \frac{1}{6}Q_{\rm sym}x^3 + \mathcal{O}(x^4)$, where $x = (\rho - \rho_o)/3\rho_o$ and ρ_o is the nuclear matter saturation density. The coefficients of this expansion, namely, J, L, $K_{\rm sym}$ and $Q_{\rm sym}$ are, at the saturation energy, the symmetry energy at the saturation density, the slope, curvature, and third derivative (skewness) of $\mathcal{S}$ respectively. In this work we use a non-relativistic limit from point-coupling versions of finite range nuclear relativistic mean field models for establish a correlations between L and J.

2. The Model

The relativistic nonlinear point-coupling (NLPC) versions of the Boguta-Bodmer models are described by the following Lagrangian density

$$\begin{aligned}\mathcal{L}_{\rm NLPC} &= \bar{\psi}(i\gamma^\mu\partial_\mu - M)\psi - \frac{1}{2}G_{\rm V}^2(\bar{\psi}\gamma^\mu\psi)^2 + \frac{1}{2}G_{\rm S}^2(\bar{\psi}\psi)^2 \\ &+ \frac{A}{3}(\bar{\psi}\psi)^3 + \frac{B}{4}(\bar{\psi}\psi)^4 - \frac{1}{2}G_{\rm TV}^2(\bar{\psi}\gamma^\mu\vec{\tau}\psi)^2, \qquad (1)\end{aligned}$$

that mimics the two, three and four body point-like interactions with the fermionic spinor field ψ associated to the nucleon of mass M. In this equation, the last term was included in order to take into account the asymmetry of the system (different number of protons and neutrons). In the nonrelativistic limit of the NLPC model, and using the mean-field approach, the energy density functional at zero temperature for asymmetric nuclear matter is written as

$$\varepsilon^{(\rm NR)}(\rho,y) = (G_{\rm V}^2-G_{\rm S}^2)\rho^2 - A\rho^3 - B\rho^4 + G_{\rm TV}^2\rho^2(2y-1)^2 + \frac{3}{10M^*(\rho,y)}\lambda\rho^{\frac{5}{3}}, \quad (2)$$

where the effective mass is

$$M^*(\rho,y) = \frac{M^2}{(M+G_{\rm S}^2\rho+2A\rho^2+3B\rho^3)H_{\frac{5}{3}}}, \quad (3)$$

with $H_{\frac{5}{3}} = 2^{\frac{2}{3}}[y^{\frac{5}{3}}+(1-y)^{\frac{5}{3}}]$, $\lambda=(3\pi^2/2)^{\frac{2}{3}}$, and $y=\rho_p/\rho$ being the proton fraction of the system. The proton density is ρ_p. For a detailed derivation of Eq. (2) from Eq. (1) in the $y=1/2$ case, we address the reader to Ref. [1].

The coupling constants of the model are $G_{\rm S}^2$, $G_{\rm V}^2$, A, B and $G_{\rm TV}^2$. The first four of them are adjusted in order to fix ρ_o, B_o, K_o and M_o^*. This is done by solving a system of four equations, namely, $\varepsilon^{(\rm NR)}(\rho_o,1/2)=-B_o$, $K^{(\rm NR)}(\rho_o,1/2)=K_o$, $P^{(\rm NR)}(\rho_o,1/2)=0$ (nuclear matter saturation), and $M^*(\rho_o,1/2)=M_o^*$. The pressure and incompressibility are defined, respectively, by $P^{(\rm NR)}(\rho,y)=\rho^2\frac{\partial(\varepsilon^{(\rm NR)}/\rho)}{\partial\rho}$ and $K^{(\rm NR)}(\rho,y)=9\frac{\partial P^{(\rm NR)}}{\partial\rho}$.

An advantage of this approach is to obtain simple analytical expressions for the equations of state (EOS) of the model, in comparison to those calculated in the exact finite range (FR) relativistic mean-field (RMF) models. Thus, the study of the correlation between the symmetry energy and its slope can be performed analytically. For this purpose, we first use Eq. (2) to write $\mathcal{S}(\rho)=\frac{1}{8}\left[\frac{\partial^2(\varepsilon^{(\rm NR)}/\rho)}{\partial y^2}\right]_{y=\frac{1}{2}}$. Then, $J=\mathcal{S}(\rho_o)$ is given by

$$J = \frac{\lambda\rho_o^{\frac{2}{3}}}{6M} + \left(G_{\rm S}^2+2A\rho_o+3B\rho_o^2\right)\frac{\lambda\rho_o^{\frac{5}{3}}}{6M^2} + G_{\rm TV}^2\rho_o. \quad (4)$$

The symmetry energy $\mathcal{S}(\rho)$ is used again in order to obtain $L=3\rho_o\left(\frac{\partial\mathcal{S}}{\partial\rho}\right)_{\rho=\rho_o}$. The result is

$$L = \frac{\lambda\rho_o^{\frac{2}{3}}}{3M} + \left(5G_{\rm S}^2+16A\rho_o+33B\rho_o^2\right)\frac{\lambda\rho_o^{\frac{5}{3}}}{6M^2} + 3G_{\rm TV}^2\rho_o. \quad (5)$$

From Eq. (4) it is possible to determine the last coupling constant $G_{\rm TV}^2$, by imposing the model to present a particular value for J.

At this point, we rewrite the coupling constants of the model, namely, $G_{\rm S}^2$, $G_{\rm V}^2$, A, and B, in terms of the bulk parameters m^*, ρ_o, B_o, and K_o. Therefore, it is possible to write L explicitly as $L=L(m^*,\rho_o,B_o,K_o)$. By doing so, and subtracting $3J$ from L, we finally find a clear correlation between J and L in the following form

$$L = 3J + f(m^*,\rho_o,B_o,K_o), \quad (6)$$

where the function

$$f(m^*,\rho_o,B_o,K_o) = \frac{5E_F^o}{(3M^2-19E_{\rm F}^oM+18E_{\rm F}^{o2})}\times$$
$$\times\left\{\frac{2M}{9m^*}(3M-14E_{\rm F}^o) - M\left(M+K_o/9\right) + E_F^o(5M+6B_o)\right\} \quad (7)$$

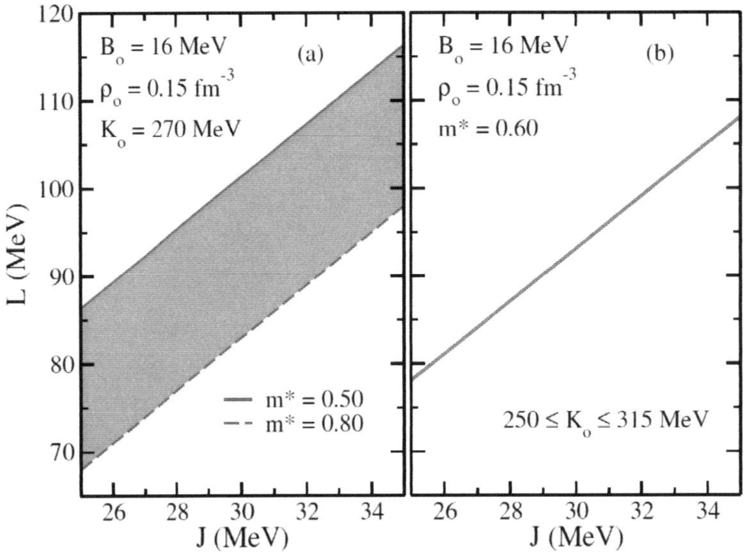

Figure 1. Effect of Δf in the $L-J$ correlation of Eq. (6) for (a) $0.50 \leq m^* \leq 0.80$, and (b) $250 \leq K_o \leq 315$ MeV.

exhibits a dependence with the inverse of the effective mass, with $E^o_{\rm F} = 3\lambda\rho_o^{\frac{2}{3}}/10M$.

Usually, in nuclear mean-field models, the binding energy and the saturation density are well established close around the values of $B_o = 16$ MeV and $\rho_o = 0.15$ fm^{-3}. The same assumption does not apply to the incompressibility and effective mass. Thus, from Eq. (7), it is straightforward to check that for a fixed value of m^*, the variation in f will be given by

$$(\Delta f)_{K_o} = -\frac{5ME^o_F}{9(3M^2 - 19E^o_{\rm F}M + 18E^{o2}_{\rm F})}\Delta K_o. \tag{8}$$

For the range of $250 \leq K_o \leq 315$ MeV, recently proposed in Ref. [2], one can verify that $|(\Delta f)_{K_o}| = 0.32$ MeV. On the other hand, by choosing two different models presenting the same incompressibility K_o but with two different effective masses m^*_1 and m^*_2, the f variation can be inferred by

$$(\Delta f)_{m^*} = \frac{5ME^o_F(3M - 14E^o_{\rm F})}{9(3M^2 - 19E^o_{\rm F}M + 18E^{o2}_{\rm F})}\frac{\Delta m^*}{m^*_1 m^*_2}, \tag{9}$$

where $\Delta m^* = m^*_2 - m^*_1$. For a typical range of $0.50 \leq m^* \leq 0.80$, presented by FR-RMF models, one has $|(\Delta f)_{m^*}| = 18$ MeV.

Fig. 1 shows how such variations affect the correlation given in Eq. (6). From this figure we can conclude that different models presenting the same effective mass, will exhibit points in a $L \times J$ graph situated very close to a same line, since in this case the variation of the linear coefficient in Eq. (6) is very small compared to that one of the case in which K_o is fixed. This leads us to draw the conclusion that in the NR limit of the NLPC models described by Eq. (1), the linear correlation between J and L in Eq. (6) is achieved for the more distinct models under the condition that their effective masses are equal.

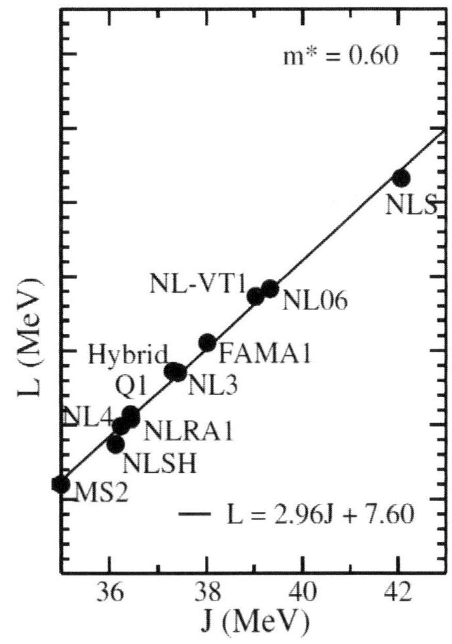

Figure 2. L versus J for FR parametrizations in which m^* is the same.

3. PREDICTIONS ON FR-RMF MODELS

3.1. Symmetry energy slope

Now, we pose the question whether the NR correlation obtained in Eq. (6), and the results showed in Fig. 1 with the subsequently conclusions, still remain valid for exact FR models. The answer is given by the study we have done for a set of representative FR models, whose results are displayed in Fig. 2.

Confident on the correlation between L and J, found in the NR approximation and confirmed for the relativistic calculations, we have constructed a region of possible L values as a function of J, and according the FRS constraint [3], that FR-RMF Boguta-Bodmer models must satisfy in order to give values for the finite nuclei spin-orbit splitting compatible with well established experimental values. In Fig. 3, we present our prediction for the lowest and highest values for L in comparison with other values found in the literature, by taking into account the region in a range of $25 \leq J \leq 35$ MeV for the symmetry energy. Notice that our limits for L are comparable with other models.

3.2. Symmetry energy curvature

In the NR framework it is also possible to find an analytical expression for $K_{\text{sym}} = 9\rho_o^2\left(\frac{\partial^2 \mathcal{S}}{\partial\rho^2}\right)_{\rho=\rho_o}$. It reads

$$K_{\text{sym}} = \frac{5E_F^o}{(3M^2 - 19E_{\text{F}}^o M + 18E_{\text{F}}^{o2})}\left\{\frac{2M}{3m^*}(5M - 18E_{\text{F}}^o)\right.$$
$$\left. + M(18B_o - 4M + 18E_{\text{F}}^o) - \frac{K_o}{9}(19M - 18E_{\text{F}}^o)\right\} \quad (10)$$

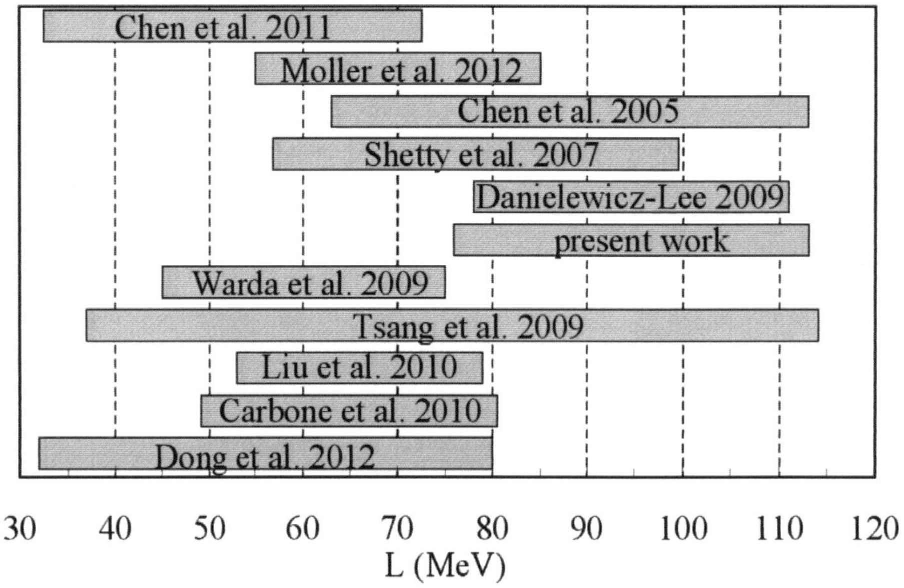

Figure 3. Comparison between the limits of L obtained in this work and those from Dong et al., Carbone et al., Liu et al., Tsang et al., Warda et al., Danielewicz and Lee, Shetty et al., Chen et al., Mόller et al., and Chen; See reference [4].

Note than the symmetry energy curvature K_{sym} depends on m^*, scaling as $1/m^*$, and is linearly correlated with K_o in the NR approach, see Eq. (10). Such dependences are not negligible. Using the Eqs. (6) and (10), we can rewrite K_{sym} as function L (or J) and establish the correlation.

4. Conclusion

In this work, by studying the nonrelativistic limit of the NLPC models, we have seen that the symmetry energy slope L is linearly correlated with J. This correlation depends explicitly on the effective nucleon mass m^* and the incompressibility K_o. However, the K_o dependence of L has been verified to be negligible. In the same way, we have also shown that it is possible to establish correlation between the symmetry energy curvature K_{sym} as function L (or J). However, a more detailed study regarding this correlation is still underway.

References

[1] O. Lourenço, M. Dutra, A. Delfino, and J. S. Sá Martins, Phys. Rev. C **81**, 038201 (2010).
[2] J. R. Stone, N. J. Stone, and S. A. Moszkowski, Phys. Rev. C **89**, 044316 (2014).
[3] R. J. Furnstahl, J. J. Rusnak, B.D. Serot, Nucl. Phys. **A632**, 607 (1998).
[4] J. Dong, W. Zuo, J. Gu, and U. Lombardo, Phys. Rev. C **85**, 034308 (2012).

Journal of Physics: Conference Series **630** (2015) 012034 doi:10.1088/1742-6596/630/1/012034

Analysis of the vertices B_sBK^* and D_sDK^* with QCD Sum Rules

A Cerqueira Jr[1], B Osório Rodrigues[2], M E Bracco[3]

[1] Centro Universitário de Volta Redonda, UniFOA, Av. Paulo Erlei Alves Abrantes 1325, 27240-560, Volta Redonda, RJ, Brazil

[2] Centro Brasileiro de Pesquisas Físicas, Rua Xavier Sigaud 150, 22290-180, Rio de Janeiro, RJ, Brazil

[3] Faculdade de Tecnologia, Universidade do Estado do Rio de Janeiro, Rod. Presidente Dutra Km 298, Pólo Industrial, 27537-000, Resende, RJ, Brazil

E-mail: angelo.cerqueira.cunha@gmail.com

Abstract. In this work, we calculated the form factors and coupling constant of the vertex D_sDK^* using the QCD Sum Rules. The calculation of the form factors were performed for the cases D, D_s and K^* off shell. Using the similarities between B and D mesons, we also calculated the form factors and coupling constant of the vertex B_sBK^*. The coupling constants of the vertices B_sBK^* and D_sDK^* were compared with each other through the Heavy Hadron Chiral Perturbation Theory(HHChPT). We found a difference of 19% between B_sBK^* and D_sDK^* coupling constants using the HHChPT relation.

1. Introduction

In previous works, we calculated the form factors and coupling constants for the beauty vertices B_s^*BK[1] and B_sB^*K[2]. The calculation of these vertices was made in comparison with the results of the vertices D_s^*DK and D_sD^*K[3]. Through the relations of Heavy Hadron Chiral Perturbation Theory(HHChPT) between coupling constants of the vertices with D mesons and B, we verified the consistency of our results for the vertices B_s^*BK and B_sB^*K when compared with the vertices D_s^*DK and D_sD^*K, respectively. The coupling constants of these vertices were calculated in order to contribute to the study of hadronic molecules. These hadronic molecules are considered as weakly-bound states of two or more hadrons. In Ref [4], the author uses effective lagrangians to describe the molecular state. This treatment was extensively applied in decays of D_{s0}^* and D_{s1}[5] mesons as molecular structure, respectively, D^*K [6] and DK[7]. In these studies, use was made of a quantum field approach based on effective lagrangians that depend on the coupling constants of vertices like B_s^*BK, B_sB^*K and B_sBK^*. A precise knowledge about their values is of great importance.

The determination of the coupling constant, $g_{D_sDK^*}$, using QCD Sum Rules is important to compare with the result obtained using the HHChPT relation. The coupling $g_{D_sDK^*}$ was used to estimate the coupling $g_{B_sBK^*}$ using HHChPT in Ref [2]. This estimated value using HHChPT was useful to compare with the coupling $g_{B_sBK^*}$ obtained from QCD Sum Rules. Both results of the coupling $g_{B_sBK^*}$ are in agreement in Ref [2].

Journal of Physics: Conference Series **630** (2015) 012034 doi:10.1088/1742-6596/630/1/012034

2. The QCD Sum Rules

The three-point correlation functions can be evaluated by two different approaches, one known as the OPE side, or QCD side, and the other known as the phenomenological side. In the OPE side the hadrons are described by their interpolating currents. In the phenomenological side the hadrons are described by hadronic states. The starting point of the calculation is the three point correlation function,

$$\Pi(p,p') = \int d^4x d^4y \, \langle 0| \, T\{j_1(x) {j_2}^\dagger(y) {j_3}^\dagger(0)\} \, |0\rangle \, e^{ip'\cdot x} e^{-iq\cdot y}. \tag{1}$$

In the OPE side, a Wilson's operator product expansion is applied, the OPE side is obtained from Eq. 1. The OPE side is described as a sum of the perturbative and non-perturbative contributions to the correlation function. The non-perturbative term includes the contribution of all the QCD condensates. In this work, the only non-perturbative term we calculate is the contribution from quark condensates, i.e, $\Pi^{non-pert} = \Pi^{\langle qq\rangle}$.

In the phenomenological side, we use the same correlation function of Eq. 1, where we perform a sum over a set of hadronic states. Applying the quark-hadron duality, the quantities of interest (masses, form factors, coupling constants) are obtained matching both sides before a Borel Transform.

3. The vertex D_sDK^*

The coupling constant of the vertex D_sDK^* was calculated using the three mesons off shell. In the case where the D meson is off shell in Fig. 1(a), there is non-perturbative contribution from the strange quark condensate (Fig. 1(d)). The D_s off shell diagram(fig. ??(b)) has contribution from the light quark condensate (Fig. 1(e)), the K^* off shell(Fig. 1(c)) is the only diagram that has no contribution from quark condensate.

3.1. D off shell

Using in Eq. 1 the interpolating currents written in terms of the quark fields: $j_3 = j_5^{D_s} = i\bar{c}\gamma_5 s$, $j_2 = j_5^D = \bar{q}\gamma_5 c$ and $j_1 = j_\mu^{K^*} = \bar{q}\gamma_\mu s$, we obtain the OPE side, given by:

$$\begin{aligned} B_{M^2}B_{M'^2}[\Pi_\mu^{QCD}(p,p')] = &-\frac{1}{4\pi^2}\int_{s_{min}}^{\infty}\int_{u_{min}}^{\infty} dsdu \frac{3}{2\pi\sqrt{\lambda}}\{p_\mu[A(2p\cdot k - p\cdot p\prime + m_c m_s - m_s^2) \\ &-k\cdot p\prime + m_s^2] + p\prime_\mu[B(2p\cdot k - p\cdot p\prime + m_c m_s - m_s^2) + k\cdot p - m_s^2 + m_s m_c]\} e^{-s/M^2} e^{-u/M'^2}. \end{aligned} \tag{2}$$

The Borel transform gives the exponentials in Eq. 2. The Borel Transform is useful to eliminate the contribution of higher moments of excited states. The contribution of the strange quark condensate, is given by:

$$B_{M^2}B_{M'^2}[\Pi^{\langle ss\rangle}] = -m_c\langle s\bar{s}\rangle e^{-m_c^2/M^2} p'_\mu. \tag{3}$$

The phenomenological side yields the following equation,

$$\begin{aligned} B_{M^2}B_{M'^2}[\Pi_\mu^{phen}(p,p')] = &-i\frac{f_{K^*} f_D f_{D_s} m_{K^*} m_{D_s}^2 m_D^2 g_{D_sDK^*}^{(D)}(Q^2)}{(m_c m_s + m_c^2)(Q^2 + m_D^2)} \\ &\times \left[-2p_\mu + p\prime_\mu \left(\frac{m_D^2 - m_{D_s}^2}{m_{K^*}^2} + 1\right)\right] e^{-m_{D_s}^2/M^2} e^{-m_{K^*}^2/M'^2}, \end{aligned} \tag{4}$$

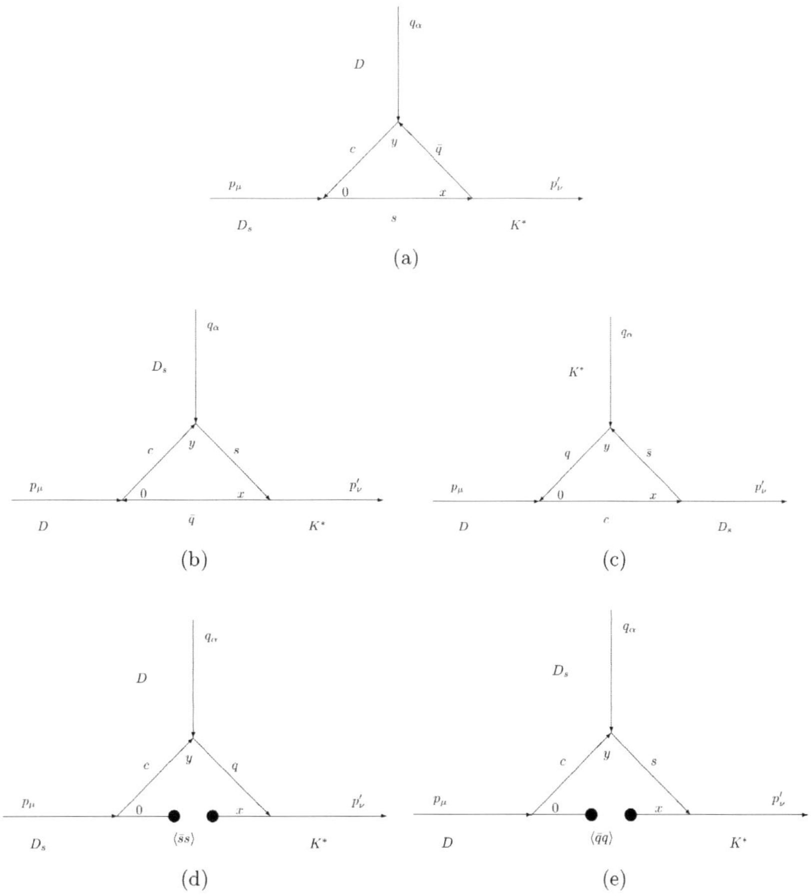

Figure 1: The diagram **(a)** corresponds to the perturbative OPE term of D meson off shell, The diagrams **(b)** and **(c)** correspond to the perturbative OPE term of D_s and K^* mesons off shell, respectively; and the diagrams **(d)** and **(e)** correspond to the strange quark and light quark condensates contributions, respectively.

where f_D, f_{D_s} and f_{K^*} are the decay constants of the mesons, the term Q^2 is the transferred momentum and $g^{(D)}_{D_sDK^*}(Q^2)$ is the form factor.

Figure 2(a) shows the pole and continuum contribution and figure 2(b) the Borel mass stability of the D off shell case. The tensor structure chosen is p' and it has a good sum rule and contribution of strange quarks condensates. The Borel mass is stable for values greater than $4.0GeV^2$ and the contribution of the pole is greater than the continuum contribution. The values used for the continuum thresholds are $\Delta s = \Delta u = 0.5GeV$. The stability of the Borel mass allowed us to choose its value equal to $M^2 = 6.0GeV^2$. The plots shown in Fig.2 were made fixing $Q^2 = 1.0GeV^2$. The data of the form factors were fitted by an exponential function, given by:

$$g^{(D)}_{D_sDK^*} = 1.32e^{-Q^2/18.16}. \tag{5}$$

So, the value of the coupling constant is given by:

$$g^{(D)}_{D_sDK^*} = 1.67. \tag{6}$$

This value is obtained extrapolating the form factor to the negative value of the squared mass of the meson off shell. The value of the function in this point, where $Q^2 = -m^2$, gives the value of the coupling constant.

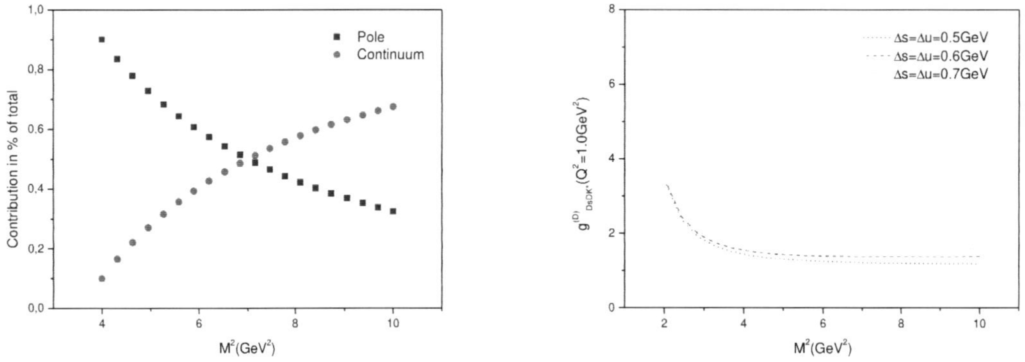

(a) (b)

Figure 2: The pole and continuum conditions **(a)** and the Borel mass stability **(b)**.

3.2. D_s *off shell*

Starting from Eq. 1 with D_s off shell(fig. 1(b)) and using the same interpolating currents, we obtain the OPE side:

$$B_{M^2}B_{M'^2}[\Pi^{QCD}_\mu(p,p')] = -\frac{1}{4\pi^2}\int_{s_{min}}^{\infty}\int_{u_{min}}^{\infty} dsdu\frac{3}{2\pi\sqrt{\lambda}}\{p_\mu[A(p\cdot p\prime - m_c m_s - 2p\cdot k) - k\cdot p\prime] \\ + p\prime_\mu[B(p\cdot p\prime - m_c m_s - 2p\cdot k) + k\cdot p]\}e^{-s/M^2}e^{-u/M'^2}. \tag{7}$$

The diagram of the D_s off shell has a contribution of the light quark condensate $\langle q\bar{q}\rangle$, given by:

$$B_{M^2}B_{M'^2}[\Pi^{\langle q\bar{q}\rangle}_\mu(p,p')] = \langle q\bar{q}\rangle\left[m_c p'_\mu - m_s p_\mu\right]e^{-m_c{}^2/M^2}e^{-m_s{}^2/M'^2}. \tag{8}$$

Starting from the correlation function Eq. 1, through the calculation of hadronic states, we obtain the following expression for the phenomenological side:

$$B_{M^2}B_{M'^2}[\Pi^{phen}_\mu(p,p')] = -i\frac{f_{K^*}f_D f_{D_s} m_{K^*} m^2_{D_s} m^2_D g^{(D_s)}_{D_sDK^*}(Q^2)}{(m_c^2 + m_s m_c)(Q^2 + m^2_{D_s})} \\ \times\left[-2p_\mu + p\prime_\mu\left(1 - \frac{m_D^2 - m^2_{D_s}}{m^2_{K^*}}\right)\right]e^{-m_D^2/M^2}e^{-m^2_{K^*}/M'^2}. \tag{9}$$

In the case of the D_s off shell, there is a contribution of light quark condensate. The tensor structure chosen to calculate the form factors is the p_μ structure. This structure shows Borel mass stability (Fig. 3(b)) and conditions of pole and continuum (Fig. 3(a)). The values of the continuum thresholds are $\Delta s = \Delta u = 0.5GeV$ and the value fixed for the Borel mass is $M^2 = 3.5GeV^2$. The plots shown in Fig. 3 were made fixing $Q^2 = 1.0GeV^2$. The form factors were fitted by an exponential function,

$$g^{(D_s)}_{D_s D K^*} = 1.15e^{-Q^2/27.08}. \quad (10)$$

The value of the constant is given by:

$$g^{(D_s)}_{D_s D K^*} = 1.35. \quad (11)$$

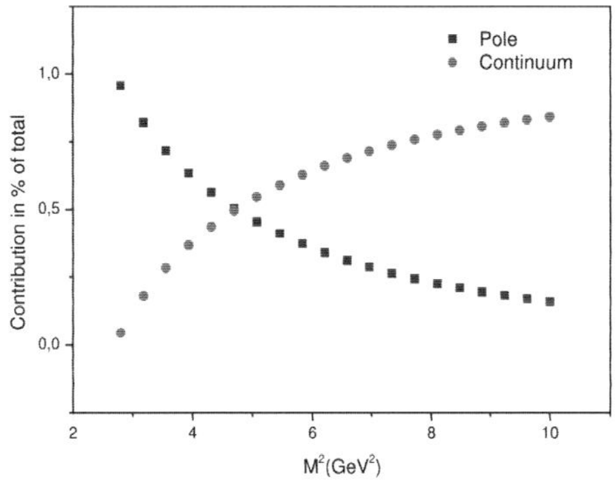

(a)

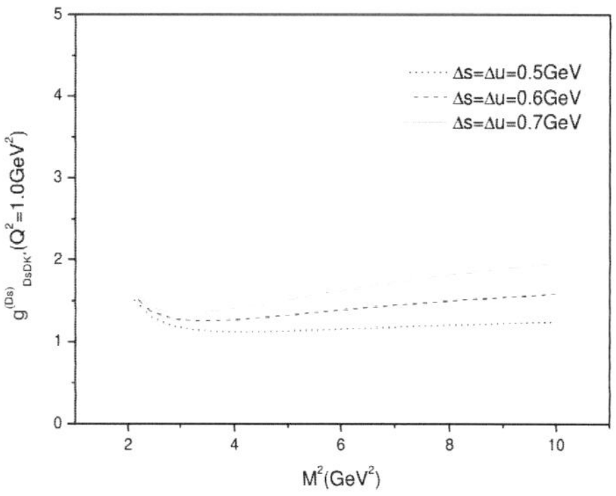

(b)

Figure 3: The pole and continuum conditions **(a)** and the Borel mass stability **(b)**.

3.3. K^* *off shell*

Starting from the correlation function of the K^* off shell(Fig. 1(c)), we obtain the OPE side:

$$B_{M^2}B_{M'^2}[\Pi^{QCD}_\mu(p,p')] = -\frac{1}{4\pi^2}\int_{s_{min}}^{\infty}\int_{u_{min}}^{\infty} dsdu\frac{3}{\sqrt{\lambda}}\{p_\mu[A(p\cdot p\prime + m_c m_s - m_c^2) + m_c^2 - k\cdot p\prime - m_c m_s] + p\prime_\mu[B(p\cdot p\prime + m_c m_s - m_c^2) - k\cdot p + m_c^2]\}e^{-s/M^2}e^{-u/M'^2}. \quad (12)$$

This diagram does not have contribution of quark condensate. The phenomenology gives the following expression:

$$B_{M^2}B_{M'^2}[\Pi^{phen}_\mu(p,p')] = -\frac{f_D f_{K^*} f_{D_s} m_{K^*} m^2_{D_s} m^2_D g^{(K^*)}_{D_s^* DK}(Q^2)}{(m_s m_c + m_c^2)(Q^2 + m^2_{K^*})}$$
$$\times\left[p_\mu\left(1 - \frac{(m_D^2 - m_{D_s}^2)}{m^2_{K^*}}\right) + p\prime_\mu\left(1 - \frac{(m^2_{D_s} - m^2_D)}{m^2_{K^*}}\right)\right] e^{-m_D^2/M^2}e^{-m^2_{D_s}/M'^2} \quad (13)$$

Figure 4 shows the pole and continuum condition (a) and the Borel mass stability (b). Our result has only the contribution of the tensor structure p_μ. The other tensor structure does not yield the necessary conditions for a good sum rule. For a good sum rule, the Borel mass must show stability and the pole must have a contribution greater the the continuum. The values of the continuum thresholds are $\Delta s = \Delta u = 0.5GeV$ and the Borel mass is $M^2 = 2.0GeV^2$. The plots shown in Fig. 4 were made fixing $Q^2 = 1.0GeV^2$. The data of this diagram were fitted also by an exponential function,

$$g_{D_s DK^*}^{(K^*)} = 1.13e^{-Q^2/2.41}. \tag{14}$$

Then, the value of the coupling constant is given by:

$$g_{D_s DK^*}^{(K^*)} = 1.63. \tag{15}$$

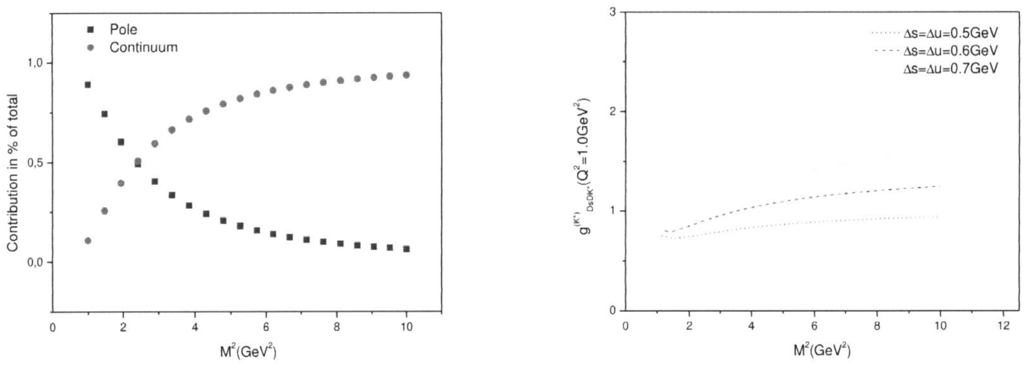

(a) (b)

Figure 4: The pole and continuum conditions **(a)** and the Borel mass stability **(b)**.

3.4. Results

The table 1 shows the values of masses and decay constants used in this work.

Table 1: Numerical values used in the calculation of $g_{D_s DK^*}(Q^2)$ and $g_{B_s BK^*}(Q^2)$.

	Particle							
Quantity	c	s	b	D_s	D	K^*	B	B_s
m (GeV) [8]	1.20	0.13	4.20	1.97	1.87	0.89	5.20	5.40
f (MeV)	-	-	-	280	200	220	191	227

The value of the K^* decay constant is in [9]. The other decay constants were taken from [3]. The results of the form factors are shown in the Fig. 5. The value of the light quark condensate is, $-0.23GeV^3$ and the value of strange quark condensate is, $0.8(0.23GeV)^3$

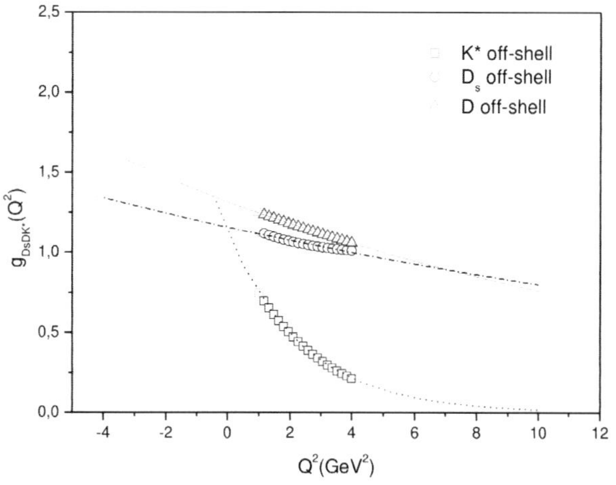

Figure 5: The QCDSR form factors for the three mesons off-shell of the D_sDK^* vertex.

The result of the coupling constant of the vertex D_sDK^* is given by:

$$g_{D_sDK^*} = 1.56. \tag{16}$$

This value is an average value for this constant considering the three values obtained for each meson off shell. We performed the same calculation substituting the D meson by the B meson and the value found for the coupling constant is $g_{B_sBK^*} = 3.30 \pm 0.10$.

4. Conclusions

The coupling constant for the vertex B_sBK^* estimated by the Heavy Hadron Chiral Perturbation Theory(HHChPT)

$$g_{B_1B_2L} = g_{D_1D_2L}\frac{m_B}{m_D},$$

using the values the coupling constant $g_{D_sDK^*}$, is given by

$$g_{B_sBK^*} = 4.10. \tag{17}$$

The result of the B_sBK^* from QCD Sum Rules is

$$g_{B_sBK^*} = 3.3. \tag{18}$$

There is a difference about 19%. But, these are preliminaries results. The next step is to perform an error analysis of these results.

Acknowledgments
This work has been partially supported by CNPq and CAPES.

Journal of Physics: Conference Series **630** (2015) 012034 doi:10.1088/1742-6596/630/1/012034

References

[1] Cerqueira Jr A, Osório Rodrigues B, Bracco M E 2012 *Nuc. Phys. A* **874** 130–142
[2] Cunha Júnior A C. Fatores de forma em processos com mésons *B*. 2013. 133 f. Tese(Doutorado em física)-Instituto de Física Armando Dias Tavares, Universidade do Estado do Rio de Janeiro, Rio de Janeiro, 2013.
[3] Bracco M E, Cerqueira Jr A, Chiapparini M, Lozeá A, Nielsen M 2006 *Phys. Lett. B* **641** 286–293
[4] Faessler A, Gutsche T, Lyubovitskij V E, Ma Y L 2008 *Phys. Rev. D* **077** 114013
[5] Faessler A, Gutsche T, Kovalenko S, Lyubovitskij V E 2007 *Phys. Rev. D* **076** 014003
[6] Faessler A, Gutsche T, Lyubovitskij V E, Ma Y L 2007 *Phys. Rev. D* **076** 114008
[7] Faessler A, Gutsche T, Lyubovitskij V E, Ma Y L 2007 *Phys. Rev. D* **076** 114005
[8] Particle Data Group, Review of Particle Physics 2014 Chin. Phys. C **38**
[9] Fang Su, Yue-Liang Wu , Ci Zhuang, Yi-Bo Yang 2011 Eur. Phys. J **C72** 1914

A study of the $g_{\eta_c D^* D}$ coupling constant from the QCD sum rules

B Osório Rodrigues[1,3], M E Bracco[2] and C M Zanetti[2]

[1] Instituto de Física, Universidade do Estado do Rio de Janeiro, Rua São Francisco Xavier 524, 20550-900, Rio de Janeiro, RJ, Brazil

[2] Faculdade de Tecnologia, Universidade do Estado do Rio de Janeiro, Rod. Presidente Dutra Km 298, Pólo Industrial, 27537-000, Resende, RJ, Brazil

E-mail: bosorio@cbpf.br

Abstract. In this work, we calculate the form factor and the coupling constant of the charmed vertex $\eta_c D^* D$ using the QCD sum rules (QCDSR) for the case where the D^* meson is considered off-shell. This is accomplished by using not only the perturbative contribution of the OPE series but also the non-perturbative effects of the gluon condensates. The form factor of this vertex presents two tensor structures, both of which were analysed. The coupling constant is obtained by the extrapolation of the exponentially parametrized form factor curve to the pole of the D^* meson.

1. Introduction

Our QCD sum rules (QCDSR) group has been working extensively for the past years in the charm sector, with the calculation of a vast number of coupling constants as the $D^* D\pi$ [1], $D^* D\rho$ [2], $J/\psi D^{(*)} D^{(*)}$ [3, 4] and more recently, the $J/\psi D_s D_s$ [5]. Continuing this line of research, we started the study of charmed vertices that include the pseudo-scalar meson η_c, in this particular work, we focus our attention on the $\eta_c D^* D$ vertex.

This vertex is related to physical problems of current interest, as the η_c decay in two light vector mesons ($\eta_c \to VV$). Experimental data indicate that this type of decay, as the $\eta_c \to \rho\rho$ and $\eta_c \to K^*(892)\bar{K}^*(892)$, is among the most contributing decay channels for the η_c, with branching ratios [6] of $(1.8 \pm 0.5) \times 10^{-2}$ and $(6.8 \pm 1.3) \times 10^{-3}$ respectively for the mentioned channels. However, these decays should be suppressed by the helicity selection rule (HSR). This is an intriguing problem that shares some similarities with the notorious $\rho\pi$ puzzle, in which the decay $J/\psi \to \rho\pi$ also presents a branching ratio of the order 10^{-2} even though being a suppressed decay according to the OZI rule. One way to circumvent the HSR suppression is by considering the $\eta_c \to VV$ decay with an intermediate step through [7] the mesons D and D^*, i.e. $\eta_c \to D\bar{D}^* \to VV$. In this picture, the coupling constant $g_{\eta_c D^* D}$ will be necessary to the calculation of these decay amplitudes, therefore, the results of this work can be useful to this analysis.

[3] Present address: Centro Brasileiro de Pesquisas Físicas, Rua Xavier Sigaud 150, 22290-180, Rio de Janeiro, RJ, Brazil.

Following our previous works, we calculate the $g_{\eta_c D^* D}$ coupling constant using the QCDSR framework for the case where the D^* meson is off-shell. In the following sections, we present our development, results and conclusion for this work.

2. Development

The starting point of the QCDSR calculation is the three-point correlation function [8]. For the case where the D^* meson is considered off-shell, the correlation function can be written as:

$$\Pi_\mu^{(D^*)}(p,p') = \int d^4x d^4y \, e^{ip'x} e^{-iqy} \langle 0'|T\{j_5^{\eta_c}(x) j_\mu^{D^{*+}\dagger}(y) j_5^{D^-\dagger}(0)\}|0'\rangle \,, \tag{1}$$

where $q = p' - p$ is the transferred momentum, T is the time ordered product, j_μ^M is the interpolating current of the meson M and $|0'\rangle$ is the non-perturbative QCD vacuum.

Each correlation function can be studied in two different ways: using quark degrees of freedom, called the *OPE side* or using hadron degrees of freedom, called the *phenomenological side*. Both representations are equivalent and can be equated invoking the quark-hadron duality after applying a double Borel transform on each side. Then, the form factor and the coupling constant can be extracted.

The OPE side corresponds to a Wilson's operator product expansion and it can be obtained from Eq. 1 by using the interpolating currents written in terms of the quark fields: $j_5^{\eta_c} = i\bar{c}\gamma_5 c$, $j_\mu^{D^{*+}} = \bar{d}\gamma_\mu c$ and $j_5^{D^-} = \bar{c}\gamma_5 d$. The OPE side can be written as a sum of the perturbative contribution and the non-perturbative contributions to the correlation function:

$$\Pi_\mu^{OPE(D^*)} = \Pi_\mu^{pert(D^*)} + \Pi_\mu^{non\text{-}pert(D^*)} \,, \tag{2}$$

where the non-perturbative term of Eq. 2 includes all the contributions from the QCD condensates ($\langle\bar{q}q\rangle$, $\langle g^2G^2\rangle$, $\langle\bar{q}g\sigma Gq\rangle$, ...). In this D^* off-shell calculation, we will only calculate non-perturbative contributions from gluon condensates:

$$\Pi_\mu^{OPE(D^*)} = \Pi_\mu^{pert(D^*)} + \Pi_\mu^{\langle g^2G^2\rangle(D^*)} \,. \tag{3}$$

This is motivated by the later use of the double Borel transform, which will suppress the contributions of quark condensates and mixed quark-gluon condensates when considering an OPE series up to dimension five operators. Calculating Eq. 3 corresponds to the calculation of the sum of the seven diagrams in Fig. 1, where the first diagram is the perturbative term and the following six diagrams are the $\langle g^2G^2\rangle$ contributions to the correlation function.

The perturbative contribution of Fig. 1a can be calculated by using dispersion relations and the Cutkosky rule, which leads to:

$$\Pi_\mu^{pert(D^*)}(p,p') = -\frac{1}{4\pi^2}\int_{s_{inf}}^{\infty}\int_{u_{inf}}^{\infty} dsdu \frac{\left[F_{p_\mu}(s,u,t)p_\mu + F_{p'_\mu}(s,u,t)p'_\mu\right]}{(s-p^2)(u-p'^2)} \,, \tag{4}$$

where $s = p^2$, $u = p'^2$ and $t = q^2$ are the Mandelstam variables and $F_{p_\mu}(s,u,t)$ and $F_{p'_\mu}(s,u,t)$ are given by

$$F_{p_\mu}(s,u,t) = \frac{3[su(s-u-t) + um_c^2(s+u-t)]}{(s^2+u^2+t^2-2tu-2st-2su)^{3/2}} \,, \tag{5}$$

$$F_{p'_\mu}(s,u,t) = \frac{3[su(u-s-t) + m_c^2(2tu+2st-s^2-u^2-t^2)]}{(s^2+u^2+t^2-2tu-2st-2su)^{3/2}} \,, \tag{6}$$

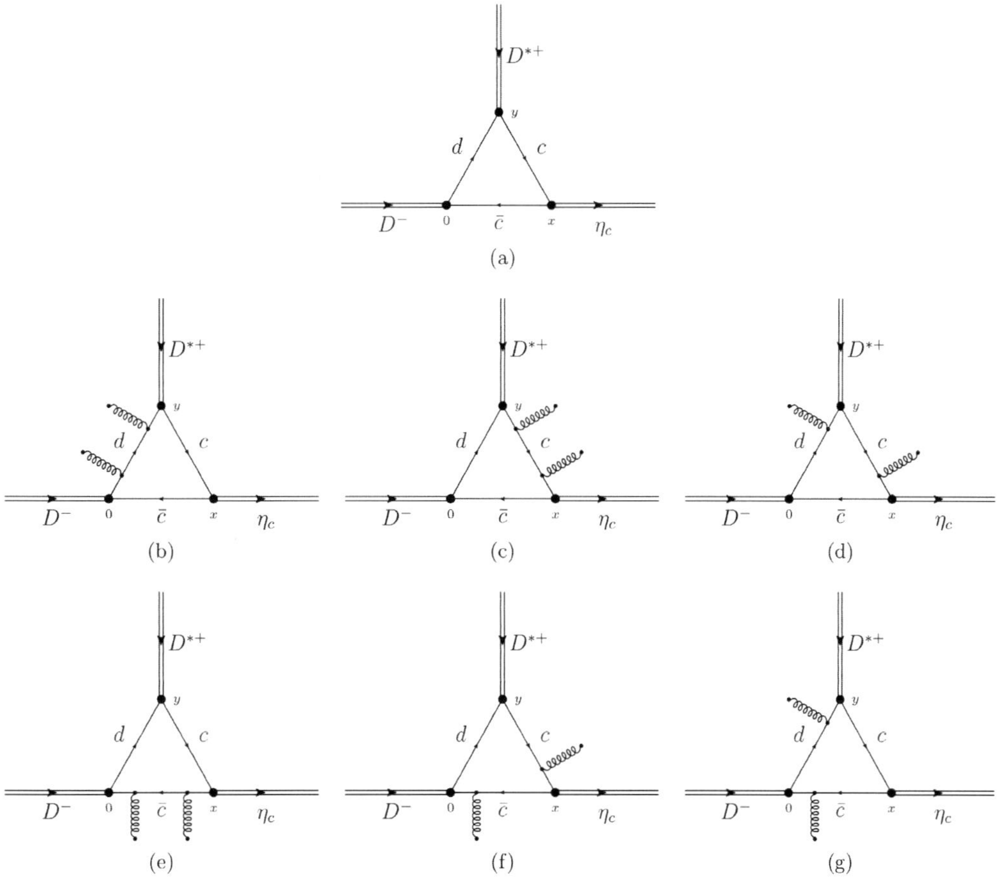

Figure 1: OPE diagrams up to dimension 5 operators for D^* off-shell. The diagram **(a)** corresponds to the perturbative OPE term, while diagrams **(b–g)** correspond to the gluon condensate contributions to the OPE series.

where the limit $m_d \to 0$ has been used.

Making the variable changes $p^2 \to -P^2$, $p'^2 \to -P'^2$, $q^2 \to -Q^2$ and applying the double Borel transform, we have the final expression for the QCDSR contribution of the perturbative term:

$$\mathcal{B}_{M^2}\mathcal{B}_{M'^2}\left[\Pi_\mu^{pert(D^*)}\right] = -\frac{1}{4\pi^2}\int_{s_{inf}}^{\infty}\int_{u_{inf}}^{\infty} ds du\, e^{-s/M^2} e^{-u/M'^2} \times \left[F_{p_\mu}(s,u,t)p_\mu + F_{p'_\mu}(s,u,t)p'_\mu\right], \quad (7)$$

where M^2 and M'^2 are new variables introduced by the double Borel transform called Borel masses. In this work, the Borel masses are related to each other by the following ansatz:

$$M'^2 = \frac{m_{\eta_c}^2}{m_D^2} M^2, \quad (8)$$

with m_{η_c} and m_D being the masses of the mesons η_c and D respectively.

After making the same variable changes and applying the double Borel transform, the gluon condensate contribution can be expressed as:

$$\mathcal{BB}\left[\Pi_\mu^{\langle g^2G^2\rangle}\right] = \frac{\langle g^2G^2\rangle}{96\pi^2}\int_{\frac{1}{M'^2}}^{\infty} d\alpha e^{\frac{\alpha t M'^2 - t}{M^2+M'^2} - \frac{\alpha m_c^2 M'^2(1+\alpha^2 M^2)}{\alpha M^2 M'^2 - M^2}} \left(F_{p_\mu}^{\langle g^2G^2\rangle} p_\mu + F_{p'_\mu}^{\langle g^2G^2\rangle} p'_\mu\right), \quad (9)$$

where the full expression for the function $F_{p_\mu}^{\langle g^2G^2\rangle}$ is presented in appendix A.

Having calculated the OPE side, the next step is to calculate the correlation function of the phenomenological side. This is achieved by expanding Eq. 1 using hadron degrees of freedom. Such expansion lead to the following correlation function:

$$\Pi_\mu^{phen} = \frac{f_{\eta_c} f_{D^*} f_D m_{\eta_c}^2 m_{D^*} m_D^2 \epsilon_\mu^*(q,\lambda)\Gamma(p,p')}{2m_c^2(p^2 - m_D^2)(p'^2 - m_{\eta_c}^2)(q^2 - m_{D^*}^2)} + h.r., \quad (10)$$

where $h.r.$ stands for the contributions of higher resonances, ϵ_μ is the polarization vector and f_{η_c}, f_{D^*} and f_D are the decay constants of the mesons η_c, D^* and D respectively.

The $\Gamma(p,p')$ term in Eq. 10 is related to the Feynman's scattering amplitude, is directly proportional to the form factor and can be obtained from the effective Lagrangian:

$$\mathcal{L}_{\eta_c D^* D} = -i g_{\eta_c D^* D}^{(D^*)}(Q^2) D^{*\pm\alpha}\left(\pm\eta_c\partial_\alpha D^\mp \mp D^\mp\partial_\alpha\eta_c\right),$$

where $g_{\eta_c D^* D}^{(D^*)}(Q^2)$ is the form factor for the case where the D^* meson is off-shell and is related to the coupling constant by the limit

$$g_{\eta_c D^* D} = \lim_{Q^2\to -m_{D^*}^2} g_{\eta_c D^* D}^{(D^*)}(Q^2). \quad (11)$$

The double Borel transform of the phenomenological side is:

$$\mathcal{BB}\left[\Pi^{phen}\right] = \frac{g_{\eta_c D^* D}^{(D^*)}(Q^2) f_{\eta_c} f_{D^*} f_D m_{\eta_c}^2 m_D^2}{2m_c^2 m_{D^*}(Q^2 + m_{D^*}^2)} e^{-\frac{m_D^2}{M^2}} e^{-\frac{m_{\eta_c}^2}{M'^2}} \times$$
$$\left[(m_{D^*}^2 + m_{\eta_c}^2 - m_D^2)p_\mu + (m_{D^*}^2 + m_D^2 - m_{\eta_c}^2)p'_\mu\right] + \mathcal{BB}\left[h.r.\right], \quad (12)$$

where once more the variable changes $p^2 \to -P^2$, $p'^2 \to -P'^2$, $q^2 \to -Q^2$ were performed.

Before obtaining the QCDSR expression for $g_{\eta_c D^* D}^{(D^*)}(Q^2)$, we need to remove the $\mathcal{BB}\left[h.r.\right]$ contribution from Eq. 12. This can be achieved by the introduction of the continuum threshold parameters $s_0 = (m_D + \Delta_s)^2$ and $u_0 = (m_{\eta_c} + \Delta_u)^2$ in Eq. 7 as the upper limits of the double integration.

Finally, applying the QCDSR, we are able to obtain the D^* off-shell form factor of the vertex $\eta_c D^* D$:

$$g_{\eta_c D^* D}^{(D^*)}(Q^2)_{p_\mu} = \frac{-\frac{1}{4\pi^2}\int_{s_{inf}}^{s_0}\int_{u_{inf}}^{u_0} ds du\, e^{-s/M^2} e^{-u/M'^2} F_{p_\mu}(s,u,t) + \mathcal{BB}\left[\Pi_{p_\mu}^{\langle g^2G^2\rangle}\right]}{\frac{f_{\eta_c} f_{D^*} f_D m_{\eta_c}^2 m_D^2}{2m_c^2 m_{D^*}(Q^2 + m_{D^*}^2)} e^{-\frac{m_D^2}{M^2}} e^{-\frac{m_{\eta_c}^2}{M'^2}}(m_{D^*}^2 + m_{\eta_c}^2 - m_D^2)}, \quad (13)$$

and

$$g_{\eta_c D^* D}^{(D^*)}(Q^2)_{p'_\mu} = \frac{-\frac{1}{4\pi^2}\int_{s_{inf}}^{s_0}\int_{u_{inf}}^{u_0} ds du\, e^{-s/M^2} e^{-u/M'^2} F_{p'_\mu}(s,u,t) + \mathcal{BB}\left[\Pi_{p'_\mu}^{\langle g^2G^2\rangle}\right]}{\frac{f_{\eta_c} f_{D^*} f_D m_{\eta_c}^2 m_D^2}{2m_c^2 m_{D^*}(Q^2 + m_{D^*}^2)} e^{-\frac{m_D^2}{M^2}} e^{-\frac{m_{\eta_c}^2}{M'^2}}(m_{D^*}^2 + m_D^2 - m_{\eta_c}^2)}, \quad (14)$$

where F_{p_μ} and $F_{p'_\mu}$ are given by eqs. 5 and 6 respectively and $\mathcal{BB}\left[\Pi^{\langle g^2G^2\rangle}\right]$ is given by Eq. 9.

3. Results

In order to obtain numerical results from eqs. 13 and 14, it is necessary to know the numerical values of all the quantities presented in these equations. The values of the masses (hadrons and quarks) and decay constants are presented in table 1. The value of the gluon condensate is $\langle g^2G^2\rangle = 0.88\ GeV^4$ [9].

Table 1: Numerical values necessary to the $g^{(D^*)}_{\eta_c D^* D}(Q^2)$ QCDSR calculation.

	Particle			
Quantity	c	η_c	D^*	D
m (MeV) [6]	1275	2983.6	2010.26	1869.61
f (MeV)	-	394.7 [10]	242[11]	204.6 [6]

The Borel mass and continuum threshold parameters are quantities introduced by the QCDSR and do not have a specific value, however, each of these quantities is confined to a particular range.

In the case of the Borel mass, it is allowed to have any value that is contained in the so called Borel window. The Borel window is a region limited by two constrains regarding the correlation function: the pole contribution must be bigger than the continuum contribution and the perturbative contribution must be bigger than the sum of the non-perturbative contributions. Since our latest work [5], we do not use an arbitrary value for the Borel mass. Instead, we use the average of the form factor on the Borel mass within their allowed values of the Borel window.

The values of the continuum threshold parameters Δ_s and Δ_u are determined by using the stability criteria, where the preferred combination of Δ_s and Δ_u is the one that presents the most stable form factor regarding the Borel window. The possible values of Δ_s and Δ_u are confined in the ranges $m_D^2 < (m_D + \Delta_s)^2 < m_D^{'2}$ and $m_{\eta_c}^2 < (m_{\eta_c} + \Delta_u)^2 < m_{\eta_c}^{'2}$ respectively, where m'_D and m'_{η_c} are the masses of the first radial resonances of the mesons.

From eqs. 13 and 14, it is evident that the $g^{(D^*)}_{\eta_c D^* D}(Q^2)$ form factor can be calculated from two different sum rules, one for each tensor structure available (p_μ and p'_μ), but in this work, just the p_μ structure presented a valid and stable Borel window. In this case, the most stable Borel window is $M^2 = [1.195–2.095]$ GeV^2, obtained with the continuum thresholds $\Delta_s = 0.6$ GeV and $\Delta_u = 0.4$ GeV in the $Q^2 = [1.0–3.0]$ GeV^2 window.

Using these parameters, we obtain the $g^{(D^*)}_{\eta_c D^* D}(Q^2)$ QCDSR form factor, represented by the squares in Fig. 2. This form factor is well fitted by an exponential curve (full line in Fig. 2):

$$g^{(D^*)}_{\eta_c D^* D}(Q^2) = 1.79e^{-0.282Q^2}\,. \tag{15}$$

Using Eq. 11, we extrapolate Eq. 15 to the pole of the D^* meson and obtain the $g_{\eta_c D^* D}$ coupling constant (black circle in Fig. 2):

$$g_{\eta_c D^* D} = 5.60\,. \tag{16}$$

4. Conclusion

In this work, we used the QCDSR to calculate the form factor and the coupling constant of the $\eta_c D^* D$ vertex for the D^* off-shell case. The OPE series used in this QCDSR considered operators

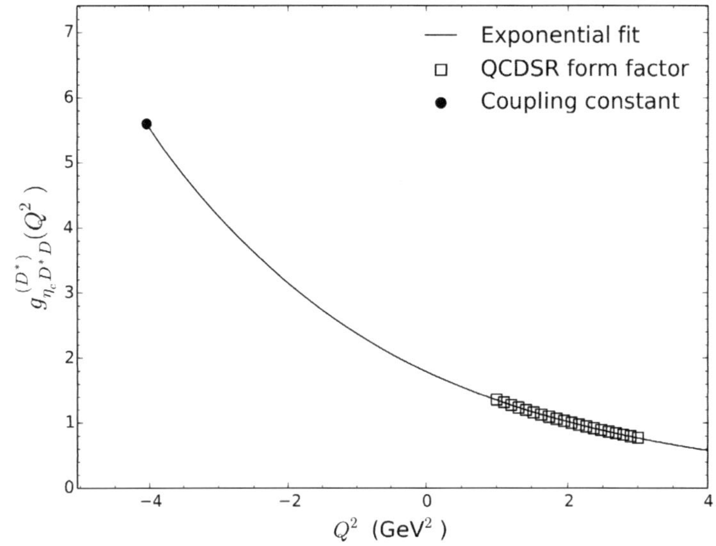

Figure 2: The QCDSR form factor for the D^* off-shell case of the $\eta_c D^* D$ vertex.

up to dimension 5 with non-perturbative contribution exclusively from the gluon condensates while quark condensates and mixed quark-gluon condensates were suppressed by the use of a double Borel transform.

The $g^{(D^*)}_{\eta_c D^* D}(Q^2)$ form factor can be calculated by two different QCDSR, given by eqs. 13 and 14, one for each tensor structure presented in the correlation function. However, just the p_μ structure could be used in this work as the p'_μ did not presented a valid Borel window.

The result for this QCDSR is an exponential form factor for the D^* off-shell case (Eq. 15) which leads to the coupling constant $g_{\eta_c D^* D} = 5.60$ (Eq. 16).

The next step in the study of the $\eta_c D^* D$ vertex is to calculated the two other possible form factors: $g^{(D)}_{\eta_c D^* D}(Q^2)$ and $g^{(\eta_c)}_{\eta_c D^* D}(Q^2)$. From Eq. 11, we can expect the coupling constant obtained from each of these two form factors to be close to the value obtained for the D^* off-shell case. This fact will be used in the analysis of the uncertainties of our QCDSR calculation.

Acknowledgments
This work has been partially supported by CNPq and CAPES.

Appendix A. Full expressions for $F^{\langle g^2G^2 \rangle}_{p_\mu}$

$F^{\langle g^2G^2 \rangle}_{p_\mu}$ and $F^{\langle g^2G^2 \rangle}_{p'_\mu}$ are functions of α_s used in Eq. 9 for the contribution of the gluon condensates to the correlation function. These functions can be defined as:

$$F^{\langle g^2G^2 \rangle}_{p_\mu,p'_\mu} = m_c F^{\langle g^2G^2 \rangle}_{p_\mu,p'_\mu(c)} - F^{\langle g^2G^2 \rangle}_{p_\mu,p'_\mu(d)} + m_c F^{\langle g^2G^2 \rangle}_{p_\mu,p'_\mu(e)} - F^{\langle g^2G^2 \rangle}_{p_\mu,p'_\mu(f)} - F^{\langle g^2G^2 \rangle}_{p_\mu,p'_\mu(g)} \tag{A.1}$$

where $F^{\langle g^2G^2 \rangle}_{(x)}$, with $x = c$–g, is related to each gluon condensate diagram in figs.1c–1g. The term $m_d F^{\langle g^2G^2 \rangle}_{p_\mu,p'_\mu(b)}$ that correspond to the diagram of Fig.1b is omitted from Eq. A.1 as we are using the limit $m_d \to 0$. We will also omit full expressions for the $F_{p'_\mu}^{\langle g^2G^2\rangle}$ function as the structure p'_μ was not used in this QCDSR.

In the next lines, we will show the full expressions for the $F_{p_\mu(x)}^{\langle g^2G^2\rangle}$ functions.

$$F_{p_\mu(c)}^{\langle g^2G^2\rangle} = -(m_cM'^2(\alpha^5M'^8M^6Q^2-2\alpha^4M'^6M^6Q^2+\alpha^3M'^4M^6Q^2+2\alpha^4M'^8M^4Q^2-4\alpha^3M'^6M^4Q^2$$
$$+2\alpha^2M'^4M^4Q^2+\alpha^3M'^8M^2Q^2-2\alpha^2M'^6M^2Q^2+\alpha M'^4M^2Q^2-\alpha^4M'^6M^8+3\alpha^3M'^4M^8-\alpha^3m_c^2M'^2M^8$$
$$-3\alpha^2M'^2M^8+\alpha M^8-2\alpha^3m_c^2M'^4M^6+2\alpha^2M'^4M^6-2\alpha^2m_c^2M'^2M^6-4\alpha M'^2M^6+2M^6+\alpha^4M'^{10}M^4$$
$$-3\alpha^3M'^8M^4-\alpha^3m_c^2M'^6M^4+7\alpha^2M'^6M^4-4\alpha^2m_c^2M'^4M^4-9\alpha M'^4M^4-\alpha m_c^2M'^2M^4+4M'^2M^4$$
$$+2\alpha^2M'^8M^2-2\alpha^2m_c^2M'^6M^2-4\alpha M'^6M^2-2\alpha m_c^2M'^4M^2+2M'^4M^2-\alpha m_c^2M'^6))$$
$$/((\alpha M'^2-1)^2(M^2+M'^2)^4). \quad (A.2)$$

$$F_{p_\mu(d)}^{\langle g^2G^2\rangle} = -(M'^2(\alpha^4M'^8M^6Q^2-3\alpha^3M'^6M^6Q^2+3\alpha^2M'^4M^6Q^2-\alpha M'^2M^6Q^2-\alpha^4M'^{10}M^4Q^2$$
$$+2\alpha^3M'^8M^4Q^2-\alpha^2M'^6M^4Q^2-\alpha^3M'^{10}M^2Q^2+2\alpha^2M'^8M^2Q^2-\alpha M'^6M^2Q^2-2\alpha^3M'^6M^8$$
$$-3\alpha^3m_c^2M'^4M^8+4\alpha^2M'^4M^8+2\alpha^2m_c^2M'^2M^8-3\alpha M'^2M^8+M^8+\alpha^3M'^8M^6-5\alpha^3m_c^2M'^6M^6$$
$$-\alpha^2M'^6M^6+\alpha^2m_c^2M'^4M^6-3\alpha M'^4M^6+\alpha m_c^2M'^2M^6+3M'^2M^6+4\alpha^3M'^{10}M^4-\alpha^3m_c^2M'^8M^4$$
$$-6\alpha^2M'^8M^4-3\alpha^2m_c^2M'^6M^4-\alpha M'^6M^4+\alpha m_c^2M'^4M^4+3M'^4M^4+\alpha^3M'^12M^2+\alpha^3m_c^2M'^{10}M^2$$
$$-\alpha^2M'^{10}M^2-\alpha^2m_c^2M'^8M^2-\alpha M'^8M^2-\alpha m_c^2M'^6M^2+M'^6M^2+\alpha^2m_c^2M'^{10}-\alpha m_c^2M'^8))$$
$$/(\alpha(\alpha M'^2-1)M^2(M^2+M'^2)^4) \quad (A.3)$$

$$F_{p_\mu(e)}^{\langle g^2G^2\rangle} = -(\alpha m_cM'^4(\alpha^3M'^4M^4Q^2-2\alpha^2M'^2M^4Q^2+\alpha M^4Q^2+\alpha^2M'^4M^2Q^2-2\alpha M'^2M^2Q^2$$
$$+M^2Q^2+2\alpha^2M'^2M^6-\alpha^2m_c^2M^6-2\alpha M^6+3\alpha^2M'^4M^4-2\alpha^2m_c^2M'^2M^4-2\alpha M'^2M^4-\alpha m_c^2M^4-M^4$$
$$+\alpha^2M'^6M^2-\alpha^2m_c^2M'^4M^2-2\alpha m_c^2M'^2M^2-M'^2M^2-\alpha m_c^2M'^4))$$
$$/((\alpha M'^2-1)^4M^4(M^2+M'^2)) \quad (A.4)$$

$$F_{p_\mu(f)}^{\langle g^2G^2\rangle} = -(M'^2(4\alpha^4M'^6M^6Q^2-7\alpha^3M'^4M^6Q^2+2\alpha^2M'^2M^6Q^2+\alpha M^6Q^2+6\alpha^3M'^6M^4Q^2$$
$$-12\alpha^2M'^4M^4Q^2+6\alpha M'^2M^4Q^2+\alpha^3M'^8M^2Q^2-2\alpha^2M'^6M^2Q^2+\alpha M'^4M^2Q^2-4\alpha^3M'^4M^8$$
$$-12\alpha^3m_c^2M'^2M^8+8\alpha^2M'^2M^8+6\alpha^2m_c^2M^8-4\alpha M^8-24\alpha^3m_c^2M'^4M^6-2\alpha^2M'^4M^6-2\alpha^2m_c^2M'^2M^6$$
$$+4\alpha M'^2M^6+6\alpha m_c^2M^6-2M^6+4\alpha^3M'^8M^4-12\alpha^3m_c^2M'^6M^4-8\alpha^2M'^6M^4-22\alpha^2m_c^2M'^4M^4$$
$$+4\alpha M'^4M^4+10\alpha m_c^2M'^2M^4+2\alpha^2M'^8M^2-14\alpha^2m_c^2M'^6M^2-4\alpha M'^6M^2+2\alpha m_c^2M'^4M^2+2M'^4M^2$$
$$-2\alpha m_c^2M'^6))/(2\alpha(\alpha M'^2-1)^2M^2(M^2+M'^2)^3) \quad (A.5)$$

$$\begin{aligned}F^{\langle g^2G^2\rangle}_{p_\mu(g)} &= -(M'^2(2\alpha^4M'^8M^4Q^2-5\alpha^3M'^6M^4Q^2+4\alpha^2M'^4M^4Q^2-\alpha M'^2M^4Q^2+\alpha^3M'^8M^2Q^2\\&-2\alpha^2M'^6M^2Q^2+\alpha M'^4M^2Q^2-4\alpha^3M'^6M^6-3\alpha^3m_c^2M'^4M^6+9\alpha^2M'^4M^6+2\alpha^2m_c^2M'^2M^6\\&-6\alpha M'^2M^6+M^6-4\alpha^3M'^8M^4-6\alpha^3m_c^2M'^6M^4+8\alpha^2M'^6M^4+2\alpha^2m_c^2M'^4M^4-4\alpha M'^4M^4\\&+\alpha m_c^2M'^2M^4-3\alpha^3m_c^2M'^8M^2-\alpha^2M'^8M^2-2\alpha^2m_c^2M'^6M^2+2\alpha M'^6M^2+2\alpha m_c^2M'^4M^2-M'^4M^2\\&-2\alpha^2m_c^2M'^8+\alpha m_c^2M'^6))/(\alpha(\alpha M'^2-1)^2M^2(M^2+M'^2)^3)\end{aligned}\quad\text{(A.6)}$$

References

[1] Navarra F S, Nielsen M, Bracco M E, Chiapparini M and Schat C L 2000 *Phys. Lett. B* **489** 319.
[2] Rodrigues B O, Bracco M E, Nielsen M and Navarra F S 2011 *Nucl. Phys. A* **852** 127.
[3] Bracco M E, Chiapparini M, Navarra F S and Nielsen M 2005 *Phys. Lett. B* **605** 326.
[4] Matheus R D, Navarra F S, Nielsen M and Rodrigues da Silva R 2005 *Int. J. Mod. Phys. A* **14** 555.
[5] Osório Rodrigues B., Bracco M E and Chiapparini M 2014 *Nucl. Phys. A* **929** 143.
[6] Olive K A et al 2014 *Chin. Phys. C* **38** 0900001.
[7] Wang Q, Liu X and Zhao Q 2012 *Phys. Lett. B* **711** 364.
[8] Bracco M E, Chiapparini M, Navarra F S and Nielsen M 2012 *Prog. Part. Nucl. Phys.* **67** 1019.
[9] Narison S 2012 *Phys. Lett. B* **706** 412.
[10] Davies C T H, McNeile C, Follana E, Lepage G P, Na H and Shigemitsu J 2010 *Phys. Rev. D* **83** 114504.
[11] Gelhausen P, Khodjamirian A, Pivovarov A A and Rosenthal D 2013 *Phys. Rev. D* **88** 014015.

Journal of Physics: Conference Series **630** (2015) 012036 doi:10.1088/1742-6596/630/1/012036

Unitary neutron matter in the on-shell limit *

E. Ruiz Arriola[1], **S. Szpigel**[2] **and V. S. Timóteo**[3]
[1]Departamento de Fisica Atómica, Molecular y Nuclear, Universidad de Granada
E-18071, Granada, Andalucia, España
[2]Centro de Rádio-Astronomia, Universidade Presbiteriana Mackenzie
01302-907, Sao Paulo, SP, Brasil
[3]Grupo de Óptica e Modelagem Numérica - GOMNI, Universidade Estadual de Campinas
13484-332, Limeira, SP , Brasil

E-mail: earriola@ugr.es (ERA)

E-mail: szpigel@mackenzie.br (SS)

E-mail: varese@ft.unicamp.br (VST)

Abstract. We compute the Bertsch parameter for neutron matter by using nucleon-nucleon interactions that are fully diagonal in momentum space. We analyze the on-shell limit with the similarity renormalization group and compare the results for a simple separable toy model to realistic calculations with high precision NN potentials.

1. Introduction

About fifteen years ago George Bertsch proposed the following problem [1]: what would be the ground state properties of a many-body system composed of spin-$\frac{1}{2}$ fermions interacting via a short-range contact interaction with an infinitely large scattering length?

The two-body scattering amplitude can be written as

$$\mathcal{T}_2(k,k) \propto \frac{1}{[k\cot\delta - ik]} , \tag{1}$$

and at low energies, $k\cot\delta$ can be described by the effective range expansion

$$k\cot\delta = -\frac{1}{\alpha_0} + \frac{1}{2}r_0k^2 + \cdots , \tag{2}$$

where α_0 is the scattering length and r_0 is the effective range. The unitary limit corresponds to $\alpha_0 \to \infty$ and $r_0 \to 0$. In this limit, $k\cot\delta \to 0$ and the scattering amplitude is then reduced to

$$\mathcal{T}_2(k,k) \propto \frac{i}{k} . \tag{3}$$

and thus the cross section in the S-wave saturates the unitarity bound $\sigma = 4\pi/k^2$. Therefore the two-body scattering amplitude becomes completely scale invariant and as a consequence the energy per particle for the ground state of such a system would be given by

$$\varepsilon = \xi \times \varepsilon_{FG} , \tag{4}$$

*Talk given by V.S.T. at XXXVII RTFNB, 37th Brazilian Workshop on Nuclear Physics, 8-12 September 2014, Maresias, SP, Brazil

where

$$\varepsilon_{FG} = \frac{3}{5}\frac{\hbar^2 k_F^2}{2m} \tag{5}$$

is the energy per particle of the free (non-interacting) Fermi Gas and ξ has been called Bertsch parameter. In fact, Bertsch wanted to know the sign of ξ. If $\xi > 0$, the system is a gas with positive pressure, and if $\xi < 0$ the system collapses since the pressure is negative.

Interesting many-body physics emerges when the interaction is close to the unitarity limit. In particular, many-body fermionic systems behave almost like a perfect fluid and may exhibit both BCS crossover and Bose-Einstein condensation which are, in fact, distinct limits of a common phenomenon. If the scattering length is small and negative, the interaction is weakly attractive and the system is in the BCS limit and there are overlapping loosely bound pairs. If the scattering length is large and negative, the interaction is strongly attractive and the system is in the BEC limit where the fermions form deeply bound pairs [2] (see Fig. 1).

Both neutron matter at low densities and ultra-cold atoms close to a Feshbach resonance are strongly interacting fermionic systems and present large pairing gaps when measured in units of the Fermi energy [3]. Monte Carlo simulations of a superfluid Fermi gas with an attractive short-range two-body interaction in the unitary limit (infinite scattering length and zero effective range) estimate the energy per particle of neutron matter in 44% of the energy of the free Fermi gas and the pairing gap to be about twice the energy per particle [4].

The S-wave interaction between two neutrons is very attractive (almost enough to produce a nn bound state) and has a significantly large (and negative) scattering length, $-\alpha_0^{nn} = 18.5\text{fm} \gg r_0^{nn} = 2.7\text{fm}$. Hence, neutron matter at low densities is a system with features close to the unitary limit. Another reason to study neutron matter at low densities is that both superconductivity and superfluidity in fermonic systems are manifestations of quantum coherence at a macroscopic level. An *ab initio* calculation of a Fermi gas in the unitary limit shows that, at $T = 0.2\ E_F$, the viscosity is close to the lower limit for a perfect fluid [5].

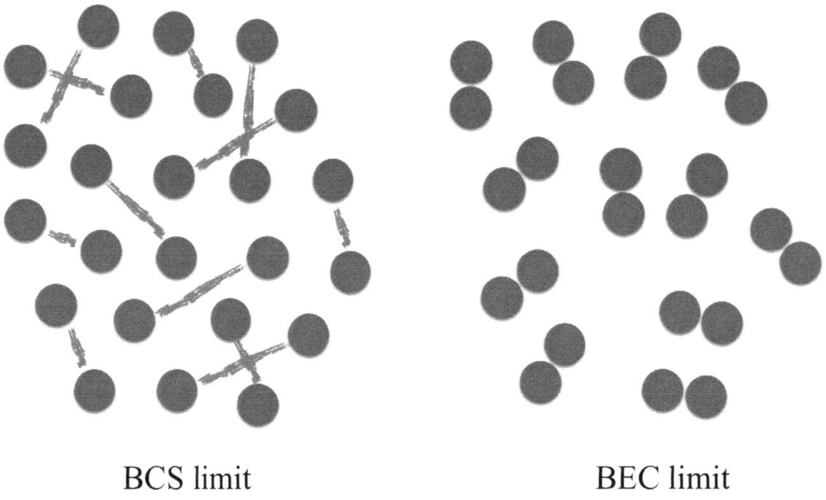

Figure 1. Pictorical representation of the BCS and BEC limits of a many-fermion system. The blurred red lines indicate the weak pairing. Adapted from Living Rev. Relativity **11**, 10 (2008).

Journal of Physics: Conference Series **630** (2015) 012036 doi:10.1088/1742-6596/630/1/012036

2. On-shell interactions from the SRG and the Bertsch parameter

In this work we want to study neutron matter at the unitary limit with on-shell interactions obtained by evolving the nuclear force with the similarity renormalization group (SRG) towards the infrared region of the similarity cutoff λ. The SRG has been widely applied to nuclear structure calculations and the nuclear many-body problem [10, 11]. The technique is based on a flow equation which for the nucleon-nucleon interaction and the Wilson generator reads

$$\frac{dV_s(p,p')}{ds} = -(p^2 - p'^2)^2 \; V_s(p,p') + \int_0^\infty dq \; q^2(p^2 + p'^2 - 2q^2)V_s(p,q) \; V_s(q,p') \; , \tag{6}$$

where the flow parameter s is usually written in terms of the so-called similarity cutoff λ (which has dimension of momentum) as $s = \lambda^{-4}$. The potential before evolution (initial) corresponds to $s = 0$ or $\lambda = \infty$ and the matrix elements of the evolved potential are denoted as $V_\lambda(p,p')$. The unitarity of the transformation ensures that all observables computed with $V_\lambda(p,p')$ are exactly the same as the observables computed with the non-evolved initial potential $V_\infty(p,p')$. This applies in particular to phase-shifts, which do not depend on λ.

Recently, we have developed techniques in nuclear physics in order to study the infrared fixed-point of the SRG by pushing the evolution towards the on-shell limit $\lambda \to 0$ [6, 7, 8] and have found an elegant an simple way to determine phase shifts from fully diagonal interactions in momentum space complying with isospectrality and Levinson's theorem [9].

A simple S-wave gaussian separable potential toy model allows to carry studies with a moderate numerical effort

$$V(p,p') = C \;\exp\left[-\frac{1}{L^2}\left(p^2 + p'^2\right)\right] \; , \tag{7}$$

where the parameters C and L are obtained by fitting the scattering length and the effective range. The nn interaction cannot be measured directly, but since the nn and np interaction in the $^1S_0$ channel have similar (and large) scattering lengths, $\alpha_0^{nn} = -18.5$ fm and $\alpha_0^{np} = -23.7$ fm, we use the np phase-shifts to access how well the toy model describes the nuclear force in the S-waves. This gives $C = -1.916$ fm and $L = 1.2$ fm^{-1}.

At similarity cutoffs close to $\lambda \sim 1$ fm^{-1} the flow equation becomes extremely stiff so that it is nearly impossible to study the infrared limit of the similarity cutoff, $\lambda \to 0$, if the potential has a long tail in momentum space, which is the case for high precision nucleon-nucleon potentials. This is the reason why we have constructed the toy model since it gives good qualitative results for the nucleon-nucleon S-waves but has a short tail in momentum space, allowing the SRG evolution towards the infrared region of the similarity cutoff with a moderate numerical effort. The fully diagonal on-shell interaction at $\lambda = 0$ is obtained by using the energy shift prescription of Ref. [9].

3. Bertsch parameter

Here we apply the toy model to compute the Bertsch parameter in the infrared region of the similarity cutoff with different grid sizes. We also compute the Bertsch parameter with high precision nucleon-nucleon potentials to compare them to the results from the toy potential.

The Bertsch parameter is the ratio between the total energy of a system of interacting fermions in the unitary limit and the energy of a free Fermi gas:

$$\xi_\lambda(k_F) = \frac{T(k_F) + V_\lambda(k_F)}{T(k_F)} = 1 + \frac{V_\lambda(k_F)}{T(k_F)} \; . \tag{8}$$

The kinetic energy in neutron matter is given by

$$T(k_F) = \frac{3k_F^2}{10m_n} \; , \tag{9}$$

where m_n is the neutron mass and k_F is the Fermi momentum. The potential energy can be obtained in the Hartree-Fock approximation and is due to the toy interaction in the $^1S_0$ channel:

$$V_\lambda(k_F) = \frac{4}{m_n}\frac{2}{\pi}\int_0^{k_F} dk\, k^2 \left(1 - \frac{3k}{2k_F} + \frac{k^3}{2k_F^3}\right) V_\lambda^{^1S_0}(k,k)\,. \tag{10}$$

For realistic high precision interactions, there are also contributions from higher partial waves and the energy per particle can be written as

$$\begin{aligned}\varepsilon(k_F) = \frac{3k_F^2}{10m_n} &+ \frac{4}{m_n}\frac{2}{\pi}\int_0^{k_F} dk\, k^2 \left(1 - \frac{3k}{2k_F} + \frac{k^3}{2k_F^3}\right) \\ &\times \left[V_\lambda^{^1S_0}(k,k) + 9V_\lambda^{^3P_c}(k,k) + 5V_\lambda^{^1D_2}(k,k) + 21V_\lambda^{^3F_c}(k,k) + 9V_\lambda^{^1G_4}(k,k)\right]\,,\end{aligned} \tag{11}$$

where ^3P_c and ^3F_c are linear combinations of the P and F waves, which are given explicitly in Ref. [12].

4. Numerical results

In Fig. 2 we show the Bertsch parameter as a function of the Fermi momentum computed with the toy model for several values of the similarity cutoff, mostly in the infrared region, for different number of grid points. For similarity cutoffs from infinity down to 0.5 fm^{-1} the number of grid points almost do not affect the results. However, for smaller values of λ the number of grid points start to change the results and in the limit $\lambda = 0$ the difference is huge as can be seen in the last panel of Fig. 2.

At $\lambda = 1$ fm^{-1} the results obtained with the toy model are very close to the results that come out if we consider only S-waves from high precision nucleon-nucleon potentials. This can be observed in Fig. 3 where we display the Bertsch parameter computed with Argonne v18 [13], Nijmegen II [14], N3LO (2003) [15] and N3LO (2005) [16] nucleon-nucleon interactions. The minimum value of the Bertsch parameter lies in between $\xi = 0.42$ and $\xi = 0.45$ for both the toy model and high precision potentials at k_F between 1.1 fm^{-1} and 1.3 fm^{-1}. Sophisticated quantum Monte Carlo calculations give $\xi = 0.44$ [4], so it is quite impressive that a simple separable potential can provide results that are so close to more accurate approaches.

While the contribution from the S-waves is rather independent of the nucleon-nucleon potential, when higher partial waves are included the results depend on whether one uses phenomenological potentials (Av18 or NijII) or chiral potentials (2003 N3LO or 2005 N3LO). This can be seen in Fig. 4 where the Bertsch parameter is computed summing up to G-waves. Also, the minimum of ξ gets much smaller and is displaced towards larger k_F.

When the similarity cutoff reaches the limit $\lambda = 0$ the interaction becomes fully diagonal and all off-shell ambiguities are then eliminated. In Fig. 5 we show a comparison of the Bertsch parameter at $\lambda = 0$ for the toy model and the high precision potentials with only S-waves (left panel) and summing up to G-waves (right panel). In the region $0 < k_F < 0.5$ fm^{-1}, the toy model matches the calculation with high precision potentials with only S-waves. For larger k_F the result is also very reasonable if one considers the extreme simplicity of the separable potential.

The λ-dependence of the Bertsch parameter is rather strong and shows that it is not determined just by two-body scattering information unless the extreme on-shell limit $\lambda \to 0$ is taken. Of course, the real limit $a_0 \to -\infty$ and $\lambda \to 0$ remains an interesting challenge.

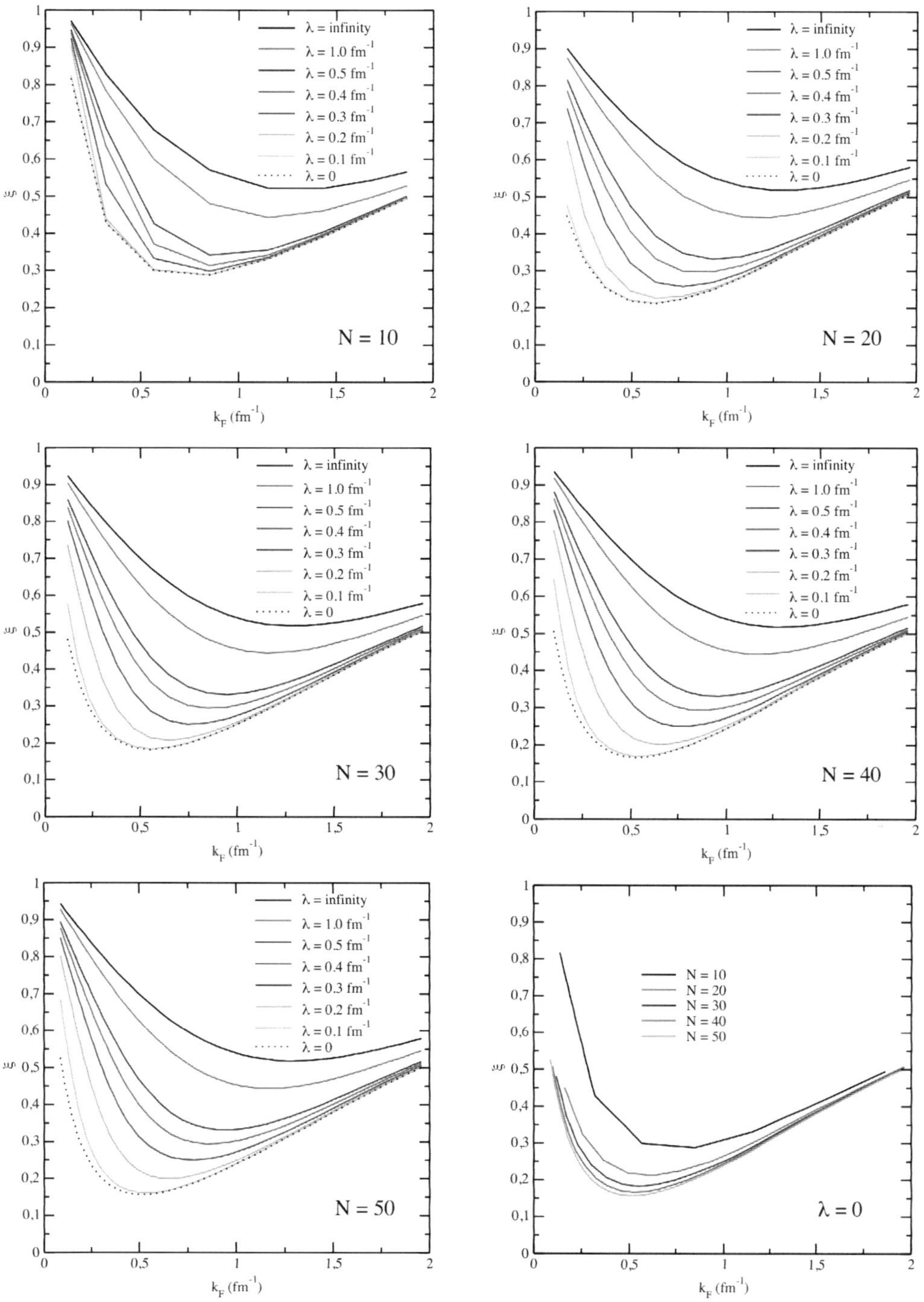

Figure 2. Bertsch parameter from the S-wave toy model for several values of the similarity cutoff λ (Wilson generator) and different number of grid points N. The right panel in the last row shows the grid dependence in the limit $\lambda = 0$. The on-shell interaction $V_{\lambda=0}(k)$ was obtained by applying the eigenvalue method of Ref. [9].

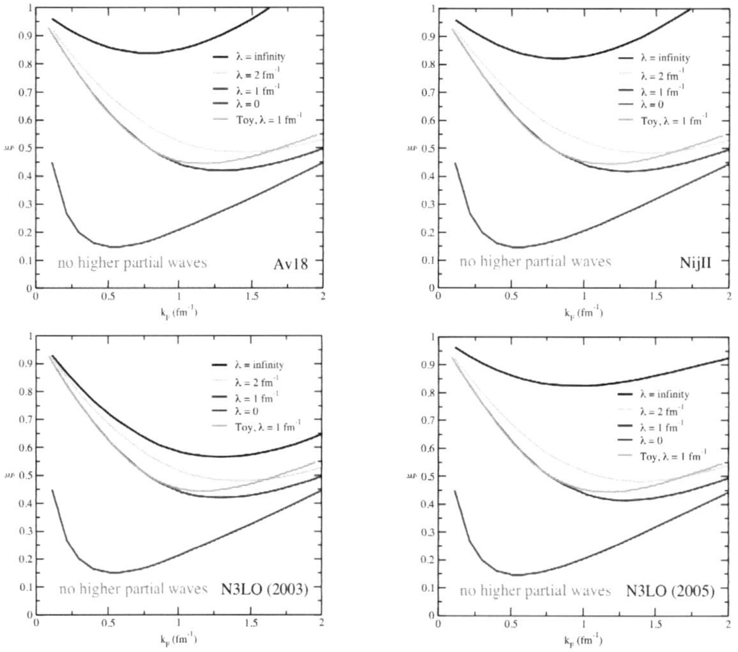

Figure 3. Bertsch parameter from high precision potentials for some values of the similarity cutoff λ (Wilson generator) with $N = 200$ grid points, taking only S-waves into account.

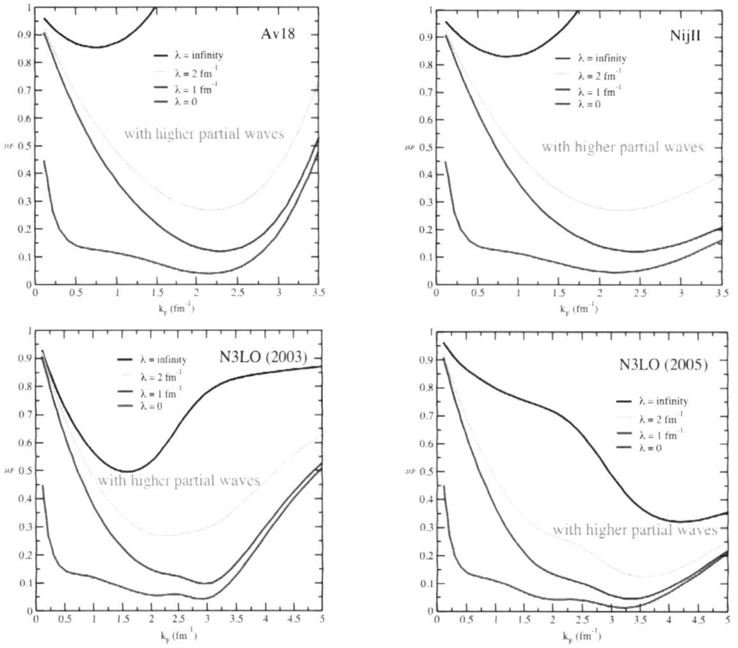

Figure 4. Bertsch parameter from high precision potentials for some values of the similarity cutoff λ (Wilson generator) with $N = 200$ grid points, but summing up to G-waves.

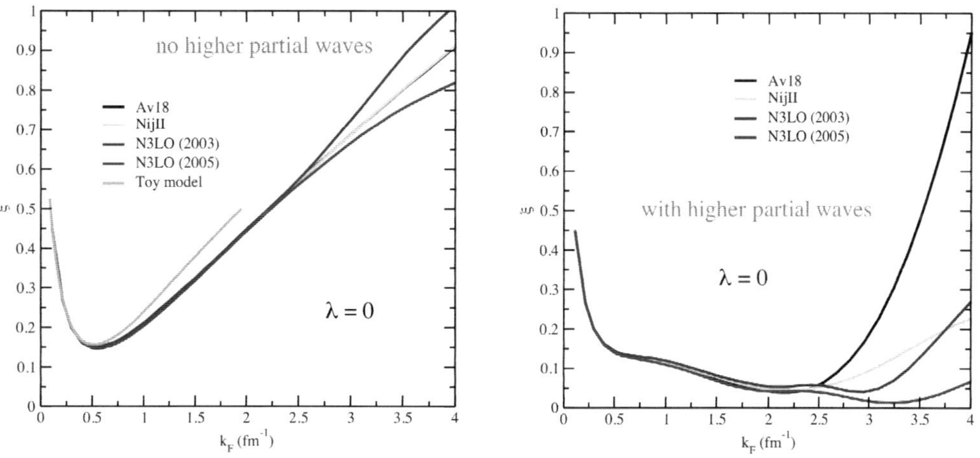

Figure 5. Bertsch parameter in the limit $\lambda = 0$ for some high precision potentials with $N = 200$ grid points with only S-waves (left) and summing up to G-waves (right).

5. Conclusions

So far we have studied the ground state of neutron matter in the unitary limit with a two-nucleon interaction given by a simple separable potential and compared the results to what is obtained with high precision nucleon-nucleon interactions. We also extended the study to the infrared region of the similarity cutoff by evolving the interactions with the SRG flow equation for the Wilson generator. The limit $\lambda = 0$ was obtained by using our energy shift prescription. Our results for the toy model provide a good estimate for the Bertsch parameter at $\lambda = 1$ fm^{-1}. In the limit $\lambda = 0$, since the two-body force gets small, ξ lowers dramatically at intermediate k_F.

Acknowledgements

E.R.A. would like to thank the Spanish DGI (Grant FIS2011-24149) and Junta de Andalucia (Grant FQM225) for support. S.S. is partially supported by FAPESP and V.S.T. thanks FAEPEX, FAPESP and CNPq for financial support.

References

[1] T. Papenbrock, G. F. Bertsch, Phys. Rev. C **59**, 2052 (1999).
[2] N. Chamel and P. Haensel, Living Rev. Relativity **11**, 10 (2008).
[3] J. Carlson, S. Gandolfi and A. Gezerlis, Prog. Theor. Exp. Phys., 01A209 (2012).
[4] J. Carlson, S. Y. Chang, V. R. Pandharipande and K. E. Schmidt, Phys. Rev. Lett. **91**, 050401 (2003).
[5] G. Wlazlowski, P. Magierski and J. E. Drut, Phys. Rev. Lett. **109**, 020406 (2012).
[6] E. Ruiz Arriola, S. Szpigel and V. S. Timóteo, Phys. Lett. B **728**, 596 (2014).
[7] E. Ruiz Arriola, S. Szpigel and V. S. Timóteo, Few Body Syst. **55**, 989 (2014).
[8] E. Ruiz Arriola, S. Szpigel and V. S. Timóteo, arXiv: 1407.8449 (2014).
[9] E. Ruiz Arriola, S. Szpigel and V. S. Timóteo, Phys. Lett. B **735**, 149 (2014).
[10] S.K. Bogner, R.J. Furnstahl and A. Schwenk, Prog. Part. Nucl. Phys. **65**, 94 (2010).
[11] R.J. Furnstahl and K. Hebeler, Rept. Prog. Phys. **76**, 126301 (2013).
[12] V. S. Timóteo, S. Szpigel and E. Ruiz Arriola, Phys. Rev. C **86**, 034002 (2012).
[13] R. B. Wiringa, V. G. J. Stoks, and R. Schiavilla, Phys. Rev. C **51**, 38 (1995).
[14] V. G. J. Stoks, R. A. M. Klomp, C. P. F. Terheggen, and J. J. de Swart, Phys. Rev. C **49**, 2950 (1994).
[15] D. R. Entem and R. Machleidt, Phys. Rev. C **68**, 041001 (2003).
[16] E. Epelbaum, W. Glöckle and U. -G. Meissner, Nucl. Phys. A **747**, 362 (2005).

$X(3872)$ production from reactions involving D and D^* mesons

A. Martínez Torres[1], K. P. Khemchandani[1], F. S. Navarra[1], M. Nielsen[1], Luciano M. Abreu[2]

[1] Instituto de Física, Universidade de São Paulo, C.P. 66318, 05389-970 São Paulo, SP, Brazil.
[2] Instituto de Física, Universidade Federal da Bahia, 40210-340, Salvador, BA, Brazil.

Abstract. In this proceeding we show the results found for the cross sections of the processes $\bar{D}D \to \pi X(3872)$, $\bar{D}^*D \to \pi X(3872)$ and $\bar{D}^*D^* \to \pi X(3872)$, information needed for calculations of the $X(3872)$ abundance in heavy ion collisions. Our formalism is based on the generation of $X(3872)$ from the interaction of the hadrons $\bar{D}^0D^{*0}$ − c.c, D^-D^{*+} − c.c and $D_s^-D_s^{*+}$ − c.c. The evaluation of the cross section associated with processes having D^* meson(s) involves an anomalous vertex, $X\bar{D}^*D^*$, which we have determined by considering triangular loops motivated by the molecular nature of $X(3872)$. We find that the contribution of this vertex is important. Encouraged by this finding we estimate the $X\bar{D}^*D^*$ coupling, which turns out to be 1.95 ± 0.22. We then use it to obtain the cross section for the reaction $\bar{D}^*D^* \to \pi X$ and find that the $X\bar{D}^*D^*$ vertex is also relevant in this case. We also discuss the role of the charged components of X in the determination of the production cross sections.

1. Introduction

Since the development of B factories like BELLE and BES a wealth of data on new hadronic states has been produced [1, 2], information which is crucial to understand the nature and properties of such states. Particularly interesting are the data on the so called exotic charmonium states. One member of this family, and probably the most widely studied theoretically, is the $X(3872)$ (from now on simply X), reported a decade ago by the Belle collaboration in the decay $B^\pm \to K^\pm\pi^+\pi^-J/\psi$ [3]. After this finding, several other collaborations [4–6] confirmed this state and its existence is now established beyond any doubt. However, it has only been very recently when the spin-parity quantum numbers of X have been confirmed to be 1^{++} [7].

During these years, several theoretical models have been proposed to describe the properties of this state, considering it as a charmonium state, a tetraquark, a $D - \bar{D}^*$ hadron molecule and a mixture between a charmonium and a molecular component [8–21]. In spite of the effort of these numerous groups, the properties of this particle are not yet well understood and represent a challenge both for theorists and experimentalists.

In line with the information on new charmonium states brought by the B factories, in a different frontier of physics, collaborations like RHIC and LHC has devoted a significant part of their physics program to study the Quark Gluon Plasma (QGP). It is now a well accepted fact that in high energy heavy ion collisions a deconfined medium is created: the quark gluon plasma (QGP) [22, 23]. In a high energy heavy ion collision the QGP is formed, expands, cools, hadronizes and is converted into a hadron gas, which lives up to 10 fm/c and then freezes out.

During this evolution, an increasing (with the reaction energy) number of charm quarks and anti-quarks move freely. The initially formed charmonium bound states are dissolved (the famous "charmonium suppression") but c's and $\bar{c}$'s, coming now from different parent gluons, can pick up light quarks and anti-quarks from the rich environment and form multiquark bound states. This is called quark coalescence and it happens during the phase transition to the hadronic gas [24,25]. Therefore, the formation of the quark gluon plasma phase increases the number of produced X's [24, 25]. Interestingly, the coalescence formalism is based on the overlap of the Wigner functions of the quarks and of the bound state, being thus sensitive to the spatial configuration of the charmonium state and hence being able to distinguish between a compact, $\simeq 1$ fm long, tetraquark configuration and a large $\simeq 10$ fm long, molecular configuration. A big difference between the predicted abundancies could be used as a tool to discriminate between different X structures and to help us to decide whether it is a molecule or a tetraquark [26]. In this way, heavy ion collisions can be used to obtain information about exotic charmonium states as X. However, there is an additional complication. Due to the rich hadronic environment present in the plasma, the X's can be destroyed in collisions with ordinary hadrons, such as $X+\pi \to D+\bar{D}^*$, and can also be produced through the inverse reactions, such as $D + \bar{D}^* \to X + \pi$. A proper determination of the abundance of X in heavy ion collisions requires a precise calculation of the cross sections of these kind of processes. In Ref. [26], the hadronic absorption cross section of the X by mesons like π and ρ was evaluated for the processes $\pi X \to D\bar{D}$, $\pi X \to D^*\bar{D}^*$, $\rho X \to D\bar{D}$, $\rho X \to D\bar{D}^*$, and $\rho X \to D^*\bar{D}^*$. Using these cross sections, the variation of the X meson abundance during the expansion of the hadronic matter was computed with the help of a kinetic equation with gain and loss terms. The results turned out to be strongly dependent on the quantum numbers of the X and on its structure.

The present work is devoted to introduce two improvements in the calculation of cross sections performed in Ref. [26]. The first and most important one is the inclusion of the anomalous vertices $\pi D^* D^*$ and $X\bar{D}^* D^*$, which were neglected before. With these vertices new reaction channels become possible, such as $\pi X \to D\bar{D}^*$, and the inverse process $D\bar{D}^* \to \pi X$. As will be seen, this reaction is the most important one for X in the hadron gas. The relevance of anomalous couplings has also been shown earlier in different contexts, for example in the J/ψ absorption cross sections by π and ρ mesons [27], radiative decays of scalar resonances and axial vector mesons [28, 29] and in kaon photoproduction [30].

The second improvement is the inclusions of the charged components of the D and D^* mesons which couple to the X [17].

2. Formalism

2.1. Determination of the cross sections

To calculate the cross section for the processes (1) $\bar{D}D \to \pi X$, (2) $\bar{D}^*D \to \pi X$ and (3) $\bar{D}^*D^* \to \pi X$ we consider the model of Refs. [16, 17, 31] in which X is generated from the interaction of $\bar{D}^0 D^{*0} -$c.c, $D^- D^{*+} -$c.c and $D_s^- D_s^{*+} -$c.c. The isospin-spin averaged production cross section for the processes $\bar{D}D, \bar{D}^*D, \bar{D}^*D^* \to \pi X$, in the center of mas (CM) frame can be determined as

$$\sigma_r(s) = \frac{1}{16\pi\lambda(s, m_{1i,r}^2, m_{2i,r}^2)} \int_{t_{\rm min,r}}^{t_{\rm max,r}} dt \, \overline{\sum_{\rm Isos,spin} |\mathcal{M}_r(s,t)|^2}\,, \tag{1}$$

where $r = 1, 2, 3$ is an index indicating the reaction considered, $\sqrt{s}$ is the CM energy, and $m_{1i,r}$ and $m_{2i,r}$ represent the masses of the two particles present in the initial state i of the reaction r. We follow the convention of associating the index 1 (2) with the particle with charm -1 ($+1$) present in the initial state. The function $\lambda(a, b, c)$ in Eq. (1) is the Källen function, $t_{\rm min,r}$ and $t_{\rm max,r}$ correspond to the minimum and maximum values, respectively, of the Mandelstam variable t and $\mathcal{M}_r$ is the reduced matrix element for the process r. The symbol $\overline{\sum_{\text{spin,Isos}}}$ represents the sum over the isospins and spins of the particles in the initial and final state, weighted by the isospin and spin degeneracy factors of the two particles forming the initial state for the reaction r, i.e.,

$$\overline{\sum_{\text{spin,Isos}}} |\mathcal{M}_r|^2 \to \frac{1}{(2I_{1i,r}+1)(2I_{2i,r}+1)} \frac{1}{(2s_{1i,r}+1)(2s_{2i,r}+1)} \sum_{\text{spin,Isos}} |\mathcal{M}_r|^2, \tag{2}$$

where,

$$\sum_{\text{spin,Isos}} |\mathcal{M}_r|^2 = \sum_{Q_{1i},Q_{2i}} \left[\sum_{\text{spin}} \left| \mathcal{M}_r^{(Q_{1i},Q_{2i})} \right|^2 \right]. \tag{3}$$

In Eq. (3), Q_{1i} and Q_{2i} represent the charges for each of the two particles forming the initial state i of the reaction r, which are combined to obtain total charge $Q_r = Q_{1i} + Q_{2i} = 0, +1, -1$. In this way, we have four possibilities: $(0,0)$, $(-,+)$, $(-,0)$ and $(0,+)$ and thus,

$$\sum_{\text{spin,Isos}} |\mathcal{M}_r|^2 = \sum_{\text{spin}} \left(\left| \mathcal{M}_r^{(0,0)} \right|^2 + \left| \mathcal{M}_r^{(-,+)} \right|^2 + \left| \mathcal{M}_r^{(-,0)} \right|^2 + \left| \mathcal{M}_r^{(0,+)} \right|^2 \right). \tag{4}$$

.

Each of the amplitudes $\mathcal{M}_r^{(Q_{1i},Q_{2i})}$ of Eq. (3) can be written as

$$\mathcal{M}_r^{(Q_{1i},Q_{2i})} = T_r^{(Q_{1i},Q_{2i})} + U_r^{(Q_{1i},Q_{2i})}, \tag{5}$$

where $T_r^{(Q_{1i},Q_{2i})}$ and $U_r^{(Q_{1i},Q_{2i})}$ are the contributions related to the t and u channel diagrams contributing to each process.

2.2. The $\bar{D}D \to \pi X$ reaction

In Fig. 1 we show the different diagrams contributing to $\bar{D}D \to \pi X$ (without specifying the charge of the reaction).

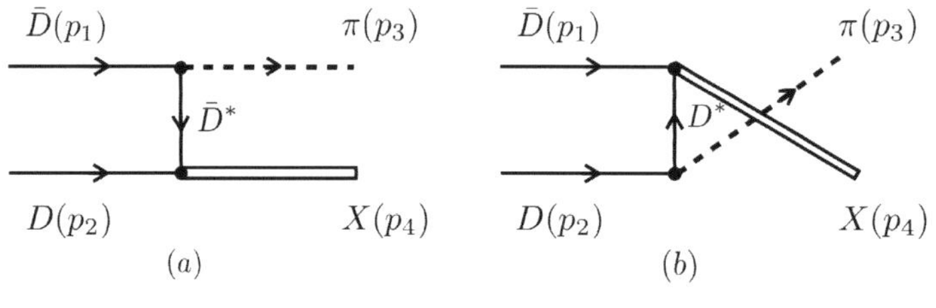

Figure 1. Diagrams contributing to the process $\bar{D}D \to \pi X$.

The t-channel amplitude for the process in Fig. 1a can be written as

$$T_1^{(Q_{1i},Q_{2i})} = W_1^{(Q_{1i},Q_{2i})} g_{PPV}\, g_X \frac{1}{t - m_{\bar{D}^*}^2} \left[(p_1 + p_3)_\mu + \frac{m_{\bar{D}}^2 - m_\pi^2}{m_{\bar{D}^*}^2} p_{2\mu} \right] \epsilon_X^\mu(p_4), \tag{6}$$

Table 1. Coefficients $W_r^{(Q_{1i},Q_{2i})}$ and couplings g_X for the amplitude given in Eq. (6). We have defined $g_n \equiv g_{X\bar{D}^0 D^{*0}}$ and $g_c \equiv g_{XD^- D^{*+}}$, whose numerical values can be found in Table 3.

r	(Q_{1i}, Q_{2i})	W_r	g_X
1	$(0,0)$	$-1/\sqrt{2}$	$-g_n$
	$(-,+)$	$1/\sqrt{2}$	$-g_c$
	$(-,0)$	-1	$-g_n$
	$(0,+)$	-1	$-g_c$
2	$(0,0)$	$-1/2$	$-g_n$
	$(-,+)$	$1/2$	$-g_c$
	$(-,0)$	$-1/\sqrt{2}$	$-g_n$
	$(0,+)$	$-1/\sqrt{2}$	$-g_c$

Table 2. Coefficients $Z^{(Q_{1i},Q_{2i})}$ and couplings g_X for the amplitude given in Eq. (7). We have defined $g_n \equiv g_{X\bar{D}^0 D^{*0}}$ and $g_c \equiv g_{XD^- D^{*+}}$, whose numerical values can be found in Table 3.

(Q_{1i}, Q_{2i})	Z_r	g_X
$(0,0)$	$1/\sqrt{2}$	g_n
$(-,+)$	$-1/\sqrt{2}$	g_c
$(-,0)$	1	g_c
$(0,+)$	1	g_n

Table 3. Couplings of X to the different pseudoscalar-vector components constituting the state $(\bar{P}_X V_X)$. The couplings for the complex conjugate components bear a minus sign.

$\bar{P}_X V_X$	$g_{X\bar{P}_X V_X}$ (MeV)
$D^- D^{*+}$	$3638/\sqrt{2}$
$\bar{D}^0 D^{*0}$	$3663/\sqrt{2}$
$D_s^- D_s^{*+}$	$3395/\sqrt{2}$

while for the u-channel amplitude (Fig. 1b) we have

$$U_1^{(Q_{1i},Q_{2i})} = Z^{(Q_{1i},Q_{2i})} g_{PPV} g_X \frac{1}{u - m_{D^*}^2} \left[(p_2 + p_3)_\mu + \frac{m_D^2 - m_\pi^2}{m_{D^*}^2} p_{1\mu} \right] \epsilon_X^\mu(p_4). \tag{7}$$

The coefficients $W_r^{(Q_{1i},Q_{2i})}$ and $Z^{(Q_{1i},Q_{2i})}$ and couplings g_X are given in Tables 1 and 2.

The coupling g_{PPV} in Eqs. (6) and (7) is the strong coupling of the D^* meson to $D\pi$. As shown in Refs. [32, 33], consideration of heavy quark symmetry gives a value of

$$g_{PPV} = \frac{m_\rho}{2 f_\pi} \frac{m_{D^*}}{m_{K^*}} \sim 9, \tag{8}$$

where we have use the pion decay constant value $f_\pi = 93$ MeV. Using this coupling, the decay width for the process $D^{*+} \to D^0 \pi^+$ is 71 KeV, in agreement with the recent experimental result of (65 ± 15) KeV [34] and compatible with the coupling found in Ref. [35] using QCD sum rules.

2.3. The $\bar{D}^* D \to \pi X$ reaction

We show the relevant diagrams contributing to this process in Figs. 2 and 3, which involve anomalous vertices, $\bar{D}^* \bar{D}^* \pi$ in the t-channel and $X \bar{D}^* D^*$ in the u-channel. The t channel

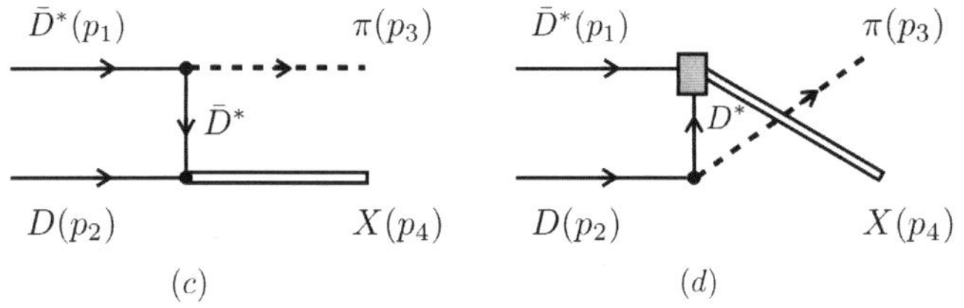

Figure 2. Diagrams contributing to the process $\bar{D}^*D \to \pi X$. The diagram containing a filled box is calculated by summing the set of diagrams shown in Fig. 3, as explained in the text.

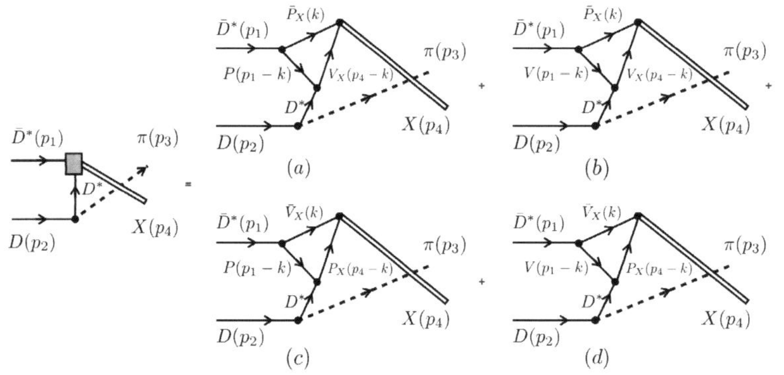

Figure 3. Diagrams considered for the determination of the filled box shown in Fig. 2. The hadrons P_X and V_X represent the pseudoscalars and vectors coupling to the state X, while P and V are any pseudoscalar and vector meson which can be exchanged conserving different quantum numbers. For a list of the different exchanged hadrons considered here see Ref. [36].

contribution is directly obtained as

$$T_2^{(Q_{1i},Q_{2i})} = W_2^{(Q_{1i},Q_{2i})} g_{VVP}\, g_X \frac{1}{t-m^2_{\bar{D}^*}} \epsilon^{\mu\nu\alpha\beta} p_{1\mu} p_{3\alpha} \epsilon_{\bar{D}^*\nu}(p_1) \epsilon_{X\beta}(p_4). \tag{9}$$

where the coefficients $W^{(Q_{1i},Q_{2i})}$ and the corresponding g_X coupling are listed in Table 1. The amplitudes for the u-channel diagram shown in Fig. 2d can be calculated as

$$U_2^{(Q_{1i},Q_{2i})} = \sum_{p=a}^{d} U_{2p}^{(Q_{1i},Q_{2i})}, \tag{10}$$

with $U_{2p}^{(Q_{1i},Q_{2i})}$ $(p = a, b, \cdots, d)$ being the amplitudes associated with the diagrams depicted in Fig. 3. As can be seen, these amplitudes depend on the hadrons present in the triangular loops (P, V_X, etc.), since the couplings, propagators, etc., depend on them. The final result for the amplitude of each diagram in Fig. 3 can be obtained by summing over the amplitudes for the different intermediate states

$$U_{2p}^{(Q_{1i},Q_{2i})} = \sum_{P,P_X,V_X,V} \mathcal{U}_{2p}^{(Q_{1i},Q_{2i})}, \tag{11}$$

where $\mathcal{U}_{2p}^{(Q_{1i},Q_{2i})}$, $p = a, b,$ etc., is the amplitude for the diagram in Fig. 3p for a particular set of hadrons in the triangular loop.

The evaluation of the amplitudes in Eqs. (9) and (11) involves PPV, VVP and VVV vertices (with P and V representing a pseudoscalar and a vector meson, respectively). To calculate them we have made used of effective Lagrangians [37–40]

$$\begin{aligned} \mathcal{L}_{PPV} &= -ig_{PPV}\langle V^\mu[P,\partial_\mu P]\rangle, \\ \mathcal{L}_{VVP} &= \frac{g_{VVP}}{\sqrt{2}}\epsilon^{\mu\nu\alpha\beta}\langle\partial_\mu V_\nu\partial_\alpha V_\beta P\rangle \\ \mathcal{L}_{VVV} &= ig_{VVV}\langle(V^\mu\partial_\nu V_\mu - \partial_\nu V_\mu V^\mu)V^\nu)\rangle. \end{aligned} \tag{12}$$

with

$$g_{VVP} = \frac{3m_V^2}{16\pi^2 f_\pi^3}, \quad g_{VVV} = \frac{m_V}{2f_\pi}, \tag{13}$$

The symbol $\langle\rangle$ in Eq. (12) indicates the trace in the isospin space.

The determination of the amplitudes in Eq. (11) is quite tedious and we refer to the reader to Ref. [36] for more details on the calculations and for a list of the different intermediate channels considered.

A different way to proceed in the determination of the u-channel diagram in Fig. 2d is to construct an effective Lagrangian of the type [41]

$$\mathcal{L}_{X\bar{D}^*D^*} = ig_{X\bar{D}^*D^*}\epsilon^{\mu\nu\alpha\beta}\partial_\mu X_\nu \bar{D}^*_\alpha D^*_\beta, \tag{14}$$

and try to estimate somehow the unknown coupling $g_{X\bar{D}^*D^*}$. However, a model like this would lose its predictive power in the absence of any reasonable constrain on the value of the coupling $g_{X\bar{D}^*D^*}$. The strategy followed in this paper consists of first determining the $\bar{D}^*D \to \pi X$ cross section by calculating the $X\bar{D}^*D^*$ vertex in terms of the loops shown in Fig. 3. After this is done, we obtain the cross section for the same process but using the Lagrangian in Eq. (14) to evaluate the diagram in Fig. 2d and compare both results. In this way, we get a reliable estimation of the $g_{X\bar{D}^*D^*}$ coupling.

2.4. The $\bar{D}^*D^* \to \pi X$ reaction

As shown in Fig. 4, the cross section for the process $\bar{D}^*D^* \to \pi X$ can get contributions from the anomalous $X\bar{D}^*D^*$ vertex. To determine the diagrams in Figs. 4b and 4d we are going to make use of the method explained in the previous section to estimate the coupling $g_{X\bar{D}^*D^*}$ and consider the $X\bar{D}^*D^*$ vertex as a point-like one. Considering the Lagrangian in Eq. (14) for the $X\bar{D}^*D^*$ vertex, we find the following amplitudes for the t and u channel diagrams:

$$\begin{aligned} T_{3a}^{(Q_{1i},Q_{2i})} &= -2g_{PPV}\, g_X^a\, \mathcal{Y}^{(Q_{1i},Q_{2i})}\frac{1}{t-m_{\bar{D}}^2}p_{3\mu}\epsilon^\mu_{\bar{D}^*}(p_1)\epsilon^\nu_{D^*}(p_2)\epsilon_{X\nu}(p_4) \\ T_{3b}^{(Q_{1i},Q_{2i})} &= -\frac{g_{VVP}}{\sqrt{2}}g_{X\bar{D}^*D^*}\,\mathcal{Y}^{(Q_{1i},Q_{2i})}\frac{1}{t-m_{\bar{D}^*}^2}\,\epsilon^{\mu\nu\alpha\beta}\,\epsilon^{\mu'\nu'\alpha'}{}_\beta\, p_{1\mu}\, p_{3\alpha}\, p_{4\mu'}\epsilon_{\bar{D}^*\nu}(p_1)\epsilon_{D^*\alpha'}(p_2)\epsilon_{X\nu'}(p_4) \\ U_{3c}^{(Q_{1i},Q_{2i})} &= -2g_{PPV}\, g_X^c\, \mathcal{Y}^{(Q_{1i},Q_{2i})}\frac{1}{u-m_D^2}\, p_{3\nu}\,\epsilon^\mu_{\bar{D}^*}(p_1)\epsilon^\nu_{D^*}(p_2)\epsilon_{X\mu}(p_4) \\ U_{3d}^{(Q_{1i},Q_{2i})} &= -\frac{g_{VVP}}{\sqrt{2}}g_{X\bar{D}^*D^*}\,\mathcal{Y}^{(Q_{1i},Q_{2i})}\frac{1}{u-m_{D^*}^2}\epsilon^{\mu\nu\alpha\beta}\epsilon^{\mu'\nu'\alpha'\beta'}\, g_{\nu'\alpha}p_{2\alpha'}p_{3\mu'}p_{4\mu}\epsilon_{\bar{D}^*\beta}(p_1)\epsilon_{D^*\beta'}(p_2)\epsilon_{X\nu}(p_4), \end{aligned} \tag{15}$$

where the values of g_X^a, g_X^c, and $\mathcal{Y}^{(Q_{1i},Q_{2i})}$ are those given in Table 4.

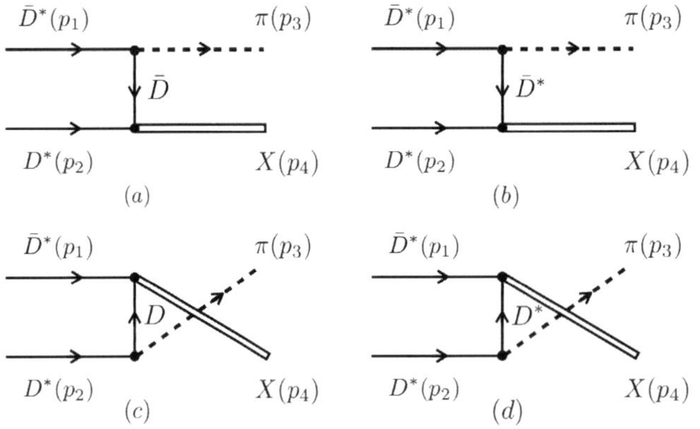

Figure 4. Different diagrams contributing to the reaction $\bar{D}^* D^* \to \pi X$.

Table 4. Values for the coupling $g_X^{a,c}$ and the coefficients $\mathcal{Y}^{(Q_{1i},Q_{2i})}$ of Eq. (15). The numerical values of g_n and g_c can be found in Table 3.

(Q_{1i}, Q_{2i})	g_X^a	g_X^c	$\mathcal{Y}^{(Q_{1i},Q_{2i})}$
$(0, 0)$	g_n	g_n	$\frac{1}{\sqrt{2}}$
$(-, +)$	g_c	g_c	$-\frac{1}{\sqrt{2}}$
$(-, 0)$	g_n	g_c	1
$(0, +)$	g_c	g_n	1

3. Results

3.1. The $\bar{D}D \to \pi X$ reaction

In Fig. 5 we show the results obtained for the production cross section of X from the reaction $\bar{D}D \to \pi X$ as a function of the center of mass energy, $\sqrt{s}$. The dashed line corresponds to the case where only the neutral components of X, i.e, $\bar{D}^0 D^{*0} -$c.c, are considered in the calculations, as in Ref. [26]. The solid line is the result for the cross section when all components of X are taken into account (using the couplings shown in Table 3). As can be seen from Fig. 5, the difference between the two curves is around a factor 2-3, depending on the energy. Thus, in a model in which X is considered as a molecular state of $\bar{D}D^* -$ c.c, a precise determination of the magnitude of the production cross section for X necessarily implies the consideration of all the components, neutral as well as charged.

3.2. The $\bar{D}^* D \to \pi X$ reaction considering triangular loops

Next, we determine the cross section related to the process $\bar{D}^* D \to \pi X$. The diagrams considered for this process (see Figs. 2c and 2d) involve anomalous vertices, $\bar{D}^*\bar{D}^*\pi$ in the t-channel and $X\bar{D}^*D^*$ in the u-channel. We find it interesting to compare the contributions arising form these vertices. We show the results in Fig. 6. The solid line, as in Fig. 5, continues representing the final result for the $\bar{D}D \to \pi X$ cross section. The dashed line is the cross section for the $\bar{D}^*D \to \pi X$ process without considering the diagrams involving the anomalous vertex $X\bar{D}^*D^*$, i.e., only with the t channel diagram shown in Fig. 2c. The shaded region represents the result found with both t and u channel diagrams shown in Figs. 2c and 3 (with the latter ones

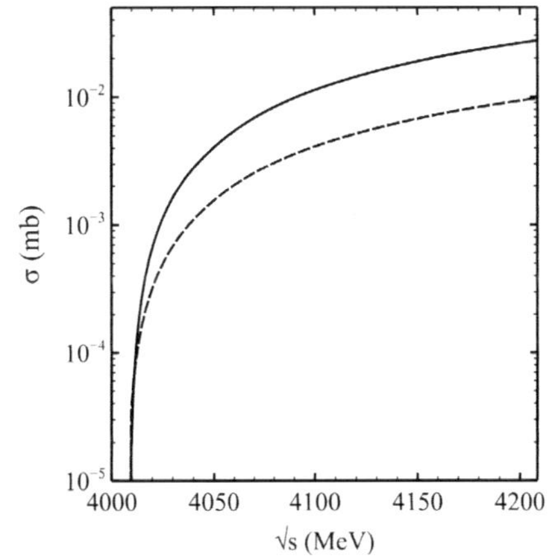

Figure 5. Cross section for the reaction $\bar{D}D \to \pi X$ considering only the neutral components of X (dashed-line) and adding the charged components (solid line).

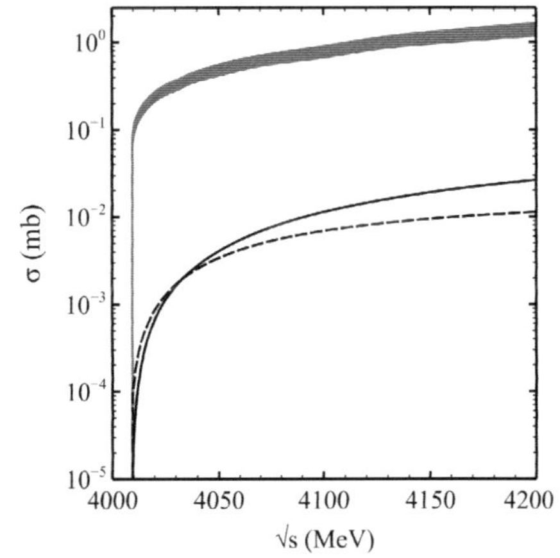

Figure 6. Cross section for the reaction $\bar{D}^*D \to \pi X$. The solid line has the same meaning as in Fig. 5, and we have shown it for the purpose of comparison. The dashed line represents the result for the cross section of the process $\bar{D}^*D \to \pi X$ considering only the t channel diagram in Fig. 2. The shaded region is the result obtained with both t and u channel diagrams of Fig. 2 considering cut-offs in the range 700-1000 MeV.

involving the $X\bar{D}^*D^*$ vertex) when changing the cut-off needed to regularize the loop integrals in the range 700-1000 MeV. As can be seen, the results do not get very affected by a reasonable change in the cut-off. Clearly, the vertex $X\bar{D}^*D^*$ plays an important role in the determination of the $\bar{D}^*D \to \pi X$ cross section, raising it by around a factor 100-150.

The importance of the anomalous vertices has been earlier mentioned in different contexts. For example, in Ref. [27] the J/ψ absorption cross sections by π and ρ mesons were evaluated for several processes producing D and D^* mesons in the final state. The authors found that the $J/\psi\,\pi \to D^*\bar{D}$ cross section obtained with the exchange of a D^* meson in the t-channel, which involves the anomalous $D^*D^*\pi$ coupling, was around 80 times bigger than the one obtained with a D meson exchange in the t-channel. In Ref. [28] the authors studied the radiative decay modes

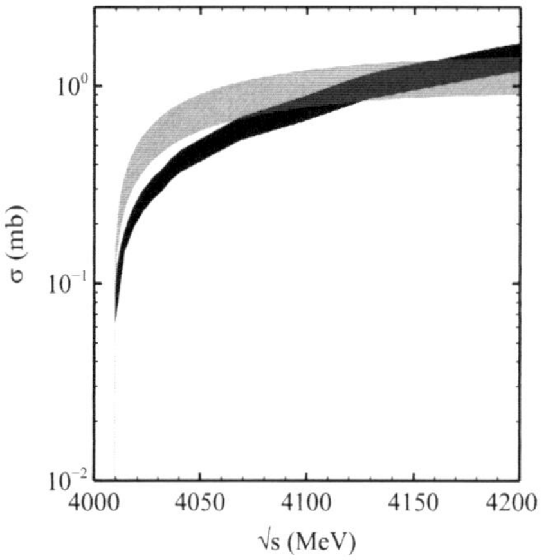

Figure 7. Cross section for the reaction $\bar{D}^*D \to \pi X$. The dark color shaded region has the same meaning as the shaded region in Fig. 6. The light color shaded region represents the result for the cross section when considering the Lagrangian in Eq. (14) to determine the $X\bar{D}^*D^*$ vertex with the value of the coupling given in Eq. (16).

of the $f_0(980)$ and $a_0(980)$ resonances, finding that the diagrams involving anomalous couplings were quite important for most of the decays, particularly for the $f_0(980) \to \rho^0\gamma$, $a_0(980) \to \rho\gamma$ and $a_0(980) \to \omega\gamma$.

Summarizing this subsection, we have shown that the cross section for the reaction $\bar{D}^*D \to \pi X$ is larger than that for $\bar{D}D \to \pi X$ and, thus, the consideration of this reaction in a calculation of the abundance of the X meson in heavy ion collisions could be important.

3.3. Estimating the $g_{X\bar{D}^*D^*}$ coupling

Having determined the contribution from the anomalous vertex $X\bar{D}^*D^*$ calculating the loops shown in Fig. 3, we could now obtain the cross section for the $\bar{D}^*D \to \pi X$ reaction using the Lagrangian of Eq. (14) to determine the amplitude for the diagram shown in Fig. 2d, which results in Eq. (15). In this way we can fix the $X\bar{D}^*D^*$ coupling to that value which gives similar results to the shaded region shown in Fig. 6. From Eq. (14), it can be seen that the coupling $g_{X\bar{D}^*D^*}$ should be dimensionless. In Fig. 7 we show the results found for the cross section of the reaction $\bar{D}^*D \to \pi X$ for $g_{X\bar{D}^*D^*}$ in the range 1.95 ± 0.22 (light color shaded region). The dark shaded region in the figure corresponds to the result for the cross section obtained by evaluating the vertex $X\bar{D}^*D^*$ using the diagrams in Fig. 3, where the loops have been regularized with a cut-off in the range $700 - 1000$ MeV. It can be seen that, although the energy dependence obtained by using the Lagrangian in Eq. (14) is not exactly the same as the one found by considering the triangular loops of Fig. 3, the two results are compatible in some energy range. Thus, the usage of the Lagrangian of Eq. (14) with the value

$$g_{X\bar{D}^*D^*} \sim 1.95 \pm 0.22, \tag{16}$$

can be considered as a reasonable approximation for describing processes involving the anomalous vertex $X\bar{D}^*D^*$, simplifying in this way the calculation of this vertex to a great extend.

3.4. The $\bar{D}^*D^* \to \pi X$ reaction

After estimating the coupling $g_{X\bar{D}^*D^*}$, we can use this value to determine the cross section for the process $\bar{D}^*D^* \to \pi X$, which could also get a contribution from the anomalous $X\bar{D}^*D^*$ vertex, that was neglected in Ref. [26]. The different Feynman diagrams considered for this process are depicted in Fig. 4.

In Fig. 8 we show the results for the cross section of the reaction $\bar{D}^*D^* \to \pi X$. The solid line corresponds to the result found without the anomalous $X\bar{D}^*D^*$ contribution, while the shaded region is the result considering the diagrams involving this anomalous vertex with the value for the $g_{X\bar{D}^*D^*}$ coupling given in Eq. (16). The first observation to be made is that the cross section

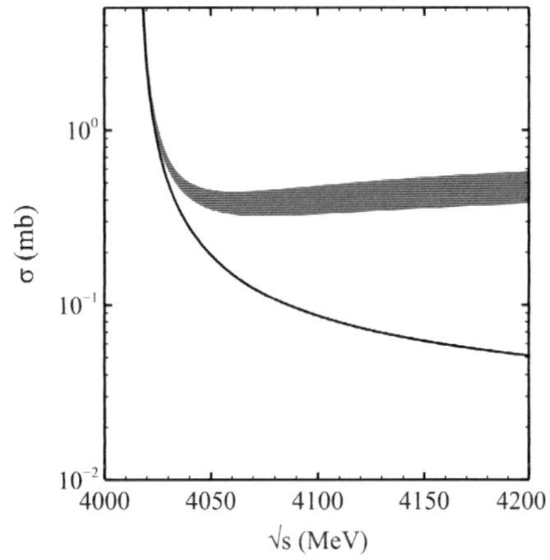

Figure 8. Cross section for the reaction $\bar{D}^*D^* \to \pi X$. The solid line represents the cross section without the contribution from the diagrams in Fig. 4b and 4d, which contain the vertex $X\bar{D}^*D^*$. The shaded region represents the result for the cross section when including the contribution of all the diagrams in Fig. 4, with the vertex $X\bar{D}^*D^*$ obtained using the Lagrangian in Eq. (14) with the value of the coupling given in Eq. (16).

for $\bar{D}^*D^* \to \pi X$ diverges close to the threshold of the reaction. This behavior is different to the cross sections of the processes studied in the previous sections. This is because the reaction $\bar{D}^*D^* \to \pi X$ is exothermic, while $\bar{D}D, \bar{D}^*D \to \pi X$ are endothermic. The second observation is that the contribution from the diagrams involving the $X\bar{D}^*D^*$ vertex is important, raising the cross section about a factor 8-10.

Therefore, as in case of the $\bar{D}^*D \to \pi X$ reaction, the consideration of the anomalous vertices could play an important role when determining the X abundance in heavy ion collisions.

4. Summary

In this work we have obtained the production cross sections of the reactions $\bar{D}D \to \pi X$, $\bar{D}^*D \to \pi X$ and $\bar{D}^*D^* \to \pi X$, considering $X(3872)$ as a molecular state of $\bar{D}D^*$ – c.c. We have shown that the consideration of the neutral as well as the charged hadrons coupling to X is important for the evaluation of the cross sections. Next, to obtain the cross section for the process $\bar{D}^*D \to \pi X$ we have included the contribution of the anomalous vertex $X\bar{D}^*D^*$. With this result, we have estimated the $X\bar{D}^*D^*$ coupling and used it to calculate the cross section for the reaction $\bar{D}^*D^* \to \pi X$. The contribution to the cross section from the vertex $X\bar{D}^*D^*$ turns out to be important and could play an important role in the determination of the abundance of the X meson in heavy ion collisions.

Our results pave the way for a new round of calculations of X abundancies in a hadron gas, as outlined in Ref. [26]. We emphasize that we expect to find some significant differences with respect to the results found in Ref. [26], because the processes $\bar{D}D \to \pi X$ and $\bar{D}^*D^* \to \pi X$ have been recalculated and, more importantly, the process $\bar{D}^*D \to \pi X$ has been included. This latter was found to give the most important contribution of all the three processes considered.

5. Acknowledgements

The authors would like to thank the Brazilian funding agencies FAPESP and CNPq for the financial support.

[1] M. Uchida [Belle Collaboration], Few Body Syst. **54**, 947 (2013).
[2] Y. p. Guo [BESIII Collaboration], EPJ Web Conf. **72**, 00009 (2014).
[3] S. K. Choi *et al.* [Belle Collaboration], Phys. Rev. Lett. **91**, 262001 (2003).
[4] D. Acosta *et al.* [CDF Collaboration], Phys. Rev. Lett. **93**, 072001 (2004).
[5] V. M. Abazov *et al.* [D0 Collaboration], Phys. Rev. Lett. **93**, 162002 (2004).
[6] B. Aubert *et al.* [BaBar Collaboration], Phys. Rev. D **71**, 071103 (2005).
[7] R. Aaij *et al.* [LHCb Collaboration], Phys. Rev. Lett. **110**, no. 22, 222001 (2013).
[8] N. A. Tornqvist, Phys. Lett. B **590**, 209 (2004).
[9] F. E. Close and P. R. Page, Phys. Lett. B **578**, 119 (2004).
[10] E. S. Swanson, Phys. Lett. B **588**, 189 (2004).
[11] E. Braaten and M. Kusunoki, Phys. Rev. D **72**, 054022 (2005).
[12] D. Gamermann and E. Oset, Eur. Phys. J. A **33**, 119 (2007).
[13] R. D'E. Matheus, S. Narison, M. Nielsen and J. M. Richard, Phys. Rev. D **75**, 014005 (2007).
[14] R. D'E. Matheus, F. S. Navarra, M. Nielsen and C. M. Zanetti, Phys. Rev. D **80**, 056002 (2009).
[15] Y. Dong, A. Faessler, T. Gutsche, S. Kovalenko and V. E. Lyubovitskij, Phys. Rev. D **79**, 094013 (2009).
[16] D. Gamermann and E. Oset, Phys. Rev. D **80**, 014003 (2009).
[17] D. Gamermann, J. Nieves, E. Oset and E. Ruiz Arriola, Phys. Rev. D **81**, 014029 (2010).
[18] S. Dubnicka, A. Z. Dubnickova, M. A. Ivanov, J. G. Koerner, P. Santorelli and G. G. Saidullaeva, Phys. Rev. D **84**, 014006 (2011).
[19] A. M. Badalian, V. D. Orlovsky, Y. .A. Simonov and B. L. G. Bakker, Phys. Rev. D **85**, 114002 (2012).
[20] S. Coito, G. Rupp and E. van Beveren, Eur. Phys. J. C **71**, 1762 (2011), *idem* Eur. Phys. J. C **73**, 2351 (2013).
[21] A. L. Guerrieri, F. Piccinini, A. Pilloni and A. D. Polosa, Phys. Rev. D **90**, 034003 (2014).
[22] I. Arsene *et al.* [BRAHMS Collaboration], Nucl. Phys. A **757**, 1 (2005).
[23] J. Adams *et al.* [STAR Collaboration], Nucl. Phys. A **757**, 102 (2005).
[24] S. Cho *et al.* [ExHIC Collaboration], Phys. Rev. Lett. **106**, 212001 (2011).
[25] S. Cho *et al.* [ExHIC Collaboration], Phys. Rev. C **84**, 064910 (2011).
[26] S. Cho and S. H. Lee, Phys. Rev. C **88**, 054901 (2013).
[27] Y. S. Oh, T. Song and S. H. Lee, Phys. Rev. C **63**, 034901 (2001).
[28] H. Nagahiro, L. Roca and E. Oset, Eur. Phys. J. A **36**, 73 (2008).
[29] H. Nagahiro, L. Roca, A. Hosaka and E. Oset, Phys. Rev. D **79**, 014015 (2009).
[30] S. Ozaki, H. Nagahiro and A. Hosaka, Phys. Lett. B **665**, 178 (2008).
[31] F. Aceti, R. Molina and E. Oset, Phys. Rev. D **86**, 113007 (2012).
[32] W. H. Liang, C. W. Xiao and E. Oset, Phys. Rev. D **89**, 054023 (2014).
[33] F. Aceti, M. Bayar and E. Oset, Eur. Phys. J. A **50**, 103 (2014).
[34] A. Anastassov *et al.* [CLEO Collaboration], Phys. Rev. D **65**, 032003 (2002).
[35] M. E. Bracco, M. Chiapparini, F. S. Navarra and M. Nielsen, Prog. Part. Nucl. Phys. **67**, 1019 (2012).
[36] A. Martinez Torres, K. P. Khemchandani, F. S. Navarra, M. Nielsen and L. M. Abreu, arXiv:1405.7583 [hep-ph].
[37] M. Bando, T. Kugo, S. Uehara, K. Yamawaki and T. Yanagida, Phys. Rev. Lett. **54**, 1215 (1985).
[38] M. Bando, T. Kugo and K. Yamawaki, Phys. Rept. **164**, 217 (1988).
[39] U. G. Meissner, Phys. Rept. **161**, 213 (1988).
[40] M. Harada and K. Yamawaki, Phys. Rept. **381**, 1 (2003).
[41] L. Maiani, F. Piccinini, A. D. Polosa and V. Riquer, Phys. Rev. D **71**, 014028 (2005).

Journal of Physics: Conference Series **630** (2015) 012038 doi:10.1088/1742-6596/630/1/012038

Differences Between The $^3P_0$ and C$^3P_0$ model in the Charming Strange Sector

D. T. da Silva

Instituto de Física e Matemática, Universidade Federal de Pelotas
Campus Universitário UFPel, CP 354, 96010-900, Capao do Leao - RS, Brazil

E-mail: neodts@gmail.com

Nathan Roschildt

Instituto de Física e Matemática, Universidade Federal de Pelotas
Campus Universitário UFPel, CP 354, 96010-900, Capao do Leao - RS, Brazil

E-mail: nathan.roschildtt@gmail.com

D. Hadjimichef

Instituto de Física, Universidade Federal do Rio Grande do Sul
Av. Bento Gonçalves, 9500, Porto Alegre, Rio Grande do Sul, CEP 91501-970, Brazil

E-mail: dimihadj@gmail.com

Hércules B. Rodrigues

CINAT - FÍSICA, Instituto Federal de Educaçao, Ciencia e Tecnologia Sul-rio-grandense - IFSUL
Praça inte de Setembro, 455, centro, CEP 96015-360, Pelotas - RS, Brazil

E-mail: herborge@gmail.com

Abstract.

The goal of this work is to establish a comparison between the very well studied $^3P_0$ model and a bound-state corrected version, the C^3P_0 model, obtained from applying the Fock-Tani transformation to the $^3P_0$ model, in the context of the charmed-strange meson sector (D_{SJ} meson). In particular, we shall calculate the decay amplitudes and decay rates of the $D_{s1}(2460)^+ \rightarrow D_s^{*+}\pi^0$ and $D_{s1}(2536)^+ \rightarrow D^*(2010)^+K^0$, showing the differences between the two models.

1. Introduction

The Fock-Tani formalism is a field theoretic method appropriated for the simultaneous treatment of composite particles and their constituents. This technique was originally used in atomic physics [1] and later in hadron physics to describe hadron-hadron scattering interactions [2, 3, 4] and meson decay [5, 6].

The $^3P_0$ model is a typical decay model which considers only OZI-allowed decay processes. The model considers a quark-antiquark pair created with the vacuum quantum numbers which interact with a meson in the initial state. It is described as the non-relativistic limit of a pair creation Hamiltonian [7].

The Fock-Tani transformation is applied to a qq pair creation Hamiltonian, producing a characteristic expansion in powers of the wave function, where the ${}^3P_0$ model is the lowest order term in the expansion. The corrected ${}^3P_0$ model (C^3P_0) is obtained from higher orders terms in this expansion, where terms contaning the bound state kernel appear [5].

Both the ${}^3P_0$ model and C^3P_0 model have been widely used in the study of meson spectroscopy. Our motivation for this work is in a comparison of these models for mesons in the charmed-strange sector. In particular, we shall calculate the decay amplitudes and decay rates of $D_{s1}(2460)^+ \to D_s^{*+}\pi^0$ and $D_{s1}(2536)^+ \to D^*(2010)^+K^0$.

2. Mesons in the Fock-Tani formalism: a brief outline

In the Fock-Tani formalism (FTf) we can write the meson creation operators in the following form: $M^\dagger_\alpha = \ {}^{\mu\nu}_{\alpha} q^\dagger_\mu q^\dagger_\nu$. A single particle state in second-quantization is

$$|\alpha\rangle = M^\dagger_\alpha |0\rangle ,$$

where ${}^{\mu\nu}_{\alpha}$ is the bound-state wave-function for two-quarks. The quark and antiquark operators obey the usual anticommutation relations. The composite meson operators satisfy non-canonical commutation relations

$$[M_\alpha, M_\beta] = 0 \quad ; \quad [M_\alpha, M^\dagger_\beta] = \delta_{\alpha\beta} - \Delta_{\alpha\beta},$$

where

$$\Delta_{\alpha\beta} = \ {}^{*\mu\gamma}_{\alpha} \ {}^{\gamma\rho}_{\beta} q^\dagger_\rho q_\mu + \ {}^{*\mu\gamma}_{\alpha} \ {}^{\gamma\rho}_{\beta} q^\dagger_\rho q_\mu \, .$$

The idea of the FTf is to make a representation change, where the composite particle operators are described by ideal particle operators that satisfy canonical commutation relations, i.e.,

$$[m_\alpha, m_\beta] = 0 \quad ; \quad [m_\alpha, m^\dagger_\beta] = \delta_{\alpha\beta} \, .$$

To implement this change of representation one can define a unitary transformation U that maps the composite state $|\alpha\rangle$ into an ideal state $|\alpha)$. In the meson case, for example, we have

$$U^{-1}\, M^\dagger_\alpha|0\rangle = m^\dagger_\alpha|0) \equiv |\alpha),$$

where $U = \exp(tF)$ and F is the generator of the meson transformation given by

$$F = m^\dagger_\alpha \tilde{M}_\alpha - \tilde{M}^\dagger_\alpha m_\alpha \, ,$$

with $\tilde{M}_\alpha$ defined up to third order

$$\tilde{M}_\alpha = M_\alpha + \frac{1}{2}\Delta_{\alpha\beta}M_\beta + \frac{1}{2}M^\dagger_\beta[\Delta_{\beta\gamma}, M_\alpha]M_\gamma \, .$$

3. The Microscopic Model

The Hamiltonian used in this model is inspirated in the ${}^3P_0$ model, deduced in [7]:

$$H_I = g\int d^3x \Psi^\dagger(\vec{x})\gamma^0\Psi(\vec{x}) \qquad (1)$$

where $\Psi(\vec{x})$ is the Dirac quark field, one should note that the bilinear $\Psi^{\dagger}\gamma^{0}\Psi$ leads to the decay $(qq)_A \to (qq)_B + (qq)_C$ through the $b^{\dagger}d^{\dagger}$ term. Introducing the following notation $b \to q$; $d \to q$; $\mu = (\vec{p}\,', s')$ e $\nu = (\vec{p}, s)$, after the expansion in the momentum representation, one obtains a compact notation for H_I:

$$H_I = V_{\mu\nu}\;q^{\dagger}_{\mu}\,q^{\dagger}_{\nu}$$

where the sum (integration) is applied over repeated indeces and

$$V_{\mu\nu} \;\;\equiv\;\; -\gamma\;\delta_{f_\mu f_\nu}\delta_{c_\mu c_\nu}\delta\left(\vec{p}_\mu + \vec{p}_\nu\right)\chi^{*}_{s_\mu}\left[\vec{\sigma}\cdot\left(\vec{p}_\mu - \vec{p}_\nu\right)\right]\chi^{c}_{s_\nu}\,. \tag{2}$$

In Eq. (2) γ is the free parameter pair production strength with $\gamma = g/2m_q$, where m_q is the quark mass of the pair creation. Applying the Fock-Tani transformation to H_I one obtains the effective Hamiltonian

$$H^{C3P0}_{FT} = U^{-1}\,H_I\,U = H_0 + \delta H_1$$

The decay amplitude h_{fi} for $m_\gamma \to m_\alpha + m_\beta$, is given by

$$\langle f \mid H^{C3P0}_{FT} \mid i\rangle = \delta(P_\gamma - P_\alpha - P_\beta)\,h_{fi} \tag{3}$$

where $|i\rangle = m^{\dagger}_{\gamma}|0\rangle$, $|f\rangle = m^{\dagger}_{\alpha}m^{\dagger}_{\beta}|0\rangle$ and

$$\begin{aligned} h_{fi} \;\;=\;\; & -V_{\mu\nu}\left\{ {}^{*\rho\nu}_{\beta}\;{}^{*\mu\eta}_{\alpha} + {}^{*\mu\eta}_{\beta}\;{}^{*\rho\nu}_{\alpha}\right\}\;{}^{\rho\eta}_{\gamma} \\ & -\frac{1}{4}V_{\mu\nu}\left\{ {}^{*\rho\tau}_{\alpha}\;{}^{*\mu\eta}_{\beta} + {}^{*\rho\tau}_{\beta}\;{}^{*\mu\eta}_{\alpha}\right\}\Delta\left(\rho\eta;\lambda\nu\right)\;{}^{\lambda\tau}_{\gamma} \\ & -\frac{1}{4}V_{\mu\nu}\left\{ {}^{*\rho\nu}_{\alpha}\;{}^{*\sigma\eta}_{\beta} + {}^{*\rho\nu}_{\beta}\;{}^{*\sigma\eta}_{\alpha}\right\}\Delta\left(\rho\eta;\mu\xi\right)\;{}^{\sigma\xi}_{\gamma} \\ & +\frac{1}{2}V_{\mu\nu}\left\{ {}^{*\rho\tau}_{\alpha}\;{}^{*\sigma\eta}_{\beta} + {}^{*\sigma\eta}_{\alpha}\;{}^{*\rho\tau}_{\beta}\right\}\Delta\left(\rho\eta;\mu\nu\right)\;{}^{\sigma\tau}_{\gamma} \end{aligned} \tag{4}$$

In the Eq. (4), the terms dependent on Δ are the bound state corrections, where the kernel represents an intermediate state of transition of the particle from the initial state to the final state. In this kernel a sum is performed over mesons with the quantum numbers of the final state. The meson wave function is defined as

$${}^{\mu\nu}_{\alpha} = \chi^{s_1 s_2}_{S_\alpha}\,f^{f_1 f_2}_{f_\alpha}\,C^{c_1 c_2}\;\;{}^{\vec{P}_\alpha - \vec{p}_1 - \vec{p}_2}_{nl}\;,$$

where χ is spin; f is flavor and C are color coefficients. The spatial part is given by the SHO wave-functions [8].

4. Applications and Results

Now we shall consider some specific processes for a comparative study between the $^3P_0$ and C$^3P_0$ models. In particular, the decay processes studied are: $D_{s1}(2460)^+ \to D_s^{*+}\pi^0$ and $D_{s1}(2536)^+ \to D^*(2010)^+K^0$. The full expressions for the decay amplitudes h_{fi} has the following form

$$h_{fi} = \left[\frac{\gamma}{\pi^{1/4}(\rho+1)^2}\right]\sum_{LS}^{N}\mathcal{C}_{LS}Y_{LM}\left(\Omega\right)\,,$$

where the coefficients $\mathcal{C}_{LS}$ are polynomials which have a dependence on the momentum P and in β gaussian width of the mesons involved in the processes. The decay amplitude h_{fi} can be combined with relativistic phase space to give the decay rate [5, 7]:

$$\Gamma_{A\rightarrow BC} = 2\pi\, P\,\frac{E_B E_C}{M_A}\left(\frac{\gamma}{\pi^{1/4}(\rho+1)^2}\right)^2 \sum_{LS}^{N} (\mathcal{C}_{LS})^2 .$$

The experimental values are extracted from Particle Data Group 2010 (PDG) [9] and the theoretical values obtained with $^3P_0$ and C$^3P_0$ model for these processes are shown in tables 1 and 2. In Tab. 1 we can see that the $^3P_0$ model is zero for the decay processes with final state $D_{SJ}\,\pi$. In Tab. 2 the two models obtain the equal results. For the theoretical results presented in tables the γ and β (in GeV) are $\gamma = 0.420$, $\beta_{\pi^0} = 0.410$, $\beta_{K^0} = 0.399$, $\beta_{D^*(2010)^+} = 0.280$, $\beta_{D_s^{*+}} = 0.200$ and for the intermediate state $\beta_6 = 0.100$ and $\beta_7 = 0.630$ is the state 1^1S_0 and the intermediate state $\beta_8 = 0.300$ and $\beta_9 = 0.200$ is the state 1^3S_1.

Table 1. Experimental values of the total decay rates and branching ratios for the meson $D_{s1}(2460)^+$.

$D_{s1}(2460)^+$ Branching ratios			
Process	Exp. (PDG)	$^3P_0$	C^3P_0
$\Gamma_{D_s^{*+}\pi^0}/\Gamma_{tot}$	0.48 ± 0.11	0	0.014

Table 2. Experimental values of the total decay rates and branching ratios for the meson $D_{s1}(2536)^+$.

$D_{s1}(2536)^+$ Branching ratios			
Process	Exp. (PDG)	$^3P_0$	C^3P_0
$\Gamma^{S-wave}_{D^*(2010)^+K^0}/\Gamma_{D^*(2010)^+K^0}$	$0.72 \pm 0.05 \pm 0.01$	0.995	0.995

5. Conclusions

Briefly we presented a comparsion of the $^3P_0$ model and the Corrected $^3P_0$ model applied for two meson decay processes of the charmed-strange sector. In this sector the decay processes are of two forms: $D^*_{SJ} \rightarrow D_{SJ}\pi$ and $D^*_{SJ} \rightarrow D^*K$. The first can not be obtained in the $^3P_0$ model, but the C$^3P_0$ model can be applied for both decay processes. The next step will be to consider the other D decay channels in the Corrected $^3P_0$ model.

6. Acknowledgements

This work is partially financed by CNPq, Brazil.

7. References

[1] Girardeau M D 1971 *Phys. Rev. Lett.* **27**, 1416, ibid. 1975 *J. Math. Phys.* **16**, 1901 1982 *Phys Rev.* A **26**, 217.
[2] Hadjimichef D, Krein G, Szpigel S and da eiga J S 1998 *Ann. of Phys.* **268**, 105 *ibid.* 1996 *Phys. Lett.* B **367**, 317.
[3] Hadjimichef D, Haidenbauer J, Krein G 2001 *Phys. Rev.* C **63**, 035204 *ibid.* 2002 *Phys. Rev.* C **66**, 055214.
[4] da Silva D T, Hadjimichef D 2004 *J. Phys.* G **30**, 191.
[5] da Silva D T, da Silva M L L, de Quadros J N and Hadjimichef D 2008 *Phys. Rev.* D **78**, 076004.
[6] da Silva M L L, Hadjimichef D and asconcellos C A Z and Bodmann B E J 2006 *J. Phys.* G **32**, 475.
[7] Ackleh E S, Barnes T and Swanson E S 1996 *Phys. Rev.* D **54**, 6811.
[8] de Quadros J N, da Silva D T, da Silva M L L and Hadjimichef D 2010 *Mod. Phys. Rev. Lett.* A **25**, p.2973.
[9] Nakamura K et al (Particle Data Group) 2010 *J. Phys.* G **37**, 7A.

Fermionic matter under the effects of high magnetic fields and its consequences in white dwarfs

E. Otoniel [a†] **M. Malheiro** [a] **J. G. Coelho** [b]

[a]Department of Physics, Institute Technological of Aeronautics - ITA - DCTA, Vila das Acácias, São José dos Campos, SP, 12228-900, Brazil.
[b] Department of Physics and ICRA, University of Rome - Sapienza P.le Aldo Moro 5, Rome, I-00185, Italy

E-mail: †otoniel@ita.br

Abstract. We investigate a recently proposed effect of strong magnetic fields in Fermionic matter that is important to the structure of magnetic white dwarfs. This work is highly relevant in view of the recent observations of magnetized white dwarfs ($B \sim 10^{8-9}$ G), and possible candidates for white dwarfs pulsars as an alternative descriptions for SGRs and AXPs. Here, we consider the matter inside white dwarfs composed by ions surrounded by an electron degenerate Fermi gas subject to a strong magnetic field. We investigate the effect of the Landau levels due to the huge magnetic field on the equation of state (EoS). We see that the behaviour of the equation of state as a function of the mass and energy density is much stiffer when only one Landau level is occupied. We also investigate the regime of lower magnetic fields where many Landau levels are occupied.

1. Introduction

The study of white dwarf allows to improve our understanding of nuclear matter under extreme densities and high magnetic fields [1] [2]. The interior of these stars offers an unique point of encounter among astrophysics and atomic physics, since the macroscopic properties of compact stars, such as mass, radius, rotation and thermal evolution depend on the microscopic composition of the stellar matter. This composition can change under the presence of strong magnetic fields, affecting the equation of state, and as a consequence the structure of the star.

Recently, Coelho et al. discussed some basic equilibrium properties of magnetized white dwarfs, in particular the condition for dynamical instability of the star in the presence of an extremely large magnetic field [3]. This analysis was done in the context of the virial theorem extended to include a magnetic term. Following the work of Chandrasekhar & Fermi of 1953, when the star magnetic energy W_B exceeds its gravitational potential energy $|W_G|$ ($W_B > |W_G|$), the system becomes dynamically unstable [4]. In light of this, the new mass limit for very magnetized and spherical white dwarf of $2.58 M_{\odot}$, recently calculated by [5], should be considered carefully, since these objects are unstable and unbound. Furthermore, it was showed that the new mass limit was obtained neglecting several macro and micro physical aspects such as gravitational, dynamical stability, breaking of spherical symmetry, general relativity, inverse β decay, and pycnonuclear fusion reactions. These effects are relevant for the self-consistent description of the structure and assessment of stability of these objects. When accounted for, they lead to the conclusion that the existence of such ultramagnetized white dwarfs in nature is very unlikely due to violation of minimal requirements of stability, and therefore the canonical Chandrasekhar mass limit of white dwarfs has to be still applied.

2. Fermion matter

Our starting point will be the microscopic energy-momentum tensor obtained from the system Lagrangian. The Lagrangian density of a fermionic system in the presence of a magnetic field is given by

$$\mathcal{L} = \overline{\psi}(\imath \not{D} - m)\psi + \frac{1}{4}F^{\mu\nu}F_{\mu\nu}, \tag{1}$$

where we used the notation usual $\not{a} = \gamma^{\mu}a_{\mu}$ and $D_{\mu} = \frac{1}{2}(\overrightarrow{\partial} - \overleftarrow{\partial}) + \Gamma_{\mu} + \imath|q|A_{\mu}$ with Γ_{μ} being the spin connection which is zero in flat space, and $F^{\mu\nu}$ is the field strength tensor of the electromagnetic field We have still the vector potential chosen as

$$A^{\mu}(x) = \delta_{\mu 2}x_1 B$$

that produces a constant magnetic field in the Z direction.

Solving the Dirac equation from the Lagrangian density we determined the dispersion relation:

$$E_{\nu} = \sqrt{p_z^2c^2 + m_e^2c^4 + 2\nu|q|\hbar Bc}. \tag{2}$$

where p_z is the longitudinal momentum, q electric charge, m_e the electron mass, c the speed light, and ν the Landau level given by

$$\nu = (l + \frac{1}{2} + s)$$

being $s = \pm 1$.

2.1. Zero Temperature

The integration in momentum for a electron gas with charge $|q|$ immersed in magnetic field, restricted to discrete Landau levels of magnetic field is [6]:

$$\int_{p_z(\nu)} \rightarrow \frac{2eB}{(2\pi\hbar)^2c}\sum_{\nu} g_{\nu 0}\int_{-\infty}^{\infty} f(E)dp_z(\nu)$$

where $f(E) = \frac{1}{e^{\beta(E-\epsilon_F(0))}+1}$ is the Fermi-Dirac function, $\beta = 1/K_BT$. and $g_{\nu 0} = (2 - \delta_{\nu 0})$ is the degeneracy in each Landau level. At zero temperature the distribution function is given by a theta function to one-particle

$$f(E) = \Theta(\mu - E)$$

where μ is the chemical potential.

2.2. Basic Equations for Landau Level Systems of Degenerate Electrons

In term of the chemical potential μ, the maximum p_z is defined

$$p_z = \sqrt{p_F^2 - \frac{2\nu|q|B\hbar}{c}} \tag{3}$$

Rewriting the equation above,introducting of the dimensionless parameter $x_F = p_F/m_e c$ we have still $x_F^2 = \mu^2 - 1$ where $\mu^2 = \epsilon_e/mc^2$ this leads to

$$p_z/m_e c = \chi(\nu) = \sqrt{\mu^2 - 1 - 2\nu\gamma}$$

being $B_c = \frac{|q|\hbar}{m_e^2 c^3} \sim 4.414 \times 10^{13} G$, $\gamma = B/B_c$, and ν_{max} is the maximum number of Landau level given by [7],

$$\nu \leq \nu_{max} = \frac{\mu^2 - 1}{2\gamma} \tag{4}$$

If the lowest Landau level $\nu = 0$ is occupied, $\nu_{max} = 1$. Similarly, for two level system, when gthe lowest $\nu = 0$, and first, $\nu = 1$, levels are occupied, $\nu_{max} = 2$, and so on [2] .

3. Equation of State

3.1. Number Density Equation

The number density is given by

$$n_e = \frac{2\gamma}{(2\pi)^2 \lambda^3} \sum_\nu g_{\nu 0} \int d\chi(\nu)$$

where $\lambda = \frac{\hbar}{m_e c}$. Integrating the equation above we have:

$$n_e = \frac{2\gamma}{(2\pi)^2 \lambda^3} \sum_\nu g_{\nu 0} \sqrt{\mu^2 - 1 - 2\nu\gamma} \tag{5}$$

3.2. Mass Density

We can relate the matter density with the electron number density n_e from

$$\rho = m_n \frac{A}{Z} n_e$$

where Z is the atomic number, A the mass number, and m_n the neutron mass. The matter density can be rewritten in the following way,

$$\rho = K_1 \sum_\nu g_{\nu 0} \sqrt{\mu^2 - 1 - 2\nu\gamma} \tag{6}$$

being $K_1 = \frac{2\gamma m_N \mu_e}{(2\pi)^2 \lambda^3}$, However, we see that $K1$ changes with the magnetic field due to the term γ.

3.3. Energy Equation and Longitudinal Pressure

The total energy density to zero temperature is given by

$$\epsilon = \rho c^2 + \epsilon_e$$

where

$$\epsilon_e = K_2 \sum_\nu g_{\nu 0} \int E_\nu d\chi(\nu)$$

is the electron energy density of the magnetized fermi gas and $k_2 = \frac{\gamma m_e c^2}{(2\pi)^2 \lambda^3}$. Then, by integration in the zero temperature limit

$$\epsilon_e = K_2 \sum_\nu g_{\nu 0}(1 + 2\nu\gamma)\xi_+ \left[\frac{\chi(x)}{\sqrt{1 + 2\nu\gamma}}\right] \quad (7)$$

As the previous case, we obtain the equation of state for the pressure

$$P_\| = K_2 \sum_\nu g_{\nu 0}(1 + 2\nu\gamma)\xi_- \left[\frac{\chi(x)}{\sqrt{1 + 2\nu\gamma}}\right] \quad (8)$$

where the function ξ is given by

$$\xi_\pm = \mu\sqrt{1 + \mu^2} \pm ln\left(\mu + \sqrt{1 + \mu^2}\right) \quad (9)$$

4. Results

In this section we present the numerical values for the longitudinal pressure, mass density and the mass-radius relations.

In Figure 1. we see the longitudinal pressure as a function of the density, we observe in graphs (a), (b), and (c) that for magnetic fields up to $B = 10^{13}G$ the equation of state with magnetic field behaves like non-magnetic equation of state, with no significant effect caused by Landau levels.

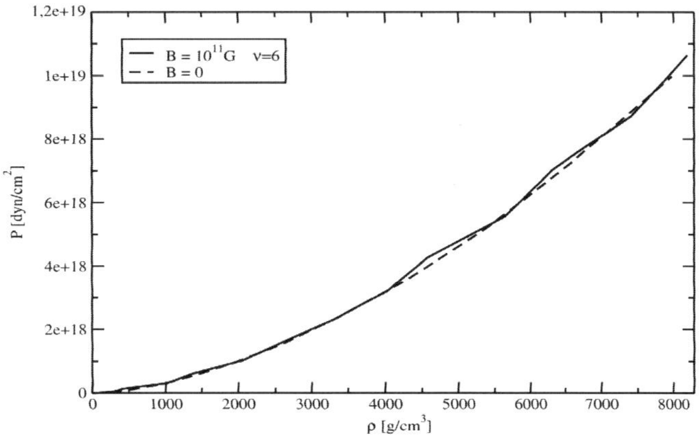

(a)

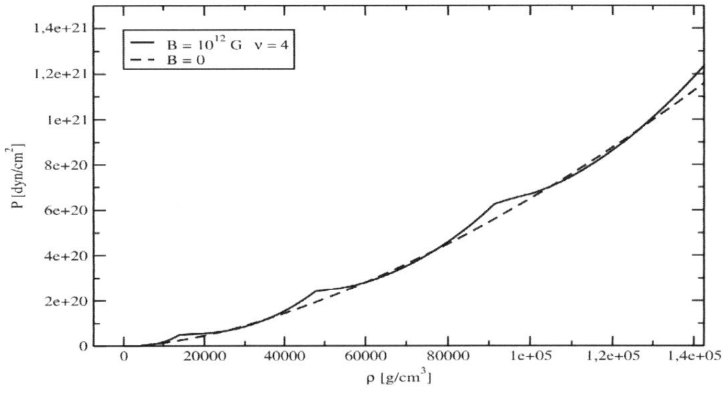

(b)

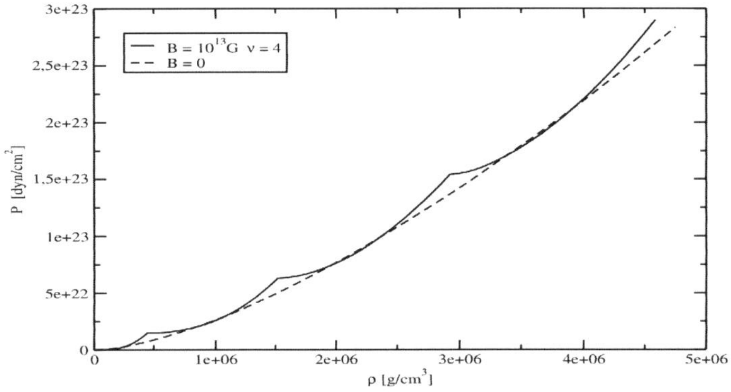

(c)

Figure 1. The figures above show the calculation of the longitudinal pressure versus mass density for a gas of strongly magnetized electron. All cases are plotted with solid line for magnetic fields below $B_c = 4.414 \times 10^{13}G$ (a) $B = 10^{11}G$, (b)$B = 10^{12}G$ and (c)$B = 10^{13}G$. The dashed line represent the EOS without magnetic field ($B = 0$).

The next figure shows the effects caused by the Landau levels in the equations of state with magnetic fields up to $8.8 \times 10^{15}G$. We see that for fields of this magnitude the EOS becomes stiffer than when we have only the the lowest state occupied. As we increase the number of occupied levels, as in Figure 2., the equation of state approaches the nonmagnetic case as can be seen in Figure 1..

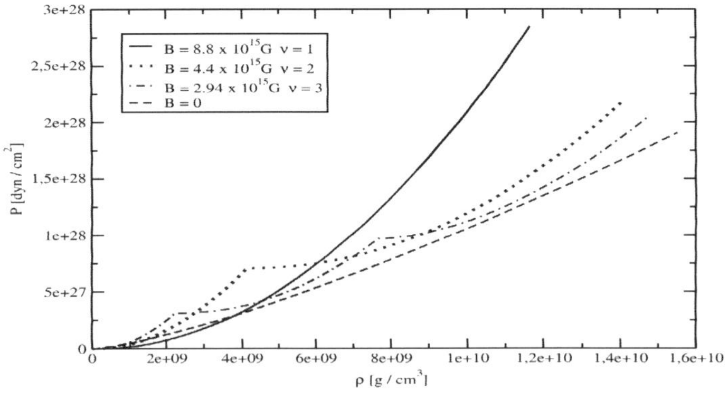

Figure 2. The graph 2 shows the pressure versus mass density for the Fermi energy $E_F = 20 m_e c^2$, The black line $B = 8.8 \times 10^{15}$, the dash point line for $B = 4.4 \times 10^{15}$, and the dashed and point line $B = 2.94 \times 10^{15}$. Each of these curves are for one, two, and three Landau levels, respectively. The dashed line is the EOS without magnetic field ($B = 0$).

In the Fig. 3 show the solution of a Tolman-Oppenheimer-Volkoff equation (TOV), In this figure we observe that for magnetic fields of values up to $B_c = 4.414 \times 10^{13} G$ the radius-mass curve behaves similarly to non-magnetic case, but for value $B = 8.8 \times 10^{15}$ the mass becomes independent of the radius reaching values of $M = 2.58 M\odot$.

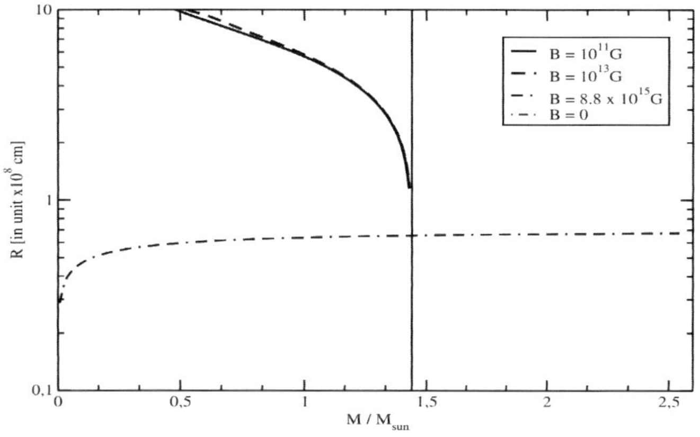

Figure 3. The figure shows the star radius R as a function of the mass with $E_F = 20 m_e c^2$ as the maximum Fermi energy. The vertical solid line marks the $1.44 M_\odot$ Chandrasekhar limit. The solid line $B = 10^{11} G$, and dashed $B = 10^{13} G$ are obtained with the EOS magnetized. The pointed-dashed $B = 0$ is obtained with the EOS non-magnetized. The dotted-dashed represent the one-level to magnetic field $B = 8.8 \times 10^{15} G$.

5. Conclusions

In this work we solve the equations of state for an degenerate electron Fermi gas under the presence of strong magnetic fields. We investigate the effect of the Landau levels due to a strong magnetic field in the equation of state (EOS) for several values of the magnetic field. We conclude that for strong magnetic fields the separation among Landau levels is large, so electrons with lower energy (non-relativistic) can only occupy the ground state. As the magnetic field decreases, the separation between Landau levels decreases, as shown in Fig. 2. Hence it becomes energetically favorable for electrons to jump to a higher level, so the number of occupied Landau levels increases accordingly. Likewise, in the case of relativistic electrons, if the magnetic field is low, the separation of Landau levels is comparable to the rest energy of the electrons and hence the electrons can pass freely between the highest Landau levels making the EOS similar to nonmagnetic case as we can see in the Fig. 1. This case is also shown in Fig. 1, which for low values of the magnetic field the Landau levels behave almost continuously and it is no longer possible to see the peaks shown in Fig. 2, that represent precisely the critical density for each Landau level.

We also conclude that for magnetic fields below B_c the mass radius relation of the star is similar to the nonmagnetic with $M = 1.44 M\odot$ as the mass limit case. Furthermore, for fields of the order $B \sim 10^{15} G$ the mass exceeds the well-known limit for a white dwarf, may reach masses $M = 2.58M\odot$. However, as has been discussed in [3], these white dwarfs are unstable and not reliable.

6. Acknowledgments

EO acknowledges the support by CAPES - Brazilian Federal Agency for Support and Evaluation of Graduate Education within the Ministry of Education of Brazil and MM acknowledges the financial support of CNPq and FAPESP (São Paulo state agency, thematic project #2013/26258-4).

References

[1] Das U and Mukhopadhyay B 2012 *Physical Review D* **86** 042001
[2] Kundu A and Mukhopadhyay B 2012 18 (*Preprint* `arXiv:1204.146v1 [astro-ph.SR] 6 Apr 2012`)
[3] Coelho J G, Marinho R M, Malheiro M, Negreiros R, Cáceres D L, Rueda J A and Ruffini R 2014 *The Astrophysical Journal* **794** 86
[4] Chandrasekhar S and Fermi E 1953 *Astrophysical Journal* **118** 116
[5] Das U and Mukhopadhyay B 2013 *Phys. Rev. Lett.* **110** 071102
[6] Strickland M, Dexheimer V and Menezes D P 2012 *Phys. Rev. D* **86** 125032
[7] Lai D and Shapiro S L 1991 *The Astrophysical Journal* **383** 745–751

Journal of Physics: Conference Series **630** (2015) 012040 doi:10.1088/1742-6596/630/1/012040

Net baryon production in the CGC formalism

F.O. Durães[1], A.V. Giannini[2], V.P. Gonçalves[3] and F.S. Navarra[2]

[1] Curso de Física, Escola de Engenharia, Universidade Presbiteriana Mackenzie, CEP 01302-907, São Paulo, Brazil

[2] Instituto de Física, Universidade de São Paulo, CEP 05315-970 São Paulo, SP, Brazil

[3] Instituto de Física e Matemática, Universidade Federal de Pelotas, CEP 96010-900, Pelotas, RS, Brazil

E-mail: duraes@mackenzie.br

E-mail: avgiannini@usp.br

E-mail: victorpbg@gmail.com

E-mail: navarra@if.usp.br

Abstract. In this work the net-baryon production at forward rapidities is investigated considering the CGC formalism. We assume that at large energies the coherence of the projectile quarks is lost and that the leading baryon production mechanism changes from recombination to independent fragmentation. The phenomenological implications for net-baryon production in $pp/pA/AA$ collisions are analyzed and predictions for LHC energies are presented. We also extend previous studies to pp and pA collisions and estimate for the first time the nuclear modification ratio, equal to the ratio of the net-proton production cross section in pA collisions over the one in pp collisions, scaled by the number of binary collisions, for leading baryon production.

In high energy hadronic collisions, baryons are produced both in the central and in the forward rapidity region. In the first case baryons are produced together with antibaryons and the net baryon number (baryons minus antibaryons) is small. In contrast, in the large rapidity region there are almost only baryons and no antibaryons. These experimental facts suggest that the forward baryons are produced from the valence quarks of the projectile, whereas low rapidity baryons are produced mainly from gluons and sea quarks.

At higher energies new phenomena are expected to affect forward baryon production and at large rapidities, baryon production requires the interaction of valence quark with a relatively large momentum fraction (x_1) of the projectile with low fractional momentum (x_2) partons in the target.

In the low x regime the target is a dense system of partons (predominantly gluons) which may form the Color Glass Condensate (CGC), a state of very high partonic densities in which the nonlinear effects of QCD change the parton distributions and hence the cross sections. The CGC is characterized by a momentum scale (Q_s) which marks the onset of nonlinear (or saturation) effects and grows with the reaction energy. At increasing projectile energies the valence quarks receive a transverse momentum kick of the order of Q_s and hence above a certain energy the coherence of the projectile quarks is lost and the leading baryon production mechanism changes from recombination to independent fragmentation (see [1] and references there in).

In the CGC formalism the differential cross section for the forward production of a hadron of transverse momentum p_T and mass m, at rapidity y, reads [1]:

$$\frac{dN}{d^2 p_T dy} = \frac{1}{(2\pi)^2} \int_{x_F}^{1} \frac{dz}{z^2} \, D(z) \frac{1}{q_T^2} \, x_1 \, q_v(x_1) \, \varphi\left(x_2, q_T\right) \,, \tag{1}$$

where $D(z) \equiv D_{B/q}(z) - D_{\bar{B}/q}(z)$ is the net-baryon fragmentation function, $z = E_B/E_q$ is the fraction of the energy of the fragmenting quark (E_q) taken by the emerging baryon B, the variable q_T is the quark transverse momentum and x_F represents the Feynman-x momentum of the produced baryon.

Moreover, $x_1 \, q_v(x_1)$ is the valence quark distribution of the projectile hadron and the function $\varphi(x_2, q_T)$ is the unintegrated gluon distribution of the hadron target, which is given by:

$$\varphi(x_2, q_T) = 2\pi q_T^2 \int dr_T \, r_T \, \mathcal{N}(x_2, r_T) J_0(r_T q_T) \,, \tag{2}$$

where J_0 is a Bessel function and $\mathcal{N}(x_2, r_T)$ is the forward scattering amplitude of a color dipole of radius r_T off a hadron target.

The evolution of $\mathcal{N}(x_2, r_T)$ is described in the mean field approximation of the CGC formalism [2] by the BK equation [3]. This quantity encodes the information about the hadronic scattering and then about the non-linear and quantum effects in the hadron wave function. In the last years, several groups have constructed phenomenological models which satisfy the asymptotic behavior of the leading order BK equation in order to fit the HERA and RHIC data.

In general, it is assumed that it can be modelled through a simple Glauber-like formula, which reads

$$\mathcal{N}(x, r_T) = 1 - \exp\left[-\frac{1}{4} (r_T^2 Q_s^2)^{\gamma(x, r_T)} \right] \,, \tag{3}$$

where $\gamma(x, r_T)$ is the anomalous dimension of the target gluon distribution.

The main difference among the distinct phenomenological models comes from the behavior predicted for the anomalous dimension, which determines the transition from the nonlinear to the extended geometric scaling regime, as well as from the extended geometric scaling to the DGLAP regime.

In this paper we restrict our analyses to the so called BUW model, which is able to describe the ep HERA data for the proton structure function and the hadron spectra measured in pp and dAu collisions at RHIC energies [4]. Another feature of the BUW model which motivates this analysis is that it explicitly satisfies the property of geometric scaling [5], which is predicted for the solutions of the BK equation in the asymptotic regime of large energies.

In the BUW model, the anomalous dimension is given by

$$\begin{aligned} \gamma(x, r_T) &= \gamma_s + \Delta\gamma(x, r_T) \\ &= \gamma_s + (1 - \gamma_s)(\omega^a - 1) \left[(\omega^a - 1) + b \right]^{-1} \,, \end{aligned} \tag{4}$$

where $\omega \equiv 1/(r_T Q_s(x))$, $\gamma_s = 0.628$ and the two free parameters $a = 2.82$ and $b = 168$ are fitted in such a way to describe the RHIC data on hadron production [4].

Besides, in comparison with other phenomenological parameterizations, in the BUW model, the behavior expected for the unintegrated gluon distribution in the large p_T limit (linear regime) is recovered: $\varphi(x_2, q_T) \propto 1/q_T^4$ at large q_T. In contrast, in Ref. [6] the nonlinear effects were taken into account considering the model (called here the MTW model) proposed long ago by Golec-Biernat and Wusthoff [7], where the forward dipole scattering amplitude is given by

Eq. (3) with $\gamma = 1$. Although this model satisfactorily describes the nonlinear regime (small-q_T), it clearly does not contain the expected behavior for large-q_T. Consequently, the resulting predictions are not valid at large values of the transverse momentum of the hadron. This model implies that the r_T integration in Eq. (2) can be carried out analytically and a simple expression for the unintegrated gluon distribution can be obtained:

$$\varphi(x_2, q_T) = 4\pi \frac{q_T^2}{Q_s^2(x_2)} \exp\left(-\frac{q_T^2}{Q_s^2(x_2)}\right) . \tag{5}$$

The net-baryon rapidity distribution is obtained integrating Eq. (1) in p_T between $p_{T_{min}} = 0$ and $p_{T_{max}} = \sqrt{s}\, e^{-y}$. The upper limit $p_{T_{max}}$ comes from the kinematical condition $x_F < 1$. Following Ref. [6] we assume that the nuclear valence quark distribution is given by $x\, q_v^A(x, Q^2) = N_{part}\, x\, q_v^p(x, Q^2)$, with N_{part} being the number of participants. The proton valence quark distribution is described by the MRST01-LO parametrization [8]. For the fragmentation function we use the KKP parametrization [9].

Substituting (5) by the unintegrated gluon distribution derived from the BUW model does not lead to significant changes in the final rapidity distributions, except at the lowest energies. In this energy region the rapidity distribution obtained with the BUW model does not show a dip at $y = 0$, in contradiction with the data (see references and discussions in [1]). This disagreement is an indication of the limitation of this approach at lower energies. On the other hand, at increasing energies, the MTW and BUW predictions for the net-baryon rapidity distributions in nucleus-nucleus collisions are essentially equivalent, even for central *PbPb* collisions at LHC energies. The same is true for proton-nucleus collisions.

As an illustration, in Fig. 1 we present the predictions for *pPb* collisions at LHC energies (the GBW curves below represent the predictions resulting from the substitution of the phenomenological model for the fragmentation function by the KKP parametrization).

In Fig. 2 we present our predictions for the net-baryon transverse momentum spectra in central *AuAu* collisions at RHIC energies. As in Ref. [6] we have assumed $N_{part} = 315$ and 357 for $\sqrt{s} = 62.4$ and 200 GeV, respectively. These plots show two striking features. First, we observe a very good agreement between data and the spectra obtained with Eq. (1) and the BUW dipole amplitude and, at the same time, a disagreement between data points and the spectra obtained with the GBW dipole amplitude, specially when $p_T > 1$ GeV. This happens because the GBW dipole amplitude has no DGLAP evolution and should not be able to reproduce data with large p_T. The BUW amplitude has the correct behavior at larger p_T and is able to describe the data in this region. Another interesting feature of these plots is the failure of the formalism at the largest rapidity and lowest energy. This may be an indication that here the baryons are not produced by independent quark fragmentation.

In order to quantify the magnitude of the nuclear effects in the net-proton production we introduce the nuclear modification ratio, equal to the ratio of the net-proton production cross section in *pA* collisions over the one in *pp* collisions scaled by the number of binary collisions and defined by

$$R_{pA} = \frac{d^2 N_{pA}}{dy d^2 p_T} / A \frac{d^2 N_{pp}}{dy d^2 p_T} . \tag{6}$$

In Fig. 3 we present our predictions. We observe a suppression at small values of p_T which increases at larger energies and rapidities, as expected from nonlinear effects. In contrast to the approach discussed in Ref. [10], which focus on the production of soft valence quarks far away (in rapidity) from the fragmentation region, here we consider the production of hard valence quarks which experience no recoil and are produced in the fragmentation region as proposed in Ref. [11]. As already emphasized in Ref. [10], both approaches are complementary. However, the behavior of R_{pA} in the latter approach is still an open issue.

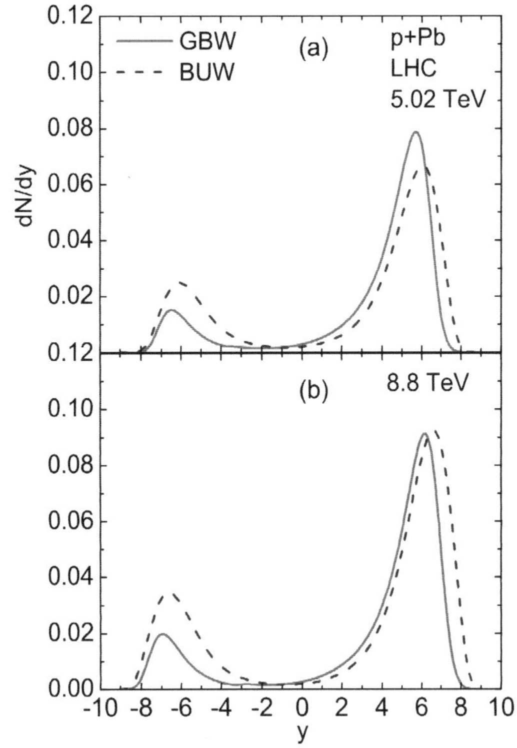

Figure 1. Net-baryon rapidity distributions in central $PbPb$ collisions at LHC energies.

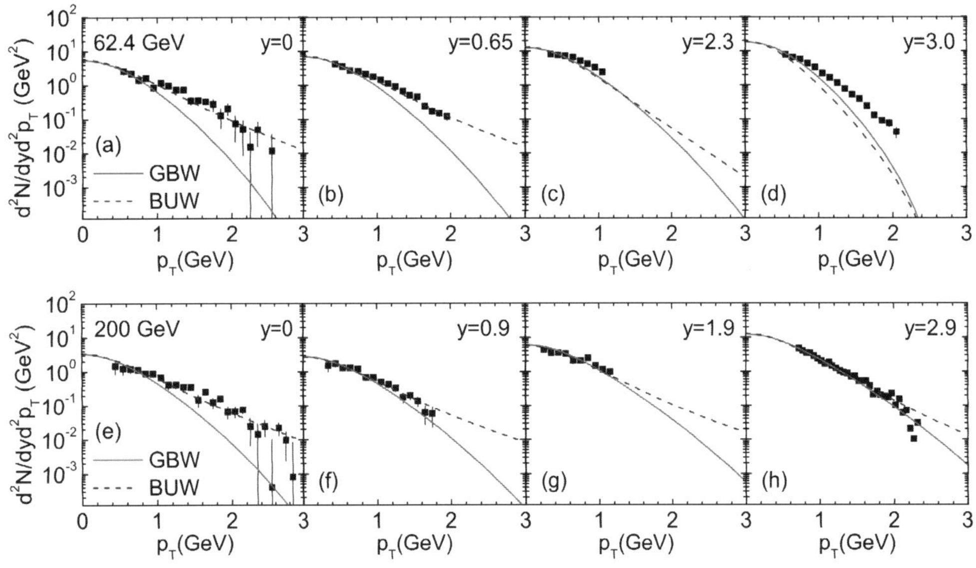

Figure 2. Net-baryon transverse momentum spectra in central $AuAu$ collisions at RHIC.

In this work we have improved the CGC formalism of forward particle production developed

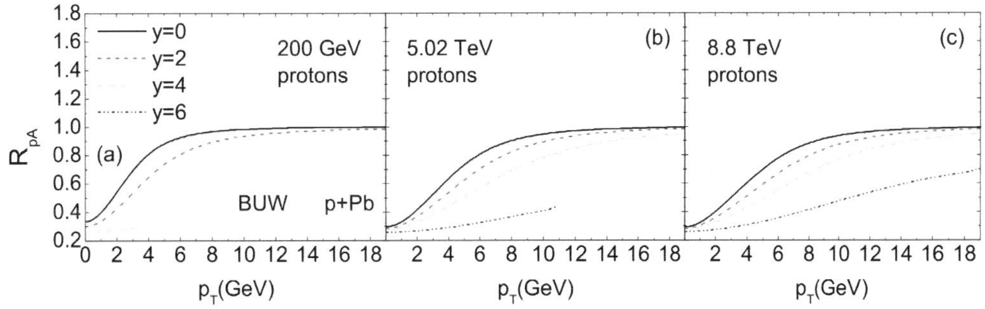

Figure 3. Nuclear modification ratio, R_{pA}, for net-proton production in pPb collisions at RHIC and LHC energies.

in [4, 6, 11] (the study of rapidity distributions, p_T and x_F spectra of forward protons and pions can be found in [1]). We obtain a good agreement with existing data and show predictions for the forthcoming LHC data.

Concerning forward proton production, our results suggest that at energies around $\sqrt{s} = 62.4$ GeV there is a transition from quark recombination to independent quark fragmentation. The independent fragmentation dynamics underpredicts the data at large rapidities and lower energies but starts to describe the data very well at higher energies. A solid conclusion about this change of mechanism still requires further theoretical and experimental work.

Finally, in the CGC formalism we do not observe any baryon anomaly. This suggests that this phenomenon is related to the central region dynamics of gluons and sea quarks.

Acknowledgments

This work was partially financed by the Brazilian funding agencies CAPES, CNPq, FAPESP and FAPERGS.

References

[1] F. O. Durães, A. V. Giannini, F. S. Navarra and V. P. Gonçalves 2014 *Phys. Rev.* C **89** 035205
[2] J. Jalilian-Marian, A. Kovner, L. McLerran and H. Weigert 1997 *Phys. Rev.* D **55** 5414
A. Kovner, J. G. Milhano and H. Weigert 2000 *Phys. Rev.* D **62** 114005
E. Ferreiro, E. Iancu, A. Leonidov, L. McLerran 2002 *Nucl. Phys.* A **703** 489
[3] I. Balitsky 1996 *Nucl. Phys.* B **463** 99
Y. V. Kovchegov 1999 *Phys. Rev.* D **60** 034008
Y. V. Kovchegov 2000 *Phys. Rev.* D **61** 074018
[4] D. Boer, A. Utermann and E. Wessels 2008 *Phys. Rev.* D **77** 054014
V. P. B. Gonçalves and M. L. L. da Silva 2013 *Nucl. Phys.* A **906** 28
[5] A. M. Staśto, K. Golec-Biernat and J. Kwieciński 2001 *Phys. Rev. Lett.* **86** 596
[6] Y. Mehtar-Tani and G. Wolschin 2009 *Phys. Rev. Lett.* **102** 182301
Y. Mehtar-Tani and G. Wolschin 2009 *Phys. Rev.* C **80** 054905
Y. Mehtar-Tani and G. Wolschin 2010 *Phys. Lett.* B **688** 174
Y. Mehtar-Tani and G. Wolschin 2011 *Europhys. Lett.* **94** 62003
[7] K. Golec-Biernat and M. Wusthoff 1998 *Phys. Rev.* D **59** 014017
K. Golec-Biernat and M. Wusthoff 1999 *Phys. Rev.* D **60** 114023
[8] A. D. Martin, W. J. Stirling and R. S. Thorne 2006 *Phys. Lett.* B **636** 259
[9] B. A. Kniehl, G. Kramer and B. Pötter 2000 *Nucl. Phys.* B **582** 514
[10] J. L. Albacete and Y. V. Kovchegov 2007 *Nucl. Phys.* A **781** 122
[11] A. Dumitru and J. Jalilian-Marian 2002 *Phys. Rev. Lett.* **89** 022301

Finite N corrections to the second order cumulant of pressure via the Optimized Perturbation Theory on the Nambu–Jona-Lasinio model

Tulio E. Restrepo*, **Juan C. M. Ramirez**†, **Gabriel N. Ferrari**‡

Departamento de Fsica, Universidade Federal de Santa Catarina, 88040-900, Florianópolis, Santa Catarina, Brazil

E-mail:
*tulioeduardo@gmail.com, †jcamilomacias@gmail.com, ‡gabriel.ferrari@ufsc.br

Abstract.
We show preliminary results for the second order cumulant of pressure at zero chemical potential, using the Nambu-Jona-Lasinio (NJL) model beyond traditional large-N_c (LN) or mean-field approximation (MFA). The main physical feature of this quantity lies on the determination of the critical point (CP) of the QCD phase diagram on the $T - \mu$ plane. Our calculations are done with the Optimized Perturbation Theory (OPT), which provides finite N corrections even in the first non-trivial order of perturbation. Also, the OPT generates dynamically the repulsive vector contribution, whose imposition in MFA has shown some weaknesses. Nevertheless, the results from both, once compared with those obtained by lattice QCD, suggest the need of a reformulation of the implementations near and above the critical temperature.

1. Introduction

Effective theories used to describe compressed strongly interacting matter should, at first, include vector channels [1, 2]. In NJL model, this can be easily implemented by adding a term such as $-G_V(\bar{\psi}\gamma^\mu\psi)^2$ to the original lagrangian density. However, G_V should be fixed using the ρ meson mass which, in general, is higher than the maximum energy scale set by Λ. At present, the vector term coupling G_V cannot be determined from experiments and lattice QCD simulations and, theoretically, there is absolutely no consensus about its fixation.

On the experiments with ultrarelativistic heavy ion collisions at SPS, RHIC and LHC energies, there are some quantities that contribute to probe the phase structure of strongly interacting matter and, in particular, to identify the deconfinement and chiral restoration transitions [3]. One of them is the second order quark number susceptibility at zero chemical potential, which theoretical evaluation depends on the determination of the second order cumulant of pressure. Very recently, Schramm and Steinheimer [4] have used the Polyakov-loop extended Nambu-Jona-Lasinio (PNJL) model, obtaining unsatisfactory results for these cumulants above the critical temperature. The main conclusion is that there should be a strong vector repulsion in the hadronic phase and near-zero repulsion in the deconfined phase.

However, due to the Fierz identities, when going beyond MFA level one may induce quantum (loop) corrections which mimic the physical effects caused by a classical (tree) term such as

$-G_V(\bar{\psi}\gamma^\mu\psi)^2$. This is precisely what has been observed in an application of the nonperturbative OPT method to the two flavor NJL model with vanishing G_V [5]. The OPT results for the NJL phase diagram show that $1/N_c$ corrections induced by this approximation reproduce the same qualitative features obtained by considering the model at large-N_c with an explicit repulsive vector channel. The reason is that the OPT two loop contributions add a term like $-G_S(\bar{\psi}\gamma^\mu\psi)^2/(N_cN_f)$ to the pressure. In order to compare OPT results with those from large-N approach with $G_V = 0$ and $G_V \neq 0$, we evaluated these cumulants and concluded that the problem at high temperatures persist. Then we point out an analytic feature which both implementations have in common as a responsible of this behavior.

2. The Nambu–Jona-Lasinio Model and its effective potential

Here we discuss the standard two flavor NJL model for quark matter using two different non perturbative methods: the large-N_c approximation and the Optimized Perturbation Theory. The respective lagrangian density can be written as [6]

$$\mathcal{L}_{NJL} = \bar{\psi}\left(i\partial\!\!\!/ - \hat{m}_0\right)\psi + G_S\left[\left(\bar{\psi}\psi\right)^2 + \left(\bar{\psi}i\gamma_5\boldsymbol{\tau}\psi\right)^2\right], \tag{1}$$

where ψ (sum over flavor and colour degrees of freedom is implicit) represents a isodoublet in flavor (u, d) and a N_c-plet quark field (N_c is the number of colors), $\boldsymbol{\tau}$ are the Pauli matrices and G_S represents the coupling strength. The quark mass matrix is $\hat{m}_0 = \mathrm{diag}\,(m_u, m_d)$, and we work in the isospin-symmetric limit $m_u = m_d = m_0$.

In order to investigate how symmetry breaking occurs in a particular theory, one can analyze its Landau's free energy, or the effective potential. To calculate it, one can use the path intgral formalism. Within this approach the partition function, $\mathcal{Z}$, can be written in terms of the effective potential as follows (for details see Refs. [7, 8]):

$$\mathcal{Z} = \exp\left[-i\int d^4x\mathcal{F}\right]. \tag{2}$$

Then, as within statistical mechanics, all the relevant thermodynamical quantities, such as the pressure, can easily be obtained once the free energy (or the partition function) is known. However, the complete evaluation of $\mathcal{F}$ is an impossible task within most interacting theories so that one needs to invoke an approximation scheme, such as the OPT or the LN approximation.

Before solving Eq. (2), for the NJL model case it is convenient to write down the Lagrangian density in terms of the auxiliary fields, σ and $\boldsymbol{\pi}$, by using the Hubbard-Stratanovich transformation, which consists in adding the following quadratic terms to the Lagrangian density:

$$\Theta^2 = -\frac{1}{G_S}\left[\frac{\sigma}{2} + G_S\left(\bar{\psi}\psi\right)\right]^2, \quad \Omega^2 = -\frac{1}{G_S}\left[\frac{\boldsymbol{\pi}}{2} + G_S\left(\bar{\psi}i\gamma_5\boldsymbol{\tau}\psi\right)\right]^2. \tag{3}$$

Then, the the original theory can be written in a bosonized fashion as

$$\mathcal{L}_{NJL} = \bar{\psi}\left(i\partial\!\!\!/ - m_0\right)\psi - \frac{1}{4G_S}\left(\sigma^2 + \boldsymbol{\pi}^2\right) - \bar{\psi}\left[\sigma + i\gamma_5\boldsymbol{\pi}\cdot\boldsymbol{\tau}\right]\psi. \tag{4}$$

2.1. The interpolated model

To implement the OPT approximation (also known as *linear delta expansion*, LDE, see [9, 10] for earlier references) in the NJL model we follow Ref. [11].

Basically, this method consists in modifying the Lagrangian of a particular theory by introducing a dummy expansion parameter, δ.

Let $\mathcal{L}$ be the original Lagrangian density we want to solve, and $\mathcal{L}_0$ a Lagrangian density of a free theory that we know how to solve exactly. Then, the Lagrangian density in terms of δ can be written as

$$\mathcal{L}(\delta) = (1-\delta)\mathcal{L}_0 + \delta\mathcal{L} = \mathcal{L}_0 + \delta(\mathcal{L} - \mathcal{L}_0). \tag{5}$$

As one can easily see, $\mathcal{L}(\delta)$ interpolates between the original theory ($\delta = 1$) and a free (exactly solvable) theory ($\delta = 0$). Then, the relevant physical quantities are evaluated as a power series in δ, which is formally treated as a small number.

It is is crucial to note that for dimensional balance, the Lagrangian density $\mathcal{L}_0$ must have at least one arbitrary mass parameter (η). This is a welcome fact since η will act as an infra red regulator if the original theory is massless. One can then fix the arbitrary η by requiring that any physical quantity $\mathcal{P}(\eta)$, be at least locally η-independent. This optimization criterion translates into the following variational condition

$$\left.\frac{\partial \mathcal{P}(\eta)}{\partial \eta}\right|_{\bar{\eta}} = 0, \tag{6}$$

which is known as the *principle of minimal sensitivity* (PMS) [12]. The solution of the PMS equation gives $\bar{\eta}$ as a function of the original parameters of the theory. The convergence of the OPT in the case of critical theories is given in Ref. [13].

The OPT has already established itself as a powerful method in dealing with critical theories. For example, in the Bose-Einstein condensation case this method and its different variations have provided some of the most precise analytical results for the shift in the critical temperature for weakly interacting homogeneous Bose gases [14, 15]. Other applications to condensed matter situations include a precise evaluation of the critical density for polyacetylene [16]. Also, when extended by hard-thermal loops, the method was successful in predicting QCD thermodynamical properties at the three-loop level [17]. Improved by the renormalization group (RG), and inspired by similar properties [18] in the Gross-Neveu model, a variation of the OPT has been recently used in the evaluation of $\Lambda_{\overline{\rm MS}}^{\rm QCD}$ [19] and α_S [20], where the stability and convergence at higher orders of this RG-OPT form was demonstrated.

Now, in order to apply this method to the NJL Lagrangian density, Eq. (4), one can define the quadratic term, $\mathcal{L}_0$, as

$$\mathcal{L}_0 = \bar{\psi}\left(i\partial\!\!\!/ - m_0 - \eta\right)\psi.$$

Then, following the OPT interpolation prescription, Eq. (5), one gets

$$\begin{aligned}\mathcal{L}_{NJL}(\delta) =& \bar{\psi}\left(i\partial\!\!\!/ - m_0 - \eta\right)\psi \\ &+ \delta\left[\eta\bar{\psi}\psi - \frac{1}{4G_S}\left(\sigma^2 + \boldsymbol{\pi}^2\right) - \left(\sigma\bar{\psi}\psi + \bar{\psi}i\gamma_5\boldsymbol{\pi}\cdot\boldsymbol{\tau}\psi\right)\right].\end{aligned} \tag{7}$$

It is easy to see that Eq. (7) can be written as

$$\mathcal{L}_{NJL}(\delta) = \bar{\psi}\left[i\partial\!\!\!/ - m_0 - \delta\left(\sigma + i\gamma_5\boldsymbol{\pi}\cdot\tau\right) - \eta(1-\delta)\right]\psi - \delta\frac{N_c}{2\lambda}\left(\sigma^2 + \boldsymbol{\pi}^2\right), \tag{8}$$

where we have done $G_S \to \lambda/(2N_c)$. To order δ, the effective potential is given by the Feynman diagrams shown in Fig. 2.1, where the mass term appearing in the OPT dressed propagators is defined by [28]

$$\hat{\eta} = \eta + m_0 - \delta\left[\eta - \left(\sigma + i\gamma_5\boldsymbol{\pi}\cdot\boldsymbol{\tau}\right)\right]. \tag{9}$$

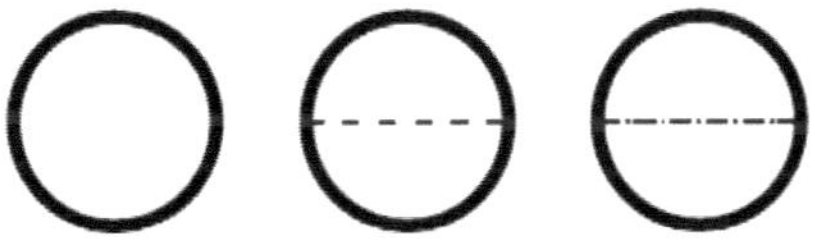

Figure 1. Diagrams contributing to $\mathcal{F}(\hat{\eta})$ at order δ. The thick continuous lines represent the OPT dressed fermionic propagators, the dashed line represents the σ propagator and the dashed-dotted line represents the π propagator. (Figure taken from Ref. [28]).

so that all the η, σ and $\boldsymbol{\pi}$ insertions are automatically accounted for.

Applying the Feynman rules one then obtains [28]

$$\begin{aligned}\mathcal{F} =& \frac{N_c}{2\lambda}\left(\sigma^2+\boldsymbol{\pi}^2\right) + i\int\frac{d^4p}{(2\pi)^4}\mathrm{Tr}\ln\left(\not{p}-\hat{\eta}\right)\\ &+\delta\frac{\lambda}{2N_c}\int\frac{d^4p}{(2\pi)^4}\int\frac{d^4q}{(2\pi)^4}\mathrm{Tr}\left[\left(\frac{\not{p}+\hat{\eta}}{p^2-\hat{\eta}^2}\right)\left(\gamma_5\tau_i\frac{\not{q}+\hat{\eta}}{q^2-\hat{\eta}^2}\gamma_5\tau_i\right)\right]\\ &-\delta\frac{\lambda}{2N_c}\int\frac{d^4p}{(2\pi)^4}\int\frac{d^4q}{(2\pi)^4}\mathrm{Tr}\left[\left(\frac{\not{p}+\hat{\eta}}{p^2-\hat{\eta}^2}\right)\left(\frac{\not{q}+\hat{\eta}}{q^2-\hat{\eta}^2}\right)\right].\end{aligned} \tag{10}$$

It is clear that (after taking the trace over colors) the first two terms are of order N_c while the last two, which are of order N_c^0, would not contribute to a LN type of calculation. This important remark emphasizes the fact that already at the first non-trivial order the OPT considers terms which belong to the NLO (next to the leading order) order in a $1/N_c$ type of expansion. Here, the first $1/N_c$ corrections due to the OPT type of expansion have the topology of exchange (Fock) type of diagrams as Fig. 2.1 explicitly shows.

The mass term $\hat{\eta}$ is dressed by the scalar field σ and the pseudoescalar field $\boldsymbol{\pi}$ and, in order to maintain the Goldstone structure of the theory [10, 21], it must be defined as [10]

$$\eta = \alpha + i\gamma_5\boldsymbol{\nu}\cdot\boldsymbol{\tau}, \tag{11}$$

where $\boldsymbol{\nu}$ is a three-vector.

Now the PMS conditions to be satisfield at $\delta = 1$ are

$$\left.\frac{\partial\mathcal{F}}{\partial\alpha}\right|_{\bar{\alpha},\bar{\nu}_i} = 0 \quad\text{and}\quad \left.\frac{\partial\mathcal{F}}{\partial\nu_i}\right|_{\bar{\alpha},\bar{\nu}_i} = 0. \tag{12}$$

One can use the identity $\mathrm{Tr}\ln = \ln\det$, to expand the logarithm to order δ. Then, by taking the trace in each one of the integrals in Eq. (10) one gets

$$\begin{aligned}\frac{\mathcal{F}}{N_c} =& \frac{1}{2\lambda}\left(\sigma^2+\boldsymbol{\pi}^2\right) + 2iN_f\int\frac{d^4p}{(2\pi)^4}\ln\left[-p^2+(\alpha+m_0)^2+\boldsymbol{\nu}^2\right]\\ &+4i\delta N_f\int\frac{d^4p}{(2\pi)^4}\frac{(\alpha+m_0)(\alpha-\sigma)+\boldsymbol{\nu}(\boldsymbol{\nu}-\boldsymbol{\pi})}{-p^2+\hat{\eta}^2}\\ &-2(n_\pi+1)\delta\frac{\lambda N_f}{N_c}\int\frac{d^4p}{(2\pi)^4}\frac{d^4q}{(2\pi)^4}\frac{p_\mu q^\mu}{(-p^2+\hat{\eta})(-q^2+\hat{\eta})}\\ &+2(n_\pi-1)\delta\frac{\lambda N_f}{N_c}\int\frac{d^4p}{(2\pi)^4}\frac{d^4q}{(2\pi)^4}\frac{(\alpha+m_0)^2-\boldsymbol{\nu}^2}{(-p^2+\hat{\eta})(-q^2+\hat{\eta})},\end{aligned} \tag{13}$$

where n_π represents the number of pseudoescalars. In the $U(1)$ version of the model, where $N_f = 1$ and $n_\pi = 1$, so that the last term does not contribute and the OPT will bring $1/N_c$ corrections only at finite chemical potential, since the other correction term is related to the quark number density, which is zero at vanishing chemical potential. In our case, $n_\pi = 3$ and finite N_c corrections are expected to occur at any regime of density and/or temperature.

Applying the PMS equation for α and ν_i leads to a cumbersome problem, which can be circumvented by noticing the symmetry between α and ν_i, which allows us to select the σ direction and set $\pi_i = 0$, implying that $\bar{\nu}_i = 0$ [10] and $\bar{\alpha} = \bar{\eta}$. We also note that in the third term of Eq. (13) only the zeroth component will eventually survive (when $\mu \neq 0$). Then, one finally gets

$$\begin{aligned}\frac{\mathcal{F}}{N_c} &= \frac{\sigma^2}{2\lambda} + 2iN_f \int \frac{d^4p}{(2\pi)^4} \ln\left[-p^2 + (\eta - m_0)^2\right] \\ &+ 4i\delta Nf \int \frac{d^4p}{(2\pi)^4} \frac{(\eta + m_0)(\eta - \sigma)}{-p^2 + (\eta + m_0)^2} \\ &- 8\delta \frac{\lambda N_f}{N_c} \left[\int \frac{d^4p}{(2\pi)^4} \frac{p_0}{-p^2 + (\eta + m_0)^2}\right]^2 \\ &+ 4\delta \frac{\lambda N_f}{N_c} (\eta + m_0)^2 \left[\int \frac{d^4p}{(2\pi)^4} \frac{1}{-p^2 + (\eta + m_0)^2}\right]^2 . \end{aligned} \tag{14}$$

Using the Matsubara's (imaginary time) formalism in order to introduce the control parameters T and μ, the time component is replaced by the temperature since the evolution of the system follows a path where each point of the path is a equilibrium state. Therefore, the effective potential in terms of temperature and chemical potential becomes

$$\begin{aligned}\mathcal{F}(\eta, \sigma, \mu, T) =& \frac{\sigma^2}{4G_S} - 2N_f N_c I_1(\mu, T) + 2\delta N_f N_c (\eta + m_0)(\eta - \sigma) I_2(\mu, T) \\ &+ 4\delta G_S N_f N_c I_3^2(\mu, T) - 2\delta G_S N_f N_c (\eta + m_0)^2 I_2^2(\mu, T), \end{aligned} \tag{15}$$

where now we have replaced $\lambda \to 2G_S N_c$. In the above equation we have also defined, for convenience, the integrals

$$I_1(\mu, T) = \int \frac{d^3p}{(2\pi)^3} \left\{E_p + \frac{1}{\beta} \ln\left[1 + e^{-\beta(E_p + \mu)}\right] + \frac{1}{\beta} \ln\left[1 + e^{-\beta(E_p - \mu)}\right]\right\}, \tag{16}$$

$$I_2(\mu, T) = \int \frac{d^3p}{(2\pi)^3} \frac{1}{E_p} \left[1 - f^+ - f^-\right] \tag{17}$$

and

$$I_3(\mu, T) = \int \frac{d^3p}{(2\pi)^3} \left[f^+ - f^-\right], \tag{18}$$

where $E_p^2 = \boldsymbol{p}^2 + (\eta + m_0)^2$ is the dispersion while

$$f^+ = \frac{1}{e^{\beta(E_p - \mu)} + 1}, \tag{19}$$

(20)

and

$$f^{-} = \frac{1}{e^{\beta(E_p + \mu)} + 1}, \tag{21}$$

represent the fermion distribution functions for particles and antiparticles respectively.

The physical interpretation of the OPT free energy, Eq. (15), becomes clear by analyzing the physical meaning of each term. The first term represents the classical potential while the second is similar to the standard result obtained in the case of free fermionic gas whose masses are given by $m_0 + \eta$ as I_1 suggests. The terms proportional to $I_2 \sim \partial I_1/\partial \eta$ are reminiscent of the one loop scalar density, $\rho_s = \langle \bar{\psi}\psi \rangle$. At the same time, the terms proportional to $I_3 \sim \partial I_1/\partial \mu$ only survive when $\mu \neq 0$ as Eq. (18) shows. This can be easily understood by recalling that, to one loop, the quark number density $\rho = \langle \psi^{+}\psi \rangle$ is given by I_3. Then, by noting that I_3 is $1/N_c$ suppressed one can readily draw the basic physical differences between the OPT and LN (or MF) approximations at this first non-trivial order. Namely, the OPT free energy is written in terms of scalar and vector condensates while only the scalar density contributes to the later. This observation is crucial for the discussions to be carried out in the sequel.

Within the two flavor version of the NJL model there are three parameters to be fixed: the current mass m_0, the cut-off Λ and the coupling constant G_S. Usually they are adjusted so as to reproduce physical observables such as the pion mass ($m_\pi \simeq 135$ MeV), the pion decay constant ($f_\pi \simeq 92.4$ MeV) and the quark condensate ($-\langle \bar{\psi}\psi \rangle^{1/3} \simeq 250$ MeV) [22]. The (non-observable) current quark mass is related to the observables via the Gell–Mann-Oakes-Renner (GMOR) relation [23],

$$m_0 = \frac{-f_\pi^2 m_\pi^2}{2\langle \bar{\psi}\psi \rangle}. \tag{22}$$

There is a family of parameter sets tailored to reproduce the numerical values of these physical observables and, in general, one chooses a set which also reproduces the vacuum effective quark mass, $M = m_0 + \Sigma \simeq 330$ MeV, where Σ represents the self energy.

The parameter sets adopted in this work are given in table 2.1 (see Ref. [28] for other possibilities).

	Λ [MeV]	m_0 [MeV]	$G_S\Lambda^2$
OPT	640	4.9	1.99
LN	640	5.2	2.14

Table 1. Parameter set for the OPT and for the LN approximation as given in Ref. [28]. These values were obtained to reproduce $m_\pi = 135$ MeV, $f_\pi = 92.4$ MeV and $-\langle \bar{\psi}\psi \rangle^{1/3} = 250$ MeV.

2.2. Repulsive vector interaction in the NJL model

In the Introduction we have emphasized the importance of a vector interaction in the studies of compressed quark matter. Within the NJL model such a term can be of the form $-G_V\left(\bar{\psi}\gamma_\mu\psi\right)^2$ with $G_V > 0$ describing repulsion which is the case here and $G_V < 0$ describing attraction. Then the standard NJL lagrangian density becomes

$$\mathcal{L}_V = \bar{\psi}\left(i\partial\!\!\!/ - \hat{m}_0\right)\psi + G_S\left[\left(\bar{\psi}\psi\right)^2 + \left(\bar{\psi}i\gamma_5\boldsymbol{\tau}\psi\right)^2\right] - G_V\left(\bar{\psi}\gamma^\mu\psi\right)^2, \tag{23}$$

Journal of Physics: Conference Series **630** (2015) 012041 doi:10.1088/1742-6596/630/1/012041

and the effective potential in the LN approximation reads [1]

$$\mathcal{F}_{LN} = \frac{\sigma^2}{4G_S} - 2N_f N_c I_1(\tilde{\mu}, T) - 4G_V N_f^2 N_c^2 I_3^2(\tilde{\mu}, T), \tag{24}$$

where $I_1(\tilde{\mu}, T)$ and $I_3(\tilde{\mu}, T)$ can be readily obtained from Eqs. (16) and (18) upon replacing $\mu \to \tilde{\mu} = \mu - 2G_V\rho$, with $\rho = 2N_cN_fI_3$, and $m_0 + \eta \to m_0 + \sigma$. Fukushima [24] has shown that the *combined* effect of $\tilde{\mu}$ and $-4G_V N_c^2 N_f^2 I_3^2$ in the above equation is to produce a *net* effect similar to $+4G_V N_c^2 N_f^2 I_3^2$. This interesting result allows us to better understand the type of $1/N_c$ contributions radiatively generated by the OPT. An inspection of Eq. (15) reveals that this approximation generates a $+4G_V N_c^2 N_f^2 I_3^2$ term which is similar to the $+4G_S N_c N_f I_3^2$ term appearing in the OPT result.

Note that, in order to obtain thermodynamical results with the LN approximation, at $G_V \neq 0$, one needs to solve the following set of equations

$$\frac{\partial \mathcal{F}}{\partial \sigma} = 0, \quad \text{and} \quad \frac{\partial \mathcal{F}}{\partial \tilde{\mu}} = 0. \tag{25}$$

Fig. 2.2 shows the predictions for the chiral transition at low temperatures and high chemical potential values, where one has a first order phase transition. This transition line starts at $T = 0$ and terminates at a critical end point (when $m_0 \neq 0$), since at supercritical temperatures a crossover takes place. This low-T/high-μ portion of the QCD phase diagram is very important for astrophysical applications and the fact that the transition is of the first kind has non-negligible consequences concerning the structure of compact stellar objects. For example, within this kind of phase transition one may have two substances, with distinct densities, coexisting at the same P, T and μ. In the case of strongly interacting matter these two substances may represent hadronic and quark matter which could lead to the formation a hybrid star instead of a pure neutron star. Our results show that both, the OPT and the LN approximation with $G_V \neq 0$ (LNG_V) reduce the first order transition in relation to the LN case with $G_V = 0$. Also, for a given temperature, the coexistence chemical potential value at which the transition occurs is shifted to higher values within the OPT and the LNG_V indicating that these two different model approximations produce a similar type of physics as one could infer by comparing their free energies.

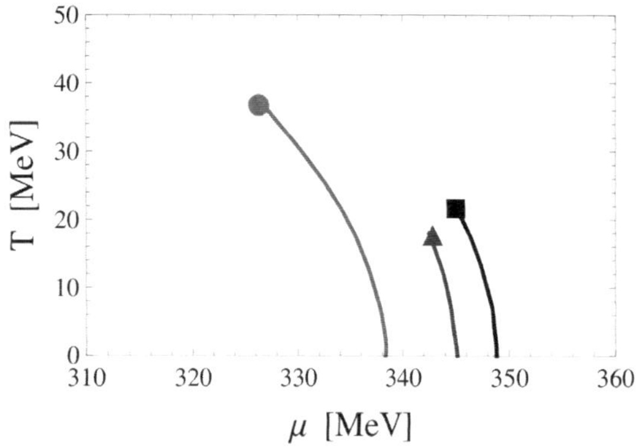

Figure 2. Phase diagram in the T–μ plane for the NJL model showing the first order transition lines and the critical end points. The LN with $G_V = 0$ is denoted by the circle, the LNG_V with $G_V = G_S/(N_f N_c)$ is denoted by the triangle and the OPT is denoted by the square.

We have already emphasized the important role played by a repulsive vector channel in the description of compressed quark matter and, as we have just shown, the OPT seems to reproduce the same effects without the need to explicitly introduce such a term (and one more parameter) at the tree level. This occurs as a consequence of the Fierz identities, which manifest themselves when exchange (two loop) diagrams are considered. This identities make, in the OPT case, the radiative corrections ($1/N_c$ suppressed) account for vector type of channel. So far, in view of our comparisons between this method, the LN and the LNG_V, one may say that at least for temperatures not much higher than the one which signals the chiral transition the OPT represents an improvement over the LN (at $G_V = 0$). So, the OPT can be seen as a powerful alternative to investigate the low-T/high-μ part of the QCD phase diagram which is currently non-accessible to the LQCD simulations. However, our evaluation of the important c_2 cumulant has revealed that at temperatures higher than T_c the OPT presents a problem similar to the one detected within the LNG_V approximation: the failure to attain the SB limit.

3. Results at finite temperature and zero density for the second order cumulant of the pressure

The statistical moments, commonly determined from a measured multiplicity distribution, are the mean (M), the variance (σ^2) (not be confused with the scalar field), the skewness (S) and the kurtosis (κ). They are related to the (net-)number N of interest by $M = \langle N \rangle$, $\sigma^2 = \langle (\Delta N)^2 \rangle$, $S = \langle (\Delta N)^3 \rangle / \sigma^3$ and $\kappa = \langle (\Delta N)^4 \rangle / \sigma^4 - 3$ where $\Delta N = N - \langle N \rangle$ is the fluctuation of N around its mean value. The cumulants c_n of the distribution are defined as $c_1 = M$, $c_2 = \sigma^2$, $c_3 = S\sigma^3$, and $c_4 = \kappa\sigma^4$ and are, for an equilibrated system, related to generalized susceptibilities given by appropriate derivatives of the pressure. Studying fluctuations in statistical systems is important from several points of view. For instance, fluctuations define the stability of the system and its way of reaching the state of thermodynamic equilibrium [25] The number-of-particle fluctuations are characterized by the dispersion

$$\langle (\Delta N)^2 \rangle = \langle N^2 \rangle - \langle N \rangle^2 \tag{26}$$

for the number-of-particles operator $N = \int \psi^+(r)\psi(r)dr$ with ψ being a field operator. The average number of particles, N, is a statistical average $N = \langle N \rangle = \int \rho(r)dr$ where $\rho(r) = \langle \psi^+(r)\psi(r) \rangle$ is the density of particles. Here we keep in mind an equilibrium state, because of which $\rho(r)$ does not depend on time. But $\rho(r)$ depends on the spatial variable r, when the system is non-uniform (which is not the case under study in this work).

To study QCD at finite densities using Monte Carlo simulations one can Taylor-expand the pressure as a power series in μ/T [26],

$$\frac{P}{T^4} = \sum_{n=0}^{\infty} c_{2n}(T) (\mu/T)^{2n} . \tag{27}$$

Note that there are only even powers of μ/T due to the reflexion symmetry, which means that the pressure is an even function with respect to μ, $P(\mu) = P(-\mu)$ [27]. The cumulants, which are the coefficients (c_n) of the series, can be identified with the quark number susceptibilities [4]:

$$\frac{\chi_n}{T^2} = n! c_n(T) . \tag{28}$$

Once the pressure has been evaluated within a given model approximation, the coefficients can be obtained from

$$c_n(T) = \frac{1}{n!} \left. \frac{\partial^n P(T,\mu)/T^4}{\partial (\mu/T)^n} \right|_{\mu=0} . \tag{29}$$

Having determined some of these coefficients one can then calculate other thermodynamical quantities like the quark number density or the quark number susceptibility, which are respectively given by

$$\frac{\rho}{T^3} = \frac{\partial P(T,\mu)/T^4}{\partial(\mu/T)} = 2c_2\frac{\mu}{T} + 4c_4\left(\frac{\mu}{T}\right)^3 + \cdots, \tag{30}$$

$$\frac{\chi_q}{T^2} = \frac{\partial^2 P(T,\mu)/T^4}{\partial(\mu/T)^2} = 2c_2 + 12c_4\left(\frac{\mu}{T}\right)^2 + \cdots \tag{31}$$

Figure 3 and 4 show the coefficient c_2 obtained respectively with the OPT and with $G_V \neq 0$, both compared to the pure large-N approximation for the NJL model. We can see from the figures that the improvements do not behave as expected, since the value of c_2 is decreasing for $T > T_\sigma$ and is moving away from the Stefan–Boltzmann limit, which takes place at sufficiently high temperatures when the thermal fluctuations overcome the interparticle interactions and the system behaves as a free gas.

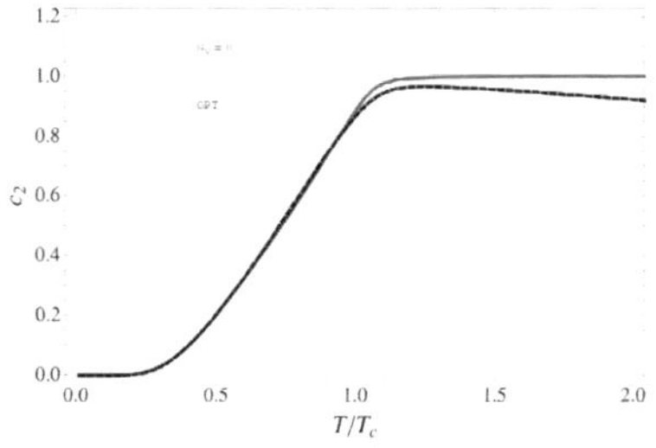

Figure 3. Taylor expansion coefficient c_2 as a function of T/T_σ obtained with the LN approximation (continuous line) and with the OPT (dashed line).

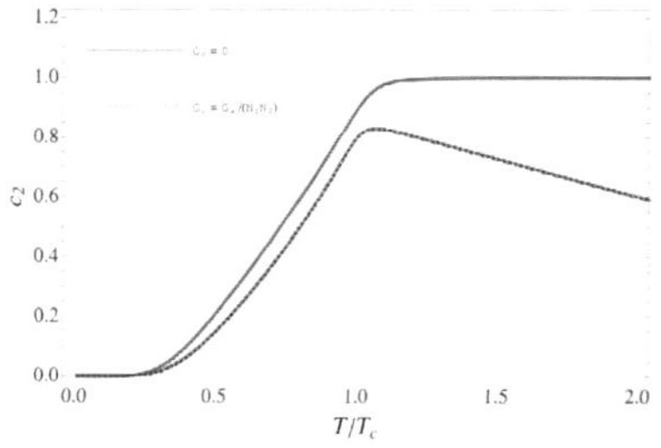

Figure 4. Taylor expansion coefficient c_2 as a function of T/T_σ obtained for the LN approximation with $G_V = 0$ (continuous line) and $G_V \neq 0$ (dotted line).

$$\frac{P_{SB}}{T^4} = \frac{8\pi^2}{45}\left[1 + \frac{21}{32}N_f\left(1 + \frac{120}{7}\hat{\mu}^2 + \frac{240}{7}\hat{\mu}^4\right)\right]. \tag{32}$$

Then, using Eq. (29) we find the Stefan–Boltzmann limit for c_2

$$c_2^{SB} = 1. \tag{33}$$

4. Conclusions

The evaluation of such cumulants has received a lot of attention from LQCD researchers and the considerable amount of data already available can be used to check the reliability of other non-perturbative techniques such as the OPT and the LN approximation. At this stage it seems puzzling that the OPT, which supposedly should produce results more accurate than the LN approximation, seems to perform rather poorly at high temperatures. Moreover, none of the LQCD evaluations for this coefficient reproduce the maximum for $c_2(T)$ observed in the OPT results. Nevertheless, this is the result that has been obtained by Steinheimer and Schramm [4] who used the LN approximation to investigate the PNJL model in the presence of this repulsive vector channel. The authors concluded that G_V should drop to zero after T_c so that the SB limit at high T while reproducing the correct physics at $T < T_c$.

However, the main source of this behavior at $T > T_\sigma$ is the medium term that appears in Eqs. (15, 18) for the OPT and in Eq. (24) for the $G_V \neq 0$ case. This integral goes to large negative values at high temperatures, and because of the pure large-N limit does not take it into account, the Stefan-Boltzmann limit is reached in this approximation. Now, the task is to fix the OPT method in order to get more coherent results for c_2, keeping in mind this question: Why a more sophisticated approximation in which more corrections diagrams are calculated would give, a priori, less satisfactory data than a simplest one? Where does this discrepance lie and how to eliminate it?

References

[1] M. Buballa, *Phys. Rept.* **407**, 205 (2005).
[2] K. Masuda, T. Hatsuda and T. Takatsuda, Prog. Ther. Exp. Phys., 073D01 (2013).
[3] V. Skokov, B. Friman and K. Redlich, Phys. Rev. C **88**, 034911 (2013).
[4] J. Steinheimer, S. Schramm, Phys.Lett. B **736** 241-245 (2014).
[5] L. Ferroni, V. Koch and M.B. Pinto; AIP Conf. Proc. 1441, 786 (2012).
[6] Nambu, Y. & Jona-Lasinio, Phys. Rev. 122, 345-358 (1961).
[7] Kapusta, J. I., Finite-temperature field theory, Cambridge University Press (1989).
[8] Bailin, D. & Love, A. Brewer, D. F. Introduction to Gauge Field Theory, Taylor & Francis Group (1993).
[9] Gandhi, S. K.; Jones, H. & Pinto, Nucl. Phys. B **359**, 429 - 440 (1991).
[10] Gandhi, S. K. & Pinto, M. B., Phys. Rev. D **46**, 2570-2577 (1992).
[11] Krein, G.; Marques de Carvalho, R.; Menezes, D.; Nielsen, M. & Pinto, M., Eur.Phys.J. A1, 45-53 (1998).
[12] Stevenson, P. M., Phys. Rev. D **23**, 2916-2944 (1981).
[13] Buckley, I. R. C.; Duncan, A. & Jones, H. F., Phys. Rev. D **47**, 2554-2559 (1993).
[14] de Souza Cruz, F. F.; Pinto, M. B. & Ramos, R. O., Phys. Rev. B **64**, 014515 (2001).
[15] Kneur, J.-L.; Neveu, A. & Pinto, M. B., Phys. Rev. A **69**, 053624 (2004).
[16] Article (PhysRevB.77.205109) Caldas, H.; Kneur, J.-L.; Pinto, M. B. & Ramos, R. O., Phys. Rev. B **77**, 205109 (2008).
[17] Andersen, J. O.; Leganger, L. E.; Strickland, M. & Su, N., JHEP **053**, 1108 (2011).
[18] Kneur, J.-L.& Neveu, A., Phys. Rev. D **81**, 125012 (2010).
[19] Kneur, J.-L. & Neveu, A., Phys. Rev. D **85**, 014005 (2012).
[20] Kneur, J.-L. & Neveu, A., Phys. Rev. D **88**, 074025 (2013).
[21] Staudt, E. Teorias Fermiônicas Efetivas Sob Condições Extremas, Universidade Federal de Santa Catarina (2009).
[22] Hatsuda, T. & Kunihiro, T. QCD phenomenology based on a chiral effective Lagrangian Phys.Rept. **247**, 221-367 (1994).
[23] Gell-Mann, M.; Oakes, R. J. & Renner, Phys. Rev. **175**, 2195-2199 (1968).
[24] Fukushima, K., Phys. Rev. D **78**, 114019 (2008).
[25] Bluhm, M.; Alba, P.; Alberico, W.; Bellwied, R.; Sarti, V. M.; Nahrgang, M. & Ratti, C., arXiv:1403.4903.
[26] C. R. Allton et al., Phys. Rev. D **66**, 074507 (2002).
[27] Philipsen, O. Lattice QCD at non-zero temperature and baryon density, arXiv:1009.4089v2.
[28] Kneur, J.-L.; Pinto, M. B. & Ramos, R. O., Phys. Rev. C **81**, 065205 (2010).

Hydrodynamics from Landau initial conditions

Abhisek Sen[1],Jochen Gerhard[2],Giorgio Torrieri[3],Kenneth Read[4,1],Cheuk-Yin Wong[4]

[1]University of Tennessee, Knoxville, TN 37996
[2]Frankfurt Institute for Advanced Studies (FIAS), Frankfurt am Main, Germany
[3] IFGW, Universidade Estadual de Campinas, Campinas, São Paulo, Brazil
[4]Oak Ridge National Laboratory, Oak Ridge, TN 37831

E-mail: torrieri@ifi.unicamp.br

Abstract. We investigate ideal hydrodynamic evolution, with Landau initial conditions, both in a semi-analytical 1+1D approach and in a numerical code incorporating event-by-event variation with many events and transverse density inhomogeneities. The object of the calculation is to test how fast would a Landau initial condition transition to a commonly used boost-invariant expansion. We show that the transition to boost-invariant flow occurs too late for realistic setups, with corrections of $\mathcal{O}(20-30\%)$ expected at freezeout for most scenarios. Moreover, the deviation from boost-invariance is correlated with both transverse flow and elliptic flow, with the more highly transversely flowing regions also showing the most violation of boost invariance. Therefore, if longitudinal flow is not fully developed at the early stages of heavy ion collisions, 2+1 dimensional hydrodynamics is inadequate to extract transport coefficients of the quark-gluon plasma. Based on [1, 2]

1. Introduction

The quantitative modeling of matter produced in high energy heavy ion collisions with relativistic hydrodynamics is now a well-established field, following the widely cited announcement that matter produced at the relativistic heavy ion collider (RHIC), behaves as a "perfect fluid" [3]. The evidence for this behavior comes from the successful modeling of RHIC anisotropic flow by boost-invariant hydrodynamics. It is now clear that the same fluid-like behavior persists at the LHC [5, 6, 7]. It is commonly argued [8, 9, 10] that, given precise enough data on soft physics, chiefly momentum spectra and their azimuthal anisotropy, the transport coefficients of matter created in ultrarelativistic heavy ion collisions can be quantitatively constrained.

These models are all based on the reduction, either exact or approximate, of the problem to a 2+1 dimensional system [11], based on the symmetry of boost-invariance. Essentially, the system is assumed to have *as an initial condition* a longitudinal flow that is Hubble-like in the beam direction (usually associated with the "z" coordinate) *only*. This means that, initially,

$$v_z = \frac{z}{t}, \qquad y_s = y_f = \langle y \rangle_p , \tag{1}$$

where y_s and y_f are respectively the spacetime and flow rapidities

$$y_s = \frac{1}{2}\ln\left(\frac{t+z}{t-z}\right), \qquad y_f = \frac{1}{2}\ln\left(\frac{1+v_z}{1-v_z}\right),$$

$$\langle y \rangle_p = \frac{1}{2}\left\langle \ln\left(\frac{E+p_z}{E-p_z}\right)\right\rangle \simeq \frac{1}{2}\ln\cot\frac{p_z}{|p|}$$

with y_p being usually referred to as pseudo-rapidity. A further simplification comes from assuming that all initial dynamics does not depend on y

$$\frac{d}{dy}\frac{dN}{dy} = 0, \qquad \frac{dv_T}{dy} = 0 \tag{2}$$

or, equivalently not on t, z separately, but just on

$$\tau = \sqrt{t^2 - z^2} \tag{3}$$

(evolved from an initial time τ_0) and transverse degrees of freedom. An initial condition that respects Eq. (1) but not Eq. (2) will slowly degrade the constraints of Eq. (1), as shown in [12, 13, 14].

Initially, a different model has originally been advocated as the obvious initial state for the hydrodynamic evolution of the fluid: Landau hydrodynamics [15, 16, 19, 20]. In this picture, the energy that forms the bulk of the expanding fireball "stops" at midrapidity at time zero (in the collider frame). The initial distribution of matter is therefore a "pancake", of thickness 2Δ related to the boosted charge radius R of the nuclei with nucleon mass m_N at center of mass energy of $\sqrt{s_{NN}}$ where

$$\Delta \to \Delta_{lim} \simeq \frac{R}{\gamma} = \frac{Rm_N}{(\sqrt{s_{NN}}/2)}. \tag{4}$$

In a more general implementation, Δ need not be defined by Eq. (4) and can be a free parameter, reflecting the spread in configuration space of low x gluons. The initial Landau condition is defined by the assumption that the initial "pancake" has no existing longitudinal flow at all, unless there are initial inhomogeneities which lead to a net momentum in local transverse space. (This is known as the "firestreak model" [21, 24]). Boost invariance is badly broken at the beginning of the fireball evolution and such a pancake has very little in common with the scenario used in [11, 13]. One can consider Bjorken and Landau as two extremes: In the Bjorken scenario, the nuclei originally pass through each other with minimal reinteraction and strings that stretch between colliding gluons arise in parallel to other strings. In the Landau scenario, they "stick together" or at least leave some energy in the middle.

While computationally reducing one dimension is highly desirable, a physics justification would be needed for the approximation of Eq. (1). A direct measurement of dN/dy is inconclusive. On the one hand, the Landau model fits a Gaussian well at all energies, with universal limiting fragmentation, as expected in [15, 16, 19]; moreover, strong violations of boost-invariance considerably lessen the HBT puzzle [25]. However, the *multiplicity* dependence on $\sqrt{s_{NN}}$ is not exactly that predicted in [15, 16, 19]. This by itself does not rule out the Landau scenario, as it can be accounted for by treating the initial thickness evolution with $\sqrt{s_{NN}}$ as a free parameter, as done in the Bjorken scenario.

There are two main arguments one can give for the Bjorken limit being more appropriate: The first one is that the perturbative partonic picture of the system [26, 27] makes this initial condition natural. However, even in the weakly-coupled limit, low x partons could lead to a breakdown of Eq. (2) (see for example, [28, 29]). Moreover, if the initial state is *strongly coupled* from the beginning, one could indeed expect that it would appear much more Landau-like [30, 31] than Bjorken-like, although the degree of stopping might also depend strongly on energy and system size [32, 33]. Stopping is therefore not determined *a priori*, as the interaction strength at the beginning of the system's evolution is currently a controversial topic.

The second reason is that, for mid-rapidity data, it is widely believed that the distinction between Bjorken and Landau evolution is irrelevant. As is clear from [15], Landau evolution

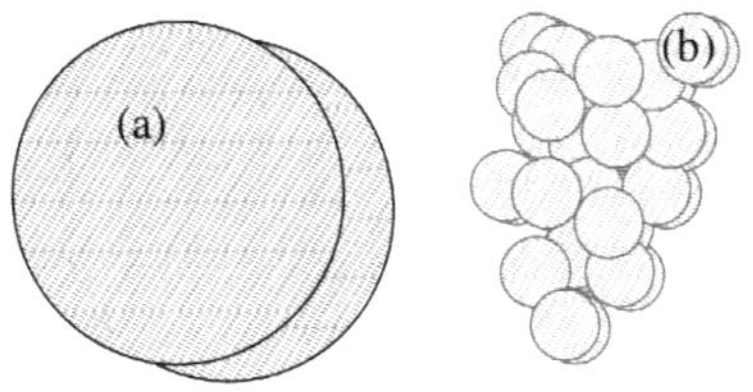

Figure 1. Landau initial condition (panel (a)), and actual Glauber initial conditions (panel (b)) for a typical event.

converges to Bjorken evolution after some sufficient time. The reason for this behavior is that longitudinal flow forms on the scale of $\sim \Delta/c_s$, while transverse flow forms on a much larger scale $\sim R/c_s$ where c_s is the speed of sound. Hence, since $\Delta \ll R$ for $\sqrt{s_{NN}} \gg 1$ GeV, initially the system can be considered, as indeed it is in [15], to be a purely 1D expanding "sharp step." As again shown in [15, 19, 20], the long-term longitudinal evolution of such a system at mid-rapidity is indistinguishable from that of [11]. Hence, 2+1 dimensional hydrodynamics can be safely used even if, at the very initial stage [13], the system is very far from boost invariance. Landau evolution at mid-rapidity can be treated as Bjorken with $\tau_0 \sim 1$ fm $\times$ GeV$/\sqrt{s_{NN}}$. Perhaps, this scaling will give unrealistically low initial proper times at the LHC, but since 2+1 simulations are only weakly sensitive to time, this might not be a fatal issue.

This idea, however, has two flaws: First of all, for a non-central collision, where anisotropic flow is most expected, locality and longitudinal momentum conservation imply that the system develops an additional initial longitudinal momentum imbalance, with extra longitudinal momentum due to the local (in transverse space) imbalance between the target and projectile $\rho_{part}^{P,T}(x_T) = d^2N_{part}^{P,T}/dx_T^2$ transverse participant density. Momentum conservation and the Landau condition (no transparency) constrain the initial $\gamma_z v_z$ to

$$\frac{v_z(x_T)}{\sqrt{1-v_z^2(x_T)}} = \frac{\rho_{part}^P(x_T) - \rho_{part}^T(x_T)}{\rho_{part}^P(x_T) + \rho_{part}^T(x_T)} K \tag{5}$$

K here is a free parameter, but it is clear that $K = \sqrt{s_{NN}}/m_N$ when $\Delta = \Delta_{lim}$ in Eq. (4). In general, $\Delta > \Delta_{\rm lim}$ reflects a picture where the partons carrying the dominant fraction of the nucleon's energy are parametrically much softer than the nucleon. This is equivalent to the "wee parton" picture, and implies they also carry less momentum. Assuming a linear dependence, the net momentum in an off-central collision is related to Δ by

$$K \simeq \frac{\sqrt{s_{NN}}}{m_N}\frac{\Delta_{lim}}{\Delta} = \frac{2R}{\Delta}. \tag{6}$$

This initial flow is trivially not boost invariant and it is not clear it disappears at *any* finite time for a general system evolving from a Landau initial condition.

Additionally, the "Landau→Bjorken" reasoning assumes that the longitudinal timescale is much larger than the transverse one. This is certainly true if the transverse scale is given as a radius of a homogeneous "pancake" of radius R given by an *average* of many events as in Fig. 1 (a). It is however less clear that such a hierarchy holds for a *typical* event as in Fig. 1(b). The inclusion of subnucleonic strong QCD fields [9] make this hierarchy even more dubious as the events with the strongest anisotropic coefficients would also have the most prominent "hotspots." Potentially, this effect makes the boost-invariant picture irrelevant *even for late-time hydrodynamics*: The more homogeneous regions will be more similar in their longitudinal expansion to [11], while the more inhomogeneous regions would, on their own, evolve to a *3D Hubble* expansion [35]. The interplay between regions of different symmetry, and local instabilities [35, 36] makes any symmetry dubious.

In this work, we examine these issues systematically by both analytical and numerical methods. In the next section, we shall use a semi-analytical 1+1D calculation to examine the assumption that Landau hydrodynamics becomes bjorken-like. Then, we use a numerical code to investigate the role in initial-state anisotropies and transverse flow from a Landau initial condition.

2. Analytical calculation

2.1. The Khalatnikov solution

For the Landau initial condition of a reflectively symmetric slab of a relativistic hot, dense matter initially at rest, the Khalatnikov solution is an analytical solution of the hydrodynamical equation that describes the space-time evolution of the system. The solution is obtained by introducing a hydrodynamical potential χ that is a function of the hydrodynamical variables of the energy density field ϵ and the velocity field v, which can be alternatively represented by the energy density logarithm y and the flow rapidity α,

$$\epsilon/\epsilon_0 = (s/s_0)^{4/3} = (T/T_0)^4 = e^{4y}, v = \tanh\alpha, \tag{7}$$

where ϵ_0 is the initial energy density and the flow rapidity α is also represented by the symbol y_f. The Khalatnikov solution consists of writing the space-time coordinates (z, t) as functions of the hydrodynamical variables (y, α), given in Eq. (4.30)-(4.12a) of Ref. [18], as

$$t(y,\alpha) = e^{-y}\left(\frac{\partial\chi}{\partial y}\cosh\alpha - \frac{\partial\chi}{\partial\alpha}\sinh\alpha\right), \tag{8}$$

$$x(y,\alpha) = e^{-y}\left(\frac{\partial\chi}{\partial y}\sinh\alpha - \frac{\partial\chi}{\partial\alpha}\cosh\alpha\right). \tag{9}$$

Belenkij and Landau considered the initial slab of width $2l$ and chose the origin of the longitudinal z-coordinate at x=$-l$. The longitudinal coordinates z is therefore related to the quantity x in Eq. (9) by $z = x + l$. As we are considering a system possessing a reflection symmetry with respect to $z = 0$, we need to examine only the region of $z \geq 0$.

The Khalatnikov solution is uniquely specified by the requirement to satisfy two boundary conditions: (i) zero velocity (v=0 and α=0) at the center of the symmetric slab at z=0 (and x=$-l$), and (ii) the matching to the Riemann simple wave solution when y=$-\alpha c_s$ at the boundary of the slab. It is given in terms of the hydrodynamic potential $\chi(y,\alpha)$ in Eq. (4.30) of [18] by

$$\chi(y,\alpha) = -l\sqrt{3}e^{y}\int_{\alpha/\sqrt{3}}^{-y} e^{2y'} I_0\left[\sqrt{y'^2 - \frac{1}{3}\alpha^2}\right] dy' \tag{10}$$

The above solution (10) has been obtained for the ideal equation of state, $p = e/3$ with speed of sound $1/\sqrt{3}$.

In the Khalatnikov solution, there are two hydrodynamical degrees of freedom which have been chosen to be the energy density ϵ and the velocity field v or alternately (y,α). There is however another Riemann wave simple wave solution of the one-dimensional relativistic hydrodynamical equations in which the energy density field represented by y and the velocity field represented by α can be expressed as a function of each other in which the space-time coordinates x and t do not explicitly appear. In the presence of a disturbance, the simple wave propagation can be visualized as the superposition of (i) the propagation of a sound wave with the speed of sound c_s and (ii) the propagation of the fluid element itself with a flow velocity $v = \tanh\alpha$. They occur at the boundary regions where the energy density decreases monotonically until the energy density vanishes when the matter is in contact with the vacuum. As the two boundaries of the slab are always in contact with the vacuum, the Riemann simple wave solution are always present on these slab boundaries.

Because of this mutual dependencies between y and α for the Riemann simple wave, there is then only a single independent hydrodynamical degree of freedom in the Riemann simple wave solution. The hydrodynamics is described by $y(\alpha(x,t))$ or vice-versa $\alpha(y(x,t))$ in the form of a running wave whose profile can change with time. In relativistic hydrodynamics, this becomes [23]:

$$\alpha = \tanh^{-1} v \quad = \quad \pm \int \frac{c_s d\epsilon}{(\epsilon + p)} = \pm \frac{c_s}{c_s^2} \ln(\epsilon/\epsilon_0)^{1/4},$$

which leads to $\alpha = -\ln(T/T_0)/c_s$, with the sign chosen since energy density decreases with time.

In terms of the potential χ in Eqs. (8) and (9), we have

$$\frac{x}{t} \quad = \quad \frac{\tanh\alpha - \frac{\partial\chi}{\partial\alpha}/\frac{\partial\chi}{\partial y}}{1 - \tanh\alpha \ (\frac{\partial\chi}{\partial\alpha}/\frac{\partial\chi}{\partial y})} = \frac{\tanh(-y/c_s) - c_s}{1 - \tanh(-y/c_s)c_s}. \tag{11}$$

since or simple waves with a center at the origin, the total derivative of the potential function $\chi(y,\alpha)$ is zero [23].

After the time $t \geq l/c_s$, the rarefaction wave that starts from the edge of the slab at z=l reaches the center of the slab at z=0 (Fig. 1). Subsequent expansion of the fluid in the central region will proceed through the Khalatnikov solution of Eqs. (8), (9), and (10). From Equation 10 we get, after some algebra

$$\frac{\partial\chi}{\partial\alpha}(y,\alpha) \quad = \quad l(\alpha/\sqrt{3})e^y \int_{\alpha/\sqrt{3}}^{-y} e^{2y'} \frac{I_1\left[\sqrt{y'^2 - \frac{1}{3}\alpha^2}\right]}{\sqrt{y'^2 - \frac{1}{3}\alpha^2}} dy' + l e^y e^{2\alpha/\sqrt{3}}. \tag{12}$$

With the knowledge of $\partial\chi/\partial y$ and $\partial\chi/\partial\alpha$ given by Eq. (12), the right-hand sides of Eqs. (8) and (9) give (x,t) as explicit functions of (y,α). The integral in Eq. (12) can be evaluated numerically as the limits of the integration and the integrands are known functions of y and α.

Not all portions of the Khalatnikov solution shown as the solid curves can be used to describe the evolution of the system as the dynamics at the edge is described by the propagation of a disturbance arising from the presence of the edge boundary, and the accompanying hydrodynamical motion is a simple wave propagating from the edge toward the center. The solution at the edge of the slab is governed by the Riemann simple wave solution. The Khalatnikov solution that is applicable in the interior of the slab needs to be matched with the simple wave solution when the energy density logarithm y matches the rapidity α with $\alpha = -y/c_s$.

We can carry out the matching in the following way. We study the hydrodynamical solution for a fixed value of t $(\geq l/c_s)$ and increase stepwise the value of α, starting from α=0. We calculate y and z as a function of t and α. After determining y for the pair of (t,α) values, we test whether $-\sqrt{3}y$ remains greater than α or not. If $-\sqrt{3}y$ remains greater than α, we proceed to the next incremented value of α and look for the Khalatnikov solution for the next set of (t,α) pair. On the other hand, when $-\sqrt{3}y$ is equal to or just begin to be greater than α, the solution will be switched to the Riemann simple wave solutions for subsequent α values. In the simple wave boundary region for a fixed value of t, we increase stepwise the value of α and the energy density logarithm variable y is given by $y = -c_s\alpha$. Knowing and the values of t, α and

y, the spatial coordinate x is given by Eq. (11). This stepwise increase of α allows us to trace the energy density as a function of the longitudinal coordinates, for a fixed value of t.

The space-time coordinates (x, t) or (t, z) can be represented alternatively by the proper-time τ and spatial rapidity y_s given by Eq. 1 and Eq. 3 The initial simple wave propagation cannot be adequately represented in terms of real τ and y_s because $t \leq z$ for the initial slab. Real proper time and spatial rapidity commence at $t = l/c_s$ at the proper time of $\tau = l/c_s$, at which the rarefaction wave reaches the center of the slab and the Khalatnikov solution begin to come into play.

We need to express the Khalatnikov solution and the Riemann wave solution in terms of the proper time τ and spatial rapidity y_s. The Khalatnikov solution are given by Eq. (23) where t and x are expressed as functions of y and α by Eqs. (8), (9), with the hydrodynamical potential χ determined by Eq. (10). With the knowledge of $\partial\chi/\partial y$ and $\partial\chi/\partial\alpha$ given by Eq. (12) the (τ, y_s) variables are explicit functions of (y, α).

In terms of the (τ, y_s) coordinates, the Riemann simple wave solution can be obtained by considering the effective velocity a as

$$a = \frac{v - c_s}{1 - vc_s}. \tag{13}$$

The simple wave solution in terms of (τ, y_s) becomes

$$e^{y_s} = \frac{1 + \sqrt{1 + (\tau/l)^2(1 - a^2)}}{(\tau/l)(1 - a)}, \tag{14}$$

which describes the hydrodynamical motion of disturbances in the boundary regions of the slab. The Khalatnikov solution needs to match to the Riemann wave solution when the energy logarithm y and the rapidity α is related by the speed of sound, $y = -\alpha c_s$.

We consider a fixed value of $\tau \geq l/c_s$ and stepwise increase the value of α, starting from α=0. We can obtain the hydrodynamical description of (y, α) as a function of (τ, y_s) by inverting Eq. (3,2). For each pair of (τ, α) values, Eq. (3) with supplementary equations (8) and (9) presents itself as an equation for the unknown quantity y. We can solve this equation with only one unknown numerically. After the solution of y is obtained, Eqs. (12) and (12) are then used with Eq. (9) to calculate the value of x, z and y_s. The newly determined y can be used as the trail value for the next α to get the new solution of y.

Using the method we have just outlined for $\tau \geq l/c_s$, we can determine the Khalatnikov solution as a function of the spatial rapidity y_s for a fixed value of τ shown as solid curves in Fig. 4. However, it is necessary to match the Khalatnikov solution to the Riemann simple wave when y is equal to $-\alpha c_s$. Starting from $y_s = 0$ at $\alpha = 0$, we can carry out the simple wave matching by testing $-y$ against $c_s\alpha$. When $-y$ is equal to or just begin to be greater than $c_s\alpha$, the solution will be switched to simple wave solutions for subsequent α values. For the simple wave solution at a fixed value of τ, the energy density logarithm variable y is given by $y = -c_s\alpha$ and the spatial rapidity y_s is given by Eq. (14). The hydrodynamical solution with properly matched Khalatnikov solution and Riemann simple waves are shown as the dashed curves in Fig. 4. They are the hydrodynamical solutions satisfying the boundary conditions.

We show in Fig. 5 the dynamics of the system for later times of τ = 10 30, 50, and 70 l. The energy density is relatively flat but the flow rapidity α is slightly greater than the spatial rapidity y_s.

The solution of (y, α) as a function of (t, z) or (τ, y_s) allows one to extract other hydrodynamical quantity of interest. In addition to the energy density, one can calculate the spatial profiles of the temperature or entropy density at different times t or proper times τ.

In Fig. 8, we show the ratio ϵ/ϵ_0 at the center of the slab at $z = 0$ as a function of the proper time τ/l. The energy density decreases with τ/l but the decrease does not follow the Bjorken

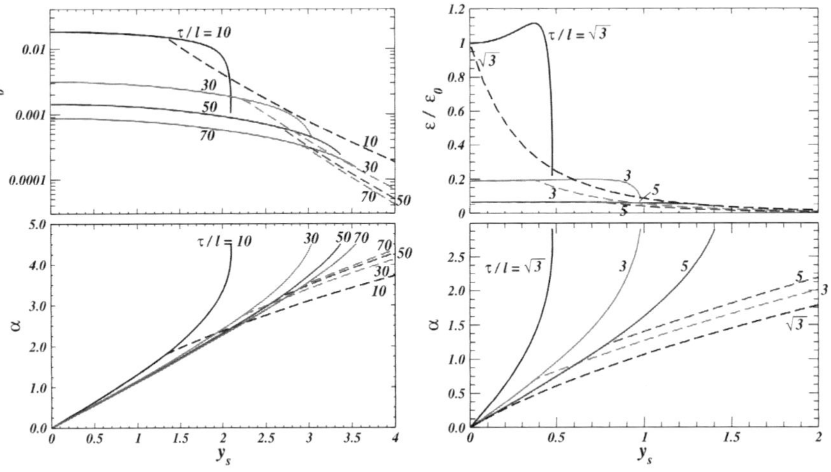

Figure 2. (Color online) right panel: The quantities $(\epsilon/\epsilon_0, \alpha)$ as a function of the spatial rapidity y_s for different values of τ/l. Left panel: The quantities $(\epsilon/\epsilon_0, \alpha)$ as a function of the spatial rapidity y_s for different values of τ/l. In both cases The solid curves give the Khalatnikov solutions which must be matched on to Riemann simple wave solutions shown as dashed curves.

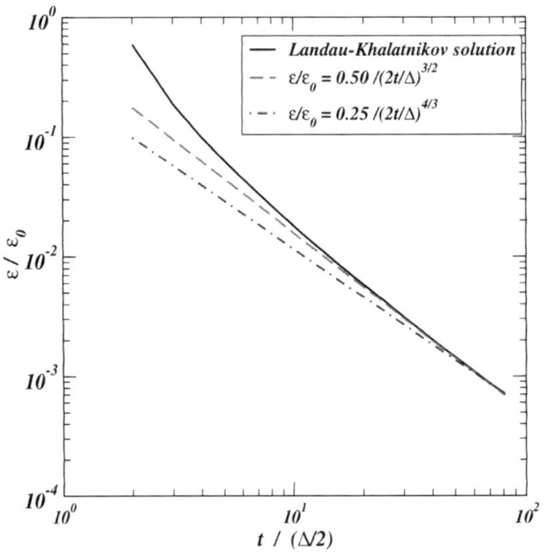

Figure 3. (Color online) The ratio ϵ/ϵ_0 at $z = 0$ as a function of τ/l. The solid curves represent the solutions from the Khalatnikov solution with the Landau initial condition, and the dashed curve is the behavior expected from Bjorken hydrodynamics.

limit of $\epsilon/\epsilon \propto 1/\tau^{4/3}$. Bjorken-like behavior of $\epsilon \propto 1/\tau^{4/3}$ behavior occurs only at the very late stage of $\tau/l \sim 80$, useless for any phenomenological study.

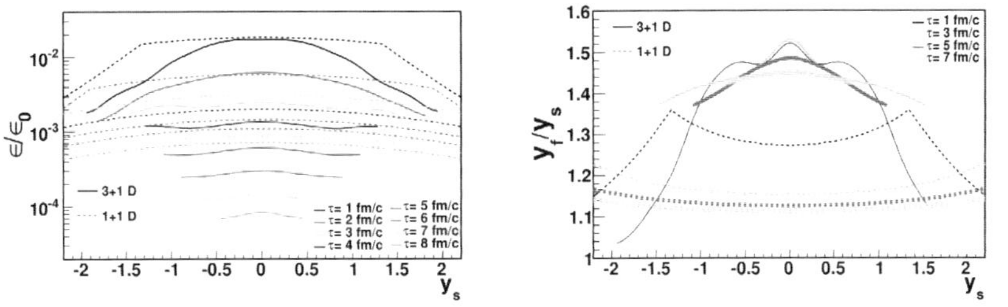

Figure 4. (color online) Left panel: Ratio of energy density at the indicated time to the initial energy density as a function of rapidity. The dashed lines show the analytical solution [22] while the solid lines show the numerical calculation, including net momentum, transverse expansion and inhomogeneities. Right panel: The y_f/y_s ratio (solid lines) as a function of proper time and rapidity, averaged over all events, at $r_\perp = 0$. The analytical 1D solution [22] is also plotted (dashed lines).

3. Numerical calculation

The Glauber Monte Carlo description of two colliding Au^{197} nuclei at 200 GeV was used to generate the initial condition relevant to RHIC. Nucleons were distributed as per a Wood-Saxon distribution with radius 6.38 fm, and diffuseness 0.535 fm. The impact parameters were simulated randomly following a distribution of $d\sigma/db = 2\pi b$. The nucleons were assumed to have no hard-core and the condition for nucleon-nucleon collision is that the inter-nucleon distance d should satisfy $\pi d^2 < \sigma_{NN}$, where $\sigma_{NN} = 42$ mb is the nucleon-nucleon cross section.

We then use the CL-SHASTA code developed in [37] to evolve this configuration according to ideal hydrodynamics, $\partial_\mu T^{\mu\nu} = 0$ with

$$T_{\mu\nu} = (\rho + p) u_\mu u_\nu + p g_{\mu\nu} \tag{15}$$

$$u_\mu = \frac{1}{\sqrt{1 - v_z^2 - v_T^2}} (1, v_T \sin(\theta), v_T \cos(\theta), v_z) \tag{16}$$

where $v_z = \tanh y_L$. With an ideal gas equation of state, $p = \rho/3, c_s = 1/\sqrt{3}$, and $\Delta = 0.1$ fm, and longitudinal flow given by Eq. (5). Our results do not qualitatively change if the longitudinal thickness if changed by $\mathcal{O}(50 - 100\%)$.

The high-statistics 3D calculations were performed at the Oak Ridge National Laboratory using the code in [37].

After simulating 10000 events for a given configuration and initial conditions, each of which evolves the millions of grid cells over 300 small time steps in the lab frame covering an expansion until 10 fm/c, we divide them into spacetime rapidity slices. The energy density evolution as a function of spacetime rapidity is shown in Fig. 4 left panel, which follows the trend in [15] to $\sim$20% precision, as expected from correction due to transverse and elliptic flow. Hence, the left panel of Fig. 4 show a decreasing bump in e (which correlates with transverse multiplicity$dN/dy \sim S\tau_0 e^{3/4}$ and transverse energy $dE_T/dy \sim S\tau_0 e$ in the Bjorken picture [11]. Here $S \sim R^2 \sim N_{part}^{2/3}$ is the transverse overlap area), making these similar to boost-invariant results [11]. Self-quenching variables (v_2, and to a lesser extent the average transverse momentum $\langle p_T \rangle$), however, will show such rapidity dependence independently of freezeout.

We then calculate the longitudinal flow rapidity y_f as well as the transverse flow for each slice of rapidity, averaged over the entire transverse volume, to explore boost invariance. Fig. 4 right

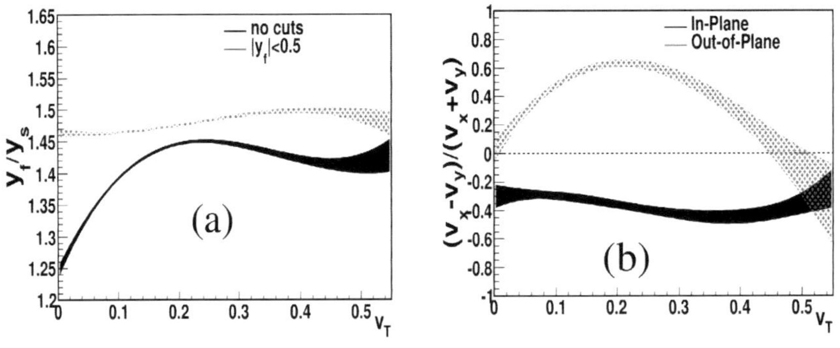

Figure 5. (color online) Panel (a) y_f/y_s as a function of mean transverse velocity for $|y_f| < 0.5$ (hashed red) and with no restriction on y_f (solid black). Panel (b) Distribution of transverse flow anisotropy versus mean total transverse velocity for in-plane (solid black) and out-of-plane (hashed red) flow. Both are averaged over fireball volume and event sample, with the bands representing the variance over the average across events.

panel shows the ratio y_f/y_s as a function of y_s at various relevant times in the evolution. If the system were exactly boost invariant, y_f/y_s would be strictly unity. As expected, the ratio does evolve towards unity as the system cools; however, it would be a gross oversimplification to treat the ratio as a constant or unity, even at significantly later times. At freezeout provided initial temperature $\simeq$ 300 MeV, we predict y_f/y_s to be above unity by about 40% around midrapidity. At earlier stages, relevant for the formation of transverse and elliptic flow ($t \sim \epsilon R/c_s$), these corrections are of order 50%. For comparison, we superimpose the same distributions for the 1D expansion calculated in an analytical work [22]. It can be seen that, unlike what was presumed in [16, 19], transverse expansion and local dynamics make a qualitative, and not just a quantitative effect: deviation from boost invariance oscillates and stays nearly constant rather than decreases in time when transverse expansion and anisotropies are taken into account.

The relevance of this dynamics for transverse degrees of freedom is elucidated in Fig. 5, which shows the dependence of y_f/y_s, an indicator of the degree of violation of boost invariance, on transverse flow. Thus, if the Landau initial condition is more appropriate, transverse flow and its azimuthal anisotropies form, to a certain extent, in strongly non-boost invariant regions. This is readily understood, as such regions are precisely the places where transverse gradients are larger w.r.t. longitudinal ones. Hotspots can also have a non-zero longitudinal momentum and vorticity [17] (the "firestreak"), further invalidating local boost-invariance. As Fig. 5 panel (a) however shows, this result somewhat depends on the rapidity region being explored. A restriction in flow rapidity, approximately tracking the pseudorapidity, will ensure v_T is independent of the degree of boost invariance. Such a cut, however, does nothing to make the evolution examined more boost-invariant, since y_f/y_s remains very well away from unity.

Fig. 5 panel (b) shows the anisotropy of the in-plane and out of plane flow, as a function of transverse flow. This plot indicates that dynamics relevant for transverse and anisotropic flow significantly violates boost invariance if Landau initial conditions are assumed.

The viability of the computations performed here depends, of course, in the longitudinal structure of the event really being close to the Landau limit. Because we do not know this from first principles, and given the many undetermined parameters in a typical hydrodynamic simulation, we suggest that experimental tests specifically probing boost invariance should be performed.

In conclusion, we have shown that, provided the system is Landau-like in its initial stages, it will not, as commonly expected, evolve to a Bjorken-like stage within realistic timescales. Furthermore, the deviation from boost invariance is directly correlated with the development of transverse and elliptic flow, the characteristic signatures used to demonstrate and quantitatively study the hydrodynamics of the quark-gluon plasma. In view of these results, the transport properties of the medium created in heavy ion collisions could be considerably different from those usually assumed.

This research used resources of the Oak Ridge Leadership Computing Facility at the Oak Ridge National Laboratory, which is supported by the Office of Science of the U.S. Department of Energy under Contract No. DE-AC05-00OR22725. We wish to thank Sean Gavin and Peter Steinberg for useful discussions.

[1] A. Sen, J. Gerhard, G. Torrieri, K. Read and C. Y. Wong, arXiv:1403.7990 [nucl-th].
[2] C. Y. Wong, A. Sen, J. Gerhard, G. Torrieri and K. Read, arXiv:1408.3343 [nucl-th].
[3] M. Gyulassy and L. McLerran, Nucl. Phys. A **750**, 30 (2005).
[4] E. Shuryak, Prog. Part. Nucl. Phys. **53**, 273 (2004).
[5] K. Aamodt *et al.* [ALICE Collaboration], Phys. Rev. Lett. **105**, 252302 (2010).
[6] S. Chatrchyan *et al.* [CMS Collaboration], Phys. Rev. C **87**, 014902 (2013).
[7] G. Aad *et al.* [ATLAS Collaboration], Phys. Rev. C **86**, 014907 (2012).
[8] H. Song, S. A. Bass and U. Heinz, Phys. Rev. C **83**, 054912 (2011) [Erratum-ibid. C **87**, 019902 (2013)].
[9] C. Gale, S. Jeon, B. Schenke, P. Tribedy and R. Venugopalan, Nucl. Phys. A904-905 **2013**, 409c (2013).
[10] R. A. Soltz, I. Garishvili, M. Cheng, B. Abelev, A. Glenn, J. Newby, L. A. Linden Levy and S. Pratt, Phys. Rev. C **87**, 044901 (2013).
[11] J. D. Bjorken, Phys. Rev. D **27**, 140 (1983).
[12] L. M. Satarov, A. V. Merdeev, I. N. Mishustin and H. Stoecker, Phys. Rev. C **75**, 024903 (2007).
[13] P. Bozek, Phys. Rev. C **77**, 034911 (2008).
[14] G. S. Denicol, T. Koide, P. Mota and T. Kodama, J. Phys. G **35**, 104130 (2008).
[15] L. D. Landau, Izv. Akad. Nauk Ser. Fiz. **17**, 51 (1953).
[16] P. Steinberg, Acta Phys. Hung. A **24**, 51 (2005) [nucl-ex/0405022].
[17] B. Betz, M. Gyulassy and G. Torrieri, Phys. Rev. C **76**, 044901 (2007) [arXiv:0708.0035 [nucl-th]].
[18] S. Z. Belenkij and L. D. Landau, Nuovo Cimento, Suppl. **3**, 15 (1956).
[19] C. -Y. Wong, EPJ Web Conf. **7**, 01006 (2010).
[20] C. -Y. Wong, Phys. Rev. C **78**, 054902 (2008).
[21] J. Gosset, J. I. Kapusta and G. D. Westfall, Phys. Rev. C **18**, 844 (1978).
[22] C.Y. Wong, A. Sen, J. Gerhard, G. Torrieri, and K.F. Read C. Y. Wong, A. Sen, J. Gerhard, G. Torrieri and K. Read, arXiv:1408.3343 [nucl-th].
[23] L. D. Landau and E. M. Lifshitz, *Fluid mechanics*, Pergamon Press, 1958.
[24] L.Cserai, Introduction to Relativistic Heavy Ion Collisions.
[25] M. Gyulassy and D. H. Rischke, Heavy Ion Phys. **17**, 261 (2003).
[26] R. P. Feynman, Phys. Rev. Lett. **23**, 1415 (1969).
[27] S. J. Brodsky, J. F. Gunion and J. H. Kuhn, Phys. Rev. Lett. **39**, 1120 (1977).
[28] W. Busza, Nucl. Phys. A **854**, 57 (2011).
[29] A. Capella, U. Sukhatme, C-ITan and J. Tran Thanh Van, Phys. Rept. **236**, 225 (1994).
[30] J. L. Albacete, Y. V. Kovchegov and A. Taliotis, JHEP **0905**, 060 (2009).
[31] D. Grumiller and P. Romatschke, JHEP **0808**, 027 (2008).
[32] J. Casalderrey-Solana, M. P. Heller, D. Mateos and W. van der Schee, Phys. Rev. Lett. **111**, 181601 (2013) [arXiv:1305.4919 [hep-th]].
[33] J. Casalderrey-Solana, M. P. Heller, D. Mateos and W. van der Schee, Phys. Rev. Lett. **112**, 221602 (2014) [arXiv:1312.2956 [hep-th]].
[34] C.E. Coleman-Smith and B. Müller, Phys. Rev. D **89**, 025019 (2014).
[35] H. Kouno, M. Maruyama, F. Takagi and K. Saito, Phys. Rev. D **41**, 2903 (1990).
[36] G. Torrieri and I. Mishustin, Phys. Rev. C **78**, 021901 (2008).
[37] J. Gerhard, V. Lindenstruth and M. Bleicher, Comput. Phys. Commun. **184**, 311 (2013).

Wave function for two-neutron halo states

Lucas A. Souza[1†], **Filipe F. Bellotti**[1], **Tobias Frederico**[1]

[1]Departamento de Física, Instituto Tecnológico de Aeronáutica, DCTA, 12228-900, São José dos Campos, SP, Brazil

E-mail: [†]lucasufsj@gmail.com

Abstract. In this work we investigate the theoretical aspects associated with the few-body universal properties of the weakly-bound two neutron halo in light nuclei. We emphasize that our study of wave functions can be used as inputs in calculations of reaction processes with light exotic nuclei. Starting from the Faddeev decomposition of the wave function for 3-body systems interacting through zero range forces, we calculate two-neutron halo states for large two-body scattering lengths coming closer to what is know today as Efimov's Physics. We report our findings for the wave function for a three-body $n - n - C$ system, where we consider two neutrons as a halo around a compact core C. We focus our study on halo structure of Borromean systems like ^{11}Li, ^{14}Be, and ^{22}C.

1. Introduction

Exotic neutron-rich light nuclei can be explored theoretically by considering few-body degrees of freedom. It forms a few-body system when it has a core nucleus surrounded by a halo of orbiting neutrons in large excess which are called halo nuclei or exotic nucleus. The halo is defined by neutrons which have a high probability to be found far away from the core. The core can be described by a more stable isotope of the element, which is surrounded by the halo neutrons. The small separation energy of the halo neutrons allows them to be distributed over a spatial region extending beyond the region occupied by the core, building a diffuse cloud that contains the neutrons, forming a halo. The concept of a short-range interaction between the particles and its implications are useful in understanding the quantum few-body physics of the halo [1]. The curious large size of the halo has motivated several studies, both experimental [2, 3] and theoretical [4], devoted to clarify the new aspects of the structure, stability and reaction of these nuclei and also the astrophysical implications.

In general, these light three-body halo nuclei systems have unusual properties in respect to the nuclear size: the radius of the neutron halo is much greater than the radius of the core, and it is assumed that the core has no structure [5, 6, 7]. We discuss here a situation which allows the use of concepts coming from short-range interactions. The two-neutron halo bound state has its properties dominated by the Efimov's effect [8, 9, 10], which is associated with the Thomas collapse [11] that corresponds to infinite value of three body binding energy in the limit of a zero-range potential. The Efimov effect [12] corresponds to the appearance in the three-boson spectrum S-wave of states with binding energy geometrically separated and condensing at zero energy, when the scattering length is let to infinity and the dimer is bound at the scattering threshold. As the scattering is large with respect to the interaction range, the limit of infinite scattering length with respect to the interaction range, is physically the same as the range

Journal of Physics: Conference Series **630** (2015) 012043 doi:10.1088/1742-6596/630/1/012043

decreasing towards zero with a finite scattering length. This relates the Thomas collapse and the Efimov effect (for more discussions see [4]).

Here we will describe a three-body halo-nuclei with a neutron-neutron-core structure (we use the notation $n-n-C$). Examples of weakly-bound three-body systems are ^{11}Li, ^{14}Be, ^{20}C, and ^{22}C, where the universal aspects and Efimov physics can be explored [13]. In figure 1 (adapted from ref. [13]), we schematically represent the large sizes of the two neutron halo in ^{11}Li and ^{22}C. The two neutron binding energy in the ^{11}Li isotope is the limit of the pairing energy at the nuclear surface. Coming closer to the neutron drip line the behavior of the separation energy of the neutrons is expected to decrease, but the interesting aspect of this nucleus is that while in ^{10}Li the neutron is unbound and escapes, the two valence neutrons are bound to ^{11}Li, an effect due to energy pairing. Thus, ^{11}Li is divided into three parts, core plus two neutrons. Cores of this type, like ^{9}Li which does not capture an extra neutron, but binds two are called Borromean. The heraldic symbol of the Renaissance Borromeo family represented by the three intertwined rings, such that if one ring is removed the others are too. Another important consequence of the large neutron halo is that it can be easily excited in a vibration mode against the core known as pigmy resonances (see e. g.[14]).

These isotopes along the neutron and proton driplines are important for understanding the limits of the nuclear stability. The study of nuclear structure and reactions of unstable weakly bound halo nuclei presents a challenge to standard methods in nuclear physics as the degrees of freedom of the neutron or proton halo should be taken into account explicitly.

In our work we model a class of halo nuclei described as a three-body system, composed by two neutrons and a core, where the pairwise interactions is modelled by Dirac $\delta-$potentials. The model is illustrated by computing a hypothetical highly excited Efimov state of the neutron halo. The present study brings to the nuclear context the previous work of Yamashita and co-workers [15] performed for cold atoms. The use of the contact interaction to study halo physics is justified by the large size of the halo compared with the range of the interaction. This simple model retains the essential physics of the weakly bound and large two-neutron halo systems. Furthermore, such structures are close to the Efimov limit and as we have discussed are associated to the collapse of the three-body state in the limit of a zero-range interaction. Due to the Thomas collapse, the low-energy properties of the three-body system are well defined if one three-body short range information is given. In addition the low-energy two-body physical informations should be supplied to the system. The ultraviolet divergences from the short range singularity coming from the contact potential can be regularised and renormalized [16].

2. The Wave Function

In this section we work out the wave function of the two-neutron halo, which is our goal . We use the notation n for neutron and C for core, but our approach is applicable to any three-particle system, which interacts via s-wave short-range interactions, where two of the particles are identical. The neutrons are supposed to be in a spin singlet state.

2.1. The Formalism

In a three-body system there are three two-body subsystems and pairwise potentials. We start by writing the Schroedinger equation for the three-body system as,

$$(E - H_0)|\psi\rangle = V|\psi\rangle, \tag{1}$$

where ψ is the bound state wave function. The potential V is the sum of three-body potential and H_0 is the free hamiltonian, which should be written in terms of the Jacobi relative momenta. For a zero-range potential the three-body wave function for a system composed by two neutrons

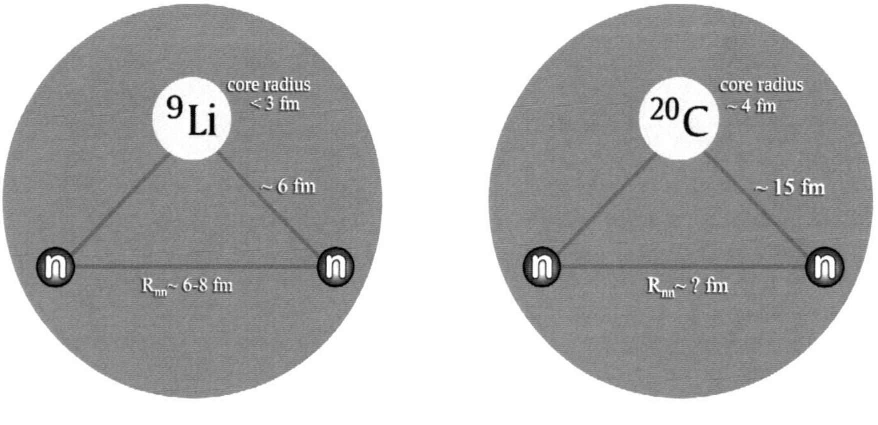

Figure 1. The sizes of the Borromean nuclei ^{11}Li and ^{22}C. Adapted from [13].

and a core, can be written in terms of the spectator functions in the basis $|\vec{q}_c\vec{p}_c\rangle$, where q_c is the relative momentum of the center of mass of the neutron in respect to the core, and p_c is the relative momentum of the two neutrons. The two-neutron halo wave function is:

$$\langle \vec{q}_c\vec{p}_c|\Psi\rangle = \frac{f_{nn}(q_c) + f_{nc}(|\vec{p}_c - \frac{\vec{q}_c}{2}|) + f_{nc}(|\vec{p}_c + \frac{\vec{q}_c}{2}|)}{E_3 + H_0} \tag{2}$$

where the functions f_{nn} and f_{nc} are the spectator functions computed in the next section. The free hamiltonian is

$$H_0 = \frac{p_c^2}{2\mu_{nn}} + \frac{q_c^2}{2\mu_{nn,c}} \tag{3}$$

and the reduced masses are given by,

$$\mu_{nn,c} = \frac{2m_c m_n}{(2m_n + m_c)}; \quad \mu_{nc,n} = \frac{m_n(m_n + m_c)}{(2m_n + m_c)}; \quad \mu_{nn} = \frac{m_n}{2}; \quad \mu_{nc} = \frac{m_c m_n}{(m_c + m_n)}.$$

We can treat the $n - n - C$ wave function (2) in configuration space, and thus can get spatial distribution of the halo neutrons. As a matter of fact, the neutrons in the halo have a large probability to be found in the classically forbidden region where the wave-function is an eigenstate of the free Hamiltonian (1), and therefore the configuration space representation of the wave function (2) is written as,

$$\Psi(\vec{r}_n, \vec{r}_{n'}) = \int d\vec{q}\,\frac{e^{-\kappa_{nn}|\vec{R}_{nn}|}}{|\vec{R}_{nn}|} e^{i\vec{q}.\vec{R}_c} f_{nn}(\vec{q}) + \int d\vec{q}\,\frac{e^{-\kappa_{nc}|\vec{R}_{nc}|}}{|\vec{R}_{nc}|} e^{i\vec{q}.\vec{R}_n} f_{nc}(\vec{q}) + \ldots, \tag{4}$$

where $\kappa_{nn} = \sqrt{2\mu_{nn}\left(S_{2n} + \frac{q^2}{2\mu_{nn,c}}\right)}$, $\kappa_{nc} = \sqrt{2\mu_{nc}\left(S_{2n} + \frac{q^2}{2\mu_{nc,n}}\right)}$ and $\vec{r}_n$, $\vec{r}_{n'}$ are neutron positions with respect to the center of mass as depicted in figure 2. (S_{2n}) is two-neutron separation energy of ^{11}Li, which in the nuclear scale is fairly small with a value of S_{2n} = 369.15(65) keV. There are another contribution to the wave function obtained from the last term by exchanging n with n' that is represented by dots in equation (4). $|\vec{R}_{nn}|$ and $|\vec{R}_{nC}|$ are relative distances for the $n - n$ and $n - core$ respectively. The relative position of the neutron to the centre of mass of the $n - C$ system is R_n. The relative positions are illustrated in fig.2.

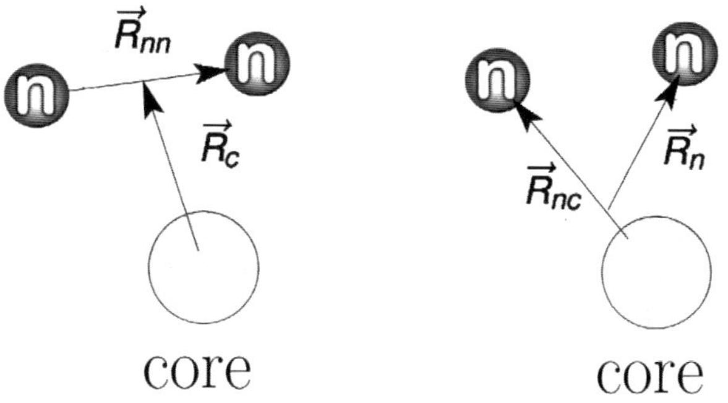

Figure 2. Relative positions of neutrons and core to the centre of mass of the subsystems $n-n$ and $n-C$.

2.2. Subtracted Faddeev Equations - Excited Bound States

The subtracted s-wave coupled integral equations for the spectator functions, f of the $n-n-C$ for the zero-range interaction, generalizes the Skorniakov and Ter-Martirosyan [18], making them regularised, and for the case of ^{11}Li with two neutrons and a core, are written as (see e.g. [4]):

$$f_{nn}(q) = 2\tau_{nn}(q,E_3)\int_0^\infty dq'\frac{q'}{q}G_1(q,q',E_3)f_{nc}(q'), \tag{5}$$

$$f_{nc}(q) = \tau_{nc}(q,E_3)\int_0^\infty dq'\left[\frac{q'}{q}G_1(q,q',E_3)f_{nn}(q') + \mathcal{A}G_2(q,q',E_3)f_{nc}(q')\right], \tag{6}$$

where the absolute value of the three-body binding energy is E_3 and,

$$\tau_{nn}(q,E_3) = \frac{1}{\pi}\left[\sqrt{E_3 + \frac{\mathcal{A}+2}{4\mathcal{A}}q^2} \pm \sqrt{E_{nc}}\right]^{-1} \tag{7}$$

$$\tau_{nc}(q,E_3) = \frac{1}{\pi}\left(\frac{\mathcal{A}+1}{2\mathcal{A}}\right)^{3/2}\left[\sqrt{E_3 + \frac{\mathcal{A}+2}{4\mathcal{A}}q^2} \pm \sqrt{E_{nc}}\right]^{-1}, \tag{8}$$

$$G_1(q,q',E_3) = \log\frac{2\mathcal{A}(E_3+q'^2+q'q)+q^2(\mathcal{A}+1)}{2\mathcal{A}(E_3+q'^2-q'q)+q^2(\mathcal{A}+1)}, \tag{9}$$

$$G_2(q,q',E_3) = \log\frac{2\mathcal{A}(E_3+q'^2+q'q)+q^2(\mathcal{A}+1)}{2\mathcal{A}(E_3+q^2-q'q)+q^2(\mathcal{A}+1)} - \log\frac{2\mathcal{A}(q+q'^2+q'q)+q^2(\mathcal{A}+1)}{2\mathcal{A}(q+q'^2-q'q)+q^2(\mathcal{A}+1)}, \tag{10}$$

we assume $\hbar = 1$ and $\mathcal{A} = m_c/m_n$ defines the mass ratio and m_n is the neutron mass. For the subsystems $n-n$ and $n-C$ we have the two-body binding energies as E_{nn} and E_{nc}.

The numerical solution of the set of subtracted coupled equations (5) for the spectator function $f_{nn}(q)$ of the $n-n-C$ system in momentum space for a highly excited state is compared to the respective asymptotic formula in fig.3. The asymptotic form for large momentum is given by

$$f_{nn}(q)|_{q\to\infty} \to C_{nn}q^2\sin(s\,\log(q/q_0)). \tag{11}$$

For more details about this form, as well as how to calculate parameter s, see reference [15]. The spectator function from the numerical solutions and the asymptotic one should coincide in the limit of an infinitely high excited state. But in practice both should be compared in the

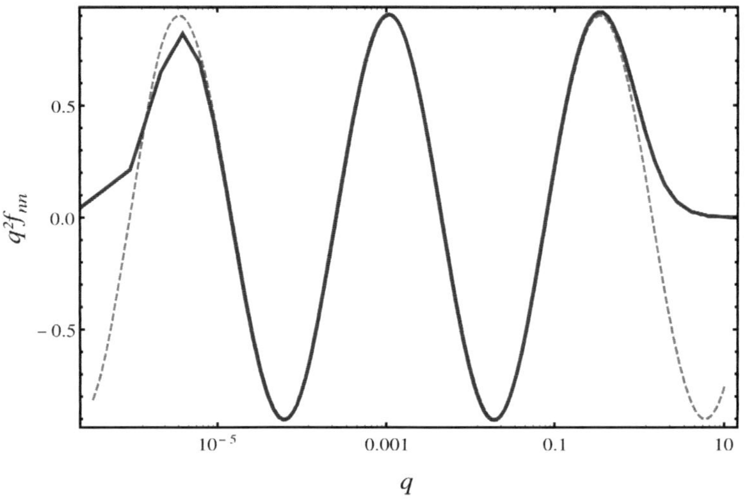

Figure 3. Spectator function $f_{nn}(q)$ of a highly excited state for subsystem energies $E_{nn} = E_{nC} = 0$ and A=9, with $E_3 = 1.478 \times 10^{-12} E_0$, solution of the coupled equations (5) (blue line), compared with the asymptotic form (11) (red line). We have introduced $E_0 = \hbar^2\mu^2/m_n$ and we work in units where $\hbar = m_n = \mu = 1$.

window $k_0 << q << \mu$, where $k_0 = \sqrt{E_3}$, where the momentum is larger then the characteristic momentum of the state, K_0, and smaller than the subtraction scale μ. We use q to denote momentum and the boundary condition $f(q_0) = 0$ is chosen to match the numerical solution of couple subtracted equation, as we see in fig. 3.

3. Conclusions and Outlook

Weakly bound systems were addressed in the context of unstable neutron-rich nuclei. In the regime of a zero-range interaction, few scales determine the low-energy properties of two-neutron halo nuclei. The low-energy properties of such systems are dominated by the Efimov physics, which is universal and model independent. The integral equations for the bound state of three-body systems are easily solved numerically. After the regularization and renormalization of the model is done by fixing the three-body binding energy, by a convenient choice of the subtraction scale. We study one example of a highly excited Efimov state and the spectator function compared to the asymptotic form. The next step is to obtain the full wave function of the s-wave two neutron halo state in configuration space and study the one and two-body neutron densities.

4. Acknowledgments

Our thanks to the Brazilian agencies CAPES, FAPESP and CNPq for partial financial support.

5. References

[1] E. Nielsen, D.V. Fedorov, A.S. Jensen and E. Garrido,*Phys. Rep.* **347** (2001) 373.
[2] I. Tanihata, H. Savajols, R. Kanungo, *Prog. Part. Nucl. Phys.* **68**, 215-313 (2013).
[3] Lu, Z.-T., Mueller, P., Drake, G.W.F., Nörtershäuser Pieper, S.C., Yan, Z.-C.: Colloquium; *Rev. Mod. Phys.* **85**, 1383 (2013).

[4] T. Frederico, M. T. Yamashita, A. Delfino, L. Tomio, *Prog. Part. Nucl. Phys.* **67**, 939-994 (2012).
[5] A.E.A. Amorim, L. Tomio and T. Frederico, *Phys. Rev. C* **56** (1997) R2378.
[6] D.V. Fedorov and A.S. Jensen, *Phys. Rev. Lett.* **25**, 4103 (1993).
[7] D.V. Fedorov, E. Garrido and A.S. Jensen, *Phys. Rev. C* **51**, 3052 (1995).
[8] I. Tanihata, *J.Phys.* G**22**, 157 (1996).
[9] C.A.Bertulani, L.F. Canto and M.S.Hussein, *Phys. Rep.* *226*, 281 (1993).
[10] P.G. Hansen, A.S. Jensen, and B. Jonson, *Annu. Rev. Nucl. Part. Sci.* **45**, 591 (1995); M.V. Zhukov, B.V. Danilin, D.V.Fedorov, J.M. Bang, I.S. Thompson, J.S. Vaagen, *Phys. Rep.* **231**, 151 (1993).
[11] L .H. Thomas, *Phys. Rev.* **47**, 903 (1935).
[12] V. Efimov, *Phys. Lett. B* **33**, 563 (1970).
[13] T. Frederico, *Few-Body Syst* **55**, 651 (2014).
[14] M. Cubero et al., *Phys. Rev. Lett.* **109**, 262701 (2012).
[15] M. T. Yamashita, F. F. Bellotti, T. Frederico, D. V. Fedorov, A. S. Jensen, and N. T. Zinner *Phys. Rev. A* **87**, 062702 (2013).
[16] S. K. Adhikari, T.Frederico and I.D. Goldman, *Phys. Rev. Lett.* **74**, 487 (1995); S .K. Adhikari and T. Frederico *Phys. Rev. Lett.* **74**, 4572 (1995).
[17] S. K. Adhikari, A. Delfino, T. Frederico, I. D. Goldman and L. Tomio, *Phys. Rev. A* **37**, 3666 (1988).
[18] G. Skorniakov and K. A. Ter-Martirosian, *Sov. Phys. JETP* **4**, 648 (1957).

Mixed Charmonium-Molecule Interpretation for the $Y(3940)$ State

R.M. Albuquerque[1], **J.M. Dias**[2], **M. Nielsen**[2] **and C.M. Zanetti**[3]
[1]Institute for Theoretical Physics, São Paulo State University (IFT-UNESP)
R. Dr. Bento Teobaldo Ferraz, 271 - Bl. II Sala 207, 01140-070 São Paulo/SP - Brasil
[2]Instituto de Física, Universidade de São Paulo,
Caixa Postal 66318, 05389-970 São Paulo, SP, Brazil
[3] Faculdade de Tecnologia, Universidade do Estado do Rio de Janeiro
Rod. Presidente Dutra Km 298, Pólo Industrial, 27537-000 , Resende, RJ, Brasil

E-mail: raphael@ift.unesp.br, jdias@if.usp.br, mnielsen@if.usp.br, carina.zanetti@gmail.com

Abstract. QCD sum rules are used to study the mass and the decay widths of the $Y(3940)$ state. We assume that it can be described by a mixed charmonium-molecule scalar current with $J^{PC} = 0^{++}$. Using a mixing angle $\theta = (76.0 \pm 5.0)^\circ$, we obtain $M_Y = (3.95 \pm 0.11)$ GeV, which is in good agreement with the experimental mass of the $Y(3940)$ state. We also evaluate the decay width in the channels $Y \to J/\psi\,\omega$ and $Y \to \gamma\gamma$ obtaining the values $\Gamma_{Y\to J/\psi\,\omega} \approx (1.7 \pm 0.6)$ MeV and $\Gamma_{Y\to\gamma\gamma} \approx (1.6 \pm 1.3)$ KeV, respectively. We also study the decay process of this state into channels containing $D\bar{D}$ mesons in the final state.

1. Introduction

Most of the new charmonium states discovered in recent years at the B factories, called X, Y, Z particles, do not seem to have a simple $c\bar{c}$ structure. Their production mechanism, masses, decay widths, spin-parity assignments and decay modes have been discussed in some reviews [1, 2, 3]. Although the masses of these states are above the corresponding thresholds of decays into a pair of open charm mesons, they decay into J/ψ or ψ' plus pions, which is unusual for $c\bar{c}$ states. Besides, their masses and decay modes are not in agreement with the predictions of potential models, which, in general, describe very well $c\bar{c}$ states. For these reasons, they are considered as candidates for exotic states such as hybrid, molecular or tetraquark states, with a more complex structure than the simple quark-antiquark states.

In particular, several states in the region of mass of about 3940 MeV has been observed in different processes of production and decay. The $Y(3940)$, observed by BELLE Collaboration in the decay $B \to (J/\psi\,\omega)K$ [4], was the first of these states. It has been also observed by BABAR Collaboration, with a slightly smaller mass and decay width[5]. BELLE Collaboration has also reported the observation of a state in the process $\gamma\gamma \to D\bar{D}$, which was called $Z(3930)$ [6, 7]. However, this state is generally linked to the charmonium state $\chi_{c2}(2P)$. Following these observations, the state $X(3915)$ was reported by BELLE Collaboration in the process $\gamma\gamma \to J/\psi\,\omega$ [8]. The mass and total decay width of these states are shown in in Table 1.

Although all these states could be connected to the same particle observed in different processes, there are evidences that, at least, two of these reported states, $Y(3940)$ and $X(3915)$,

Table 1. Mass and decay width of the states in the mass region 3940 GeV

state	mass (MeV)	total decay width (MeV)	ref.
$Y(3940)$	$3943 \pm 11 \pm 13$	$87 \pm 22 \pm 26$	[4]
$Y(3940)$	$3914.6^{+3.9}_{-3.4} \pm 2.0$	$34^{+12}_{-8} \pm 5.0$	[5]
$Z(3930)$	$3929 \pm 5 \pm 2$	$29 \pm 10 \pm 2$	[6, 7]
$X(3915)$	$3915 \pm 3 \pm 2$	$17 \pm 10 \pm 3$	[8]

could be interpreted as molecular states. In Ref. [1] these two states were interpreted as the same state and it was called $X(3915)$. However, the Particle Data Group [9] associates the label $X(3915)$ to the charmonium state $\chi_{c0}(2P)$. Therefore, to avoid misinterpretation, here we use the label $Y(3940)$ to identify the state observed in the decay mode $J/\psi\,\omega$.

In Ref. [10], it was proposed that the $Y(3940)$ could be a molecular state $D^*\bar{D}^*$, with quantum numbers $J^{PC} = 0^{++}$ or 2^{++}. In Ref. [11], the Y state was studied with QCD Sum Rules (QCDSR) method [12, 13, 14] as a $D^*\bar{D}^*$ molecule with quantum numbers 0^{++} and the mass obtained was $m_{D^*\bar{D}^*} = (4.13 \pm 0.10)$ MeV, failing to reproduce the experimental mass of the state. Here we study the $Y(3940)$, within QCDSR approach, using a mixed charmonium-molecule current, following the works in Refs. [15, 16, 17, 18] where the states $X(3872)$ and $Y(4260)$ were considered as molecule-charmonium states.

2. Mixed Hadronic Current

We use the QCDSR approach to describe the $Y(3940)$ as a mixing between the χ_{c0} charmonium and the $D^*\bar{D}^*$ molecule, with $J^{PC} = 0^{++}$. We use the following hadronic current

$$j \;=\; a\cos\theta\; j_{\chi_{c0}} \;+\; \sin\theta\; j_{D^*D^*} \tag{1}$$

where θ is an arbitrary mixing angle and the meson and molecule currents are given by:

$$j_{\chi_{c0}} = \bar{c}_k c_k, \quad j_{D^*D^*} = (\bar{q}_i\gamma_\mu c_i)\,(\bar{c}_j\gamma^\mu q_j) \quad . \tag{2}$$

The factor a is introduced in Eq. (2) in such way that the mixed current can be evaluated at the same Fock space. We use [15, 16, 17]

$$a = -\frac{\langle\bar{q}q\rangle}{\sqrt{2}} \; . \tag{3}$$

2.1. Two-Point Correlation Function

The mass of a hadronic state is obtained, in the QCDSR approach, by considering the two-point correlation function

$$\Pi(q) \;=\; i\int d^4x\; e^{iq\cdot x}\langle 0|\; T[j(x)j^\dagger(0)]\;|0\rangle \tag{4}$$

In the QCDSR approach, Eq. (4) is evaluated in two ways: the phenomenological side and the QCD side. In the QCD side, one uses the Wilson's operator product expansion (OPE), so the QCD side is also called the OPE side. Inserting Eq. (1) into the above equation, one obtains:

$$\Pi^{OPE}(q) = i\int d^4x\, e^{iq\cdot x}\left\{\frac{1}{2}\langle\bar{q}q\rangle^2\cos^2\theta\;\Pi_{\chi_{c0}} + \sin^2\theta\;\Pi_{D^*D^*} - \frac{\langle\bar{q}q\rangle}{\sqrt{2}}\sin\theta\cos\theta\left[\Pi_{mix} + \Pi^*_{mix}\right]\right\} \tag{5}$$

where the $\Pi_{\chi_{c0}}(x)$ and $\Pi_{D^*D^*}(x)$ functions are, respectively, the correlation functions of the χ_{c0} meson and the D^*D^* (0^{++}) molecular state, which have been calculated in other works [13, 19]. Thus, one only has to calculate the $\Pi_{mix}(x)$ and $\Pi^*_{mix}(x)$ functions defined as follows:

$$\Pi_{mix}(x) = \langle 0|\, T[j_{\chi_{c0}}(x) j^\dagger_{D^*D^*}(0)]\, |0\rangle = -\mathrm{Tr}\Big[S^q_{ji}(0)\, \gamma_\mu\, S^c_{ik}(-x) S^c_{kj}(x)\, \gamma^\mu\Big] \quad (6)$$

$$\Pi^*_{mix}(x) = \langle 0|\, T[j_{D^*D^*}(x) j^\dagger_{\chi_{c0}}(0)]\, |0\rangle = -\mathrm{Tr}\Big[S^q_{ji}(0)\, \gamma_\mu\, S^c_{ik}(x) S^c_{kj}(-x)\, \gamma^\mu\Big] \quad (7)$$

where $S^c(x)$ and $S^q(x)$ are the charm- and light-quark propagators, respectively. The correlation function can be written in terms of a dispersion relation, such that

$$\Pi^{OPE}(q^2) = \int_{4m_c^2}^{\infty} ds\, \frac{\rho^{OPE}(s)}{s - q^2}, \quad (8)$$

where $\rho^{OPE}(s)$ is given by the imaginary part of the correlation function: $\pi\ \rho^{OPE}(s) = \mathrm{Im}[\,\Pi^{OPE}(q^2 = s)\,]$. According to Eq. (5), the expression for the spectral density is:

$$\rho^{OPE}(s) = \frac{1}{2}\langle\bar{q}q\rangle^2 \cos^2\theta\ \rho_{\chi_{c0}}(s) + \sin^2\theta\ \rho_{D^*D^*}(s) - \frac{\langle\bar{q}q\rangle}{\sqrt{2}} \sin\theta\cos\theta\ \rho_{mix}(s)\ . \quad (9)$$

We calculate the sum rule at leading order in α_s considering the contributions from the condensates up to dimension-8 in the OPE. The expressions for the spectral density are given in ref. [20].

To evaluate the phenomenological side we insert, in Eq. (4), a complete set of intermediate states, Y, which couple to the hadronic current in Eq. (1). We get

$$\Pi^{PHEN}(q) = \frac{\lambda_Y^2}{M_Y^2 - q^2} + \int_0^{\infty} ds\, \frac{\rho^{cont}(s)}{s - q^2}, \quad (10)$$

where M_Y is the ground state mass, λ_Y, gives the coupling between the state and the current:

$$\langle 0|\, j\, |Y\rangle = \lambda_Y\ , \quad (11)$$

and the second term in the RHS of Eq. (10) denotes the continuum (or higher resonance) contributions. Assuming that the continuum contribution to the spectral density, $\rho^{cont}(s)$ in Eq. (10), coincides with the one obtained in the OPE side above a certain threshold s_0, we get [21]:

$$\rho^{cont}(s) = \rho^{OPE}(s)\, \Theta(s - s_0) \quad (12)$$

where $\Theta(s - s_0)$ is the Heaviside step function.

To improve the matching between the two sides of the sum rule, one performs the Borel transform. After transferring the continuum contributions to the OPE side and perfoming the Borel transform to the two sides of the sum rule, one gets:

$$\lambda_Y^2\, e^{-M_Y^2/M_B^2} = \int_{4m_c^2}^{s_0} ds e^{-s/M_B^2}\ \rho^{OPE}(s)\ . \quad (13)$$

The ground state mass can be obtained from the following ratio:

$$M_B^2 = \frac{\int_{4m_c^2}^{s_0} ds\, s\, e^{-s/M_B^2}\, \rho^{OPE}(s)}{\int_{4m_c^2}^{s_0} ds\, e^{-s/M_B^2}\, \rho^{OPE}(s)}. \tag{14}$$

2.2. Three-Point Correlation Function

To compute the form factor associated with the vertex $Y\, J/\psi\, \omega$ and to estimate the width of the channel $Y(3940) \to J/\psi\, \omega$ we consider the three-point function:

$$\Pi_{\mu\nu}(p, p', q) = \int d^4x\, d^4y\, e^{ip'\cdot x}\, e^{iq\cdot y}\, \Pi_{\mu\nu}(x, y), \tag{15}$$

where $p = p' + q$ and $\Pi_{\mu\nu}(x, y)$ is given by

$$\Pi_{\mu\nu}(x, y) = \langle 0|T\{j_\mu^\psi(x) j_\nu^\omega(y) j^\dagger(0)\}|0\rangle. \tag{16}$$

The interpolating currents for the J/ψ and ω mesons are given by:

$$j_\mu^\psi = \left(\bar{c}_a \gamma_\mu c_a\right), \quad j_\nu^\omega = \frac{1}{6}\left(\bar{u}_a \gamma_\nu u_a + \bar{d}_a \gamma_\nu d_a\right), \tag{17}$$

and the mixed $(\chi_{c0})-(D^*\bar{D}^*)$ state is described by the current in Eq. (1).

In the OPE side, we calculate the correlation function at leading order in α_s and we consider condensates up to dimension 7. The three-point function includes a number of different Lorentz structures and we choose to work with the $q_\mu p'_\nu$ structure [20]. In general, for any given structure, the sum rule method is inapplicable at large Q^2 where the power corrections become large and uncontrollable. At small Q^2, the situation is even worse since when approaching the physical region the operator expansion stops working. In this sense, one has to consider that the sum rule is valid up to a rather small Q^2 and the extrapolation from the values of Q^2 to the physical region can be obtained with a good accuracy.

In the phenomenological side we consider intermediate states for the J/ψ, ω and $Y(3940)$ mesons. We get:

$$\Pi_{\mu\nu}^{PHEN}(p, p', q) = \frac{\lambda_Y\, M_\psi f_\psi\, M_\omega f_\omega\, g_{Y\psi\omega}(q^2)}{(p^2 - M_Y^2)(p'^2 - M_\psi^2)(q^2 - M_\omega^2)} \times \left[q_\mu p'_\nu - (p' \cdot q) g_{\mu\nu}\right] + \cdots, \tag{18}$$

where we have used the following relations:

$$\langle 0|\, j_\mu^\psi\, |J/\psi(p')\rangle = M_\psi f_\psi\, \epsilon_\mu(p'), \quad \langle 0|\, j_\nu^\omega\, |\omega(q)\rangle = M_\omega f_\omega\, \epsilon_\nu(q), \quad \langle Y(p)|\, j\, |0\rangle = \lambda_Y.$$

The dots in Eq. (18) stand for the contribution of all possible excited states.

The form factor, $g_{Y\psi\omega}(q^2)$, is defined by the generalization of the on-shell mass matrix element, $\langle J/\psi\, \omega\, |\, Y\rangle$, for an off-shell ω meson:

$$\langle J/\psi\, \omega\, |\, Y\rangle = g_{Y\psi\omega}(q^2)\left[\left(p' \cdot \epsilon^*(q)\right)\left(q \cdot \epsilon^*(p')\right) - (p' \cdot q)\left(\epsilon^*(p') \cdot \epsilon^*(q)\right)\right], \tag{19}$$

which can be extracted from the effective Lagrangian that describes the coupling between two vector mesons and one scalar meson:

$$\mathcal{L} = \frac{i}{2}\, g_{Y\psi\omega} V_{\alpha\beta} \Psi^{\alpha\beta}\, Y \tag{20}$$

where $V_{\alpha\beta} = \partial_\alpha\omega_\beta - \partial_\beta\omega_\alpha$ and $\Psi^{\alpha\beta} = \partial^\alpha\psi^\beta - \partial^\beta\psi^\alpha$, are the tensor fields of the ω and ψ fields respectively.

Matching both side of the sum rule, taking the approximation $p^2 \simeq p'^2 = -P^2$ and doing the Borel transform to $P^2 \to M_B^2$, we get the following expression in the $q_\mu p'_\nu$ structure:

$$\frac{\lambda_Y M_\omega f_\omega M_\psi f_\psi \; g_{Y\psi\omega}(Q^2)}{(M_Y^2 - M_\psi^2)(Q^2 + M_\omega^2)} \left(e^{-M_Y^2/M_B^2} - e^{-M_\psi^2/M_B^2}\right) + H(Q^2)\, e^{-s_0/M_B^2} = \Pi^{OPE}(M_B^2, Q^2), \quad (21)$$

where $Q^2 = -q^2$, and the function $H(Q^2)$ represents the contribution of the pole-continuum transitions [15, 22, 23, 24]. The $\Pi^{OPE}(M_B^2, Q^2)$ function is

$$\Pi^{OPE}(M_B^2, Q^2) = \sin\theta \int_{4m_c^2}^{+\infty} ds\; e^{-s/M_B^2}\, \rho(s, Q^2)\ , \quad (22)$$

where $\rho = \rho^{pert} + \rho^{\langle\bar{q}q\rangle} + \rho^{\langle G^2\rangle} + \rho^{\langle\bar{q}Gq\rangle} + \rho^{\langle\bar{q}q\rangle\langle G^2\rangle}$ is given in ref. [20].

As in previous calculations [15, 16, 17, 18], only the molecule part of the mixed current defined in Eq. (1) contributes to the decay channel $Y(3940) \to J/\psi\,\omega$. This can be seen by the presence of the sine function in Eq. (22).

To extract the value of the coupling constant associated with the $Y \to J/\psi\,\omega$ vertex we first determine the form factor $g_{Y\psi\omega}(Q^2)$ in Eq. (21). To do that we follow the procedure described in [20].

3. Numerical Analysis

The numerical values for the quark masses and condensates are listed in Table 2.

Table 2. QCD input parameters.

Parameters	Values
$\overline{m}_c$	$(1.23 - 1.47)$ GeV
$\langle\bar{q}q\rangle$	$-(0.23 \pm 0.03)^3$ GeV^3
$\langle g_s^2 G^2\rangle$	(0.88 ± 0.25) GeV^4
$\langle g_s^3 G^3\rangle$	(0.58 ± 0.18) GeV^6
$m_0^2 \equiv \langle\bar{q}Gq\rangle/\langle\bar{q}q\rangle$	(0.8 ± 0.1) GeV^2
$\rho \equiv \langle\bar{q}q\bar{q}q\rangle/\langle\bar{q}q\rangle^2$	$(0.5 - 2.0)$

We start our analysis discussing the possible values of the continuum threshold s_0 and the mixing angle θ. A reasonable initial value for the continuum threshold would be $\sqrt{s_0} \sim M_Y + 0.5\,\text{GeV} \sim 4.40\,\text{GeV}$, since we are interested in a state with a mass $M_Y \sim 3.9\,\text{GeV}$. To fix the value of the mixing angle θ, we search for a value which allows us to determine the best M_B^2-stability inside of a valid Borel window. We find that the optimal choice is

$$\sqrt{s_0} = (4.40 \pm 0.10)\,\text{GeV} \quad (23)$$
$$\theta = (76.0 \pm 5.0)^\circ\ . \quad (24)$$

We notice that the OPE does not converge for θ values outside this range.

Using these values, we analyze the relative contributions of the terms in the OPE. As one can see in Fig. 1, the contribution of the dimension-8 condensate is smaller than 20% of the total contribution for values of $M_B^2 \geq 2.40\,\text{GeV}^2$, which indicates the starting point for a good OPE

Journal of Physics: Conference Series **630** (2015) 012044 doi:10.1088/1742-6596/630/1/012044

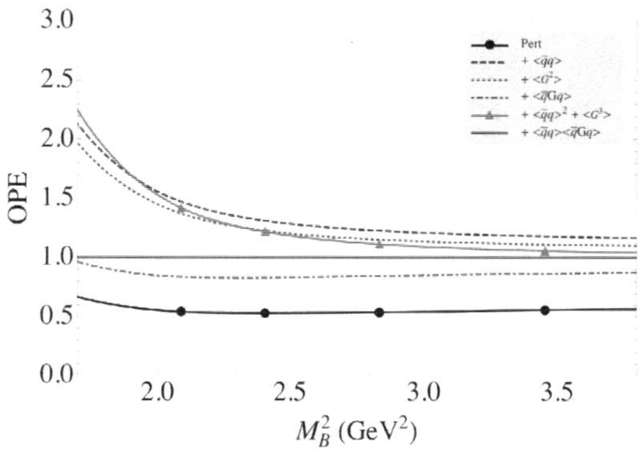

Figure 1. OPE convergence in the region $1.7 \leq M_B^2 \leq 3.8$ GeV^2 for $\sqrt{s_0} = 4.40\,\text{GeV}$ and $\theta = 76.0°$. One plots the relative contributions starting with the perturbative contribution (line with circles), and each other line represents the relative contribution after adding of one extra condensate in the expansion: + $\langle \bar{q}q \rangle$ (dashed line), + $\langle G^2 \rangle$ (dotted line), + $\langle \bar{q}Gq \rangle$ (dot-dashed line), + $\langle \bar{q}q \rangle^2$ + $\langle G^3 \rangle$ (line with triangles) and $\langle \bar{q}q \rangle \cdot \langle \bar{q}Gq \rangle$ (solid line). Extracted from ref. [20]

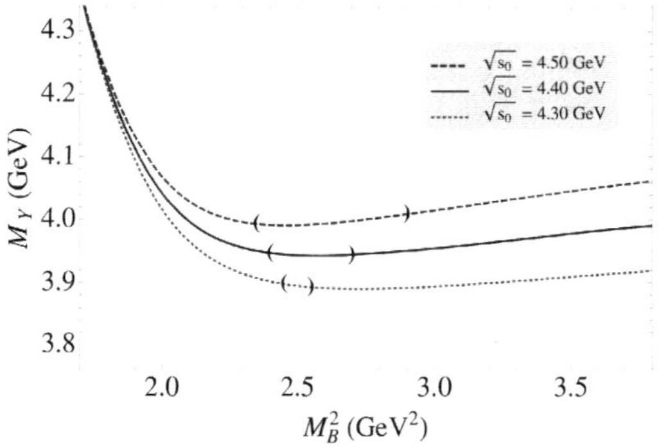

Figure 2. The mass as a function of the sum rule parameter M_B^2 for $\sqrt{s_0} = 4.30\,\text{GeV}$ (dotted line), $\sqrt{s_0} = 4.40\,\text{GeV}$ (solid line) and $\sqrt{s_0} = 4.50\,\text{GeV}$ (dashed line). The respective parentheses indicate the valid Borel window. Extracted from ref. [20]

convergence. We fix the maximum value of the Borel mass parameter as the value for which the pole contribution is greater than or equal to the continuum contribution. This condition is satisfied when $M_B^2 \leq 2.70\,\text{GeV}^2$ [20]. Therefore, the Borel window is set as $2.40 \leq M_B^2 \leq 2.70$ GeV^2.

The ground state mass, as a function of M_B^2, is shown in Fig. 2, for three different values of $\sqrt{s_0}$. We conclude that there is a good M_B^2-stability in the determined Borel window.

Varying the value of the continuum threshold in the range $\sqrt{s_0} = (4.40 \pm 0.10)\,\text{GeV}$, the mixing angle in the range $\theta = (76.0 \pm 5.0)°$, and the other parameters as indicated in Table 2, we get:

$$M_Y = (3.95 \pm 0.11)\ \text{GeV}\ . \tag{25}$$

This mass is compatible with the experimental mass of the $Y(3940)$ state observed by BELLE Collaboration [4]. Therefore, from a QCD sum rule point of view, a mixed scalar (χ_{c0})–$(D^*\bar{D}^*)$ state could be a good candidate to explain the $Y(3940)$ state.

The coupling parameter, defined in Eq. (11), can be estimates by using the result obtained for the mass in Eq. (13). We obtain:

$$\lambda_Y = (2.1 \pm 0.6) \times 10^{-2}\ \mathrm{GeV}^5. \tag{26}$$

In the numerical analysis of the three-point function we use the experimental values of the meson masses and decay constants: $M_\psi = 3.10\,\mathrm{GeV}$, $f_\psi = 0.405$ GeV, $M_\omega = 0.782$ GeV, $f_\omega = 0.046$ GeV. For the Y mass, we use the experimental value in Ref. [4] and the meson-current parameter λ_Y given in Eq. (26).

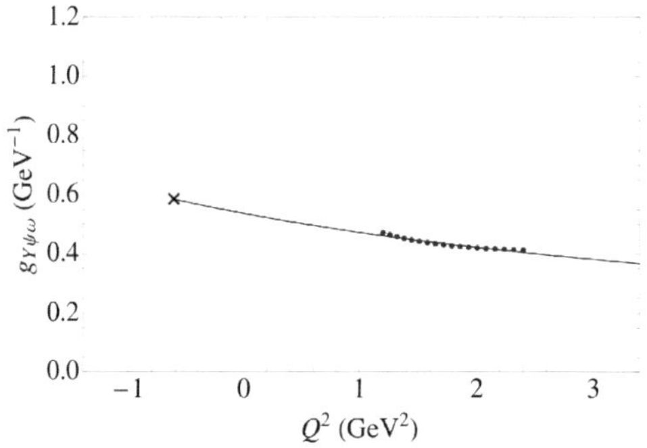

Figure 3. QCDSR results for the form factor $g_{Y\psi\omega}(Q^2)$, for $\sqrt{s_0} = 4.40$ GeV (circles). The solid line gives the parametrization of the QCDSR results through Eq. (27). The cross is the value of the coupling constant. Extraced from ref. [20]

We find that the form factor $g_{Y\psi\omega}(Q^2)$ is a very stable function of M_B^2 in Borel mass region $1.8\ \mathrm{GeV}^2 \le M_B^2 \le 4.0\ \mathrm{GeV}^2$ [20]. Therefore, the form factor dependence in Q^2 can be evaluated by taking the average of the M_B^2 values inside this stability region. The results are shown in Fig. 3.

The coupling constant is defined by the value of the form factor at the ω meson pole, $Q^2 = -M_\omega^2$. Therefore, to determine the coupling constant we have to extrapolate the form factor to the region of Q^2 where the QCDSR is not valid. This extrapolation can be done by parametrizing the QCDSR results, shown in Fig. 3 for $g_{Y\psi\omega}(Q^2)$, using a monopolar function:

$$g_{Y\psi\omega}(Q^2) = \frac{g_1}{g_2 + Q^2}, \tag{27}$$

and the results for the fitting parameters are:

$$g_1 = (4.0\ \pm\ 1.0)\ \mathrm{GeV};$$
$$g_2 = (7.4\ \pm\ 0.2)\ \mathrm{GeV}^2. \tag{28}$$

The theoretical errors are evaluated considering errors on the following parameters: $\sqrt{s_0} = 4.40 \pm 0.10$ GeV, $\theta = 76.0^\circ \pm 5.0^\circ$, and also the error on the meson coupling parameter λ_Y, given by Eq. (26). We notice that the results do not depend much on the parameters $\sqrt{s_0}$ and θ, while the theoretical errors are mainly affected by the meson coupling λ_Y.

The solid line in Fig. 3 represents Eq. (27) with values given in Eq. (28). The coupling constant, $g_{Y\psi\omega}$, indicated by the cross in Fig. 3, is obtained by using $Q^2 = -M_\omega^2$ in Eq. (27). We get:

$$g_{Y\psi\omega} = g_{Y\psi\omega}(-M_\omega^2) = (0.58\ \pm\ 0.14)\ \ \mathrm{GeV}^{-1}. \tag{29}$$

The decay width for this process $Y(3940) \to J/\psi\,\omega$ is given by

$$\Gamma_{Y(3940)\to J/\psi\,\omega} = \frac{g^2_{Y\psi\omega}}{3}\frac{p(M_Y, M_\omega, M_\psi)}{8\pi M_Y^2} \times \left(M_\psi^2 M_\omega^2 + \frac{1}{2}(M_Y^2 - M_\psi^2 - M_\omega^2)^2\right), \tag{30}$$

where

$$p(a,b,c) \equiv \frac{\sqrt{a^4+b^4+c^4-2a^2b^2-2a^2c^2-2b^2c^2}}{2a}. \tag{31}$$

Inserting the value obtained for the coupling constant (29) in (30) we get:

$$\Gamma_{Y(3940)\to J/\psi\,\omega} = (1.7\ \pm\ 0.6)\ \text{MeV}. \tag{32}$$

This result is consistent with the experimental width of the state and the lower limit for the process $Y \to J/\psi\,\omega$ [4, 5, 25, 26]. It is also of the same order as other available theoretical evaluations [27, 28].

We can use the same technique discussed above to determine the $Y(3940) \to \gamma\gamma$ decay width. This was done in ref. [20] and the result obtained was:

$$\Gamma_{Y(3940)\to\gamma\gamma} = (1.6 \pm 1.3)\,\text{KeV}. \tag{33}$$

Based on this decay width value and the one obtained for the channel $J/\psi\omega$ in Eq. (32), the product of the two partial widths of the $Y(3940)$ is given by

$$\Gamma_{\gamma\gamma} \times \Gamma_{J/\psi\,\omega} \sim \mathcal{O}(10^3)\,\text{KeV}^2\ , \tag{34}$$

which is in an excellent agreement with the measurements from BELLE and BABAR Collaborations in Refs. [8, 29].

4. Conclusions

We have used the QCDSR approach to study the two-point and three-point functions of the $Y(3940)$ state, considering it as a mixed charmonium-molecule state. We have determined the mixing angle to be $\theta = (76.0 \pm 5.0)^0$. For this mixing angle the mass obtained is in a very good agreement with the experimental value for the $Y(3940)$ state.

We have also evaluated the decay width in the channels $Y(3940) \to J/\psi\omega$ and $Y(3940) \to \gamma\gamma$. The values obtained for the width in these channels are: $\Gamma_{J/\psi\,\omega} = (1.7 \pm 0.6)$ MeV and $\Gamma_{\gamma\gamma} = (1.6 \pm 1.3)$ KeV. These results are consistent with the lower limit for the channel $Y(3940) \to J/\psi\,\omega$: $\Gamma > 1$ MeV [27, 28], and with the product $\Gamma_{\gamma\gamma} \times \Gamma_{J/\psi\,\omega} \sim \mathcal{O}(10^3)\,\text{KeV}^2$, determined by BELLE and BABAR Collaborations in Refs. [8, 29].

Thus, according to the available experimental data, we can conclude that a mixing between the χ_{c0} charmonium and the $D^*\bar{D}^*$ molecule, with $J^{PC} = 0^{++}$ quantum numbers, could be a good candidate to explain the $Y(3940)$ state.

Acknowledgments

This work has been supported by FAPESP and CNPq-Brazil.

Journal of Physics: Conference Series **630** (2015) 012044 doi:10.1088/1742-6596/630/1/012044

References

[1] N. Brambilla, *et al.*, Eur. Phys. J. C **71**, 1534 (2011) [arXiv:1010.5827].
[2] M. Nielsen, F. S. Navarra, S. H. Lee, Phys. Rept. **497**, 41-83 (2010) [arXiv:0911.1958].
[3] S. L. Olsen, Nucl. Phys. A **827**, 53C-60C (2009) [arXiv:0901.2371].
[4] K. Abe *et al.* [Belle Collaboration], Phys. Rev. Lett. **94**, 182002 (2005) [hep-ex/0408126].
[5] B. Aubert *et al.* [BaBar Collaboration], Phys. Rev. Lett. **101**, 082001 (2008) [arXiv:0711.2047].
[6] S. Uehara *et al.* [Belle Collaboration], Phys. Rev. Lett. **96**, 082003 (2006) [hep-ex/0512035].
[7] B. Aubert *et al.* [BaBar Collaboration], Phys. Rev. D **81**, 092003 (2010) [arXiv:1002.0281].
[8] S. Uehara *et al.* [Belle Collaboration], Phys. Rev. Lett. **104**, 092001 (2010) [arXiv:0912.4451].
[9] J. Beringer *et al.* (Particle Data Group), Phys. Rev. D **86**, 010001 (2012).
[10] X. Liu and S. -L. Zhu, Phys. Rev. D **80**, 017502 (2009) [Erratum-ibid. D **85**, 019902 (2012)] [arXiv:0903.2529].
[11] R. M. Albuquerque, M. E. Bracco and M. Nielsen, Phys. Lett. B **678**, 186 (2009) [arXiv:0903.5540 [hep-ph]].
[12] M.A. Shifman, A.I. and Vainshtein and V.I. Zakharov, Nucl. Phys. B **147**, 385 (1979).
[13] L.J. Reinders, H. Rubinstein and S. Yazaki, Phys. Rept. **127**, 1 (1985).
[14] For a review and references to original works, see e.g., S.Narison, *QCD as a theory of hadrons*, Cambridge Monogr. Part. Phys. Nucl. Phys. Cosmol. **17**, 1 (2002) [hep-h/0205006]; *QCD spectral sum rules*, World Sci. Lect. Notes Phys. **26**, 1 (1989); Acta Phys. Pol. **B26**, 687 (1995); Riv. Nuov. Cim. **10N2**, 1 (1987); Phys. Rept. **84**, 263 (1982).
[15] R. D. Matheus, F. S. Navarra, M. Nielsen and C. M. Zanetti, Phys. Rev. D **80**, 056002 (2009) [arXiv:0907.2683].
[16] M. Nielsen and C. M. Zanetti, Phys. Rev. D **82**, 116002 (2010) [arXiv:1006.0467].
[17] C. M. Zanetti, M. Nielsen and R. D. Matheus, Phys. Lett. B **702**, 359 (2011) [arXiv:1105.1343].
[18] J. M. Dias, R. M. Albuquerque, M. Nielsen and C. M. Zanetti, Phys. Rev. D **86**, 116012 (2012) [arXiv:1209.6592].
[19] R.M. Albuquerque, M.E. Bracco and M. Nielsen, Phys. Lett. B **678**, 186 (2009); arXiv:0903.5540.
[20] R. M. Albuquerque, J. M. Dias, M. Nielsen and C. M. Zanetti, Phys. Rev. D **89**, 076007 (2014) [arXiv:1311.6411].
[21] B.L. Ioffe, Nucl. Phys. B **188**, 317 (1981); *ibid.*, Nucl. Phys. B **191**, 591(E) (1981).
[22] B.L. Ioffe and A.V. Smilga, Nucl. Phys. **B232**, 109 (1984).
[23] F.S. Navarra, M. Nielsen, Phys. Lett. B **639**, 272 (2006); [arXiv:hep-ph/0605038].
[24] M. Nielsen, Phys. Lett. B **634**, 35 (2006); [arXiv:hep-ph/0510277].
[25] S. Godfrey and S. L. Olsen, Ann. Rev. Nucl. Part. Sci. **58**, 51 (2008) [arXiv:0801.3867].
[26] E. Eichten, S. Godfrey, H. Mahlke and J. L. Rosner, Rev. Mod. Phys. **80** (2008) 1161 [hep-ph/0701208].
[27] T. Branz, R. Molina and E. Oset, Phys. Rev. D **83**, 114015 (2011) [arXiv:1010.0587].
[28] T. Branz, T. Gutsche and V. E. Lyubovitskij, Phys. Rev. D **80**, 054019 (2009) [arXiv:0903.5424].
[29] J. P. Lees *et al.* [BaBar Collaboration], Phys. Rev. D **86**, 072002 (2012) [arXiv:1207.2651].

Journal of Physics: Conference Series **630** (2015) 012045 doi:10.1088/1742-6596/630/1/012045

Cold quark matter under intense magnetic elds: the role of flavor mixing and vector interactions

P. G. Allen[a], V. Pagura[b,c] and N. N. Scoccola[a,b,d]

[a] Department of Theoretical Physics, Comisión Nacional de Energía Atómica, Av.Libertador 8250, 1429 Buenos Aires, Argentina

[b]CONICET, Rivadavia 1917, 1033 Buenos Aires, Argentina

[c]IFLP, CONICET – Dpto. de Física, Universidad Nacional de La Plata, C.C. 67, (1900) La Plata, Argentina

[d] Universidad Favaloro, Solís 453, 1078 Buenos Aires, Argentina

Abstract. We study the effect of intense magnetic fields on the phase diagram of cold, strongly interacting matter within the Nambu-Jona-Lasinio model. Model extensions that include flavor mixing effects and vector interactions were analyzed, varying all relevant model parameters in acceptable ranges. Charge neutrality and beta equilibrium effects, which are specially relevant to the study of compact stars, were also considered.

1. Introduction

The influence of intense magnetic fields on the properties of strongly interacting matter has become an issue of increasing interest in recent years [1]. This is mostly motivated by the realization that in some relevant physical situations, like high energy non-central heavy ion collisions [2] and compact stellar objects called magnetars [3], very strong magnetic fields may be produced. Since in these systems extreme temperatures and/or densities may be found, it is interesting to investigate which modifications are induced by the presence of strong magnetic fields on the whole CD phase diagram. Unfortunately, even in the absence of those fields, the present knowledge of such phase diagram is only schematic due to the well-known difficulty given by the so-called sign problem which affects lattice calculations at finite chemical potential [4]. Of course, the presence of strong magnetic fields makes the situation even more complex. Thus, most of our present knowledge of their effect comes from investigations performed in the framework of effective models (see e.g. [5] and refs. therein). In this contribution we present some results of a study of the phase diagram of cold quark matter subject to intense magnetic fields in the framework of a generalized Nambu-Jona-Lasinio (NJL) model. The NJL-type models are effective relativistic quark models for non perturbative CD, where gluon degrees of freedom are integrated out and interactions are modelled through point like interactions. In its simplest version [6] it only includes scalar and pseudo scalar interactions that describe chiral symmetry breaking effectively. As well known, however, a more detailed description of the low-energy quark dynamics requires that other channels like flavor mixing and vector meson interactions are taken into account [7]. In fact, some aspects of the effect of those interactions on the magnetized quark matter have already been investigated [8, 9]. The purpose of the present work is to extend those analyses putting special emphasis on the way in which the so-called inverse magnetic catalysis [10] is affected. Moreover, the phase diagram of cold magnetized quark matter under conditions relevant for the physics of compact stars will be also analyzed. This contribution is organized as follows. In Sec. 2 we provide some details of the model and its parametrizations as well as the way to deal with an external constant magnetic field. In Sec. 3 we present and discuss our results for symmetric quark matter. The situation for stellar matter is analyzed in Sec. 4. Finally, our conclusions are given in Sec. 5.

2. Formalism

We consider the most general SU(2) Lagrangian density which includes a scalar-pseudoscalar interaction, vector-axial vector and the t´Hooft determinant interaction [7]. In the presence of an external magnetic field and chemical potential it reads:

$$\mathcal{L} = \bar{\psi}\left(i \not{D} - m_c + \mu\,\gamma^0\right)\psi + \mathcal{L}_{int} \tag{1}$$

where

$$\begin{aligned}\mathcal{L}_{int} \; = \; & G_1 \sum_{a=0}^{3}\left[\left(\bar{\psi}\tau_a\psi\right)^2 + \left(\bar{\psi}i\gamma_5\tau_a\psi\right)^2\right] + G_2 \sum_{a=0}^{3}\left[\left(\bar{\psi}\gamma_\mu\tau_a\psi\right)^2 + \left(\bar{\psi}\gamma_\mu\gamma_5\tau_a\psi\right)^2\right] \\ & + G_3\left[\left(\bar{\psi}\gamma_\mu\psi\right)^2 + \left(\bar{\psi}\gamma_\mu\gamma_5\psi\right)^2\right] + G_4\left[\left(\bar{\psi}\gamma_\mu\psi\right)^2 - \left(\bar{\psi}\gamma_\mu\gamma_5\psi\right)^2\right] + 2G_D\left(d_+ + d_-\right)\end{aligned} \tag{2}$$

Here, G_i with $i = 1, 4$ and G_D are coupling constants, $\psi = (u, d)^T$ represents a quark field with two flavors, $d_\pm = det\left[\bar{\psi}\left(1 \pm \gamma_5\right)\psi\right]$, $\mu = \text{diag}\left(\mu_u, \mu_d\right)$ the quark chemical potentials, m_c is the (current) mass matrix that we take to be the same for both flavors, $\tau_0 = I$, where I is the unit matrix in the two flavor space, and τ_a, $0 < a \leq 3$ denote the Pauli matrices. The coupling of the quarks to the electromagnetic field $\mathcal{A}_\mu$ is implemented through the covariant derivative $D_\mu = \partial_\mu - iq\mathcal{A}_\mu$ where q represents the quark electric charge matrix $q = \text{diag}\left(q_u, q_d\right)$ where $q_u/2 = -q_d = e/3$. In the present work we consider a static and constant magnetic field in the 3-direction, $\mathcal{A}_\mu = \delta_{\mu 2} x_1 B$. In the mean-field approximation the associated grand-canonical thermodynamical potential for cold and dense quark matter reads

$$\Omega^{\text{MFA}} = -\sum_{f=u,d}\theta_f + \Omega_{pot} \tag{3}$$

where θ_f gives the contribution from the gas of quasi-particles of each flavor $f = u, d$ and can be written as the sum of 3 contributions [11]

$$\begin{aligned}\theta_f^{\text{vac}} &= \frac{N_c}{8\pi^2}\left\{\Lambda\left(2\Lambda^2 + M_f^2\right)\sqrt{\Lambda^2 + M_f^2} - M_f^4 \ln\left[\frac{(\Lambda + \sqrt{\Lambda^2 + M_f^2})}{M_f}\right]\right\}, \\ \theta_f^{\text{mag}} &= \frac{N_c}{2\pi^2}\left(|q_f|B\right)^2\left[\zeta^{(1,0)}(-1, x_f) - \frac{1}{2}(x_f^2 - x_f)\ln x_f + \frac{x_f^2}{4}\right], \\ \theta_f^{\text{med}} &= \frac{N_c}{4\pi^2}|q_f|B\sum_{\nu=0}^{\nu_f^{max}}\alpha_\nu\left[\tilde{\mu}_f\sqrt{\tilde{\mu}_f^2 - s_f(\nu, B)^2}\right. \\ & \qquad \left. - s_f(\nu, B)^2 \ln\left(\frac{\tilde{\mu}_f + \sqrt{\tilde{\mu}_f^2 - s_f(\nu, B)^2}}{s_f(\nu, B)}\right)\right],\end{aligned} \tag{4}$$

where $M_f = m_c + \sigma_f$ and $\tilde{\mu}_f = \mu_f - \omega_f$, with σ_f and ω_f being the mean field values of the scalar and vector meson fields, respectively. Λ represents a non covariant ultraviolet cutoff and

$\zeta^{(1,0)}(-1, x_f) = d\zeta(z, x_f)/dz|_{z=-1}$ where $\zeta(z, x_f)$ is the Riemann-Hurwitz zeta function. In addition, $s_f(\nu, B) = \sqrt{M_f^2 + 2|q_f|B\nu}$ while $x_f = M_f^2/(2|q_f|B)$. In θ_f^{med}, the sum is over the Landau levels (LLs), represented by ν, while $\alpha_\nu = 2 - \delta_{\nu 0}$ is a degeneracy factor and ν_f^{max} is the largest integer that satisfies $\nu_f^{max} \leq (\tilde{\mu}_f^2 - M_f^2)/(2|q_f|B)$.

The Ω_{pot} contribution reads

$$\Omega_{pot} = \frac{(1-c_s)(\sigma_u^2 + \sigma_d^2) - 2c_s\,\sigma_u\,\sigma_d}{8g_s(1-2c_s)} - \frac{(1-2c_v)(\omega_u^2 + \omega_d^2) + c_v\,\omega_u\,\omega_d}{8g_v(1-2c_v)} \tag{5}$$

where we have introduced a convenient parametrization of the coupling constants in terms of the quantities g_s, c_s, g_v and c_v given by

$$\begin{aligned} g_s &= G_1 + G_D \quad &;& \quad g_v = G_2 + G_3 + G_4 \\ c_s &= \frac{G_D}{G_1 + G_D} \quad &;& \quad c_v = \frac{G_3 + G_4}{2(G_2 + G_3 + G_4)} \end{aligned} \tag{6}$$

The relevant gap equations are given by

$$\frac{\partial \Omega^{\rm MFA}(\sigma_u, \sigma_d, \omega_u, \omega_d)}{\partial(\sigma_u, \sigma_d, \omega_u, \omega_d)} = 0\ . \tag{7}$$

In our calculations we will consider first the simpler case of symmetric matter where both quarks carry the same chemical potential μ. Afterwards, we will analyze the case of stellar matter in which leptons are also present and β-equilibrium and charge neutrality are imposed. In this case the chemical potential for each quark, μ_f, is a function of μ and the lepton chemical potentials which have to be selfconsistently determined.

In order to analyze the dependence of the results on the model parameters, we will consider two SU(2) NJL model parametrizations. Set 1 corresponds to that leading to $M_0 = 340$ MeV while Set 2 to that leading to $M_0 = 400$ MeV. Here, M_0 represents the vacuum quark effective mass in the absence of external magnetic fields. The corresponding model parameters are listed in Table 1.

Table 1. Parameter sets for the NJL SU(2) model.

Parameter set	M_0 MeV	m MeV	$g_s\Lambda^2$	Λ MeV	$-<uu>^{1/3}$ MeV
Set 1	340	5.595	2.212	620.9	244.3
Set 2	400	5.833	2.440	587.9	240.9

The presence of the t'Hooft determinant interaction is very important since it reflects the $U_A(1)-$anomaly of CD. Its strength, and consequently the amount of flavor mixing induced by this term, is controlled by the parameter c_s. An estimate for its value can be obtained from the $\eta - \eta'$ mass splitting within the 3−flavor NJL model [12]. This leads to $c_s \simeq 0.2$ [13]. In any case, to obtain a full understanding of the effects of flavor mixing we will vary the value of c_s in a range going from 0, which corresponds to a situation in which the two flavors are completely decoupled, to 0.5, being this the case of maximum flavor mixing described for example in Ref. [14]. Regarding the vector term coupling g_v it is important to recall that one can obtain naturally the vector axial-vector interaction if one starts from a CD-inspired color current-current interaction and then performs a Fierz transform into color-singlet channels, and

Journal of Physics: Conference Series **630** (2015) 012045 doi:10.1088/1742-6596/630/1/012045

that in this case the relation between coupling strengths is $g_v = g_s/2$ [7]. Yet, the value of g_v cannot be accurately determined from experiments nor from lattice CD simulations and this is why it has been taken as a free parameter in most works. In the present work we take $0 < g_v/g_s < 0.5$. It is worth mentioning that due to the mixing of pseudoscalar and longitudinal axial vector interaction terms pseudoscalar meson properties depend on g_v. Thus, strictly speaking the parameters given in Table 1 only lead to the empirical values of f_π and m_π when $g_v = 0$. However, as shown in Ref. [15], m_π and f_π only change by $\sim 10\%$ when g_v/g_s increases from 0 to 1. Thus, for simplicity, we will keep the model parameter values fixed when varying g_v. A last comment regarding c_v, i.e. the parameter that regulates the ratio between the singlet and octet vector-axial vector interaction strengths: we will take it as a free parameter in the range $0 \leq c_v \leq 1/2$. Note that for $c_v = 1/2$ only singlet vector-axial interactions are present and, thus, there is no mixing between the pseudoscalar and longitudinal axial vector channels.

We end this section by describing the way in which the different phases of the magnetized quark matter will be denoted as well as the procedure used to identify the boundaries between them. For the phases we adopt the notation of Refs. [14, 16]. Thus, the vacuum (i.e. fully chirally broken) phase is denoted by B, the massive phases in which M_f depends on the chemical potential by C_α and, finally, the chirally restored phases by A_α. Here, α is a set of two numbers indicating the highest LL populated for each flavor. To obtain the critical chemical potential at a given eB we proceed as follows. In the case of first order phase transitions we calculate the thermodynamical potential for each of the neighboring phases and then searched the chemical potential at which these values become degenerate. In the case of crossover transitions the critical value is identified by the peak of the chiral susceptibility corresponding to each quark flavor.

3. Results for symmetric matter

In this section we present the results obtained for the case of symmetric matter. These results were obtained solving the set of coupled gap equations , Eqs. (7), for different values of magnetic field and chemical potential. In order to understand the effect of the different terms in the Lagrangian we consider in Sec. 3.1 the effect of flavor mixing on quark behavior within the model in the absence of vector interactions. The role of those interactions will be subsequently addressed in Sec. 3.2.

3.1. Effect of the flavor mixing interactions

To neglect vector interactions implies taking $G_2 = G_3 = G_4 = 0$ in Eq. (2), while $G_1 \neq 0$ and $G_D \neq 0$. Therefore, Eqs. (7), will become a set of two coupled equations that must be solved for the independent variables M_u and M_d. The parameter c_s acts as a coupling between both flavors and we study how the phase transitions are modified as we vary this parameter in the range $0 \leq c_s \leq 0.5$.

The case $c_s = 0.5$ corresponds to ordinary NJL where flavor mixing is maximum and, thus, both flavors have identical behavior. In fact, the first term in Eq. (5) will tend to infinity as c_s goes to 0.5 unless $M_u = M_d$, which leaves only one equation to be solved. But if $c_s < 0.5$, then both masses will be independent variables, and transitions for each flavor might occur simultaneously or not in different regions of the phase diagrams.

It is common to all these cases that chiral symmetry is completely broken for chemical potentials well below M_0 and that restoration occurs for high enough chemical potentials, usually accompanied by a large drop in the dressed mass. The inclusion of a constant magnetic field modifies the quark dispersion relations, introducing Landau levels (LL's) into its spectrum. A consequence of this is that chiral symmetry restoration occurs in several steps, each of which is a transition where quark population appears on previously unoccupied LL's. This gives rise to a potentially complex phase diagram, whose precise form depends on the parameter set and

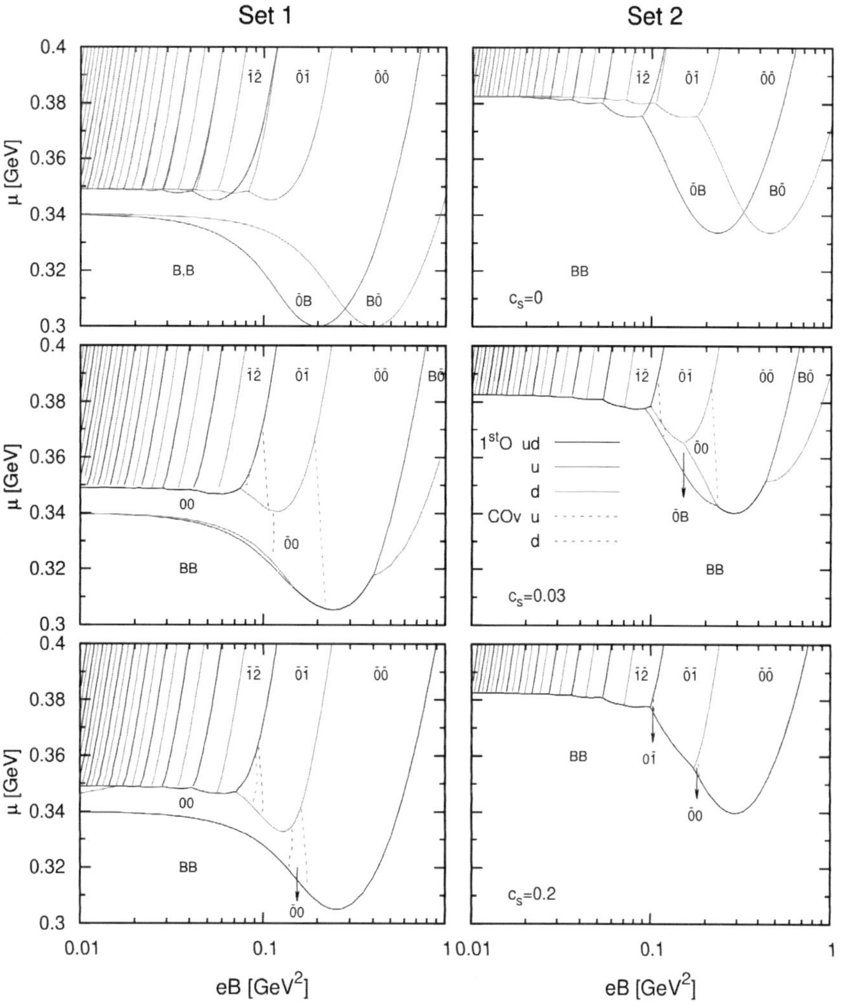

Figure 1. (Color online) Phase diagrams in the $eB - \mu$ plane for different values of flavor-mixing parameter c_s. To simplify the figure we have introduced a compact notation to indicate the phases. The pair of integers mn corresponds to the C_{mn} phase and the pair mn to the A_{mn} phase. The case in which one quark is in a C-type phase and the other in the A-type phase is indicated by putting a bar on top of the integer associated with the A-type phase.

magnetic field (The $B = 0$ case is simple yet depending on the parameter set there can be a few differences).

We will start by analyzing the phase transitions for the case with no flavor mixing, i.e. $c_s = 0$, previously studied in Ref. [8]. The phase diagrams for both parameter sets can be seen in the upper two panels of Fig. 1. In a sense, each flavor will have its own independent phase diagram because the gap equations are decoupled. Here, each line corresponds to a transition where LL population of a single quark flavor occurs (red lines for down quarks, blue lines for up quarks). Both flavors will coincide for $B = 0$ where $SU(2)$ symmetry is recovered and behave differently as B increases, due to their different electric charges. Since this is the only difference between

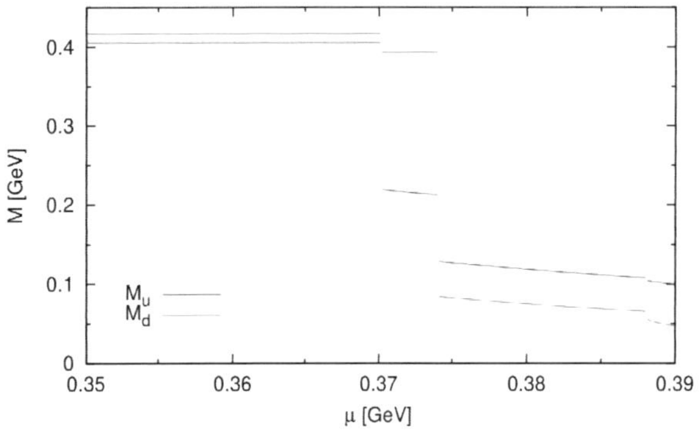

Figure 2. (Color online) Dressed masses for both flavors for Set 2, $c_s = 0.03$, for $eB = 0.11$ GeV^2.

the equations for both flavors and since it only appears in the product $q_f B$, one phase diagram may be obtained from the other one through the replacement $q_u B = (2q_d)B = q_d(2B)$ which amounts to a stretching along the B axis of its phase diagram with respect to the up quark one.

When c_s becomes finite (second row onwards in Fig. 1), flavors become coupled and a complex pattern is created, where transitions move closer together in some regions and even merge together in others, that is, the LL population changes simultaneously for both flavors (these are represented by black lines). We can see that already for $c_s = 0.03$, the transitions in both parameter sets occur together for low magnetic fields, and then separate for $B = 0.02$ GeV^2 (in Set 1) and $B = 0.1$ GeV^2 in Set 2.

In order to make the meaning of the transition lines clearer, we present on Fig. 2 the dressed masses for both flavors for Set 2, $c_s = 0.03$, for $eB = 0.11$ GeV^2. When the first discontinuity is encountered, at $\mu = 370$ MeV, M_u jumps to half its value and its lowest LL (LLL) becomes populated. On the other hand, the down flavor remains in vacuum. Actually, its mass presents a small discontinuity caused by the weak coupling to the up quark which is not to be interpreted as a down transition. The down quark LLL is occupied at $\mu = 374$ MeV. When both flavors are in the vacuum phase, $M_u > M_d$, while in the populated phase it happens the other way around, which is consistent with the phenomena of catalysis and anticatalysis discussed in Ref. [14].

For $c_s = 0.2$, the transitions for both flavors occur together, so the qualitative behavior is very similar to the full flavor mixing case $c_s = 0.5$. In fact, the model tends to full mixing quite quickly and only for $c_s \lesssim 0.1$ are relevant mixture effects (or more precisely, the absence of it) actually seen. On the other hand, cross over transitions do not have this property and tend much more slowly to the $c_s = 0.5$ behavior.

3.2. Effect of the vector interactions

In this section we analyze the effect of vector interaction terms. As discussed in the previous section, for $c_s \gtrsim 0.1$ the phase diagrams do not present qualitative variations. Here, therefore, we consider $c_s = 0.2$ which also is in the range of realistic values suggested in Ref. [13]. Although these results correspond to $c_v = 0$, our studies show that only small quantitative differences occur when varying this parameter from 0 to 1.

In Fig. 3 we present a series of phase diagrams obtained for different values of the ratio g_v/g_s. The left column corresponds to Set 1 while the right column was obtained with parameter Set 2.

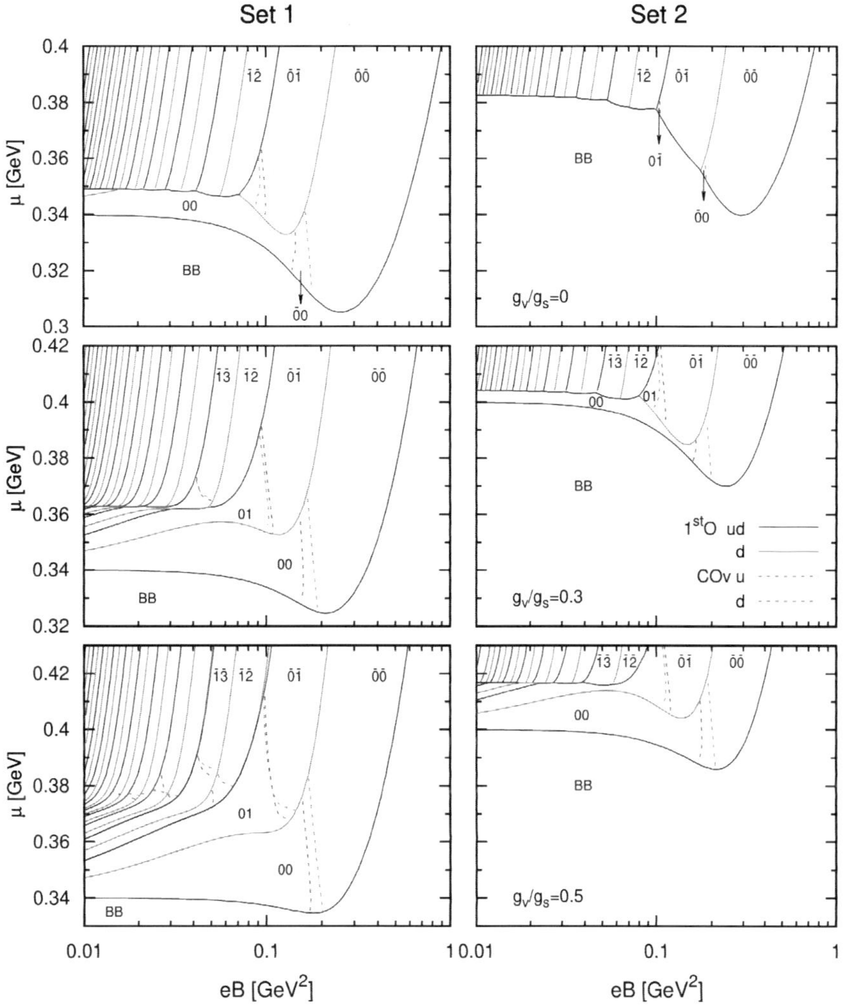

Figure 3. (Color Online) Phase diagrams in the $eB - \mu$ plane for different values of g_v/g_s. Different phases are denoted as in Fig. 1

For Set 1 we observe that as g_v/g_s increases the two lower and main transitions separate and several new transitions appear in the low eB region of the diagram. For Set 2 at some small $g_v/g_s \simeq 0.1$ the lower transition first splits into two, leaving a 00 phase in between which was not present before. For larger values of g_v/g_s the behavior is similar to that of Set 1. The appearance of new transitions at low eB as g_v/g_s increases is clearly seen in the left panel of Fig. 4, where we show d quark density normalized to nuclear matter density ($\rho_0 = 1.30$ MeV^3) as a function of chemical potential for $eB = 0.016$ GeV in Set 1. In the absence of vector interaction, the density jumps from close to 0 to 3 times nuclear matter density, while quark population jumps from 00 to high LLs occupied. As g_v increases, the amount of transitions increases too. In fact, increasing the vector interaction coupling has the same effect as going to a parameter set that reproduces a lower value of current mass M_0, described in Ref. [14]. Notice that the phase diagram for Set 2 and $g_v/g_s = 0.3$ is in this sense very similar to Set 1 and $g_v/g_s = 0$.

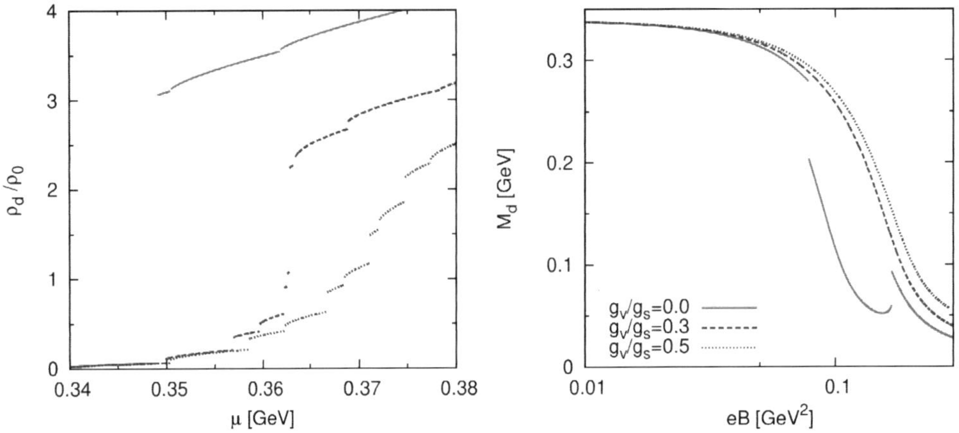

Figure 4. (Color online) uark density over nuclear matter density as function of the chemical potential for $eB = 0.016$ GeV^2 (left) and current mass as function of magnetic field for $\mu = 0.345$ GeV (right) for different values of g_v/g_s. Results correspond to d flavor and were obtained with parameter Set 1.

It was also found that as vector coupling is increased, the inverse magnetic catalysis effect becomes attenuated. This can be observed in the right panel of Fig. 4 where we plot the current quark mass for d quarks as a function of the magnetic field for $\mu = 345$ MeV, for all g_v/g_s values considered here. For low magnetic fields, the mass remains constant and after $eB \sim 0.02$ GeV^2 it starts to decrease smoothly. This decrease, however, is less abrupt for higher g_v values. The first discontinuity found in the current mass for $g_v/g_s = 0$ correlates to the d quark transition from the phase 00 to 01. For a higher eB there is a second discontinuity, in this case a jump to a higher mass value, corresponding to a return to phase 00. Moreover, if we measure the depth of the IMC well as the difference in μ between the lowest critical chemical potential at $eB = 0$ MeV and the lowest possible one in the whole diagram, we find that this $\Delta\mu$ is reduced by an 84% for Set 1 and by a 67% for Set 2 when going from $g_v = 0$ to $g_v = 0.5$.

4. Results for stellar matter

We now turn our attention to stellar matter. Following the discussions in the previous section only results for $c_s = 0.2$ will be presented. If charge neutrality is imposed on our system, there will be a fixed relation between the densities of the up and down quarks, which will be necessarily different unless they are both zero. So, even though the value of $c_s = 0.2$ was close enough to the full mixing case according to what was established in previous sections, charge neutrality will cause the flavors to behave differently among themselves.

The phase diagrams in the $\mu - B$ plane for both parameter sets and for increasing values of g_v are plotted in Fig. 5. The chemical potential μ on the horizontal axis is now the quark number chemical potential, which is related to flavor chemical potentials through $\mu = \mu_u + 2\mu_e/3$ and $\mu = \mu_d - \mu_e/3$. The introduction of charge neutrality has a few effects similar to those of vector interaction, in the sense that diagrams become similar to the ones corresponding to lower M_0 sets: main transitions separate for low eB and several transitions appear in the region between them. The magnetic anticatalysis effect is reduced also: The depth of the anticatalysis well, as defined in the previous section, is reduced from 35 MeV to 9 MeV in Set 1 and from 43 MeV to 22 MeV in Set 2. As expected, the introduction of vector interaction will further enhance these effects, so that anticatalysis has completely disappeared for $g_v = 0.5$.

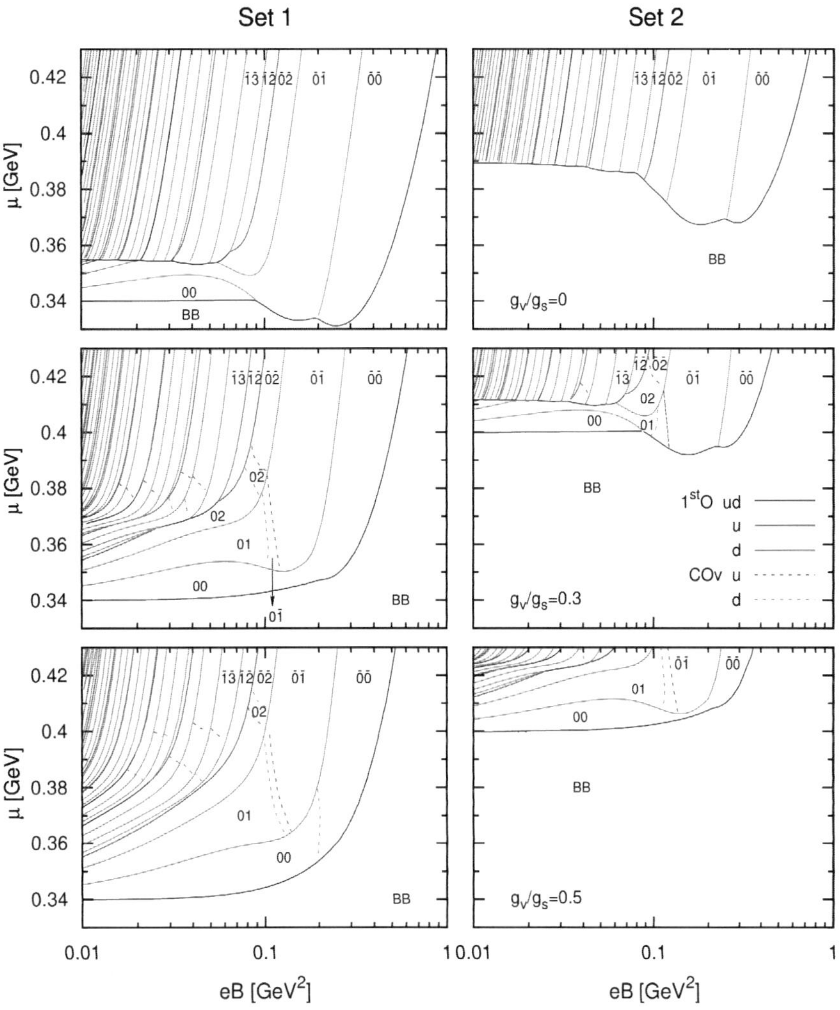

Figure 5. (Color online) Phase diagrams in the $eB - \mu$ plane for stellar matter and different values of g_v/g_s. Different phases are denoted as in Fig. 1

The Van Alphen - de Haas (VA-dH) transitions acquire relatively independent behaviors when charge neutrality is introduced. In particular, for intermediate $eB\,(\simeq 0.1\ \mathrm{GeV}^2)$ the transition between $k_u = 0$ and $k_u = 1$ is well separated from the down transition. For symmetric matter, the VA-dH transitions occurred to a very good level of approximation for a critical chemical potential $\mu_c = \sqrt{2kq_fB}$ (being this condition exact for the chiral case), so the relation between the charges produced one up transition every two down transitions. Although this relation is still satisfied for each flavor chemical potential, these are now different between themselves and related through $\mu_u - \mu_d + \mu_e = 0$. As a result of this, in Fig. 5 we roughly have one up transition every four down transitions.

5. Conclusions

We have investigated how the phase diagram of cold strongly interacting matter is modified in the presence of intense magnetic fields in the context of NJL-type models which include flavor mixing and vector interaction. The whole range of possible flavor mixing values was swept through. For symmetric matter it was found that for rather small values of the flavor mixing parameter, $c_s = 0.1$, phase diagrams already display a behavior that is qualitatively very similar to the full mixing case. The role of the vector interactions and stellar matter conditions were studied for $c_s = 0.2$, which is deduced from $SU(3)$ estimates. The relevant effects are similar to those occurring when changing to a parameter set that fits to a smaller dressed mass. Particularly interesting is the reduction of the Inverse Magnetic Catalysis phenomenon, which can completely disappear by the joint effect of the vector interactions and the stellar matter conditions.

Acknowledgments

This work was partially supported by CONICET (Argentina) under grant PIP 00682 and by ANPCyT (Argentina) under grant PICT-2011-0113.

References

[1] D.E. Kharzeev, K. Landsteiner, A. Schmitt and H.-U. Yee, Lect. Notes Phys. 871 (2013) 1.
[2] D.E. Kharzeev, L.D. McLerran and H.J. Warringa, Nucl. Phys. A803 (2008) 227.
[3] R.C. Duncan and C. Thompson, Astrophys J. 392 (1992) L9.
[4] F. Karsch, E. Laermann, 2004 *Quark Gluon Plasma 3*, edited by R.C.Hwa et al. (World Scientific Singapore), p.1.
[5] E.S. Fraga, Lect. Notes Phys. 871 (2013) 121 R. Gatto and M. Ruggieri, Lect. Notes Phys. 871 (2013) 87.
[6] Y. Nambu and G. Jona-Lasinio, Phys. Rev. **122** (1961) 345 **124** (1961) 246.
[7] U. ogl and W. Weise, Prog. Part. Nucl. Phys. **27** (1991) 195 S. Klevansky, Rev. Mod. Phys. **64** (1992) 649 T. Hatsuda and T. Kunihiro, Phys. Rep. **247** 221 (1994) 221.
[8] J. K. Boomsma and D. Boer, Phys. Rev. D **81** (2010) 074005.
[9] R. Z. Denke and M. B. Pinto, Phys. Rev. D **88** (2013) 056008 D. P. Menezes, M. B. Pinto, L. B. Castro, P. Costa and C. Providencia, Phys. Rev. C **89** (2014) 055207.
[10] F. Preis, A. Rebhan and A. Schmitt, JHEP **1103** (2011) 033 F. Preis, A. Rebhan and A. Schmitt, Lect. Notes Phys. **871** (2013) 51.
[11] D.P. Menezes, M.B. Pinto, S.S. Avancini, A. Pérez Martínez and C. Providencia, Phys. Rev. **C 79** (2009) 035807 D.P. Menezes, M.B. Pinto, S.S. Avancini and C. Providencia, Phys. Rev. **C 80** (2009) 065805 S.S. Avancini, D.P. Menezes and C. Providencia, Phys. Rev. **C 83** (2011) 065805.
[12] T. Kunihiro, Phys. Lett. B **219** (1989) 363 [Erratum-ibid. B **245** (1990) 687].
[13] M. Frank, M. Buballa and M. Oertel, Phys. Lett. B **562** (2003) 221.
[14] P. G. Allen and N. N. Scoccola, Phys. Rev. D **88** (2013) 094005.
[15] M. Hanauske, L. M. Satarov, I. N. Mishustin, H. Stoecker and W. Greiner, Phys. Rev. D **64** (2001) 043005.
[16] D. Ebert, K. G. Klimenko, M. A. dovichenko, and A. S. shivtsev, Phys. Rev. D **61** (1999) 025005.

The role of asymptotic freedom for the pseudocritical temperature in magnetized quark matter

R.L.S. Farias[1,2], K.P. Gomes[1], G. Krein[3], M.B. Pinto[4]
[1]Departamento de Ciências Naturais, Universidade Federal de São João Del Rei, 36301-000, São João Del Rei, MG, Brazil
[2]Departamento de Física, Universidade Federal de Santa Maria, 97105-900, Santa Maria, RS, Brazil
[3]Instituto de Física Teórica, Universidade Estadual Paulista, Rua Dr. Bento Teobaldo Ferraz, 271 - Bloco II, 01140-070 São Paulo, SP, Brazil
[4]Departamento de Física, Universidade Federal de Santa Catarina, 88040-900 Florianópolis, Santa Catarina, Brazil

E-mail: ricardosonegofarias@gmail.com

Abstract.
Motivated by discrepancies observed between lattice QCD simulations and quark models regarding the behavior of the pseudo critical temperature for chiral symmetry restoration as a function of the magnetic field B, we investigate the effects of a running the quark coupling constant G with temperature T and the magnetic field B in the context of the Nambu–Jona-Lasinio model(NJL). Our point that when asymptotic freedom, an essential feature of QCD and absent in the model, is included through a running of G with T and B results from the NJL model can be brought in qualitative agreement with lattice QCD simulations.

1. Introduction

Recently much effort has been devoted to the understanding of the effects produced by a magnetic field in the phase diagram of Quantum Chromodynamics (QCD). The motivation is due to the fact that strong magnetic fields may be produced in non central heavy ion collisions [1, 2, 3, 4]. Other scenarios that strong magnetic fields are also important are in the phenomenology of stars, in particular in magnetars [5, 6], and in the physics of the early universe [7]. Lattice QCD simulations [8, 9] predict that at zero baryon density and zero magnetic field there is a crossover transition at a pseudo critical temperature T_{pc}. More recent lattice simulations show that this crossover persists if we include the effects of external strong magnetic fields [10, 11, 12, 13]. At zero temperature the lattice results confirm the existence of the phenomenon of magnetic catalysis (MC) [14, 15, 16], where the chiral order parameter is enhanced with the increase of the magnetic field. At finite temperature, lattice results are in agreement with effective models [17, 18, 19, 20, 21]. However, recent lattice results of Refs. [12, 13], that consider physical values of quark masses, predict an inverse magnetic catalysis (IMC), in that the pseudo critical temperature T_{pc} for the crossover decrease with B, in total disagreement with all effective model calculations in the literature [17, 18, 19, 22, 23, 24].

Many efforts have been dedicated to understand the disagreement between lattice and models in the behavior of T_{pc} with B; see e.g. Refs. [25, 26, 34, 27]. Very sophisticated versions of quark models were implemented, as entangled PNJL model [28, 29, 30] where the quark coupling constant of the NJL model is dependent on the Polyakov loop, but no qualitative changes have been observed. Another tentative to produce inverse catalysis was implemented in Ref. [31] where the authors fitted the lattice data assuming that pure-gauge critical temperature T_0 has a dependence with B, and the model gives IMC. In the context of quark-meson (QM) models no qualitative agreement with the lattice is obtained [32], even if we include the Polyakov loop (PQM) with $T_0 = T_0(B)$. More refinements in QM where implemented and the model has no IMC [33].

There is an interesting sugestion [34] that the phenomenon of the IMC is the result of the back-reaction of the gluons due to the coupling of the magnetic field to the sea quarks. It seems therefore possible that the failure of models in obtaining IMC is due to the fact that the coupling constants in effective models do not depend on the magnetic field. Miransky and Shovkovy have shown [35] that the QCD coupling constant decreases for large B, an effect due to asymptotic freedom of QCD. In a recent publication we [36] have introduced an ansatz for G in way it decreases with B and T, similar to the running of the strong coupling in QCD due to asymptotic freedom. Our results show that at $T = 0$ there is magnetic catalysis, and at $T \neq 0$ the model realizes IMC in qualitative agreement with the recent lattice QCD simulations of Refs. [12, 13]. The idea of using a running coupling constant for the NJL coupling with the temperature is not new, it was used in Ref. [37] to understand the the decoupling of the pion and recently it was used to study QCD Dyson-Schwinger equations at finite temperature and density [38, 39].

The usefulness of implementing a running coupling in the NJL model can be understood examining the relation between the quark condensate and the constituent quark mass: $M \sim -G\langle\bar{\psi}_f\psi_f\rangle$. For a fixed value for G, one has that $\langle\bar{\psi}_f\psi_f\rangle$ and M (and consequently T_{pc}) growing with B, signalling MC. We circumvent this problem assuming that the coupling constant G decreases with B and T. Our results are in qualitative agreement with the lattice; the agreement is quite impressive, given the simplicity of the model. Our results are in the direction that IMC can be result of the back reaction of the sea quarks to the magnetic field [34].

2. Running Coupling in the NJL Model at Finite Temperature and Magnetic Field

The $SU(2)$ version of the NJL model can be defined by a fermionic Lagrangian density given by [40]

$$\mathcal{L}_{\rm NJL} = \bar{\psi}\left(i\partial\!\!\!/ - m\right)\psi + G\left[(\bar{\psi}\psi)^2 - (\bar{\psi}\gamma_5\vec{\tau}\psi)^2\right], \tag{1}$$

We need to introduce a sharp cut off Λ due the nonrenormalizability of the NJL model; the value of Λ is chosen to fit physical quantities. In this work we use $\Lambda = 650\,\text{MeV}$. The other parameters of the model are taken $G = 5.022\,\text{GeV}^{-2}$, $m = 5.5\,\text{MeV}$. These choices reproduce the vacuum values of the pion decay constant $f_\pi = 93\,\text{MeV}$, the pion mass $m_\pi = 140\,\text{MeV}$, and the quark condensate $\langle\bar{\psi}_f\psi_f\rangle^{1/3} = -250\,\text{MeV}$.

The evaluation of thermodynamic potential $SU(2)$ NJL model in the mean field approximation (MFA) is well documented in the literature [41, 22] and results beyond the MFA can be found in Ref. [42]. Here we use the MFA; the gap equation is given by

$$M_f = m_f - 2G\sum_f \langle\bar{\psi}_f\psi_f\rangle, \tag{2}$$

where $\langle\bar{\psi}_f\psi_f\rangle$ represents the quark condensate of flavor f

$$\langle\bar{\psi}_f\psi_f\rangle \;=\; -\frac{N_c M_f}{2\pi^2}\left\{\Lambda\sqrt{\Lambda^2 + M_f^2} - \frac{M_f^2}{2}\ln\left[\frac{(\Lambda + \sqrt{\Lambda^2 + M_f^2})^2}{{M_f}^2}\right]\right\}$$

$$- \frac{N_c M_f}{2\pi^2}|q_f|B\left\{ \ln[\Gamma(x_f)] - \frac{1}{2}\ln(2\pi) + x_f - \frac{1}{2}(2x_f - 1)\ln(x_f)\right\}$$
$$+ \frac{N_c M_f}{2\pi^2}\sum_{k=0}^{\infty}\alpha_k|q_f|B\int_{-\infty}^{\infty}\frac{dp_z}{E_{p,k}(B)}\left\{\frac{1}{e^{[E_{p,k}(B)]/T}+1}\right\}, \tag{3}$$

where $E_{p,k}(B) = (p_z^2 + 2k|q_f|B + M_f^2)^{1/2}$, $x_f = M_f^2/(2|q_f|B)$, and $\alpha_k = 2 - \delta_{0k}$. In addition, $|q_f|$ is the absolute value of the quark electric charge; $|q_u| = 2e/3$, $|q_d| = e/3$, with $e = 1/\sqrt{137}$ representing the electron charge – we use Gaussian natural units where $1\,\mathrm{GeV}^2 = 1.44 \times 10^{19}\ G$. We consider the condition of chemical equilibrium assuming $\mu_u = \mu_d = \mu$. More details of the manipulations to obtain this equation can be found in Ref. [22]. It is important to note that the condensates for the flavors u and d are different due to the different electric charges of the quarks. In principle, one would have two coupled gap equations, one for each flavor: $M_u = m_u - 2G(\langle\bar\psi_u\psi_u\rangle + \langle\bar\psi_d\psi_d\rangle)$ and $M_d = m_d - 2G(\langle\bar\psi_u\psi_u\rangle + \langle\bar\psi_d\psi_d\rangle)$. However, in the $SU(2)$ version of the NJL model, when $m_u = m_d = m$, the different condensates contribute to M_u and M_d in a symmetric way and we can write $M_u = M_d = M$.

The lattice results for $(\Sigma_u + \Sigma_d)/2$ and $\Sigma_u - \Sigma_d$ are presented in Ref. [13]. The definition of $\Sigma_f = \Sigma_f(B,T)$ is given by

$$\Sigma_f(B,T) = \frac{2m_f}{m_\pi^2 f_\pi^2}\left[\langle\bar\psi_f\psi_f\rangle - \langle\bar\psi_f\psi_f\rangle_0\right] + 1, \tag{4}$$

where $\langle\bar\psi_f\psi_f\rangle_0$ is the quark condensate evaluated at $T = 0$ and $B = 0$.

The result derived by Miransky and Shovkovy [35] for the QCD running coupling α_s in the regime of sufficiently strong magnetic fields $eB \gg \Lambda_{QCD}^2$:

$$\frac{1}{\alpha_s} \sim b\ \ln\frac{eB}{\Lambda_{QCD}^2}, \tag{5}$$

where $b = (11N_c - 2N_f)/12\pi$. This behavior of α_s with the magnetic field was our motivation in Ref. [36] to propose an NJL coupling, at $T = 0$, of the form

$$G(B) = \frac{G_0}{1 + \alpha\ln\left(1 + \beta\frac{eB}{\Lambda_{QCD}^2}\right)}, \tag{6}$$

where $G_0 = 5.022\ \mathrm{GeV}^{-2}$ is the value of the coupling at $B = 0$. α and β are free parameters that were fixed requiring a reasonable description of the lattice average $(\Sigma_u + \Sigma_d)/2$ at $T = 0$. We choose this parametrization because we believe that similar physics of that of asymptotic freedom is responsible for the running of the coupling with magnetic field.

In the high temperature limit, α_s also runs as the inverse of $\ln(T/\Lambda_{QCD})$. However, the values of T used in the lattice QCD simulations of Refs. [12, 13] are such that $T \le \Lambda_{QCD}$ and one cannot use a G running with T like in QCD. Since we do not know the running of α_s with B *and* T, we propose the following ansatz

$$G(B,T) = G(B)\left(1 - \gamma\frac{|eB|}{\Lambda_{QCD}^2}\frac{T}{\Lambda_{QCD}}\right). \tag{7}$$

The parameter γ is fixed when we require the lattice behavior for the temperature dependence of the lattice average $(\Sigma_u + \Sigma_d)/2$.

3. Numerical Results

We next present our numerical results, but before it is relevant to mention that Refs. [12, 13] used $m_u = m_d = 5.5\,\text{MeV}$, $m_\pi = 135$ MeV and $f_\pi = 86$ MeV in the multiplicative factor $m_f/m_\pi^2 f_\pi^2$ in Eq. (4) to make $\Sigma_f(B,T)$ dimensionless. We use this set of parameters when we compared our NJL results with the lattice simulations. In this work we consider $\Lambda_{QCD} = 200\,\text{MeV}$.

Fig. 1 displays the quark condensate average $(\Sigma_u + \Sigma_d)/2$. We determine the values $\alpha = 2$ and $\beta = 0.000327$ in Eq. (6) with a good fit of the lattice data at $T = 0$. In Fig. 1 we can see the phenomena of the magnetic catalysis reproduced at $T = 0$. The T dependence is also reasonably well reproduced (we used $\gamma = 0.0175$ in Eq. (7)). With this set of parameters fixed we proceed in the calculation of the other quantities.

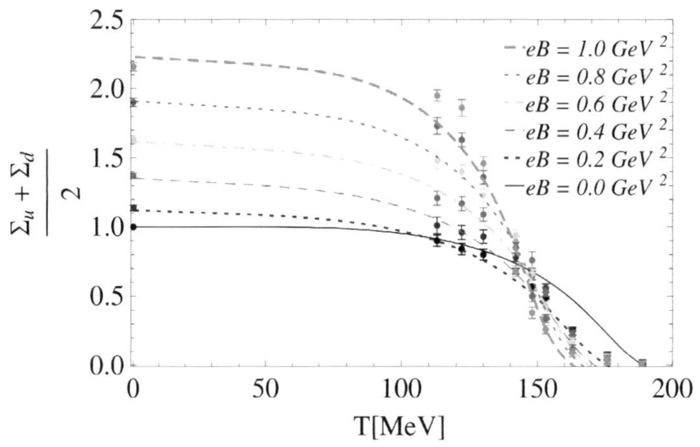

Figure 1. The quark condensate average as a function of the temperature. Data points are from the lattice QCD simulations of Ref. [13].

In Fig. 2 we show the results for the difference $\Sigma_u - \Sigma_d$. There is a small deviation from the lattice results for some values of B, but the agreement is quite impressive, given the simplicity of the version of the NJL model we are using, with a minimum number of parameters.

The model displays a crossover at high temperatures, since we consider the physical point with nonzero current quark masses, and we can establish only a pseudo critical temperature for the partial restoration of chiral symmetry. The value of T_{pc} depends on the observable that we use to define it. In this work we use the location of the peak for the vacuum normalized quark condensates. The thermal susceptibilities is given by:

$$\chi_T = -m_\pi \frac{\partial \sigma}{\partial T}\,, \tag{8}$$

where σ is defined by

$$\sigma = \frac{\langle \bar{\psi}_u \psi_u \rangle(B,T) + \langle \bar{\psi}_d \psi_d \rangle(B,T)}{\langle \bar{\psi}_u \psi_u \rangle(B,0) + \langle \bar{\psi}_d \psi_d \rangle(B,0)}\,. \tag{9}$$

In Fig. 3 we show our results for the thermal susceptibility as a function of the temperature for different values of the magnetic field. We can see clearly that $T_{\rm pc}$ decrease for increasing values of the magnetic field. It is clear that the diminishing of G with B and T leads to results that are in good agreement with lattice simulations.

In Fig. 4 we show the NJL results for the pseudocritical $T_{\rm pc}$ temperature as a function of the magnetic field. This figure shows clearly that the T_{pc} decreases as B increases. These results are

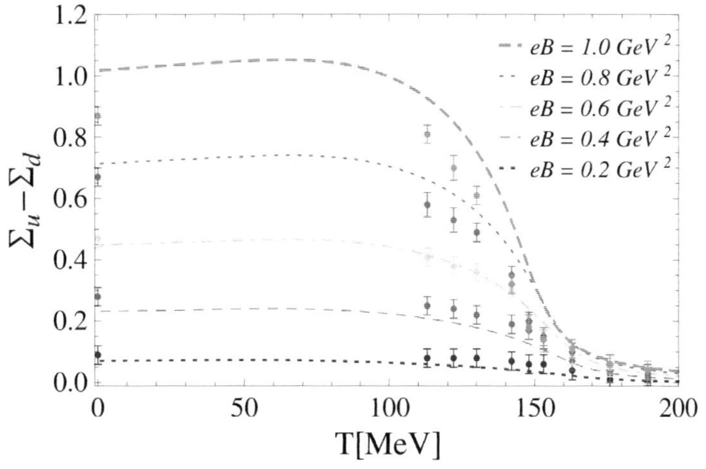

Figure 2. The quark condensate difference as a function of the temperature. Data points are from the lattice QCD simulations of Ref. [13].

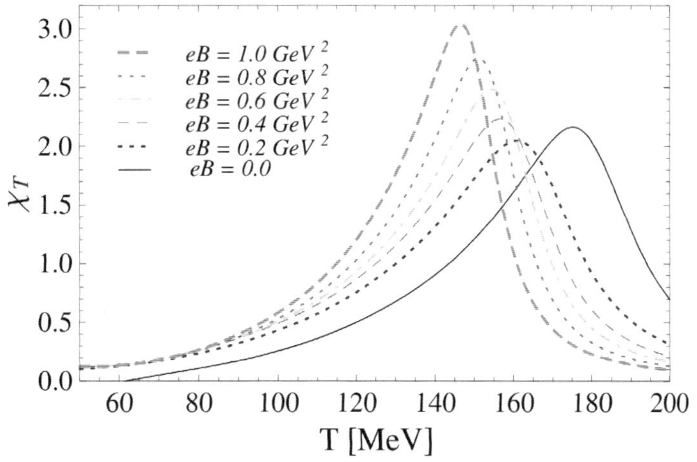

Figure 3. The thermal susceptibility (normalized) as a function of the temperature for different values of the magnetic field B.

in qualitative agreement with recent lattice QCD results of Refs. [12, 13] that consider physical values of quark masses.

Fig. 4 shows our phase diagram; the lattice results are reproduced qualitatively. We can obtain a better quantitative agreement if we include more fitting parameters in our ansatz for $G(B)$. One very important point here is that our results avoid the undesired "turn over" effect in NJL models (with B independent couplings) where for some values of B, after a initial decrease, T_{pc} starts increase with B (this can be found in Refs. [31, 32]).

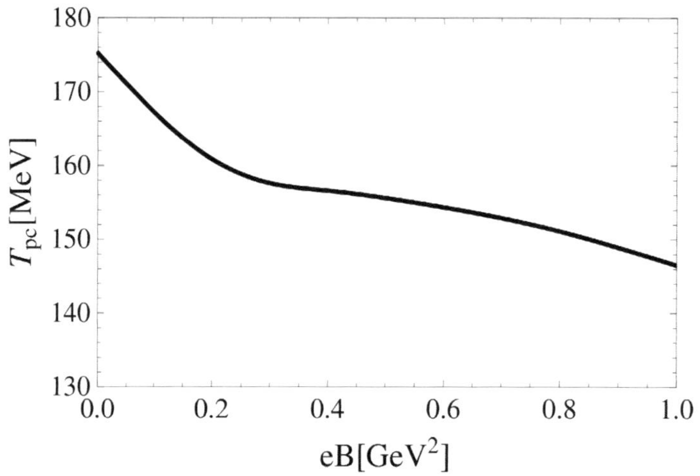

Figure 4. The pseudo critical temperature T_{pc} for the chiral transition of quark matter as a function of the magnetic field.

4. Conclusions and Perspectives

In this work we presented results for several physical quantities associated with the partial restoration of chiral symmetry at finite temperature and in the presence of an external constant magnetic field. They were calculated within the mean field approximation in two flavor NJL model. Our aim was to understand discrepancies between effective quark models and recent lattice QCD results regarding the behavior of the pseudo critical temperature for the chiral symmetry restoration as a function of magnetic field. Motivated by the property of asymptotic freedom in QCD to introduce an ansatz that implements a running coupling with T and B. It is well know in the literature [37] that if G decreases with the temperature, general features of the chiral phase transition do not change considerably. But this is not the case in the regime of strong magnetic fields. The quark condensates $\langle\bar{\psi}_u\psi_u\rangle$ and $\langle\bar{\psi}_d\psi_d\rangle$ increase with B, due to the magnetic catalysis. The constituent quark mass M in the NJL model also increases with B, as it is given by the gap equation $M_f \sim -G\langle\bar{\psi}_f\psi_f\rangle$. In our work in Ref. [36] we have shown that the lattice result that T_{pc} decreases with B can be explained when the coupling G decreases with B and T.

There are several directions for future work. We can improve the ansatz to get a quantitative agreement with the lattice. The extension of NJL model including asymptotic freedom proposed in Ref. [36] can be used to study the behavior of other quantities such as meson-quark couplings and decay constants. We can improve the ansatz including dependence with chemical potential, working in a region that is inaccessible to the lattice. At finite density and in the presence of B we can study the phenomenology of stars [43], in particular magnetars.

The parametrization $G(B,T)$ mimics asymptotic freedom of QCD and it can be understood as the back reaction of the sea quarks. First principles, nonperturbative calculations of the effects of an external magnetic field are provided by Dyson-Schwinger equations; for calculations at finite T and μ, see Ref. [39]. The coupling of sea quarks with the magnetic field appear when we solve the system of integral equations for gluon and quark propagators. Such an approach was recently applied to study fermion mass generation at finite B in QED [44, 45] and QCD [26, 46, 47]. The extension of DSE studies to finite T and B should shed light on the phenomenon of IMC beyond the scope of effective models for QCD.

Acknowledgments
We thank Gergely Endrodi for discussions and also for providing the lattice data of the up and down quark condensates. This work was partially financed by Fundação de Amparo à Pesquisa do Estado de São Paulo (FAPESP), Grant No. 2009/50180-0 (G.K.) and 2013/01907-0 (G.K.); Coordenação de Aperfeiçoamento de Pessoal de Nível Superior (CAPES), (K.P.G); Conselho Nacional de Desenvolvimento Científico e Tecnológico (CNPQ), Grants No. 305894/2009-9 (G.K.), 475110/2013-7 (R.L.S.F.), 303592/2013-3 (M.B.P); Fundação de Amparo à Pesquisa do estado de Minas Gerais (FAPEMIG), Grant No CEX APQ 01415/10 (R.L.S.F); Fundação de Amparo à Pesquisa e Inovação do Estado de Santa Catarina (FAPESC), Grant No 3448/2012 (M.B.P.).

References

[1] Fukushima K, Kharzeev D E, and Warringa H J 2008 *Phys. Rev.* D **78** 074033
[2] Kharzeev D E and Warringa H J 2009 *Phys. Rev.* D **80** 0304028
[3] Kharzeev D E 2009 *Nucl. Phys.* A **830** 543c
[4] Tuchin K 2013 *Adv. High Energy Phys.* **2013** 490495
[5] Duncan R and Thompson C 1992 *Astron. J* **32** L9
[6] Kouveliotou C et al. 1998 *Nature* **393** 235.
[7] Vaschapati T 1991 *Phys. Lett.* B **265** 258
[8] Aoki Y, Endrodi G, Fodor Z, Katz S D and Szabo K K 2006 *Nature* **443** 675
[9] Aoki Y, Fodor Z, Katz S D and Szabo K K 2006 *Phys. Lett.* B **643** 46
[10] D'Elia M, Mukherjee S and Sanfilippo F 2010 *Phys. Rev.* D **82** 051501
[11] Ilgenfritz E M, Kalinowski M, Müller-Preussker M, Petersson N and Schreiber A 2012 *Phys. Rev.* D **85** 114504
[12] Bali G S, Bruckmann F, Endrödi G, Fodor Z, Katz S D, Krieg S, Schäfer A and Szabó K K 2012 *JHEP* **1202** 044
[13] Bali G S, Bruckmann F, Endrödi G, Fodor Z, Katz S D, Krieg S and Schäfer A 2012 *Phys. Rev.* D **86** 071502(R)
[14] Klevansky S P and Lemmer R H 1989 *Phys. Rev.* D **39** 3478
[15] Gusynin V P, Miransky V A and Shovkovy I A 1995 *Phys. Lett.* B **349** 477
[16] Shushpanov I A and Smilga A V 1997 *Phys. Lett.* B **402** 351
[17] Fraga E S and Mizher A J 2008 *Phys. Rev.* D **78** 025016
[18] Mizher A J, Chernoub M N and Fraga E S 2010 *Phys. Rev.* D **82** 105016
[19] Avancini S S, Menezes D P, Pinto M B and Providencia C 2012 *Phys. Rev.* D **85** 091901
[20] Ferrari G N, Garcia A F and Pinto M B 2012 *Phys. Rev.* D **86** 096005
[21] Andersen J O and Khan R 2012 *Phys. Rev.* D **85** 065026
[22] Menezes D P, Pinto M B, Avancini S. S., Pérez Martínez A and Providência C 2009 *Phys. Rev.* C **79** 035807
[23] Andersen J O and Tranberg A *JHEP* **08** 002
[24] Skokov V 2012 *Phys. Rev.* D **85** 034026
[25] Fukushima K and Hidaka Y 2013 *Phys. Rev. Lett.* **110** 031601
[26] Kojo T and Su N 2013 *Phys. Lett.* B **720** 192
[27] Fraga E S, Noronha J and Palhares L F 2013 *Phys. Rev.* D **87** 114014
[28] Sakai Y, Sasaki T, Kouno H and Yahiro M 2010 *Phys. Rev.* D **82** 076003; *ibid* 2012 *J. Phys. G* **39** 035004
[29] Sasaki T, Sakai Y, Kouno H and Yahiro M 2011 *Phys. Rev.* D **84** 091901(R)
[30] Fukushima K, Ruggieri M and Gatto R 2010 *Phys. Rev.* D **81**, 114031
[31] Ferreira M, Costa P, Menezes D P, Providência C and Scoccola N 2014 *Phys. Rev.* D **89**, 016002
[32] Fraga E S, Mintz B W and Schaffner-Bielich J 2014 *Phys. Lett.* B **731** 154
[33] Andersen J O, Naylor W R and Tranberg A 2014 arXiv:1410.5247
[34] Bruckmann F, Endrodi G and Kovacs T G 2013 *JHEP* **1304** 112
[35] Miransky V A and Shovkovy I A 2002 *Phys. Rev.* D **66** 045006
[36] Farias R L S, Gomes K P , Krein G and Pinto M B 2014 *Phys. Rev.* C **90** 025203
[37] Bernard V, Meissner U G and Zahed I 1987 *Phys. Rev.* D **36**, 819
[38] Qin S, Chang L, Li Y and Roberts C D 2011 *Phys. Rev.* D **84** 014017
[39] Roberts C D and Schmidt S M 2000 *Prog. Part. Nucl. Phys.* **45**, S1
[40] Nambu Y and Jona-Lasinio G 1961 *Phys. Rev.* **122** 345
[41] Buballa M 2005 *Phys. Rep.* **407** 205
[42] Kneur J L, Pinto M B and Ramos R O 2010*Phys. Rev.* C **81** 065205

[43] Preis F, Rebhan A and Schmitt A 2011 *JHEP* **1103** 033; Preis F, Rebhan A and Schmitt A 2013 *Lect. Notes Phys.* **871** 51
[44] Ayala A, Bashir A, Raya A and Rojas E 2006 *Phys. Rev.* D **73** 105009
[45] Rojas E, Ayala A, Bashir A and Raya A 2008 *Phys. Rev.* D **77** 093004
[46] Watson P and Reinhardt H 2014 *Phys. Rev.* D **89** 045008
[47] Mueller N, Bonnet J A and Fischer C S, arXiv:1401.1647 [hep-ph].

Charm tetraquarks in a non-relativistic quark model

V.R. Debastiani, F.S. Navarra

Instituto de Física, Universidade de Sao Paulo, Rua do Matao Travessa R, 187, 05508-090 Sao Paulo, SP, Brazil

E-mail: vinicius.debastiani@gmail.com, navarra@if.usp.br

Abstract.
New charmonium states have been discovered since 2003. Although some of them can be explained as $c\bar{c}$ states, others do not fit in the predictions of that spectrum. Exotic models have been proposed to explain these new states, regarded now as multiquark states in the charm sector. In this work we focus on two of the most discussed models: the *tetraquark*, which consists on a diquark-antidiquark system, and the *meson molecule*, which is a weakly bound state of two mesons. We have developed a non-relativistic approach to study exotic heavy hadron spectroscopy, where the four body system is considered as three subsequent two-body systems. We solve numerically the Schroedinger equation using an effective Cornell-like potential to model the two-body interaction in each step of the tetraquark calculation, and a Yukawa-like potential to describe the meson molecule.

1. Introduction

Due to the B-meson factories there has been great progress on the study of the charmonium states. These machines are e^+e^- colliders operating on a center of mass energy of 10.600 MeV, just enough to produce B meson pairs. The decay of B produces charmonium. Over the past decade, the BABAR, BELLE and BESIII collaborations, working in different B factories, produced a large amount of data on charmonium spectra. Since 2003, new charmonium states have been continuously discovered [1, 2, 3]. While some of them, like the $X(3943)$, can be explained as conventional $c\bar{c}$ states, others, such as the $X(3872)$, do not fit in the predictions of the spectrum of $c\bar{c}$ states and are called exotic charmonium states [1, 2, 3]. Models have been proposed to explain these new states, considering them as four quark states in the charm sector ($cq\bar{c}\bar{q}$, $q \in u, d$). The most discussed models are the tetraquark, the meson molecule, hadro-charmonium, hibrid states and mixtures. Here, the tetraquark model consists on a diquark-antidiquark system ($Qq - \overline{Q}\overline{q}$), in which the diquarks interact through the exchange of colored objects, and the attraction force can be intense. In contrast, the meson molecule ($Q\bar{q} - \overline{Q}q$) is a weakly bound state of two mesons.

1.1. Meson molecules

The molecular picture has been used to understand the well established $X(3872)$, which is considered to be a DD^* molecule. Of course, the actual physical state is probably rather complex, with a short-range component of (cc) nature and a long-range component with a charmed meson and an anticharmed meson almost bound by attractive Yukawa forces. The idea of molecules with hidden charm has been proposed long ago by Okun and Voloshin [4]. A molecular interpretation was proposed for some high-lying 1^{--} charmonium resonances [5], due to puzzling branching ratios into DD, DD^* and D^*D^*, which turned out to be due to the node structure of these states as radial excitations of the J/ψ [6]. The possibility of hidden-charm meson molecules has been revisited in the 90s by Törnqvist [7], Ericson and Karl [8] and Manohar and Wise [9]. More recently, after the discovery of the $X(3872)$, the meson molecule picture was further developed by several authors [10]. The main idea is that the Yukawa interaction, that successfully binds nuclei, is not restricted to the nucleon-nucleon interaction. The exchange of light mesons also generates a potential between flavored mesons, which is sometimes attractive. Although usually weaker than the proton-neutron interaction that binds the deuteron, it is probed by much heavier particles, and thus can lead to bound states with a binding energy of a few MeV, or even a few tens of MeV, as shown by Törnqvist [7].

1.2. Tetraquarks

The first papers discussing the existence of tetraquark configurations were all based on the MIT bag model and were published in the seventies [11, 12]. In the beginning, only light tetraquark states were considered. Later on, Weinstein and Isgur [13, 14] extended the tetraquark picture to other quark models and tetraquarks with heavy quarks were included. After the discovery of the $X(3872)$ the multiquark picture has been revisited by Jaffe and Wilczek [15] and more recently by other authors [16, 17, 18, 19].

A very popular representation of the pictures above discussed can be seen in Fig. 1.

In this work, following the steps of Ref. [19], we have developed a non-relativistic approach to study exotic heavy hadron spectroscopy, where the heavy hadrons are tetraquarks. The four-body system is considered as three two-body systems. Our work consists on numerically solving the Schroedinger equation using an effective Cornell-like potential ($V(r) = -\alpha/r + \sigma r$) to model the two-body interactions among the hadron constituents. More specifically we wish to answer two questions: 1) Can the $Z(4430)$ be the first radial excitation of the $X(3872)$? 2) Is there a T_{cc} $(ccqq)$? If so with which mass ?

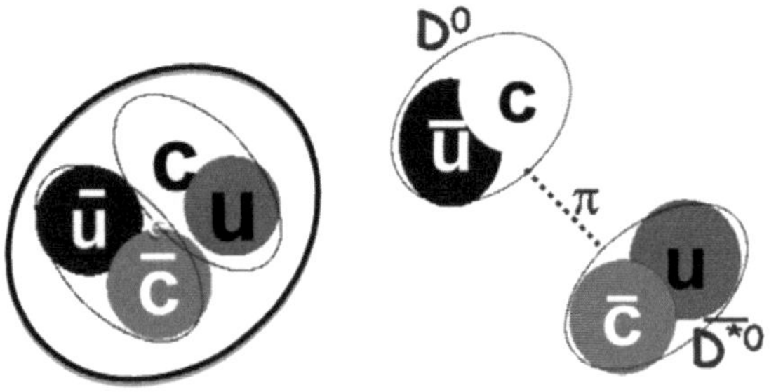

Figure 1. $X(3872)$ as a tetraquark (left panel) and as a meson molecule (right panel).

2. A non-relativistic model

2.1. The Schroedinger equation

In the charmonium states the kinetic energy is small compared to the rest energy of the constituents hence and we can develop a non-relativistic approach. In order to obtain the energy spectrum we only need to solve the time independent Schroedinger equation:

$$\left[\frac{1}{2\mu}\left(-\frac{d^2}{dr^2}+\frac{l(l+1)}{r^2}\right)+V(r)\right]y_{n,l}(r)=E_{n,l}\,y_{n,l}(r) \tag{1}$$

where $y_{n,l}(r) = rR(r)$ is the reduced radial wave function and $\mu \equiv \frac{m_1 m_2}{m_1+m_2}$ is the reduced mass (we use natural units $h = c = 1$). By solving the above eigenvalue equation, we find the interaction energy of the two body system, $E_{n,l}$, which depends on the number of nodes and on the orbital angular momentum. We are also interested in the wave function itself, which will be used to calculate the normalization factor, the mean square radius and the spin-dependent corrections. With the Cornell potential the above equation has no analytical solution and we need to solve it numerically. We have used the program written by the authors of [20], built specially to solve Schroedinger equation for bound states.

2.2. The Cornell Potential

The Cornell Potential [21] has a Coulomb term, representing the one gluon exchange in CD, and a linear term responsible for confinement. It is given by:

$$V(r) = V_{Coul} + V_{conf} = \kappa_s \frac{\alpha_s}{r} + \sigma r \tag{2}$$

In the above equation the constant κ_s (the color factor) is related to the color configuration of the quarks. The constant α_s is the CD analogue to the fine structure constant of the eletromagnetic interactions. The constant σ is called the string tension.

In our calculation we first use the potential (2) to calculate the mass of the diquark, solving the two body problem with the masses of a charm and a light quark. Then, the diquark mass is calculated by summing the masses of its constituents plus the interaction energy, i.e., $m_d = m_c + m_q + E_{cq}$, where E_{cq} is the solution of (1). Spin-dependent corrections can be included perturbatively, but their contribution is much smaller than the ones coming from a small change in the Cornell potential parameters, so they have been neglected at this stage.

The antidiquark mass is calculated in the same way (without spin-corrections it is essentially the same as the diquark mass). Then the tetraquark (treated now as a diquark-antidiquark two-body system) interaction energy is calculated, using different parameters in the potential, and the diquark masses as inputs. The final state mass is the sum of diquark and antidiquark masses and the interaction energy between them.

3. Results

3.1. $Z(4430)$ as the first excitation of $X(3872)$

Relations between exotic states have been suggested in recent works on the field [3]. Here we explore the relation between the $X(3872)$ and $Z(4430)$, assumed to be tetraquarks. We adjust the parameters of (2) to reproduce the mass of the $X(3872)$ and then, with all the parameters fixed, we calculate the energy of the first excitation. Using $m_c = 1.486$ GeV, $m_u = m_d = 0.323$ GeV, as suggested in [19], we find the diquark mass to be: $m_d = 1.949$ GeV. Then we use this mass as input to calculate the tetraquark state with constants chosen based on a parametrization of the usual $c\bar{c}$ meson [21], where the final state energy is calculated subtracting a constant. Here we use $\Delta E = 0.312$ GeV and hence:

$$m_{4q} = m_d + m_{\bar{d}} + E_{d\bar{d}} - \Delta E \quad (3)$$

For the ground state we find $m_{4q} = m_X = 3.872$ GeV, and the mass of the first radial excitation (one node) turns out to be:

$$m_{4q} = m_Z = 4.431\text{GeV} \quad (4)$$

The values of the parameters used in the calculation are listed in Table 1 and the wave functions are plotted in Fig. 2 and 3, where $R(r) = y(r)/r$ (see equation 1).

Table 1. Parameters and results for X(3872) as a ground state tetraquark and Z(4430) as its first radial excitation ($n = 1$). Units are GeV^2 for σ, GeV for energies and masses ($c = 1$) and fermi (fm) for the radius.

Object	κ_s	α_s	σ	m_1	m_2	n	l	E_{int}	ΔE	E_{total}	$< r^2 >^{1/2}$
diquark	$-2/3$	0.30	0.015	1.486	0.323	0	0	0.140	0	1.949	1.538
tetraquark	$-4/3$	0.34	0.180	1.949	1.949	0	0	0.286	0.312	3.872	0.379
tetraquark	$-4/3$	0.34	0.180	1.949	1.949	1	0	0.845	0.312	4.431	0.756

This result strongly favors the assignment proposed first in [17] and then in [3] and also in [22].

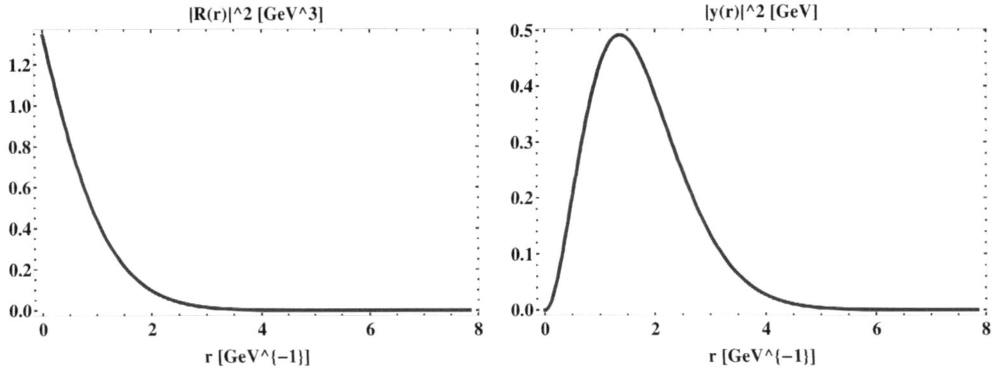

Figure 2. Radial squared normalized wave function $|R(r)|^2$ and reduced radial squared normalized wave function $|y(r)|^2$ for $X(3872)$ as the ground state in the tetraquark model.

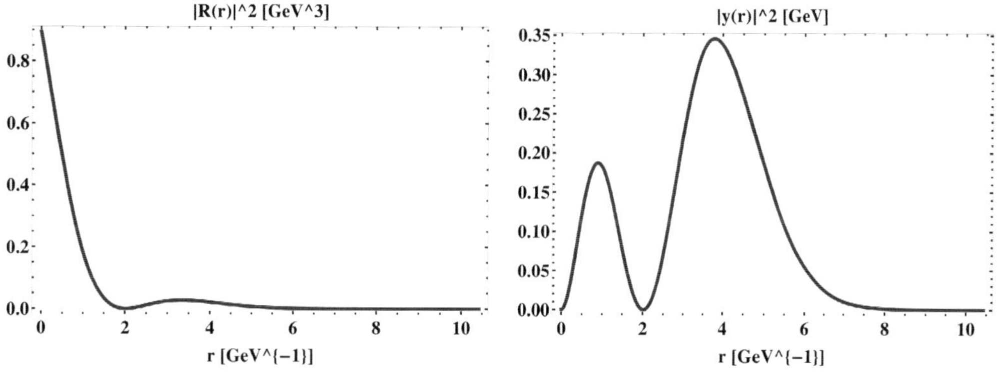

Figure 3. Radial squared normalized wave function $|R(r)|^2$ and reduced radial squared normalized wave function $|y(r)|^2$ for $Z(4430)$ as the first radial excitation of $X(3872)$ in the tetraquark model.

3.2. The meson molecule model

The meson molecule, shown in Fig. 1, is a weakly bound state of two mesons $(Q\overline{q} - \overline{Q}q)$, usually described as $D^0 - \overline{D}^{*0}$. Its mass can be calculated in the same way as in the tetraquark model, solving Eq. 1 for the interaction of the quarks inside the meson, and then for the interaction between the two mesons inside the molecule. However, instead of this three steps approach, here we already adopt the experimentally measured meson masses $m_{D^0} = 1.864$ GeV and $m_{\overline{D}^{*0}} = 2.008$ GeV in the molecule calculation.

The interaction between mesons bound in a molecule can be described by a Yukawa potential, representing the exchange of light mesons like the pion. Because of the exponential decrease in this binding force, the meson molecule is expected to have a weaker coupling and to be bigger in size compared to the tetraquark. Also, it may not be able to support radial excitations, which would destroy the bound state. The Yukawa potential adopted is the following:

$$V_Y = -g^2 \frac{e^{-kr}}{r} \tag{5}$$

First we choose $k = m_{\pi^\pm} = 0.139$ GeV and $g^2 = 1$. Solving Eq. 1, we obtain negative interaction energies for the ground state and its first radial excitation, as shown in Table 2. The wavefunctions are shown in Figs. 4 and 5:

Table 2. Parameters and results for the $D^0 - \overline{D}^{*0}$ meson molecule bound by the Yukawa potential with $k = m_{\pi^\pm} = 0.139$ GeV. Units are GeV for energies and masses ($c = 1$) and fermi (fm) for the radius.

Object	g^2	k	m_{D^0}	$m_{D^{*0}}$	n	l	E_{int}	E_{total}	$< r^2 >^{1/2}$
molecule	1	0.139	1.864	2.008	0	0	−0.358	3.514	0.360
molecule	1	0.139	1.864	2.008	1	0	−0.029	3.843	1.585

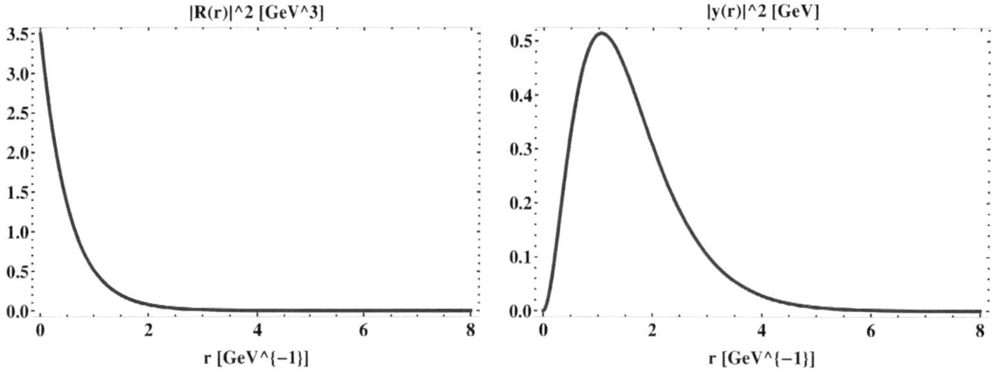

Figure 4. Radial squared normalized wave function $|R(r)|^2$ and reduced radial squared normalized wave function $|y(r)|^2$ for the $D^0 - \overline{D}^{*0}$ meson molecule bound by the Yukawa potential with $k = m_{\pi^{\pm}} = 0.139$ GeV.

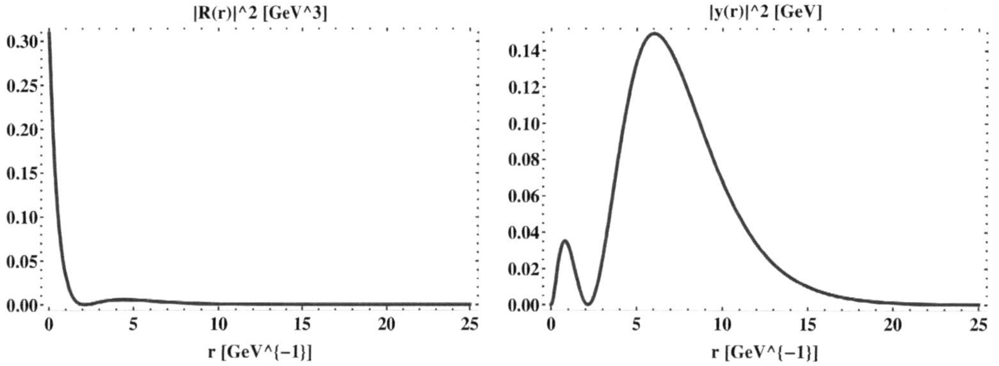

Figure 5. Radial squared normalized wave function $|R(r)|^2$ and reduced radial squared normalized wave function $|y(r)|^2$ for the first radial excitation of the $D^0 - \overline{D}^{*0}$ meson molecule in Yukawa-like potential with $k = m_{\pi^{\pm}} = 0.139$ GeV.

If we choose $k = 1$, we obtain a much more weakly bound ground state, as shown in Table 3:

Table 3. Parameters and results for the $D^0 - \overline{D}^{*0}$ meson molecule bound by the Yukawa potential with $k = 1$ GeV. Units are GeV for energies and masses ($c = 1$) and fermi (fm) for the radius.

Object	g^2	k	m_{D^0}	$m_{D^{*0}}$	n	l	E_{int}	E_{total}	$< r^2 >^{1/2}$
molecule	1	1	1.864	2.008	0	0	-0.007	3.865	1.373
molecule	1	1	1.864	2.008	1	0	4×10^{-6}	3.872	110.189

Now the first radial excitation has an unrealistically large radius with negligible interaction energy. This is an indication that this excited state cannot exist, supporting our hypothesis that the meson molecule is not able to hold radial excitations. The wavefunctions are shown in Figs. 6, 7 and 8. The $|y(r)|^2$ wavefunction acctually has a wide range. A zoom around the origin is shown in Fig. 7 and the whole range is shown in 8.

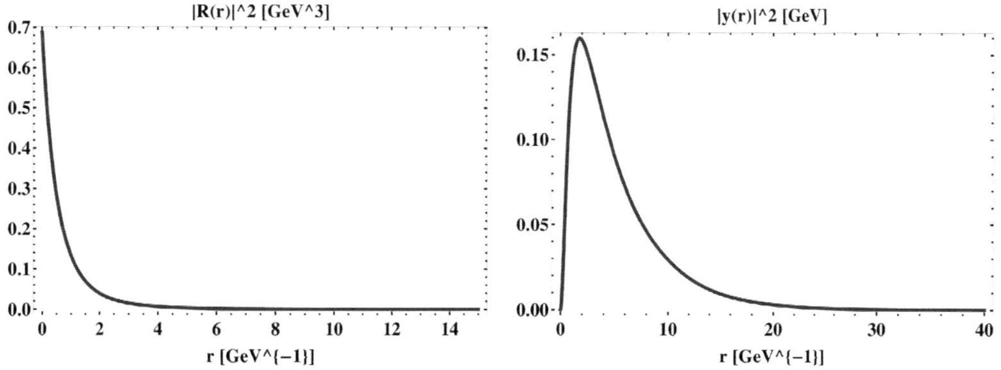

Figure 6. Radial squared normalized wave function $|R(r)|^2$ and reduced radial squared normalized wave function $|y(r)|^2$ for the $D^0-\overline{D}^{*0}$ meson molecule bound by the Yukawa potential with $k = 1$ GeV.

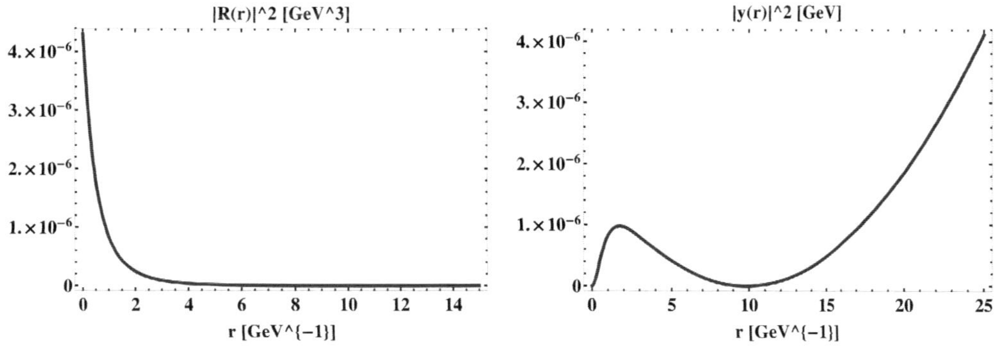

Figure 7. Radial squared normalized wave function $|R(r)|^2$ and zoom around the origin of the reduced radial squared normalized wave function $|y(r)|^2$ for the first radial excitation of the $D^0 - \overline{D}^{*0}$ meson molecule bound by the Yukawa potential with $k = 1$ GeV.

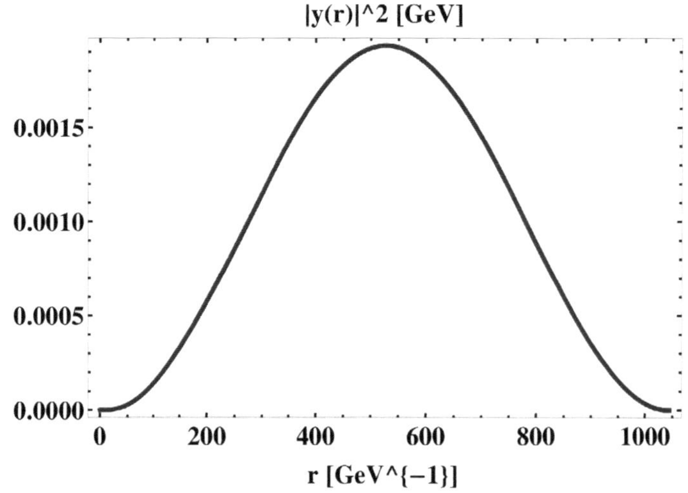

Figure 8. Complete range of the reduced radial squared normalized wave function $|y(r)|^2$ for the first radial excitation of the $D^0 - \overline{D}^{*0}$ meson molecule in Yukawa-like potential with $k = 1$ GeV.

3.3. The T_{cc} tetraquark

In this subsection we present our results for the tetraquark configuration known as T_{cc} [23]. In this configuration, instead of a diquark composed by a charm and a light quark, we consider a diquark as being composed by two charm quarks, whereas the two light antiquarks compose the antidiquark, forming a $cc - \overline{qq}$ tetraquark structure, where $q \in u, d$, (or $\overline{cc} - qq$). Using the same parameters as in [19], we calculate the T_{cc} ground state energy and the first radial excitation ($n = 1$). In these calculations we have adopted $\Delta E = 0$. Parameters and results are listed in Table 4 and wavefunctions in Figs. 9 and 10.

Table 4. Parameters and results for Tcc tetraquark model in ground state tetraquark and its first radial excitation ($n = 1$). Units are GeV^2 for σ, GeV for energies and masses ($c = 1$) and fermi (fm) for the radius.

Object	κ_s	α_s	σ	m_1	m_2	n	l	E_{int}	E_{total}	$< r^2 >^{1/2}$
diquark cc	$-2/3$	0.3	0.015	1.486	1.486	0	0	0.072	3.044	0.990
antidiquark $\overline{dd}$	$-2/3$	0.3	0.015	0.323	0.323	0	0	0.178	0.824	1.864
tetraquark	$-4/3$	0.3	0.030	3.044	0.824	0	0	0.072	3.940	0.749
tetraquark	$-4/3$	0.3	0.030	3.044	0.824	1	0	0.278	4.146	1.539

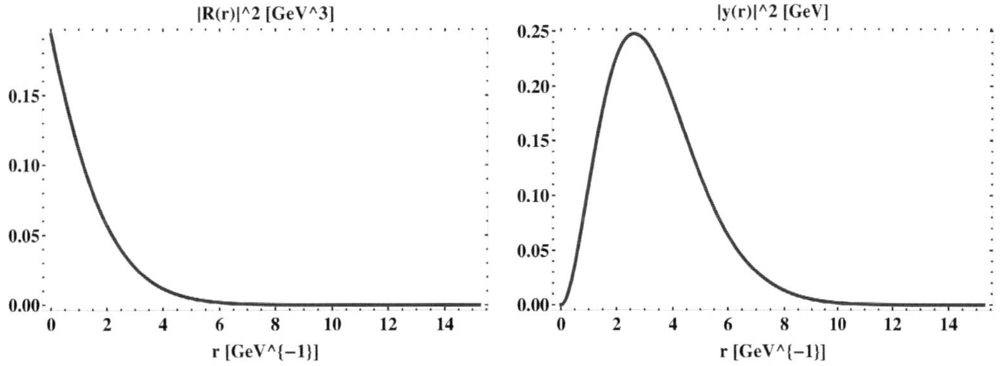

Figure 9. Normalized squared reduced radial wavefunction $|y(r)|^2$ and normalized squared radial wavefunction $|R(r)|^2$ for the T_{cc} ground state.

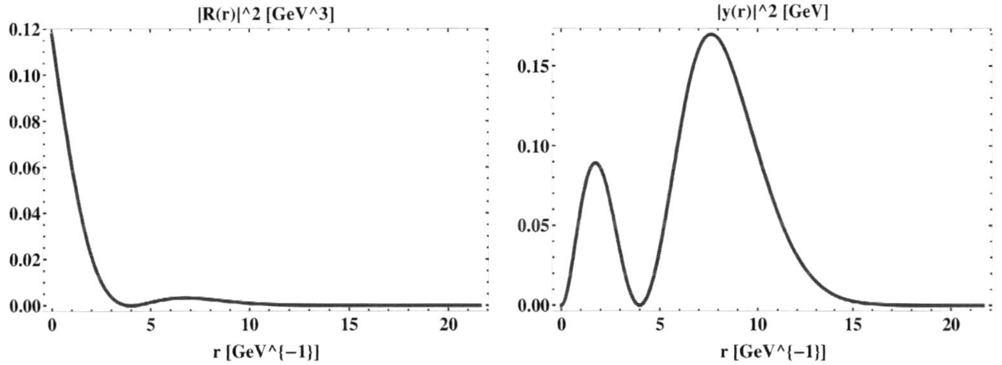

Figure 10. Normalized squared reduced radial wavefunction $|y(r)|^2$ and normalized squared radial wavefunction $|R(r)|^2$ for the first radial excitation of the T_{cc} ($n = 1$).

4. Conclusions

As it could be seen, in quark models, as the one used here, it is difficult to justify the choice of some parameters. Finding reliable theoretical justification for those choices is an important task yet to be done. In spite of the limitations of our approach, our results are rather suggestive. We can say that we could give preliminary answers to the questions raised in the introduction. It is very interesting to observe how, fixing the parameters so as to reproduce the $X(3872)$ mass, we can obtain the first radial excitation with the mass so close to the $Z(4430)$ mass. Assuming that the $X(3872)$ is a meson molecule interacting through a Yukawa potential, we can reproduce its mass. It remains to be checked that the radius of the $X(3872)$ is larger than 1 fm. Moreover it seems quite clear that the meson molecule does not have radial excitations. Finally, we find that the T_{cc} mass is very close to estimates made with other methods, such as in Ref. [23].

5. Acknowledgments

We would like to thank Prof. Dr. Franz F. Schoeberl for sending us the Mathematica notebook. This work was partially supported by FAPESP, CNPq, Pró-Reitoria de Pesquisa da Universidade de São Paulo and by the Santander Bank.

6. References

[1] For a comprehensive review on charmonium physics, see N. Brambilla, *et al*, Eur. Phys. J. C **71**, 1534 (2011).
[2] For a review on exotic charmonium states, see M. Nielsen, F.S. Navarra, S.H. Lee, Phys. Rept. **497**, 41 (2010).
[3] For a review on charged exotic charmonium states, see M. Nielsen and F.S. Navarra, Mod. Phys. Lett. A **29**, 1430005 (2014).
[4] M. B. oloshin and L. B. Okun, JETP Lett. **23**, 333 (1976) [Pisma Zh. Eksp. Teor. Fiz. **23**, 369 (1976)].
[5] A. De Rujula, H. Georgi and S. L. Glashow, Phys. Rev. Lett. **38**, 317 (1977).
[6] A. Le Yaouanc, L. Oliver, O. Pene and J. C. Raynal, Phys. Lett. B **71**, 397 (1977) **72**, 57 (1977) E. Eichten, K. Gottfried, T. Kinoshita, K. D. Lane and T. M. Yan, Phys. Rev. D **21**, 203 (1980).
[7] N. A. Tornqvist, Z. Phys. C **61**, 525 (1994).
[8] T. E. O. Ericson and G. Karl, Phys. Lett. B **309**, 426 (1993).
[9] A. . Manohar and M. B. Wise, Nucl. Phys. B **399**, 17 (1993).
[10] E. S. Swanson, Phys. Rept. **429**, 243 (2006).
[11] R.L. Jaffe, Phys. Rev. D **15**, 281 (1977).
[12] R.L. Jaffe, Phys. Rev. D **15**, 267 (1977).
[13] J.D. Weinstein and N. Isgur, Phys. Rev. D **27**, 588 (1983).
[14] J.D. Weinstein and N. Isgur, Phys. Rev. D **41**, 2236 (1990).
[15] R.L. Jaffe, F. Willczek, Phys. Rev. Lett. **91**, 232003 (2003).
[16] L. Maiani, F. Piccinini, A.D. Polosa and . Riquer, Phys. Rev. D **71**, 014028 (2005).
[17] L. Maiani, A.D. Polosa and . Riquer, ar iv:0708.3997, and New Journal of Physics, **10** 073004 (2008).
[18] J. ijande, A. alcarce, J.M. Richard and N. Barnea, Few Body Syst. **45**, 99 (2009).
[19] S. Patel, M. Shah and P.C. inodkumar, Eur. Phys. J. A **50**, 131 (2014).
[20] W. Lucha, F.F. Schoeberl, Int. J. Mod. Phys. C **10**, 607 (1999).
[21] E. Eichten, K. Gottfried, T. Kinoshita, K.D. Lane, T.M. Yan, Phys Rev. D **21**, 203 (1980).
[22] F.S. Navarra, M. Nielsen and J.M. Richard, J. Phys. Conf. Ser. **348**, 012007 (2012).
[23] F.S. Navarra, M. Nielsen and S. H. Lee, Phys. Lett. B **649**, 166 (2007).

On improvements of Double Beta Decay using FQTDA Model

L. de Oliveira[†,a] **A.R. Samana**[a], **F. Krmpotić**[b,c], **A.E. Mariano**[b,c] **C.A. Barbero**[b,c]

[a] Dep. de Cs. Exatas e Tecnológicas, Universidade Estadual de Santa Cruz, Ilhus, Brazil
[b]Departamento de Fsica, Universidad Nacional de La Plata, C. C. 67, 1900 La Plata, Argentina [c] Instituto de Física La Plata,CONICET, 1900 La Plata, Argentina

E-mail: †leandroliv@yahoo.com.br

Abstract.
The Quasiparticle Tamm-Dancoff Approximation (QTDA) is applied to describe the nuclear double beta decay with two neutrinos. Several serious inconveniences found in the Quasiparticle Random Phase Approximation (QRPA) are not present in the QTDA, as such as the ambiguity in treating the intermediary states, and further approximations necessary for evaluation of the nuclear matrix elements (NMEs) or, the extreme sensitivity of NME with the ratio between the pn and $pp + nn$ pairings. Some years ago, the decay ^{48}Ca $\rightarrow^{48}$Ti was discussed within the particle-hole limit of QTDA. We found some mismatch in the numerical calculations when the full QTDA was being implemented, and a new performance in the particle-hole limit of QTDA is required to guarantee the fidelity of the approximation.

1. INTRODUCTION

At the end of the 20th century the neutrino physics was born. This new era in neutrino physics was opened by the experimental evidence of neutrino oscillations obtained from the results of atmospheric, solar, reactor, and accelerator neutrino experiments. Nevertheless, we do not until this moment which is the absolute scale mass, and whether the neutrino is a Majorana or Dirac particle. The atomic nuclei are the detectors of the evasive neutrinos and the key of this puzzle is the neutrinoless double beta decay. The next generation of experiment for many different nuclei are searching for this rare decay mode, including ^{48}Ca, ^{76}Ge, ^{100}Mo, ^{116}Cd,^{130}Te, ^{124,126,134}Xe, ^{136}Ce, ^{150}Nd, and ^{160}Gd. A summary of the experiments with the above nuclei is well explained in recent review works as such as in Barabash [1], or Tosi [2]. In particular ^{48}Ca is studied at NEMO-III [3], CANDLES [4] and CARVEL [5], because the $Q_{\beta\beta}$-value lies well above the energy of naturally occurring background, a good signal-to-noise ratio is ensured, while the large phase-space factor enhances the $0\nu\beta\beta$-decay rate as it is remarked in Ref. [6].

The single β-decay is energetically forbidden from about 50 nuclear systems in nature, then the $\beta\beta$-decay turns out to be the only possible mode of disintegration. This kind of anomaly is due to the nuclear pairing force, by making the odd-odd isobar, within the isobaric triplet (N, Z), $(N-1, Z+1)$, $(N-2, Z+2)$, to have a higher mass than the even-even neighbors. The usual modes of $\beta\beta$ disintegrations are: (i) the two-neutrino double beta ($2\nu\beta\beta$) decay, that can occur by two successive β decays, passing through the intermediate virtual states of the

$(N-1, Z+1)$ nucleus, and (ii) the neutrinoless $\beta\beta$ $(0\nu\beta\beta)$ decay, where there are not neutrinos in the final state. The half-lives of these decays are written as:

$$T_{2\nu}^{-1} = \mathcal{G}_{2\nu}\mathcal{M}_{2\nu}^2, \quad T_{0\nu}^{-1} = \mathcal{G}_{0\nu}\mathcal{M}_{0\nu}^2\langle m_\nu\rangle^2, \tag{1}$$

respectively, where $\mathcal{G}'s$ are geometrical phase space factors, and $\mathcal{M}'s$ are nuclear matrix elements (NME's) which present many similar features. There is consensus in the scientific community that we shall not understand the $0\nu\beta\beta$-decay unless we understand the $2\nu\beta\beta$-decay. So that, our goal will be describe in a consistent way the $2\nu\beta\beta$-decay.

Independently of the nuclear model which is used, and when only the allowed transitions are considered, the $2\nu\beta\beta$ matrix element for the $|0_f^+\rangle$ final state [8] reads

$$\mathcal{M}_{2\nu}(f) = \sum_{\lambda=0,1}(-)^\lambda\sum_\alpha\left[\frac{\langle 0_f^+||\mathcal{O}_\lambda^{\beta^-}||\lambda_\alpha^+\rangle\langle\lambda_\alpha^+||\mathcal{O}_\lambda^{\beta^-}||0^+\rangle}{\mathcal{D}_{\lambda_\alpha^+,f}}\right] \equiv \mathcal{M}_{2\nu}^F(f) + \mathcal{M}_{2\nu}^{GT}(f), \tag{2}$$

where the summation goes over all intermediate virtual states $||\lambda_\alpha^+\rangle$, and

$$\mathcal{O}_\lambda^{\beta^-} = (2\lambda+1)^{-1/2}\sum_{pn}\langle p||\mathrm{O}_\lambda||n\rangle\left(c_p^\dagger c_{\bar{n}}\right)_\lambda, \text{ where } \begin{cases} \mathrm{O}_0 = 1 & \text{, for F} \\ \mathrm{O}_1 = \sigma & \text{, for GT} \end{cases}, \tag{3}$$

We assume that Fermi (F) and Gamow-Teller (GT) operators for β^--decay, and $c^\dagger$ (c) are the particle creation (annihilation) operators. The corresponding β^+-decay operators are $\mathcal{O}_\lambda^{\beta^+} = \left(\mathcal{O}_\lambda^{\beta^-}\right)^\dagger$, and

$$\mathcal{D}_{\lambda_\alpha^+,f} = E_{\lambda_\alpha^+} - \frac{E_0 + E_{0_f^+}}{2} = E_{\lambda_\alpha^+} - E_0 - \frac{E_{0_f^+} - E_0}{2}, \tag{4}$$

is the energy denominator. E_0 and $E_{0_f^+}$ are, respectively, the energy of the initial state $|0^+\rangle$ and of the final states $|0_f^+\rangle$. There is a divergent issue about the existence of the Fermi transition in double beta decay. Some authors [9] claim that there is not Fermi transition in double beta decay due the isospin conservation. For other hand some authors this transition could be exist, assuming the non-isospin conservation and after a partial restoration of isospin and SU(4) symmetries [13]. We leave this discussion to a future work.

The $\beta\beta$ decays occur in medium-mass nuclei that are often far from closed shells, and as a consequence most of the recent attempts to evaluate $\mathcal{M}_{2\nu}$ and $\mathcal{M}_{0\nu}$ are: (i) the RPA makes a large fraction of nucleons to take part in a large single-particle space, but within a modest configuration space, so it is by far simpler computationally, and (ii) the SM (Shell Model) deals with a small fraction of the nucleons in a limited single-particle space, but allows them to correlate in arbitrary ways within a large configuration space. Already it was noted in Ref. [8] that the kinds of correlations that these methods include are not the same. Also, various and different RPA calculations have been solved, but that none of the amendments of the RPA were able to change qualitatively the behavior of the amplitude $\mathcal{M}_{2\nu}$ (the model collapses in the physical region of the coupling constant for the particle-particle channel) unless we agree to assume the violation of the Ikeda's Sum Rule. The RPA $\beta\beta$-decay amplitudes shown to be very sensitivity to the ratio between the pn and $pp+nn$ pairings. These amplitudes before goes to zero, meaning that the isospin and Wigner SU(4) (broken by the mean field) are restored by RPA correlations. After this, the lower RPA energies become complex, and the amplitudes $\mathcal{M}_{2\nu}$ grow, developing a pole and the entire method breaks down [8].

In Ref. [8] the particle hole (ph-) limit of F TDA was employed to describe the quenching

Journal of Physics: Conference Series **630** (2015) 012048 doi:10.1088/1742-6596/630/1/012048

mechanism in the ^{48}Ca $\rightarrow^{48}$ Ti decay within the 1fp-shell. In a second work [12], the authors extended this study to a complete F TDA, still within the 1fp-shell, but some mismatch in the numerical evaluations of the NME were performed. This fact requires a new analysis and the present work is developed to correct this issue and to extent the space $1fp$-shell to a two major shell ($sdfp$-shell consisting of sd and pf shells). On the other side, new efforts and advances were obtained in SM calculations with ^{48}Ca [7]. In that work, the authors developed a new fast algorithm and computing code to calculate the two-body matrix elements of the neutrinoless double-beta decay transition operator in the closure approximation reducing considerably the evaluation computer time.

2. FOUR QUASIPARTICLE TAMM-DANCOFF APPROXIMATION

A simple nuclear model for evaluating the $\beta\beta$ decay rates, based on the well-known Four uasiparticle Tamm-Dancoff Approximation - F TDA [10] is briefly sketched here. As in the RPA, we conveniently express the total Hamiltonian as

$$H = H_p + H_n + H_{pn} + H_{pp} + H_{nn} \equiv H_0 + H_{res}, \tag{5}$$

where H_p and H_n are, respectively, the effective proton and neutron single-quasiparticle Hamiltonians (with eigenvalues ϵ_p and ϵ_n) , and H_{pn}, H_{pp}, and H_{nn} are the matching effective two-quasiparticle interaction Hamiltonians for the valence quasiparticles. Assuming that:

(1) the initial state is the BCS vacuum in the (N, Z) nucleus $|0^+\rangle = |BCS\rangle$,

(2) the intermediate states are two quasiparticle excitations on this vacuum:

$$|\lambda^+_\alpha\rangle = \sum_{pn} X_{pn;\lambda^+_\alpha}|pn;\lambda^+\rangle, \ with \ |pn;\lambda^+\rangle = [a^\dagger_p a^\dagger_n]_{\lambda^+}|BCS\rangle, \tag{6}$$

(3) the final states are four quasiparticle excitations on this vacuum:

$$|0^+_f\rangle = \sum_{p_1p_2n_1n_2J} Y_{p_1p_2n_1n_2J;0^+_f}|p_1p_2, n_1n_2; J\rangle, \tag{7}$$

with

$$|p_1p_2, n_1n_2; J\rangle = N(p_1p_2)N(n_1n_2)\{[a^\dagger_{p_1}a^\dagger_{p_2}]_J[a^\dagger_{n_1}a^\dagger_{n_2}]_J\}_0|BCS\rangle, \tag{8}$$

and $N(ab) = (1 + \delta_{ab})^{-1/2}$. Here $a^\dagger$ (a) is the quasiparticle creation (annihilation) operator relative to the BCS vacuum.

The energies in the denominator in (4) are

$$E_{\lambda^+_\alpha} = E_0 + \omega_{\lambda^+_\alpha} + \lambda_p - \lambda_n, \quad E_{0^+_f} = E_0 + \omega_{0^+_f} + 2\lambda_p - 2\lambda_n, \tag{9}$$

and the denominator energy is written as

$$\mathcal{D}_{\lambda^+_\alpha,f} = \omega_{\lambda^+_\alpha} - \frac{\omega_{0^+_f}}{2}, \tag{10}$$

where $\omega_{\lambda^+_\alpha}$ and $\omega_{0^+_f}$ are the eigenvalues of the Hamiltonian for the intermediate states $|\lambda^+_\alpha\rangle$ and of the final states $|0^+_f\rangle$, respectively, and λ_p and λ_n are the chemical potentials. A residual interaction kind δ-force: $V = -4\pi(v_sP_s + v_tP_t)\delta(r)$ is employed in the calculations as in the previous work [8].

3. NUMERICAL RESULTS

The choice of the suitable effective single-particle energies is a delicate issue. We used three sets of s.p.e labeled as:

SET 1 - s.p.e of neutrons and protons (corrected by the Coulomb interaction) of ^{40}Ca employed to describe ^{48}Ca using a core of ^{40}Ca . The ϵ^0_{jn} are from the experimental SPE for ^{40}Ca, taken from Ref. [8], namely: $\epsilon^0_{f_{7/2}} = 0$, $\epsilon^0_{f_{5/2}} = 6.5$, $\epsilon^0_{p_{3/2}} = 2.1$, and $\epsilon^0_{p_{1/2}} = 4.1$, in units of MeV. They were corrected using the self-energies that account for 8 neutrons in the $f_{7/2}$ shell, and using the Coulomb displacement energy in ^{48}Ca for the protons. This set was used in [8].

SET 2 - experimental s.p.e of neutrons and protons ^{48}Ca. They were used in the previous RPA calculations of Refs. [11, 13]).

SET 3 - modified s.p.e from usual harmonic oscillator using the inverse gap equations procedure [15] to reproduce the experimental s.p.e in ^{48}Ca.

These s.p.e. SETS are shown in Table 1. The underlined s.p.e. of Set 1 were employed to make the *ph*-limit of TDA previously performed in Refs. [8, 12] for ^{48}Ca. For *ph*-limit we understand that the occupation probabilities satisfy the conditions: $v_p \to 0$ and $v_n \to 1$. An extension of the underlined s.p.e was necessary to complete an available space where the BCS equations must be consistently solved. As it was done in Ref.[8], these additional s.p.e are taken from experimental energies for neutrons in ^{40}Ca and for protons these s.p.e are corrected by the Coulomb energy. The resulting energies complete the Set 1.

Table 1. Single-particle energies (s.p.e) e_j (in Mev) for the SET 1, SET 2 and SET 3 of s.p.e. used for ^{48}Ca according the text. The pairing strength v_s^{pair} (dimensionless) for neutrons and proton within the BCS for $N = 28$ and $P = 20$ are also sketched. The underlined s.p.e. are the energies used in the preliminary *ph*-limit of TDA.

	Neutrons			Protons		
Shell	*SET*1	*SET*2	*SET*3	*SET*1	*SET*2	*SET*3
$1f_{5/2}$	6.50	−1.54	−2.05	$\underline{12.89}$	−5.80	−6.27
$2p_{1/2}$	4.10	−3.11	−3.79	$\underline{10.49}$	−4.58	−4.86
$2p_{3/2}$	2.10	−5.14	−8.18	$\underline{8.49}$	−6.93	−7.39
$1f_{7/2}$	$\underline{0.00}$	−9.94	−8.44	$\underline{6.39}$	−9.62	−10.87
$1d_{5/2}$	−12.73	−12.52	−11.71	−6.34	−15.69	−14.10
$2s_{1/2}$	−10.50	−12.55	−11.68	−4.11	−15.30	−12.48
$1d_{3/2}$	−7.24	−14.60	−14.05	−0.85	−26.40	−18.58
v_s^{pair}	31.00	32.00	23.87	41.90	34.43	24.40

Figure 1 shows the energy dependence of double GT strengths $S^{\beta\beta}_{GT}$, measured from the ground state of ^{48}Ca. The unperturbed strength (residual interaction off) is compared with the next perturbed results (residual interaction on). The perturbed results are obtained with the *ph*-parameters $v_s = 35, v_t = 65$ and *pp*-parameters $s = \frac{v_s^{pp}}{v_s^{pair}} = 1$ and $t = \frac{v_t^{pp}}{v_s^{pair}} = 0$, where $v_s^{pair} = (v_s^{pair}(p) + v_s^{pair}(n))/2$ [13]. According the choice set of s.p.e of Table 1, these results are separated in: (a) SET 2, (b) SET 3 and, (c) SET 1. In (a),(b) and (c) the BCS equations were solved consistently in a complete space with the v_s^{pair} shown in Table 1. It means that these calculations are a complete F TDA calculations. We can observe that the unperturbed calculation for double GT strength, $S^{\beta\beta}_{GT}$, has been focused in three energy regions. On the other hand, the perturbed cases (a) SET 2 and (b) SET 3, lead a shift of strength to the left, instead

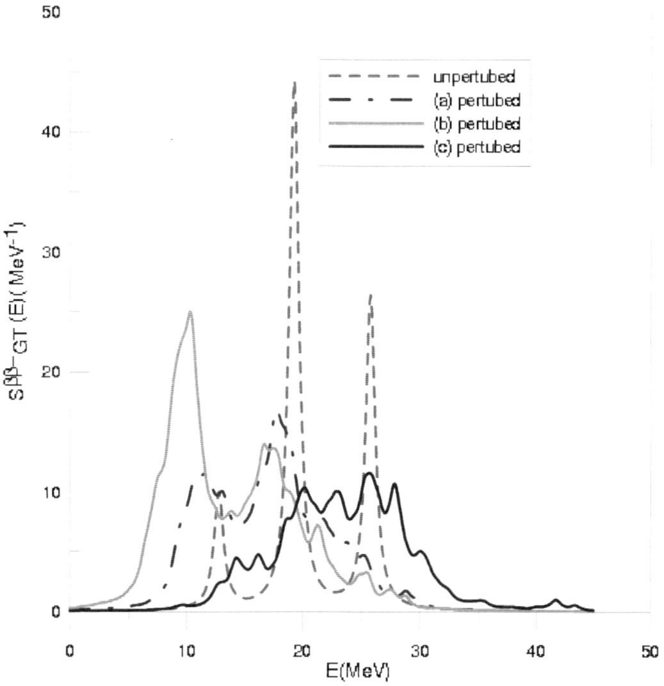

Figure 1. Double GT strength, $S^{\beta\beta}_{GT}$, for ^{48}Ca as function of the energy of daughter nucleus. The unperturbed case (red dashed line) is compared with the perturbed cases (a) SET 2 in blue dash-dot line, (b) SET 3 in green solid line, and (c) SET 1 in black solid line.

that the case (c) SET 1 shown the shift to the other side. This is an effect of the choice of the suitable effective single-particle energies. We see below how the choice of the suitable effective s.p.e. affect in the same way the behavior of the NME, because the $S^{\beta\beta}_{GT}$ are the numerators in Eq. 2.

One of the improvements on F TDA in this work was to separate the *pp* and *ph*-channels in the residual interaction. In this way, we shown in Figure 2 the $\mathcal{M}^{GT}_{2\nu}$ as function of t, *pp*-strength parameter, with three different values of (v_s, v_t) *ph*-parameters. For SET 1, the results in the *ph*-limit of F TDA (dashed lines), are compared with those obtained in the reduced F TDA (solid lines). Reduced F TDA means that the BCS equations are solved in the complete space of SET 1, and after a reduction to the *ph*-space of 1n-4p is employed. Here, we noted the effect of the probability occupation numbers v^2 have over the NME. We note in both cases that the behavior of $\mathcal{M}^{GT}_{2\nu}$ as function of t is continuous until to very high values of t parameter (values may be without physical meaning) and there is not a collapse of the F TDA. A brief discussion of this issue is given below when we will analyze the behavior of NME. The effect of inclusion of pairing is notable moving from constant values of $\mathcal{M}^{GT}_{2\nu}$ at each *ph* parametrization to smooth decreasing curve as function of t. This effect appears in the reduced and single *ph*-space and it could be washed when the complete space will be used.

The results of Figures 1 and 2 were obtained with different strengths in the *pp* and *ph*-channels of δ-residual interaction. But in previous calculations of Ref. [8], only one pair of parameters was adopted in the residual interaction :$v^{PH}_s = v^{PP}_s = v_s$ and $v^{PH}_t = v^{PP}_t = v_t$. This adoption, required for simplicity, have been implemented for two main reasons: (i) to avoid a fine fitting of the residual interaction and, (ii) to show how the TDA quenching-mechanism works. Then

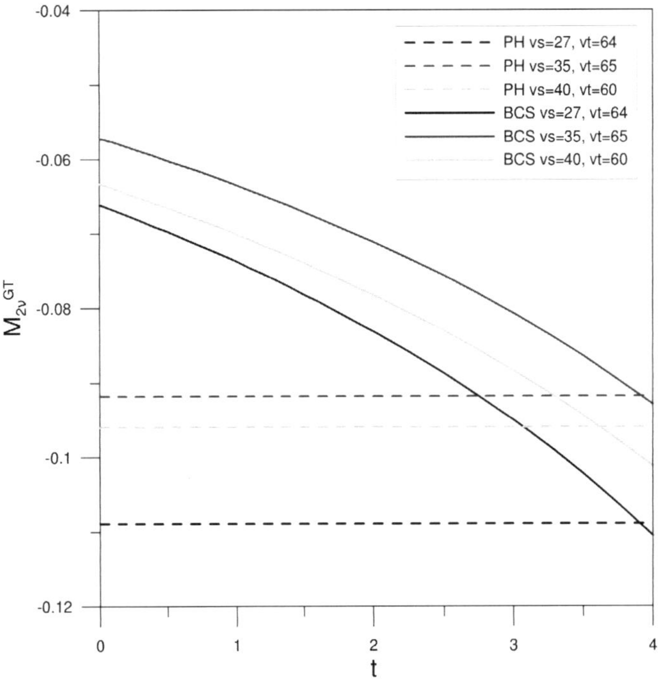

Figure 2. $\mathcal{M}_{2\nu}^{GT}$ as function of t for the three different ph parameters with SET 1. The ph-limit is shown with dashed lines, while that those from reduced calculations are shown with solid lines.

we can learn, using this recipe, which is the effect when the BCS is on, without worrying of the parametrization in the residual interaction. Figure 3 shows a comparison of $\mathcal{M}_{2\nu}^{GT}$ as function of v_t for different v_s values with the SET 1. We noted the ph-limit in the complete space of SET 1 with dashed lines. When the BCS calculations are included in a complete F TDA (solid lines) the $\mathcal{M}_{2\nu}^{GT}$ reduces its values, like a quenching effect in the NME. It is necessary to remark that the ph-limit is obtained when $v_p \to 0, v_n \to 1$, and with BCS-on means that the $v_{p(n)}^2$, merged from BCS equations, give the occupation probability for the quasiparticle in the p (n) level.

Among the simplicity introduced when one chooses the same parameters in pp and ph-channels of residual interaction, this could lead to non-physical values when both channels work separately. Already, we mentioned that one improvement was to separate these channels and then the parameters of δ-interaction must be selected. We adopted three different set of $(v_s^{PH}, v_s^{PH}) = \{(27,64),(35,65),(40,60)\}$ (in units of MeV.fm^3) parameters adopted to reproduce reasonably well the experimental data of energy levels of ^{48}Sc (Figure 5, [12]). The pp-channel parameters are fixed on the basis of the SU(4) and isospin symmetry, as $v_s^{pp} = v_s^{pair}$, and $v_t^{pp} \gtrsim v_s^{pp}$ [11]. Then $s = 1$ and, t is variable to be responsible for the known collapse in RPA calculations.

Figure 4 shows a comparison of $\mathcal{M}_{2\nu}^{GT}$ as a function of t, for several ph parameters in F TDA calculations. In the left side are the results for complete SET 1, while that in the right side those for complete SET 2. We previously obtained that the behavior $\mathcal{M}_{2\nu}^{GT}$ as function of t in a reduced space yielded high values of t and this effect could be washed when a complete space is implemented. Nevertheless, when the complete SET 1 is used the $\mathcal{M}_{2\nu}^{GT}$ are still weakly dependent of t. On the right side we shown the $\mathcal{M}_{2\nu}^{GT}$ for complete SET 2. They are most

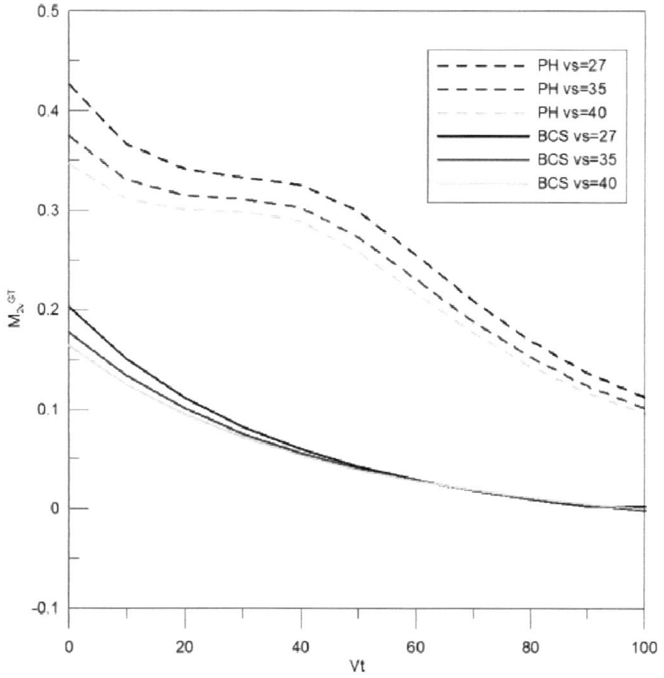

Figure 3. $\mathcal{M}_{2\nu}^{GT}$ as a function of v_t for complete SET 1: (dashed lines) in the *ph*-limit of F TDA and, (solid lines) with complete F TDA (BCS on).

sensitivity to the *ph*-parameters ($v_s = 27, v_t = 64$) doing that the F TDA results in $t \approx 0.4$ assuming a discontinuity and , as well as with the other parameters ($v_s = 35, v_t = 65$) and ($v_s = 40, v_t = 60$) where the F TDA collapses in $t \approx 0.54$ and $t \approx 0.6$ respectively. The main difference between the $\mathcal{M}_{2\nu}^{GT}$ was the choice os s.p.e.

We compared the $\mathcal{M}_{2\nu}^{GT}$ obtained with another set of s.p.e. In Figure 5 is shown that comparison. At right-hand, we can observe that the behavior of $\mathcal{M}_{2\nu}^{GT}$ of SET 3 is similar to SET 2, the values are decreasing when t increases with a smooth behavior. The spread relative to the different values of *ph*-strength is observed in $t = 0$. A similar spread was appeared with the another SET 2, but in SET 1 this phenomena does not appear. For the set 3, the $\mathcal{M}_{2\nu}^{GT}$ values according to the ($v_s = 27, v_t = 64$) *ph*-parameters are most sensitive at the t parameter up to $t \approx 1$ where appears a discontinuity. With ($v_s = 35, v_t = 65$) and ($v_s = 40, v_t = 60$) the t-values are extended up to $t \approx 1.3$. The irregularities presented in the NME for the SET 2 and SET 3 are not a collapse type such in RPA calculations. There are not complex values in the TDA equations and the discontinuities in the NME, according to the different sets of s.p.e, obey to the ratio between the of the numerator (product of single beta decay NME) and the denominator energies. In these discontinuities, the energy denominators are not zero to be a pole of $\mathcal{M}_{2\nu}^{GT}$.

Journal of Physics: Conference Series **630** (2015) 012048 doi:10.1088/1742-6596/630/1/012048

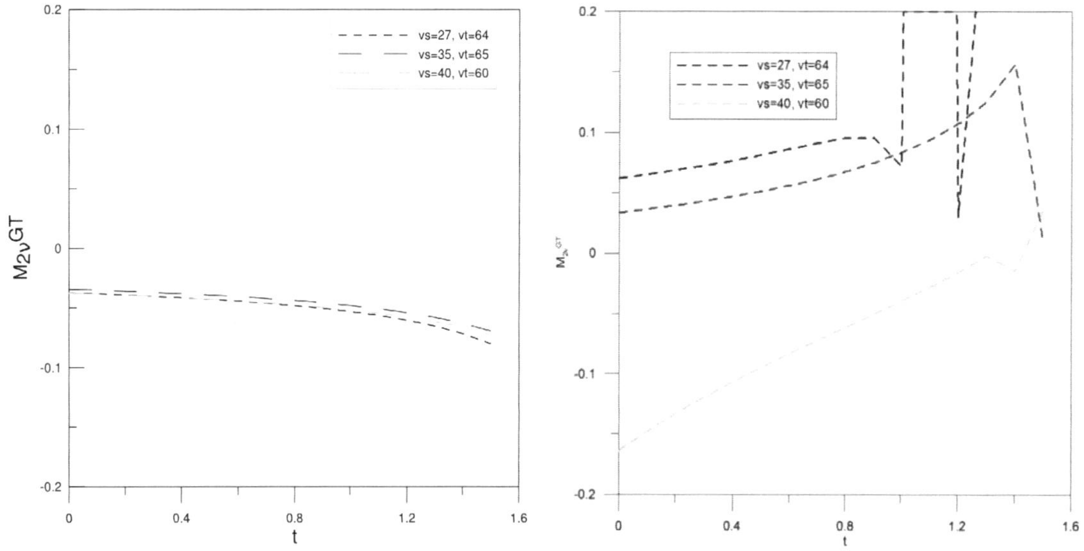

Figure 4. $\mathcal{M}_{2\nu}^{GT}$ as a function of t, pp-strength parameter and several ph parameters in a complete F TDA. Left side: for SET 1. Right side: $-\mathcal{M}_{2\nu}^{GT}$ for SET 2.

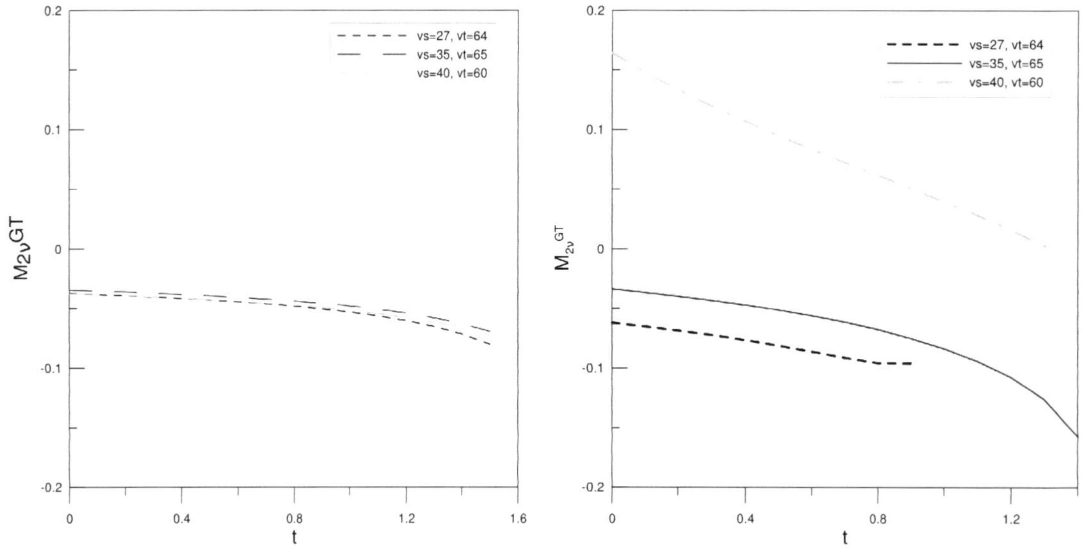

Figure 5. $\mathcal{M}_{2\nu}^{GT}$ as a function of t, pp-strength parameter and several ph parameters in a complete F TDA. Left side: for SET 1. Right side: for SET 3.

4. DISCUSSION AND BRIEF CONCLUSIONS

The use of F TDA model around the problem of collapse of RPA, where the elements of the nuclear matrix $M_{2\nu}$ goes to zero and after to infinity for certain values of the interaction strength of pp. A combination of the method of F TDA, treating the initial and final 0^+ states with the same BCS wave function as vacuum and as a four quasiparticle excitation on this vacuum, respectively, the intermediary states as two quasiparticle excitations on this vacuum, leads to avoid the known collapse of RPA. The collapse in RPA is related with the appearance of complex eigenvalues in the RPA equations at certain values of *pp* strength. This leads to obtain a non-physical wave function of the ground state with zero energy (or complex) values with a subsequent zero and latter one pole in the expression of the 2ν NMEs [13]. Also the collapse of the RPA equation indicates that the 1^+ component is ambiguous and thus also limits somewhat the reliability of the result on the neutrinoless decay [14]. We noted from Figures (4) and (5) the extreme sensitivity of $M_{2\nu}$ with interaction strength of *pp*, here represented by the values of t parameter. In particular for two different set of values (SET 2 and SET3) of s.p.e., this phenomena is reflected in a discontinuity of the $M_{2\nu}$ values when t is increased, more notable in SET2. At difference of the weakly decreasing behavior of $M_{2\nu}$ as function of t for SET1. We do not note that the indications of a kind 'collapse of RPA' appears in F TDA beyond the discontinuity appearing in $M_{2\nu}$ for some t-values with some set of s.p.e. The $M_{2\nu}$ increases significatively in this point. Neither the ground state energy of wave function of the 0^+ final state have been presented some kind of irregularity, either the intermediary 1^+ states. A detailed study of the collapse of RPA was performed by Civitarese *et al.* [16], where the interpretation of the collapse was pointed out as a signature of a phase transition. This important issue in the F TDA will be studied in detail in future work , anyway we enforced the idea that the choice of the suitable effective single-particle energies is a delicate issue, that leads to singularities in the NME, non interpreted at all with the known collapse of RPA.

As final conclusions, with the experimental half-life ($(4.3 \pm 2.3)10^{19}$yr [?] in Eq.?? the NME for two-neutrino double-beta decay of ^{48}Ca is $M_{2\nu}^{GT} = 0.0560 \pm 0.0162$ MeV^{-1}. It is possible to reproduce this value with the parametrizations employed for the *ph*-chanel as well as for some physical parameters of the t parameter in *pp*-chanel. A fine-tune to choice the *ph*-parameters must be performed comparing the experimental strength of single β-decay with our theoretical results. After this procedure, one needs to fit the best t-value of *pp* channel to reproduce the experimental $M_{2\nu}^{GT}$.

Our future work is to expand the study of the model, testing the decay $2\nu\beta\beta$ in ^{76}Ge and ^{100}Mo to check the reliability of F TDA evaluating double beta decay NME.

5. ACKNOWLEDGMENT

L.O and A.R.S thank the financial support of FAPESB, UESC and CAPES/PROAP. C.A. Barbero, F. Krmpotić and A.E. Mariano fellow to CONICET (Argentina) and CCT La Plata, Argentina

6. REFERENCES

[1] A. Barabash, ar iv:1403.2870v1 [nucl-ex], 12 Mar 2014.
[2] D. Tosi on behalf of the E O Collaboration, ar iv:1402.1170v1 [nucl-ex] 5 Feb 2014.
[3] R. Arnold et al. (NEMO Collaboration), Phys. Rev. Lett. **95**, 182302 (2005).
[4] S. Umehara *et al.*, J. Phys.: Conf. Ser. **39**, 356 (2006).
[5] Y. G. Zdesenko *et al.*, Astropart. Phys. **23**, 249 (2005).
[6] A. A. Kwiatkowski *et al.*, ar iv:1308.3815v1 [nucl-ex].
[7] A. Neacsu *et al.*, Phys. Rev. C **86**, 067304 (2012).
[8] F. Krmpotić, *Fizica* **B** 14, (2005).
[9] J. Suhonen and O. Civitarese, Phys. Rev. C **49**, 3055 (1994).
[10] Y. K. Pal, M. K. Gambir and Ram Raj. *Phys. Rev.* 155, 1144 (1966).
[11] J. Hirsch and F. Krmpotić, Phys. Rev. C **41**, 304 (1990), and references therein.
[12] A.R. Samana and F. Krmpotić, Proocedings of II Workshop of Nucler Physixs in Brasi - Malulhy Co 2006, 151 (2006).
[13] F. Krmptotić and S. Shelly Sharma, Nuc. Phys. **A572**, 329 (1984) F. Krmptotić, Revista Mexicana de Fsica **40**, Sup. 1, 285 (1994).
[14] A. Staudt, K. Muto b and H. . Klapdor-Kleingrothaus, Phys. Lett. **B** 268, 312 (1991).
[15] A.R. Samana, F. Krmpotić and C.A. Bertulani, Comp. Phys. Comm. **181**, 1123 (2010).
[16] O. Civitarese, P.O. Hess and J.G. Hirsch, Phys. Lett. **B 412**, 1 (1997).

Meson phenomenology and phase transitions in nonlocal chiral quark models

J P Carlomagno[a,b], **D Gómez Dumm**[a,b], **V Pagura**[a,b] **and N N Scoccola**[b,c,d]

[a] IFLP, CONICET – Dpto. de Física, FCE, Universidad Nacional de La Plata, C.C. 67, 1900 La Plata, Argentina
[b] CONICET, Rivadavia 1917, 1033 Buenos Aires, Argentina
[c] Physics Department, Comisión Nacional de Energía Atmica, Av.Libertador 8250, 1429 Buenos Aires, Argentina
[d] Universidad Favaloro, Solís 453, 1078 Buenos Aires, Argentina

E-mail: dumm@fisica.unlp.edu.ar

Abstract. We study the features of nonlocal chiral quark models that include wave function renormalization. Model parameters are determined from meson phenomenology, considering different nonlocal form factor shapes. In this context we analyze the characteristics of the deconfinement and chiral restoration transitions at finite temperature and chemical potential, introducing the couplings of fermions to the Polyakov loop for different Polyakov potentials. The results for various thermodynamical quantities are compared with data obtained from lattice QCD calculations.

1. Introduction

The detailed understanding of the behavior of strongly interacting matter under extreme conditions of temperature and/or density has become an issue of great interest in recent years. It is widely believed that as the temperature and/or density increase, one finds a transition from a hadronic phase to a partonic one, in which chiral symmetry is restored. From the theoretical point of view, one way to address the study of these phase transitions is through lattice CD calculations [1, 2, 3]. However, this ab-initio approach shows difficulties when dealing with small current quark masses and finite real chemical potentials. Thus it is worth to develop effective models that show consistency with lattice results, and can be extrapolated into regions not accessible by lattice calculation techniques. Here we will concentrate on one particular class of effective theories, namely the so-called nonlocal Polyakov–Nambu–Jona-Lasinio (nlPNJL) models [4, 5, 6, 7, 8], in which quarks move in a background color field and interact through covariant nonlocal chirally symmetric four-point couplings.

The aim of this work is to study the chiral restoration and deconfinement transitions at nonzero temperature and both real and imaginary chemical potentials, taking into account the effect of dynamical quarks on the Polyakov loop (PL) effective potential [9, 10, 11]. We consider both SU(2) chiral models and the corresponding extension to three flavors, including flavor mixing through a nonlocal 't Hooft-like six-fermion interaction [12, 13, 14]. Special attention is paid to models in which the nonlocal form factors are chosen on the basis of lattice CD results for the effective quark propagators. This requires the inclusion of a nonlocal interaction that leads to wave function renormalization (WFR) of the quark fields [15].

2. Meson phenomenology

We start by considering a chiral quark model in which the Lagrangian includes the coupling between quark-antiquark nonlocal currents. The explicit expression for this Lagrangian for the three-flavor case (including WFR) can be found in Ref. [14]. Now, in order to work with mesonic degrees of freedom, we proceed to perform a standard bosonization of the fermionic theory, introducing scalar and pseudoscalar fields. After integrating out the fermionic degrees of freedom, we expand the bosonic fields around their vacuum expectation values. At this mean field level the Euclidean effective action reduces to

$$\frac{S_E^{\rm MFA}}{V^{(4)}} = -2\ {\rm Tr} \int \frac{d^4p}{(2\pi)^4} \log\left[\frac{M(p)^2+p^2}{Z(p)^2}\right] - \sigma_a S_a - \zeta R - \frac{G}{2}(S_a S_a + R^2) - \frac{H}{4} A_{abc} S_a S_b S_c\ , \quad (1)$$

where G and H are the coupling constants corresponding to the four- and six-fermion nonlocal interactions, respectively, and the trace acts over color and flavor spaces. The mean field values of the scalar fields σ_a and ζ, and the auxiliary fields S_a, P_a and R are obtained through the minimization of the partition function and the usage of the random phase approximation, in the path integral formalism. This leads to a set of coupled gap equations that can be found in Ref. [14]. The functions $M_i(p)$ and $Z(p)$ correspond to momentum-dependent effective masses and WFR of the quark propagators. In terms of the model parameters and form factors, these are given by

$$M_i(p) = Z(p)\ [m_i + \sigma_i\, g(p)]\ , \quad Z(p) = \left[1 - \kappa^{-1}\zeta\, f(p)\right]^{-1}\ , \quad i = u, d, s\ , \quad (2)$$

where κ is a parameter that controls the relative weight of the coupling leading to the WFR, m_i are the current quark masses, and $f(p)$ and $g(p)$ are the fourier transforms of the nonlocal form factors in the quark-antiquark currents.

In order to analyze the properties of meson fields it is necessary to go beyond the mean field level, considering quadratic fluctuations in the Euclidean action:

$$S_E^{\rm quad} = \frac{1}{2}\int \frac{d^4p}{(2\pi)^4} \sum_M r_M\ G_M(p^2)\ \phi_M(p)\, \phi_M(-p)\ , \quad (3)$$

where ϕ_M stand for the meson fields in the charge basis, M being the scalar and pseudoscalar mesons in the lowest mass nonets, plus the ζ field. The coefficient r_M is 1 for charge eigenstates $M = {\rm a}_0^0, \sigma, f_0, \zeta, \pi^0, \eta, \eta'$, and 2 for $M = {\rm a}_0^+, \kappa, K_0^*, \pi^+, K^+, K^0$. Thus meson masses are given by the equations

$$G_M(-m_M^2)\ =\ 0\ . \quad (4)$$

Full expressions for the one-loop functions $G_M(q)$ can be found in Refs. [13, 14].

The model includes five parameters, namely the current quark masses $m_{u,s}$ and the coupling constants G, H and κ. In addition, one has to specify the form factors $g(p)$ and $f(p)$. We will consider here two parametrizations, corresponding to different functional forms for the form factors: the first one (PI) involves the often used exponential functions,

$$g(p) = \exp\left(-p^2/\Lambda_0^2\right)\ , \qquad f(p) = \exp\left(-p^2/\Lambda_1^2\right)\ , \quad (5)$$

while for the second one (PII) we use the functions [15]

$$g(p) = \frac{1+\alpha_z}{1+\alpha_z\ f_z(p)} \frac{\alpha_m\ f_m(p) - m_u\ \alpha_z f_z(p)}{\alpha_m - m_u\ \alpha_z}\ , \qquad f(p) = \frac{1+\alpha_z}{1+\alpha_z\ f_z(p)} f_z(p)\ , \quad (6)$$

where

$$f_m(p) = \left[1 + \left(p^2/\Lambda_0^2\right)^{3/2}\right]^{-1}, \qquad f_z(p) = \left[1 + \left(p^2/\Lambda_1^2\right)\right]^{-5/2}, \tag{7}$$

with $\sigma_u = (\alpha_m - m_u\alpha_z)/(1+\alpha_z)$, $\zeta = \kappa\alpha_z/(1+\alpha_z)$.

Given the form factor functions, one can fix the model parameters (see Ref. [14]) so as to reproduce the observed meson phenomenology. To the above mentioned parameters $m_{u,s}$, G, H and κ one has to add the cutoff scales Λ_0 and Λ_1, introduced through the form factors. We have taken as input some selected observables, such as the pion and kaon masses. As it is shown in Fig. 1, the functions defined by the parametrization PII allow to fit adequately the momentum dependence of mass and WFR in the light quark propagators to the results obtained in lattice CD calculations [16].

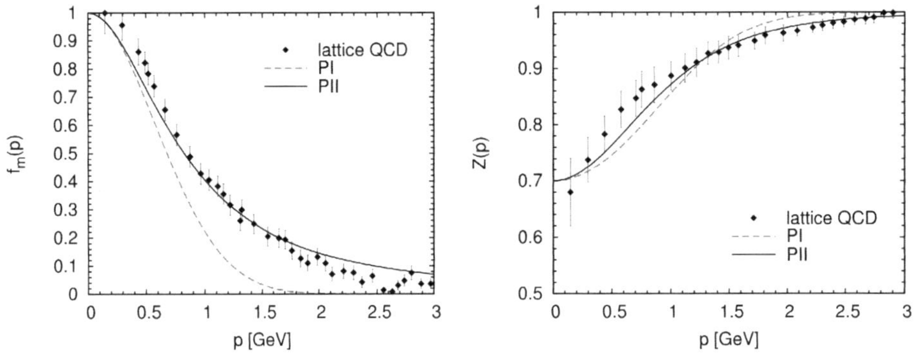

Figure 1. Momentum dependences for mass and WFR for our parameterizations PI and PII, in comparison with lattice results from Ref. [16].

Once the parameters have been determined, we can calculate the values of several meson properties for the scalar and pseudoscalar sector. The parameter values for PI and PII are given in Table 1 and our numerical results are quoted in Table 2, where input observables are indicated. In general, it is seen that the meson masses, mixing angles and weak decay constants predicted by the model are in a reasonable agreement with phenomenological expectations. Moreover, the results for PI do not differ significantly from those obtained in Ref. [6] for a nlPNJL model with a Gaussian form factor $g(p)$ and no WFR (we call this parametrization PIII).

	PI	PII
m_u [MeV]	5.7	2.6
m_s [MeV]	136	65
Λ_0	814	799
Λ_1 [MeV]	1033	1570
$G\,\Lambda_0^2$	23.6	16.6
$H\,\Lambda_0^5$	−526	−202
κ/Λ_0	5.36	11.1

Table 1. Parameter values for PI and PII.

Journal of Physics: Conference Series **630** (2015) 012049 doi:10.1088/1742-6596/630/1/012049

σ_u [MeV]	529	469	-
σ_s [MeV]	702	707	-
ζ/κ	-0.429	-0.429	-
$-\langle\bar{u}u\rangle^{1/3}$ [MeV]	240	316	-
$-\langle\bar{s}s\rangle^{1/3}$ [MeV]	198	340	-
m_π [MeV]	input	input	139
m_K [MeV]	input	input	495
m_η [MeV]	527	522	547
$m_{\eta'}$ [MeV]	input	input	958
m_σ	601	561	$400 - 550$
m_κ [MeV]	810	732	$650 - 710$
f_π [MeV]	input	input	92.4
f_K/f_π	1.17	1.17	1.22
f_η^0/f_π	0.17	0.21	(0.11 - 0.507)
f_η^8/f_π	1.12	1.08	(1.17 - 1.22)
$f_{\eta'}^0/f_\pi$	1.09	1.54	(0.98 - 1.16)
$f_{\eta'}^8/f_\pi$	-0.48	-0.48	$-(0.42$ - $0.46)$
θ_0	$-8.63°$	$-7.91°$	$-(0°$ - $10°)$
θ_8	$-22.9°$	$-23.7°$	$-(19°$ - $22°)$

Table 2. Results for various phenomenological quantities within nonlocal models for parametrizations PI and PII.

3. Nonzero temperature

In order to investigate the phase transitions and the temperature dependence of thermodynamical quantities within our model, we consider the thermodynamical potential per unit volume at the mean field level. In order to deal with the interaction between quarks and color gauge fields we assume that quarks move on a constant background field $\phi = A_4 = iA_0 = ig\,\delta_{\mu 0}\,G_a^\mu \lambda^a/2$, where G_a^μ are the SU(3) color gauge fields and λ_a are the Gell-Mann matrices. Then the traced Polyakov loop, which in the infinite quark mass limit can be taken as order parameter of confinement, is given by $\;= \frac{1}{3}\mathrm{Tr}\,\exp(i\phi/T)$.

We proceed by using the standard Matsubara formalism. The mean field thermodynamical potential will be given by

$$\Omega^{\rm MFA} \;=\; \Omega^{\rm reg} + \Omega^{\rm free} + \mathcal{U}(\;,T)\;, \tag{8}$$

where

$$\begin{aligned}
\Omega^{\rm reg} \;&=\; -2T \sum_{n=-\infty}^{\infty} \sum_{c,f} \int \frac{d^3p}{(2\pi)^3} \log\left[\frac{p_{nc}^2 + M_f^2(p_{nc})}{Z^2(p_{nc})\,(p_{nc}^2 + m_f^2)}\right] \\
&\quad - \left(\zeta R + \frac{G}{2} R^2 + \frac{H}{4} S_u S_d S_s\right) - \frac{1}{2}\sum_f \left(\sigma_f S_f + \frac{G}{2} S_f^2\right)\;, \\
\Omega^{\rm free} \;&=\; -2T \sum_{c,f} \sum_{s=\pm 1} \int \frac{d^3p}{(2\pi)^3}\, \mathrm{Re}\,\log\left[1 + \exp\left(-\frac{\epsilon_{fp} + \imath\, s\, \phi_c}{T}\right)\right]\;.
\end{aligned} \tag{9}$$

Here we have defined $p_{nc}^2 = [(2n+1)\pi T + \phi_c]^2 + \vec{p}^{\,2}$, $\epsilon_{fp} = \sqrt{\vec{p}^{\,2} + m_f^2}$. The sums over color and flavor indices run over $c = r, g, b$ and $f = u, d, s$, respectively, and the color background fields are $\phi_r = -\phi_g = \phi_3$, $\phi_b = 0$. The term $\mathcal{U}(\;,T)$ is known as the Polyakov-loop potential, which accounts for effective gauge field self-interactions. It is usual to take for this potential a functional form based on properties of pure gauge CD. One possible ansatz is that based on the

Journal of Physics: Conference Series **630** (2015) 012049 doi:10.1088/1742-6596/630/1/012049

logarithmic expression of the Haar measure associated with the SU(3) color group integration. This yields [17]

$$\frac{\mathcal{U}_{\log}(\ \ ,T)}{T^4} = -\frac{1}{2}a(T)\ \ ^2 + b(T)\log\left(1-6\ \ ^2+8\ \ ^3-3\ \ ^4\right)\,, \tag{10}$$

where

$$a(T) = a_0 + a_1\left(\frac{T_0}{T}\right) + a_2\left(\frac{T_0}{T}\right)^2\,, \qquad b(T) = b_3\left(\frac{T_0}{T}\right)^3\,. \tag{11}$$

The parameters in the above expression can be fitted to pure gauge lattice CD data so as to properly reproduce the corresponding equation of state and Polyakov loop behavior [17]. In absence of dynamical quarks T_0 is the deconfinement temperature, which is expected to be about 270 MeV from lattice calculations. However, it has been argued that in the presence of light dynamical quarks this temperature scale should be adequately reduced to about 210 and 190 MeV for the case of two and three flavors, respectively, with an uncertainty of about 30 MeV [18]. Besides the logarithmic function in Eq. (10), other possible forms for the Polyakov-loop potential can be found in Ref. [14].

In order to analyze the phase transitions it is useful to define a subtracted quark condensate

$$\langle qq\rangle_{\rm sub} = \frac{\langle uu\rangle - \frac{m_u}{m_s}\langle ss\rangle}{\langle uu\rangle_0 - \frac{m_u}{m_s}\langle ss\rangle_0}\,, \tag{12}$$

where $\langle qq\rangle_0$ are the values of the chiral condensates at zero temperature. In Fig. 2 we quote our results for the subtracted quark condensate and the Polyakov loop, together with the corresponding susceptibilities, as functions of the temperature. The values correspond to PII and the PL potential given by Eq. (10), taking $T_0 = 270$ and 200 MeV. For comparison we include lattice CD results from Refs. [19, 20]. As expected, it is found that when the temperature is increased the system undergoes both a chiral restoration and deconfinement transitions, which proceed as smooth crossovers. It is seen that both transitions occur essentially at the same critical temperatures, in agreement with lattice CD results. The numerical values are $T_c \simeq 200$ MeV and $T_c \simeq 165$ MeV for $T_0 = 270$ and 200 MeV, respectively, while lattice CD analyses lead to a transition temperature of about 160 MeV [19, 20]. Thus, the agreement with lattice CD data favors the suggested rescaling of the reference temperature T_0 from the pure gauge transition temperature (270 MeV) towards values around 200 MeV.

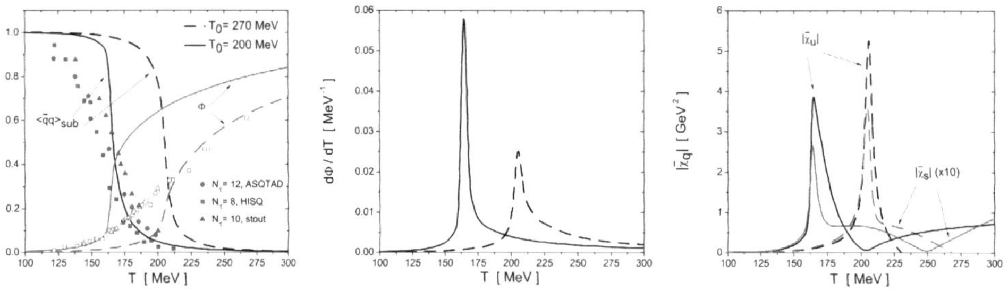

Figure 2. Subtracted chiral condensate, Polyakov loop, chiral susceptibilities and PL susceptibility $d\ \ /dT$ as functions of the temperature. Solid (dashed) curves correspond to PII, for a logarithmic PL potential with $T_0 = 200$ (270) MeV. Triangles, circles and squares stand for lattice CD results from Refs. [19, 20].

In order to compare the features of parameterizations PI and PII, it is useful to consider some thermodynamical quantities such as the energy and the entropy densities, which can be obtained from the thermodynamical potential:

$$\varepsilon = \Omega + Ts \ , \qquad s = -\frac{\partial \Omega}{\partial T} \ . \tag{13}$$

Our results are shown in Fig. 3, where we plot the normalized interaction energy and entropy. It can be seen that the curves for PI show a pronounced dip at about $T \sim 300$ MeV, which is not found for PII. The comparison with lattice data turn out to favor the lattice-inspired parametrization PII, as one would have expected for consistency. We have taken here the logarithmic PL potential given by Eq. (10), with $T_0 = 200$ MeV.

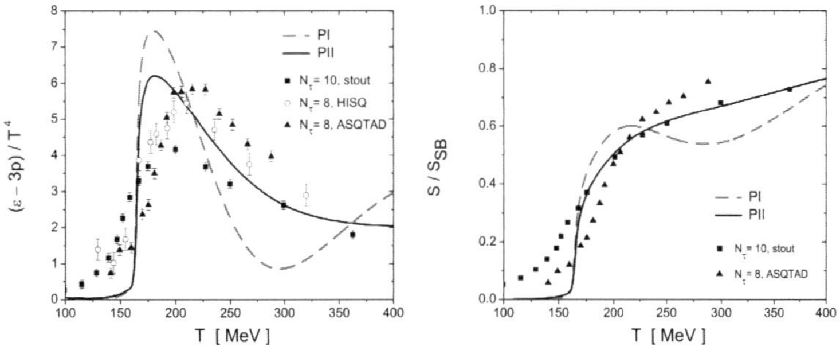

Figure 3. Normalized interaction energy (left) and entropy density (right) as functions of the temperature, for parameterizations PI and PII and a logarithmic PL potential with $T_0 = 200$ MeV. Squares, circles and triangles stand for lattice data from Refs. [20], [21] and [22], respectively.

Finally, in Fig. 4 we show the behavior of the (normalized) interaction energy, entropy density and energy density as functions of the temperature. We compare the results obtained for different PL potentials, considering the previously defined logarithmic potential as well as a widely used polynomic form (see e.g. Refs. [23, 24]) and the so-called improved polynomic potential proposed in Ref. [25]. The curves correspond to parametrization PII. It is seen that the improved polynomic potential shows a better agreement with lattice results up to the critical temperature, while for higher temperatures the agreement is better for the usual logarithmic and polynomic potentials.

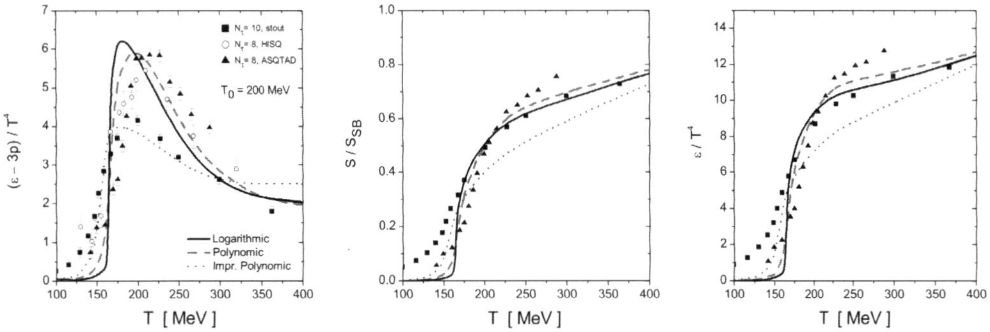

Figure 4. Normalized interaction energy, entropy density and energy density as functions of the temperature, for parameterization PII and three different PL potentials. Squares, circles and triangles stand for lattice data from Refs. [20, 21, 22].

4. Nonzero chemical potential

We have studied the phase transitions at finite chemical potential in the framework of the SU(2) version of nonlocal PNJL models. The corresponding Euclidean mean field action can be obtained from that in Eq. (1) in the limit of two flavors. Once again we consider a system at finite temperature and use the Matsubara formalism, working in the Polyakov gauge. In order to compare with lattice calculations we consider in general the case of a complex chemical potential μ. The mean field thermodynamical potential is then given by

$$\Omega^{\rm MFA}_{\rm SU(2)} = -4T \sum_{n=-\infty}^{\infty} \sum_{c} \int \frac{d^3\vec{p}}{(2\pi)^3} \log \left[\frac{(\rho^c_{n,\vec{p}})^2 + M(\rho^c_{n,\vec{p}})^2}{Z(\rho^c_{n,\vec{p}})^2} \right] + \frac{\sigma^2 + \zeta^2}{2G} + \mathcal{U}(\ \ , \ \ ^*, T) \; , \tag{14}$$

where we have defined

$$\left(\rho^c_{n,\vec{p}}\right)^2 = \Big[(2n+1)\pi T - i\mu + \phi_c \Big]^2 + \vec{p}^{\,2} \; . \tag{15}$$

The sum over color indices runs over $c = r, g, b$, with $\phi_r = \phi_3 + \phi_8/\sqrt{3}$, $\phi_g = -\phi_3 + \phi_8/\sqrt{3}$, $\phi_b = -2 + \phi_8/\sqrt{3}$. Here ϕ_3 and ϕ_8 parametrize the traced PL according to

$$= \frac{1}{3} \left[\exp\left(-\frac{2i}{\sqrt{3}} \frac{\phi_8}{T} \right) + 2 \exp\left(\frac{i}{\sqrt{3}} \frac{\phi_8}{T} \right) \cos(\phi_3/T) \right] . \tag{16}$$

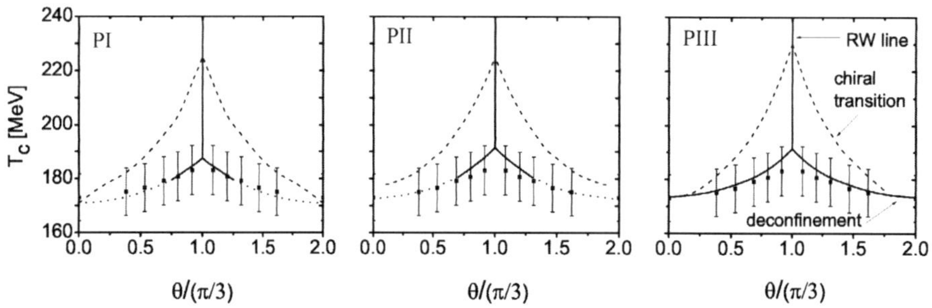

Figure 5. Critical temperature as a function of $\theta/(\pi/3)$ for parametrizations PI (left), PII (center) and PIII (right). Solid and dashed lines stand for first order and crossover-like transitions, respectively, while dots correspond to lattice CD results [28]. Vertical solid lines correspond to the first order RW transition.

Let us first consider a purely imaginary chemical potential parametrized as $\mu = i\theta T$. In this case the CD thermodynamical potential is invariant under the so-called extended Z_3 symmetry [26], which is a combination of a Z_3 transformation of the quark and gauge fields and a shift $\theta \to \theta + 2k\pi/3$, with integer k. It turns out that this symmetry is also present in our nlPNJL model. In CD with dynamical quarks, if the temperature is larger than a certain value T_{RW} it can be seen that three Z_3 vacua appear. Roberge and Weiss showed that there is a first order phase transition at $\theta = \pi/3 \bmod 2\pi/3$, in which the vacuum jumps to one of its Z_3 images. This is known as the Roberge-Weiss transition , and the point at the end of the RW transition line in the (T, θ) plane, i.e. $(T, \theta) = (T_{RW}, \pi/3)$, is known as the RW endpoint . The order of the RW transition at the RW endpoint has been subject of considerable interest recently in the framework of lattice CD [27] due to the implications it might have on the CD phase diagram at finite real μ. According to lattice calculations, it appears that for two light flavors the RW endpoint is first order for realistically small values of the current quark mass. In Fig. 5 we show the dependence of the chiral restoration and deconfinement transition temperatures with θ, together with lattice data. We consider gaussian and lattice inspired parameterizations PI and PII, and also a gaussian parametrization PIII with no WFR, whereas for the PL potential we take the logarithmic form with $T_0 = 200$ MeV. We observe that, while for PIII both transitions are always first order, in the case of PI and PII there is a value $\theta_{\rm CEP} \sim 0.7 \times \pi/3$ below which the transitions become crossover-like. Thus, we find that for all three parametrizations the RW endpoint is a triple point, being the RW transition first order there. If we choose $T_0 = 270$ MeV instead, we observe qualitatively the same behavior, however the critical temperature turns out to be too high compared to lattice CD results. Regarding the effect of different PL potentials, the main difference we find is that for the polynomic potential there are no first order deconfinement transitions, which implies a second order RW transition at the RW endpoint, in contradiction with lattice CD predictions.

We turn now to consider the case of finite real chemical potential. In order to account for both the case of real and imaginary μ in a single plot, in Fig. 6 we show the results for the phase transition curves in the $T - \mu^2$ plane. The curves correspond to PII and a logarithmic PL potential with $T_0 = 200$ MeV. Finally, in Fig. 7 we include various $T - \mu$ phase diagrams for real μ in order to present a more detailed comparison between the results obtained for different form factors (parametrizations PI, PII and PIII in upper, center and lower panel, respectively) and PL potentials (polynomic and logarithmic, in the left and right panels respectively). Full lines correspond to first order phase transitions, dashed lines to pure chiral restoration crossover transitions, dotted lines to pure deconfinement crossover transitions, and dash-dotted lines to simultaneous chiral restoration and deconfinement crossover transitions. By increasing the chemical potential, it is seen that for both PI and PII the chiral restoration and deconfinement transitions occur at approximately the same temperature, until they reach a critical end point. The position of this point depends on the parametrization, ranging from $\mu = 150$ to 250 MeV, and from $T = 120$ to 170 MeV for $T_0 = 200$ MeV. In the case of the model with no WFR, the assumption of a logarithmic PL potential with $T_0 = 200$ MeV leads to a first order chiral phase transition at $\mu = 0$ in contradiction with lattice CD results.

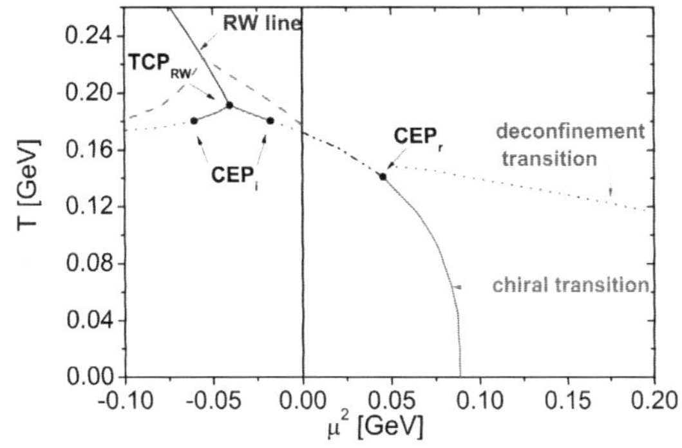

Figure 6. Phase diagram in $T - \mu^2$ space. Full lines correspond to first order phase transitions, dashed lines to pure chiral restoration crossover transitions, dotted lines to pure deconfinement crossover transitions, and dash-dotted lines to simultaneous chiral restoration and deconfinement crossover transitions. Curves correspond to PII and logarithmic PL potential with $T_0 = 200$ MeV.

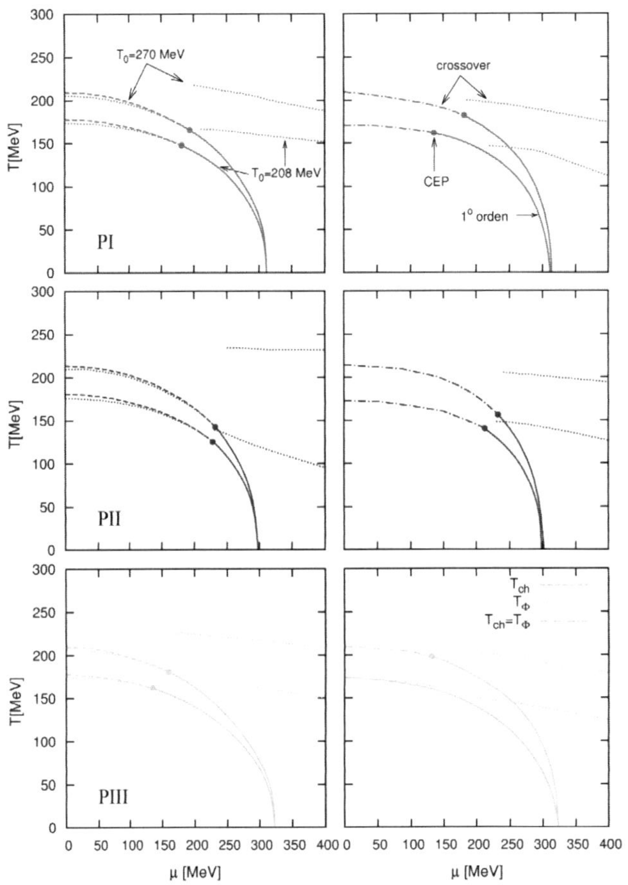

Figure 7. Phase diagrams for the three parameterizations considered. PI and PIII correspond to exponential form factors while PII to lattice motivated form factors. Full (dotted) lines correspond to first (second) order chiral restoration transitions, while dashed lines correspond to deconfinement transitions.

5. Summary

We have analyzed the features of three-flavor nlPNJL models that include wave function renormalization in the effective quark propagators. In this framework, we have obtained a parametrization (PII) of the model that reproduces lattice CD results for the momentum dependence of the effective quark mass and WFR, and at the same time leads to an acceptable phenomenological pattern for particle masses and several meson properties in both the scalar and pseudoscalar meson sectors.

As a second step we have analyzed the characteristics of the deconfinement and chiral restoration transitions at finite temperature. In general it has been found that both transitions occur at the same critical temperature, in agreement with lattice CD results. A critical temperature of about 170 MeV, consistent with that arising from lattice CD calculations, has been obtained for $T_0 \simeq 200$ MeV, in agreement with theoretical expectations for a model with two/three light dynamical quarks. It is seen that the lattice-inspired power-like parameterization

Journal of Physics: Conference Series **630** (2015) 012049 doi:10.1088/1742-6596/630/1/012049

PII shows indeed the best agreement with lattice CD results.

Finally, we have considered the case of nonzero chemical potential. For imaginary chemical potentials we have analyzed the Roberge Weiss transition, which has been found to be first order at the RW endpoint for the considered parametrizations. In the case of real chemical potentials we have shown the features of the corresponding phase diagrams. At low μ, for PI and PII it is seen that chiral restoration and deconfinement transitions proceed approximately at the same critical temperatures, up to a critical end point whose location is rather dependent on the model parametrization.

Acknowledgments

This work has been partially funded by CONICET (Argentina) under grants PIP 00682 and PIP 00449, and by ANPCyT (Argentina) under grant PICT11-03-00113.

References

[1] Allton C R *et al.* 2003 *Phys. Rev.* D **68** 014507 *Phys. Rev.* D **71** 054508.
[2] Fodor Z and Katz S D 2004 *JHEP* **0404** 050 Aoki Y, Fodor Z, Katz S D and Szabo K K 2006 *JHEP* **0601** 089
[3] Karsch F and Laermann E, in *Quark Gluon Plasma III*, edited by R.C. Hwa and . N. Wang (World Scientific, Singapore, 2004), ar iv:hep-lat/0305025
[4] Blaschke D, Buballa M, Radzhabov A E and olkov M K 2008 *Yad. Fiz.* **71** 2012 [*Phys. Atom. Nucl.* **71** 1981]
[5] Contrera G A, Gomez Dumm D and Scoccola N N 2008 *Phys. Lett.* B **661** 113
[6] Contrera G A, Gomez Dumm D and Scoccola N N 2010 *Phys. Rev.* D **81** 054005
[7] Hell T, Roessner S, Cristoforetti M and Weise W 2009 *Phys. Rev.* D **79** 014022
[8] Hell T, Rossner S, Cristoforetti M and Weise W 2010 *Phys. Rev.* D **81** 074034
[9] Contrera G A, Orsaria M and Scoccola N N 2010 *Phys. Rev.* D **82** 054026
[10] Pagura , Gomez Dumm D and Scoccola N N 2012 *Phys. Lett.* B **707** 76
[11] Pagura , Gomez Dumm D and Scoccola N N 2013 *Phys. Rev.* D **87** 014027
[12] Scarpettini A, Gomez Dumm D and Scoccola N N 2004 *Phys. Rev.* D **69** 114018
[13] Hell T, Kashiwa K and Weise W 2011 *Phys. Rev.* D **83** 114008
[14] Carlomagno J P, Gomez Dumm D and Scoccola N N 2013 *Phys. Rev.* D **88** 074034
[15] Noguera S and Scoccola N N 2008 *Phys. Rev.* D **78** 114002
[16] Parappilly M B, Bowman P O, Heller U M, Leinweber D B, Williams A G and Zhang J B 2006 *Phys. Rev.* D **73** 054504
[17] Roessner S, Ratti C and Weise W 2007 *Phys. Rev.* D **75** 034007
[18] Schaefer B -J, Pawlowski J M and Wambach J 2007 *Phys. Rev.* D **76** 074023, Schaefer B -J, Wagner M and Wambach J 2010 *Phys. Rev.* D **81** 074013
[19] Borsanyi S *et al.* [Wuppertal-Budapest Collaboration] 2010 *JHEP* **1009** 073
[20] Bazavov A *et al.* [HotQCD Collaboration] 2010 *J. Phys. Conf. Ser.* **230** 012014
[21] Bazavov A *et al.* 2009 *Phys. Rev.* D **80** 014504
[22] Borsanyi S, Endrodi G, Fodor Z, Jakovac A, Katz S D, Krieg S, Ratti C and Szabo K K 2010 *JHEP* **1011** 077
[23] Scavenius O, Dumitru A and Lenaghan J T 2002 *Phys. Rev.* C **66** 034903
[24] Ratti D, Thaler M A and Weise W 2006 *Phys. Rev.* D **73** 014019
[25] Haas L M, Stiele R, Braun J, Pawlowski J M and Schaffner-Bielich J 2013 *Phys. Rev.* D **87** 076004
[26] Roberge A, Weiss N 1986 *Nucl. Phys.* B **275** 734
[27] D'Elia M, Di Renzo F, Lombardo M P 2007 *Phys. Rev.* D **76** 114509
D'Elia M, Sanfilippo F 2009 *Phys. Rev.* D **80** 111501 de Forcrand P, Philipsen O 2010 *Phys. Rev. Lett.* **105** 152001 Bonati C, Cossu G, D'Elia M and Sanfilippo F 2011 *Phys. Rev.* D **83** 054505
[28] Wu L -K, Luo -Q, Chen H -S 2007 *Phys. Rev.* D **76** 034505

Journal of Physics: Conference Series **630** (2015) 012050 doi:10.1088/1742-6596/630/1/012050

Ward identities, $B \to V$ transition form factors and applications

M. Ali Paracha

Laboratório de Física Teórica e Computacional, Universidade Cruzeiro do Sul, 01506-000 São Paulo SP, Brazil; Department of Physics, School of Natural Sciences, National University of Science and Technology, Islamabad, Pakistan

E-mail: paracha@phys.qau.edu.pk

Bruno El-Bennich

Laboratório de Física Teórica e Computacional, Universidade Cruzeiro do Sul, 01506-000 São Paulo, SP, Brazil; Instituto de Física Teórica, Universidade Estadual Paulista, 01140-070 São Paulo, SP, Brazil

E-mail: bruno.bennich@cruzeirodosul.edu.br

M. Jamil Aslam

Department of Physics, Quaid-i-Azam University, Islamabad 45320, Pakistan

E-mail: jamil@phys.qau.edu.pk

Ishtiaq Ahmed

Laboratório de Física Teórica e Computacional, Universidade Cruzeiro do Sul, 01506-000 São Paulo, SP, Brazil; National Centre for Physics, Quaid-i-Azam University Campus, Islamabad 45320, Pakistan

E-mail: ishitaq@ncp.edu.pk

Abstract. Long distance effects are studied in the rare exclusive semileptonic $B_{(d,s)} \to V\ell^+\ell^-$ decays, where V denotes a K^* or ϕ meson. The form factors, which describe the meson transition amplitudes in the effective Hamiltonian approach, are calculated by means of Ward identities, experimental constraints and extrapolated within a general vector meson dominance framework. These form factors are then compared to the ones obtained in Lattice QCD simulations, with Light Cone Sum Rules and a Dyson-Schwinger equation approach. Additionally, the $B_d \to K^*\ell^+\ell^-$ and $B_s \to \phi\ell^+\ell^-$ branching ratios are computed and the differential branching fractions are given as a function of the squared-momentum transfer.

1. Introduction

Rare B meson decays have been a subject of great interest for the past two decades. These decays not only provide a stringent tests of the Standard Model (SM), but also serve as a tool to extract physics beyond the SM in the flavor sector. We remind that in the SM rare decays do not occur at tree order but proceed at loop level through the Glashow-Iliopoulos-Maiani

(GIM) mechanism [1]. Attempts to unravel the imprints of physics in and beyond the SM in the flavor sector imply the study of inclusive [2] and exclusive [3] B-meson decays. From an experimental point of view, exclusive decays are easier to observe than inclusive decays. However, theoretically, exclusive decays are more challenging due to the uncertainties in the calculation of hadronic transition form factors which, so far, are model-dependent quantities when computed for the entire range of squared momentum transfer q^2. Different frameworks have been used to compute transition form factors, namely within the constituent quark model (CQM), light cone sum rules (LCSR), QCD sum rules (QCDSR), Dyson-Schwinger equation (DSE) approaches and lattice QCD (LQCD) amongst others. These form factors are the ingredients of physical observables, such as branching ratios and different asymmetries related to final particle states, especially in $B \to K(K^*)\ell^+\ell^-$ [4, 5, 6], $B \to \gamma\ell^+\ell^-$ [7] decays.

Nowadays, rare decays, and in particular the decays $B_d \to K^*\mu^+\mu^-$ and $B_s \to \phi\mu^+\mu^-$, are of particular interest as they are the object of intense experimental scrutiny by the LHCb Collaboration [8]. Updated values of experimental branching ratios for these decays have been published [9] and their numerical values are given by,

$$\mathrm{Br}(B_d \to K^*\mu^+\mu^-) = (1.06 \pm 0.09) \times 10^{-6}, \quad (1)$$
$$\mathrm{Br}(B_s \to \phi\mu^+\mu^-) = (7.6 \pm 1.5) \times 10^{-7}\ . \quad (2)$$

Here, we study the transition form factors, $B_d \to K^*$ and $B_s \to \phi$, using Ward identities to compute their values at $q^2 = 0$ and then extrapolate them in a general vector meson dominance model to larger q^2 values. We compare these form factors with those obtained in Refs. [4, 10, 11]. In addition, for the decay $B_d \to K^*\mu^+\mu^-$, we also compare the transition form factors with the results of a DSE model of QCD [12].

The vector dominance approach has been successfully applied to different heavy-to-light semileptonic decays, such as $B \to \rho$ [13], $B \to \gamma$ [14], $B \to K_1$[15] and $B_c \to D_s^*$ [16]. The main purpose here is to apply the same technique to $B \to K^*(\phi)$ decays, where we relate the form various factors of the transition in a model-independent way via Ward identities. This allows us to make a clear separation between pole and non-pole type contributions [17, 18]. The residue of the pole is then determined in a self-consistent manner in terms of $\xi_\perp(0)$ or $g_+(0)$ which in turn defines the couplings of vector and axialvector B mesons in the $B \to V$ channel. The form factors in the $q^2 \to 0$ limit are expressed in terms of a universal function, $\xi_\perp(0) \equiv g_+(0)$, that is introduced in Large Energy Effective Theory (LEET) [17, 18] of heavy-to-light form factors, and the masses of the particles. The value of $g_+(0)$ will be extracted from the experimental values of the branching ratios of corresponding $B_{d,s} \to V\gamma$ decays. Eventually, using the above form factors, we compute branching ratios and compare them with their experimental values in Eqs. (1) and (2) as well as with the predictions of the LCSR, Lattice QCD and DSE approaches.

2. Theoretical Framework

2.1. *Effective Hamiltonian*

At the quark level, the decays $B \to V\ell^+\ell^-$ ($V = K^*, \phi$) are governed by the transition $b \to s\ell^+\ell^-$. The phenomenology of the rare decays is commonly described by the heavy-quark effective Hamiltonian,

$$H_{\text{eff}} = -\frac{4G_F}{\sqrt{2}} V_{tb}V_{ts}^* \sum_{i=1}^{10} C_i(\mu)O_i(\mu)\ , \quad (3)$$

where the Wilson coefficients $C_i(\mu)$ encode contributions from energy scales above the renormalization point μ; due to asymptotic freedom property in QCD, these can be computed in perturbation theory, for instance in the naive dimensional regularization scheme [19]. The local quark operators, O_i, describe via their matrix elements the strong and electromagnetic contributions from *all* scales below μ and thus require a nonperturbative approach to their determination. Additional SM/NP operators can be inserted in the above effective Hamiltonian. The operators which describe the decay $B \to V\ell^+\ell^-$ in the SM are given by,

$$\begin{aligned} O_7 &= \frac{e^2}{16\pi^2} m_b \left(\bar{s}\sigma_{\mu\nu} P_R b\right) F^{\mu\nu}, \\ O_9 &= \frac{e^2}{16\pi^2} (\bar{s}\gamma_\mu P_L b)(\bar{l}\gamma^\mu l), \\ O_{10} &= \frac{e^2}{16\pi^2} (\bar{s}\gamma_\mu P_L b)(\bar{l}\gamma^\mu \gamma_5 l), \end{aligned} \tag{4}$$

where $P_{L,R} = (1 \mp \gamma_5)/2$ are left- and right-handed projectors. The effective Hamiltonian given in Eq. (3) yields the following matrix element for the decay $B \to V\ell^+\ell^-$,

$$\begin{aligned} \mathcal{M}(B \to V\ell^+\ell^-) &= \frac{\alpha_{\rm em} G_F}{2\sqrt{2}\pi} V_{tb} V_{ts}^* \Big[\langle V(k,\varepsilon)|\bar{s}\gamma^\mu(1-\gamma^5)b|B(p)\rangle \left\{ C_9^{\rm eff}(\bar{l}\gamma^\mu l) + C_{10}(\bar{l}\gamma^\mu\gamma^5 l)\right\} \\ &\quad -2\, C_7^{\rm eff} m_b \langle V(k,\varepsilon)|\bar{s}\, i\sigma_{\mu\nu}\tfrac{q^\nu}{q^2}(1+\gamma^5)b|B(p)\rangle(\bar{l}\gamma^\mu l)\Big], \end{aligned} \tag{5}$$

where $\alpha_{\rm em}$ is the electromagnetic coupling constant while p and k are the momenta of B and V mesons, respectively. Moreover, ε is the polarization vector of final state vector meson and $q^2 = (p-k)^2$ is the squared momentum transfer. The explicit expressions of Wilson Coefficients $C_{7,9}^{\rm eff}$ and C_{10} are given in Ref. [20] and we do not reproduce them here.

2.2. Form Factors and Ward Identities

Both decays, $B_d \to K^*\ell^+\ell^-$ and $B_s \to \phi\ell^+\ell^-$, involve the hadronic matrix elements of the quark operators introduced in Eq. (4) between the B and V ($V = K^*, \phi$) meson states. These matrix elements can be parameterized in terms of transition form factors which are functions of the square of momentum transfer, q^2. The different matrix elements can be written as

$$\langle V(k,\varepsilon)\,|\bar{s}\gamma_\mu b|\,B(p)\rangle = \frac{2\epsilon_{\mu\nu\alpha\beta}}{M_B+M_V}\varepsilon^{*\nu}p^\alpha k^\beta V(q^2), \tag{6}$$

$$\begin{aligned} \langle V(k,\varepsilon)\,|\bar{s}\gamma_\mu\gamma_5 b|\,B(p)\rangle &= i\,(M_B+M_V)\,\varepsilon^{*\mu}A_1(q^2) \\ &\quad -i\,\frac{(\varepsilon^*\cdot q)}{M_B+M_V}\,(p+k)^\mu\,A_2(q^2) \\ &\quad -2iM_V\,\tfrac{q^\mu}{q^2}\,(\varepsilon^*\cdot q)\left[A_3(q^2)-A_0(q^2)\right], \end{aligned} \tag{7}$$

where

$$A_3(q^2) = \frac{M_B+M_V}{2M_V}A_1(q^2) - \frac{M_B-M_V}{2M_V}A_2(q^2), \tag{8}$$

and where the relation $A_3(0) = A_0(0)$ holds. Likewise,

$$\langle V(k,\varepsilon)\,|\bar{s}\sigma_{\mu\nu}q^\nu b|\,B(p)\rangle = 2\,F_1(q^2)\epsilon_{\mu\nu\alpha\beta}\,\varepsilon^{*\nu}p^\alpha k^\beta \tag{9}$$

$$\begin{aligned} \left\langle V(k,\varepsilon)\,|\bar{s}\sigma_{\mu\nu}q^\nu\gamma^5 b|\,B(p)\right\rangle &= i\left[\left(M_B^2-M_V^2\right)\varepsilon_\mu^* - (\varepsilon^*\cdot q)(p+k)_\mu\right]F_2(q^2) \\ &\quad +\; i\,(\varepsilon^*\cdot q)\left[\,q_\mu - \tfrac{q^2}{M_B^2-M_V^2}(p+k)_\mu\,\right]F_3(q^2)\,, \end{aligned} \tag{10}$$

where $F_1(0) = F_2(0)$. Now, the form factors appearing in Eqs. (6) to (10) can be related to each other by means of Ward identities [13, 26], i.e.,

$$\langle V(k,\varepsilon)\,|\bar{s}\sigma_{\mu\nu}q^\nu b|\,B(p)\rangle = -(m_b+m_s)\,\langle V(k,\varepsilon)\,|\bar{s}\gamma_\mu b|\,B(p)\rangle \tag{11}$$
$$\left\langle V(k,\varepsilon)\,\left|\bar{s}\sigma_{\mu\nu}q^\nu\gamma^5 b\right|\,B(p)\right\rangle = (m_b-m_s)\,\langle V(k,\varepsilon)\,|\bar{s}\gamma_\mu\gamma_5 b|\,B(p)\rangle\ . \tag{12}$$

Making use of Eqs. (6) to (10) in Eqs. (11) and (12), one can relate the transition form factors as follows:

$$F_1(q^2) \;=\; \frac{m_b+m_s}{M_B+M_V}\,V(q^2), \tag{13}$$
$$F_2(q^2) \;=\; \frac{m_b-m_s}{M_B-M_V}\,A_1(q^2), \tag{14}$$
$$F_3(q^2) \;=\; -(m_b-m_s)\tfrac{2M_V}{q^2}[A_3(q^2)-A_0(q^2)]\ . \tag{15}$$

Note that the form factor relations as stated in Eqs. (13), (14) and (15) are model independent. Yet, in deriving the above Ward identities [13], one assumes that the light-quark degrees of freedom can be neglected and hence takes $p = p_b$ as suggested by heavy quark effective theory in the limit $m_b \to \infty$ with p_b/m_b constant. This is a sensible approach and justified in the case of B mesons, though nonperturbative $\Lambda_{\rm QCD}/m_c$ corrections due to the light quarks may be important in charmed mesons. As known from a series of nonperturbative studies, heavy quark effective theory is not generally a good guide to charm physics [21, 22, 23, 24, 25] since the charm quark is neither light nor really heavy. Eventually, a reliable calculation of heavy-light form factors, couplings and decay constants requires the correct description of dynamical chiral symmetry breaking, that is the effect of the light degrees of freedom.

The normalization of the form factors at $q^2 = 0$ can be expressed in terms of $g_+(q^2), g_-(q^2)$, $h(q^2)$ and $h_1(q^2)$ which are discussed in detail in Refs. [13, 16]. Namely, we me make use of

$$\begin{aligned}\langle V(k,\varepsilon^*)|\bar{s}i\sigma_{\alpha\beta}b|B(p)\rangle \;&=\; -i\,\epsilon_{\alpha\beta\rho\sigma}\varepsilon^{*\rho}[(p+k)^\sigma g_+ + q^\sigma g_-] - (\varepsilon^*\cdot q)\,\epsilon_{\alpha\beta\rho\sigma}(p+k)^\rho q^\sigma h\\ &\quad -i[(p+k)^\alpha\epsilon_{\beta\rho\sigma\tau}\varepsilon^{*\rho}(p+k)^\sigma q^\tau - \alpha\leftrightarrow\beta]\,h_1\ ,\end{aligned} \tag{16}$$

so the form factors $V(q^2), A_{1,2}(q^2), F_{1,2,3}(q^2)$ become,

$$F_1(q^2) \;=\; g_+(q^2) - q^2h(q^2)\ , \tag{17}$$
$$F_2(q^2) \;=\; g_+(q^2) + \frac{q^2}{M_B^2-M_V^2}\,g_-(q^2)\ , \tag{18}$$
$$F_3(q^2) \;=\; -g_-(q^2) - (M_B^2-M_V^2)h(q^2)\ , \tag{19}$$

and

$$V(q^2) \;=\; \frac{M_B+M_V}{m_b+m_s}\left[g_+(q^2)-q^2h_1(q^2)\right]\ , \tag{20}$$
$$A_1(q^2) \;=\; \frac{M_B-M_V}{m_b-m_s}\left[g_+(q^2)+\frac{q^2}{M_B^2-M_V^2}\,g_-(q^2)\right]\ , \tag{21}$$
$$A_2(q^2) \;=\; \frac{M_B-M_V}{m_b-m_s}\left[g_+(q^2)-q^2h(q^2)\right] - \frac{2M_V}{M_B-M_V}\,A_0(q^2)\ . \tag{22}$$

At $q^2 = 0$, the form factors F_1, F_2, V and A_1 in Eqs. (17), (18), (20) and (21) are thus parameterized by a single constant $g_+(0)$ whereas A_2 and A_0 in Eq. (22) can be expressed in

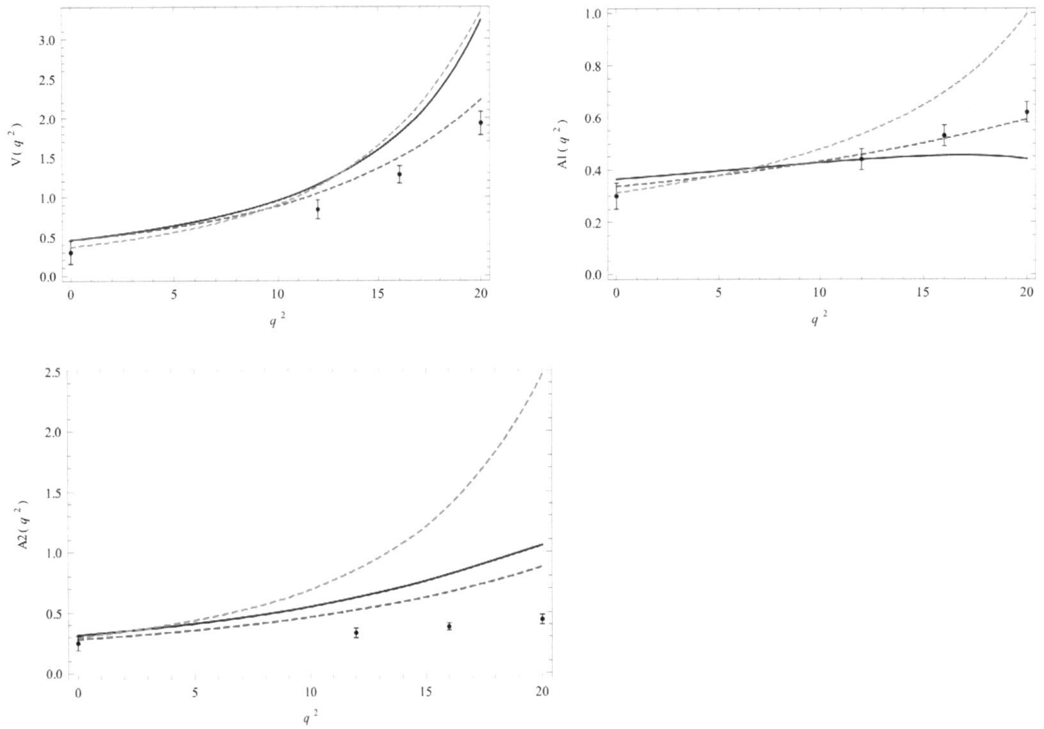

Figure 1. Comparison of the $B_d \to K^*$ transition form factors, Eqs. (30), (31) and (32) (black solid curve) with predictions from LQCD [10] (points with error bars), LCSR [4] (red dashed curve) and DSE [12] (green dashed curve).

terms of $g_+(0)$ and $A_0(0)$. It is well known that in case of real photon emission in $B_{d,s} \to V\gamma$ decays, the decay rate is a function of $g_+(0)$ [18]:

$$\Gamma(B \to V\gamma) = \frac{G_F^2\, \alpha_{\rm em}}{32\pi^4} |V_{tb}V_{ts}^*|^2 m_b^2 M_B^3 \left(1 - \frac{M_V^2}{M_B^2}\right)^3 |C_7^{\rm eff}|^2 |g_+(0)|^2. \tag{23}$$

Using the experimental values of branching ratios of $B_d \to K^*\gamma$ and $B_s \to \phi\gamma$ decays [9],

$$\mathrm{Br}(B_d \to K^*\gamma) = (4.33 \pm 0.15) \times 10^{-5}, \tag{24}$$

$$\mathrm{Br}(B_s \to \phi\gamma) = (3.6 \pm 0.4) \times 10^{-5}\ , \tag{25}$$

the extracted $g_+(0)$ for these decays are,

$$g_+(0)^{B_d \to K^*} = 0.365^{+0.025}_{-0.025}, \tag{26}$$

$$g_+(0)^{B_s \to \phi} = 0.335^{+0.02}_{-0.02}\ . \tag{27}$$

With $g_+(0)$ in hand, the other unknown, i.e. $A_0(0)$, can be expressed in terms of it as [13],

$$A_0(0) = \left(\frac{1 - M_V^2/M_B^2}{1 + M_V^2/M_B^2} + \frac{M_B}{M_V}\right) g_+(0)\ . \tag{28}$$

Journal of Physics: Conference Series **630** (2015) 012050 doi:10.1088/1742-6596/630/1/012050

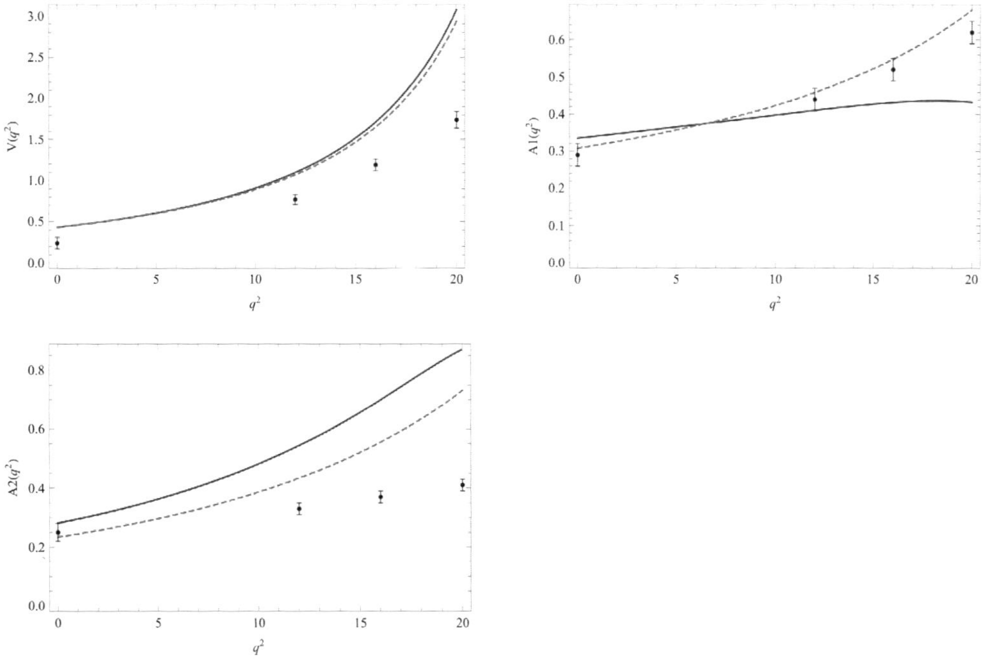

Figure 2. Comparison of form factors for $B_s \to \phi$; curve description as in Fig. 1 though no DSE results available in this transition.

It is worth remembering that the form factor relations derived from Ward identities do not hold for the entire physical momentum region. Therefore, we employ the following parametrization between $q^2 = 0$ and near the poles:

$$F(q^2) = \frac{F(0)}{(1 - q^2/M^2)(1 - q^2/M'^2)} , \tag{29}$$

where M is $M_{B^*}(1^-)$ or $M_{B_A^*}(1^+)$ and M' is the radial excitation of M. The parametrization given in Eq. (29) not only takes into account the correction to the single pole dominance, as suggested by dispersion relations [26], but also help us to determine the couplings of B^* or B_A^* to the BV channel [13, 14], which is not the aim of this work. Finally, using the above parameterization, the form factors can be expressed as:

$$V(q^2) = \frac{V(0)}{(1 - q^2/M_{B^*}^2)(1 - q^2/M_{B^*}'^2)}, \tag{30}$$

$$A_1(q^2) = \frac{A_1(0)}{(1 - q^2/M_{B^*}^2)(1 - q^2/M_{B^*}'^2)} \left(1 - \frac{q^2}{M_B^2 - M_V^2}\right), \tag{31}$$

$$A_2(q^2) = \frac{\tilde{A}_2(0)}{(1 - q^2/M_{B^*}^2)(1 - q^2/M_{B^*}'^2)} - \frac{2M_V}{M_B - M_V} \frac{A_0(0)}{(1 - q^2/M_B^2)(1 - q^2/M_B'^2)}, \tag{32}$$

where $\tilde{A}_2(0)$ is defined as [13]:

$$\tilde{A}_2(0) = \left(\frac{M_B - M_V}{m_b - m_s}\right) g_+(0) . \tag{33}$$

	V	A_1	A_2	$\frac{V(q^2_{\rm max})}{V(0)}$	$\frac{A_1(q^2_{\rm max})}{A_1(0)}$	$\frac{A_2(q^2_{\rm max})}{A_2(0)}$
Present Work	0.456 (2.858)	0.365 (0.4476)	0.316 (1.007)	6.267	1.226	3.186
LCSR	0.457 (2.064)	0.337 (0.577)	0.282 (0.830)	4.516	1.712	2.943
LQCD	0.30 (1.93)	0.30 (0.62)	0.25 (0.44)	6.433	2.06	1.76
DSE	0.37 (2.98)	0.29 (1.28)	0.30 (2.177)	8.054	4.431	7.256

Table 1. Variation of form factors for the transition $B_d \to K^*$ at $q^2 = 0$ and $q^2 = q^2_{\rm max}$ in different approaches. The first value represents the form factors at $q^2 = 0$ whereas the value at maximum q^2 is in parentheses. The ratios of the form factors at $q^2_{\rm max}$ and $q^2 = 0$ are listed in the last three columns.

	V	A_1	A_2	$\frac{V(q^2_{\rm max})}{V(0)}$	$\frac{A_1(q^2_{\rm max})}{A_1(0)}$	$\frac{A_2(q^2_{max})}{A_2(0)}$
Present Work	0.433 (2.566)	0.335 (0.436)	0.282 (0.828)	5.926	1.301	2.936
LCSR	0.434 (2.455)	0.311 (0.637)	0.234 (0.676)	5.656	2.048	2.888
Lattice	0.24 (1.74)	0.29 (0.62)	0.25 (0.41)	7.25	2.137	1.64

Table 2. Variation of form factors for the transition $B_s \to \phi$ at $q^2 = 0$ and $q^2 = q^2_{\rm max}$ in different approaches. Table entries are as in Table 1.

The transition form factors in Eqs. (30), (31) and (32) are plotted as a function of q^2 in Figs. 1 and 2 for $B_d \to K^*$ and $B_s \to \phi$, respectively, where we use the Particle Data Group values for the masses of B, B^*, B^*_A and their excited states [9]. In the same figures, we also compare our form factors with those obtained in the Lattice QCD and LCSR approaches and in case of the $B_d \to K^*$ transition the DSE form factors [12] are also included. The trend of the $V(q^2)$ evolution we obtain parallels that of the DSE form factor, whereas our $A_1(q^2)$ and $A_2(q^2)$ compare more favorably with the LCSR results, as becomes clear from Fig. 1. Similarly, for the $B_s \to \phi\mu^+\mu^-$ decay, our form factors are mostly in agreement with LCSR predictions except for larger momenta, $q^2 \gtrsim 10$ GeV^2; see Fig. 2. It is also apparent that lattice QCD simulations produce a softer slope of the $A_1(q^2)$ and $V(q^2)$ form factors than the other calculations. For reference, we also list the form factor values at $q^2 = 0$ and $q^2 = q^2_{\rm max}$ as well as their ratios in Tables 1 and 2.

3. Applications of Transition Form Factors: Branching Fractions

To conclude, we present the calculated branching ratios of the $B \to K^*\mu^+\mu^-$ and $B_s \to \phi\mu^+\mu^-$ decays, where we remind that the transition form factors are the major hadronic input as well as source of uncertainties. Here we use the form factors introduced in Section 2.2 and extrapolated via Eqs. (30), (31) and (32) to compute the differential branching fractions and compare them to those obtained with form factors of the corresponding LCSR, Lattice QCD and DSE approaches.

The formula for the differential decay rate is given by,

$$\frac{d^2\Gamma(B \to V\ell^+\ell^-)}{d\cos\theta\, dq^2} = \frac{1}{2M_V^3}\frac{2\beta\sqrt{\lambda}}{(8\pi)^3}|\mathcal{M}|^2 \tag{34}$$

with $\mathcal{M}$ from Eq. (5) and where $\beta \equiv \sqrt{1-4m_\ell^2/s}$, $\lambda \equiv \lambda(M_B, M_V, s) = M_B^4 + M_V^4 + q^4 - 2M_B^2M_V^2 - 2q^2M_B^2 - 2q^2M_V^2$. The explicit form of differential decay rates for the decay $B_d \to K^*\mu^+\mu^-$ and $B_s \to \phi\mu^+\mu^-$ can be found in Ref. [27]. The plot of the differential branching fractions, $d\Gamma(B \to V\ell^+\ell^-)/dq^2$, for the above mentioned decays in different approaches are depicted in Figs. 3a and 3b. The functional form of the differential branching fractions we

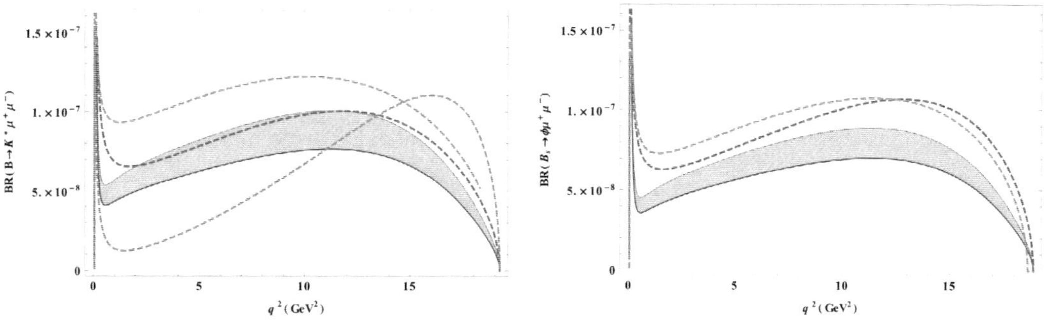

Figure 3. Differential branching fractions, $d\Gamma/dq^2$, for the decays $B_d \to K^*\mu^+\mu^-$ (left panel) and $B_s \to \phi\mu^+\mu^-$ (right panel). The gray-shaded band represents the estimated error of the differential branching fraction in the present approach; the same branching fractions computed with the LCSR, Lattice QCD and DSE form factors are plotted with red-dashed, green-dashed and orange-dashed curves, respectively.

	$\text{Br}(B_d \to K^*\mu^+\mu^-)$	$\text{Br}(B_s \to \phi\mu^+\mu^-)$
Present Work	$(1.37 \pm 0.2) \times 10^{-6}$	$(1.21 \pm 0.15) \times 10^{-6}$
LCSR	1.31×10^{-6}	1.70×10^{-6}
Lattice	1.97×10^{-6}	1.67×10^{-6}
DSE	1.18×10^{-6}	—
PDG	$(1.06 \pm 0.09) \times 10^{-6}$	$(7.6 \pm 1.5) \times 10^{-7}$

Table 3. Numerical values of branching ratios for the semileptonic decays $B_d \to K^*\mu^+\mu^-$ and $B_s \to \phi\mu^+\mu^-$ in different form factors approaches.

obtain is comparable with the one using other form factor calculations, albeit with variations in magnitude. Note, however, that the DSE approach leads to a stronger variation of $d\Gamma/dq^2$ with q^2 in the $B \to K^*$ transition.

Furthermore, numerical values of branching ratios of $B_d \to K^*\mu^+\mu^-$ and $B_s \to \phi\mu^+\mu^-$ decays are listed in Table 3. The $B_d \to K^*\mu^+\mu^-$ branching ratio obtained with transition form factors of the DSE based model lies close to the experimental value. Within theoretical and experimental errors, both of which may be underestimated, our numerical results for both decays do agree with the PDG average. We cannot make a statement in case of the LQCD and LCSR results as no error estimate is given, yet the LQCD branching ratios are considerably larger than experimental values. This could be due to the fact that LQCD does not produce as strongly rising form factors, hence might not describe a VMD behavior. In the case of the $B_s \to \phi\mu^+\mu^-$ decay, the discrepancy between the experimental average and LCSR and LQCD values is equally large. It is worthwhile to remind that the values of $g_+(0)$, Eqs. (26) and (27), are very similar. Given that the functional form of the $V(q^2)$, $A_1(q^2)$ and $A_2(q^2)$ form factors of the $B_d \to K^*$ transition is not noticeable different from the one in $B_s \to \phi$ and that the phase space difference is minor, we do not expect very different branching ratio values.

Acknowledgments
M. A. P., B. E. and I. A. are grateful to the organizers of the *XXXVII Reunião de Trabalho sobre Física Nuclear no Brasil* in Maresias, São Paulo, for their kind invitation and acknowledge support by the São Paulo Research Foundation (FAPESP) under grant nos. 2012/13047-2, 2013/23177-3 and 2013/16088-4. M. A. P. and I. A. would also like to thank the Physics Department at Quaid-i-Azam University for the kind hospitality during the last stages of the work. M. J. A would like to thank the support by Quaid-i-Azam University through the University Research Fund. B. E. is partially supported by a CNPq fellowship no. 301190/2014-3.

[1] S. L. Glashow, J. Iliopoulos, and L. Maiani, Phys. Rev. D **2**, 1285 (1970).
[2] M. S. Alam et al. Phys. Rev. Lett. **74**, 2885 (1995).
[3] R. Ammar et al., Phys. Rev. Lett. **71**, 674 (1993); CLEO CONF 96-05 (1996).
[4] A. Ali, P. Ball, L. T. Handoko and G. Hiller, Phys. Rev. D **61**, 074024 (2000).
[5] T. M. Aliev, M. K. Cakmak and M. Savci, Nucl. Phys. B **607**, 305 (2001); T. M. Aliev, A. Ozpineci, M. Savci and C. Yuce, Phys. Rev. D **66**, 115006 (2002); T. M. Aliev, A. Ozpineci and M. Savci, Phys. Lett. B **511**, 49 (2001); T. M. Aliev and M. Savci, Phys. Lett. B **481**, 275 (2000); T. M. Aliev, D. A. Demir and M. Savci, Phys. Rev. D **62**, 074016 (2000); T. M. Aliev, C. S. Kim and Y. G. Kim, Phys. Rev. D **62**, 014026 (2000); T. M. Aliev and E. O. Iltan, Phys. Lett. B **451**, 175 (1999); C. H. Chen and C. Q. Geng, Phys. Rev. D **66**, 034006 (2002); C. H. Chen and C. Q. Geng, Phys. Rev. D **66**, 014007 (2002); G. Erkol and G. Turan, Nucl. Phys. B **635**, 286 (2002); E. O. Iltan, G. Turan and I. Turan, J. Phys. G **28**, 307 (2002); T. M. Aliev, V. Bashiry and M. Savci, JHEP **0405**, 037 (2004); W. J. Li, Y. B. Dai and C. S. Huang, Eur. Phys. J. C **40**, 565 (2005); Q. S. Yan, C. S. Huang, W. Liao and S. H. Zhu, Phys. Rev. D **62**, 094023 (2000); S. R. Choudhury, N. Gaur, A. S. Cornell and G. C. Joshi, Phys. Rev. D **68**, 054016 (2003); S. R. Choudhury, A. S. Cornell, N. Gaur and G. C. Joshi, Phys. Rev. D **69**, 054018 (2004).
[6] A. Ali, E. Lunghi, C. Greub and G. Hiller, Phys. Rev. D **66**, 034002 (2002); F. Kruger and E. Lunghi, Phys. Rev. D **63**, 014013 (2001).
[7] S. Rai Choudhury , N. Gaur and N. Mahajan, Phys. Rev. D **66**, 054003 (2002); T. M. Aliev, V. Bashiry and M. Savci, Phys. Rev. D **71**, 035013 (2005); U. O. Yilmaz, B. B. Sirvanli and G. Turan, Nucl. Phys. **692**, 249 (2004); U. O. Yilmaz, B. B. Sirvanli and G. Turan, Eur. Phys. J. C **30**, 197 (2003).
[8] The LHCb Collaboration, R. Aaij *et al.* JHEP **08**, 034 (2011); The LHCb Collaboration, R. Aaij *et al.* Preprint CERN-PH-EP-2011-106.
[9] K. A. Olive et al. (Particle Data Group), Chin. Phys. C **38**, 090001 (2014).
[10] Ronald R. Horgan, Zhaofeng Liu, Stefan Meinel, Matthew Wingate, Phys. Rev. D **89**, 094501 (2014).
[11] P. Ball and R. Zwicky, Phys. Rev. D **71**, 014029 (2005).
[12] Mikhail A. Ivanov, Jürgen G. Körner, Sergey G. Kovalenko, and Craig D. Roberts Phys. Rev. D **76**, 034018 (2007).
[13] A. H. S. Gilani, Riazuddin and T. Al-Aithan, JHEP **0309**, 065 (2003).
[14] M. Saleh Khan, M. Jamil Aslam, A. H. S Gilani and Riazuddin, Eur. Phys. J.C **49**, 665 (2007).
[15] M. Ali Paracha, Ishtiaq Ahmed and M. Jamil Aslam, Eur. Phys. J.C **52**, 967 (2007).
[16] M. Ali Paracha, Ishtiaq Ahmed and M. Jamil Aslam, Phys. Rev. D **84**, 035003 (2011).
[17] J. Charles, A. Le Yaouanc, L. Oliver, O. Pène, J.-C. Raynal, Phys. Rev. D **60**, 014001 (1999).
[18] M. Jamil Aslam and Riazuddin, Phys. Rev. D **72**, 094019 (2005); M. Jamil Aslam, Eur. Phys. J. C **49**, 651 (2007).
[19] A. J. Buras et al., Nucl. Phys. B **424**, 374 (1994).
[20] C.S. Kim, T. Morozumi, A.I. Sanda, Phys. Lett. B **218**, 343 (1989); X. G. He, T. D. Nguyen and R. R. Volkas, Phys. Rev. D **38**, 814 (1988); B. Grinstein, M. J. Savage, M. B. Wise, Nucl. Phys. B **319**, 271 (1989); N. G. Deshpande, J. Trampetic and K. Panose, Phys. Rev. D **39**, 1461 (1989); P. J. O' Donnell and H. K. K. Tung, Phys. Rev. D **43**, 2067 (1991); N. Paver and Riazuddin, Phys. Rev. D **45**, 978 (1992); A. Ali, T. Mannel and T. Morozumi, Phys. Lett. B **273**, 505 (1991); D. Melikhov, N. Nikitin and S. Simula, Phys. Lett. B **430**, 332 (1998) [arXiv:hep-ph/9803343]; J. M. Soares, Nucl. Phys. B **367**, 575 (1991); G. M. Asatrian and A. Ioannisian, Phys. Rev. D **54**, 5642 (1996) [arXiv:hep-ph/9603318]; J. M. Soares, Phys. Rev. D **53**, 24 (1996); C. H. Chen and C. Q. Geng, Phys. Rev. D **64**, 074001(2001) [arXiv:hep-ph/0106193].
[21] B. El-Bennich, M. A. Ivanov and C. D. Roberts, Nucl. Phys. Proc. Suppl. **199**, 184 (2010).
[22] B. El-Bennich, C. D. Roberts and M. A. Ivanov, arXiv:1202.0454 [nucl-th].
[23] B. El-Bennich, G. Krein, L. Chang, C. D. Roberts and D. J. Wilson, Phys. Rev. D **85**, 031502 (2012).

[24] A. Bashir, L. Chang, I. C. Cloët, B. El-Bennich, Y. X. Liu, C. D. Roberts and P. C. Tandy, Commun. Theor. Phys. **58**, 79 (2012).
[25] E. Rojas, B. El-Bennich and J. P. B. C. de Melo, Phys. Rev. D **90**, no. 7, 074025 (2014).
[26] C. A. Dominguez, N. Paver and Riazuddin, Z. Physik C **48**, 55 (1990).
[27] Ishtiaq Ahmed, M. Jamil Aslam and M. Ali Paracha, Phys. Rev. D **89**, 015006 (2014).

Hadron Physics and QCD: Just the Basic Facts.

Crag D. Roberts
Physics Division, Argonne National Laboratory, Argonne, Illinois 60439, USA
E-mail: cdroberts@anl.gov

Abstract. With discovery of the Higgs boson, the Standard Model of Particle Physics became complete. Its formulation is a remarkable story; and the process of verification is continuing, with the most important chapter being the least well understood. Quantum Chromodynamics (QCD) is that part of the Standard Model which is supposed to describe all of nuclear physics and yet, almost fifty years after the discovery of quarks, we are only just beginning to understand how QCD moulds the basic bricks for nuclei: pious, neutrons, protons. QCD is characterized by two emergent phenomena: confinement and dynamical chiral symmetry breaking (DCSB), whose implications are extraordinary. This contribution describes how DCSB, not the Higgs boson, generates more than 98% of the visible mass in the Universe, explains why confinement guarantees that condensates, those quantities that were commonly viewed as constant mass-scales that fill all spacetime, are instead wholly contained within hadrons, and elucidates a range of observable consequences of confinement and DCSB whose measurement is the focus of a vast international experimental programme.

1. Introduction

A context for this report is provided by the experimental programmes underway and planned at the Thomas Jefferson National Accelerator Facility (JLab), Newport News, Virginia. Funds for the development of research plans and designs for the Continuous Electron Beam Accelerator Facility (CEBAF) at JLab were initially provided in 1984, and construction began in 1987. Seven years later, in 1994, the facility achieved its design capability, delivering 4 GeV electron beams on targets in the three associated experimental halls. The goal was to "write the book" about the strongest known force in Nature – the force that holds nuclei together – and determine how that force can be explained in terms of the gluons and quarks of QCD.

An aim of the original JLab programme was verification of the following prediction: at energy-scales greater than some minimum value, Q_0, whose value was not determined by the proof, cross-sections and form factors involving hadrons should behave as follows: [1, 2, 3, 4, 5, 6]

$$\mathcal{A}(\sigma^2) \overset{\sigma \gg Q_0}{\propto} \left[\frac{Q_0^2}{\sigma^2}\right]^N \ln\left[\frac{Q_0^2}{\sigma^2}\right]^{\gamma_{\mathcal{A}}} \tag{1}$$

where the integer $N = n_{\rm quarks} - 1 + \Delta\lambda$, with $n_{\rm quarks}$ counting the lowest possible number of quarks and/or antiquark in the target and $\Delta\lambda = 0, 1$, depending upon whether a spin-flip occurs in the scattering process. The power-law term gives rise to parton-model scaling and can be explained on dimensional grounds. The distinctive signature of QCD lies in the logarithmic factor, the exponent on which, $\gamma_{\mathcal{A}}$, can be computed and whose appearance signals scaling violation, which is well documented in deep inelastic scattering [7]. It was optimistically imagined that Q_0 might be as small as 1 GeV and so CEBAF was designed to reach 4 GeV.

In the ten years following achievement of design capacity, numerous fascinating experimental results were obtained at JLab, including an empirical demonstration that the distribution of charge and magnetization within the proton are completely different [8, 9, 10, 11, 12, 13], *viz.*

$$\mu_p G_E^p(Q^2)/G_M^p(Q^2) \neq 1\,. \tag{2}$$

This fact, which overturned a longstanding particle physics paradigm, along with a range of related observations, can be explained by the presence of strong, nonpointlike, electromagnetically-active scalar and axial-vector diquark correlations within the nucleon [14, 15, 16, 17, 18, 19, 20, 21]. However, notwithstanding the breadth of JLab's programme and the excitement it has generated, no experiment has yet produced an unambiguous signal for the realisation of Eq. (1), *i.e.* as yet, we have no clear sign of parton model scaling and certainly no sign of scaling violations. Partly owing to this but also with a vast array of new experimental tests of the Standard Model's strong interaction sector in mind [22], it was decided in 2004 that CEBAF should be upgraded so that it could deliver 12 GeV electrons on targets. That upgrade is now complete and commissioning beams are being sent to the experimental halls.

The JLab 12 facility and, indeed, an array of modern accelerators worldwide, will confront a diverse range of scientific challenges. In the foreseeable future, we will know the results of a search for hybrid and exotic hadrons, the discovery of which would force a dramatic reassessment of the distinction between matter and force fields in Nature. Opportunities provided by new data on hadron elastic and transition form factors will be exploited, yielding insights into the infrared running of QCD's coupling and dressed-masses, revealing correlations that are key to hadron structure, exposing the facts or fallacies in contemporary descriptions of hadron structure, and seeking for verification of Eq. (1) in numerous processes. Precision experimental studies of the valence-quark region will proceed, with the results being used to confront predictions from theoretical computations of distribution functions and distribution amplitudes – computation is critical here because without it no amount of data will reveal anything about the theory underlying the phenomena of strong interaction physics. In addition, the international community will seek and exploit opportunities to use precision-QCD as a probe for physics beyond the Standard Model. Each of the pieces in this body of exploration, however, can be viewed as steps aimed toward understanding a single overarching puzzle within the Standard Model, *viz.* what is confinement and how is it related to DCSB – the origin of the vast bulk of visible mass.

In order to match the rate at which progress is being made experimentally and, better, to guide and enhance that programme, flexible, responsive theoretical approaches are needed: methods that are capable both of rapidly providing an intuitive understanding of complex problems and illuminating a path toward answers and new discoveries. In this milieu, notwithstanding its steady progress toward results with input parameters that approximate the real world, the numerical simulation of lattice-regularised QCD will not suffice. Approaches formulated in the continuum and inspired by, based upon, or connected directly with QCD are necessary. Prominent amongst such tools are QCD Sum Rules [23] and Dyson-Schwinger equations (DSEs) [24], both of which were canvassed at this Workshop.

2. Quantum Chromodynamics

QCD is a quantum gauge field theory based on the group $SU(3)$ [25]; and following around forty years of study, there is no confirmed breakdown over an enormous energy domain: $0 < E < 8$ TeV. Consequently, QCD is plausibly the only known instance of a quantum field theory that can rigorously be defined nonperturbatively [26], in which case it is truly a field theory *not merely* an effective field theory. The possibility that QCD might be rigorously well

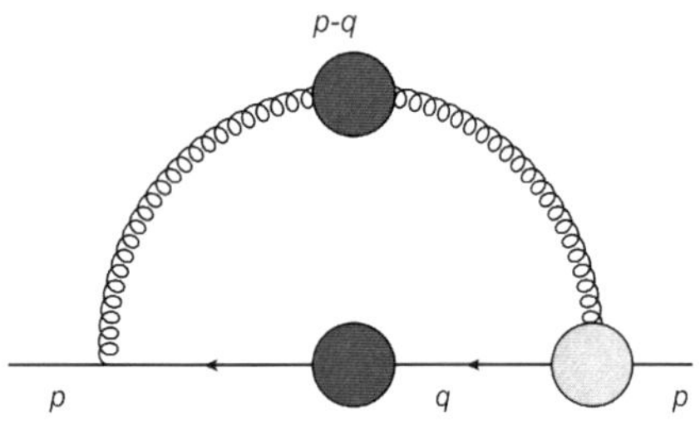

Figure 1. Dressed-quark self energy, which is the dynamical content of QCD's most basic fermion gap equation. The kernel is composed from the dressed-gluon propagator (spring with dark circle) and the dressed-quark-gluon vertex (light-circle). The equation is nonlinear owing to the appearance of the dressed-quark propagator (line with dark circle). This image encodes every valid Feynman diagram relevant to the quark dressing process. (Momentum flows from right-to-left.)

defined is one of its deepest fascinations. In that case, QCD could stand alone as an archetype – the only internally consistent quantum field theory which is defined at all energy scales. This is a remarkable possibility with wide-ranging consequences and opportunities, *e.g.* it means that QCD-like theories provide a viable paradigm for extending the Standard Model to greater scales than those already probed. Contemporary research in this direction is typified by the notion of extended technicolour [27, 28], in which electroweak symmetry breaks via a fermion bilinear operator in a strongly-interacting non-Abelian theory; and the Higgs sector of the Standard Model becomes an effective description of a more fundamental fermionic theory, similar to the Ginzburg-Landau theory of superconductivity.

2.1. Confinement

This notion has already been mentioned. However, in order to consider the concept further it is actually crucial to *define* the subject. That problem is canvassed in Sec. 2.2 of Ref. [24], which explains that the potential between infinitely-heavy quarks measured in numerical simulations of quenched lattice-regularised QCD – the so-called static potential [29] – is *irrelevant* to the question of confinement in our Universe, in which light quarks are ubiquitous and the pion is unnaturally light. This is because light-particle creation and annihilation effects are essentially nonperturbative in QCD and so it is impossible in principle to compute a quantum mechanical potential between two light quarks [30, 31]. This means there is no flux tube in a Universe with light quarks and consequently that the flux tube is not the correct paradigm for confinement.

An alternative perspective associates confinement with dramatic, dynamically-driven changes in the analytic structure of QCD's propagators and vertices. In this realisation, confinement is a dynamical process. In fact, as will subsequently be explained, contemporary theory predicts that both quarks and gluons acquire running mass distributions in QCD, which are large at infrared momenta (see, *e.g.* Refs. [32, 33, 34, 35, 36, 37]). The generation of these masses leads to the emergence of a length-scale $\varsigma \approx 0.5\,\text{fm}$, whose existence and magnitude is evident in all existing studies of dressed-gluon and -quark propagators and which characterizes a dramatic change in their analytic structure. In models based on such features [38], once a gluon or quark is produced, it begins to propagate in spacetime; but after each "step" of length ς, on average, an interaction occurs so that the parton loses its identity, sharing it with others. Finally a cloud of partons is produced, which coalesces into colour-singlet final states. Such pictures of parton propagation, hadronisation and confinement can be tested in experiments at modern and planned facilities.

2.2. Dynamical chiral symmetry breaking

Whilst the nature and realisation of confinement in empirical QCD is still being explored, DCSB; namely, the generation of *mass from nothing*, is a theoretically-established nonperturbative feature of QCD. It is important to insist on the term "dynamical," as distinct from spontaneous,

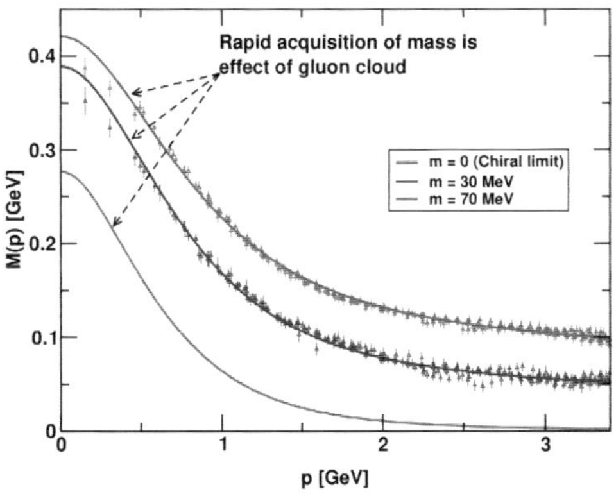

Figure 2. Dressed-quark mass function, $M(p)$ in Eq. (3): *solid curves* – DSE results, explained in Refs. [32, 34], "data" – numerical simulations of lattice-regularised QCD [33]. (NB. $m = 70\,\text{MeV}$ is the uppermost curve and current-quark mass decreases from top to bottom.) The current-quark of perturbative QCD evolves into a constituent-quark as its momentum becomes smaller. The constituent-quark mass arises from a cloud of low-momentum gluons attaching themselves to the current-quark. This is DCSB: an essentially nonperturbative effect that generates a quark *mass from nothing*; namely, it occurs even in the chiral limit.

because nothing is added to QCD in order to effect this remarkable outcome and there is no simple change of variables in the QCD action that will make it apparent. Instead, through the act of quantising the classical chromodynamics of massless gluons and quarks, a large mass-scale is generated. DCSB is the most important mass generating mechanism for visible matter in the Universe, being responsible for approximately 98% of the proton's mass.

A fundamental expression of DCSB is the behaviour of the quark mass-function, $M(p)$, which is a basic element in the dressed-quark propagator

$$S(p) = 1/[i\gamma \cdot pA(p^2) + B(p^2)] = Z(p^2)/[i\gamma \cdot p + M(p^2)]\,, \tag{3}$$

which may be obtained as a solution to QCD's most basic fermion gap equation (see Fig. 1). The highly nontrivial behaviour of the mass function, depicted in Fig. 2, arises primarily because a dense cloud of gluons comes to clothe a low-momentum quark; and explains how an almost-massless parton-like quark at high energies transforms, at low energies, into a constituent-like quark, which possesses an effective "spectrum mass" $M_D \sim 350\,\text{MeV}$. Consequently, the proton's mass is two orders-of-magnitude larger than the sum of the current-masses of its three valence-quarks.

2.3. Whence the mass?

One might ask just how the self-energy depicted in Fig. 1 is capable of generating *mass from nothing*, *viz.* the $m = 0$ curve in Fig. 2, which cannot arise in the classical theory. The answer lies in the fact that Fig. 1 is a deceptively simply picture. It actually corresponds to a countable infinity of diagrams, all of which can potentially contribute. To provide a context, quantum electrodynamics, an Abelian gauge theory, has 12 672 diagrams at order α^5 in the computation of the electron's anomalous magnetic moment [39]. Owing to its foundation in the non-Abelian group $SU(3)$, the analogous perturbative computation of a quark's anomalous chromomagnetic moment has many more diagrams at this order in the strong coupling. The number of diagrams represented by the self energy in Fig. 1 grows equally rapidly, *i.e.* combinatorially with the number of propagators and vertices used at a given order. Indeed, proceeding systematically, a computer will very quickly generate the first diagram in which the number of loops is so great that it is simply impossible to calculate in perturbation theory: impossible in the sense that we don't yet have the mathematical capacity to solve the problem.

Each of the diagrams which contributes to $M(p^2)$ in a weak-coupling expansion of Fig. 1 is multiplied by the current-quark mass, $\hat{m}$. Plainly, any finite sum of diagrams must therefore vanish as $\hat{m} \to 0$. However, with *infinitely many* diagrams the situation might be very different:

Journal of Physics: Conference Series **630** (2015) 012051 doi:10.1088/1742-6596/630/1/012051

one has "$0 \times \infty$," a product whose limiting value is contingent upon the cumulative magnitude of each term in the sum. Consider therefore the behaviour of $M(p^2)$ at large p^2. QCD is asymptotically free [40, 41, 42]. Hence, on this domain, each of the regularised loop diagrams must individually evaluate to a small number whose value depends on just how large is the coupling. It will not be surprising, therefore, to learn that for a monotonically-decreasing running-coupling, $\alpha_S(k^2)$, there is a critical value of $\alpha_S(0)$ above which the magnitude of the sum of infinitely many diagrams is sufficient to balance the linear decrease of $\hat{m} \to 0$, so that the answer is nonzero and finite in this limit, *viz.*,

$$\exists \alpha_S^c(0) \mid \forall \alpha_S(0) > \alpha_S^c(0), M_0(p^2) := \lim_{\hat{m} \to 0} M(p^2; \hat{m}) \neq 0 . \tag{4}$$

The internal consistency of QCD appears to guarantee that the limit is always finite. (The case of Abelian theories is more complicated [43] because they are not asymptotically free.)

2.4. Enigma of mass

The pion is Nature's lightest hadron. In fact, it is peculiarly light, with a mass just one-fifth of that which quantum mechanics would lead one to expect. This remarkable feature has its origin in DCSB. In quantum field theory the pion's structure is described by a Bethe-Salpeter amplitude (here k is the relative momentum between the valence-quark and -antiquark constituents, and P is their total momentum):

$$\Gamma_\pi(k;P) = \gamma_5 \left[i E_\pi(k;P) + \gamma \cdot P F_\pi(k;P) + \gamma \cdot k \, G_\pi(k;P) - \sigma_{\mu\nu} k_\mu P_\nu H_\pi(k;P) \right], \tag{5}$$

which is simply related to an object that would be its Schrödinger wave function if a nonrelativistic limit were appropriate. In QCD if, and only if, chiral symmetry is dynamically broken, then for $\hat{m} = 0$ [44, 45]:

$$f_\pi E_\pi(k;0) = B(k^2) , \tag{6}$$

where the right-hand-side is a scalar function in the dressed-quark propagator, Eq. (3). This identity is miraculous. It means that the two-body problem is solved, almost completely, without lifting a finger, once the solution to the one body problem is known. Eq. (6) is a quark-level Goldberger-Treiman relation. It is also the most basic expression of Goldstone's theorem in QCD, *viz.*

Goldstone's theorem is fundamentally an expression of equivalence between the one-body problem and the two-body problem in QCD's colour-singlet pseudoscalar channel.

Eq. (6) emphasises that Goldstone's theorem has a pointwise expression in QCD; and, furthermore, that pion properties are an almost direct measure of the mass function depicted in Fig. 2. Thus, enigmatically, properties of the (nearly-)massless pion are the cleanest expression of the mechanism that is responsible for almost all the visible mass in the Universe. Plainly, DCSB has a very deep and far-reaching impact on physics within the Standard Model.

2.5. Gluon cannibalism

It is not just the propagation of quarks that is affected by strong interactions in QCD. The propagation of gluons, too, is described by a gap equation; and its solution shows that gluons are cannibals: they are a particle species whose members become massive by eating each other! The associated gluon mass function, $m_g(k^2)$, is monotonically decreasing with increasing k^2 and recent work [37] has established that

$$m_g(k^2 = 0) \approx 0.5 \, \text{GeV}. \tag{7}$$

The value of the mass-scale in Eq. (7) is *natural* in the sense that it is commensurate with but larger than the value of the dressed light-quark mass function at far infrared momenta: $M(0) \approx 0.3\,\text{GeV}$ (see Fig. 2). Moreover, the mass term appears in the transverse part of the gluon propagator, hence gauge-invariance is not tampered with; and the mass function falls as $1/k^2$ for $k^2 \gg m_g(0)$ (up to logarithmic corrections), so the gluon mass is invisible in perturbative applications of QCD.

Gauge boson cannibalism presents a new physics frontier within the Standard Model. Asymptotic freedom means that the ultraviolet behaviour of QCD is controllable. At the other extreme, dynamically generated masses for gluons and quarks entail that QCD creates its own infrared cutoffs. Together, these effects eliminate both the infrared and ultraviolet problems that typically plague quantum field theories and thereby make reasonable the hope that QCD is nonperturbatively well defined.

The dynamical generation of gluon and quark masses provides a basis for understanding the notion of a maximum wavelength for gluons and quarks in QCD [46]. Indeed, given the magnitudes of the gluon and quark mass-scales, it is apparent that field modes with wavelengths $\lambda > \varsigma \approx 2/m_g(0) \approx 0.5\,\text{fm}$ decouple from the dynamics. They are screened in the sense described in Sec. 2.1. This is just one consequence of the appearance of a dynamically generated gluon mass-scale.

There are many more. For example, the exceptionally light pion degree-of-freedom becomes dominant in QCD at those length-scales above which dressed-gluons and -quarks decouple from the theory owing to the large magnitudes of their dynamically generated masses. It is therefore conceivable that Gribov copies have no measurable impact on observables within the Standard Model because they affect only those gluonic modes whose wavelengths lie in the far infrared; and such modes are dynamically screened, by an exponential damping factor $\sim \exp(-\lambda/\varsigma)$, so that their role in hadron physics is superseded by the dynamics of light-hadrons. This conjecture is consistent with the insensitivity to Gribov copies of the dressed-gluon and -quark two-point Schwinger functions observed in numerical simulations of QCD on fine lattices [47, 48]. Another plausible conjecture is that dynamical generation of an infrared gluon mass-scale leads to saturation of the gluon parton distribution function at small Bjorken-x within hadrons. The possible emergence of this phenomenon stirs great scientific interest and curiosity. It is a key motivation in plans to construct an electron ion collider (EIC) that would be capable of producing a precise understanding of collective behaviour amongst gluons [49].

3. Continuum-QCD and ab initio predictions of hadron observables

Within contemporary hadron physics there are two common methods for determining the momentum-dependence of the interaction between quarks: the top-down approach, which works toward an *ab initio* computation of the interaction via direct analysis of the gauge-sector gap equations; and the bottom-up scheme, which aims to infer the interaction by fitting data within a well-defined truncation of those equations in the matter sector that are relevant to bound-state properties. These two approaches have recently been united [37] by a demonstration that the renormalisation-group-invariant (RGI) running-interaction predicted by contemporary analyses of QCD's gauge sector coincides with that required in order to describe ground-state hadron observables using a nonperturbative truncation of QCD's Dyson-Schwinger equations in the matter sector, *i.e.* the DCSB-improved (DSE-DB) kernel elucidated in Refs. [50, 51, 52].

The unification is illustrated in Fig. 3: the right panel presents a comparison between the top-down RGI interaction (solid-black curve, derived from the left panel) and the DB-truncation bottom-up interaction (green band containing dashed curve). Plainly, the interaction predicted by modern analyses of QCD's gauge sector is in near precise agreement with that required for a veracious description of measurable hadron properties using the most sophisticated matter-sector gap and Bethe-Salpeter kernels available today. This is a remarkable result, given that

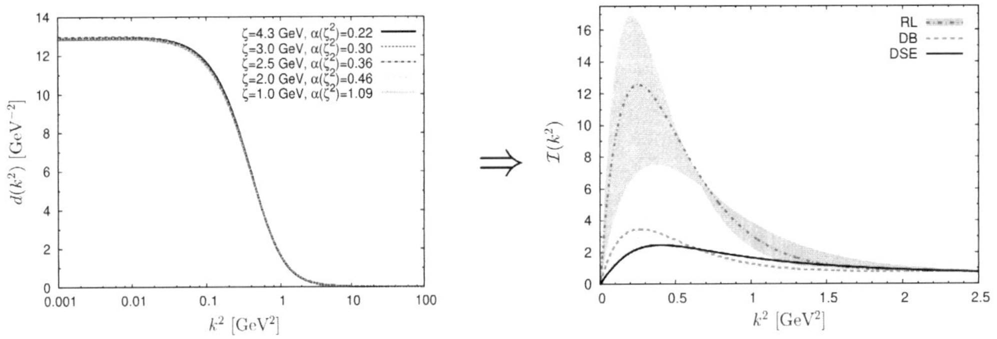

Figure 3. *Left panel* – RGI running interaction strength computed via a combination of DSE- and lattice-QCD analyses, as explained in Ref. [53]. The function obtained with five different values of the renormalisation point is depicted in order to highlight that the result is RGI. The interaction is characterized by a value $\alpha_s(0) \approx 0.9\,\pi$ and the gluon mass-scale in Eq. (7). *Right panel* – Comparison between top-down results for the gauge-sector interaction (derived from the left-panel) with those obtained using the bottom-up approach based on hadron physics observables. *Solid curve* – top-down result for the RGI running interaction; *dot-dashed curve* within *pale-red band* – bottom-up result obtained in the RL truncation; and *dashed curve* within *pale-green band* – advanced bottom-up result obtained in the DB truncation. The bands denote the domain of uncertainty in the bottom-up determinations of the interaction. All curves are identical on the perturbative domain: $k^2 > 2.5\,\mathrm{GeV}^2$.

there had previously been no serious attempt at communication between practitioners from the top-down and bottom-up hemispheres of continuum-QCD. It bridges a gap that had lain between nonperturbative continuum-QCD and the *ab initio* prediction of bound-state properties.

A comparison between the top-down prediction and that inferred using the simple DSE-RL kernel (red band containing dot-dashed curve) is also important. One observes that the DSE-RL result has the correct shape but is too large in the infrared. This is readily explained [37]; and it follows that whilst the RL truncation supplies a useful computational link between QCD's gauge sector and measurable hadron properties, the model interaction it delivers is *not a pointwise-accurate* representation of ghost-gluon dynamics. Notwithstanding this, it remains true that the judicious use of RL truncation can yield reliable predictions for a known range of hadron observables, with an error that may be estimated and whose origin is understood.

4. A Physical Vacuum

As the preceding discussion indicates, DCSB is a crucial emergent feature of the Standard Model. It is very clearly expressed in the dressed-quark mass function of Fig. 2. However, this understanding is relatively recent. DCSB was historically conflated with the existence of a spacetime-independent quark-antiquark condensate, $\langle\bar{q}q\rangle$, that permeates the Universe. This notion was born with the introduction of QCD sum rules as a theoretical artifice to estimate nonperturbative strong-interaction matrix elements [54, 55] and is typically tied to a belief that the QCD vacuum is characterized by numerous distinct, spacetime-independent condensates, as illustrated in the left panel of Fig. 4.

This belief is harmless unless one imagines that the theory of gravity is understood well enough so that it may reliably be coupled to quantum field theory. Subscribers to this view argue [60, 61] that the energy-density of the Universe must receive a contribution from such vacuum condensates and that the only possible covariant form for the energy of the (quantum) vacuum, *viz.*

$$T_{\mu\nu}^{\rm VAC} = \rho_{\rm VAC}\,\delta_{\mu\nu}\,, \tag{8}$$

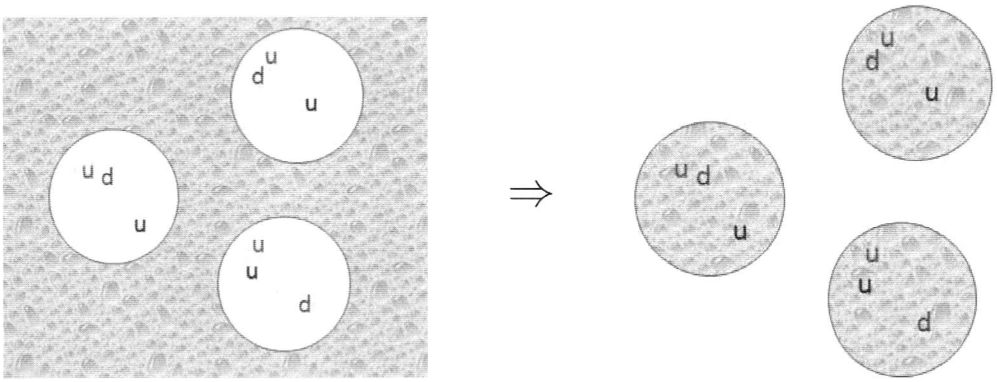

Figure 4. *Left panel* – The QCD vacuum was historically imagined to be a "frothing sea," with hadrons being merely bubbles of foam containing nothing but quarks and gluons interacting perturbatively throughout. Only near the bubble's boundary did some sort of strong interaction occur, enforcing quark and gluon confinement. *Right panel* – A modern view [56, 57, 58, 59] flips the picture completely. The space between hadrons is "empty," except for the perturbative quantum fluctuations that characterise all quantum field theories. The interior of hadrons, however, is extremely complex, with nonperturbative dynamics dominating in $\sim 98\%$ of the interior: the so-called condensates are spacetime-dependent and confined within the hadrons.

is mathematically equivalent to the cosmological constant. From this perspective, the quantum vacuum is [60] "...a perfect fluid and precisely spatially uniform ..." so that "Vacuum energy is almost the perfect candidate for dark energy." Now, if the ground state of QCD is really expressed in a nonzero spacetime-independent expectation value $\langle \bar{q}q \rangle$, then the energy difference between the symmetric and broken phases is roughly $M_{\rm QCD} \sim 0.3\,{\rm GeV}$, as indicated by Fig. 2. One obtains therefrom:

$$\rho_\Lambda^{\rm QCD} = 10^{46} \rho_\Lambda^{\rm obs}, \tag{9}$$

i.e. the contribution from the QCD vacuum to the energy density associated with the cosmological constant exceeds the observed value by forty-six orders-of-magnitude. In fact, the discrepancy is far greater if the Higgs vacuum expectation value is treated in a similar manner.

This mismatch has been called [62] "...one of the gravest puzzles of theoretical physics." However, it vanishes if one discards the notion that condensates have a physical existence, which is independent of the hadrons that express QCD's asymptotically realisable degrees of freedom [56]; namely, if one accepts that such condensates are merely mass-dimensioned parameters in one or another theoretical computation and truncation scheme. This appears mandatory in a confining theory [58, 57, 59], a perspective one may embed in a broader context by considering just what is observable in quantum field theory [63]: "...although individual quantum field theories have of course a good deal of content, quantum field theory itself has no content beyond analyticity, unitarity, cluster decomposition and symmetry." If QCD is a confining theory, then the principle of cluster decomposition is only realised for colour singlet states [64] and all observable consequences of the theory, including its ground state, can be expressed via a hadronic basis. This is quark-hadron duality.

The new hypothesis [56, 57, 58, 59] can therefore be stated succinctly as follows: "If quark-hadron duality is a reality in QCD, then condensates, those quantities that have commonly been viewed as constant empirical mass-scales that fill all spacetime, are instead wholly contained within hadrons, *i.e.* they are a property of hadrons themselves and expressed, for example, in their Bethe-Salpeter or light-front wave functions." This view is depicted in the right panel of Fig. 4 and canvassed fully in Sec. 4 of Ref. [24]. It presents the reasonable perspective that the understanding of hadrons requires that one explain what lies within those hadrons in contrast to the historical alternative, which suggested that hadrons could only be understood by explaining the properties of the vast spacetime domains that contain no hadrons at all.

The shift in perspective highlighted by the right panel of Fig. 4 does not undermine the utility of the QCD sum rules approach to the estimation of hadron observables. Instead, it tames the condensates so that they return to being merely mass-dimensioned parameters in a valuable computation scheme. Its implications are, however, significant and wide-ranging. For example, in connection with the cosmological constant, putting QCD condensates back into hadrons reduces the mismatch between experiment and theory by a factor of 10^{46}. Furthermore, if technicolour-like theories [27, 28] are the correct scheme for explaining electroweak symmetry breaking, then the impact of the notion of in-hadron condensates is far greater still.

5. Hadron interiors

Since the advent of the parton model and the first deep inelastic scattering experiments there has been a determined effort to deduce the parton distribution functions (PDFs) of the most stable hadrons [65]. The behaviour of such distributions on the valence domain (Bjorken-$x > 0.5$) is of particular interest because this domain is definitive of hadrons, *e.g.* quark content on the valence domain is how one distinguishes between a neutron and a proton: a neutron possesses one valence u-quark plus two valence d-quarks whereas the proton possesses two valence u-quarks plus one valence d-quark. Indeed, all Poincaré-invariant properties of a hadron: baryon number, charge, total spin, *etc.*, are determined by the PDFs which dominate on the valence domain. Moreover, via QCD evolution [66, 67, 68, 69], PDFs on the valence-quark domain determine backgrounds at the large hadron collider. There are also other questions, *e.g.* regarding flavour content of a hadron's sea and whether that sea possesses an intrinsic component [70, 71]. The answers to all these questions are essentially nonperturbative properties of QCD.

Recognising the significance of the valence domain, a new generation of experiments, focused on $x \gtrsim 0.5$, is planned at JLab, and under examination in connection with Drell-Yan studies at Fermilab and a possible EIC. Consideration is also being given to experiments aimed at measuring parton distribution functions in mesons at J-PARC. Furthermore, at FAIR it would be possible to directly measure the Drell-Yan process from high-x antiquarks in the antiproton annihilating with quarks in the proton.

A concentration on such measurements requires theory to move beyond merely parametrising distribution functions and amplitudes. Computation within QCD-connected frameworks becomes critical because without it, no amount of data will reveal anything about the theory underlying strong interaction phenomena. This is made clear by the example of the pion's valence-quark PDF, $u_v^\pi(x)$, in connection with which a failure of QCD was suggested following a leading-order analysis of πN Drell-Yan measurements [72]. As explained in Ref. [65], this confusion was fostered by the application of a diverse range of models. On the other hand, a series of QCD-connected calculations [73, 74, 75, 76] subsequently established that the leading-order analysis was misleading, so that $u_v^\pi(x)$ may now be seen as a success for the unification of nonperturbative and perturbative studies in QCD.

A framework that provides access to the pion's valence-quark PDF can also be employed to compute its PDAs. For example, the pion's leading-twist two-particle PDA is given by the following projection of the pion's Bethe-Salpeter wave function onto the light-front [77]

$$f_\pi\, \varphi_\pi(x;\zeta) = \mathrm{tr}_{\mathrm{CD}} Z_2 \int_{dq}^{\Lambda} \delta(n\cdot q_\eta - x\, n\cdot P)\, \gamma_5\gamma\cdot n\, S(q_\eta)\Gamma(q;P)S(q_{\bar\eta})\,, \tag{10}$$

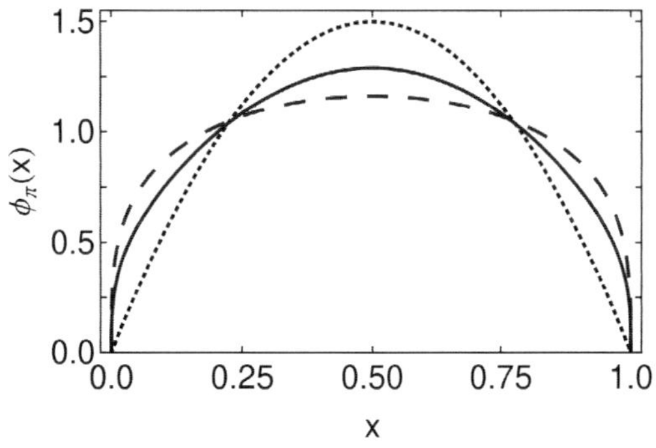

Figure 5. Twist-two pion PDA computed using two very different DSE truncations at a scale $\zeta = 2\,\text{GeV}$. *Dashed curve* – rainbow-ladder (RL), the leading-order in a systematic, symmetry-preserving scheme [78, 79]; and *solid curve* – the most sophisticated kernel that is currently available; namely, a DCSB-improved (DB) kernel that incorporates nonperturbative effects generated by DCSB that are omitted in RL truncation and any stepwise improvement thereof [50, 51, 52]. These results are consistent with contemporary lattice-QCD [80, 81, 82]. The dotted curve is $\varphi_\pi^{\rm asy}(x) = 6x(1-x)$, the result obtained in conformal QCD [4, 5].

where: f_π is the pion's leptonic decay constant; $\int_{dq}^{\Lambda}$ is a Poincaré-invariant regularization of the four-dimensional integral, with Λ the ultraviolet regularization mass-scale; $Z_2(\zeta,\Lambda)$ is the quark wave-function renormalisation constant, with ζ the renormalisation scale; n is a light-like four-vector, $n^2 = 0$; and P is the pion's four-momentum, $P^2 = -m_\pi^2$ and $n \cdot P = -m_\pi$, with m_π being the pion's mass.

The amplitude in Eq. (10) has been computed using two very different truncations of QCD's DSEs [77], with the result depicted in Fig. 5. Both kernels agree: compared with the asymptotic form, which is valid when ζ is extremely large, there is a marked broadening of $\varphi_\pi(x;\zeta)$, which owes exclusively to DCSB. This causal connection may be claimed because the PDA is computed at a low renormalisation scale in the chiral limit, whereat the quark mass function owes entirely to DCSB via Eq. (6). Moreover, the dilation measures the rate at which a dressed-quark approaches the asymptotic bare-parton limit. It can be verified empirically at JLab12, *e.g.* in measurements of the pion's electromagnetic form factor, the ratio of the proton's electric and magnetic form factors, and the behaviour of the form factors that characterise transitions between the nucleon and its excited states.

A question of more than thirty-years standing can be answered using Fig. 5; namely, when does $\varphi_\pi^{\rm asy}(x)$ provide a good approximation to the pion PDA? Plainly, not at $\zeta = 2\,\text{GeV}$. The ERBL evolution equation [4, 5] describes the ζ-evolution of $\varphi_\pi(x;\zeta)$; and applied to $\varphi_\pi(x;\zeta)$ in Fig. 5, one finds [80, 81, 82] that $\varphi_\pi^{\rm asy}(x)$ is a poor approximation to the true result even at $\zeta = 200\,\text{GeV}$. Thus at empirically accessible energy scales, the PDAs of ground-state hadrons are "squat and fat". Evidence supporting this picture had long been accumulating [83, 84, 85, 86]; and the dilation is verified by simulations of lattice-QCD [80, 81, 82].

6. Electromagnetic properties of hadrons

It is now possible to add flesh to the bones of Eq. (1). In the theory of strong interactions, the cross-sections for many hard exclusive hadronic reactions can be expressed in terms of the PDAs of the hadrons involved. An example is the pion's elastic electromagnetic form factor, for which the prediction can be stated succinctly [3, 4, 5, 6]:

$$\exists Q_0 > \Lambda_{\rm QCD} \mid Q^2 F_\pi(Q^2) \overset{Q^2>Q_0^2}{\approx} 16\pi\alpha_s(Q^2) f_\pi^2 w_\varphi^2, \quad w_\varphi = \frac{1}{3}\int_0^1 dx\, \frac{1}{x}\varphi_\pi(x)\,, \tag{11}$$

where $\alpha_s(Q^2) = 4\pi/[\beta_0 \ln(Q^2/\Lambda_{\rm QCD}^2)]$, with $\beta_0 = 11 - (2/3)n_f$ (n_f is the number of active quark flavours), is the leading-order expression for the strong running coupling. As noted in the Introduction, the value of Q_0 is not predicted by perturbative QCD. Here $\Lambda_{\rm QCD} \sim 0.2\,\text{GeV}$

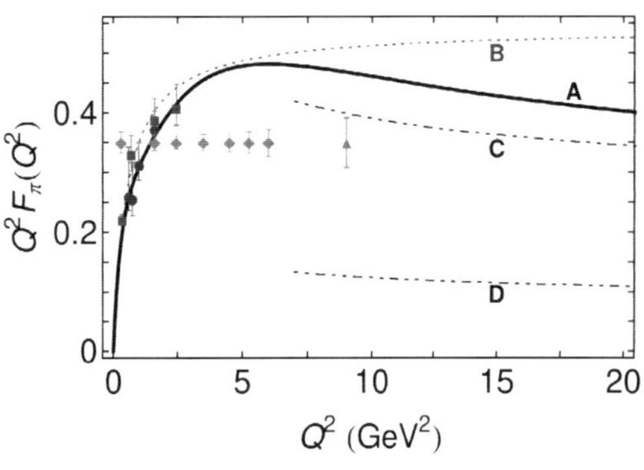

Figure 6. $Q^2F\pi(Q^2)$. Solid curve (A) - Theoretical prediction [94]; dotted curve (B) -monopole form fitted to data; dot-dot-dashed curve (C) - perturbative QCD (pQCD) prediction computed with the modern, dilated pion PDA described in Sec. 5; and dot-dot-dashed curve (D) - pQCD prediction computed with the asymptotic PDA, which had previously been used to guide expectations for the asymptotic behaviour of $Q^2F\pi(Q^2)$. The filled-circles and -squares represent existing JLab data [93]; and the filled diamonds and triangle, whose normalisation is arbitrary, indicate the projected Q^2-reach and accuracy of forthcoming experiments [95, 96].

is the natural mass-scale of QCD, whose dynamical generation through quantisation spoils the conformal invariance of the classical massless theory [87, 88, 89]: $\Lambda_{\rm QCD}$, whose value must be determined empirically within the Standard Model, sets the scale for the dynamically generated masses described above.

It was hoped that JLab would verify this fundamental Standard Model prediction, Eq. (11); and in 2001, seven years after commencing operations, the facility provided the first high-precision pion electroproduction data for $F_\pi(Q^2)$ between Q^2 values of 0.6 and 1.6 GeV^2 [90]. In 2006 and 2007, this domain was revisited [91] and a new result was reported, this time at $Q^2 = 2.45\,{\rm GeV}^2$ [92]. However, there was disappointment and surprise, with the collaboration stating that experiment is "still far from the transition to the Q^2 region where the pion looks like a simple quark-antiquark pair." This data is represented in Fig. 6 by the filled circles and squares, drawn from Ref. [93], and the collaboration were comparing with Curve-D in the figure, which is the prediction obtained when $\varphi_\pi^{\rm asy}(x)$, the asymptotic PDA appropriate to the conformal limit of QCD, is used in Eq. (11).

The confusion was compounded by the fact that JLab's data confirmed the behaviour predicted by a DSE-RL prediction for the pion form factor, computed in 2000 [97]. That prediction could not resolve the difficulty because it relied on brute numerical methods and hence could not produce a result at $Q^2 > 4\,{\rm GeV}^2$. This limit is marked by the vertical dashed-line in Fig. 6. In appearance, however, the shape of the prediction suggested to many that one might never see parton model scaling and QCD scaling violations at the momentum transfers accessible to JLab, even after its upgrade.

This conundrum was recently resolved [94]. Using a refinement of known methods [98, 99], also employed in the successful analysis of $\varphi_\pi(x;\zeta)$, described in Sec. 5, a reliable prediction of $F_\pi(Q^2)$ is now available on the entire domain of spacelike Q^2. This is Curve-A in Fig. 6. Moreover, the analysis enables correlation of that result with Eq. (11), using the modern PDA computed in precisely the same framework, which is Curve-C in the figure. This leading-order, leading-twist QCD prediction, obtained with a pion valence-quark PDA evaluated at a scale appropriate to the experiment, underestimates the full DSE-RL computation by merely an approximately uniform 15% on the domain depicted. The small mismatch is explained by a combination of higher-order, higher-twist corrections to Eq. (11) in pQCD on the one hand and, on the other hand, shortcomings in the rainbow-ladder truncation (see Sec. 3), which predicts the correct power-law behaviour for the form factor but not precisely the right anomalous dimension (exponent on the logarithm) in the strong coupling calculation. Hence, disappointment is now transformed into optimism because the comparison of Curves-A and C in Fig. 6 predicts that the upgraded JLab facility will reveal a maximum at $Q^2 \approx 6\,{\rm GeV}^2$ and an experiment at $Q^2 = 9\,{\rm GeV}^2$ will see a clear sign of parton model scaling for the first time in a hadron elastic form factor.

Journal of Physics: Conference Series **630** (2015) 012051 doi:10.1088/1742-6596/630/1/012051

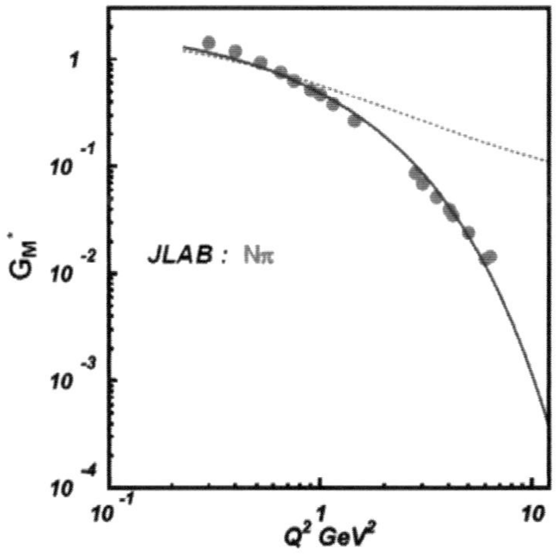

Figure 7. Comparison between CLAS data [100] on the magnetic $\gamma + N \to \Delta$ transition form factor and a recent theoretical prediction [21]. The dashed curve shows the result that would be obtained if the interaction between quarks in QCD were momentum-independent [18]. The solid curve is obtained with precisely the same QCD-based formulation as was employed in a successful analysis of nucleon elastic form factors, which explains [19] the behaviour of the ratio $\mu_p G_E^p(Q^2)/G_M^p(Q^2)$ described in the Introduction. The experiment-theory comparison establishes that experiments are sensitive to the momentum dependence of the running couplings and masses in QCD; and the theoretical unification of N and Δ properties highlights the material progress that has been made in constraining the long-range behaviour of these fundamental quantities. (Figure courtesy of V. Mokeev.)

The implications of this marked shift in perspective are manifold. In the foreseeable future one may reasonably expect empirical confirmation of the theory of factorisation in hard exclusive processes, and expose the dominance of hard contributions to the pion form factor for $Q^2 > 8\,\text{GeV}^2$. One will simultaneously find that the normalisation of $F_\pi(Q^2)$ is fixed by a pion wave-function whose dilation with respect to $\varphi_\pi^{\text{asy}}(x)$ is a definitive signature of DCSB. The experiments will thus provide a direct measurement of the strength of DCSB in the Standard Model, the origin of the vast bulk of visible mass. These will be important pages in a book on the Standard Model, in which the first lines were written forty years ago.

7. Implications for protons, neutrons and their excited states

The understanding at which we have now arrived paves the way for a dramatic reassessment of pictures of proton and neutron structure, which is already well underway [21]. One particular example shall here serve to illustrate that progress.

Given the challenges posed by non-perturbative QCD, it is insufficient to study hadron ground-states alone. Many novel perspectives and additional insights are provided by nucleon-to-resonance transition form factors, whose behaviour at large momentum transfers can reveal much about the long-range behaviour of the interactions between quarks and gluons [101]. Indeed, the properties of nucleon resonances are more sensitive to long-range effects in QCD than are those of hadron ground states. The lightest baryon resonances are the $\Delta(1232)$-states; and despite possessing a width of 120 MeV, these states are well isolated from other nucleon excitations. Hence the $\gamma + N \to \Delta$ transition form factors have long been used to probe strong interaction dynamics. They excite keen interest because of their use in probing, *inter alia*, the relevance of perturbative QCD to processes involving moderate momentum transfers [102, 103, 104]; shape deformation of hadrons [105]; and, of course, the role that resonance electroproduction experiments can play in exposing non-perturbative features of QCD [101]. Using the "CLAS" detector at JLab, precise data on the dominant $\gamma + N \to \Delta$ magnetic transition now reaches to $Q^2 = 8\,\text{GeV}^2$; an eventuality that poses both great opportunities and challenges for QCD theory, some of which have recently been met, as illustrated in Fig. 7.

8. Insights from continuum-QCD

Perhaps the most important understanding to draw from this contribution is highlighted in connection with Eq. (6); namely, dynamical chiral symmetry breaking (DCSB) in QCD – the generation of mass from nothing – is expressed fundamentally in a near-identity between the dressed-quark one-body problem and the pseudoscalar two-body problem. As illustrated in Secs. 4-6, this has enabled theory to arrive at a detailed understanding of Nature's lightest hadron, *viz.* the pion. The dressed-quark mass function (Fig. 2) is known and there is no material model dependence. Hence, continuum-QCD theory is on the threshold of drawing a complete picture of the pion's light-front valence-quark wave-function, which provides the closest thing in quantum field theory to a quantum mechanical picture of the pion.

In parallel, continuum-QCD is providing predictions of the light-front structure of other light-quark hadrons [82, 106] and embarking on analyses of generalised parton distributions for these systems [76, 107], with a view to providing a unified QCD-based and quantum mechanical picture of the light-quark meson sector of the Standard Model. Sec. 7 indicates that developments in the baryon sector are equally promising; and Sec. 3 suggests that continuum-QCD is on the verge of providing genuinely *ab initio* predictions for hadron observables.

Acknowledgments

I am grateful to the organisers of the "37th Brazilian Workshop on Nuclear Physics" and my hosts in Maresias for their support and gracious hospitality. I thank, too, D. Binosi, S. J. Brodsky, L. Chang, C. Chen, I. C. Cloët, M.-H. Ding, B. El-Bennich, F. Gao, G. Krein, T.-S. H. Lee, Y.-X. Liu, C. Mezrag, V. Mokeev, H. Moutarde, J. Papavassiliou, M. Pitschmann, S.-X. Qin, J. Rodríguez-Quintero, T. Sato, J. Segovia, S. M. Schmidt, R. Shrock, C. Shi, P. C. Tandy, S.-L. Wan, S.-S. Xu and H.-S. Zong for valuable discussions during the preparation of this contribution. This work was supported by U.S. Department of Energy, Office of Science, Office of Nuclear Physics, contract no. DE-AC02-06CH11357.

References

[1] Brodsky S J and Farrar G R 1973 *Phys. Rev. Lett.* **31** 1153–1156
[2] Brodsky S J and Farrar G R 1975 *Phys. Rev. D* **11** 1309
[3] Farrar G R and Jackson D R 1979 *Phys. Rev. Lett.* **43** 246–249
[4] Efremov A V and Radyushkin A V 1980 *Phys. Lett. B* **94** 245–250
[5] Lepage G P and Brodsky S J 1979 *Phys. Lett. B* **87** 359–365
[6] Lepage G P and Brodsky S J 1980 *Phys. Rev. D* **22** 2157–2198
[7] Ellis R K, Stirling W J and Webber B R 1996 QCD and Collider Physics (Cambridge University Press, Cambridge (UK))
[8] Jones M K *et al.* 2000 *Phys. Rev. Lett.* **84** 1398–1402
[9] Gayou O *et al.* 2002 *Phys. Rev. Lett.* **88** 092301
[10] Gayou O, Wijesooriya K, Afanasev A, Amarian M, Aniol K *et al.* 2001 *Phys. Rev. C* **64** 038202
[11] Punjabi V, Perdrisat C, Aniol K, Baker F, Berthot J *et al.* 2005 *Phys. Rev. C* **71** 055202
[12] Puckett A J R *et al.* 2010 *Phys. Rev. Lett.* **104** 242301
[13] Puckett A J R, Brash E J, Gayou O, Jones M K, Pentchev L *et al.* 2012 *Phys. Rev. C* **85** 045203
[14] Eichmann G, Cloët I C, Alkofer R, Krassnigg A and Roberts C D 2009 *Phys. Rev. C* **79** 012202(R)
[15] Cloët I C, Eichmann G, El-Bennich B, Klähn T and Roberts C D 2009 *Few Body Syst.* **46** 1–36
[16] Chang L, Cloët I C, Roberts C D and Roberts H L L 2011 *AIP Conf. Proc.* **1354** 110–117
[17] Cloët I C, Roberts C D and Wilson D J 2011 *AIP Conf. Proc.* **1388** 121–127
[18] Wilson D J, Cloët I C, Chang L and Roberts C D 2012 *Phys. Rev. C* **85** 025205
[19] Cloët I C, Roberts C D and Thomas A W 2013 *Phys. Rev. Lett.* **111** 101803
[20] Cloët I C, Bentz W and Thomas A W 2014 *Phys. Rev. C* **90** 045202
[21] Segovia J, Cloët I C, Roberts C D and Schmidt S M 2014 *Few Body Syst.* **55** 1185–1222
[22] Dudek J *et al.* 2012 *Eur. Phys. J. A* **48** 187
[23] Nielsen M, Navarra F S and Lee S H 2010 *Phys. Rept.* **497** 41–83
[24] Cloët I C and Roberts C D 2014 *Prog. Part. Nucl. Phys.* **77** 1–69
[25] Marciano W J and Pagels H 1979 *Nature* **279** 479–483

[26] *The Millenium Prize Problems*, eds. J. Carlson, A. Jaffe, and A. Wiles. (American Mathematical Society, Providence, 2006)
[27] Andersen J R *et al.* 2011 *Eur. Phys. J. Plus* **126** 81
[28] Sannino F arXiv:13066346 *Higgs Discovery: Impact on Composite Dynamics [Thinking Fast and Slow]*
[29] Wilson K G 1974 *Phys. Rev. D* **10** 2445–2459
[30] Bali G S, Neff H, Duessel T, Lippert T and Schilling K 2005 *Phys. Rev. D* **71** 114513
[31] Chang L, Cloët I C, El-Bennich B, Klähn T and Roberts C D 2009 *Chin. Phys. C* **33** 1189–1196
[32] Bhagwat M, Pichowsky M, Roberts C and Tandy P 2003 *Phys. Rev. C* **68** 015203
[33] Bowman P O *et al.* 2005 *Phys. Rev. D* **71** 054507
[34] Bhagwat M S and Tandy P C 2006 *AIP Conf. Proc.* **842** 225–227
[35] Boucaud P, Leroy J P, Le-Yaouanc A, Micheli J, Pene O and Rodríguez-Quintero J 2012 *Few Body Syst.* **53** 387–436
[36] Ayala A, Bashir A, Binosi D, Cristoforetti M and Rodriguez-Quintero J 2012 *Phys. Rev. D* **86** 074512
[37] Binosi D, Chang L, Papavassiliou J and Roberts C D 2015 *Phys. Lett. B* **742** 183–188
[38] Stingl M 1996 *Z. Phys. A* **353** 423–445
[39] Aoyama T, Hayakawa M, Kinoshita T and Nio M 2012 *Phys. Rev. Lett.* **109** 111807
[40] Politzer H 2005 *Proc. Nat. Acad. Sci.* **102** 7789–7793
[41] Gross D 2005 *Proc. Nat. Acad. Sci.* **102** 9099–9108
[42] Wilczek F 2005 *Proc. Nat. Acad. Sci.* **102** 8403–8413
[43] Roberts C D and Williams A G 1994 *Prog. Part. Nucl. Phys.* **33** 477–575
[44] Maris P, Roberts C D and Tandy P C 1998 *Phys. Lett. B* **420** 267–273
[45] Qin S X, Roberts C D and Schmidt S M 2014 *Phys. Lett. B* **733** 202–208
[46] Brodsky S J and Shrock R 2008 *Phys. Lett. B* **666** 95–99
[47] Bowman P O, Heller U M, Leinweber D B and Williams A G 2002 *Phys. Rev. D* **66** 074505
[48] Zhang J *et al.* 2005 *Phys. Rev. D* **71** 014501
[49] Accardi A *et al.* arXiv:12121701 [nucl-ex] *Electron Ion Collider: The Next QCD Frontier - Understanding the glue that binds us all*
[50] Chang L and Roberts C D 2009 *Phys. Rev. Lett.* **103** 081601
[51] Chang L, Liu Y X and Roberts C D 2011 *Phys. Rev. Lett.* **106** 072001
[52] Chang L and Roberts C D 2012 *Phys. Rev. C* **85** 052201(R)
[53] Aguilar A, Binosi D, Papavassiliou J and Rodríguez-Quintero J 2009 *Phys. Rev. D* **80** 085018
[54] Shifman M A, Vainshtein A and Zakharov V I 1979 *Nucl. Phys. B* **147** 385–447
[55] Leinweber D B 1997 *Annals Phys.* **254** 328–396
[56] Brodsky S J and Shrock R 2011 *Proc. Nat. Acad. Sci.* **108** 45–50
[57] Brodsky S J, Roberts C D, Shrock R and Tandy P C 2010 *Phys. Rev. C* **82** 022201(R)
[58] Chang L, Roberts C D and Tandy P C 2012 *Phys. Rev. C* **85** 012201(R)
[59] Brodsky S J, Roberts C D, Shrock R and Tandy P C 2012 *Phys. Rev. C* **85** 065202
[60] Turner M S astro-ph/0108103 *Dark energy and the new cosmology*
[61] Bass S D 2011 *J. Phys. G* **38** 043201
[62] Zee A 2008 *Int. J. Mod. Phys. A* **23** 1295–1308
[63] Weinberg S 1979 *Physica A* **96** 327
[64] Roberts C D, Williams A G and Krein G 1992 *Int. J. Mod. Phys. A* **7** 5607–5624
[65] Holt R J and Roberts C D 2010 *Rev. Mod. Phys.* **82** 2991–3044
[66] Dokshitzer Y L 1977 *Sov. Phys. JETP* **46** 641–653
[67] Gribov V N and Lipatov L N 1972 *Sov. J. Nucl. Phys.* **15** 438–450
[68] Lipatov L N 1975 *Sov. J. Nucl. Phys.* **20** 94–102
[69] Altarelli G and Parisi G 1977 *Nucl. Phys. B* **126** 298
[70] Brodsky S J, Hoyer P, Peterson C and Sakai N 1980 *Phys. Lett. B* **93** 451–455
[71] Signal A I and Thomas A W 1987 *Phys. Lett. B* **191** 205
[72] Conway J S *et al.* 1989 *Phys. Rev. D* **39** 92–122
[73] Hecht M B, Roberts C D and Schmidt S M 2001 *Phys. Rev. C* **63** 025213
[74] Aicher M, Schäfer A and Vogelsang W 2010 *Phys. Rev. Lett.* **105** 252003
[75] Nguyen T, Bashir A, Roberts C D and Tandy P C 2011 *Phys. Rev. C* **83** 062201(R)
[76] Chang L, Mezrag C, Moutarde H, Roberts C D, Rodríguez-Quintero J and Tandy P C 2014 *Phys. Lett. B* **737** 2329
[77] Chang L, Cloët I C, Cobos-Martinez J J, Roberts C D, Schmidt S M and Tandy P C 2013 *Phys. Rev. Lett.* **110** 132001
[78] Munczek H J 1995 *Phys. Rev. D* **52** 4736–4740
[79] Bender A, Roberts C D and von Smekal L 1996 *Phys. Lett. B* **380** 7–12

[80] Cloët I C, Chang L, Roberts C D, Schmidt S M and Tandy P C 2013 *Phys. Rev. Lett.* **111** 092001
[81] Segovia J, Chang L, Cloët I C, Roberts C D, Schmidt S M *et al.* 2014 *Phys. Lett. B* **731** 13–18
[82] Shi C, Chang L, Roberts C D, Schmidt S M, Tandy P C and Zong H S 2014 *Phys. Lett. B* **738** 512–518
[83] Mikhailov S and Radyushkin A 1986 *JETP Lett.* **43** 712–715
[84] Petrov V Y, Polyakov M V, Ruskov R, Weiss C and Goeke K 1999 *Phys. Rev. D* **59** 114018
[85] Braun V *et al.* 2006 *Phys. Rev. D* **74** 074501
[86] Brodsky S J and de Teramond G F 2006 *Phys. Rev. Lett.* **96** 201601
[87] Collins J C, Duncan A and Joglekar S D 1977 *Phys. Rev. D* **16** 438–449
[88] Nielsen N 1977 *Nucl. Phys. B* **120** 212–220
[89] Pascual P and Tarrach R 1984 *QCD: Renormalization for the Practitioner* (Springer-Verlag, Berlin) Lecture Notes in Physics **194**
[90] Volmer J *et al.* 2001 *Phys. Rev. Lett.* **86** 1713–1716
[91] Tadevosyan V *et al.* (Jefferson Lab F(pi)) 2007 *Phys. Rev.* **C75** 055205
[92] Horn T *et al.* 2006 *Phys. Rev. Lett.* **97** 192001
[93] Huber G *et al.* 2008 *Phys. Rev. C* **78** 045203
[94] Chang L, Cloët I C, Roberts C D, Schmidt S M and Tandy P C 2013 *Phys. Rev. Lett.* **111** 141802
[95] Huber G M and Gaskell D 2006 Jefferson Lab Experiment E12-06-10
[96] Horn T and Huber G M 2007 Jefferson Lab Experiment E12-07-105
[97] Maris P and Tandy P C 2000 *Phys. Rev. C* **62** 055204
[98] Nakanishi N 1963 *Phys. Rev.* **130** 1230–1235
[99] Nakanishi N 1969 *Prog. Theor. Phys. Suppl.* **43** 1–81
[100] Aznauryan I *et al.* 2009 *Phys. Rev. C* **80** 055203
[101] Aznauryan I *et al.* 2013 *Int. J. Mod. Phys. E* **22** 1330015
[102] Carlson C E 1986 *Phys. Rev. D* **34** 2704
[103] Pascalutsa V, Vanderhaeghen M and Yang S N 2007 *Phys. Rept.* **437** 125–232
[104] Aznauryan I and Burkert V 2012 *Prog. Part. Nucl. Phys.* **67** 1–54
[105] Alexandrou C, Papanicolas C and Vanderhaeghen M 2012 *Rev. Mod. Phys.* **84** 1231
[106] Gao F, Chang L, Liu Y X, Roberts C D and Schmidt S M 2014 *Phys. Rev. D* **90**(1) 014011
[107] Mezrag C, Moutarde H, Rodrguez-Quintero J and Sabati F arXiv:14067425 [hep-ph] *Towards a Pion Generalized Parton Distribution Model from Dyson-Schwinger Equations*

Introduction to Hard Scattering Processes and Recent Results from Hard Probes at RHIC and LHC

John W. Harris
Physics Department, Yale University, New Haven CT 06520-8120 U.S.A.
E-mail: john.harris@yale.edu

Abstract. A summary is presented of recent heavy-ion collision results from hard scattering processes at RHIC and LHC. Hadrons with large transverse momentum are suppressed in heavy-ion collisions compared to proton-proton interactions, including those hadrons with heavy quarks (charm, beauty). Jets are quenched and modifications to their fragmentation are observed, the effects increasing with the centrality of the heavy-ion collision. The heavy-ion measurements are compared to those in proton-proton and proton-nucleus collisions. These results when compared to theory are consistent with the suppression and jet quenching being a product of parton energy loss in a hot QCD medium. Correlations are presented for di-jets, jets with trigger photons, and hadrons with trigger jets. The di-jet correlations exhibit a di-jet energy and momentum imbalance that increases with the centrality of the collision. The correlation results indicate that the parton energy loss is redistributed to lower momentum particles at larger angles from the axis of the hard-scattered parton.

1. Introduction

Hard or highly penetrating probes originate from hard scattering of partons (quarks and gluons) and are characterized typically by either large momentum transfer (i.e. large 4-momentum transfer squared Q^2), large transverse momentum (p_T) or a large mass (m) scale. Thus by definition, hard probes are highly penetrating observables used to explore properties of matter that cannot be viewed directly! Examples of these observables include radiation, high p_T particles, jets, and particles with heavy quarks even at low p_T.

The study of the products of hard-scattering processes between constituent partons provides a means of investigating the hot QCD medium formed in relativistic heavy ion collisions (see [1, 2]). The production rates of hard processes are calculable in perturbative Quantum Chromodynamics (QCD). After the initial hard parton-parton collisions, scattered partons traverse the colored medium. The interactions of these scattered partons with the medium and momentum exchange lead to parton energy loss, which alters the distributions of fragmentation products in the final state. By comparing hard-scattering results in heavy ion collisions with those in elementary pp collisions, modifications of the parton fragmentation products can be studied. Thorough understanding of parton energy loss in the hot QCD medium requires detailed experimental and theoretical investigation. Shown in Figure 1 are cross sections (left ordinate) predicted in perturbative QCD and rates (right ordinate) of various hard probes as a function of center-of-mass energy. The cross sections for hard probes increases as the collision energy is increased.

In this paper, heavy-ion results from the Relativistic Heavy Ion Collider (RHIC) and the Large Hadron Collider (LHC) will be presented for high p_T particles (R_{AA}) and azimuthal correlations, jet spectra (jet R_{AA} and R_{CP}), jet-shape modifications, heavy-quark jets, photon-tagged jet correlations, jet-hadron and dijet correlations, and energy balance.

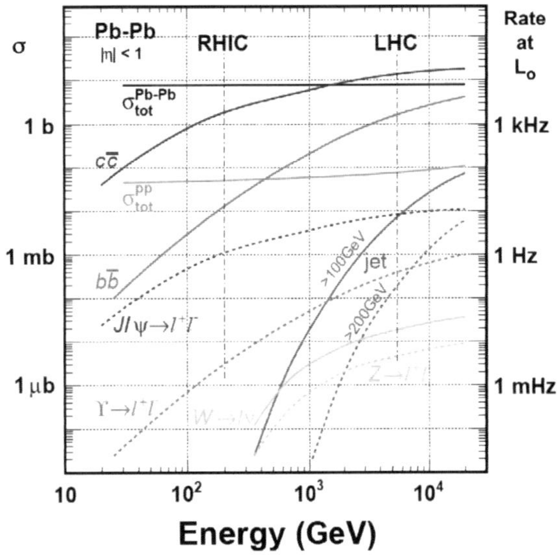

Figure 1. Cross sections (left axis) for various production processes (labeled) for Pb-Pb at mid-rapidity as a function of $\sqrt{s_{NN}}$. Rates (right axis) are derived from the cross sections using the LHC Pb-Pb design luminosity (L_0).

2. Why study proton-proton (pp), proton-nucleus (p-A) and nucleus-nucleus (A-A) collisions?

Comparisons of measurements in the pp, p-A and A-A colliding systems are expected to allow identification of final state effects due to the presence of a quark-gluon plasma (QGP) in A-A collisions. Comparing p-A to pp collisions allows determination of initial state effects, e.g. gluon saturation and effects of a possible color-glass condensate, and discrimination between different nuclear parton distribution functions. Furthermore, our overall understanding of these high energy nuclear collisions and the initial state of nuclei would be strengthened, if we could understand pp, p-A or d-A, and A-A collisions in a consistent framework. Can we unambiguously distinguish the initial state from the final state in these collisions? Is the initial state composed of gluon fields? To what extent are they saturated, and is there a color glass condensate? These questions about the initial state of nuclei still await resolution, and may require future electron-ion collisions at high energies for a final resolution.

It is important to determine the effects of cold nuclear matter on final state observables for A-A collisions. This requires the study of p-A collisions in order to determine the multiplicity and energy dependence of observables measured in p-A and A-A collisions. To extract precise information on parton energy loss mechanisms, it is instructive to compare hard scattering observables in pp, p-A, and A-A collisions at RHIC and LHC energies. For example, comparison of p-A events at the LHC with similar multiplicity A-A events at LHC and RHIC may shed light on these mechanisms as a function of the virtuality of the parton.

3. High p_T particle suppression

Leading particles and correlations provide information on hard scattering in addition to, or as a substitute for, jets. The spectra of particles produced in A-A collisions can be compared to those in pp collisions, scaled by the number of binary nucleon-nucleon collisions in the A-A geometry. This is carried out through measurements using the nuclear modification factor, which is written as

$$R^i_{AA}(p_T) = \frac{d^2N^i_{AA}/dp_T d\eta}{\langle T_{AA}\rangle d^2\sigma^i_{NN}/dp_T d\eta} \qquad (1)$$

with N^i_{AA} the yield for particle (or jet) type i in A-A collisions and σ^i_{NN} the cross section for particle type i in pp collisions. $\langle T_{AA}\rangle$ is the ratio of the number of binary nucleon-nucleon collisions (typically calculated in the Glauber Model) to the inelastic nucleon-nucleon cross section. R_{AA} is unity when A-A collisions are a superposition of pp collisions, e.g. in the absence of nuclear effects such as parton energy loss. A version of this ratio, called R_{CP}, is often used where the ratio of central to peripheral A-A collisions is taken.

The nuclear modification factor R_{AA} is investigated in heavy-ion collisions as a function of particle p_T, collision energy $\sqrt{s_{NN}}$, collision centrality and for different particle types. Shown in Figure 2 is R_{AA} for charged hadrons and pions in central collisions of heavy ions at SPS, RHIC and LHC energies, as labeled in the figure. At SPS energies there is an enhancement, commonly attributed to initial-state scattering and often called the Cronin effect [3]. A large suppression is observed at the higher energies of RHIC and the LHC at intermediate p_T (2–20 GeV) with a gradual rise for p_T beyond 20 GeV. Various model calculations are also shown, with widely diverging predictions at larger p_T. Parameters used in the models are from fits to the RHIC data and extrapolated to the LHC energy. The predictions depend on the parton density and parton energy-loss mechanisms such as scattering and radiation used in the models (for details see [4, 5, 6, 7]). In all cases the suppression results from parton-energy loss in the dense medium.

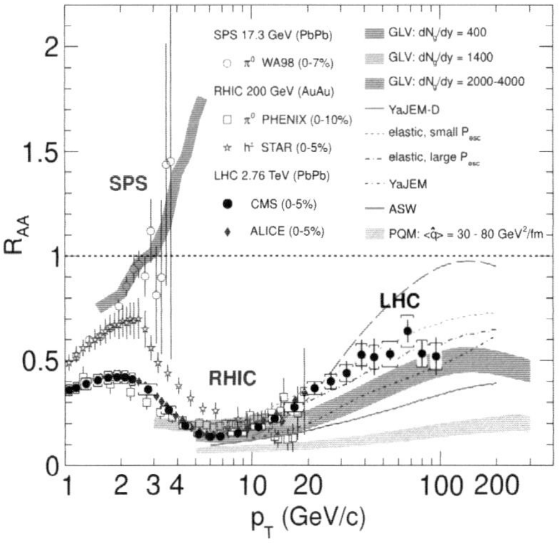

Figure 2. R_{AA} for charged hadrons and pions in central collisions of heavy ions at SPS, RHIC and LHC energies. Various model predictions (see text) are also shown. [8]

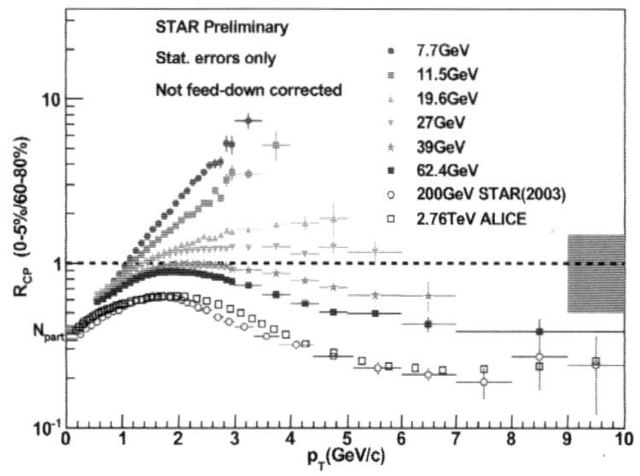

Figure 3. Preliminary R_{CP}) for charged particles in central (0-5%) relative to peripheral (60-80%) Au-Au collisions for a range of energies [9]. R_{AA} results for central Pb-Pb collisions at $\sqrt{s_{NN}}$ = 2.76 TeV from ALICE [10] at the LHC.

Nuclear modification factors (R_{AA} and R_{CP}) exhibit medium-induced suppression at $\sqrt{s_{NN}}$ = 0.2 TeV at RHIC and at $\sqrt{s_{NN}}$ = 2.76 TeV at LHC. STAR has analyzed R_{CP} for charged particles in central (0-5%) relative to peripheral (60-80%) Au-Au collisions in the RHIC Beam Energy Scan covering a range of energies $\sqrt{s_{NN}}$ = 0.008 0.2 TeV. Shown in Figure 3 are the results for seven energies in this range, as labeled in the figure. Also shown for reference R_{AA} results for central Pb-Pb collisions at $\sqrt{s_{NN}}$ = 2.76 TeV from ALICE at the

LHC [10]. These data exhibit significant suppression as already seen at the top RHIC and LHC energies to a dominant Cronin-type enhancement at the lower energies. Disentangling the competition between parton energy loss (suppression) and Cronin-type effects (enhancement) requires further theoretical investigation and additional experimental measurements such as di-hadron correlations, perhaps with identified particles, as a function of energy. The theoretical models will need further refinement to include the relative roles of various initial and final state effects including nuclear shadowing, initial state broadening via various mechanisms, multi-parton correlations, parton energy loss and collective flow as a function of energy, which is clearly no small task.

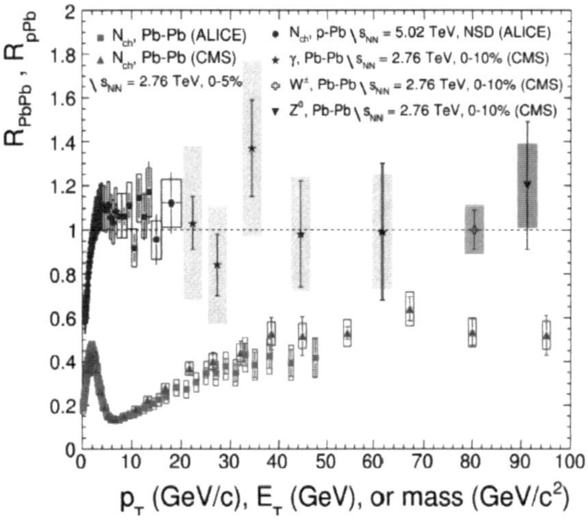

Figure 4. R_{PbPb} versus p_T, E_T or m for charged particles and for gauge bosons (γ, $W^{\pm}$ and Z^o) in central collisions [8, 11]. Also shown are R_{pPb} data for charged particles [12].

Collisions of p-Pb at the LHC and d-Au at RHIC have been studied in order to investigate potential cold nuclear effects and initial state effects. Shown in Figure 4 are R_{pPb} and R_{PbPb} for charged particles, and R_{PbPb} for photons, $W^{\pm}$, and Z^0. Note, as expected, R_{PbPb} for the non-strongly interacting gauge bosons is consistent with a value of 1. This is also the case for charged particles in p-Pb reactions, indicating no observation of initial state nuclear effects in the production of charged particles at mid-rapidity. The R_{PbPb} for charged particles for central collisions at the LHC exhibits a strong suppression, as seen in Figure 2. The R_{dAu} of neutral pions in $\sqrt{s_{NN}} = 0.2$ TeV minimum bias d-Au collisions at RHIC are also consistent with a value of 1, as seen in Figure 5. Also, shown in the figure are R_{AuAu} for central (10%) Au-Au collisions at the same energy. Clearly, for both cases p-Pb at LHC and d-Au at RHIC there appear to be no strong initial state effects, while for central A-A collisions at RHIC and LHC energies there is a strong suppression of high p_T particles. Together these data provide strong evidence that the suppression is a final state effect.

The suppression in A-A collisions is observed to increase with collision centrality. This can be seen in Figure 6 where R_{PbPb} is presented as a function of p_T for central (0-10%) and peripheral (70-80%) collisions at mid-rapidity[12]. Also shown for reference is the p-Pb data that exhibit no suppression. There is also a dependence of R_{PbPb} on the particle-type for hadrons, as seen in Figure 7. The π's, K's and protons exhibit increased suppression for the lighter hadrons over $1 < p_T < 8$ GeV. The π's and K's have similar suppression above $p_T \sim 3$ GeV, whereas protons are less suppressed over the entire $1 < p_T < 8$ GeV range. Over this p_T range there may be a combination of strong suppression and a mass-dependent radial flow or recombination in Pb-Pb that results in the observed R_{PbPb} particle-type dependence. The suppression observed

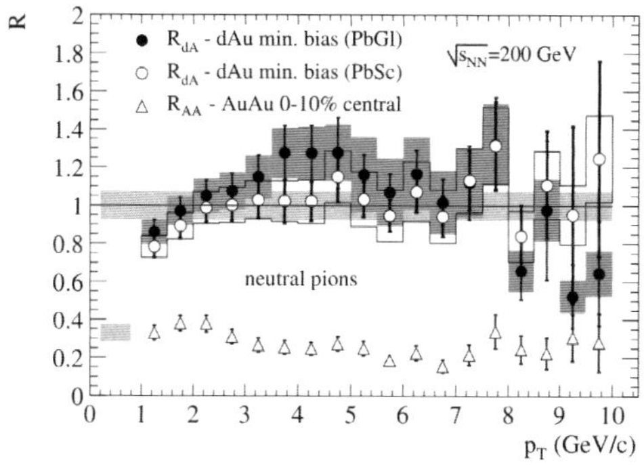

Figure 5. R_{dAu} for neutral pions in minimum bias d-Au collisions from PHENIX. Shown are two sets of data each taken with a different calorimeter system (PbGl and PbSc). Systematic errors are shown as bands, with the band around unity the uncertainty in the overall normalization. Shown for reference is R_{AuAu} for central (10%) Au-Au collisions. [13]

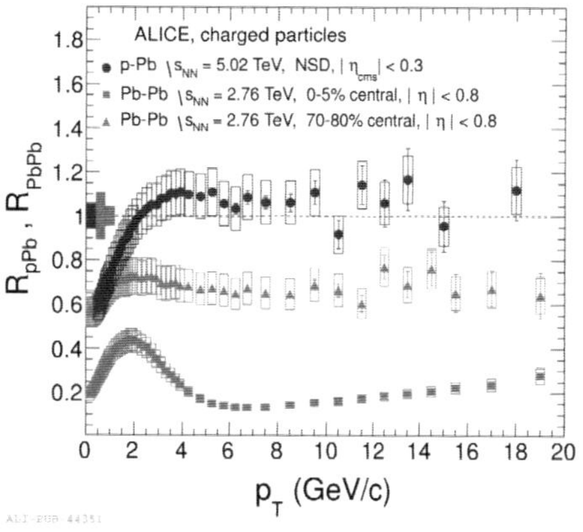

Figure 6. R_{pPb} for non-single diffractive interactions and R_{PbPb} for central and peripheral collisions as a function of p_T. [12]

for more central collisions in Pb-Pb and the R_{PbPb} mass dependence coupled with the absence of suppression in p-Pb suggests parton energy loss in the medium in A-A collisions.

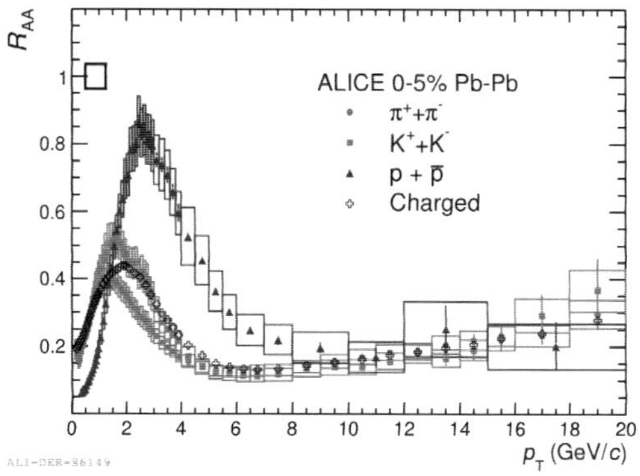

Figure 7. R_{PbPb} for π, K, proton in central (0-5%) Pb-Pb collisions at $\sqrt{s_{NN}}$ = 2.76 TeV. Adapted from [14].

4. Jets

Measurements of jets allow determination of the initial binary parton-parton scattering kinematics. Comparison of jet results in pp, p-A and A-A aids in discrimination of effects due to parton propagation in the medium, c.f. parton energy loss, from other effects that are derived from either the initial state or interactions in cold nuclear matter. Di-jet and γ-jet asymmetries relative to the collision geometry and orientation relative to the reaction plane can provide further information about parton energy loss, and in the future as a function of path length.

The technique of jet reconstruction in p-A and A-A collisions follows that established in pp collisions, taking into account also the additional complications associated with the large hadronic background from soft particles in A-A collisions that is unassociated with the jet. Background subtraction techniques have been developed to be able to correctly determine the jet energy and shape. Clustering algorithms are used to distinguish hadrons from parton fragmentation and those presumably associated with the background. For jet measurements ATLAS and CMS incorporate both electromagnetic and hadronic calorimetry, while ALICE utilizes electromagnetic calorimetry and charged particle tracking to identify the hadronic component of jets. Clustering algorithms that are both infra-red and collinear-safe have been developed and are used to identify jets and to compare directly with theory. For jet-finding the anti-k_T algorithm [15] from the FASTJET package [16] is used primarily. The soft background fluctuates event-to-event and is identified from the underlying event typically through the use of the k_T [17] or iterative-cone algorithms. Finally the background is subtracted, detector efficiency and resolution corrections implemented, and the jet energy scale adjusted via data-driven and/or Monte Carlo techniques in order to derive a final jet energy spectrum. [18] Details of the implementation of these techniques for jet-finding in A-A collisions can be found in [19].

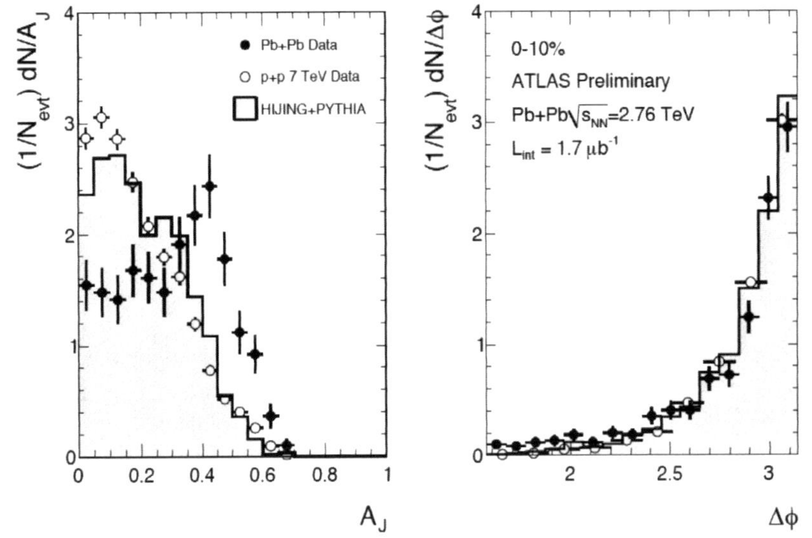

Figure 8. Left: Di-jet energy asymmetry A_J (defined in text) measured in $\sqrt{s_{NN}} = 2.76$ TeV central (0-10%) Pb-Pb events and in $\sqrt{s} = 7$ TeV pp events [20]. Right: Corresponding azimuthal angular correlation between leading and next-to-leading jet on opposite-side. Monte Carlo simulations using HIJING+PYTHIA Di-jets are also shown for comparison.

Measurements of complete jets were reported for the first time in A-A collisions by ATLAS [20] and CMS [21]. Displayed in the left panel of Figure 8 are the ATLAS measurements of the di-jet energy asymmetry A_J, where $A_J = E_{T1}E_{T2}/(E_{T1} + E_{T2})$ for $\sqrt{s_{NN}} = 2.76$ TeV Pb-Pb compared to $\sqrt{s} = 7$ TeV pp data and Monte Carlo simulations using PYTHIA jets embedded in HIJING events. T1 is defined as the highest energy jet with transverse energy $E_{T1} > 100GeV$ and T2 the highest transverse energy jet in the opposite hemisphere with $E_{T2} > 25GeV$. The Pb-Pb events exhibit a large asymmetry in the jet energies with the opposite-side jet having lost considerable energy compared to the pp data and HIJING+PYTHIA simulations. This is a clear indication of jet quenching in Pb-Pb collisions at the LHC. In the right panel of Figure 8 the azimuthal angular correlation $\Delta\phi$ between the leading jet and the opposite-side jet indicates that the jet angles, however, are not appreciably altered and exhibit the expected peak at $\Delta\phi = \pi$ as seen in the pp data and HIJING+PYTHIA simulations. A preliminary result of a similar asymmetry in momentum for di-jets was recently reported by STAR for $\sqrt{s_{NN}} = 0.2$ TeV Au-Au for jets of considerably lower energy with $p_{T1} > 20$ GeV and $p_{T2} > 10$ GeV [22].

Jets have been measured in $\sqrt{s} = 0.9$, 2.76, 7 and 8 TeV pp, $\sqrt{s_{NN}} = 5.02$ TeV p-Pb and in $\sqrt{s_{NN}} = 2.76$ TeV Pb-Pb collisions at the LHC and in $\sqrt{s_{NN}} = 0.2$ TeV Au-Au collisions at RHIC. Displayed in the left panel of Figure 9 is the R_{pPb} for fully-reconstructed jets in ATLAS [23] for R=0.4 and CMS [24] for R=0.3, and charged jets in ALICE [25] for R=0.4 at midrapidity in $\sqrt{s_{NN}} = 5.02$ TeV p-Pb collisions. The data are consistent with $R_{pPb} = 1$ over the range of jet momenta $25 < p_T < 800$ GeV/c. The R_{AA} for fully-reconstructed jets measured in ALICE [26] for R=0.2 and CMS [27] for R=0.3 in $\sqrt{s_{NN}} = 2.76$ TeV Pb-Pb collisions for various centralities are shown in the right panel of Figure 9. Note that different techniques are used in the two experiments for the full jet measurements, as described above. The results are consistent and exhibit an increased suppression of jets compared to pp and p-Pb for more central collisions ($0-5\%$ and $0-10\%$) compared to less central collisions ($10-30\%$).

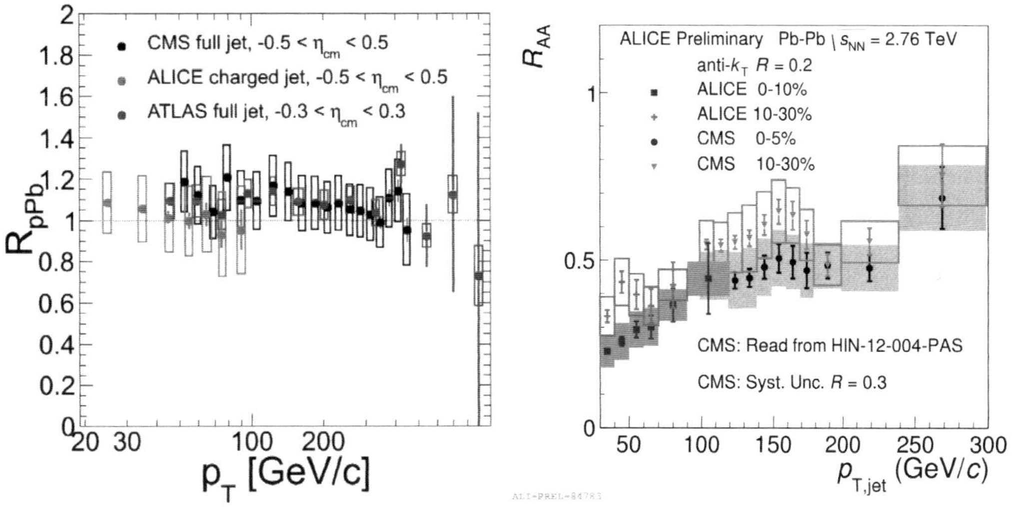

Figure 9. R_{pPb} for jets at midrapidity in each of the three experiments in $\sqrt{s_{NN}} = 5.02$ TeV p-Pb collisions [23, 24, 25] and R_{PbPb} for jets measured in ALICE and CMS for different centralities in $\sqrt{s_{NN}} = 2.76$ TeV Pb-Pb collisions [26, 27].

Investigation and measurement of differences in the suppression of parton probes of different flavor or mass is important to determine parton energy loss mechanisms and flavor or mass dependences. Shown in the left panel of Figure 10 for central collisions of Pb-Pb is the R_{AA} for D-mesons [28] and charged particles, which are dominated by pions. The suppression of D-mesons is consistent with that of charged particles, indicating that the energy loss of the heavier charm quark is not very different from that of light quarks. Also shown are the results from B-decays to J/ψ for inclusive Pb-Pb collisions [29]. The B-decays are less suppressed, but the difference in centrality between the B results and the others in the figure must be noted and make up for the difference. Thus, it is not clear yet whether b-quarks are less suppressed for a given collision centrality (which may reflect path-length, to some extent). Shown in the right panel of Figure 10 are the jet R_{AA} from b-jets [30] and inclusive jets [27] measured in central Pb-Pb collisions. This result clearly indicates that the b-jet suppression is of similar magnitude to that of inclusive jets demonstrating the lack of flavor dependence in the quenching mechanism at these large momenta.

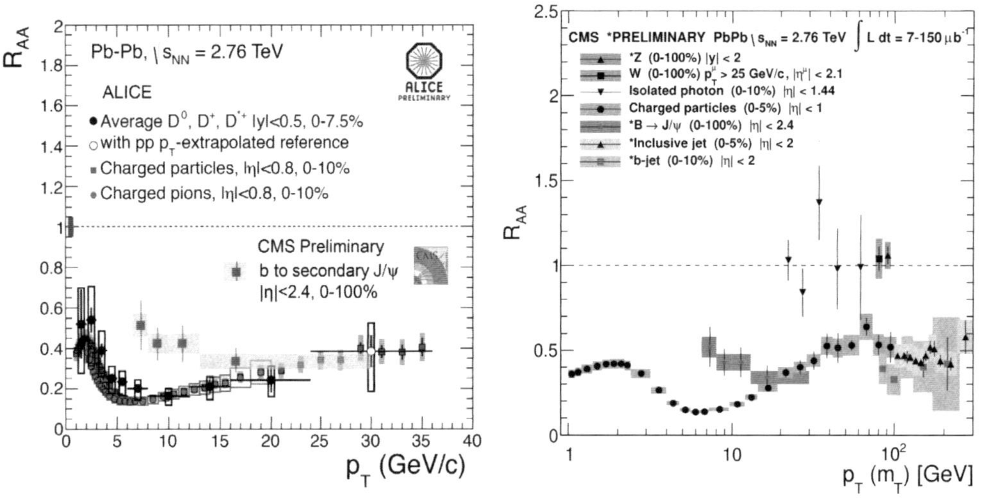

Figure 10. Left: R_{AA} for D-mesons and charged particles in central collisions of Pb-Pb from ALICE [28], and B-decays to J/ψ for inclusive Pb-Pb collisions from CMS [29]. Right: R_{AA} for various particles (see legend), inclusive jets and b-jets measured in central Pb-Pb collisions from CMS [27, 30].

Parton energy loss in the QCD medium is expected to modify jet fragmentation in A-A collisions as compared to that in pp collisions. Patterns of fragmentation can be investigated and compared to pp collisions to determine any modifications. A measurement of the fragmentation function can be determined by measuring R_{AA} or R_{CP} as a function of the momentum fraction z, of the jet with momentum p_{jet}, that is carried by the detected particle of momentum p. This is defined as $z = p/p_{jet}$ or often designated as $\xi = ln(1/z)$. The R_{AA} measured by CMS [31] and R_{CP} by ATLAS [32] are shown in the left panel of Figure 11. The two are consistent and exhibit enhancements at large ξ representing more particles with low momentum fraction z in Pb-Pb as compared to pp collisions. Likewise, there appears to be a slight enhancement observed at low ξ or large momenta. Thus, jets are significantly modified in Pb-Pb compared to pp collisions.

This is also seen in the γ-hadron correlations measured by PHENIX at RHIC. The right-top panel of Figure 11 shows the yield per trigger for 0-40% central Au-Au collisions and pp collisions as a function of ξ and z_T (top scale). The I_{AA} defined as the ratio of Au-Au to pp fragmentation functions is shown in the bottom-right panel. Also shown are pQCD-inspired model calculations, represented as curves in the figure. These incorporate parton energy loss and result in suppression at low ξ, reappearance of the lost energy in enhanced production of low p_T particles, i.e. at larger ξ. See [33] and references therein for details in the models.

Jet modification can also be investigated by studying jet shapes in Pb-Pb collisions relative to those in pp collisions. The ratio of the charged particle density in central Pb-Pb relative to that in pp is shown as a function distance r from the jet center for anti-k_T jets in a cone of radius R = 0.3 in Figure 12. An increase in the ratio is observed for large distances near the

Journal of Physics: Conference Series **630** (2015) 012052 doi:10.1088/1742-6596/630/1/012052

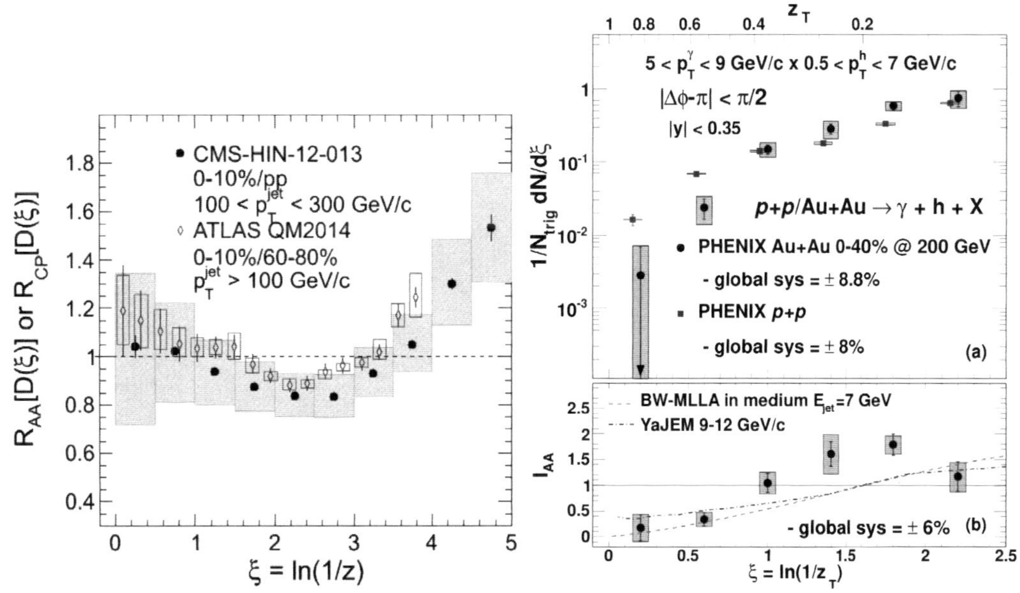

Figure 11. Left panel: Fragmentation function ratios R_{AA} from CMS [31] and R_{CP} from ATLAS [32] as a function of $\xi = ln(1/z)$ of jets as defined in legend. Right panel: (top) γ-hadron correlation yield per trigger for 0-40% central Au-Au collisions (circles) and pp collisions (squares) as a function of ξ. (bottom) I_{AA} ratio of Au-Au to pp fragmentation functions. Also shown are model calculations (curves), see text and [33] for details. The γ-hadron p_T ranges are given in the legend and the dependence on z_T is shown on the top scale.

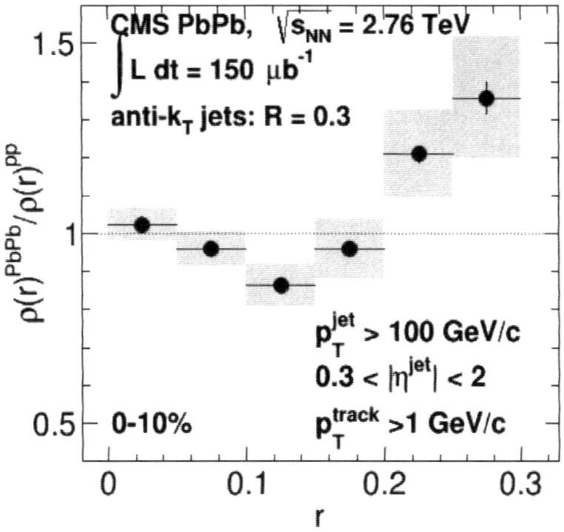

Figure 12. Ratio of the number of charged tracks with $p_T > 1$ GeV/c in Pb-Pb central collisions relative to that in pp for $p_T > 100$ GeV/c jets using anti-k_T with R = 0.3 as a function of radial distance from the center of the jet from [34].

cone "edge" (r = R) with a depletion at intermediate distances. This is indicative of broadening of the jet, and is suggestive of parton energy loss. Detailed studies of the particle densities as a function of cone size for central Pb-Pb exhibit enhanced particle densities that also increase with cone size relative to pp or peripheral Pb-Pb collisions [35].

The centrality dependence of jet R_{AA} can be seen in Figure 13. As expected jet quenching, like high p_T particle suppression, increases (i.e. jet R_{AA} decreases) for more central collisions. This reflects the increase in the size of the medium as collisions become more central and may be the reason for the slight differences in jet R_{AA} observed by the three collaborations (above) for slightly different centrality ranges.

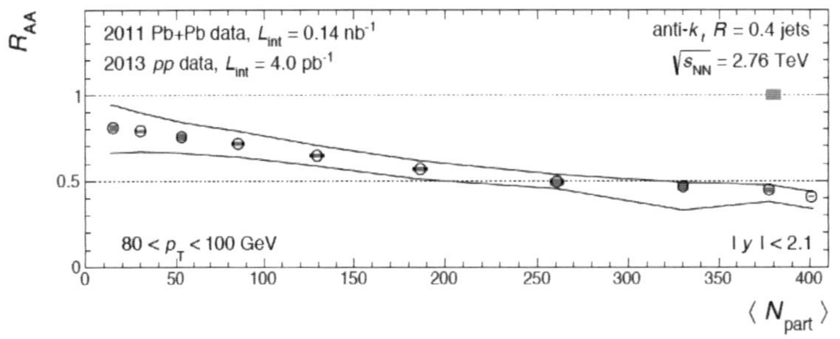

Figure 13. R_{AA} for jets with $80 < p_T < 100$ GeV and $|y| < 2.1$ as a function of $\langle N_{part} \rangle$. [36]

ATLAS, ALICE and CMS have measured finite values of v_2 for particles and jets at large transverse momenta in Pb-Pb collisions at $\sqrt{s_{NN}} = 2.76$ TeV. All three experiments agree well in the measurements of $v_2(4)$ for $< p_T < 20$ GeV over the entire range of centralities [38, 37]. At larger momenta CMS has measured a value of $v_2 \simeq 0.02$ over the range $20 < p_T < 50$ GeV for particle tracks in 0-5 % central collisions [37]. ATLAS finds $v_2^{jet} \simeq 0.02$ over the range $50 < p_T < 200$ GeV for jets in 5-10 % central collisions [38]. ALICE measures a somewhat larger v_2^{jet} than ATLAS but with lower track p_T threshold and with fairly large uncertainties for charged jets in the range $50 < p_T < 100$ GeV.[26] Since hydrodynamics is not expected to play any direct role at these larger transverse momenta, a significant value of v_2 at these momenta is highly indicative of a path length dependence of parton energy loss. The partons lose more energy in the out-of-plane direction than in-plane due to the elongation out-of-plane in the overlap region, thus resulting in a finite value of v_2 from the different path lengths in- and out-of-plane.

Direct photons are not expected to interact strongly with the medium in Pb-Pb collisions. Thus, they can be used to determine the kinematics of the initial hard scattering. This is substantiated in Pb-Pb collisions where $R_{AA}^{\gamma} = 1.0$ within uncertainties, which indicates that the yields of isolated photons in Pb-Pb is a superposition of those measured in pp data. Photons at leading order are produced back-to-back with an associated parton of nearly identical transverse momenta. Thus, measurements of back-to-back photons and jets can be used to test jet quenching and the energy lost by the parton by comparison with the photon. Shown in Figure 14 are $R_{J\gamma}$ of isolated prompt photons with jets that are back-to-back with the photon from ATLAS [39] and CMS [40]. The ATLAS data are for photon $60 < p_T^{\gamma} < 90$ GeV at mid rapidity, with $p_T^{jet} > 25$ GeV within a back-to-back azimuthal angular range $\Delta\phi_{J\gamma} > 7\pi/8$. The CMS data are for isolated photons with transverse momentum $p_T^{\gamma} > 60$ GeV and an associated jet with $p_T^{jet} > 30$ GeV in the same back-to-back range. Also shown are Monte Carlo simulations of hard scattering using PYTHIA superimposed on data in the ATLAS case, with PYTHIA alone and PYTHIA+HYDJET in the CMS case. The pp reference data are also shown for CMS. The

Journal of Physics: Conference Series **630** (2015) 012052 doi:10.1088/1742-6596/630/1/012052

$R_{J\gamma}$ is observed to decrease with increasing centrality (N_{part}) and remains below the values from the PYTHIA+Monte Carlo simulations and pp reference data, except for the least central Pb-Pb case. In addition, no angular broadening is observed (not shown) beyond that seen in the pp data and MC reference at all centralities. Similar results have been observed for Z-jet back-to-back coincidence measurements [41]. Furthermore, the average ratio of jet transverse momentum to photon transverse momentum for the most central events was found by CMS to be 0.73 ± 0.02(stat.) ± 0.04(syst.), which is lower than the value of 0.86 seen in the pp data and that predicted by PYTHIA + HYDJET at the same centrality [40].

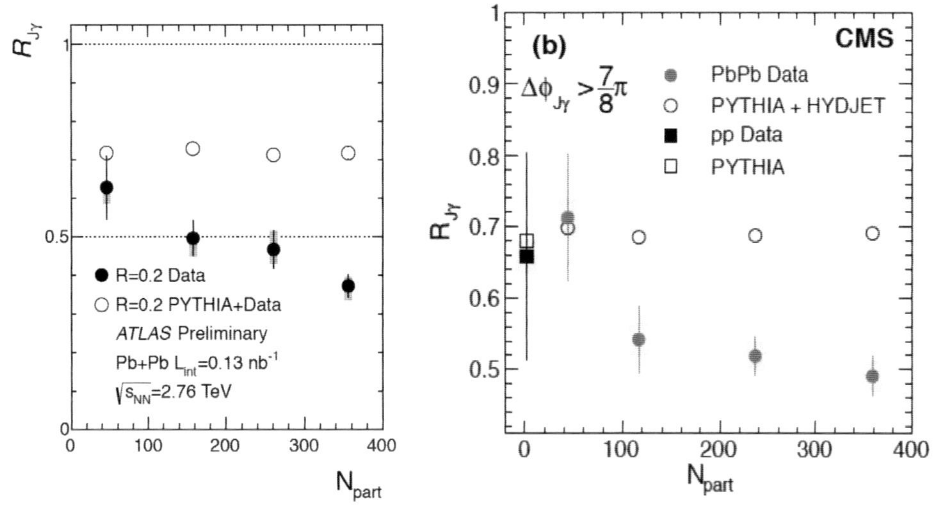

Figure 14. Left: $R_{J\gamma}$ of isolated prompt photons with back-to-back jets from ATLAS [39] with R=0.2. Right: $R_{J\gamma}$ from CMS with R=0.4 [40]. Data shown as a function of centrality as represented by N_{part} for $\sqrt{s_{NN}}$ = 2.76 TeV Pb-Pb collisions, see text for details.

It is important to understand what happens to the energy lost by partons as they traverse the medium. There is some indication from the fragmentation function measurements above that the jet suppression leads to re-distribution of the lost energy to lower momentum particles. The top-left panel of Figure 15 shows the dependence of I_{AA} on the angular distribution of particles on the away-side of the trigger γ for γ-hadron correlations in 0-40% central Au-Au collisions [33]. The γ and hadron momentum ranges are shown in the legend, with the γ-hadron azimuthal angle difference defined as Δ. Three regions of away-side angles are presented, as indicated in the figure. In going from the larger away-side hadron acceptance range $|\Delta\phi - \pi| < \pi/2$ to the smaller ranges of $|\Delta\phi - \pi| < \pi/3$ and $|\Delta\phi - \pi| < \pi/6$, the observed enhancement at larger ξ (lower momentum particles) is reduced, becoming negligible for the most restrictive range. However, the suppression at low ξ remains in the cases studied. The bottom-left panel depicts the ratio of I_{AA} for full away-side acceptance to the most limited, $|\Delta\phi - \pi| < \pi/2$ divided by $|\Delta\phi - \pi| < \pi/6$. The I_{AA} is observed to vary with significant enhancement at larger ξ as the away-side hadron acceptance becomes less limited, indicating that the enhancement at large ξ is due to hadrons emitted at the large angles associated with $|\phi - \pi| > \pi/6$.

STAR has studied jet-hadron correlations, measuring the widths of the peaks on the away-side of the jet in 0-20% central Au-Au collisions [42]. D_{AA} measures the p_T difference between Au-Au and pp collisions for each p_T^{assoc} bin with mean $\langle p_T^{assoc}\rangle$ as defined below:

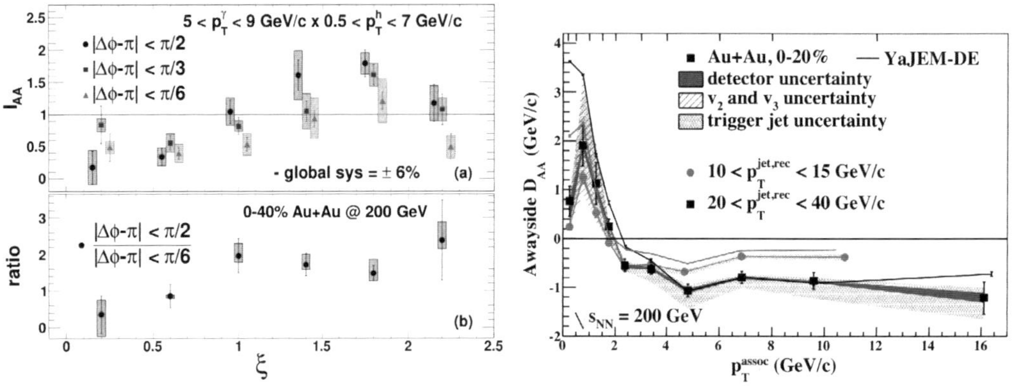

Figure 15. Left panel: γ-hadron correlations in 0-40% central Au-Au collisions with the γ-hadron azimuthal angle difference defined as Δ. (top) I_{AA} with the full away-side acceptance for hadrons $|\Delta\phi - \pi| < \pi/2$ (circles) and for two ranges of away-side acceptance $|\Delta\phi - \pi| < \pi/3$ (squares) and $|\Delta\phi - \pi| < \pi/6$ (triangles). (bottom) Ratio of I_{AA} for full away-side acceptance to the most limited, $|\Delta\phi - \pi| < \pi/2$ divided by $|\Delta\phi - \pi| < \pi/6$. [33] Right panel: Jet-hadron correlations in central (0-20%) collisions of $\sqrt{s_{NN}} = 0.2$ TeV Au-Au [42]. Plotted is the away-side D_{AA} (see text for definition) as a function of p_T^{assoc} for two different reconstructed jet energy ranges. Bands depict uncertainties as defined in the legend. YaJEM calculations are shown as the lines for the two different triggers [43].

$$D_{AA}(p_T^{assoc}) = Y_{AuAu}(p_T^{assoc}) \cdot \langle p_T^{assoc} \rangle_{AuAu} - Y_{pp}(p_T^{assoc}) \cdot \langle p_T^{assoc} \rangle_{pp} \tag{2}$$

Thus, $\mathrm{D}_{AA} \neq 0$ indicates modification of jet fragmentation in Au-Au compared to pp collisions. $\mathrm{D}_{AA} = 0$ represents the case where Au-Au collisions have identical fragmentation patterns in p_T as pp collisions for all p_T^{assoc} bins. It is expected that the near-side trigger jet has a surface bias, making it more likely that the away-side hadron traverses a larger distance in the medium and that effects of parton energy loss will be prominent for the away-side hadrons. Furthermore, the widths of the trigger jets (defined to be the near-side) are observed to be the same in Au-Au and pp collisions in this study. Shown in the right panel of Figure 15 are the awayside momentum difference D_{AA} for two ranges of jet reconstructed momentum. The measured D_{AA} show that high p_T^{assoc} hadrons on the away-side are suppressed and those at low p_T^{assoc} are enhanced. This indicates that jets in Au-Au are significantly softer that those in pp collisions. Also shown in the figure are YAJEM Monte Carlo model calculations of in-medium shower evolution which incorporate radiative and elastic energy loss. This model qualitatively describes the general trends of the jet-hadron correlation data.

Another approach to study the energy redistribution due to parton energy loss in A-A collisions is to measure and correlate charged particles with the momentum imbalance of di-jets and centrality. [44] Di-jets were reconstructed by CMS and the di-jet imbalance, as presented above for ATLAS in Figure 8, was found to increase with centrality. CMS found that as the di-jet momentum imbalanced increases, the fragmentation pattern of the away-side jet becomes softer. This can be seen in Figure 16. Plotted is the missing p_T of charged tracks projected onto the lead-jet axis as given by

$$\not{p}_T^{\parallel} = \sum_i -p_T^i \cos(\phi_i - \phi_{lead-jet}) \quad (3)$$

for tracks with momenta $p_T > 0.5$ GeV/c at mid-rapidity ($|\eta| < 2.4$) and jets with $p_T^{jet} > 120$ GeV/c at $|\eta| < 1.6$. In the left panel the momentum is balanced between the lead-jet (negative values of $\not{p}_T^{\parallel}$) and away-side jet (positive $\not{p}_T^{\parallel}$) hemispheres. When only tracks inside the jet cones are included in the calculation of $\not{p}_T^{\parallel}$, shown in the middle panel, there is an excess of track momentum in the lead-jet direction and opposite that for tracks outside the jet cones as shown in the right panel.

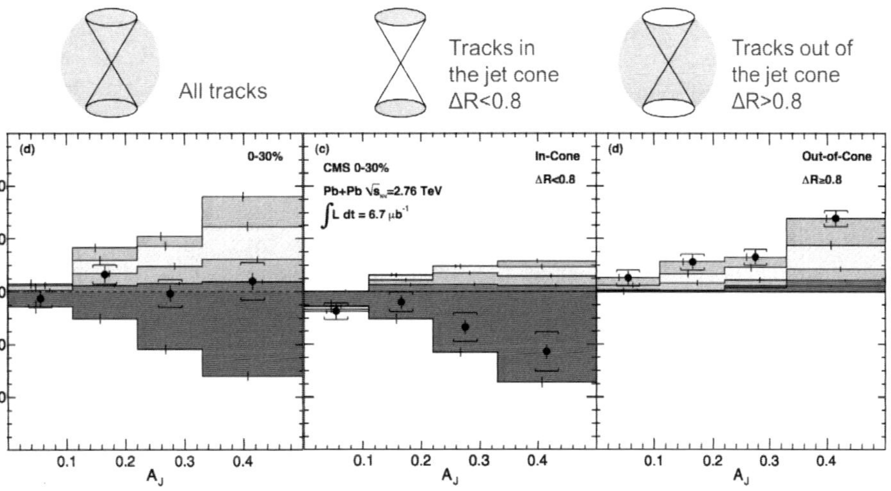

Figure 16. Missing transverse momentum $\not{p}_T^{\parallel}$ of charged tracks projected onto the leading jet axis as a function of di-jet momentum asymmetry A_J for central collision event class 0-30%. Average values are plotted as data points, with colored regions for contributions from each track p_T range as designated in the legend. Left panel: all charged tracks. Middle panel: only charged tracks inside the jet cones of $\Delta R < 0.8$. Right panel: charged tracks outside the jet cones. Track selections are depicted relative to the jet cones in the top row of diagrams for each panel. Average values are plotted as data points, with colored regions for contributions from each track p_T range (in GeV): data (> 0.5), blue (0.5-1.0), yellow (1.0-2.0), orange (2.0-4.0), green (4.0-8.0), red (> 8.0).[44]

The results on di-jets from ATLAS and CMS show that there is an increase in the di-jet energy and momentum asymmetry as the centrality of the collision increases, which is consistent with energy loss of the parton as it traverses a larger distance in the medium. CMS results indicate that this imbalance is recovered only when lower p_T tracks, down to $p_T = 0.5$ GeV/c, are included in the calculation of $\not{p}_T^{\parallel}$.

5. Summary and Conclusions

In summary, the recent heavy-ion collision results from hard scattering processes at RHIC and LHC have provided a wealth of information to address parton energy loss in the medium. Hadrons with large transverse momentum and jets are suppressed in heavy-ion collisions compared to proton-proton collisions. Hadrons and jets with heavy quarks appear to also be suppressed at about the same level as those with light quarks. The jet fragmentation is modified, with the modification increasing with the centrality of the heavy-ion collision. When a comparison of the heavy ion results is made to those in pp, p-A and theory, a consistent picture emerges. The suppression of particles and jets with large transverse momentum is a product of parton energy loss in the hot QCD medium. Di-jet correlations exhibit a di-jet energy and momentum imbalance that increases with the centrality of the collision. Correlations of di-jets, jets and trigger photons, and hadrons and trigger jets indicate that the parton energy loss is redistributed to lower momentum particles at larger angles from the axis of the hard-scattered parton.

In order to understand at a fundamental level the parton energy loss in QCD at high temperatures, its dependence on the properties of the parton and of the medium need further addressing. This leads to more specific questions. What is the dependence of parton energy loss on the properties (e.g. flavor, mass and energy) of the traversing parton? Can this be accessed experimentally in the final state, for example by incorporating identification of particles? What is the extent to which radiative and collisional energy loss mechanisms play a role? How does the energy loss depend on other properties of the medium, especially temperature? Can we say more about happens to the lost energy? How is it dependent on the virtuality of the parton and its pathlength in the medium? Answers to these questions are being pursued vigorously both theoretically and experimentally. More differential measurements with a well-determined initial state and controlled initial- and final-state geometries may provide the key to obtain these answers.

6. References

[1] Wiedemann UA, arXiv:0908.2306v1 [hep-ph].
[2] dEnterria D, arXiv:0902.2011v2 [nucl-ex].
[3] J.W. Cronin, H. J. Frisch, M. J. Shochet, J. P. Boymond, R.Mermod, P. A. Piroue and R. L. Sumner, Phys. Rev. D 11 (1975) 3105.
[4] I. Vitev and M. Gyulassy, arXiv:hep-ph/0209161, Phys. Rev. Lett. 89 (2002) 252301.
[5] T. Renk et al., arXiv:1103.5308, Phys. Rev. C 84 (2011) 014906.
[6] C. A. Salgado and U. A. Wiedemann, arXiv:hep-ph/0302184, Phys. Rev. D 68 (2003) 014008; N. Armesto, A. Dainese, C. A. Salgado et al., arXiv:hep-ph/0501225, Phys. Rev. D 71 (2005) 054027.
[7] A. Dainese, C. Loizides, and G. Paic, arXiv:hep-ph/0406201, Eur. Phys. J. C 38 (2005) 461.
[8] CMS Collaboration, Eur. Phys. J. C 72 (2012) 1945.
[9] S. Horvat for the STAR Collaboration, Quark Matter 2014, to be published in Nuc. Phys A.
[10] ALICE Collaboration, Phys. Lett. B 720 (2013) 52.
[11] CMS Collaboration, Phys. Lett. B 715 (2012) 66; Phys. Lett. B 710 (2012) 256; Phys. Rev. Lett. 106 (2011) 212301.
[12] ALICE Collaboration, arXiv:1210.4520v1 [nucl-ex], Phys. Rev. Lett. 110, 082302 (2013).
[13] PHENIX Collaboration, Phys. Rev. Lett. 91, 072303 (2003).
[14] ALICE Collaboration, arXiv:1401.1250v3 (2014).
[15] M.Cacciari, G.P.Salam and G.Soyez, JHEP, 04:063, 2008.
[16] M. Cacciari and G. P. Salam, Phys. Lett. B 641, 57 (2006); Matteo Cacciari, Gavin P. Salam, and Gregory Soyez. FastJet user manual, Eur.Phys.J., C72:1896, 2012.
[17] S. Catani, Y. Dokshitzer,M.H. Seymour and B.R.Webber, Nucl. Phys. B 406, 187 (1993).
[18] G. Cowan. A Survey of Unfolding Methods for Particle Physics. Proc. Advanced Statistical Techniques in Particle Physics, Durham, 2002.
[19] D. D'Enterria, J.Phys.G34:S53-S82,2007.
[20] ATLAS Collaboration, arXiv:1011.6182v2 [hep-ex], Phys. Rev. Lett. 105, 252303 (2010).
[21] CMS Collaboration, arXiv:1102.1957, Phys. Rev. C84 (2011) 024906.
[22] J. Putschke for the STAR Collaboration, Quark Matter 2014, Nucl. Phys. A in press.
[23] ATLAS Collaboration, ATLAS-CONF-2014-024, Quark Matter 2014, Nucl. Phys. A in press.
[24] CMS Collaboration, CMS PAS HIN-14-001, Quark Matter 2014, Nucl. Phys. A in press.
[25] R. Haake for the ALICE Collaboration, arXiv:1310.3612, EPS-HEP2013 proceedings.
[26] S. Aiola for the ALICE Collaboration, Quark Matter 2014, Nucl. Phys. A in press.
[27] CMS Collaboration, CMS PAS HIN-12-004.

Journal of Physics: Conference Series **630** (2015) 012052 doi:10.1088/1742-6596/630/1/012052

[28] ALICE Collaboration, arXiv:1203.2160v4 [nucl-ex], JHEP 9 (2012) 112.
[29] CMS Collaboration, arXiv:1201.5069, JHEP 05 (2012) 063.
[30] CMS Collaboration, arXiv:1312.4198v1 [nucl-ex].
[31] CMS Collaboration, arXiv:1406.0932v1 [nucl-ex], CMS-HIN-12-013.
[32] M. Spousta for the ATLAS Collaboration, Quark Matter 2014, Nucl. Phys. A in press.
[33] PHENIX Collaboration, Phys. Rev. Lett. 111, 032301.
[34] CMS Collaboration, arXiv:1310.0878v2 [nucl-ex] (2014), PLB 730 (2014) 243.
[35] ATLAS Collaboration, arXiv:1208.1967v2 [hep-ex], Phys. Lett. B719 (2013) 220.
[36] ATLAS Collaboration, ATLAS-CONF-2014-025, Quark Matter 2014, Nucl. Phys. A in press.
[37] CMS Collaboration, arXiv:1204.1850v2, Phys. Rev. Lett.109.022301.
[38] ATLAS Collaboration, ATLAS-CONF-2014-027, Quark Matter 2014, Nucl. Phys. A in press.
[39] ATLAS Collaboration, ATLAS-CONF-2012-121,
[40] CMS Collaboration, arXiv:1205.0206 [nucl-ex], Phys. Lett. B 718 (2013) 773.
[41] ATLAS Collaboration, ATLAS-CONF-2012-119.
[42] STAR Collaboration, Phys. Rev. Lett. 112 (2014) 122301.
[43] T. Renk, Phys. Rev. C84, 067902 (2011).
[44] CMS Collaboration, arXiv:1102.1957v2 [nucl-ex], Phys. Rev. C84:024906,2011.

Journal of Physics: Conference Series **630** (2015) 012053 doi:10.1088/1742-6596/630/1/012053

Exploring the Phase Diagram of Nuclear Matter with Relativistic Heavy Ion Collisions

Rene Bellwied[1]
[1] University of Houston, Physics Dept., 617 SR1 Building, Houston, TX 77204, USA

E-mail: bellwied@uh.edu

Abstract. I will review the present status of relativistic heavy ion collisions at RHIC and the LHC with special emphasis on the most recent discoveries, which characterize the deconfined phase of matter and the phase diagram itself.

1. Introduction

Since the discovery of the deconfined, strongly interacting phase of matter at RHIC in 2005 [1], the experiments at RHIC and the LHC have focussed on characterizing this new phase and determining the features of the QCD transition phase diagram at lower energies. The deconfined phase, which was originally termed the Quark Gluon Plasma (QGP) turned out to show quite unexpected features, in particular a strong collective behaviour which is characterized by a very small mean free path between interactions. This led to the rather dramatic announcement that the partonic matter phase behaves more like an ideal liquid than a weakly interacting plasma. Ideal hydrodynamics came close to describing the collective flow patterns, and after small viscous corrections were added, the data could be described well enough to make first estimates of state variables [2], such as the shear viscosity over entropy ratio (η/s), which showed that the produced fluid indeed features a minimal viscosity ratio, very close to the conjectured quantum limit of $\eta/s = 1/4\pi$ [3].

Since these discoveries, which are almost a decade old, the experiments have evolved by a.) increasing the collision energy range by almost a factor 30 to the LHC energies in order to characterize the deconfined in more detail and with newer detector technology, and b.) decreasing the collision energy in order to determine the exact temperature and density when the QCD transition sets in, and to determine whether the phase diagram of nuclear matter features a critical point on the deconfinement line. The nature of the phase diagram at high temperatures and low densities has come into focus due to continuum extrapolated lattice QCD results which were, for the first time, conducted with realistic quark and pion masses and with very small lattice spacings [4,5]. These first principle calculations show unambiguously that at LHC and the highest RHIC energies where the baryon density (as described by the quark chemical potential) is negligible, the QCD transition is an analytic crossover rather than a transition of first or second order [6]. Effective models such as Dyson-Schwinger chiral theory [7] or PNJL [8] have indicated that at finite density the transition should become first order, which means that the transition line will feature a critical point. These postulates have not yet been confirmed by lattice QCD calculations, since the so-called sign problem, makes it impossible to

[1] To whom any correspondence should be addressed.

cleanly calculate the transition at higher quark chemical potential. Several methods to approximate the finite density behaviour on the lattice have been attempted, either through Taylor expansion or imaginary chemical potentials, but none of them have yielded conclusive evidence for a critical point [9,10]. Still, since in particular the RHIC machine is capable to lower the collision energy to study the baryon densities in question, a significant experimental program, labelled the Beam Energy Scan (BES) program, has been initiated and its first phase was concluded in the past year.

At the higher LHC energies the main questions that have been addressed are whether one can a.) determine the shear viscosity to textbook accuracy and b.) show directly that hydrodynamics and strong interactions are indeed the proper model to describe the behaviour above the transition. This picture deviates significantly from the original belief that the deconfined phase consists, even near the phase transition, of weakly interacting, asymptotically free quark and gluons. Early lattice QCD calculations showed that from about 1.5-2 Tc on the chosen order parameters approach the Stefan Boltzmann limit and show deviations of at most 20% from the limit, which characterizes an asymptotically free phase. Since the latest calculations show more of an analytic crossover the limit is approached slightly slower. Still, the fact that in this crossover region the phase acts like a strongly coupled plasma requires us to rethink the properties of the relevant degrees of freedom in the deconfined phase near the hadronization temperature. Fluctuation measurements are sensitive to the chiral, confinement and flavour properties of the relevant particle state in the transition region and they can be directly related to susceptibility calculations on the lattice [11]. Regarding the interaction strength between the degrees of freedom, several parton based transport models (BAMPS [12], PHSD [13], non-Abelian bremsstrahling [14]) taking into account multi-parton interactions in the dense regime have been proposed to describe the collective features of the emitting system without invoking parton hydrodynamics. This has become more relevant since some of the hadron correlation structures, which have been attributed to collective flow, have now also been measured in proton-nucleus (pA) and even proton-proton (pp) collisions at the LHC. More detailed measurements are needed to sort out whether the hydrodynamical approach is also valid for even the smallest interaction system, where no thermalization is historically expected, or whether multi-parton interactions in a non-equilibrium system can explain correlation structures even in heavy ion collisions.

On a broader scale these measurements will relate to cosmology by studying the evolution of the universe during the QCD transition, which leads to correlations and flavour patterns during the hadronization, that might still be imprinted in the hadronic cosmological matter in the core of stars and cosmic rays today.

In the following I will first discuss the latest LHC measurements in the context of determining collectivity in the heavy ion systems, then I will briefly review the most dramatic measurements of the RHIC-BES program, before ending with some interesting observations from collision energy dependent fluctuation measurements. I will end with a short conclusion that relates these measurements to the QCD transition and its role in the evolution of matter in the universe.

2. LHC measurements on collective behaviour

Over the past three years the LHC has successfully completed a program that included proton-proton, proton-Pb and Pb-Pb collisions at various beam energies, all of which correspond to a regime very near zero baryon density on the nuclear phase diagram. This program enables us to study the source of collective behaviour as a function of the system size. Furthermore the high multiplicity achievable at these beam energies allows a very precise determination not only of the higher harmonics in the Fourier decomposed moment spectrum of the emitted charged particles, but also of the features of multi-particle correlations for a given harmonic.

Journal of Physics: Conference Series **630** (2015) 012053 doi:10.1088/1742-6596/630/1/012053

2.1. Heavy Ion Collision Measurements

The realization that collectivity in a heavy ion collision can be measured by determining how much an asymmetry in the original coordinate system of the produced fireball causes an anisotropy in the momentum spectrum of the emitted particles was realized during the early days of the fixed target relativistic heavy ion programs at the BEVALAC and the AGS [15,16]. As expected the ideal hydrodynamic limit was never reached at these low energies. Still, a certain level of collectivity could be shown, in particular in the lowest harmonic v_2 in the Fourier decomposed momentum spectrum (elliptic flow) in non-central collisions. During the analysis of the RHIC data it became also apparent that the energy density fluctuations in the fireball, which are largely caused by the distribution of nucleons, and subsequently quarks and gluons in the initial state, will cause higher harmonics in the Fourier series to occur on an event-by-event basis [17].

Early measurements at RHIC energies indicated that the higher harmonics, which are more sensitive to the initial conditions, might be used to constrain the viscosity ratio by distinguishing the contribution from asymmetries caused by the initial density profile. In particular the early PHENIX results showed that v_3 can be used to show that an initial profile based on a highly saturated gluon density (Color Glass Condensate) does not describe that data [18]. Since then, though, models that merge reasonable assumptions of the initial parton densities with a hydrodynamic evolution of the system have proven to be very successful in describing the higher harmonics and constraining the η/s value [19].

This has become possible due to the tremendous wealth of higher order harmonic data from all three LHC experiments. Fig.1 shows a comparison of v_2 data as a function of transverse momentum from ATLAS [20], ALICE [21], and CMS [22]. The data are in excellent agreement and reach out to almost 20 GeV/c for all experiments. Fig.2 shows the latest v_3 and v_4 results from ATLAS based on four-particle correlations [20].

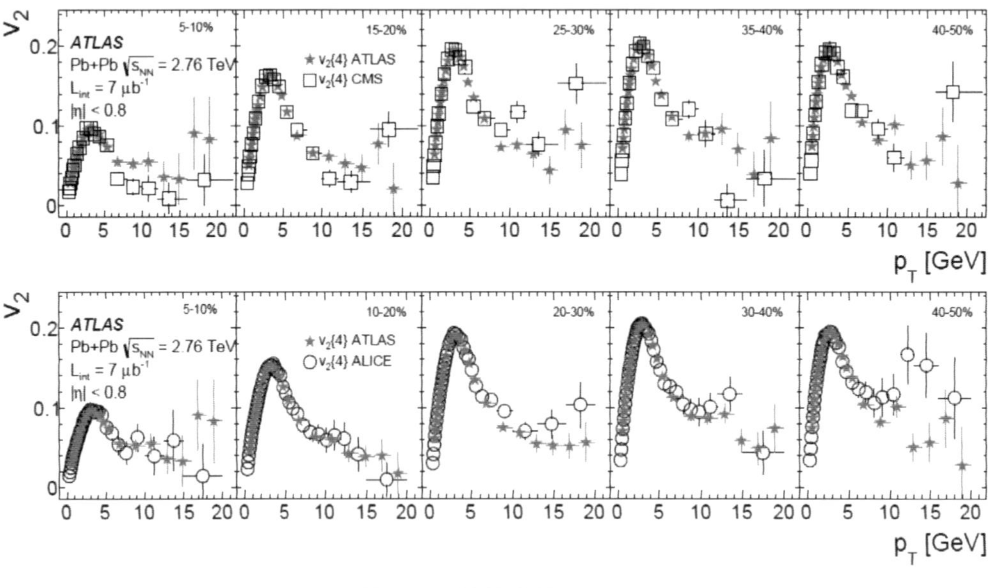

Figure 1. Comparison of ATLAS [20] and CMS [22] (top panel), and ATLAS [20] and ALICE [21] (bottom panel) measurements of v2{4} for selected centrality intervals at |h| < 0.8 (from [20]).

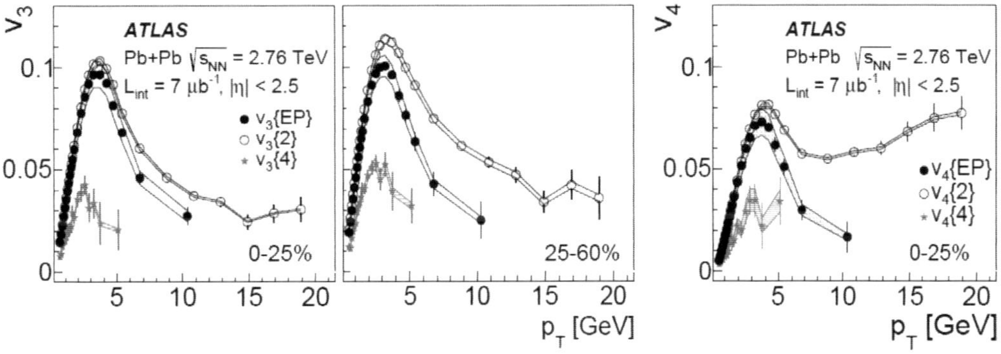

Figure 2. The transverse momentum dependence of v_3 calculated with two- and four-particle cumulants and with the event-plane method $v_3\{EP\}$ for the centrality interval 0–25% (left plot) and 25–60% (middle plot). The right plot shows the same results for v_4 for the centrality interval 0–25% (from [20]).

The argument to measure more than two-particle correlations for a given harmonic is relatively simple: the more particles contribute to a collective effect, the less likely it is that this effect can be caused by fundamental non-equilibrium gluon exchange processes. The latest transport calculations expanded the basic 2 to 2 processes to include 2 to 3 and 3 to 2 processes [12], but it is difficult to imagine that exchange processes can be extended to more than a few particles. Therefore any collective effect that includes more than four particles can be viewed as a likely sign for collective dynamics outside of multi-parton interactions. I will come back to this point when discussing the results from the small system measurements. In order to constrain the initial conditions and ultimately the precise value of the viscosity over entropy ratio, the data have been analysed up to v_5 in the context of a hybrid model taking into account initial gluon densities and hydrodynamical evolution. Fig.3 show the p_T-dependence as measured by ATLAS [23] and Fig.4 shows the centrality dependence as measured by ALICE [24] in comparison to a MUSIC model calculation which yields a $\eta/s=0.2$ or roughly 2.5 times the quantum limit [19]. This is still many times smaller than one would expect in a perturbative QCD calculation.

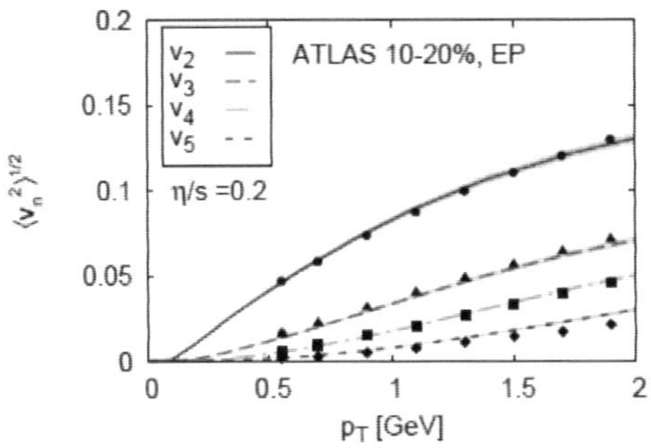

Figure 3. RMS anisotropic flow coefficients $(v^2_n)^{1/2}$, computed as a function p_T, compared to experimental data of $v_n\{2\}$, $n \in \{2, 3, 4\}$, by the ATLAS collaboration [23]. Results are for 200 events per momentum bin with bands indicating statistical errors (from [19])..

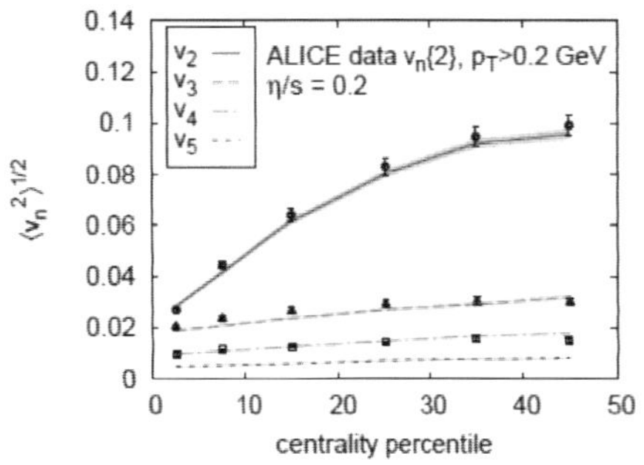

Figure 4. RMS anisotropic flow coefficients $(v^2_n)^{1/2}$, computed as a function of centrality, compared to experimental data of $v_n\{2\}$, $n \in \{2, 3, 4\}$, by the ALICE collaboration [24]. Results are for 200 events per centrality with bands indicating statistical errors (from[19]).

The viscous corrections are not negligible but they are small and well within the capabilities of existing viscous hydrodynamical models. If one applies the same model to the existing RHIC data the η/s reduces to 0.12 [25], so as expected the very strong coupling reduces as a function of collision energy but the pQCD limit is still far away and we are seemingly still dominated by strongly coupled degrees of freedom above the QCD transition.

2.2. pPb and pp Collision Measurements

The analysis of the first year of proton-proton collisions at the LHC led to a very surprising result, when CMS announced that they found long-range two particle correlations in very high multiplicity events that are consistent with elliptic flow. Historically proton-proton collisions were considered reference data obtained in a system that is distinctly non-equilibrium and shows no collective effects from emission from a thermalized fireball. Since then certain multi-parton interactions, if properly included in fragmentation models, such as PYTHIA, have given rise to long-range correlations. In particular final state effects, characterized as color reconnection, seem to be able to at least qualitatively confirm the correlation trends observed in high multiplicity proton proton collisions. A definitive test of either theory (collective or non-equilibrium) should be possible by studying nucleon-nucleus collisions, such as p-Pb or d-Au, where we expect again no collective behaviour but a significant sensitivity to the initial gluon density, since the nucleon system is isolated in the incoming channel. In this case the most relevant flow measurement is the multi-particle correlation probability for a fixed harmonic. Since the charged particle multiplicity is limited in the small systems only the lowest collective harmonic (v_2) has been fully analysed at this time. Fig.5 shows the rather dramatic result based on the preliminary CMS analysis [25]. As can be seen, although the flow in pPb is about a factor 2 smaller than in PbPb collisions, the multi-particle correlations up to $v_2\{8\}$ are finite for the small system.

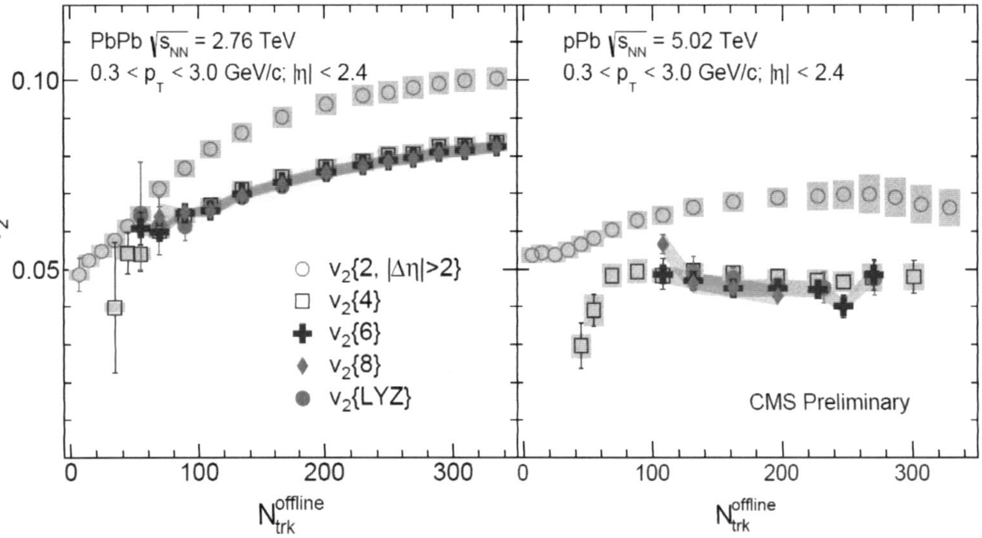

Figure 5. The v2 results obtained from mutli-particle cumulants, and LYZ method, averaged over the particle p_T range of 0.3–3.0 GeV/c, as a function of number of tracks recorded in PbPb at $\sqrt{s_{NN}}$ = 2.76 TeV (left) and pPb at $\sqrt{s_{NN}}$ = 5.02 TeV (right). Shaded areas denote systematic uncertainties (from [25]).

All v_2 measurements in both systems are consistent from $v_2\{4\}$ to $v_2\{8\}$. $v_2\{2\}$ is generally larger because non-flow effects cannot be fully suppressed when only two particles are correlated. The measurement gives rise to an interpretation based on hydrodynamics, even for very small collision systems that lead to a sufficiently large charged particle multiplicity. At this point it seems very difficult to reconcile these results with just fundamental gluon exchange processes. If small systems indeed show a high level of collectivity then it might be possible to determine potentially a sub-structure in the proton from the flow patterns since they will be, as in the large systems, dependent on the initial distribution of the energy density in the colliding system. The question on how such a small system, or even a large system, can thermalize in such a short time is still not fully answered. Several ideas have been proposed, but need to be verified with more experimental data [26,27].

3. Beam Energy Scan Results from RHIC

The discovery of a deconfined phase and the realization that at high collision energies the QCD transition is an analytic crossover makes it very interesting to study the phase diagram as a function of collision energy, i.e. as a function of finite density (μ_B). One needs to remember, though, that all previous fixed target experiments at the BEVALAC, SIS, AGS and SPS have yielded a significant wealth of data on temperature and density dependence of the measured particle properties. They are well described by statistical hadronization models (SHM), which assume emission from a chemically and kinetically equilibrated source. The models yield a chemical freeze-out curve, which is very close to QCD transition curve, at least for the small densities at which lattice can yield definitive results on the transition temperature based on Taylor expansion calculations. Fig.6 shows the long established chemical freeze-out curve based on measured particle yields and ratios [28].

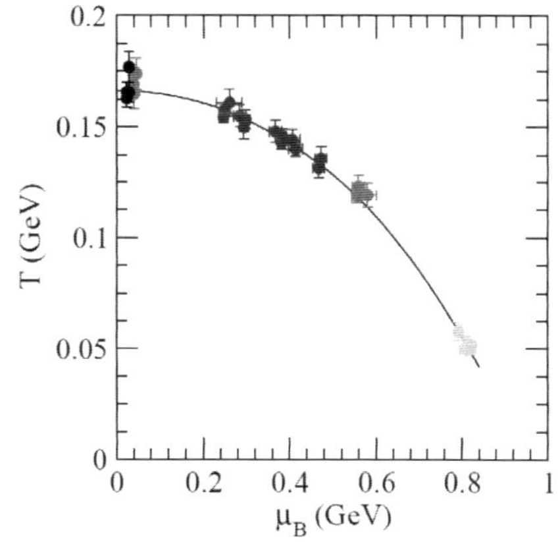

Figure 6. Chemical freeze-out curve in SHM based on data from RHIC (black and red), SPS (blue), AGS (red) and SIS (green) (from [28]).

The questions to answer with the BES are: a.) can we determine the collision energy, and thus the energy density, for which the transition is not achieved and the system stays in the hadronic phase. b.) can we determine the existence of a critical point on the nuclear phase diagram, and c.) can we show that the chemical freeze-out temperature coincides with the hadronization temperature and is common for all hadrons ?

3.1. Flow measurements as a function of energy

The discrepancy between realistic hydrodynamical calculations and the data from the AGS and SPS is contrasted by the near perfect agreement between the same models and the highest energy RHIC and

LHC data. This difference gives rise to the assumption that the onset of small viscosity over entropy ratios will coincide with the transition to the deconfined phase. STAR has measured the elliptic flow as a function of transverse momentum at five different energies ranging from $\sqrt{s_{NN}}$ = 7.7. GeV to 200 GeV, as shown in Fig.7 [29]. The results show that a decrease in the v_2 strength is measured below $\sqrt{s_{NN}}$ = 39 GeV and the flow is weakest at the lowest energy. This might indicate that the minimum collision energy required to deconfine the matter is in the $\sqrt{s_{NN}}$ =10-40 GeV range.

Another interesting flow measurement, which was suggested early on as a possible signature for the transition energy density, is directed flow (v_1). In the early models the directed flow changes from a 'bounce off' type of interaction to a 'squeeze-out' interaction [30, 31], which means v_1 would change sign, reach a minimum and then continue to grow again as a function of increasing collision energy. Such behaviour would signal a softest point in the equation of state, i.e. a system of maximum compressibility. The measurement by STAR, conducted for the net-protons emitted from the interaction, is shown in Fig.8 [32] and shows a remarkable likeness to the predicted effect. In theory the softest point coincides with the energy necessary to convert from a dense hadronic system to a deconfined partonic system. Based on this measurement this would occur at collision energies between $\sqrt{s_{NN}}$ = 10-20 GeV.

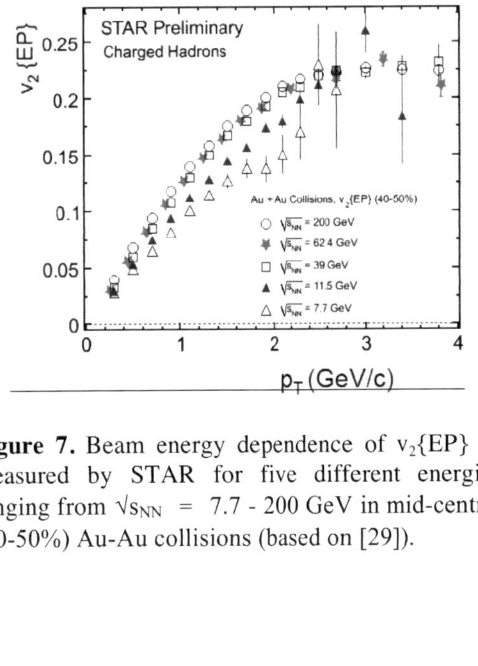

Figure 7. Beam energy dependence of $v_2\{EP\}$ as measured by STAR for five different energies ranging from $\sqrt{s_{NN}}$ = 7.7 - 200 GeV in mid-central (40-50%) Au-Au collisions (based on [29]).

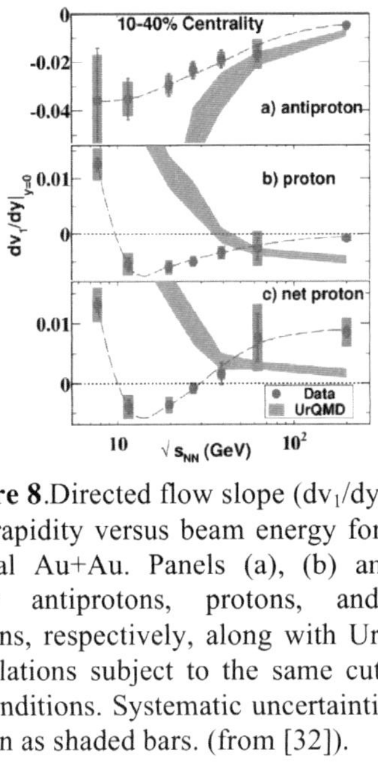

Figure 8.Directed flow slope (dv_1/dy) near mid-rapidity versus beam energy for mid-central Au+Au. Panels (a), (b) and (c) show antiprotons, protons, and net protons, respectively, along with UrQMD calculations subject to the same cuts and fit conditions. Systematic uncertainties are shown as shaded bars. (from [32]).

The two aforementioned measurements are only examples of a whole series of studies initiated to determine the transition conditions. In the final analysis one needs to assume that all measurements, which can be related to the QCD transition show a consistent collision energy dependence. The early results are intriguing but they will require additional consistency test of dynamic variables in the probed energy regime.

3.2. Fluctuation measurements as a function of energy

Dynamic fluctuation measurements have a long history of controversy in the field of relativistic heavy ion measurements, in part because they are very sensitive to critical effects near the phase transition and to quantum number dependent patterns, which might allow us to study various aspects of the transition in a first principle approach. On the flip side, fluctuations can often be artificially generated

through experimental limitations, such as geometrical acceptance and reconstruction efficiencies, and they are prone to background from trivial sources, such as detector response effects [33].
The main goal of the BES study of fluctuations is to find evidence for a critical point in the QCD phase diagram. Although the lattice QCD results on this subject are at best inconclusive, many effective models predict the existence at finite chemical potential. Semi-quantitative predictions of actual measurable quantities are largely based on non-linear sigma model simulations [34, 35]. The exact correspondence of susceptibilities on the lattice, which are the derivatives of the pressure with respect to the chemical potential of a specific quantum number, to higher moments of the measured particle multiplicity, allows, for the first time, to relate a measured quantity to a calculated quantity on the lattice as long as the measurement reflects uniquely the behaviour of a conserved quantum number [11]. For example it can be shown that net-protons are a good proxy for the net-baryon number [36] and net-charges obviously reflect the behaviour of the electric charge. Fluctuations in these moments of multiplicities are predicted near a critical point in the sigma model. Generally, the higher the moment the more pronounced the non-Poissonian fluctuation. The order of the moment also determines the required statistics for a significant measurement, though. At this time, STAR was able to determine the first four moments (mean (c_1), variance (c_2), skewness (c_3) and kurtosis (c_4)) of the net-proton and the net-charge particle distributions [37, 38].
In particular the shape and the magnitude of the kurtosis over variance ratio was proposed as a sensitive signature for critical behaviour. A moment ratio has the advantage that volume effects that are significant in a limited acceptance detector cancel out. The sigma model has predicted two unique effects near the critical point: a.) the ratio should go negative (compared to a Poissonian baseline) and b.) the magnitude should rise significantly above the Poissonian baseline at the critical point. The figures above show the collision energy dependence of moment ratios for net-protons, (Fig.9) and net-charges (Fig.10).

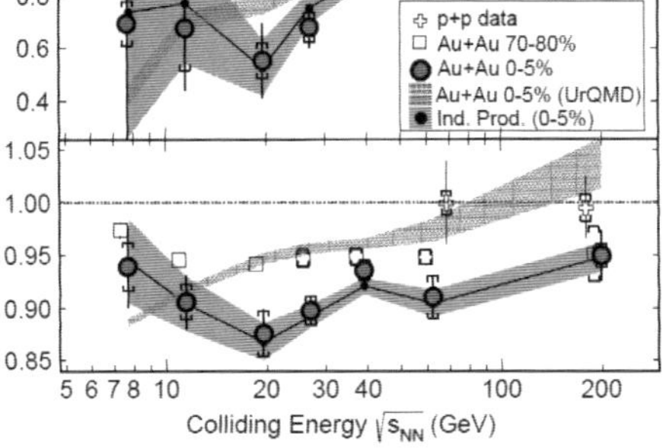

Figure 9. Beam-energy dependence of net-proton $S\sigma$ and $\kappa\sigma^2$, after all corrections, for most central (0-5%) and peripheral (70-80%) bins. The error bars are statistical and the caps represent systematic errors. Results from URQMD and independent production are superimposed (from [37]).

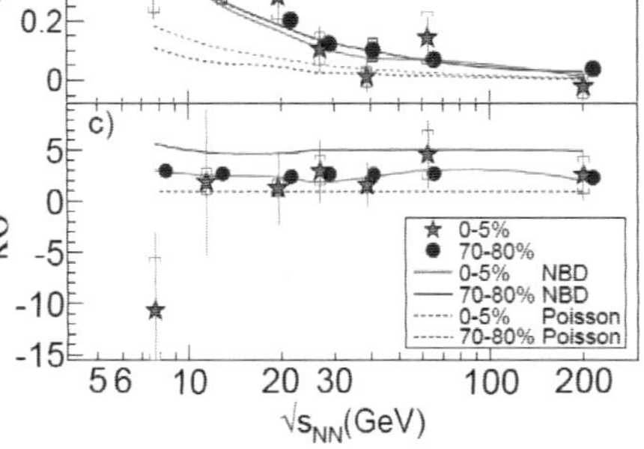

Figure 10. Beam-energy dependence of net-charge σ^2/M, $S\sigma$, and $\kappa\sigma^2$, after all corrections, for most central (0-5%) and peripheral (70-80%) bins. The error bars are statistical and the caps represent systematic errors. Results from the Poisson and the NBD baselines are superimposed (from [38]).

Journal of Physics: Conference Series **630** (2015) 012053 doi:10.1088/1742-6596/630/1/012053

A slight dip in the c_4/c_2 ($\kappa\sigma^2$) ratio is observed in the net-proton but no significant rise can yet be established. The net charges show no effect which can be understood since the charges are a combination of pions, kaons and protons, pre-dominantly, and the sigma model predicts that the strength of the critical behaviour is hadron mass dependent [39]. More measurements in particular below $\sqrt{s_{NN}}$ = 20 GeV are planned for phase II of the RHIC-BES which is scheduled for 2017/2018.

Quantum number fluctuations can also be used for a slightly more mundane, but nevertheless very important measurement, regarding the dynamic evolution of the system through the hadronization process. As was stated earlier the particle yields determined in statistical hadronization models with a single freeze-out temperature describe the measured yield extremely well over many orders of magnitude. The model is a static model, though, which can only determine the yields assuming a fixed temperature. If confirmed by more sensitive measurements, namely fluctuations, one would have to conclude that the hadron yields and fluctuations tell us very little about the hadronization mechanism and the relevant degrees of freedom above the transition temperature. If, though, specific patterns arise, depending on which quantum number is studied, then we might be able to determine a quantum number specific evolution through the phase transition. This topic has garnered considerable interest recently when it was shown that at least for some fluctuation measures on the lattice one can determine a flavour specific freeze-out temperature based on differences in susceptibility ratio between light and strange quarks [40]. It was shown that the continuum extrapolated c_4/c_2 ratio exhibits a 20 MeV temperature difference between the two quark flavors. This difference in the chemical freeze-out is not necessarily expected since the statistical fits to the particle yields describe all species reasonably well with a single temperature. A closer look, though, shows that at the highest RHIC and LHC energies the proton yields are consistently over estimated when assuming a common temperature of 164 MeV. The best fits to the proton/pion ratios yield a temperature of 148 MeV. A compromise of 152 MeV slightly overestimates the protons and underestimates the strange baryons by 2-3σ. Studies of the net-proton and net-charge fluctuations, independently performed in the framework of lattice QCD [41] and a hadron resonance gas (HRG) approach [42], shows that for fluctuations, which are dominated by light flavour particles, the preferred freeze-out temperature is less than 150 MeV.

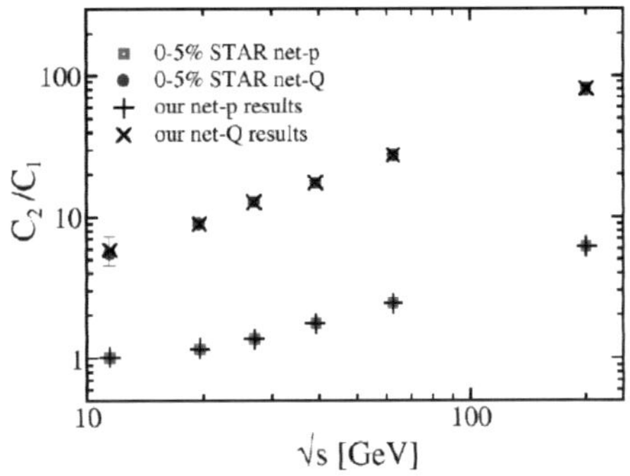

Figure 11. Comparison between HRG model results and experimental data for the most central collisions (0-5%) (from Refs. [37, 38]) for σ^2/M of net-electric charge (blue symbols) and net protons (red symbols). The experimental data have been used in the HRG model in order to extract T and μ_B for each collision energy (from [42]).

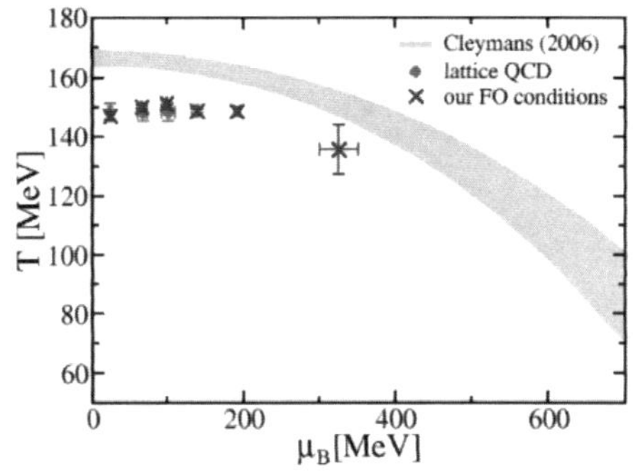

Figure 12. Freeze-out parameters in the T-μ_B plane: comparison between the curve obtained in Ref. [28] (grey band) and the values obtained from the combined analysis of σ^2/M for net-electric charge and net protons in the HRG model (blue symbols) and in lattice QCD (Ref.[41]) (from [42]).

Fig.11 shows the HRG fit to the experimental c_2/c_1 data, Fig.12 shows a comparison of the obtained freeze-out temperature from lattice and HRG with the freeze-out properties from fits to the yields.

If it can be established that strange particle fluctuations require a higher temperature then one could deduce that the chemical freeze-out is flavour dependent. In other words during the dynamic expansion of the system in the crossover region the system is likely emitting strange particles at a higher temperature than light quark particles. There will be a finite time window during which the system will predominantly produce strange matter, which could lead to higher mass strange baryon configurations as postulated by the quark model [43] or even exotic strange quark matter configurations as suggested for the core of neutron stars [44].

4. Summary

The study of the deconfined matter produced at RHIC and LHC energies in heavy ion collisions has reached a level of sophistication that allows us to determine the properties of the high temperature phase just above the phase transition. It turns out that this phase is very strongly interacting with ideal liquid like features. The viscosity over entropy ratio is near the quantum limit at RHIC and only rises slightly when reaching LHC energies. The system is still well below the pQCD limit and therefore cannot be described with weakly interacting quarks and gluons. Whether quasi-particles or color-neutral bound states dominate the system between 1-2 T_c remains to be seen, but there are first indications that the de-excitation to hadronic matter might be flavour dependent, which opens the possibility of a phase of mixed degrees of freedom which follows a flavour specific pattern during the hadronization process. Exotic states could be formed during this time in the cosmological evolution and might be found in relativistic heavy ion collisions as well as the interior of stars.

The search for a critical point is still ongoing, but it seems likely that the minimum required energy density to achieve a deconfined state is somewhere in the $\sqrt{s_{NN}}$ = 10-30 GeV range. It might be that a critical point, which should exist at finite chemical potential, is located at such a high quark density and temperature that it cannot be reached with accelerator based experiments.

Finally, the phenomena that enable us to quantify the collective behaviour of the deconfined parton phase also seem to be achievable in the collision of very small systems at energies that produce high particle multiplicities in the final state. If true, then the paradigm of proton-proton collision reference data needs to be revised and distinctly non-perturbative effects in the system, such as hadronization and equilibration, could be an explanation for effects in elementary particle collisions within the same framework that is presently used to describe large multi-particle initial state systems such as heavy ion collisions. This would also allow us to study the Quark Gluon Plasma in collisions that are generally reserved for the study of Higgs Bosons or exotic phenomena beyond the Standard Model.

5. References

[1] RHIC White Papers 2005 *Nucl. Phys* **A757** 1
[2] e.g. Song H and Heinz UW 2007 *Phys.Rev.* **C77** 064901
[3] Kovtun PK et al. 2005 *Phys.Rev.Lett.* **94** 111601
[4] Borsanyi S et al. (WB collaboration) 2010 *JHEP* **1011** 077
[5] Bazavov A et al. (Hot QCD collaboration) 2014 *Phys.Rev.* **D90** 094503
[6] Aoki Y et al. (WB collaboration) 2006 *Nature* **443** 675
[7] e.g. Fischer C et al. 2014 (*Preprint* arXiv:1410.0124)
[8] e.g. Ratti C et al. 2006 (*Preprint* nucl-th/0604025)
[9] Endrodi G et al. (WB Collaboration) 2011 *JHEP* **1104** 001
[10] De Forcrand P and Philipsen O 2007 *JHEP* **0701** 077
[11] Karsch F 2012 *Central Eur.J.Phys.* **10** 1234

[12] Xu Z and Greiner C 2007 *Phys.Rev.* **C76** 024911
[13] Cassing W and Bratkovskaya EL 2009 *Nucl. Phys.* **A831** 215
[14] Gyulassy M et al. 2014 *Phys.Rev.* **D90** 054025
[15] Sorge H 1999 *Phys.Rev.Lett.* **82** 2048
[16] Poskanzer AM and Voloshin S 1998 *Phys.Rev.* **C58** 1671
[17] Alver B and Roland G 2010 *Phys.Rev.* **C81** 054905
[18] Adare A et al. (PHENIX collaboration) 2011 *Phys.Rev.Lett.* **107** 252301
[19] Gale C et al. 2013 *Phys.Rev.Lett.* **110** 012302
[20] Aad G et al. (ATLAS Collaboration) *Preprint* arXiv:1408.4342)
[21] Aamodt K et al. (ALICE collaboration) 2013 *Phys.Lett.* **B719** 18
[22] CMS collaboration 2013 *Phys.Rev.* **C87** 014902
[23] Aad G et al. (ATLAS Collaboration) 2012 *Phys.Rev.* **C80** 014907
[24] Aamodt K et al. (ALICE collaboration) 2011 *Phys. Rev.Lett* **107** 032301
[25] CMS collaboration 2014, CMS PAS HIN-14-006
[26] Castorina P et al. 2007 *Eur.Phys.J.* **C52** 187
[27] Tomboulis ET and Velytsky A 2005 **Phys. Rev. D72** 074509
[28] Cleymans J et al. 2006 *Phys.Rev.* **C73** 034905
[29] Adamczyk L et al. (STAR collaboration) 2013 *Phys.Rev.Lett.* **110** 142301
[30] Rischke DH et al. 1995 *Heavy Ion Phys.* **1** 309
[31] Stoecker H 2005 Nucl.Phys. **A750** 121
[32] Adamczyk L et al. (STAR collaboration) 2014 *Phys.Rev.Lett.* **112** 162301
[33] Koch V. (*Preprint* arXiv:0810.2520)
[34] Stephanov MA et al. 1998 *Phys.Rev.Lett.* **81** 4816
[35] Stephanov MA 2011 *Phys.Rev.Lett.* **107** 052301
[36] Kitazawa M and Asakawa M 2012 *Phys.Rev.* **C86** 024904
[37] Adamczyk L et al. (STAR collaboration) 2014 *Phys.Rev.Lett.* **112** 032302
[38] Adamczyk L et al. (STAR collaboration) 2014 *Phys.Rev.Lett.* **113** 092301
[39] Athanasiou C et al. 2010 *Phys.Rev.* **D82** 074008
[40] Bellwied R et al. (WB collaboration) 2013 *Phys.Rev.Lett.* **111** 202302
[41] Borsanyi S et al. (WB collaboration) 2014 *Phys.Rev.Lett.* **113** 052301
[42] Alba P et al. 2014 *Phys.Lett.* **B738** 305
[43] Bazavov A et al. (HotQCD collaboration) 2014 *Phys.Rev.Lett.* **113** 072001
[44] e.g. Ellis JR et al. 1991 *Nucl.Phys.* **B348** 345

Journal of Physics: Conference Series **630** (2015) 012054 doi:10.1088/1742-6596/630/1/012054

Studying meson baryon systems with strangeness +1

[1,2†]K P Khemchandani, [1]A Martínez Torres, [1]F S Navarra, [1]M Nielsen and [3,4]L Tolos

[1]Instituto de Física, Universidade de São Paulo, C.P 66318, 05314-970 São Paulo, SP, Brazil.
[2]Faculdade de Tecnologia, Universidade do Estado do Rio de Janeiro, Rod. Presidente Dutra Km 298, Pólo Industrial, 27537-000 , Resende, RJ, Brazil.
[3]Instituto de Ciencias del Espacio (IEEC/CSIC), Campus Universitat Autònoma de Barcelona, Carrer de Can Magrans, s/n 08193 Cerdanyola del Vallès, Spain.
[4]Frankfurt Institute for Advanced Studies (FIAS), 60438 Frankfurt am Main, Germany.

E-mail: †kanchan@if.usp.br

Abstract. We discuss the results of our recent study of the s-wave KN-K^*N coupled interaction. The K^*N amplitude is obtained by calculating t- and u- channel diagrams and a contact interaction. For the KN amplitude we calculate the Weinberg-Tomozawa term obtained from the lowest order chiral Lagrangian. The $KN \leftrightarrow K^*N$ amplitudes are calculated by replacing the photon by a vector meson in the standard Kroll-Ruderman term. The subtraction constants required to calculate the loops are fixed by demanding the KN amplitudes to fit the data available on the isospin 0 and 1 s-wave phase shifts. We find that the coupling between the two channels plays an important in the isospin 0 configuration. We obtain updated amplitudes and cross sections for the KN and K^*N systems, which can be used to understand some recent findings of K, K^* production in p-p and p-A collisions studied by the Hades collaboration. We also look for resonances in these systems but find none.

1. Introduction

We study the s-wave interaction of K and K^* with nucleons with a twofold motivation: one is to obtain reliable amplitudes and cross sections for KN, K^*N systems, which can be useful for studies related to strange mesons in nuclear medium, and another is to look for possible resonance formation in these systems. Interaction of strangeness +1 mesons with nucleons is known to be relatively weak since the absence of a light (u, d) antiquark in the system does not allow formation of a three quark intermediate state, unlike systems like πN, $\bar{K}N$, $\bar{K}^*N$, etc. Still, the KN amplitudes and phase shifts show some structure in some energy range giving rise to discussions on the possibility of the existence of resonances with positive strangeness since more than four decades (see Refs. [1] for some of the oldest articles on this subject). More recently, an exploration of a narrow pentaquark baryon, $\Theta^+(1540)$, at several experimental facilities was triggered by a signal reported in Ref. [2] but the results seem to be negative (for a review on this, see Ref. [3]). Moreover, an alternative explanation has been brought forward in Refs. [4, 5] for the enhancement of the cross section seen in Ref. [2]. These findings, however, do not imply that no pentaquarks exist. It might be that one has to look for a different mass/width and/or quantum numbers. In fact, evidences for existence of other kind of exotic hadrons are being reported recently, like baryons with an anticharm quark [6, 7, 8, 9, 10], mesons with a tetra quark nature [11, 12, 13, 14], etc. It is, then, also reasonable to contemplate the existence of baryons with strangeness +1.

Besides, although finding a baryon with an antistrange (or anticharm) quark would unquestionably confirm existence of a pentaquark structure, it is important to note that some known nonexotic baryons can be essentially understood as pentaquarks. For example, we need a pentaquark structure to understand why the lowest mass nucleon resonance with spin-parity $1/2^-$, $N^*(1535)$, is lighter than the corresponding hyperon, $\Lambda(1405)$. An extra $\bar{s}s$ component in the wave function of $N^*(1535)$ and $\bar{q}q$ for $\Lambda(1405)$ plays a crucial role in explaining the properties of these states [15, 16, 17, 18, 19]. Analogously, it is possible that a state with strangeness +1 may also exist.

Another motivation for our present study is to obtain reliable amplitudes and cross sections for the KN and K^*N systems. The need for such an information arises since some recent findings of experimental investigations of K, K^* in nuclear medium do not seem to be explained within transport approach or statistical models. For example, a deep sub-threshold K^{*0} production has been reported in Ar+KCl collisions by HADES, with the experimental K^{*0} yield and K^{*0}/K^0 larger by about a factor five and two, respectively, as compared to the one obtained from the UrQMD transport approach [20]. The need for reliable information on the in-medium kaon potential has also been discussed in Ref. [21] where the K^0 production in p+Nb reactions at a beam kinetic energy of 3.5 GeV is analyzed by HADES. Also, the attenuation of the K^* and $\bar{K}^*$ states in the hadronic phase of the expanding fireball, as determined by the observation of a strong suppression of the total yield ratios $< K^* > / < K^+ >$ and $< \bar{K}^* > / < K^- >$ in central Pb+Pb collisions compared to p+p, C+C and Si+Si by the NA49 Collaboration [22], was not reproduced using UrQMD [23, 24] or statistical HQGM [25] models. Such findings thus indicate that it is important to look into KN and K^*N interactions in vacuum and in matter. In the present work, we focus on the s-wave interaction of K and K^* with nucleons.

We study the KN and K^*N systems by using a formalism based on the chiral Lagrangian and the theory of hidden local symmetry (HLS) of Ref. [26], which treats vector mesons as the gauge bosons. As discussed in the study of vector meson-baryon (VB) systems of Ref. [27], the gauge invariance of the HLS Lagrangian requires the consideration of a contact term together with s-, t- and u-channel interactions, which all turn out to give important contributions. We, thus, solve the Bethe-Salpeter equation using the sum of all these interactions as the VB kernel. We obtain KN interaction from the lowest order chiral Lagrangian. We couple KN and K^*N systems by extending the Kroll-Ruderman theorem for the pion photoproduction by replacing the photon by a vector meson, consistently with the phenomena of the vector meson dominance. This formalism has been earlier applied to strange and nonstrange meson-baryon systems [28, 29] to study dynamical generation of resonances.

In the present work we extend the model of Refs. [27, 28, 29, 30], where exchange of octet baryons was considered in the s- and u-channel, by including the exchange of hyperon resonances in the u-channel too. The couplings of the resonances to meson and octet-baryon channels required in such calculations are taken from Ref. [28].

In the following sections we first present the formalism briefly and then discuss the results obtained in the present work.

2. Formalism

Let us start the discussion by looking at the amplitude for KN, which can be obtained from the standard chiral Lagrangian [17]

$$V^I_{KN} = -\frac{C^I_{KN}}{4f_K^2}(\omega + \omega'), \tag{1}$$

where the superscript label I indicates the isospin of the meson-baryon system. A system of a kaon and a nucleon can have total isospin $I = 0$ or 1. The term C^I_{KN}, in the equation above, denotes an isospin coefficient whose value is 0 (-2) for isospin 0 (1), when using average masses for the kaons (K^0, K^+) and the nucleons (n, p). Further, $f_K = 113.46$ MeV is the kaon decay constant, and ω (ω') corresponds to the energy of the kaon in the initial (final) state. Thus, the KN amplitude is repulsive in the isospin 1 configuration and null in isospin 0. It is clear that these amplitudes cannot result in formation of a resonance. On the other hand, however, these amplitudes may be lacking some information since it is not possible to explain the non-zero KN phase shifts and scattering lengths in isospin 0 and compare them with those available from the partial wave analyses of the relevant experimental data [31, 32, 33, 34]. In the present work, we explore the possibility of improving the situation by coupling KN and K^*N channel.

For the K^*N interaction we use the general VB Lagrangian of Ref. [27]

$$\begin{aligned}\mathcal{L}_{VB} &= -g\Big\{\langle\bar{B}\gamma_\mu\left[V_8^\mu, B\right]\rangle + \langle\bar{B}\gamma_\mu B\rangle\langle V_8^\mu\rangle + \frac{1}{4M}\Big(F\langle\bar{B}\sigma_{\mu\nu}\left[V_8^{\mu\nu}, B\right]\rangle + D\langle\bar{B}\sigma_{\mu\nu}\left\{V_8^{\mu\nu}, B\right\}\rangle\Big)\Big\}\\ &\quad -g\Big\{\langle\bar{B}\gamma_\mu B\rangle\langle V_0^\mu\rangle + \frac{C_0}{4M}\Big(\langle\bar{B}\sigma_{\mu\nu}V_0^{\mu\nu}B\rangle\Big)\Big\},\end{aligned} \tag{2}$$

where the expressions in the parentheses (...) are related to the anomalous magnetic moment of the baryons. V in Eq. (2) denotes the SU(3) matrix for the vector mesons and $V^{\mu\nu}$ represents the tensor field of the vector mesons,

$$V^{\mu\nu} = \partial^\mu V^\nu - \partial^\nu V^\mu + ig\left[V^\mu, V^\nu\right]. \tag{3}$$

The subscripts 8 and 0 stand for the octet and the singlet part of the ω and ϕ wave functions (written within the the ideal mixing assumption). The coupling g in Eq. (2) is given by the Kawarabayashi-Suzuki-Riazuddin-Fayyazuddin relation (KSRF) [35, 36]

$$g = \frac{m_{K^*}}{\sqrt{2}f_{K^*}}. \tag{4}$$

The constants $D = 2.4$, $F = 0.82$ and $C_0 = 3F - D$ are chosen to ensure that the couplings for the ϕNN, ωNN and ρNN vertices are compatible with the known values. The values of D and F used here were also found to well reproduce the magnetic moments of the baryons in Ref. [37]. It should be mentioned here that Eq. (2) is in good agreement with the VB Lagrangians obtained within other approaches [38, 39].

With the Lagrangian given by Eq. (2) we can calculate Yukawa-type vertices which can be used to calculate t- and u-channel diagrams. Amplitudes corresponding to s-channel are trivially null in the present case since it requires exchange of a strangeness positive baryon. In addition to these contributions, one can obtain a contact term using the commutator part of $V_{\mu\nu}$ (Eq. (3)) in the Lagrangian of Eq. (2). This term turns out to be large in comparison with the other diagrams. Further motivation to consider this contact term comes from the fact that the commutator part of the tensor $V_{\mu\nu}$ is crucial to guarantee the gauge invariance of the VB Lagrangian (Eq. (2)) under the HLS transformations.

The t-channel amplitude obtained from Eq. (2) for K^*N is analogous to Eq. (1), (as also obtained in Ref. [40])

$$V^I_{t,K^*N} = -\frac{C^I_{t,K^*N}}{4f^2_{K^*}}(\omega + \omega')\vec{\epsilon}_1\cdot\vec{\epsilon}_2, \tag{5}$$

where $\omega(\omega')$ and $\epsilon_1(\epsilon_2)$ represent the energy and the polarization vector of the K^*-meson in the initial (final) state, respectively. The values of C^I_{t,K^*N} are 0 and -2 for isospin 0 and 1, respectively, and $f_{K^*} = 171.12$ MeV is the decay constant for K^* [41, 42]. Similar to the case of KN, the isospin 1 K^*N amplitude is also repulsive while isospin 0 amplitude is null. Since

K^* has spin 1 and the spin of the nucleon N is 1/2, we can have total spin $S = 1/2$ and 3/2 for the system in s-wave. Notice that Eq. (5) is spin independent, giving identical amplitudes for spin 1/2 and 3/2.

Next, the amplitude for exchange of an octet baryon in the u-channel and the contact term are obtained as

$$V^I_{u,K^*N} = C^I_{u,K^*N}\left(\frac{g^2}{2\bar{M}-m}\right)\vec{\epsilon}_1\cdot\vec{\sigma}\,\vec{\epsilon}_2\cdot\vec{\sigma}, \tag{6}$$

$$V^I_{\rm CT,K^*N} = iC^I_{\rm CT,K^*N}\frac{g^2}{2M}\vec{\sigma}\cdot\vec{\epsilon}_2\times\vec{\epsilon}_1, \tag{7}$$

where m (M) is the mass of the K^* (nucleon) and $\bar{M}$ represents an average mass for the baryons involved in the process. The isospin coefficients C^I_{u,K^*N} and $C^I_{\rm CT,K^*N}$ are given in Table 1. It can be seen that the spin structure of these amplitudes leads to different amplitudes for total spin 1/2 and 3/2. We obtain $V^I_{\rm CT,K^*N}$ and V^I_{u,K^*N} to be attractive in nature for total spin 3/2 in isospin 0, and for spin 1/2 in isospin 1. They are repulsive in the remaining two configurations.

Table 1. Isospin coefficients for the u-channel amplitude and contact term given by Eqs. (6) and (7).

	$I=0$	$I=1$
C^I_u	$\frac{Dm[(D-3F)m-6\bar{M}]}{6\bar{M}^2}$	$\frac{12\bar{M}^2+12Fm\bar{M}+(D^2+3F^2)m^2}{12\bar{M}^2}$
$C^I_{\rm CT}$	D	$-F$

Further, we include exchange of those resonances in the u-channel for which the couplings to meson-baryon channels have been obtained in Ref. [28] using the same formalism as the present one. To begin with, we do so to restrict the number of unknown parameters to be minimum. But later we find this contribution to be negligibly small (see Ref. [43] for more details) and, thus, we do not consider exchange of more resonances.

To determine the contribution from the exchange of resonances, we write the following phenomenological effective field Lagrangians [44, 45]

$$\mathcal{L}_{NKH^*} = ig_{NKH^*}\bar{H}^*K^\dagger N + \text{h.c},$$
$$\mathcal{L}_{NK^*H^*} = g_{NK^*H^*}\bar{H}^*\gamma^\mu\gamma_5 K^{*\dagger}_\mu N + \text{h.c}, \tag{8}$$

where H^* stands for Λ^* or Σ^* and $K^\dagger(K^{*\dagger})$ creates a K (K^*) meson. The couplings $g_{NK\Lambda^*}$ and $g_{NK^*\Lambda^*}$ have been obtained (in Ref. [28]) by studying VB interaction using the Lagrangian given in Eq. (2). Thus, although we write phenomenological Lagrangians in Eq. (8), the contributions from both vector and tensor VB interaction are embedded in $g_{NK\Lambda^*}$ and $g_{NK^*\Lambda^*}$.

Keeping in mind the extended structure of the hadrons and the fact that only the negative energy solution of the Dirac equation contributes to the u-channel diagrams for the s-wave K^*N interaction, the following form factor is multiplied to all the u-channel amplitudes,

$$F(\Lambda,u) = \frac{\Lambda^4}{\Lambda^4+(u-M_u^2)^2}, \tag{9}$$

where u is the usual Mandelstam variable, M_u is the mass of the baryon exchanged and Λ is a cut-off which we vary in the range 650-1000 MeV.

We sum the K^*N amplitudes obtained from the contact term, t-channel and u-channel diagrams and use it as the kernel when solving scattering equations.

Finally, we obtain the $KN \leftrightarrow K^*N$ amplitudes using the Lagrangian from Ref. [30]

$$\mathcal{L}_{\rm PBVB} = \frac{-ig_{KR}}{2f_\pi}\left(F'\langle\bar{B}\gamma_\mu\gamma_5\left[\left[P,V^\mu\right],B\right]\rangle + D'\langle\bar{B}\gamma_\mu\gamma_5\left\{\left[P,V^\mu\right],B\right\}\rangle\right), \tag{10}$$

where the subscript PBVB stands for the pseudoscalar meson-baryon–vector meson-baryon Lagrangian and $F' = 0.46$, $D' = 0.8$ ensure that $F' + D' \simeq g_A = 1.26$ [30]. Using Eq. (10) we obtain

$$V^I_{KNK^*N} = i\sqrt{3}\frac{g_{\rm KR}}{2\sqrt{f_K f_{K^*}}}C^I_{KNK^*N}. \tag{11}$$

where $g_{\rm KR}$ is obtained using the KSRF relation [30]

$$g_{\rm KR} = \frac{m_{K^*}}{\sqrt{2 f_K f_{K^*}}} \sim 4.53. \tag{12}$$

The coefficient $C^I_{KNK^*N}$ is $-2D'$ for isospin 0 and $2F'$ for isospin 1.

3. Results

In the previous section we discussed the s-wave KN and K^*N interactions with which the Bethe-Salpeter equation can be solved in its on-shell factorization form [17, 46]

$$T = (1 - VG)^{-1}V, \tag{13}$$

where V is the kernel and G is the loop function which is divergent in nature. We calculate it using the dimensional regularization and fix the require subtraction constants by making a χ^2-fit to the latest data [31] on the s-wave isospin 0 and 1 KN phase shifts. The results for the K^*N system are predictions.

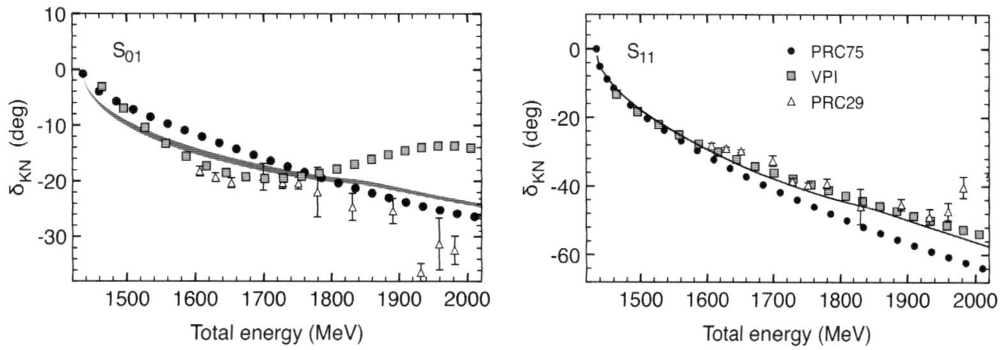

Figure 1. Scattering phase shifts for the KN system obtained by using the subtraction constants: $b^{I=0}_{KN} = -6.82$, $b^{I=0}_{K^*N} = 1.84$, $b^{I=1}_{KN} = -1.59$ and $b^{I=1}_{K^*N} = -1$ ($\mu = 630$ MeV)) and for Λ in Eq. (9) varying between 650-1000 MeV. The data from the partial wave analysis, represented by filled circles, boxes and empty triangles are taken from Refs. [31, 32, 33], respectively.

We obtain the best fit with the values of the subtraction constants: $b^{I=0}_{KN} = -6.82$, $b^{I=0}_{K^*N} = 1.84$, $b^{I=1}_{KN} = -1.59$ and $b^{I=1}_{K^*N} = -1$ for the regularization scale μ fixed to 630 MeV. We show the KN phase shifts obtained with these constants in Fig 1 for the value of the Λ, in Eq. (9), varied between 650 MeV -1000 MeV. As can be seen from these figures, there is a good agreement between our results and the partial wave analysis data of Refs. [31, 32, 33] for both isospins. However, the results for isospin 0 seem to be slightly sensitive to the the value of the Λ used in Eq. (9) while the isospin 1 phase shifts seem to be stable. This apparent stability arises due to a weak $KN \leftrightarrow K^*N$ amplitude in the isospin 1 case.

We show our predictions for the K^*N s-wave scattering phase shifts in Fig. 2 for different spin-isospin configurations. It can be seen in these figures too that the K^*N phase shifts are

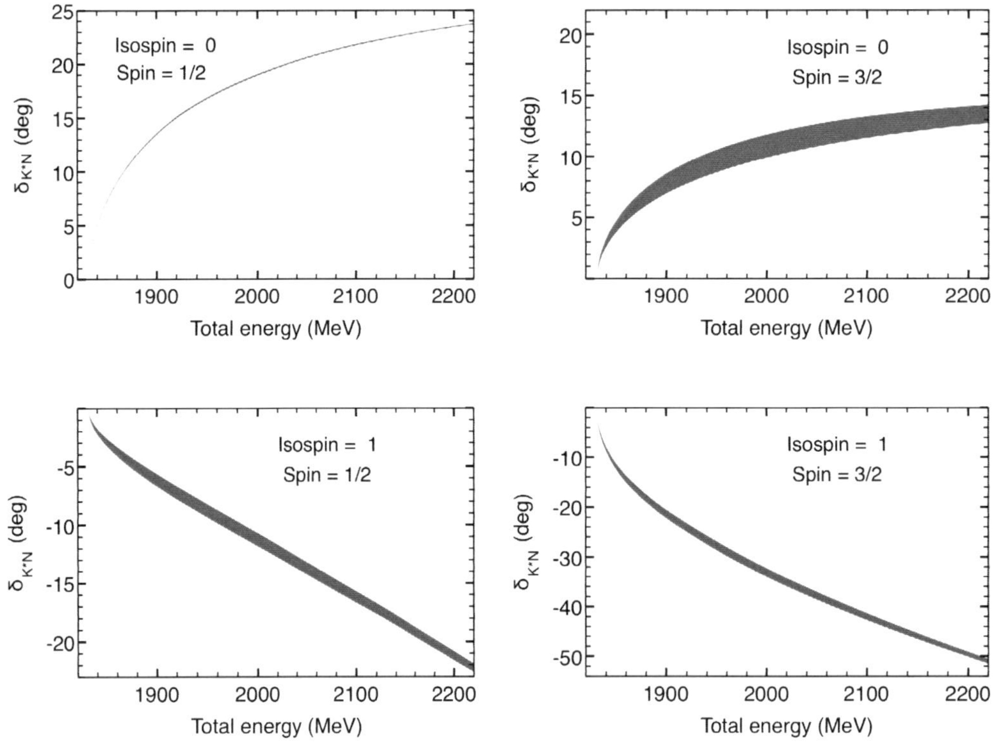

Figure 2. Scattering phase shifts for the K^*N interaction in s-wave.

not very sensitive to the changes in the cut-off Λ used in the form factor defined by Eq. (9). These results are of importance for understanding the data related to K and K^* production in $p+p$ and $p+A$ collisions, as reported by HADES [20, 21, 47], STAR [48] and NA49 [22] Collaborations.

Using the amplitudes corresponding to the phase shifts shown in Figs. 1, 2, we search for poles in the complex plane. By coupling the KN and K^*N channels involved, we have calculated the amplitudes for isospin 0 and 1 and spin-parities $1/2^-$ and $3/2^-$. We do not find any pole in none of these configurations. From our work, thus, it can be concluded that a strangeness +1 resonance is not generated from KN and K^*N coupled channel dynamics, contrary to the anticharm systems [9]. Also, our results differ from those of Ref. [49] where a narrow, isoscalar resonance with spin-parity $3/2^-$ was found within a different formalism to study KN-K^*N coupled systems.

Acknowledgements
The authors thank Profs. D. Jido, A. Hosaka and H. Nagahiro for useful discussions. The authors would also like to thank the Brazilian funding agencies FAPESP and CNPq for the financial support. L.T acknowledges support from the Ramón y Cajal Research Programme (Ministerio de Ciencia e Innovación) and from Grants No. FPA2010-16963 and No. FPA2013-43425-P (Ministerio de Ciencia e Innovación), No. FP7-PEOPLE-2011-CIG under Contract No. PCIG09-GA-2011-291679 and the European Community-Research Infrastructure Integrating Activity Study of Strongly Interacting Matter (acronym HadronPhysics3, Grant Agreement n. 283286) under the Seventh Framework Programme of EU.

References

[1] Aaron R, Silbar R R,and Amado R D 1971 *Phys. Rev. Lett.* **26** 407; Aaron R, Hogan W L, Srivastava Y N and Rich M 1973 *Phys. Rev.* D **7** 1401
[2] Nakano T *et al.* [LEPS Collaboration] 2003 *Phys. Rev. Lett.* **91** 012002
[3] Hicks K H 2012 *Eur. Phys. J.* H **37** 1
[4] Martinez Torres A and Oset E 2010 *Phys. Rev. Lett.* **105** 092001
[5] Martinez Torres A and Oset E 2010 *Phys. Rev.* C **81** 055202
[6] Aktas A *et al.* [H1 Collaboration] 2004 *Phys. Lett.* B **588** 17
[7] Kim H, Lee S H and Oh Y -s 2004 *Phys. Lett.* B **595** 293
[8] Lee S H , Kwon Y and Kwon Y 2006 *Phys. Rev. Lett.* **96** 102001
[9] Gamermann D, Garcia-Recio C, Nieves J, Salcedo L L and Tolos L 2010 *Phys. Rev.* D **81** 094016
[10] Albuquerque R M, Lee S H and Nielsen M 2013 *Phys. Rev.* D **88** 076001
[11] Nielsen M and Navarra F S 2014 *Mod. Phys. Lett.* A **29** 1430005
[12] Faccini R, Pilloni A and Polosa A D 2012 *Mod. Phys. Lett.* A **27** 1230025 [arXiv:1209.0107 [hep-ph]]
[13] Navarra F S, Nielsen M and Richard J M 2012 *J. Phys. Conf. Ser.* **348** 012007 [arXiv:1108.1230 [hep-ph]]
[14] Guo F K, Meiner U G and Wang W 2014 *Commun. Theor. Phys.* **61** 354 [arXiv:1308.0193 [hep-ph]]
[15] Yuan S G, Wei K W, He J, Xu H S and Zou B S 2012 *Eur. Phys. J.* A **48** 61
[16] Zou B S 2006 *Int. J. Mod. Phys.* A **21** 5552.
[17] Oset E and Ramos A 1998 *Nucl. Phys.* A **635** 99
[18] Jido D, Oller J A, Oset E, Ramos A and Meissner U G 2003 *Nucl. Phys.* A **725** 181
[19] Meissner U G and Oller J A 2000 *Nucl. Phys.* A **673** 311
[20] Agakishiev G *et al.* [HADES Collaboration] 2013 *Eur. Phys. J.* A **49** 34
[21] Agakishiev G *et al.* [HADES Collaboration] *Medium effects in proton-induced* K^0 *production at 3.5 GeV* arXiv:1404.7011 [nucl-ex]
[22] Anticic T *et al.* [NA49 Collaboration] 2011 *Phys. Rev.* **84** 064909
[23] Bleicher M, Zabrodin E, Spieles C, Bass S A, Ernst C, Soff S, Bravina L and Belkacem M *et al.* 1999 *J. Phys.* G **25** 1859
[24] Bass S A, Belkacem M, Bleicher M, Brandstetter M, Bravina L, Ernst C, Gerland L and Hofmann M *et al.* 1998 Prog. Part. Nucl. Phys. **41** 255
[25] Becattini F, Manninen J and Gazdzicki M 2006 *Phys. Rev.* C **73** 044905; Becattini F and Heinz U W 1997 *Z. Phys.* C **76** 269 [Erratum-ibid. C **76** (1997) 578]
[26] Bando M, Kugo T and Yamawaki K 1985 *Nucl. Phys.* B **259** 493
[27] Khemchandani K P, Kaneko H, Nagahiro H and Hosaka A 2011 *Phys. Rev.* D **83** 114041
[28] Khemchandani K P, Martinez Torres A, Nagahiro H and Hosaka A 2012 *Phys. Rev.* D **85** 114020
[29] Khemchandani K P, Martinez Torres A, Nagahiro H and Hosaka A 2013 *Phys. Rev.* D **88** 114016; Khemchandani K P, Martinez Torres A, Nagahiro H and Hosaka A 2014 *Int. J. Mod. Phys. Conf. Ser.* **26** 1460060
[30] Khemchandani K P, Martinez Torres A, Kaneko H, Nagahiro H and Hosaka A 2011 *Phys. Rev.* D **84** 094018
[31] Gibbs W R and Arceo R 2007 *Phys. Rev.* C **75** 035204
[32] Hylop J S, Arndt R A, Roper L D and Workman R L 1992 *Phys. Rev.* D **46** 961
[33] Hashimoto K 1984 *Phys. Rev.* C **29** 1377
[34] Martin B R 1978 *J. Phys.* G **4** 335
[35] Kawarabayashi K and Suzuki M 1966 *Phys. Rev. Lett.* **16** 255

[36] Riazuddin and Fayyazuddin 1966 *Phys. Rev.* **147** 1071
[37] Jido D, Hosaka A, Nacher J C, Oset E, Ramos A 2002 *Phys. Rev.* C **66**, 025203
[38] Jenkins E E, Luke M E, Manohar A V, Savage M J 1993 *Phys. Lett.* B **302** 482
[39] Meissner U G, Steininger S 1997 *Nucl. Phys.* B **499**, 349
[40] Oset E and Ramos A 2010 *Eur. Phys. J.* A **44** 445
[41] Maris P and Tandy P C 1999 *Phys. Rev.* C **60** 055214
[42] Maris P 2000 *Nucl. Phys.* A **663** 621
[43] Khemchandani K P, Martinez Torres A, Navarra F S,Nielsen M and Tolos L *Phys. Rev.* D **91**, 094008
[44] Nam S -i, Park J -H, Hosaka A and Kim H -C 2011 *J. Korean Phys. Soc.* **59** 2676
[45] Ozaki S, Nagahiro H and Hosaka A 2008 *Phys. Lett.* B **665** 178
[46] Oller J A, Oset E 1997 *Nucl. Phys.* A **620** 438 [Erratum-ibid. A **652** (1999) 407]
[47] Agakishie G *et al.* [HADES Collaboration] *Phys. Rev.* C **90**, 015202
[48] Adams J *et al.* [STAR Collaboration] 2005 *Phys. Rev.* C **71** 064902; Aggarwal M M *et al.* [STAR Collaboration] 2011 *Phys. Rev.* C **84** 034909
[49] Garcia-Recio C, Nieves J and Salcedo L L 2006 *Phys. Rev.* D **74** 034025; Garcia-Recio C, Nieves J and Salcedo L L 2007 *Eur. Phys. J.* A **31** 499

Testing Left-Right extensions of the standard model of electroweak interactions with double-beta decay and LHC measurements

O. Civitarese[a], J. Suhonen[b] and K. Zuber[c]

[a] Department of Physics, University of La Plata. 49 y 115.c.c.67 (1900) La Plata, Argentina
[b] Department of Physics. Unversity of Jyväskylä, P.O. Box 35 (YFL) FL-40014, Jyväskylä, Finland
[c] Department of Physics, TU Dresden, Zellescher 19, 01069 dresden, Germany

E-mail: osvaldo.civitarese@fisica.unlp.edu.ar

Abstract. The minimal extension of the standard model of electroweak interactions allows for massive neutrinos, a massive right-handed boson W_R, and a left-right mixing angle ζ. While an estimate of the light (electron) neutrino can be extracted from the non-observation of the neutrinoless double beta decay, the limits on the mixing angle and the mass of the right-handed (RH) boson may be extracted from a combined analysis of the double beta decay measurements (GERDA, EXO-200 and KamLAND-Zen collaborations) and ATLAS data on the two-jets two-leptons signals following the excitation of a virtual RH boson mediated by a heavy-mass neutrino. In this work we shall compare results of both types of experiments, and show that the estimates are not in tension.

1. Introduction

The theory of the electroweak interactions, in the form of the standard model (SM), assumes massless neutrinos and left-handed couplings [1]. In contrast to this, the observation of neutrino oscillations demonstrates that neutrinos are massive particles [2]. The symmetries of the SM can indeed be extended to accommodate, at least minimally, massive neutrinos and/or left-right (LR) couplings [3, 4, 5, 6]. The experimental search for signals of LR couplings is conducted at LHC [7], where the ATLAS [8] and CMS [9] collaborations have analysed the right-handed boson decay produced in the reactions p-p$\rightarrow$ $\mathrm{W_R}$ $\rightarrow$ two-jets two-leptons, mediated by a heavy-mass neutrino. From these experimental studies a lower mass limit of the order of 3 TeV, for the mass of the right-handed boson, was extracted provided the mass of the heavy neutrino is of the order (or larger than) 500 MeV. The result has significance for the construction of LR models, since it determines the range of allowed values for the mixing between left and right-handed currents [10], since the mass of the left-handed boson is already known. Neutrinoless double beta decay is a unique process which, if detected [11], will provide an answer about the nature of neutrinos (Dirac or Majorana) and their couplings [3]. Recently the GERDA [12], EXO-200 [14] and KamLAN-Zen [13] experiments have produced world-record lower limits of the half-life of the neutrinoless double beta decay of some of the double beta decay emitters (^{76}Ge and ^{136}Xe, respectively). From these values one can extract the corresponding limits for the neutrino mass, the right-handed and the LR couplings. To determine these values one needs not only the experimental information about the half-life, but also the involved nuclear matrix elements [15, 16]. The complexity of the nuclear structure calculations, due to the relatively large number of parameters entering the calculations, has been a challenging problem since the early calculations performed in the 1960's. After so many years a consensus has been reached about the order of magnitude and individual values of the participant nuclear matrix elements [16, 17], to the extent that we are in conditions to make definite predictions about the half-life of the transitions. In this work we shall discuss about the complementarity of high energy (p-p) and double bera decay data. An earlier version of the work has been published long ago [18], when the information about the high energy sector of the problem was rather limited.

2. Formalism

In this section we shall briefly present the essentials of the formalism which is currently applied to describe double beta decay transitions. It contains three main components, namely:

- Nuclear matrix elements (NME) for neutrinoless double beta decay transitions
- Neutrino mixing matrix and phase space factors from the leptonic sector of the decay
- Left and right symmetries of the electroweak currents

The neutrinoless double beta decay is a processes where a nucleus decays by emission of two electrons (or two positrons), and change its charge in (+2) (-2) units. The decay violates lepton number conservation, since the electrons are not accompanied by the corresponding antineutrinos, like in the double beta decay with neutrinos. The two-neutrino mode is allowed by the SM since it takes place *via* two independent single beta decays. The two neutrino double beta decay mode has been measured intensively and its half-life is the longer ever measured in labs, it is of the order of 10^{20} years. It is described a s a second order process in the weak interaction Hamiltonian and it is independent of any assumption about the neutrino. The neutrinoless double beta decay mode is much more interesting, since it may take place only if neutrinos are massive and it depends also on the model assumptions about the neutrino. It is a unique source of information about the scale of the mass of the neutrino [16]. It has not been observed and the information about the neutrino may be extracted from the lower limits of the half-life established experimentally. Today, the lower limit for the neutrinoless double beta decay half-life is of the order of 10^{25} years. The minimal electroweak interaction Hamiltonian which may accommodate massive neutrinos, is a current-current interaction where the currents do have left- and right-handed components, and the participant neutrinos are described as superposition of neutrino mass-eigenstates. The inclusion of right-handed currents activates, in the neutrino sector of the currents, three heavy mass mass-eigenstates in addition to the three light neutrino mass-eigenstates. This minimal extension of the model does not include CP violation and the basic factorization obeys the SU(2)×SU(2) scheme, one for the left-handed currents and the other for the right-handed currents. Each current is then written in terms of a left-handed boson and a right-handed boson depending on the symmetry. Thus, in addition to a mass sector the model lagrangian has left-right and right-right terms. To compute the half-life of the process one has to evaluate the expectation value of the Hamiltonian, between the initial and final states, an a neutrino propagator between each decay vertex [15, 3, 16],

The inverse of the half-life for neutrinoless double beta decay transitions $0\nu\beta\beta$ is written

$$\begin{aligned}
\left[T_{1/2}^{(0\nu)}\right]^{-1} &= C_{mm}^{(0\nu)}\left(\frac{\langle m_\nu\rangle}{m_e}\right)^2 + C_{m\lambda}^{(0\nu)}\langle\lambda\rangle\left(\frac{\langle m_\nu\rangle}{m_e}\right) \\
&+ C_{m\eta}^{(0\nu)}\langle\eta\rangle\left(\frac{\langle m_\nu\rangle}{m_e}\right) + C_{\lambda\lambda}^{(0\nu)}\langle\lambda\rangle^2 \\
&+ C_{\eta\eta}^{(0\nu)}\langle\eta\rangle^2 + C_{\lambda\eta}^{(0\nu)}\langle\eta\rangle\langle\lambda\rangle\ . \qquad (1)
\end{aligned}$$

In the above expression, the factors $C^{(0\nu)}_{a,b}$, where the super-indexes a and b stand for the neutrino mass (m), and RR (λ) and LR (η) couplings [3], are functions of the NME and leptonic phase space factors [15, 16]. The calculation of these coefficients, in the context of different models for the nuclear structure of the participant nuclei, yields results which may vary substantially from model to model, however, it is possible to extract from the different calculations some similarities. In the work of [17] the reader can find a systematics of the NME obtained within the quasiparticle random phase approach (QRPA), which is the set of values we have adopted for the present calculations. The results are quite comparable to those of other models of nuclear structure, where the proton-neutron interactions are explicitly accounted for in relatively large configuration spaces, as it is discussed in the review of [16].

Once the values of the NME are fixed, the expression for the half-life can be viewed as the equation of a surface in three dimensions (each of the three axis being given by each of the three variables, that is the neutrino mass and the RR and LR couplings). The procedure to extract the upper values of each of the variables is rather simple, since to extract them one should perform a diagonalization to find their values in the intrinsic frame, where the half-life becomes a diagonal function of the square of each variable. The results of such a procedure, for the cases of ^{76}Ge and ^{136}Xe, with the matrix elements extracted from the compilation of [17], are shown in Table 1.

In the space of these variables, we have determined the ratios between the RR and LR couplings which are simultaneously consistent with the neutrino mass. All these variables are intrinsic functions of the neutrino mixing parameters, since they are average values taken on the basis of neutrino mass eigenstates [3]. With the values extracted in this manner we have determined allowed values of the right-handed boson, by using the expression [3, 18]

$$\frac{\mathrm{M_L}}{\mathrm{M_R}} = \sqrt{\frac{(\alpha - \tan\zeta)\tan\zeta}{(1+\alpha\tan\zeta)}}\,, \qquad (2)$$

where $\alpha = \langle\lambda\rangle/\langle\eta\rangle$, $\mathrm{M_L}$ is the mass of the left-handed boson, and ζ is the mixing angle between the massive boson generators of the $\mathrm{SU}(2)_L \times \mathrm{SU}(2)_R$ minimal extension of the SM.

3. Results

With the elements introduced in the previous section we are in conditions to calculate the mass of the right-handed boson and the limits on the mixing angle ζ by the combined analysis of ATLAS [8] and CMS[9] results and those of GERDA[12], EXO-200[14] and KamLAND-Zen[13]. We have extracted the couplings of the RR and LR sectors of the minimal extended electroweak

Table 1. Extracted upper values of the average neutrino mass and the RR and LR couplings

Case	Half-life limit (10^{25}yr)	$\langle m_\nu\rangle_{\max}$ (eV)	$\langle\lambda\rangle_{\max}$	$\langle\eta\rangle_{\max}$
^{76}Ge	2.5	0.325	$0.431(10^{-6})$	$0.286(10^{-8})$
^{136}Xe	1.1	0.182	$0.197(10^{-6})$	$0.176(10^{-8})$

lagrangian. In order to determine exclusion regions in the parametric space we have taken, for our analysis, the data acquired in the measurement of two jets+two leptons final states in pp collisions performed at the LHC facility[7] by ATLAS [8] and CMS [9]. The data reported by CMS correspond to collisions with center of mass energies of 8 TeV and integrated luminosity of 19.7 fb^{-1} [9], while the limits determined by ATLAS [8] correspond to a luminosity of 2.1 fb^{-1}. From these results a lower limit for the mass of the right-handed boson, of 3 TeV, was established. We then proceeded to extract the same information, from the reported lower limits for the half-life of two double beta decay emitters, germanium and xenon, as measured by GERDA[12] and EXO-200[14] and KamLAND-Zen[13], respectively. In order to extract these values we have used a set of NME whose validity, as representative of the state of the art results in microscopic nuclear structure calculations, has been well established [17].

Figure 1 shows the intersection between the extracted values, for the mixing angle ζ and the right-handed boson mass M_R from the $0\nu\beta\beta$ half-life lower limits, and the regions excluded by TWIST [10] $\zeta > 10^{-2}$, and the region excluded by the CMS [9] data: $M_R < 3$ TeV. The curves constructed from the neutrinoless double beta decay half-life lower limits display the results for allowed values of the average neutrino mass, as explained before. The detail of the calculations and the corresponding discussion are presented in [19].

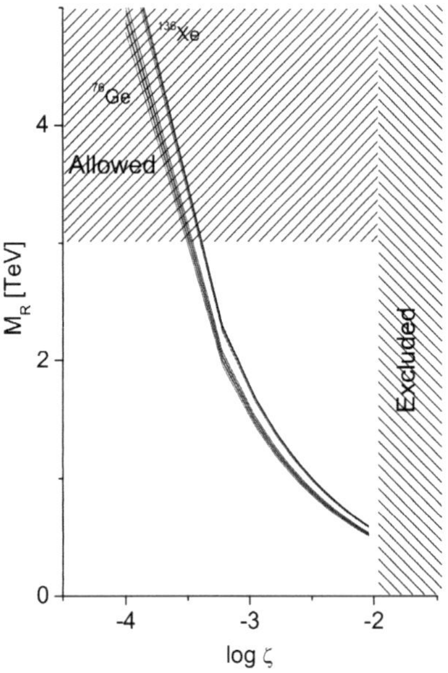

Figure 1. Allowed and excluded regions for the mass of the right-handed boson M_R and mixing angle ζ, extracted from the CMS [9] and ATLAS [8] data. The curves labelled 76Ge and 136Xe show the results extracted from the $o\nu\beta\beta$ half-life limits determined by GERDA [12], EXO-200 [14] and KamLAND-Zen [13] measurements, respectively

4. Conclusions

By the comparison between the limits determined by the experiments performed at CERN and the data obtained by the neutrinoless double beta decay experiments, we conclude that both type of experiments are complementary, since they explore the same physics. The results of our calculations show that the neutrinoless double beta decay relevant factors, like the average neutrino mass and couplings, are consistent with the limits determined by the experiments performed with the LHC facility, for channels where the decay of the right-handed boson in two jets and two leptons could be inferred from the data.

5. Acknowledgments

This work has been partially supported by the CONICET (Argentina) and by the Academy of Finland under the Finnish Centre of Excellence Programme 2012-2017 (Nuclear and Accelerator Based Programme at JYFL).

6. References

[1] Aitchinson I J R and Hey A J G 2003 *Gauge Theories in Particle Physics* Briston. IOP Publishers.
[2] Gonzales-Garcia M C et al. 2012 *JHEP* **123** 1212

[3] Vergados J D 1986 *Phys. Rep.* **133** 1
[4] Keung W-Y, and Senjanovic G 1983 *Phys. Rev. Lett.* **50** 1427
[5] Nemevsek M, Nesti F, Senjanovic G, and Zhang Y, 2011 *Phys. Rev. D* **83** 115014
[6] Tello V *et al.* 2011 *Phys. Rev. Lett.* **106** 151801,
[7] Evans L, Bryant P (Eds.), 2008 *JINST* **3** S08001.
[8] Aad G *et al.*, ATLAS collaboration 2012 *Eur. Phys. J. C* **72** 2056
[9] Chatrchyan S *et al.*, CMS collaboration 2012 *Phys. Rev. Lett.* **109** 261802; ibid. 2012 arXiv:1210.2402v3; 2014 arXiv:1407.3683v1
[10] Bayes R *et al.*, TWIST collaboration 2011 *Phys. Rev. Lett.* **106** 041804
[11] Zuber K 2012 *J. Phys. G: Nucl. Part. Phys.* **39** 124009
[12] Agostini M *et al.*, GERDA collaboration 2013 *Phys. Rev. Lett.* **111** 122503
[13] Gando A *et al.*, KamLAND-Zen collaboration 2013 *Phys. Rev. Lett.* **110** 062502
[14] Albert J B *et al.*, EXO-200 collaboration 2014 arXiv:1402.6956
[15] Doi M, Kotani T, and Takasugi E 1985 *Prog. Theor. Phys. Suppl.* **83** 1
[16] Suhonen J, and Civitarese O 1998 *Phys. Rep.* **300** 123
[17] Suhonen J, and Civitarese O 2012 *J. Phys. G: Nucl. Part. Phys.* **39** 124005
[18] Suhonen J and Civitarese O 1993 *Phys. Lett. B* **312** 367
[19] Civitarese O, Suhonen J and Zuber K 2014 (to be published).

Journal of Physics: Conference Series **630** (2015) 012056 doi:10.1088/1742-6596/630/1/012056

Running of the contact interactions in chiral N3LO potentials from subtractive renormalization *

E. F. Batista[1], **S. Szpigel**[2] **and V. S. Timóteo**[3]

[1]Departamento de Ciências Exatas e Naturais, Universidade Estadual do Sudoeste da Bahia 45700-000, Itapetinga, BA , Brazil

[2]Centro de Rádio-Astronomia e Astrofísica Mackenzie, Universidade Presbiteriana Mackenzie 01302-907, Sao Paulo, SP, Brasil

[3]Grupo de Óptica e Modelagem Numérica - GOMNI, Faculdade de Tecnologia - FT Universidade Estadual de Campinas - UNICAMP, 13484-332, Limeira, SP , Brasil

E-mail: `efbatista@uesb.edu.br` (EFB)

E-mail: `szpigel@mackenzie.br` (SS)

E-mail: `varese@ft.unicamp.br` (VST)

Abstract.
In this work a subtracted kernel renormalization procedure (SKM) is applied to the chiral NN potential up to next-to-next-to-next-to-leading-order (N^3LO) to obtain the running of the renormalized contact strengths with the subtraction scale μ and the phase shifts for all uncoupled waves with contact interaction (S, P, D). We use two potentials constructed within the framework of Weinberg's approach to ChEFT, which provide a very accurate description of NN scattering data below laboratory energies $E \sim 350\ MeV$, namely Epelbaum, Glöckle and Meissner (N3LO-EGM) and Entem and Machleidt (N3LO-EM). For both potentials, we consider a large cutoff (30 fm^{-1}) and analyze the phases and the running of the contact strengths with the subtraction point μ by making a fit of the K-matrix with five subtractions to the K-matrix from the Nijmegen II potential at low energies ($E \leq 20$ MeV).

1. Introduction

Quantum Chromodynamics (QCD) is widely accepted as the fundamental theory of strong interactions. The degrees of freedom described in the QCD Lagrangian are quarks and gluons. The strength of the interaction is strongly energy dependent and decreases as the energy increases. As a consequence, the QCD coupling constant α_s decreases for large momentum and the quarks inside the hadrons behave as free particles. Thus, in the high-energy regime α_s is very small and QCD can be treated perturbatively. On the other hand, in the low-energy regime the QCD coupling constant increases and perturbation theory can no longer be applied. Therefore, QCD is not effective to describe the strong interactions at low energies. To avoid this problem alternative approaches to treat strong interactions in the non-perturbative regime have been developed over the last two decades.

At low-energies the relevant degrees of freedom are baryons and mesons rather than quarks and gluons. Hence, it is more convenient to use Effective Fields Theories (EFT) which are

*Presented by E.F.B. at XXXVII RTFNB, 37th Brazilian Workshop on Nuclear Physics, 8-12 September 2014, Maresias, SP, Brazil

constructed with nucleons and pions as the relevant degrees of freedom. Weinberg proposed an EFT formalism for the nucleon-nucleon (NN) interaction based on a chiral expansion of an effective Lagrangian known as Chiral Perturbation Theory (ChPT) [1, 2, 3]. This approach, first applied by Ordóñez, Ray and van Kolck [4] allows one to work with the NN interaction perturbatively where the classes of the Feynman diagrams required to complete each order are organised by a power counting scheme called Weinberg's Power Counting (WPC). At leading order (LO) the NN potential consists of one-pion-exchange (OPE) plus a contact term. At next-to-leading order (NLO) two-pion-exchange (TPE) and $\mathcal{O}(p^2)$ contact interactions are added, at next-to-next-to-leading order (N2LO) there is an additional set of TPE diagrams and finally, at next-to-next-to-next-to-leading order (N3LO) corrections to both OPE and TPE are included along with $\mathcal{O}(p^4)$ contact interactions.

At any order in the chiral expansion the NN interaction contributions have ultraviolet (UV) divergencies and therefore a renormalization procedure is required. Usually a momentum cutoff (sharp or smooth) is applied to the potential so that the integral in the Lippmann-Schwinger (LS) equation becomes finite. An alternative procedure based on a kernel subtraction scheme has been proposed to modify the scattering equation and keep the potential uncut. This method was first introduced in Ref.[5] where the authors applied a single subtraction to renormalize the NN interaction at leading order.

In this formalism only one scale parameter is introduced, the subtraction point μ, which is the scale where a physical input for the two-body scattering amplitude has to be supplied in order to fix the renormalized counter-terms. The number of subtractions necessary to render the amplitude finite depends on the degree of divergence of the interaction. Each subtraction introduces a factor of p^{-2} in the kernel of the LS equation. So, if the interaction has terms with $\mathcal{O}(p^d)$, at least $d+1$ subtractions have to be performed in the LS equation kernel to generate finite results.

A generalisation of the SKM method for any number of subtractions was developed in Ref. [6] and it was shown that a non-relativistic Callan-Symanzik equation arises from the condition that the two-body amplitude has to be invariant under the change of the subtraction point. The renormalization group flow equation tell us how the driving terms have to be modified with the renormalization scale in order to keep the scattering amplitude invariant.

In previous works this formalism was applied with three [7] and four [8, 9] subtractions in the study of higher orders of NN interaction. In the latter, the Callan-Symanzik flow equation with four subtractions was numerically integrated for the ${}^1S_0$ channel and renormlization group invariance of the method was shown explicitely.

2. Chiral forces at N3LO: renormalization with five subtractions

The NN potential at N3LO can be separated in three main contributions:

$$V_{\rm N3LO} = V_{\rm OPE} + V_{\rm TPE} + V_{\rm Contact} \, . \tag{1}$$

The first term is the one-pion-exchange potential which describes the long range part of the interaction. The second is the two-pion-exchange potential that is associated with the mid-range part and the last one is the contact term which parametrizes the short range part of the interaction.

At N3LO, there are contact interactions only in the partial waves with $J \leq 2$ and for uncoupled channels we have twelve Low-Energy Constants (LEC):

$$\begin{aligned} V^{\rm 1S0}_{\rm Contact} &= C_0 + C_2\,(p^2+p'^2) + C_4\,(p^2\times p'^2) + C'_4\,(p^4+p'^4)\,, \\ V^{\rm 3P0,\,1P1,\,3P1}_{\rm Contact} &= C_1\,(p\times p') + C_3\,(p\times p')\,(p^2+p'^2)\,, \\ V^{\rm 1D2,\,3D2}_{\rm Contact} &= D_4\,(p^2\times p'^2)\,. \end{aligned} \tag{2}$$

The OPE and TPE terms are given in terms of well known physical quantities like the axial coupling constant, g_a, the pion weak-decay constant, f_π, the nucleon mass, m, and the pion mass, m_π. The LECs C_i and D_4 stand for the strengths of the contact interactions which have to be adjusted to reproduce the two-nucleon observables.

Both pion-exchange and contact interaction terms can lead to UV divergences when the effective NN potential at a given order in the chiral expansion is iterated in the LS equation, requiring a regularization and renormalization procedure in order to obtain well-defined finite solutions. Both a sharp cutoff renormalization scheme [10] or a smooth cutoff regularization function [11] can applied with very successful results.

In this work we renormalize the N3LO interactions by applying five subtractions to the kernel of the LS equation for the K-matrix with an iterative procedure at a given momentum scale μ. The LS equation for the K-matrix with five subtraction is given by

$$K_5(p,p';k,\mu) = V^{(5)}(p,p';k,\mu) + \frac{2}{\pi}\mathcal{P}\int dq\, q^2\, V^{(5)}(p,q;k,\mu)\; G^{(5)}(q,k;\mu)\; K_5(q,p';k,\mu)\;, \quad (3)$$

where $G^{(5)}$ is the five-fold subtracted Green's function

$$G^{(5)}(q,k;\mu) = \left(\frac{\mu^2+k^2}{\mu^2+q^2}\right)^5 \frac{1}{k^2-q^2}\;, \quad (4)$$

and $V^{(5)}$ is the so called fifth-order driving term

$$V^{(5)}(p,p';k,\mu) = V^{(4)}(p,p';k,\mu) + \frac{2}{\pi}\mathcal{P}\int dp\, p^2\, V^{(4)}(p,q;k,\mu)\; \frac{(\mu^2+k^2)^4}{(\mu^2+q^2)^5}\; V^{(5)}(q,p';k,\mu)\;. \quad (5)$$

Thus, for a given subtraction scale μ we determine the recursive driving terms by applying the recursive procedure described in detail in our previous works. Then we insert the driving term $V^{(5)}$ in Eq.3 and a finite solution for the K-matrix is obtained which can be used to calculate the NN phase-shift at a given energy. In the next section we will show some numerical results for all uncoupled channels up to $J = 2$.

3. Numerical results

In the following we present the numerical procedure used to fix the strengths of the contact interactions and some results for the NN phase shifts and the running of the contact strengths with the renormalization scale are shown.

3.1. Fitting Procedure

We compute the mean square function

$$\mathcal{F}(\mu) = \frac{1}{N}\sum_{j=1}^{N}\left[K_5(k_j;\;\mu) - K_{\rm ref}(k_j)\right]^2\;, \quad (6)$$

and obtain $C(\mu)$ by minimizing $\mathcal{F}(\mu)$ for each value of the renormalization scale μ. We use $N = 21$ points in the interval $0 \le k_j \le 0.5\ \mathrm{fm}^{-1}$ ($0 \le E_j \le 20$ MeV). Once the strengths of the contact interactions have been determined, the phase shifts can be obtained as

$$\delta_\mu(k_j) = \mathrm{atan}\left[-k_j \times K_5(k_j;\;\mu)\right]\;. \quad (7)$$

Note that we fit only the low-energy part of the K-matrix so that the results for higher energy will be just predictions.

Table 1. Strengths of the renormalized contact interactions, all given in fm, for the reference scale μ_0, in fm^{-1}, for the EGM potential.

	μ_0	C_0	$\sqrt[3]{C_1}$	$\sqrt[3]{C_2}$	$\sqrt[3]{C_3}$	$\sqrt[5]{C_4}$	$\sqrt[5]{C_4'}$	$\sqrt[5]{D_4}$
$^1S_0$	0.97	-1.248	-	-2.451	-	3.176	-0.646	-
$^3P_0$	3.00	-	-1.707	-	-0.288	-	-	-
$^1P_1$	1.90	-	0.429	-	0.416	-	-	-
$^3P_1$	1.50	-	-0.499	-	0.722	-	-	-
$^1D_2$	5.25	-	-	-	-	-	-	-0.346
$^3D_2$	0.47	-	-	-	-	-	-	-0.614

Table 2. Strengths of the renormalized contact interactions, all given in fm, for the reference scale μ_0, in fm^{-1}, for the EM potential.

	μ_0	C_0	$\sqrt[3]{C_1}$	$\sqrt[3]{C_2}$	$\sqrt[3]{C_3}$	$\sqrt[5]{C_4}$	$\sqrt[5]{C_4'}$	$\sqrt[5]{D_4}$
$^1S_0$	1.00	-0.838	-	-2.798	-	3.388	-0.817	-
$^3P_0$	3.05	-	3.672	-	-0.3	-	-	-
$^1P_1$	1.70	-	0.574	-	0.541	-	-	-
$^3P_1$	1.50	-	-0.584	-	0.649	-	-	-
$^1D_2$	4.00	-	-	-	-	-	-	3.094
$^3D_2$	0.47	-	-	-	-	-	-	-0.423

In Tables 1 and 2 we show the strengths of the renormalized contact interactions for the best fit in each partial wave for the N3LO-EGM and N3LO-EM respectively. We define the scale which provides the best fit as the reference scale μ_0.

In Figs. 1 and 2 we show the phase-shifts for the $^1D_2$ and $^3D_2$ waves for both N3LO-EM and N3LO-EGM chiral potentials compared to the Nijmegen partial wave analysis [12]. The magenta and yellow bands correspond to a range for the renormalization scale μ. With the exception of the $^1D_2$ with the N3LO-EGM potential which doesn't show a good fit, there is a narrow range of the renormalization scale where the fits are very good up to $E_{\text{lab}} = 100$ MeV.

The running of the contact strength D_4 for the two potentials are displayed in Fig. 3 for the uncoupled D-waves. In the case of the $^3D_2$ channel (blue line), the running is nearly identical for the two potentials and is similar to what was obtained by Epelbaum and Meissner with cutoff regularization [13]. The strengths for the $^1D_2$ channel (red line) is almost scale invariant in the range $0 < \mu < 2$ fm^{-1}.

Figs. 4 shows the phase-shifts for the $^1P_1$ and $^3P_1$ channels. In these channels a proper choice for the renormalization scale range (see figure legends) allows a good prediction for the phases up to $E_{\text{lab}} = 100$ MeV like in the D-waves. The associated running of C_1 and C_3 are displayed in Fig. 5, where we can observe that the running for C_1 is rather stable (upper panels) but for C_3 there are regions with stability followed by sharp variations (lower panels). Note that C_3 in the $^3P_1$ channel changes sign for N3LO-EM in the range $1 < \mu < 2$ fm^{-1} and for N3LO-EGM there is a maximum for C_3 at $\mu = 1.5$ fm^{-1} in the $^3P_1$ channel and at $\mu = 2.0$ fm^{-1} in the $^1P_1$ channel. In the range $2 < \mu < 4$ fm^{-1} the strengths of C_3 are nearly the same for the $^1P_1$ and $^3P_1$ channels for both N3LO-EM and N3LO-EGM.

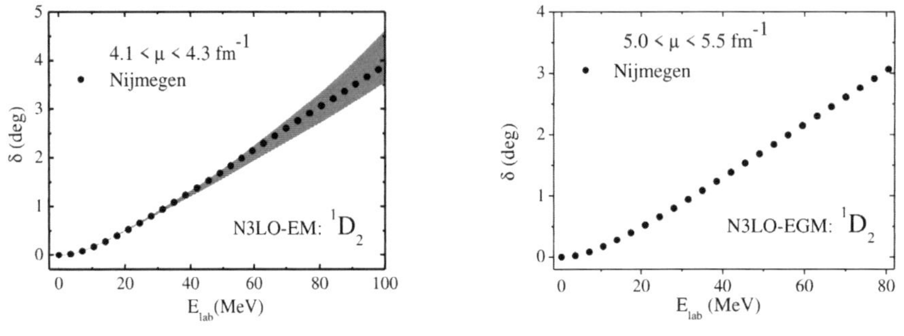

Figure 1. Phase-shifts for the $^1D_2$ channel from both N3LO-EM (left) and N3LO-EGM (right) chiral potentials compared to the Nijmegen partial wave analysis.

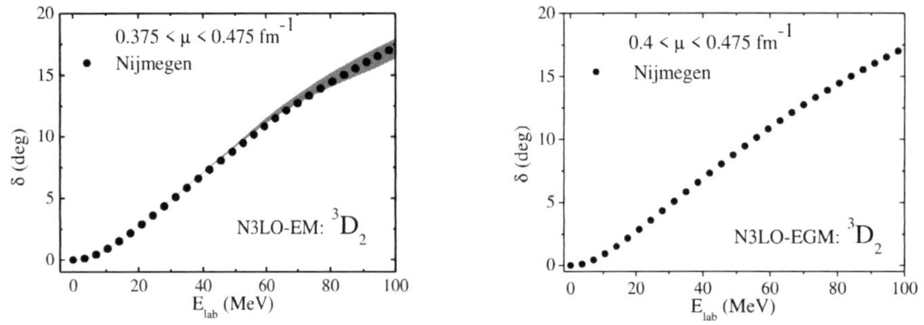

Figure 2. Phase-shifts for the $^3D_2$ channel from both N3LO-EM (left) and N3LO-EGM (right) chiral potentials compared to the Nijmegen partial wave analysis.

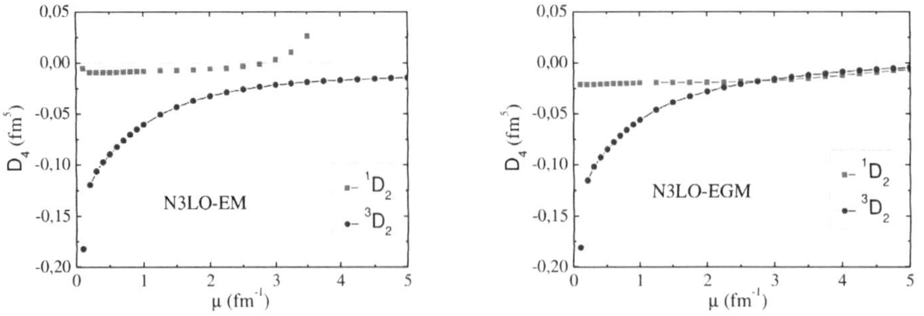

Figure 3. Running of the contact strength D_4 with the renormalization scale μ for the uncoupled D-waves with both N3LO-EM (left) and N3LO-EGM (right) chiral potentials.

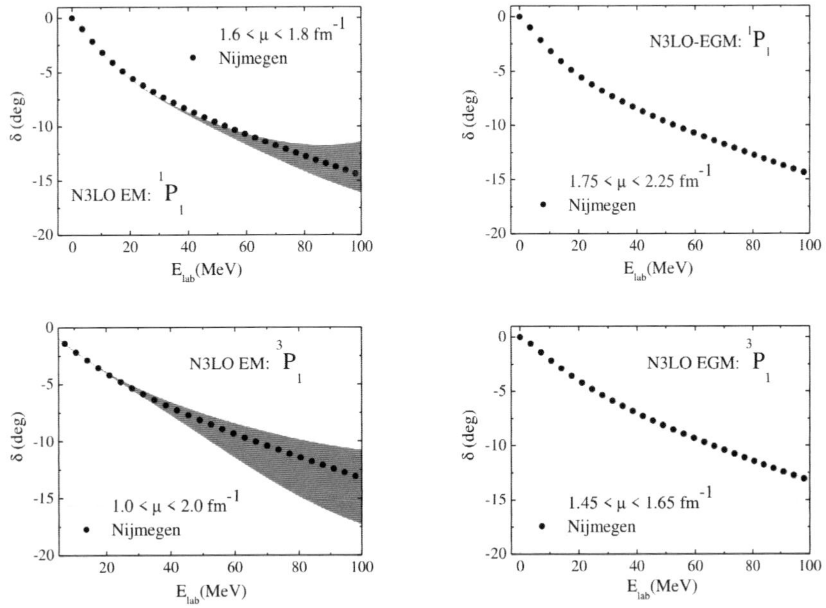

Figure 4. Phase-shifts for the $^1P_1$ (top) and $^3P_1$ (bottom) channels from both N3LO-EM and N3LO-EGM chiral potentials compared to the Nijmegen partial wave analysis.

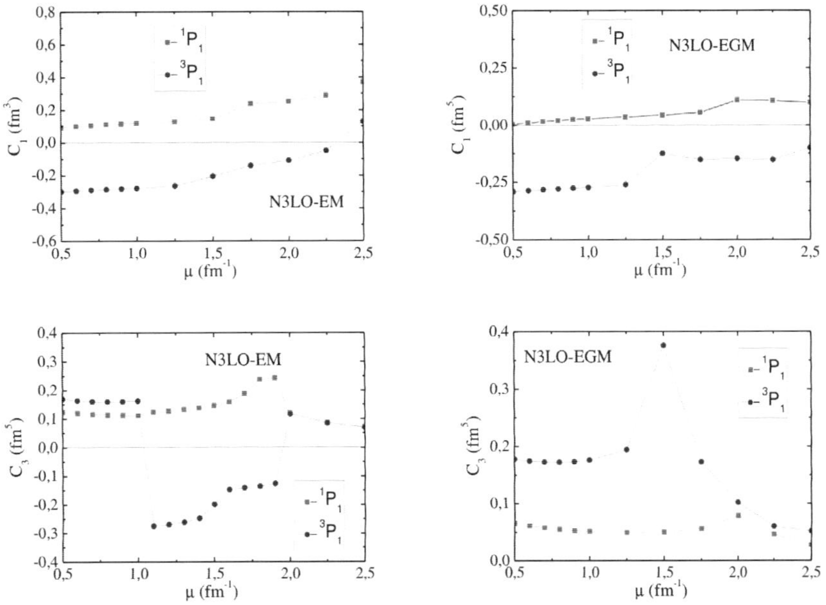

Figure 5. Running of the contact strengths C_1 and C_3 with the renormalization scale μ for the uncoupled P-waves with both N3LO-EM (left) and N3LO-EGM (right) chiral potentials.

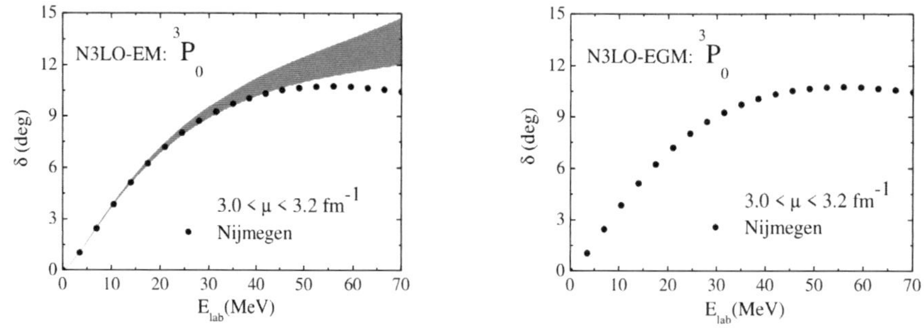

Figure 6. Phase-shifts for the $^3P_0$ channel from both N3LO-EM and N3LO-EGM chiral potentials compared to the Nijmegen partial wave analysis.

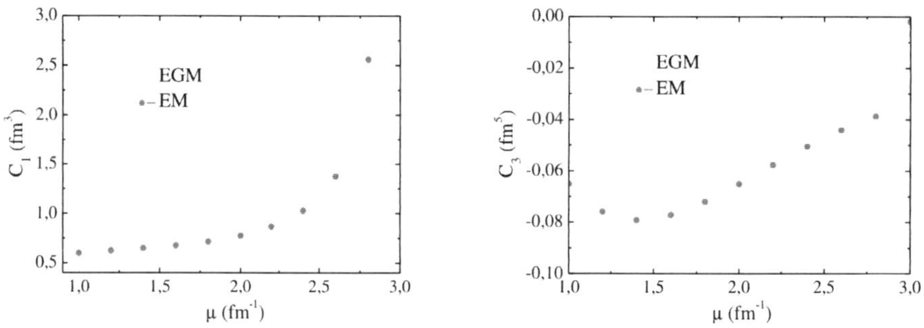

Figure 7. Running of the contact strengths C_1 and C_3 with the renormalization scale μ for the $^3P_0$ channel with both N3LO-EM and N3LO-EGM chiral potentials.

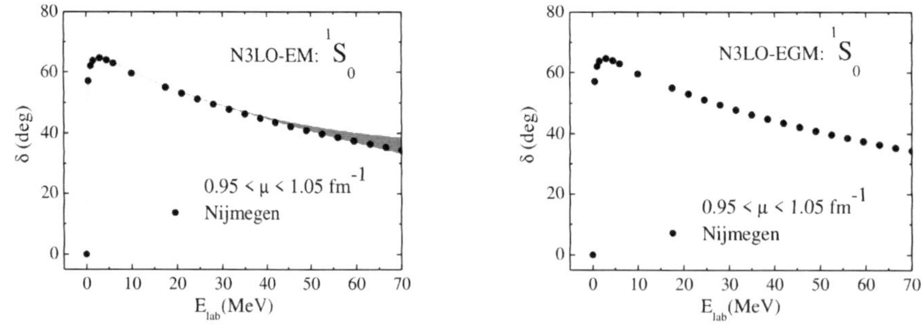

Figure 8. Phase-shifts for the $^1S_0$ channel from both N3LO-EM and N3LO-EGM chiral potentials compared to the Nijmegen partial wave analysis.

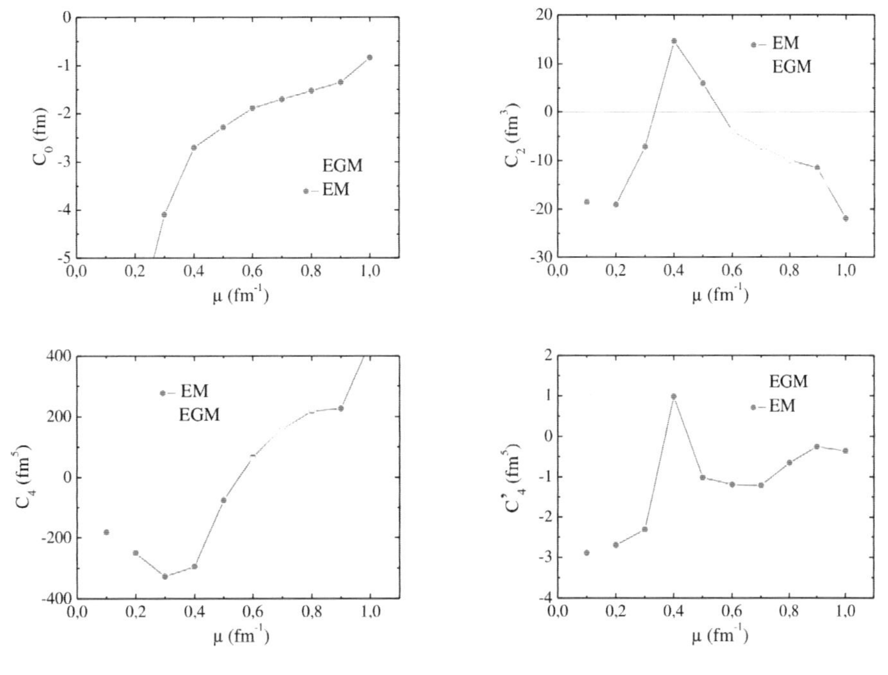

Figure 9. Running of the contact strengths C_0, C_2, C_4 and C_4' with the renormalization scale μ for the $^1S_0$ channel with both N3LO-EM and N3LO-EGM chiral potentials.

Now we turn to the channels with $J = 0$. The phase-shifts for the $^3P_0$ wave are shown in Fig. 6 where we can observe that the experimental data are described only for very low energies ($E_{\rm lab} \sim 50$ MeV) and for the same range of the renormalization scale, $3.0 < \mu < 3.2$ fm^{-1}, the N3LO-EM provides a wider band containing the at low energies.

The running for C_1 and C_3 in the $^3P_0$ channel, displayed in Fig. 7, are very similar for both N3LO-EM and N3LO-EGM. In fact, the difference between the strengths of the contact interactions for the two potentials is nearly constant.

The phase-shifts for the $^1S_0$ channel are very well described up to $E_{\rm lab} \sim 70$ MeV with renormalization scales in the range $0.95 < \mu < 1.05$ fm^{-1} for both N3LO-EM and N3LO-EGM, as can be seen in Fig. 8. This is consistent with what has been obtained with four subtractions at N2LO in Ref. [9].

Fig. 9 shows the running of the four contact terms present in the $^1S_0$ wave and we can observe that the strengths of the interactions for N3LO-EM and N3LO-EGM match for renormalization scales in the range $0.6 < \mu < 1.0$ fm^{-1}.

4. Summary and outlook

In this paper we have used the subtracted kernel renormalization method (SKM) with multiple subtractions to obtain the phase shifts and the running of the contact strengths for all uncoupled channels up to $J = 2$. We have considered two state of the art chiral potentials at N3LO and performed five subtractions while keeping a large instrumental cutoff $\Lambda = 30$ fm^{-1}.

In future works we will study the renormalization group invariance with N3LO potentials by integrating the non-relativistic Callan-Symanzik flow equation in order to obtain an exact solution (non-perturbative) for the driving terms in several channels of the NN interaction.

Acknowledgements
S.S. is partially supported by FAPESP and V.S.T. would like to thank FAEPEX, FAPESP and CNPq for financial support.

References

[1] Weinberg S 1990 *Phys. Lett.* B **251** 288.
[2] Weinberg S 1991 *Nucl. Phys.* B **363** 3.
[3] Weinberg S 1992 *Phys. Lett.* B **295** 114.
[4] Ordóñez C, Ray L and van Kolck U 1994 *Phys. Rev. Lett.* **72** 1982.
[5] Frederico T, Timóteo V S and Tomio L 1999 *Nucl. Phys.* A **653** 209.
[6] Frederico T, Delfino A and Tomio L 2000 *Phys. Lett.* B **481** 143.
[7] Timóteo V S, Frederico T, Tomio L and Delfino A 2005 *Phys. Lett.* B **621** 109.
[8] Timóteo V S, Frederico T, Delfino A and Tomio L 2011 *Phys. Rev.* C **83** 064005.
[9] S. Szpigel and V. S. Timóteo 2012 *J. Phys.* G **39** 105102.
[10] Entem D R and Machleidt R 2003 *Phys. Rev.* C **68** 041001
[11] Epelbaum E, Glöckle W, and Meißner U G 2005 *Nucl. Phys.* A **747** 362.
[12] Stoks V G J, Klomp R A M, Terheggen C P F and de Swart J J 1994 *Phys. Rev.* C **49** 2950.
[13] Epelbaum E and Meissner U -G 2013 *Few Body Systems* **54** 2175.

Journal of Physics: Conference Series **630** (2015) 012057 doi:10.1088/1742-6596/630/1/012057

Computational Analysis of *3D* Ising Model Using Metropolis Algorithms

A. F. Sonsin, M. R. Cortes, D. R. Nunes, J. V. Gomes and R. S. Costa

Universidade Federal de Rondônia
Departamento de Física de Ji-Paraná. Rua Rio Amazonas, 351
Jardim dos Migrantes, 78961970 - Ji-Paraná, RO – Brasil

arturfalqueto_sonsin@hotmail.com, mariane_cortess@hotmail.com,
danii.nunes@hotmail.com, janygovi@gmail.com and rscosta.unir@gmail.com

Abstract. We simulate the Ising Model with the Monte Carlo method and use the algorithms of Metropolis to update the distribution of spins. We found that, in the specific case of the three-dimensional Ising Model, methods of Metropolis are efficient. Studying the system near the point of phase transition, we observe that the magnetization goes to zero. In our simulations we analyzed the behavior of the magnetization and magnetic susceptibility to verify the phase transition in a paramagnetic to ferromagnetic material. The behavior of the magnetization and of the magnetic susceptibility as a function of the temperature suggest a phase transition around $KT/J \approx 4.5$ and was evidenced the problem of finite size of the lattice to work with large lattice.

1. Introduction

In this work, we used large volume of lattice compared to lattice values used in the literature for Ising model in three dimensions. Therefore we see and evidenced in our calculations the problem of finite size of the lattice.

The Ising Model, even simple, is still of current interest. The model consists of a system of atoms arranged in a regular lattice, described by spin variables that take only two values, +1 and -1. Further, each of these variables interact only with the adjacent former (first neighbors interaction), so that the energy parallel and anti parallel spins is different. Since its creation, the model became the basis for much of the research in the field of critical phenomena and phase transition. The model was proposed by Lenz theme of the doctoral thesis of E. Ising on magnetism, and the first model of the analytical solution has been published by Ising [1].

2. Simulation Computational of the Ising model in 3D

In this work, all simulated physical models are models on the lattice. We understand such models as defined not in the space-time, but the net, an approach the complex n-space-time or simply space. We use the Monte Carlo method [2] to statistical analyzes, and the lattice used were 150^3 200^3 and 250^3. The computational model Ising solution of 3-dimensions can be used to calculate the critical exponent for its corresponding universality class. In 3-dimensions, the Hamiltonian of the model can be written as

$$H = -J \sum_{i,j,k=1}^{N} (S_{i-1,j}S_{i,j} + S_{i,j}\, S_{i+1,j} + S_{i,j-1}\, S_{i,j} + S_{i,j}\, S_{i,j+1} + S_{i-1,k}S_{i,k} + S_{i,k}S_{i+1,k} + S_{i,k-1}S_{i,k}$$
$$+ S_{i,k}S_{i,k+1} + S_{j-1,k}S_{j,k} + S_{j,k}S_{j+1,k} + S_{j,k}S_{j,k-1} + S_{j,k}S_{j,k+1}) - \boldsymbol{H}\sum_{i}^{N} \boldsymbol{s_i}\,. \quad (1)$$

We do not consider any external force, so $\boldsymbol{H} = 0$. The model Ising studied here is a cube with N spins located at the corners of a cubic lattice. We obtain a magnetization system for different mesh sizes, and temperatures in order to verify the behavior near the phase transition. The magnetization M of a given configuration is simply given by the sum of all the spins in the lattes. And being written by sites (N) as

$$M = \frac{1}{N}\sum_{i}^{N} S_i\,. \quad (2)$$

The magnetic susceptibility (χ) is the greatness featuring a magnetic material according to their response to an applied magnetic field. The determination of χ can assist in the identification phase transitions. The susceptibility is given by:

$$\chi = \frac{J}{K_B T}(\langle M^2\rangle - \langle M\rangle^2), \quad (3)$$

where k_B is the constant Boltzmann, J the exchange term, T the temperature and the magnetization M.

The Cumulant of the Binder is an observational tool for estimating critical points. For different sizes of the lattices , the average magnetization curve always passes through a fixed point, which coincides with the critical point. Thus, generation of magnetization curves for several different sizes of the network can estimate the critical temperature T_c. The place where the curves intersect. Thus, it is a visual characteristic for detection of phase transition, as defined

$$U_4 = 1 - \frac{\langle M^4\rangle}{3\langle M^2\rangle^2}. \quad (4)$$

Journal of Physics: Conference Series **630** (2015) 012057 doi:10.1088/1742-6596/630/1/012057

3. Numerical Results

The number of Monte Carlo steps was equal to 10000 and with lattices of the 150^3 , 200^3 and 250^3 to verify the behavior of the magnetization as a function of temperature. And thus estimating a critical temperature of phase transition. Due to the finite size of the lattice problem use the Magnetic Susceptibility and Cumulant of the Binder as a function to estimate an accurate value of the critical temperature.

In Figure 1, we can see that the magnetization behavior is almost the same for different sizes of the lattices: 150^3 , 200^3 and 250^3. Decreasing to reach the critical temperature value and then going to zero after the critical temperature.

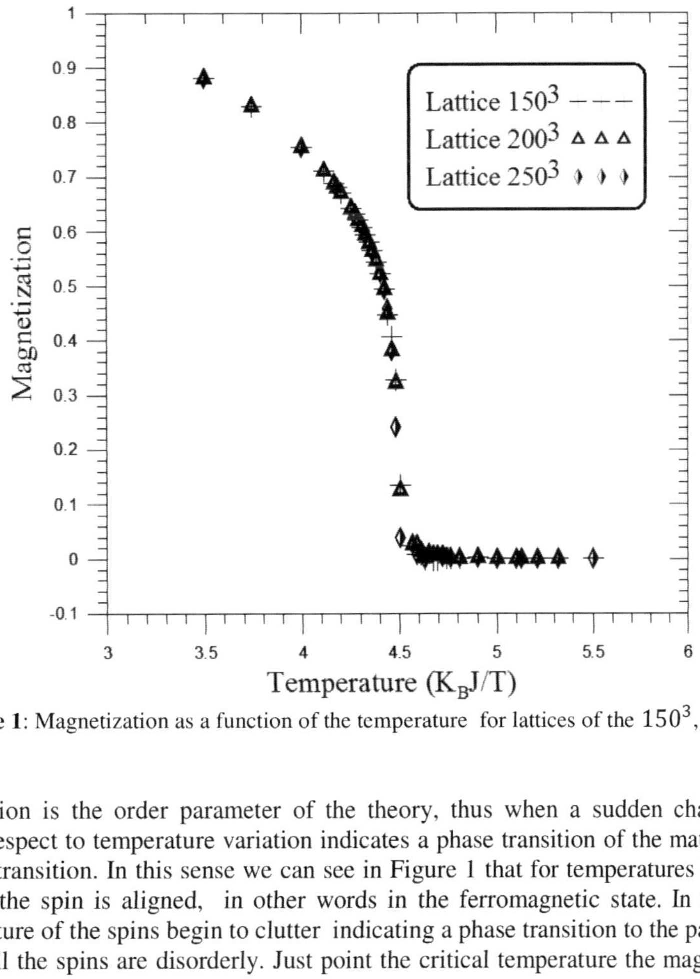

Figure 1: Magnetization as a function of the temperature for lattices of the 150^3, 200^3 and 250^3.

The magnetization is the order parameter of the theory, thus when a sudden change much of its behavior with respect to temperature variation indicates a phase transition of the material. In this case a second-order transition. In this sense we can see in Figure 1 that for temperatures below the critical temperature of the spin is aligned, in other words in the ferromagnetic state. In the region of the critical temperature of the spins begin to clutter indicating a phase transition to the paramagnetic state, that is, where all the spins are disorderly. Just point the critical temperature the magnetization should be close to zero. However, to reach such a situation we should work with infinite lattices ($L^3 \rightarrow \infty$) which would make them unviable computations. This situation is called the finite size of the lattice,

Journal of Physics: Conference Series **630** (2015) 012057 doi:10.1088/1742-6596/630/1/012057

since we can only work with finite limits. Therefore, the magnetization curve with infinite lattices will have an asymptote to reach the critical temperature, indicating a phase transition.

In Figure 2, we can see that with the size of the lattice $L \to \infty$ 1, $U_4 \to 0$ for T > Tc and $U_4 \to 2/3$ for $T < T_c$. Thus, the determination of T_c technique is to make the $U_4 \times T$ chart for different sizes of the lattice (150^3, 200^3 and 250^3). Analyzing the graphic, one can determine the point corresponding to the critical temperature Tc, as the point where all the curves cross. In Figure 2 clearly shows the meeting point of the curves.

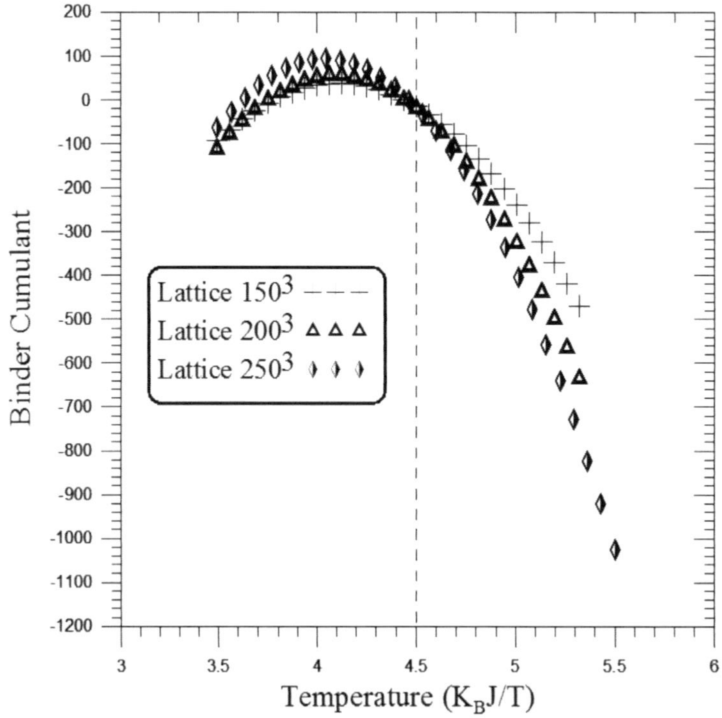

Figure 2:Binder Cumulant as a function of the temperature for lattices of the 150^3, 200^3 and 250^3.

We can see in Figure 3 we get a critical temperature $\boldsymbol{k_B T/J \approx 4.5}$ right next to temperature obtained via computer simulation [3]. To a temperature close to zero, the spins of the lattice tend to align them selves in a similar way to a ferromagnetic even when the **H** field tends to zero. However, after a certain temperature value (called the critical temperature), the spins lose this self-alignment.

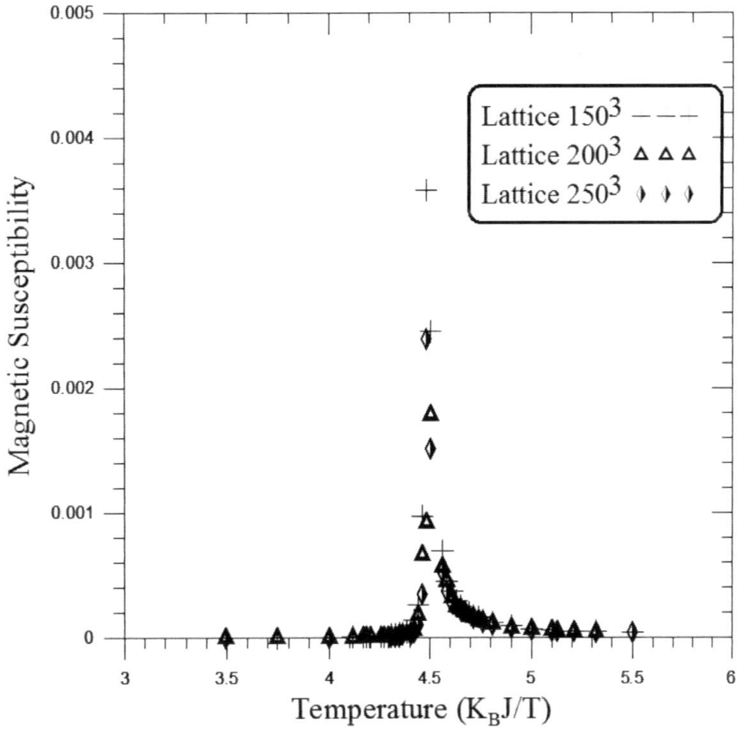

Figure 3: Magnetic Susceptibility as a function of the temperature for lattices of the 150^3, 200^3 and 250^3.

4. Conclusion

We simulate the Ising model with Monte Carlo and we use the Metropolis algorithm to update the distribution of spins. We found that, in the specific case of the three-dimensional Ising model, the Metropolis method is efficient. Studying the system near the phase transition point, we observed that the magnetization goes to zero. The behavior of the magnetization and magnetic susceptibility versus temperature suggest a phase transition around $\boldsymbol{k_B T/J \approx 4.5}$. The value of the critical temperature obtained via computer simulation with large lattice of 150^3, 200^3 and 250^3 was very close to the value found in the literature, via computer simulation for the Ising model in three dimensions with small lattices [3][4][5].

References

[1] Ising, E. 1925. *Beitrag zur Theorie des Ferromagnetismus*. Z. Phys , pp. v. 31, 253.

[2] Newman, M. E. 1999. *Monte Carlo Methods in Statistical Physics.* New York: Oxford University press.

[3] Pinto, F.S.C. 1999. *Um estudo computacional do modelode Ising* . (Dissertação submetida à Faculdade de Engenharia da Universidade do Porto para obtenção do grau).

[4] Ferrenberg, A. M., Landau, D. P. 1991. Phys. Rev. B 44, 5081 .

[5] Barber, M. N., Pearson, R. B., Toussaint, D., 1985. Phys. Rev. B 32, 1720 .

Mass-Radius diagram for compact stars

G. A. Carvalho[1]; R. M. Marinho Jr[1] and M. Malheiro[1]

[1]Departamento de Física, Instituto Tecnológico de Aeronáutica , Vila das Acácias, São José dos Campos, SP, 12228-900, Brazil

E-mail: geanderson.araujo.carvalho@gmail.com

Abstract. The compact stars represent the final stage in the evolution of ordinary stars, they are formed when a star ceases its nuclear fuel, in this point the process that sustain its stability will stop. After this, the internal pressure can no longer stand the gravitational force and the star colapses [2]. In this work we investigate the structure of these stars which are described by the equations of Tolman-Openheimer-Volkof (TOV) [1]. These equations show us how the pressure varies with the mass and radius of the star. We consider the TOV equations for both relativistic and non-relativistic cases. In the case of compact stars (white dwarfs and neutron stars) the internal pressure that balances the gravitational pressure is essentialy the pressure coming from the degeneracy of fermions. To have solved the TOV equations we need a equation of state that shows how this internal pressure is related to the energy density or mass density. Instead of using politropic equations of state we have solved the equations numericaly using the exact relativistic energy equation for the model of fermion gas at zero temperature. We obtain results for the mass-radius relation for white dwarfs and we compared with the results obtained using the politropic equations of state. In addition we discussed a good fit for the mass-radius relation.

1. Introduction

The compact stars are the final stage of the evolution of an ordinary star, and they also constitute a laboratory of tests in general relativity [2][4]. We can study the effects of general relativity on the compact stars when it is taken into account [1]. The compact stars are formed of matter in high densities, in the case of the white dwarfs they are constituted of degenerate electron gas or in the case of neutron star that are formed by neutron degenerate gas [2]. The mass of a white dwarf is tipically 0.5 to 1 $M_\odot$ ($M_\odot$ represents the solar mass) and they have a radii of the order of 10000 km, however the mass of the neutron stars is about 1.4 to 3 $M_\odot$ and the raddi is tipically 10 km [5].

2. Diferential Equations

We always assume in our accounts that the configuration of the mass of the star is static and spherically symmetric, in this case the pressure and the density of the star are functions of only the radial coordinate r [1]. In the compact star there are two forces acting on the star, one is the gravitational force and the other one came from the degeneracy pressure in which is the force that essentially balances the gravitational one. For the structure of a newtonian star we have the following equations [2]:

$$\frac{dp}{dr} = -\frac{Gm(r)\rho(r)}{r^2} \tag{1a}$$

$$\frac{dm}{dr} = 4\pi r^2\rho(r), \tag{1b}$$

where P denotes pressure, G is the gravitational constant, and ρ density, the mass $m(r)$ represents the mass inside of the radius r.
Note that while the dm/dr is positive, dp/dr must always be negative. We solve the equations numerically with the initial conditions $m(0) = 0$ and $p(0) = p_0$ in the region at the center of the star, with increasing rarius the mass also increases while the pressure decreases occasionally reaching zero in the surface of the star [2]. However for an isotropic, general relativistic, static, ideal fluid sphere in hydrostatic equilibrium we have the Tolman-Oppenheimer-Volkoff (TOV) equations that describes the structure of the star [1]

$$\frac{dp}{dr} = -\frac{Gm(r)\epsilon(r)}{c^2r^2}\left[1+\frac{p(r)}{\epsilon(r)}\right]\left[1+\frac{4\pi r^3p(r)}{m(r)c^2}\right]\left[1-\frac{2Gm(r)}{c^2r}\right]^{-1} \tag{2a}$$

$$\frac{dm}{dr} = \frac{4\pi r^2\epsilon(r)}{c^2}, \tag{2b}$$

the $\epsilon(r)$ represents the energy density, in case of the newtonian equations we can also substitute

$$\epsilon = \rho c^2, \tag{3}$$

where c is the speed of light.
The new three terms in the equilibrium equation (2a) are corrections terms. The general relativistic effects become important when the star is sufficiently compact, i.e., when the factor $2Gm(r)/c^2r$ approaches unity [4] and when the pressure is considerable compared to energy density.

3. Equation of State EoS

To solve numerically the coupled diferential equations in both cases, general relativistic and non-relativistic, we must have an equation of state (EoS). A first approach for the EoS is the polytrope approximation, at zero temperature we have [3]

$$P = K\rho^{1+\frac{1}{n}}, \tag{4}$$

where K is a constant, and n is the politropic index in which defines if the electrons in the gas are relativistic or non-relativistic. The value of the constant K depends on the politropic index. There is a more general EoS without the polytrope approximation that is given by [2]

$$P = \frac{\epsilon_0}{24}\left[(2x +^3 -3x)(1+x^2)^{1/2} + 3\sinh^{-1}(x)\right], \tag{5}$$

where x is a dimensionless variable given by $x = P_f/m_ec$, P_f is the Fermi moment, m_e is the electron mass and ϵ_0 is a constant which carries the dimension of pressure. In terms of relative parameter x the density is given by

$$\rho = 9.738 \times 10^5\mu x^3 \text{ g/cm}^3, \tag{6}$$

μ is the ratio between the mass number and atomic number. If we assume that the white dwarf is made of ^4He, ^{12}C or ^{16}O, hence $\mu = 2$.
In order to compare the equation (4) with (5) we plot these functions in figure (1).

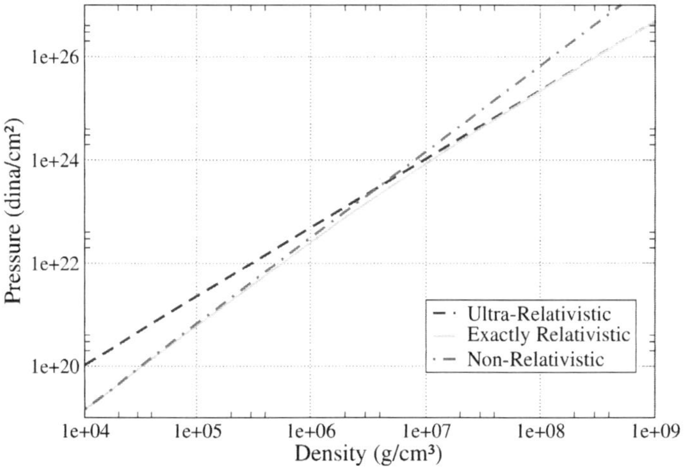

Figure 1. Graphic of the pressure in function of the density. For $n = 3$ and $n = 3/2$ we have the relativistic and non-relativistic EoS, respectively.

We can see that the politropic approaches have a density regime in which they work as an approximation of the general EoS. For low densities the non-relativistic politropic EoS will work, but for densities between $10^5 - 10^7 g/cm^3$ neither non-relativistic nor relativistic EoS correspond to the general solution for the EoS. So for the mass-radius relation it's important that we use the general EoS.

4. Mass-radius relation

Using the set of equations (1) or (2) we can calculate the mass-radius relation and to verify the effects of the general relativity on the model of white dwarfs stars.

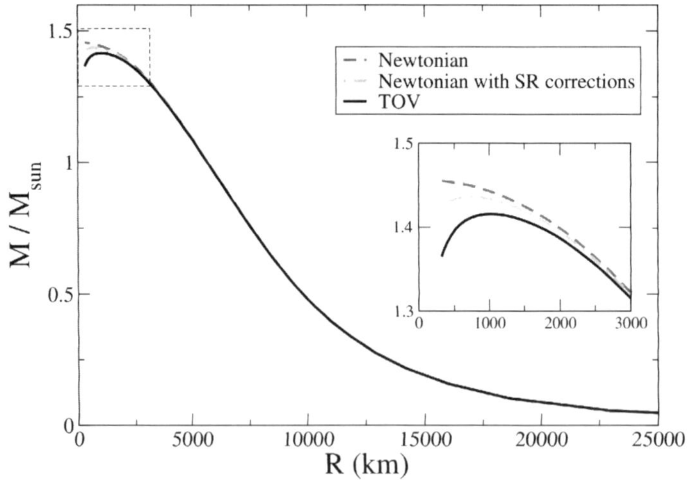

Figure 2. Mass-radius diagram for the general relativistic and newtonian ODE.

In the figure (2) we have solved the set of equations (1) and (2) using the general EoS (5). In "Newtonian with SR corrections" we solved the equations (1) and use energy density instead of mass density according to equation (3) and sum the contributions of the energy density of the electrons.
We can determine the maximum mass for the models of white dwarfs and realize that the difference between the models are small.

Table 1. Maximum mass for the models of white dwarfs stars.

Model	Maximum mass
Newton	$1.4559 \times M_{\odot}$
Newton with SR corrections	$1.4358 \times M_{\odot}$
TOV	$1.4154 \times M_{\odot}$

In figure (3) we calculated the mass-radius relation for the TOV equations (2) using the general EoS (5) and the non-relativistic EoS (equation (4) for $n = 3/2$).

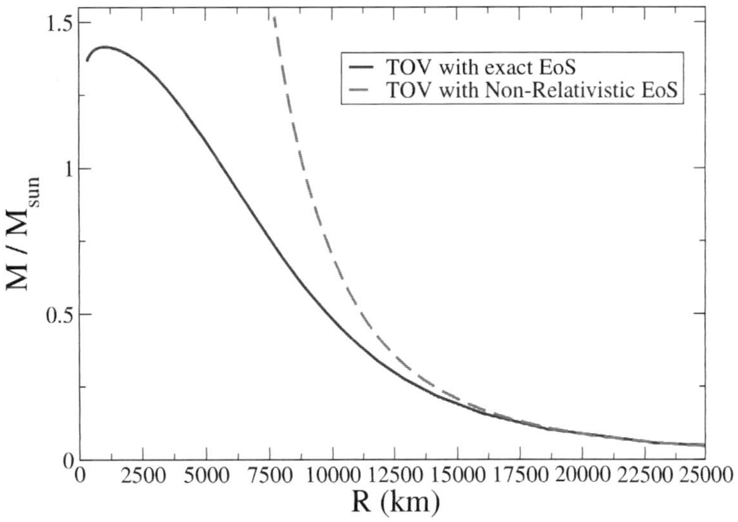

Figure 3. Mass-radius diagram using TOV equations and non-relativistic and exact EoS.

We see that the non-relativistic EoS works for the mass-radius relation for low values of mass and large radius untill approximately values of 15000 km.
In addition we try to find good fits for the mass-radius relation as a possible analytical correlation between the mass and radius of a white dwarf star. We fit the two curves in figure (3) as ploted in following graphics.

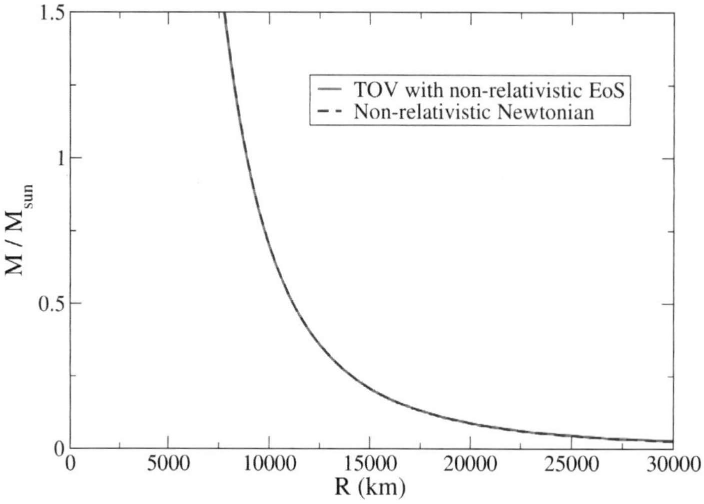

Figure 4. Non-relativistic mass-radius diagram fit.

In figure (4) we fit the mass-radius relation generated with TOV and non-relativistic EoS using,

$$\frac{M}{M_\odot} = 2.08 \times 10^{-6} \left(\frac{R}{R_\odot}\right)^{-3} . \tag{7}$$

We see that the non-relativistic EoS in general relativistic calculations doesn't modifies the well known non-relativistic Newtonian result obtained by Chandrasekhar in [6]. We also fit the general relativistic results in order to obtain an analitical expression for mass-radius relation in that case. In figure (5) we fit the mass-radius relation generated with TOV and general EoS

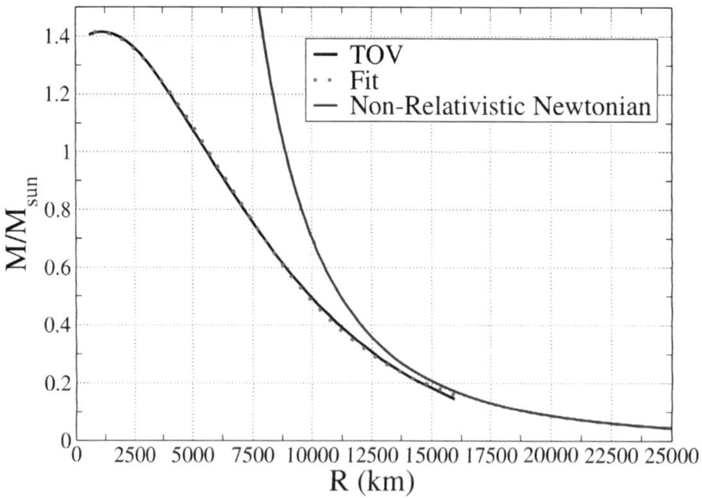

Figure 5. General relativistic mass-radius diagram fit.

using:

$$\frac{M}{M_{\odot}} = \frac{aR+b}{e^{cR^2}+d} \tag{8}$$

where a, b, c and are constants given by:

$$\begin{aligned} a &= 2.325 \times 10^{-5}\ \mathrm{km}^{-1} \\ b &= 0.4617 \\ c &= 7.277 \times 10^{-9}\ \mathrm{km}^{-2} \\ d &= -0.644 \end{aligned} \tag{9}$$

The fit (8) can be useful for white dwarfs with radius untill values approximatelly 17000 km.

5. Acknowledgments

I would like to thank CAPES (Coordenação de Aperfeiçoamento de Pessoal de Nível Superior) for the financial support.

References

[1] R. ADLER; M. BAZIN; M. SCHIFFER **Introduction to General Relativity**. second edition, 1975.
[2] I. SAGERT; M. HEMPEL; C. GREINER; J. SCHAFFNER-BIELICH. **Compact stars for undergraduates**. 2006 Eur. J. Phys. 27 577.
[3] S.L. SHAPIRO; S.A. TEUKOLSKY **Black Holes, White Dwarfs, and Neutron Stars: The Physics of Compact Objects.**. WILEY-VCH. 2004.
[4] R.R. SILBAR; S. REDDY **Neutron Stars for Undergraduates.** nucl-th/0309041v2. 2003.
[5] S.O. KEPLER; M.F.O. SARAIVA **Astronomia e Astrofsica.**. 790 p. Porto Alegre, Rio grande do Sul, 2012.
[6] S. CHANDRASEKHAR. **An Introduction to the Study of Stellar Structure**. Dover, 1967.

Stellar structures and the enigma of pulsars rotation frequency decay

H. O. de Oliveira, R. M. Marinho Jr, N. S. Maglhães

Physics Dept., Technological Institute of Aeronautics, Vila das Accias, São José dos Campos, SP, 12228-900, Brazil

E-mail: heitor.de.oliveira@hotmail.com

Abstract. Pulsars are astrophysical objects normally modelled as compact neutron stars that originated from the collapse of another star. This model, that we name canonical, assumes that pulsars are described by spherical magnetized dipoles that rotate, usually with the magnetic axis misaligned to the rotation axis. This misalignment would be responsible for the observation of radiation emitted in well-defined time intervals in a certain direction (lighthouse effect), the typical observational characteristic of this kind of star. It has been noticed that the rotation frequency of pulsars is slowly decaying with time (spin down), implying a gradual decrease of the rotational angular velocity (Ω). Such decay can be quantified by a dimensionless parameter called "braking index" ("n"), given by $n = \Omega\ddot{\Omega}/(\dot{\Omega})^2$, where a dot indicates a time derivative. The canonical model predicts that this index has one only value for all pulsars, equal to three. However, observational data indicate that actual braking indices are less than three, representing an enigma. The main goal of this research is the exploration of a more precise model for pulsars' rotation frequency decay.

1. Introduction

In July 1967, scientists at the University of Cambridge detected signals using a radio telescope. The observed signals were not only isolated peaks, but periodic pulses and, according the authors, "clearly displays the regular periodicity and also the characteristic irregular variation of pulse amplitude." [1]

Pulsar became the name of the new phenomenon, an abbreviation of pulsating stars, "The constancy of frequency in the recently discovered pulsed radio sources can be accounted for by the rotation of a neutron star." [2]

According to Gold [3], pulsars have a strong magnetic field and high rotational speeds, and the radiation is released as a pattern of a rotating beacon.

Neutron stars are the remnants of the supernova process. The compression of the remains of the star after the supernova effect also implies a compression of magnetic field leaving it stronger where the field strength is greatest at its poles due to a larger magnetic density. The gravitational attraction of the remains of the star that went through a supernova process results in an extremely compact object with radius of about 10 km, which has high-speed rotation due to conservation of angular momentum.

The discovery of the first pulsar and curiosity by this object led the analysis and discovery of more pulsars. The pulsar B0531+21, in the Crab Nebula, had an increase in its period [4] and brought more support to the theory that the period of pulsars increases.

2. The Lighthouse Model

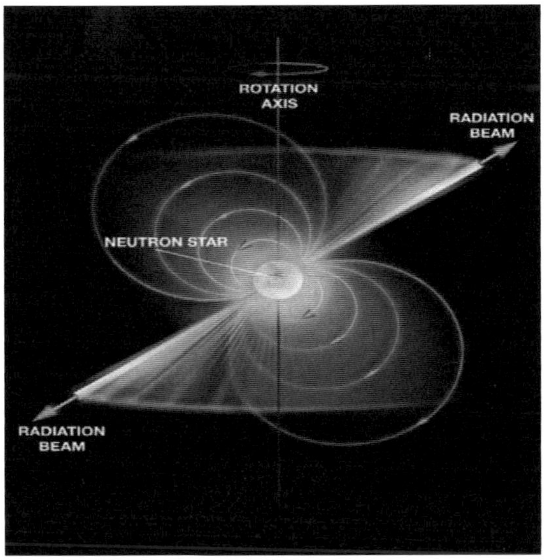

Figure 1. Artistic illustration of a Pulsar, available at: "http://www.astron.nl/about-astron/press-public/news/neutron-stars-laboratory/neutron-stars-laboratory"

This model is accepted as a way to explain the behavior of neutron stars. These neutron stars have a high magnetic field, where the intensity is greater at its poles, since the field lines are closer than at its equator, and with its high rotation speed particles are accelerated along these field lines forming a beam through which these particles emit radiation at radio frequencies.

2.1. The Model for Pulsars' Spin-down

We treat the pulsar as a rotating magnetic dipole, we can see that its rotational kinetic energy equals [5]

$$E_{rot} = \frac{1}{2} I \Omega^2 \tag{1}$$

and the time derivative of (1)

$$\frac{dE_{rot}}{dt} = \frac{1}{2} \cdot 2 \cdot I \cdot \Omega \cdot \dot{\Omega} + \frac{1}{2} \cdot \Omega^2 \cdot \dot{I} \tag{2}$$

where the second term of the above expression is very small in magnitude. Soon, we will have,

$$\dot{E}_{rot} = I \cdot \Omega \cdot \dot{\Omega} \tag{3}$$

and the magnetic energy radiated by this pulsar is the radiated energy of a magnetic dipole rotating [6],

$$\dot{E}_{mag} = -\frac{2}{3c^3} |\ddot{m}|^2 \tag{4}$$

being m the magnetic dipole moment, and $|\ddot{m}|^2$ equal to [5],

$$|\ddot{m}|^2 = m^2 \cdot \sin^2 \alpha \cdot (\Omega^4 + \dot{\Omega}^2)$$

where $\dot{\Omega}^2 \ll \Omega^4$,

$$|\ddot{m}|^2 = m^2 \cdot \sin^2 \alpha \cdot \Omega^4$$

then, the derivative of the magnetic energy has the form,

$$\dot{E}_{mag} = -\frac{2}{3c^3} m^2 \sin^2 \alpha \cdot \Omega^4. \tag{5}$$

Since the magnetic dipole $m = \frac{1}{2} B_P \cdot R^3$ [5], we get the expression for the magnetic energy,

$$\dot{E}_{mag} = -\frac{1}{6c^3} {B_P}^2 R^6 (sin\alpha)^2 \Omega^4 \tag{6}$$

where R is the radius of the sphere and B_P the magnetic field at the poles. And being k positive constant equal to,

$$k \equiv \frac{1}{6c^3} \frac{R^6}{I} (B_P \sin \alpha)^2$$

and with that, we come to a simplified expression for the rate of change of radiated energy by the dipole in time,

$$\dot{E}_{mag} = kI\Omega^4. \tag{7}$$

In the canonical model, one of the hypotheses is the conversion of rotational energy (E_{rot}) into magnetic energy (E_{mag}). From this we found that the expected value of n by the pulsars' spin-down model is equal to 3,

$$E_{rot} + E_{mag} = 0. \tag{8}$$

Substituting the values of the time derivatives of the kinetic energy of rotation (3) and magnetic energy (7) found,

$$I\Omega\dot{\Omega} + kI\Omega^4 = 0 \tag{9}$$

$$\dot{\Omega} = -k \cdot \Omega^3. \tag{10}$$

The expression in equation (10) relates the increasing of the rotation period of pulsars, i.e., the orbital velocity decay. The energy conservation leading into account the model of rotating magnetic dipole radiating its rotational energy shows that the *spin down* occurs with a power 3 of its angular velocity. If we derive the equation (10), we obtain

$$\ddot{\Omega} = -3 \cdot k \cdot \Omega^2 \cdot \dot{\Omega} \tag{11}$$

where we can organize it and replace $-k \cdot \Omega^3$ by $\dot{\Omega}$,

$$\ddot{\Omega} = 3 \cdot \frac{(-k \cdot \Omega^3)}{\Omega} \cdot \dot{\Omega} \tag{12}$$

$$\ddot{\Omega} = 3 \cdot \frac{(\dot{\Omega})}{\Omega} \cdot \dot{\Omega} \tag{13}$$

and we find that the product of $\ddot{\Omega}$ with Ω divided by $\dot{\Omega}$ should have exactly the value of 3.

$$\frac{\ddot{\Omega} \cdot \Omega}{\dot{\Omega}^2} = 3 \tag{14}$$

being, the value 3 in the equation (10) and (14) represents an intensity factor of braking. It is the called *braking index* of pulsars'. The observable Ω, $\dot{\Omega}$ and $\ddot{\Omega}$ are used to calculate the braking index and obtained experimentally, and should result in the value of 3. However, the values found for braking index of pulsars does not corresponds the expected value. We can observe that the generalization of the equation (10),

$$\dot{\Omega} = -k \cdot \Omega^n \tag{15}$$

and of (14) we have,

$$\frac{\ddot{\Omega} \cdot \Omega}{\dot{\Omega}^2} = n \tag{16}$$

even if, we generalize the equations we can see that the equation (16) would only make sense to a value of $n = 3$, but the first equation become a definition [7] where the angular deceleration is associated with decreased angular velocity raised to the exponent n.

3. Divergence in the observed values with the predicted

In the following table, we present the pulsars' with the value of braking index well observed [8] [9] [10] [11] [12].

Table 1. Experimental Data

Data	PSR B0531+21	PSR B0833-45	PSR B0540-69	PSR B1509-58	PSR J1119-6127	PSR J1846-0258
$f(Hz)$	30.2	11.2	19.2	6.63	2.45	3.06
$\frac{df}{dt}(10^{-10}s^{-2})$	−3.86	−0.15	−1.88	−0.67	−0.24	−0.66
$\frac{d^2f}{dt^2}(10^{-21}s^{-3})$	12.4	0.031	3.81	1.95	0.63	3.13
n	2.51	1.54	2.07	2.88	2.67	2.20

4. Conclusions

The current model does not corresponds the reality observations, and a possible fix is necessary to obtain correctly the values of braking index. However, the mathematical model must be able to relate any implication that may cause changes in the values of braking index.

From this, arise questions that, we lead to think, which conditions the pulsars' are subject to modify it value expected for the braking index, and hence, we realize that, we can propose an alternative way to calculate the braking index with the development a new model with related changes of variables (mass, radius, momentum of inertia, magnetic field, ...) that can to be related directly to the distortion of the predicted and observed values.

This work aims to present to the scientific community the problem which we believe to be important for nuclear astrophysics, thus to develop a model to solve correctly the problem of braking index.

5. Acknowledgments

To Coordenação de Aperfeiçoamento de Pessoal de Nível Superior CAPES, to the Technological Institute of Aeronautics and my advisors Nadja S. Magalhães e Rubens Marinho.

References

[1] Hewish, Antony; Bell, S Jocelyn; Pilkington, J D H; Scott, P F; Collins, R A; Observation of a rapidly pulsating radio source. Nature, v. 217, p. 709–713, 1968.
[2] Gold, Thomas; Rotating neutron stars as the origin of the pulsating radio sources. Nature, v.218, p. 731–732, 1968.
[3] Gold, Thomas; Rotating neutron stars and the nature of pulsars. In The Physics of Pulsars, v. 1, p. 101, 1972.
[4] Lorimer, Ducan R; Binary and millisecond pulsars. Living Rev. Relativity, v. 11, n. 8 (http://www.livingreview.org/lrr-2008-8).
[5] Shapiro, Stuart L; Teukolsky, Saul A; Black holes, white dwarfs and neutron stars: the physics of compact objects. [S.l.], John Wiley & Sons, 2008.
[6] Griffiths, David Jeffrey; Reed College; Introduction to eletrodynamics. 4 Ed. Boston, Addison-Wesley, 2012.
[7] Glendenning, Norma K; Compact stars: Nuclear physics, particle physics, and general relativity. New York, Springer, 2000.
[8] Magalhães, Nadja S; Predicting Ranges for Pulsars' Braking Indices. The Astrophysical Journal, v. 755, p. 54–57, 2012.
[9] Groth, E J; Timing of the crab Pular III. The slowing down and the nature of the random process. The Astrophysical Journal Supplement Series, v29, p. 453–465, 1975.
[10] Lyne, A G; Pritchard, R S; Graham-Smith, F; Camilo, F; Very low braking index for the Vela pulsar, v381, p. 497–498, 1996.
[11] Kaspi, V M; Manchester, R N; Siegman, B; Johnston, S; Lyne A G; On the spin-down of PSR B1509-58. The Astrophysical Journal, v. 422, p. L83–L86, 1994.
[12] Manchester, R N; Peterson, B A; A braking index for PSR 0540-69. The Astrophysical Journal, v 342, p. L23–L25, 1989.

Journal of Physics: Conference Series **630** (2015) 012060 doi:10.1088/1742-6596/630/1/012060

Low Energy Nuclear Structure from Ultrarelativistic Heavy-Light Ion collisions *

Enrique Ruiz Arriola[1] and Wojciech Broniowski[2,3]

[1] Departamento de Física Atómica, Molecular y Nuclear and Instituto Carlos I de Física Teórica y Computacional, Universidad de Granada, E-18071 Granada, Spain
[2] Institute of Physics, Jan Kochanowski University, 25-406 Kielce, Poland
[3] The H. Niewodniczański Institute of Nuclear Physics PAN, 31-342 Cracow, Poland

E-mail: earriola@ugr.es (ERA)

E-mail: Wojciech.Broniowski@ifj.edu.pl (WB)

Abstract. The search for specific signals in ultrarelativistic heavy-light ion collisions addressing intrinsic geometric features of nuclei may open a new window to low energy nuclear structure. We discuss specifically the phenomenon of α-clustering in ^{12}C when colliding with ^{208}Pb at almost the speed of light.

1. Introduction

Even before the neutron was discovered, based on the α decay explained successfully as a quantum tunneling effect across the Coulomb barrier, Gamow proposed that alpha particles (^{4}He nuclei) are the constituents of atomic nuclei [1]. Indeed, alpha clusters are present in light nuclei (such as, e.g., ^{12}C) as effective degrees of freedom and lead to large intrinsic deformation of their nuclear distributions, which in some cases lead to polyhedral symmetric structures: an equilateral triangle for ^{12}C, tetrahedron for ^{16}O, trigonal bipyramid for ^{20}Ne, octahedron pentagonal bipyramid for ^{24}Mg, hexagonal bipyramid for ^{28}Si, etc. [2, 3, 4, 5, 6], (for reviews see, e.g., [7, 8, 9, 10, 11]).

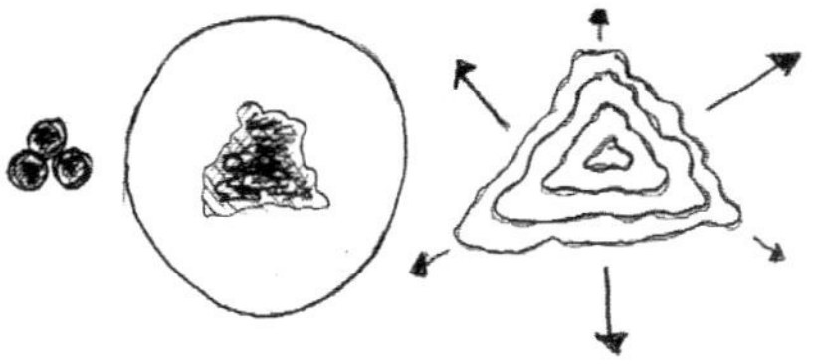

Figure 1. A triangle hitting the wall: ^{12}C pictured as a triangle of three α particles (left) colliding with a heavy ion drawn as a round flat object (middle). The triangular shape of the fireball yields a triangular pattern of particle emission after the collision (right).

*Talk by ERA at 37th Brazilian Workshop on Nuclear Physics, 8-12 September 2014, Maresias, SP, Brazil

When such a clustered geometric object, say ^{12}C, hits a large nucleus such as ^{208}Pb at almost the speed of light (see Fig. 1), the shape of the created fireball in the transverse plane reflects (up to random fluctuations) the deformation of the light nucleus and effectively implements the almost instantaneous collapse of the collective wave function [12]. As the hit nucleus is large, the fireball is abundant enough to evolve collectively, much as in collisions of two heavy nuclei at RHIC or the LHC. Because of the initial deformation, harmonic flow develops, leading to very specific and measurable signatures in the transverse momentum distributions of the detected hadrons produced in the collision. For instance, the ^{12}C+^{208}Pb system develops large triangular flow, increasing with the multiplicity of the number the produced hadrons. Traditionally, it has been assessed that only some very bulk information on the nuclear structure content could be probed by ultrarelativistic heavy-ion collisions. Our calculations show, however, that certain very distinct geometric features, such as the α-clustering in the nuclear ground state wave function, could indeed be traced significantly by properly analyzing the multiparticle spectrum in the final state [12, 13].

This talk returns to this remarkable link between the lowest energy nuclear structure and the highest-energy heavy-light-ions collisions, emphasizing some important aspects of this connection, mostly from the nuclear physics point of view. This is intended for specialists from either the $\alpha-$cluster and the relativistic heavy-ion communities who might be unfamiliar with the ideas of the other side.

In order to motivate the presentation below, let us review the basic elements needed to describe the collisions between two nuclei, A+B. Let the two nuclei A and B be characterized by wave functions satisfying $H_A\Psi_{n,A} = E_{n,A}\Psi_{n,A}$ and $H_B\Psi_{n,A} = E_{n,A}\Psi_{n,A}$. The collision at invariant CM energy $\sqrt{s}$, in their initial ground state $\Psi_{0,A}$ and $\Psi_{0,B}$, can be described by the total Hamiltonian,

$$H = H_A + H_B + V_{AB}, \tag{1}$$

where additivity of the elementary energy-dependent NN interactions is assumed,

$$V_{AB}(s;\vec{x}_1,\ldots,\vec{x}_A;\vec{y}_1,\ldots,\vec{y}_B) = \sum_{i=1}^{A}\sum_{j=1}^{B} V_{NN}(s,\vec{x}_i-\vec{y}_j), \tag{2}$$

with $V_{NN}(s,\vec{x}) = \mathrm{Re}V_{NN}(s,\vec{x}) + i\mathrm{Im}V_{NN}(s,\vec{x})$ denoting an optical potential. The basic tool of our analysis is given by the Glauber formalism (see, e.g., [14, 15]), which gives the density $P_{AB}(s,\vec{b})$ of inelastic collisions in the transverse plane, $\vec{x} = (\vec{x}_T, z)$, for the A+B reaction in terms of the elementary NN density, $P_{NN}(s,\vec{b})$. For instance, in the binary collision model (the popular wounded nucleon model applied in our studies includes the same ingredients)

$$\begin{aligned} P_{AB}(s,\vec{b}) &= \prod_{i=1}^{A}\int d^3x_i \prod_{j=1}^{B}\int d^3y_j |\psi_{0,A}(\vec{x}_1,\ldots,\vec{x}_A)|^2|\psi_{0,B}(\vec{y}_1,\ldots,\vec{y}_B)|^2 \\ &\times \frac{1}{\sigma_{AB}^{\mathrm{inel}}(s)}\left\{1-\prod_{i=1}^{A}\prod_{j=1}^{B}\left[1-\sigma_{NN}^{\mathrm{inel}}(s)P_{NN}(s,\vec{b}-\vec{x}_{i,T}+\vec{y}_{j,T})\right]\right\}, \end{aligned} \tag{3}$$

where $\sigma_{NN}^{\mathrm{inel}}(s)$ and $\sigma_{AB}^{\mathrm{inel}}(s)$ are the inelastic cross sections for NN and AB, respectively, and $\int d^2bP_{NN}(s,b) = \int d^2bP_{AB}(s,b) = 1$. This formula shows that we need a knowledge of the nuclear wave functions and the interaction between nucleons in the different nuclei. This is a purely hadronic picture of the collision process and the only partonic underlying features are hidden under $P_{NN}(s,\vec{b})$ which is usually obtained phenomenologically.

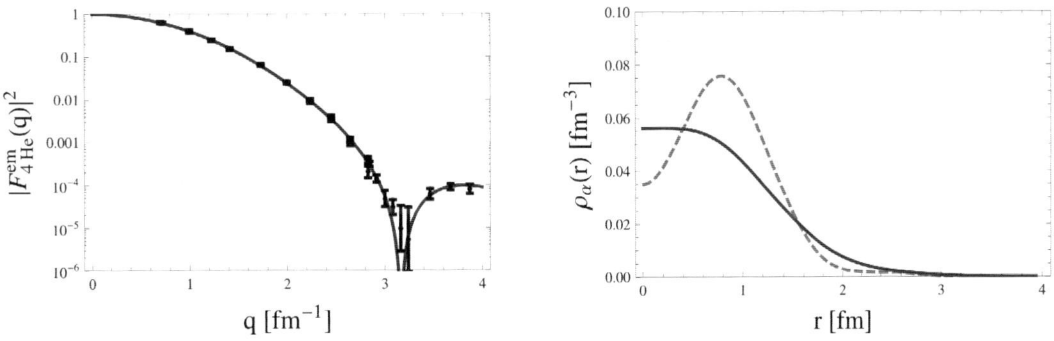

Figure 2. Left: Charge form factor for ^{4}He as a function of the momentum transfer (data from Ref. [16]). Right: Normalized charge (full line) and baryon (dashed line) density for the α particle as functions of the radial distance.

2. The α-particle

2.1. Basic properties

The α particle is a $^4\mathrm{He} = ppnn$ nucleus, which in the ground state has the quantum numbers $(J^P, T) = (0^+, 0)$. Its charge is $Z_\alpha e = +2e$ and mass $M_\alpha = 2(M_p + M_n) - B = 3727.37$ MeV with the binding energy $B = 28.2957$ MeV. There are no excited states below the $t+p$ or ^{3}He+n thresholds, and the nucleon separation energy is $S_n \sim 20$ MeV. The size information is obtained from the electron scattering, where the elastic $e +^4$ He differential cross section is given by

$$\left(\frac{d\sigma}{d\Omega}\right)_{e,\alpha} = \frac{d\sigma_{\mathrm{Ruth}}}{d\Omega}|F^{\mathrm{em}}_{^4\mathrm{He}}(q)|^2, \tag{4}$$

deviating from the point-like Rutherford result. The charge form factor can be well fitted by [16] $F^{\mathrm{em}}_{^4\mathrm{He}}(q) = \left(1 - q^{12}c^{12}\right) e^{-a^2_{em,\alpha} q^2}$, with $a_{em,\alpha} = 0.681(2)$ fm and $c = 0.316(1)$ fm, and vanishes at $q \simeq 3$ fm, see left Fig. 2. The corresponding charge mean squared radius is $\langle r^2 \rangle_{em,\alpha} = 6a^2_{em,\alpha} = (1.668(5)\ \mathrm{fm})^2$. The nucleon distribution is obtained by unfolding the charge distribution with the nucleon (proton) charge form factor $F^{\mathrm{em}}_p(q) = 1/(1 + q^2/\Lambda^2)^2$, $\Lambda^2 = 0.5\mathrm{GeV}^2$:

$$F^{\mathrm{em}}_\alpha(q) = \int d^3x e^{i\vec{q}\cdot\vec{x}} \rho^{\mathrm{em}}_\alpha(r) = Z_\alpha F^{\mathrm{em}}_p(q) \int d^3x e^{i\vec{q}\cdot\vec{x}} \rho_\alpha(r) \tag{5}$$

and assuming isospin invariance. Then the baryon density $\rho_B(r) = 4\rho_\alpha(r)$. Normalized charge and baryon densities are depicted in right Fig. 2. Thus, we have $\langle r^2 \rangle_\alpha = \langle r^2 \rangle_{em,\alpha} - \langle r^2 \rangle_{em,p} = (1.42\ \mathrm{fm})^2$.

2.2. Simple wave functions

A good and simple approximation for the wave function in the CM system, as far as the form factor is concerned for $c = 0$ in Eq. (5), is given by the shell model as a $|(1s)^4\rangle$ state,

$$\Psi_\alpha(\vec{x}_1, \vec{x}_2, \vec{x}_3, \vec{x}_4) = \prod_{i=1}^{4} \frac{e^{-\frac{(\vec{x}_i - \vec{R})^2}{2b^2}}}{(\sqrt{\pi}b)^{3/2}} \mathcal{A}(n\uparrow, n\downarrow, p\uparrow, p\downarrow), \tag{6}$$

where $\mathcal{A}$ is a normalized antisymmetrizer and $\vec{R} = \sum_{i=1}^4 \vec{x}_i/4$ is the CM. The ground state probability $|\Psi_\alpha(\vec{x}_1, \vec{x}_2, \vec{x}_3, \vec{x}_4)|^2$ can be constructed by randomly generating 12 Gaussian variables describing $\vec{x}_i$ of the four particles, and then relocating the CM $\vec{R} = \sum_{i=1}^4 x_i/4$ to the origin. More realistic wave functions can be achieved by including a short-distance correlation features, say a hard core of size d, by excluding configurations with $|\vec{x}_i - \vec{x}_j| \le d$ [17]. This simple approach will be pursued below for the ^{12}C ground state wave function. Further improvements incorporate asymptotic long-distance cluster decomposition conditions on the total wave function into ^{2}H+^{2}H, ^{3}H+n or ^{3}He+p subsystems [18], implying exponential fall-offs with proper separation wave numbers.

3. α-α interaction and effective elementarity

The study of the α-α interaction, involving eight nucleons, has been a continuous playground in cluster model studies [19, 20]. The total wave function is written as $\Psi_{\alpha,\alpha} = \Psi_\alpha(1,2,3,4)\Psi_\alpha(5,6,7,8)\chi_{\alpha\alpha}(r)$, where the relative wave function satisfies a RGM equation containing a direct term which is *local* and corresponds to a folding structure [21] very much like the nucleus-nucleus Sao Paulo potential [22], and an exchange term which is non-local and short range.

3.1. α-α potential

The direct electromagnetic interaction between the two α particles can be written as a folding potential, leading in the overlapping region to the screening of the pure Coulomb interaction below an elementarity radius, r_c,

$$V^{\rm em}_{\alpha\alpha}(r) = \int d^3\vec{r}_1 d^3\vec{r}_2 \frac{\rho^{\rm em}_\alpha(\vec{r}_1)\rho^{\rm em}_\alpha(\vec{r}_2)}{|\vec{r}_1 - \vec{r}_2 - \vec{r}|} = \int \frac{d^3\vec{q}}{(2\pi)^3}\frac{4\pi e^2}{\vec{q}^2}|F^{\rm em}_\alpha(\vec{q})|^2 e^{i\vec{q}\cdot\vec{r}} \sim \frac{Z_\alpha^2 e^2}{r}, \qquad r \gtrsim r_c.$$

The leading exchange interaction corresponds to separating one neutron or proton, leaving ^{3}H or ^{3}He behind, respectively, and is $\mathcal{O}(e^{-2\gamma_{n,t}r})$, where $S_n = \gamma_{n,t}^2/(2\mu_{n,t})$ the nucleon separation energy. Using the nuclear form factor, see Fig. 2, we immediately see from Fig. 3 that α particles interact effectively as if they were point-like for distances larger than $r \gtrsim 2.5$ fm. We call this *effective elementarity*, a feature that entitles us to treat interactions between α particles as point-like objects above this separation † and disregard the short-range exchange term. One interpretation of the α clustering is based on the fact that in $A = 4n$ nuclei α particles are on the average packed at a larger distance than r_c, and thus preserve their identity.

3.2. α-α scattering

Assuming this simple picture, we can analyze α-α scattering without much detailed knowledge of the internal α particle structure at sufficiently low energies, with an upper bound fixed by the relative CM momentum $p = \hbar/r_c \sim 200$ MeV corresponding to $T_{LAB} = 2p^2/M_\alpha \sim 15$ MeV. The scattering amplitude is written as a partial wave expansion [25]

$$f(\theta) = f_C(\theta) + \sum_{l=0}^{\infty}(2l+1)e^{2i\sigma_l(p)}\frac{e^{2i\delta_l(p)} - 1}{2ip}P_l(\cos\theta), \tag{7}$$

where $f_C(\theta)$ is the Coulomb amplitude, $\sigma_l(p)$ and $\delta_l(p)$ are the Coulomb and strong phase-shifts, respectively, obtained from solving the relative α-α wave function $\chi_{\alpha,\alpha}(r) \equiv u_{l,p}(r)/r$ with the boundary condition $u_{l,p}(r) \sim G_l(\eta, pr)\sin\delta_l(p) + F_l(\eta, pr)\cos\delta_l(p)$. Here $G_l(\eta, pr)$ and $F_l(\eta, pr)$ are Coulomb wave functions and $\eta = 1/(pa_B)$ is the Sommerfeld parameter and

†Due to the quantum numbers of the α particle, the longest range non-em interaction is of van der Waals nature and is given by Two-Pion-Exchange, a tiny effect [23, 24]. Note that this notion of elementarity has to do with the interaction; the simple overlap integral $\int d^3s\rho_\alpha(s-r)\rho_\alpha(s) \approx 0$ for $r = r_c$ is not the relevant property.

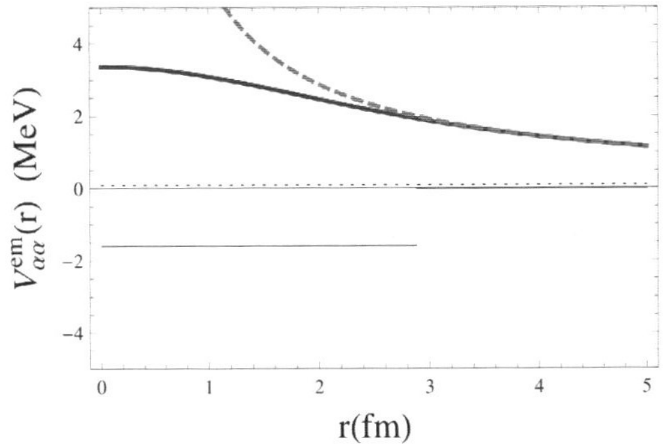

Figure 3. Direct α-α electromagnetic potential (in MeV) as a function of the distance (in fm). We compare the folding potential (solid line) with the elementary Coulomb potential (dashed line), the $Q = 92\,$KeV value for ^{8}Be (dotted line) and the $B_{\alpha\alpha} = 1.52$MeV binding when Coulomb is removed above $r_c = 2.88\,$fm (solid broken line).

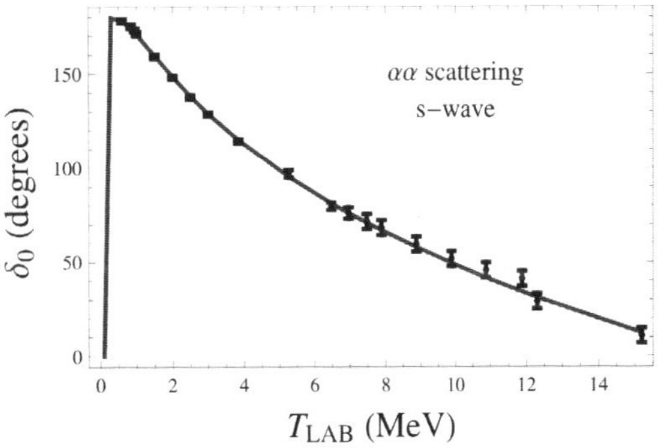

Figure 4. S-wave phase shift for the α-α scattering as a function of the LAB energy. The line is the fit using a boundary condition at a distance of $r_c = 2.88$fm and elementary Coulomb potential above r_c. The data can be traced from Ref. [25].

$a_B = 2/(M_\alpha Z_\alpha^2 e^2) = 3.63\,$fm the Bohr radius. Due to the Bose statistics, we have $l = 0, 2, 4, \dots$. The partial wave expansion, Eq. (7), is limited in practice to $l_{\rm max} \sim pa$, with a denoting the range of the non-electromagnetic interaction, since classically $L = pb$ and there is no scattering for $b > a$. Below the first inelastic ^{7}Li+p and ^{7}B+n thresholds, which take place at $T_{\rm LAB} \sim 15$MeV, we have $l_{\rm max} \sim 4$. Using an energy independent boundary condition at a given cut-off radius r_c we get for $l = 0$ (S-wave)

$$-\frac{1}{M_\alpha} u''_{0,p}(r) + \frac{e^2 Z_\alpha^2}{r} u_{0,p}(r) = \frac{p^2}{M_\alpha} u_{0,p}(r), \qquad L = \frac{u'_{0,p}(r_c)}{u_{0,p}(r_c)}, \tag{8}$$

from which we can fit the S-wave phase-shift with the parameters [23, 24]

$$L = -0.357(3)\ \text{fm}^{-1}, \qquad r_c = 2.88(3)\ \text{fm}, \qquad \chi^2/\text{DOF} = 0.5., \tag{9}$$

which is rather good, see Fig. 4. Thus, all the necessary information is encoded in these parameters, in harmony with an effective elementarity of α particles and the irrelevance of the short distance characteristics. Therefore, as long as α particles remain at distances larger than ~ 3 fm, we may ignore all the substructure details, including the exchange terms. That way one can sidestep a full *ab initio* calculation with 8 nucleons.

3.3. α-α binding and tunneling

The steep rise of the phase-shift starting at zero corresponds to the ^{8}Be resonance, which arises as a pole of the S matrix in the second Riemann sheet. The asymptotics $u_R(r) \to e^{ipr}$ with the complex momentum $p = p_R + ip_I$ gives a complex energy $E = Q - i\Gamma/2 = (p_R + ip_I)^2/(2\mu_{\alpha\alpha}) =$ 92 KeV $- \frac{i}{2}3.4(2)$ eV, so that the turning point, fulfilling $Q = V^{\rm em}_{\alpha\alpha}(r_0)$, is rather large $r_0 = 62$fm, yet tunneling through the barrier occurs. A further interesting aspect is that if the Coulomb force were removed above $r_c = 2.88$ fm, the ^{8}Be resonance would then become a stable weakly bound state with a binding energy given by $B_{\alpha\alpha} = -\gamma_B^2/M_\alpha = -1.5229$ MeV (see Fig. 3), a fairly large radius, $\langle r^2 \rangle^{\frac{1}{2}} = 4.0764$ fm, and the asymptotic coefficient of the normalized reduced wave function, $u_B(r) \to A_B e^{-\gamma_B r}$ of $A_B = 2.4163\ \text{fm}^{-1/2}$.

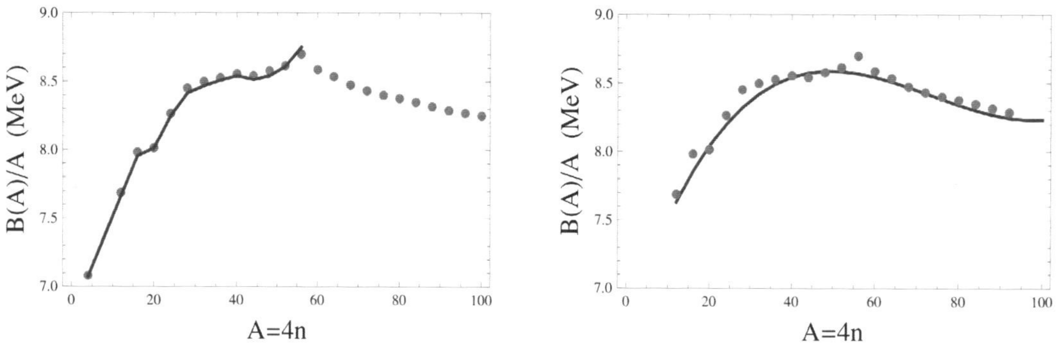

Figure 5. Binding energy per particle as a function of the mass number for $A = 4n$ nuclei. Left: Bonding geometric model. Right: Many-body model containing two-,three- and four-α forces.

4. Evidence for α-clustering

The effective elementarity of α particles has far reaching consequences. The most clear signatures for α clustering in nuclei are seen in the binding energies and form factors. Already in 1966 Harrington proposed to treat ^{12}C as a bound state of three elementary α particles [26].

4.1. Binding energies

Since the α particle is very tightly bound, a hint for clustering for $A = 4n$ nuclei was given by the bonding approximation [3, 27],

$$E(4n) = -B(4n) = -nB_\alpha + n_{\rm bonds} V_{\alpha\alpha} \qquad V_{\alpha\alpha} = -2.29 {\rm MeV}\,, \qquad n \geq 3\,, \tag{10}$$

where $n_{\rm bonds} = 3, 4, 8, 12, 16, 19, 22, 25, 27, 30$ for ^{12}C, ^{16}O, ^{20}Ne, ^{24}Mg, ^{28}Si, etc., respectively, yielding $n_{\rm bonds}/n \to 2.5$ (we stop plotting at $A = 56$ since $n_{\rm bonds}$ can always be adjusted onwards). This corresponds a next-neighbors interaction of closely packed spheres (see [7] for pictures illustrating the polyhedral configurations) ‡. The result of the fit can be seen in Fig. 5 (left). One drawback of this approach is that the fit does not reproduce the simplest $n = 2$ case, which corresponds to ^{8}Be.

In a different scheme we just count the numbers of pairs, triplets and quartets, corresponding to two-,three- and four body forces, and the energy reads

$$E(4n) = -nB_\alpha + \binom{n}{2} V_{2\alpha} + \binom{n}{3} V_{3\alpha} + \binom{n}{4} V_{4\alpha}\,, \qquad n \geq 3\,, \tag{11}$$

A fit for $A = 4n$ nuclei gives $V_{2\alpha} = -2.39$ MeV, $V_{3\alpha} = 0.47$ MeV, $V_{4\alpha} = -0.04$ MeV . The result can be seen in Fig. 5 (right), with a mean standard deviation of 0.1 MeV per nucleon. Again, the meaning of $n = 2$ is particularly subtle since, as we have shown, this corresponds to the ^{8}Be case which is unbound by ~ 0.092 MeV while direct extrapolation of the general formula for n implies ~ -2.4 MeV. The ^{8}Be is an exceptionally large system which becomes very sensitive to the long range Coulomb interaction. As we have seen elimination of $V_{\alpha\alpha}^{\rm em}(r)$ above r_c gives $B_{\alpha\alpha} \sim 1.52$ MeV, see Fig. 3, reducing partly the puzzle in the $n = 2$ case both for the bonding model, Eq. (10), and the many body model, Eq. (11).

‡The case of ^{20}Ne would require $n_{\rm bonds} = 9$ for this interpretation but that gives a worse binding.

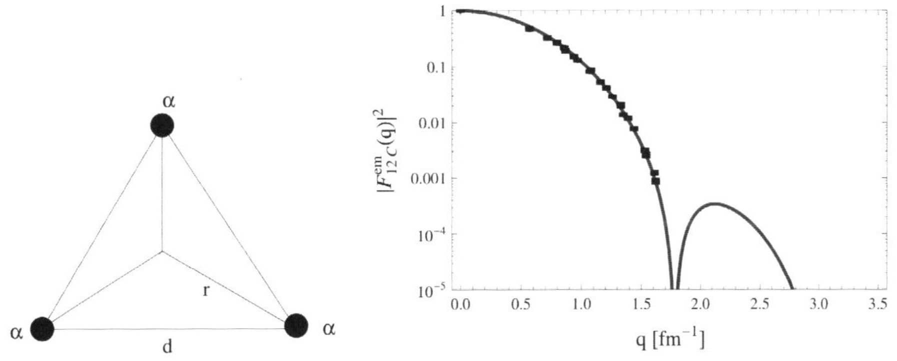

Figure 6. Left: Triangle structure in ^{12}C. Right: Corresponding charge corm factor of ^{12}C from folding the α charge form factor with the triangle $F^{\rm em}_{^{12}C}(q) = F^{\rm em}_{^4He}(q) F_\triangle(q)$ with $d = 3$fm (data from Ref. [28]).

4.2. Cluster Sizes

Charge form factors provide valuable insight on cluster geometries and sizes. If we focus on ^{12}C, we fold an equilateral triangle with side d and radius $r_\triangle = d/\sqrt{3}$ with an α particle, see Fig. 6,

$$F^{\rm em}_{^{12}C}(q) = F^{\rm em}_{^4He}(q) F_\triangle(q)\,, \qquad F_\triangle(q) = \frac{\sin(qd/\sqrt{3})}{qd/\sqrt{3}}\,, \tag{12}$$

A fit to the experimental elastic form factor of ^{12}C [28] taking the experimental $F^{\rm em}_{^4He}(q)$ (see Eq. 5 and Fig. 2) yields $d = 3.05$ fm $> r_c = 2.88$ fm, consistent with the effective elementarity.

4.3. Modeling clustering

Besides this very simple but direct evidence, a massive effort is being carried out [7, 8, 9, 10]) to probe the clusterization features of light nuclei. Clustering in ^{12}C has been described by the Bose-Einstein Condensation (BEC) model [29], the fermionic molecular dynamics (FMD) [30], antisymmetrized molecular dynamics [31], effective chiral field theory on the lattice [32], the no-core shell model [33, 34], or the variational Green's function method (VMC) [35]. The recently discovered 5^- rotational state of ^{12}C in low energy $\alpha + ^{12}C$ collisions points to the triangular $\mathcal{D}_{3h}$ symmetry of the system [36]. While it would be best to incorporate realistic calculations (see, e.g., [37, 38]), in Ref. [12] we have applied a simple and practical procedure with α clustered (or unclustered for comparison) random distributions constructed as follows. In the clustered case we randomly generate positions of the 12 nucleons, 4 in each cluster of a Gaussian shape and size r_α. The centers of the clusters are placed in an equilateral triangle of side length d, cf. Fig. 6. The distribution of the 12 nucleons is recentered such that the center of mass is placed at the origin. The short-distance NN repulsion is incorporated by precluding the centers of each pair of nucleons to be closer than the expulsion distance of 0.9 fm [17].

The single and pair distribution functions are defined as (see [40] for multiple distributions)

$$\begin{aligned} \rho(\vec{x}) &= \langle \Psi_A | \hat{\rho}(\vec{x}) | \Psi_A \rangle = \int d^3x_2 \ldots d^3x_A |\Psi_A(\vec{x}, \vec{x}_2, \ldots, \vec{x}_A)|^2, \\ \rho_2(\vec{r}) &= \int d^3R d^3x_3 \ldots d^3x_A |\Psi_A(\vec{R} + \frac{1}{2}\vec{r}, \vec{R} - \frac{1}{2}\vec{r}, \vec{x}_3, \ldots, \vec{x}_A)|^2, \end{aligned} \tag{13}$$

and are normalized to unity, i.e. $\int d^3r \rho(\vec{r}) = \int d^3r \rho_2(\vec{r}) = 1$, admitting a simple probability interpetation. These are the densities corresponding to point-like nucleons, hence the charge

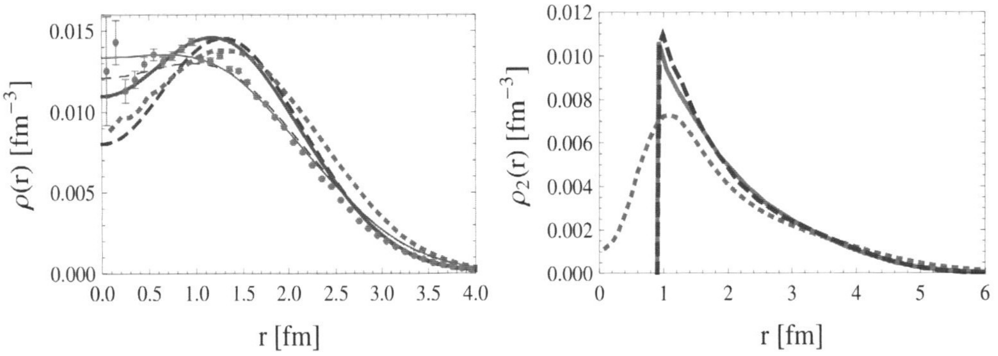

Figure 7. Left: Normalized one-particle distributions in ^{12}C. The electric charge density $\rho^{\rm em}(r)/Z$ (thin lines) and the corresponding distribution of the centers of nucleons $\rho(r)$ (thick lines) in ^{12}C for the data [39] and BEC [29] calculations (dashed lines), FMD [30] calculations (solid lines), and Jastrow correlated wave function [37] (dotted line). Right: Normalized two particle distribution $\rho_2(r)$ in ^{12}C. We show our results for the fitted $\rho(r)$ in the FMD (solid line) and BEC cases (dashed line), and compare to the Jastrow correlated wave function [37] (dotted line).

form factor reads,

$$F_A^{\rm em}(q) = ZF_p^{\rm em}(q)\int d^3x\rho(\vec{x})e^{i\vec{x}\cdot\vec{q}}, \tag{14}$$

where $F_p^{\rm em}(q)$ is the proton charge form factor (we neglect the small neutron charge form factor and assume isospin symmetry). These densities have exponential fall-offs,

$$\begin{aligned}\Psi_A(\vec{x},\vec{x}_2,\ldots,\vec{x}_A) &\xrightarrow{x\to\infty} Z_A\chi_1(\vec{x})\Psi_{A-1}(\vec{x}_2,\ldots,\vec{x}_A),\\ \Psi_A(\vec{x},\vec{x}',\vec{x}_3,\ldots,\vec{x}_A) &\xrightarrow{x,x'\to\infty} Z_A\chi_{12}(\vec{x},\vec{x}')\Psi_{A-2}(\vec{x}_3,\ldots,\vec{x}_A),\end{aligned} \tag{15}$$

because of the cluster decomposition. In Refs. [18, 41] the calculations explicitly implemented this exact asymptotic property, which would become relevant when separation energies are small or in the case of halo nuclei, which is not the case for ^{12}C in the ground state.

The parameters d and r_α are optimized such that the one particle density $\rho(r)$ of the BEC [29] or FMD [30] calculations are accurately reproduced (as explained, the unfolding of the proton charge density from the charge distribution $\rho^{\rm em}(r)$ is necessary), see Fig. 7. Note a large central depletion in the distributions, originating from the separation of the α clusters. Besides, a fair reproduction of the two-particle densities, $\rho_2(r)$, obtained from multiclustered Jastrow correlated calculations [37] is observed. The radial distribution $4\pi r^2\rho_2(r)$ peaks at the size of the triangle, $d \simeq 3$ fm [42]. Thus, in our simulations we deal with simplified but realistic nuclear distributions of ^{12}C.

5. Ultrarelativistic collisions

We now sketch the basic ingredients behind Eq. (3). We start by analyzing the NN collisions at high energies, where the Mandelstam $s = 4(p^2 + M_N^2)$ in the CM frame.

5.1. *NN collisions at high energies*

Neglecting spin effects, the scattering amplitude can be written as in Eq. (7) and in the limit $l_{\rm max} \sim pa \gg 1$ we can replace the sum over l by an integral. Using the impact parameter as the

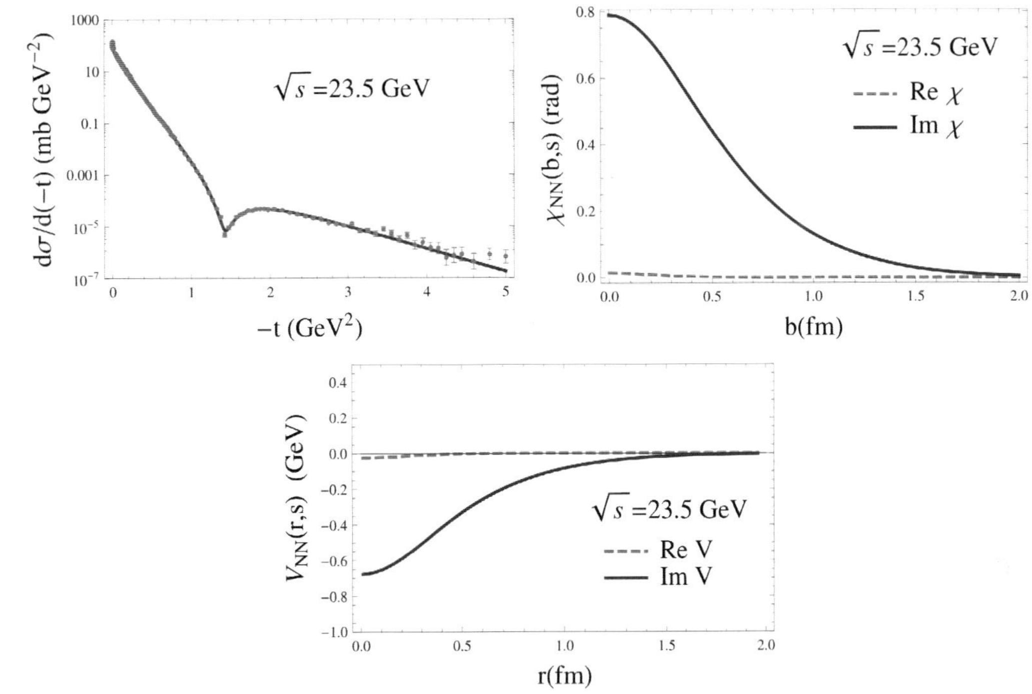

Figure 8. NN scattering at CM energy $\sqrt{s} = 23$ GeV. Top Left: Differential cross section as a function of the invariant squared momentum transfer $-t$ vs. data from [43, 44]. Top Right: Eikonal real and imaginary phase as a function of the impact parameter. Bottom: Optical potential as a function of the distance.

integration variable $bp = l + 1/2$, the formula $P_l(\cos\theta) \to J_0(bq)$, and dropping the Coulomb interaction, yields [14]

$$f_{NN\to NN}(s,t) = \frac{ip}{2\pi}\int d^2b[1 - e^{2i\delta(b)}]e^{i\vec{b}\cdot\vec{q}}. \tag{16}$$

At high energies relativistic effects and inelasticities must be taken into account. The general field theoretic approach would require a coupled channel Bethe-Salpeter equation, where the kernel would ultimately be determined phenomenologically from the NN scattering data. Under these circumstances, it is far simpler to use a minimal relativistic approach based on the squared mass operator [45]. Here $\mathcal{M}^2 = P^\mu P_\mu + W$, where W represents the (invariant) interaction determined in the CM frame by matching to the non-relativistic limit with an energy-dependent and local optical potential, $V(\vec{r},s) = \text{Re}V(\vec{r},s) + i\text{Im}V(\vec{r},s)$. This yields for NN after quantization $\hat{\mathcal{M}}^2 = 4(\hat{p}^2 + M_N^2) + 8M_N V$, with $\hat{p} = -i\nabla$, so that the relativistic equation can be written as $\hat{\mathcal{M}}^2\Psi = 4(k^2 + M_N^2)\Psi$, i.e., as a non-relativistic Schrödinger equation

$$(-\nabla^2 + M_N V)\Psi = (s/4 - M_N^2)\Psi\,. \tag{17}$$

In the eikonal approximation the (complex) phase-shift is given by

$$\delta_l(k)\Big|_{l+\frac{1}{2}=bk} \approx -\frac{2\mu}{2k}\int_b^\infty dr\frac{rV(r,s)}{\sqrt{r^2-b^2}} \equiv \frac{1}{2}\chi(b,s) = \frac{1}{2}\text{Re}\chi(b,s) + i\frac{1}{2}\text{Im}\chi(b,s), \tag{18}$$

where $\chi(b,s)$ is the eikonal phase. The optical potential $V(r,s)$ can be obtained from a fit to the experimental elastic differential cross section $d\sigma/d(-t) = |f|^2\pi/p^2$. The fits for pp scattering at $\sqrt{s} = 23.5$ GeV [43, 44] are depicted in Fig. 8 for illustration (details and further results will be presented elsewhere). As we see, both the phase shift and the optical potential are almost purely imaginary, i.e., $\mathrm{Re}\chi(b) \ll \mathrm{Im}\chi(b)$. This is a general feature above $\sqrt{s} \gtrsim 10$ GeV, which allows us to write the inelastic cross section from the optical theorem $\sigma_{\rm tot} = 4\pi \mathrm{Im} f(0)/p \sim \sigma_{\rm inel}$,

$$\sigma_{\rm inel}(s) = \int d^2b \left[1 - e^{-\mathrm{Im}\chi_{NN}(s,b)}\right] \equiv \sigma_{\rm inel}(s) \int d^2b P_{NN}(s,b), \tag{19}$$

in terms of a profile function $P_{NN}(s,b)$ interpreted as the probability of the NN inelastic collision at impact parameter b.

5.2. *Nucleus-nucleus scattering (Glauber)*

Nucleus-nucleus interactions is most conveniently studied by the Glauber theory applied to multiple scattering (for a review see e.g. [46].), where the highly asymmetric role played by the NN interactions within a given nucleus and NN interactions between nucleons of different nuclei is explicitly exploited. Actually, the interaction time is so short that only high energy states are excited, so basically both incident nuclei remain in their ground states. Under these assumptions, the initial wave function can be written as $\Psi = e^{ipz}\Phi_{AB}\psi_A\psi_B$ for a collision along the $z-$axis, with a slowly varying relative wave function Φ_{AB}. The elastic scattering amplitude becomes

$$f_{AB\to AB}(s,t) = \frac{ip}{2\pi}\int d^2b \langle \Psi_A\Psi_B|[1 - e^{i\chi_{AB}(b)}]|\Psi_A\Psi_B\rangle e^{i\vec{b}\cdot\vec{q}}. \tag{20}$$

The key simplifying aspect is that from the additivity of the NN interactions one obtains the additivity of the eikonal phases,

$$\chi_{AB}(b;\vec{x}_{1,T},\ldots,\vec{x}_{A,T};\vec{y}_{1,T},\ldots,\vec{y}_{B,T}) = \sum_{i=1}^{A}\sum_{j=1}^{B}\chi_{NN}(\vec{b} - \vec{x}_{i,T} + \vec{y}_{j,T}). \tag{21}$$

A purely imaginary NN phase, $\chi_{NN}(b) \sim i\mathrm{Im}\chi_{NN}(b)$, implies a purely imaginary AB-phase, $\chi_{AB}(b) \sim i\mathrm{Im}\chi_{AB}(b)$, and Eq. (3) follows from using again the optical theorem for the AB case.

The evaluation of the Glauber multidimensional integral in Eq. (3), or for the applied wounded nucleon model, can be done through the Monte Carlo techniques such as, e.g., in GLISSANDO [47, 48]. Within such an approach the two-dimensional density of sources can be constructed. We note that a realistic distribution that can be used for the hydrodynamic evolution requires smearing, as to make it sufficiently smooth. We refer to Ref. [12] for illustrations resembling our cartoon of Fig. 1.

Valuable information on the collision geometry and its fluctuations can be obtained from the study of eccentricity parameters (heavily used in relativistic heavy-ion analyses). One introduces the Fourier decomposition of the source density in the transverse plane *in a given event*,

$$\epsilon_n e^{in\Phi_n} = \frac{\int d^2b b^n P(b) e^{in\phi}}{\int d^2b b^n P(b)}, \tag{22}$$

with the eccentricity parameters ϵ_n encoding the information on the shape of the fireball distribution. Here n denotes the rank and Φ_n is the principal axis angle. Then ϵ_2 is called *ellipticity*, ϵ_3 *triangularity*, etc. In general, there are two sources for the ϵ_n parameters: the intrinsic mean deformation of the fireball and fluctuations.

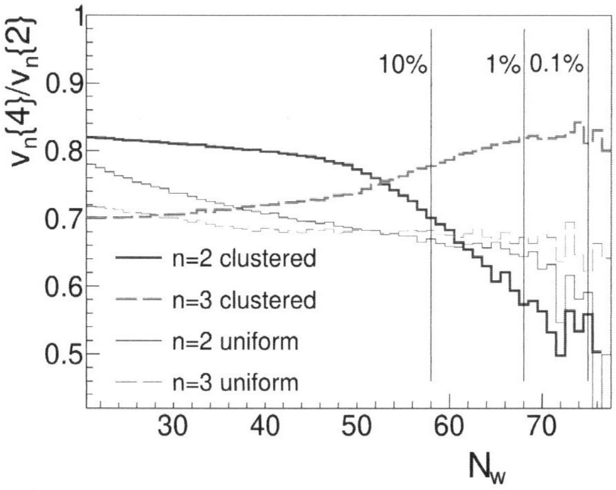

Figure 9. Ratios of harmonic flow coefficients of 4- and 2-particle cumulant moments for the elliptic ($n = 2$) and triangular ($n = 3$) case, plotted as functions of the number of wounded nucleons, and displaying the radically different behavior for the clustered and uniform ^{12}C wave functions (thick vs. thin lines). BEC case, mixed Glauber model for $^{12}C+^{208}Pb$ collisions with the SPS parameters [13]. The vertical lines indicate centralities.

5.3. Harmonic flow

The harmonic flow consequences, particularly elliptic and triangular flow have been analyzed in detail in Ref. [13, 49], where it is shown that the intrinsic deformation results in very characteristic behavior of the ratios of the so called cumulant moments as functions of the number of participant nucleons, both for the elliptic and triangular deformations. Such ratios are insensitive to the details of dynamics as long as the response of the evolution is linear in the initial deformation, which is the fact for the studied systems [50].

In Fig. 9 we show the ratio of the four- to two-particle cumulant moments for ellipticity and triangularity for the $^{12}C+^{208}Pb$ collisions at the highest the SPS energy of $\sqrt{s_{NN}} = 17$ GeV. We note a significant difference in the behavior for the clustered wave functions of ^{12}C (thick lines) and the uniform distributions (thin lines). Such effects should be the base for searches of clusterization signals in experiments.

6. Conclusions

When the experimental data on relativistic light-heavy nuclear collisions become available, our method, linking the lowest-energy nuclear phenomena (alpha clustering) and the highest-energy nuclear collisions (collective flow), will offer a novel way to study nuclear correlations in the ground state. Conversely, a detailed knowledge of the clustered light-nucleus wave functions will help to constrain the dynamics of the fireball evolution models (hydrodynamics, transport) and thus gain information on the fundamental properties of the hot and dense matter in the fireball. With this in mind, our method, inherently investigating multiparticle correlations leading to collective effects (harmonic flow), may provide new insights. We stress that since the clusterization phenomenon concerns multi-particle correlations, it is accessible directly only through observables which are many-body. Thus the typically studied one-body quantities, such as excitation spectra of the EM form factors, by definition cannot "prove" clusterization in a direct manner. This unexpected bridge between these two disparate fields of intense activity might enhance the mutual interest of researchers in nuclear structure to produce realistic wave functions which could be used to prepare the initial starting point for the hydrodynamic description of relativistic heavy-ion collisions.

Acknowledgments
This research was supported by the Polish National Science Centre, grants DEC-2011/01/D/ST2/00772 and DEC-2012/06/A/ST2/00390, Spanish DGI (grant FIS2011-24149) and Junta de Andalucía (grant FQM225).

References
[1] Gamow G 1931 *Constitution of atomic nuclei and radioactivity* (Oxford University Press)
[2] Wefelmeier W 1937 *Zeitschrift für Physik* **107** 332–346
[3] Hafstad L R and Teller E 1938 *Phys. Rev.* **54**(9) 681–692
[4] Wheeler J A 1937 *Phys. Rev.* **52**(11) 1083–1106
[5] Brink D 1965 *Proc. Int. School Enrico Fermi, Course* **36**
[6] Brink D, Friedrich H, Weiguny A and Wong C 1970 *Physics Letters B* **33** 143 – 146
[7] Freer M 2007 *Reports on Progress in Physics* **70** 2149
[8] Ikeda K, Myo T, Kato K and Toki H 2010 *Clusters in Nuclei-Vol.1* (Lecture Notes in Physics **818**, Springer)
[9] Beck C 2012 *Clusters in Nuclei-Vol.2* (Lecture Notes in Physics **848**, Springer)
[10] Okolowicz J, Nazarewicz W and Ploszajczak M 2013 *Fortsch.Phys.* **61** 66–79 (*Preprint* 1207.6225)
[11] Zarubin P I 2014 *Clusters in Nuclei-Vol.3* (Springer)
[12] Broniowski W and Arriola E R 2014 *Phys.Rev.Lett.* **112** 112501 (*Preprint* 1312.0289)
[13] Bozek P, Broniowski W, Arriola E R and Rybczynski M 2014 (*Preprint* 1410.7434)
[14] Florkowski W 2010 *Phenomenology of Ultra-Relativistic Heavy-Ion Collisions* (World Scientific Publishing Company, Singapore)
[15] Florkowski W 2014 (*Preprint* 1410.7904)
[16] Frosch R, McCarthy J, Rand R and Yearian M 1967 *Physical Review* **160** 874
[17] Broniowski W and Rybczyński M 2010 *Phys. Rev.* **C81** 064909 (*Preprint* 1003.1088)
[18] Schiavilla R, Pandharipande V and Wiringa R B 1986 *Nucl.Phys.* **A449** 219–242
[19] Tang Y, Lemere M and Thompson D 1978 *Phys.Rept.* **47** 167–223
[20] Friedrich H 1981 *Phys.Rept.* **74** 211–275
[21] Satchler G and Love W 1979 *Phys.Rept.* **55** 183–254
[22] Chamon L, Carlson B, Gasques L, Pereira D, De Conti C *et al.* 2002 *Phys.Rev.* **C66** 014610
[23] Ruiz Arriola E 2007 (*Preprint* 0709.4134)
[24] Ruiz Arriola E 2008 *AIP Conf.Proc.* **1030** 135–140 (*Preprint* 0804.2199)
[25] Afzal S, Ahmad A and Ali S 1969 *Rev.Mod.Phys.* **41** 247–273
[26] Harrington D R 1966 *Phys. Rev.* **147**(3) 685–688
[27] Goldhammer P 1963 *Rev. Mod. Phys.* **35**(1) 40–107
[28] Sick I and McCarthy J 1970 *Nuclear Physics A* **150** 631–654
[29] Funaki Y, Tohsaki A, Horiuchi H, Schuck P and Ropke G 2006 *Eur.Phys.J.* **A28** 259–263
[30] Chernykh M, Feldmeier H, Neff T, von Neumann-Cosel P and Richter A 2007 *Phys.Rev.Lett.* **98** 032501
[31] Kanada-En'yo Y 2007 *Prog.Theor.Phys.* **117** 655–680
[32] Epelbaum E, Krebs H, Lahde T A, Lee D and Meissner U G 2012 *Phys.Rev.Lett.* **109** 252501
[33] Roth R, Binder S, Vobig K, Calci A, Langhammer J *et al.* 2012 *Phys.Rev.Lett.* **109** 052501
[34] Barrett B R, Navratil P and Vary J P 2013 *Prog.Part.Nucl.Phys.* **69** 131–181
[35] Pieper S C, Varga K and Wiringa R B 2002 *Phys.Rev.* **C66** 044310
[36] Marin-Lambarri D, Bijker R, Freer M, Gai M, Kokalova T *et al.* 2014 *Phys.Rev.Lett.* **113** 012502
[37] Buendia E, Galvez F and Sarsa A 2004 *Phys.Rev.* **C70** 054315 (*Preprint* nucl-th/0405027)
[38] Wiringa R, Schiavilla R, Pieper S C and Carlson J 2014 *Physical Review C* **89** 024305 (*Preprint* 1309.3794)
[39] De Vries H, De Jager C and De Vries C 1987 *Atom.Data Nucl.Data Tabl.* **36** 495–536
[40] Viollier R and Walecka J 1977 *Acta Phys.Polon.* **B8** 25
[41] Wiringa R B 1991 *Phys.Rev.* **C43** 1585–1598
[42] Broniowski W and Ruiz Arriola E 2014 (*Preprint* 1407.8495)
[43] Amaldi U and Schubert K R 1980 *Nucl.Phys.* **B166** 301
[44] Amos N A, Block M, Bobbink G, Botje M, Favart D *et al.* 1985 *Nucl.Phys.* **B262** 689
[45] Allen T, Payne G and Polyzou W N 2000 *Phys.Rev.* **C62** 054002 (*Preprint* nucl-th/0005062)
[46] Miller M L, Reygers K, Sanders S J and Steinberg P 2007 *Ann. Rev. Nucl. Part. Sci.* **57** 205–243
[47] Broniowski W, Rybczyński M and Bożek P 2009 *Comput. Phys. Commun.* **180** 69 (*Preprint* 0710.5731)
[48] Rybczyński M, Stefanek G, Broniowski W and Bożek P 2013 (*Preprint* 1310.5475)
[49] Bożek P and Broniowski W 2014 *Phys. Lett. B* **739** 308
[50] Bzdak A, Bozek P and McLerran L 2014 *Nucl.Phys.* **A927** 15–23 (*Preprint* 1311.7325)

Diffractive photoproduction of J/ψ at LHC energies

V.P. Gonçalves[1,a], **B.D. Moreira**[2,b], **F.S. Navarra**[2,c]

[1]High and Medium Energy Group, Instituto de Física e Matemática, Universidade Federal de Pelotas, Caixa Postal 354, CEP 96010-900, Pelotas, RS, Brazil

[2]Instituto de Física, Universidade de São Paulo, Caixa Postal 66318, 05315-970, São Paulo, SP, Brazil

E-mail: [a]barros@ufpel.edu.br, [b]bduartesm@gmail.com, [c]navarra@if.usp.br

Abstract. In this work we calculate the rapidity distribution and the total cross section for the diffractive photoproduction of J/ψ in $PbPb$ collisions in LHC energies. We use the color dipole formalism and different models for the forward dipole-proton scattering amplitude ($\mathcal{N}$) which take into account nonlinear effects. Using always the same model for the wave function of the vector meson, we estimate the theoretical uncertainty related to the choice of the model for $\mathcal{N}$. We compare our results with available data from ALICE [1, 2] and present our prediction for higher enegies.

1. Introduction

With the advent of modern colliders like RHIC and specially LHC, a deeper study of high energy physics became possible. One of the most interesting topics of high energy hadron physics is the study of QCD in the low x regime. Our knowledge of the parton distributions inside the hadrons is still far from complete. In lepton-hadron and hadron-hadron collisions with fixed virtualities and low x, we expect a fast growth of gluon distributions inside the hadron, as described by the BFKL equation, making it a dense system. However an indefinite growth of gluons inside the hadron would violate the unitarity bound in a given hadronic process. In order to avoid this violation, there is a mechanism to tame the growth of the gluon distribution. This mechanism is called gluon saturation and is associated to effects of gluon recombination. The investigation of saturation effects can be best made through the study of photon-hadron collisions in the color dipole picture. In this formalism, one photon produces a $q\bar{q}$ pair (a color dipole) which scatters off the hadron, producing a given final state. The cross section of dipole-hadron interaction is obtained from the forward scattering amplitude for which there are some models that take into account saturation effects. A way to study photon-hadron collisions is through the diffractive production of vector mesons in ultraperipheral collisions, where one of the hadrons acts as a photon source that interacts with the other hadron.

Here we present our results for the diffractive photoproduction of J/ψ in $PbPb$ collisions for three different saturation models. This work is organized in the following way: in the next section we present the basic formulae of the equivalent photon approximation, the dipole formalism applied to dipole-nucleus interaction is presented in section 3, our results are showed in section 4 and, finally, we present our conclusions.

2. Equivalent photon approximation

Let us consider an ultraperipheral collision between two fast nuclei. In an ultraperipheral collision we must take $b > R_{h_1} + R_{h_2}$, where b is the impact parameter of the collision and R_{h_i} is the radius of the nucleus i. In this regime we can factorize the J/ψ production in nucleus-nucleus collisions in terms of the equivalent flux of photons of the projectile and the photon-nucleus cross section [3, 4]. This method is called equivalent photon approximation. The cross section for diffractive J/ψ photopoduction in a nucleus-nucleus collision is given by

$$\sigma(h_1 h_2 \to h_1 \otimes J/\psi \otimes h_2) = \int_{\omega_{min}}^{\infty} d\omega \frac{dN_{\gamma/h_1}}{d\omega}(\omega)\sigma_{\gamma h_2 \to J/\psi h_2}(\omega) + \int_{\omega_{min}}^{\infty} d\omega \frac{dN_{\gamma/h_2}}{d\omega}(\omega)\sigma_{\gamma h_1 \to J/\psi h_1}(\omega) \ , \tag{1}$$

where $dN/d\omega$ is the photon spectrum associated to the nucleus, given by

$$\frac{dN_{\gamma/A}(\omega)}{d\omega} = \frac{2Z^2\alpha_{em}}{\pi\omega}\left[\bar{\eta}K_0(\bar{\eta})K_1(\bar{\eta}) + \frac{\bar{\eta}^2}{2}\left(K_1^2(\bar{\eta}) - K_0^2(\bar{\eta})\right)\right] \ . \tag{2}$$

Here, $\bar{\eta} = \omega(R_{h_1} + R_{h_2})/\gamma_L$, ω is the photon energy in the center-of-mass frame (c.m.s), $\omega_{min} = M^2_{J/\psi}/(4\gamma_L m_p)$, and γ_L is the Lorentz factor. Equation (1) allows us to study the photon-nucleus scattering, which will be addressed below.

3. Diffractive J/ψ photoproduction at high energies

To compute the cross section $\sigma(h_1 h_2 \to h_1 \otimes J/\psi \otimes h_2)$ from Eq.(1), we need to know the photon-nucleus cross section. In high energy collisions we can do this in the dipole picture, where a virtual photon fluctuates into a quark pair (a color dipole) which scatters off the nucleus (Fig.1). For high energy photon-nucleus coherent scattering we have [4] :

$$\sigma^{coh}_{\gamma A \to J/\psi A} = \int d^2\mathbf{b}\left[\int d^2\mathbf{r}\int dz\left(\psi^*_{J/\psi}\psi\right)\mathcal{N}_A(x,\mathbf{r},\mathbf{b})\right]^2 \ , \tag{3}$$

where $\left(\psi^*_{J/\psi}\psi\right)$ describe the overlap between the J/ψ wave function and the virtual photon wave function. For the meson wave function, we have used the Gaus-LC model [5]. The $\mathcal{N}_A(x,\mathbf{r},\mathbf{b})$ is the forward dipole-nucleus scattering amplitude, which contains all information about the target and the strong interaction physics.

In our calculations, we use for $\mathcal{N}_A(x,\mathbf{r},\mathbf{b})$ [6]

$$\mathcal{N}_A(x,\mathbf{r},\mathbf{b}) = 1 - \exp\left[-\frac{1}{2}\sigma_{dip}(x,r^2)T_A(\mathbf{b})\right] \ , \tag{4}$$

where $T_A(\mathbf{b})$ is the nuclear profile function, which is obtained from a three-parameter Fermi distribution for the nuclear density normalized to A, and σ_{dip} is the dipole-proton cross section given by

$$\sigma_{dip}(x,r^2) = 2\int d^2\mathbf{b}\mathcal{N}_p(x,\mathbf{r},\mathbf{b}) \ . \tag{5}$$

Here, $\mathcal{N}_p$ is the forward dipole-proton scattering amplitude.

Journal of Physics: Conference Series **630** (2015) 012061 doi:10.1088/1742-6596/630/1/012061

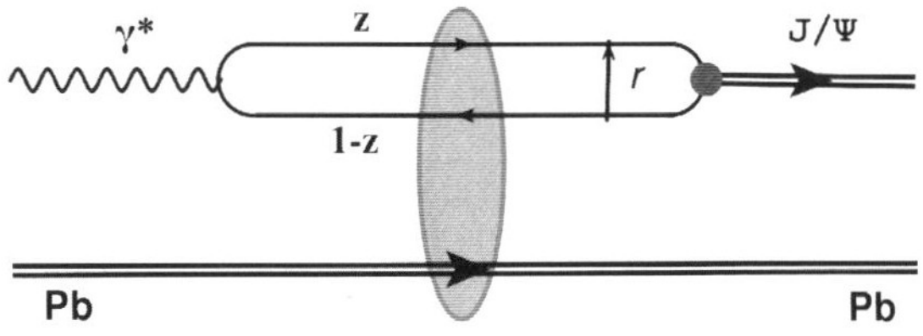

Figure 1. Production of J/ψ in the scattering $\gamma^* Pb$ in the color dipole picture.

We use three different models for $\mathcal{N}_p$, which include saturation effects. They are **rcBK** [7], which is based on the solution of the Balitsky-Kovchegov equation, and the phenomenological models **GBW** [8] and **bCGC** [5]. In the bCGC model we present a comparison between the older version and the updated version [9], in which the recently released high precision combined HERA data were included in the analysis. This last model is called **bCGC NEW**.

In the next section, we present our results for the diffractive photoproduction of J/ψ in ultraperipheral ($PbPb$) collisions. For this, we use the different models for $\mathcal{N}_p$ mentioned above.

4. Results

In this section, we present our results for the rapidity distribution and total cross section for the J/ψ production. The distribution on rapidity Y of the produced final state can be directly computed from Eq.(1), by using its relation with the photon energy ω, i.e. $Y \propto \ln(2\omega/m_p)$. Explicitly, the rapidity distribution is given by:

$$\frac{d\sigma(h_1 + h_2 \to h_1 \otimes J/\psi \otimes h_2)}{dY} = \left[\omega \left.\frac{dN}{d\omega}\right|_{h_1} \sigma_{\gamma h_2 \to J/\psi \otimes h_2}(\omega)\right]_{\omega_L} + \left[\omega \left.\frac{dN}{d\omega}\right|_{h_2} \sigma_{\gamma h_1 \to J/\psi \otimes h_1}(\omega)\right]_{\omega_R}, \quad (6)$$

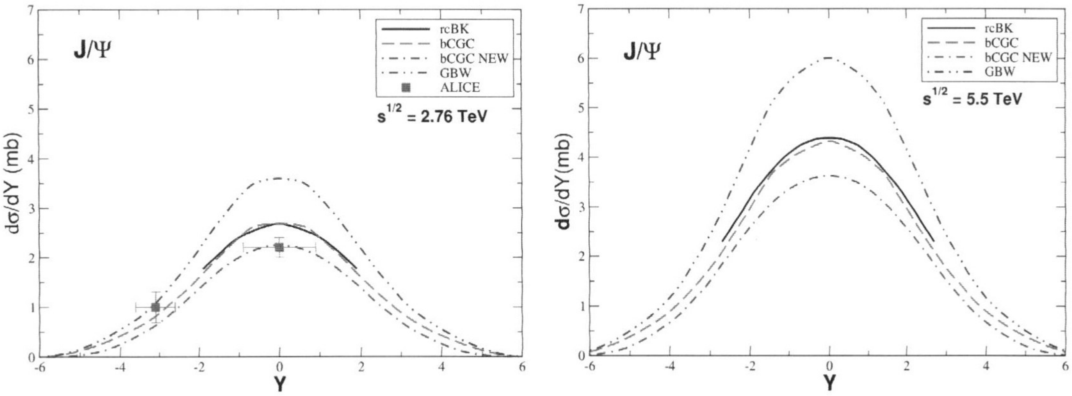

Figure 2. Rapidity distribution for the diffractive photoproduction of J/Ψ in $PbPb$ collisions at $\sqrt{s} = 2.76$ TeV (left panel) and $\sqrt{s} = 5.5$ TeV (right panel). Data from ALICE Collaboration [1, 2].

where $\omega_L \propto e^{-Y}$ and $\omega_R \propto e^{Y}$ denote photons from h_1 and h_2, respectively. Consequently, given the photon flux, the rapidity distribution is thus a direct measure of the photoproduction cross section for a given energy.

In Fig.2 we present our predictions for the rapidity distribution for the diffractive photoproduction of J/ψ in $PbPb$ collisions. For $\sqrt{s}$ = 2.76 TeV (left panel of Fig.2), we compare our results with the ALICE data [1, 2]. For forward rapidity, all the models agree among themselves, and they are in a good agreement with the experimental data. On the other hand, in the central rapidity region, the models are quite different and only the bCGC NEW model is able to describe the data. In the right panel of Fig.2, we present our predictions for the rapidity distribution of J/ψ in $PbPb$ collisions at $\sqrt{s}$ = 5.5 TeV. We can see that all the models still have similar predictions at forward rapidities, but with the growth of the energy, the difference between the models increases in the central rapidity region. In these two plots, we can observe that the rcBK model is limited because of its range of validity ($x < 0.01$).

In Table 1 we present our predictions for the total cross sections for the diffractive photoproduction of J/ψ in $PbPb$ collisions at $\sqrt{s} = 2.76$ and 5.5 TeV. Due to the discrepancies in the central rapidity region observed in Fig.2, there are large differences in the value of the total cross sections.

	GBW	bCGC	bCGC NEW
$PbPb$ ($\sqrt{s} = 2.76$ TeV)	18.2 mb	13.6 mb	11.0 mb
$PbPb$ ($\sqrt{s} = 5.5$ TeV)	33.8 mb	24.4 mb	20.3 mb

Table 1. The total cross section for diffractive photoproduction of J/Ψ in pp, pPb and $PbPb$ collisions at LHC energies.

5. Conclusions

In this work, we have calculated the rapidity distribution and total cross section of diffractive photoproduction of J/ψ in ultraperipheral $PbPb$ collisions at LHC energies. With the purpose of studying the theoretical uncertainty in the choice of models of the forward dipole-proton scattering amplitude, we have used only one model for the vector meson wave function and we have used three different models for $\mathcal{N}_p$. Among the considered models, we have found that the GBW model can be taken as a upper limit while the bCGC NEW can be taken as a lower limit. Moreover, the bCGC NEW model is able to describe the available ALICE data. Our final conclusion, is that the ultraperipheral collisions can be used to constrain the QCD dynamics at high energies.

ACKNOWLEDGMENTS
This work was partially financed by rhe Brazilian funding agencies CAPES, CNPq, FAPESP and FAPERGS.

References
[1] B. Abelev *et al.* [ALICE Collaboration], Phys. Lett. B **718**, 1273 (2013)
[2] E. Abbas *et al.* [ALICE Collaboration], Eur. Phys. J. C **73**, 2617 (2013)
[3] G. Baur, K. Hencken, D. Trautmann, S. Sadovsky, Y. Kharlov, Phys. Rep. **364**, 359 (2002); V. P. Goncalves and M. V. T. Machado, Mod. Phys. Lett. A **19**, 2525 (2004); C. A. Bertulani, S. R. Klein and J. Nystrand, Ann. Rev. Nucl. Part. Sci. **55**, 271 (2005); K. Hencken *et al.*, Phys. Rept. **458**, 1 (2008).

Journal of Physics: Conference Series **630** (2015) 012061 doi:10.1088/1742-6596/630/1/012061

[4] V. P. Goncalves, B. D. Moreira and F. S. Navarra, Phys. Rev. C **90**, 015203 (2014).
[5] H. Kowalski, L. Motyka, and G. Watt, Phys. Rev. D 74, 074016 (2006).
[6] N. Armesto, Eur. Phys. J. C 26, 35 (2002).
[7] J.L. Albacete, N. Armesto, J.G. Milhano, and C.A. Salgado, Phys. Rev. D 80, 034031 (2009).
[8] K. Golec-Biernat and M. Wusthoff, Phys. Rev. D 59, 014017 (1998).
[9] A.H. Rezaeian and I. Schmidt, Phys. Rev. D 88, 074016 (2013).